Uniform Central Limit Theorems

Second Edition

This work about probability limit theorems for empirical processes on general spaces, by one of the founders of the field, has been considerably expanded and revised from the original edition. When samples become large, laws of large numbers and central limit theorems are guaranteed to hold uniformly over wide domains. The author gives a thorough treatment of the subject, including an extended treatment of Vapnik–Červonenkis combinatorics, the Ossiander L2 bracketing central limit theorem, the Giné–Zinn bootstrap central limit theorem in probability, the Bronstein theorem on approximation of convex sets, and the Shor theorem on rates of convergence over lower layers. This new edition contains proofs of several main theorems not proved in the first edition, including the Bretagnolle–Massart theorem giving constants in the Komlós–Major–Tusnády rate of convergence for the classical empirical process, Massart's form of the Dvoretzky–Kiefer–Wolfowitz inequality with precise constant, Talagrand's generic chaining approach to boundedness of Gaussian processes, a characterization of uniform Glivenko–Cantelli classes of functions, Giné and Zinn's characterization of uniform Donsker classes of functions (i.e., classes for which the central limit theorem holds uniformly over all probability measures P), and the Bousquet–Koltchinskii–Panchenko theorem that the convex hull of a uniform Donsker class is uniform Donsker.

The book will be an essential reference for mathematicians working in infinite-dimensional central limit theorems, mathematical statisticians, and computer scientists working in computer learning theory. Problems are included at the end of each chapter so the book can also be used as an advanced text.

R. M. DUDLEY is a Professor of Mathematics at the Massachusetts Institute of Technology in Cambridge, Massachusetts.

All the titles listed below can be obtained from good booksellers or from Cambridge University Press.
For a complete series listing visit www.cambridge.org/mathematics.

Uniform Central Limit Theorems

Second Edition

R. M. DUDLEY

Massachusetts Institute of Technology

CAMBRIDGE
UNIVERSITY PRESS

CAMBRIDGE
UNIVERSITY PRESS

Shaftesbury Road, Cambridge CB2 8EA, United Kingdom

One Liberty Plaza, 20th Floor, New York, NY 10006, USA

477 Williamstown Road, Port Melbourne, VIC 3207, Australia

314–321, 3rd Floor, Plot 3, Splendor Forum, Jasola District Centre, New Delhi – 110025, India

103 Penang Road, #05–06/07, Visioncrest Commercial, Singapore 238467

Cambridge University Press is part of Cambridge University Press & Assessment, a department of the University of Cambridge.

We share the University's mission to contribute to society through the pursuit of education, learning and research at the highest international levels of excellence.

www.cambridge.org
Information on this title: www.cambridge.org/9780521738415

© R. M. Dudley 1999, 2014

First published 1999
First paperback edition 2008
Second edition 2014

A catalogue record for this publication is available from the British Library

Library of Congress Cataloging-in-Publication data
Dudley, R. M. (Richard M.)
Uniform central limit theorems / R. M. Dudley, Massachusetts Institute of Technology. – Second edition.
      pages cm – (Cambridge studies in advanced mathematics)
Includes bibliographical references and index.
ISBN 978-0-521-49884-5 (hardback) – ISBN 978-0-521-73841-5 (paperback)
1. Central limit theorem.   I. Title.
QA273.67.D84   2014
519.2–dc23   2013011303

ISBN   978-0-521-49884-5   Hardback
ISBN   978-0-521-73841-5   Paperback

To Liza

Contents

Preface to the Second Edition

This book developed out of some topics courses given at M.I.T. and my lectures at the St.-Flour probability summer school in 1982. The material of the book has been expanded and extended considerably since then. At the end of each chapter are some problems and notes on that chapter.

Starred sections are not cited later in the book except perhaps in other starred sections. The first edition had several double-starred sections in which facts were stated without proofs. This edition has no such sections.

The following, not proved in the first edition, now are: (i) for Donsker's theorem on the classical empirical process $\alpha_n := \sqrt{n}(F_n - F)$, and the Komlós–Major–Tusnády strengthening to give a rate of convergence, the Bretagnolle–Massart proof with specified constants; (ii) Massart's form of the Dvoretzky–Kiefer–Wolfowitz inequality for α_n with optimal constant; (iii) Talagrand's generic chaining approach to boundedness of Gaussian processes, which replaces the previous treatment of majorizing measures; (iv) characterization of uniform Glivenko–Cantelli classes of functions (from a paper by Dudley, Giné, and Zinn, but here with a self-contained proof); (v) Giné and Zinn's characterization of uniform Donsker classes of functions; (vi) its consequence that uniformly bounded, suitably measurable classes of functions satisfying Pollard's entropy condition are uniformly Donsker; and (vii) Bousquet, Koltchinskii, and Panchenko's theorem that a convex hull preserves the uniform Donsker property.

The first edition contained a chapter on invariance principles, based on a 1983 paper with the late Walter Philipp. Some techniques introduced in that paper, such as measurable cover functions, are still used in this book. But I have not worked on invariance principles as such since 1983. Much of the work on them treats dependent random variables, as did parts of the 1983 paper which Philipp contributed. The present book is mainly about the i.i.d. case. So I suppose the chapter is outdated, and I omit it from this edition.

For useful conversations and suggestions on topics in the book I'm glad to thank Kenneth Alexander, Niels Trolle Andersen, the late Miguel Arcones, Patrice Assouad, Erich Berger, Lucien Birgé, Igor S. Borisov, Donald Cohn, Yves Derrienic, Uwe Einmahl, Joseph Fu, Sam Gutmann, David Haussler, Jørgen Hoffmann-Jørgensen, Yen-Chin Huang, Vladimir Koltchinskii, the late Lucien Le Cam, David Mason, Pascal Massart, James Munkres, Rimas Norvaiša, the late Walter Philipp, Tom Salisbury, the late Rae Shortt, Michel Talagrand, Jon Wellner, He Sheng Wu, Joe Yukich, and Joel Zinn. I especially thank Denis Chetverikov, Peter Gaenssler and Franz Strobl, Evarist Giné, and Jinghua Qian, for providing multiple corrections and suggestions. I also thank Xavier Fernique (for the first edition), Evarist Giné (for both editions), and Xia Hua (for the second edition) for giving or sending me copies of expositions.

Notes

Throughout this book, all references to "RAP" are to the author's book *Real Analysis and Probability*, second edition, Cambridge University Press, 2002.

Also, "$A := B$" means A is defined by B, whereas "$A =: B$" means B is defined by A.

1

Donsker's Theorem, Metric Entropy, and Inequalities

Let P be a probability measure on the Borel sets of the real line $\mathbb{R}$ with distribution function $F(x) := P((-\infty, x])$. Let $X_1, X_2, \ldots$, be i.i.d. (independent, identically distributed) random variables with distribution P. For each $n = 1, 2, \ldots$, and any Borel set $A \subset \mathbb{R}$, let $P_n(A) := \frac{1}{n}\Sigma_{j=1}^n \delta_{X_j}(A)$, where $\delta_x(A) = 1_A(x)$. For any given $X_1, \ldots, X_n$, P_n is a probability measure called the *empirical measure*. Let F_n be the distribution function of P_n. Then F_n is called the *empirical distribution function*.

Let U be the $U[0, 1]$ distribution function $U(x) = \min(1, \max(0, x))$ for all x, so that $U(x) = x$ for $0 \le x \le 1$, $U(x) = 0$ for $x < 0$ and $U(x) = 1$ for $x > 1$. To relate F and U we have the following.

Proposition 1.1 *For any distribution function F on $\mathbb{R}$:*

(a) For any y with $0 < y < 1$, $F^{\leftarrow}(y) := \inf\{x : F(x) \ge y\}$ is well-defined and finite.

(b) For any real x and any y with $0 < y < 1$ we have $F(x) \ge y$ if and only if $x \ge F^{\leftarrow}(y)$.

(c) If V is a random variable having $U[0, 1]$ distribution, then $F^{\leftarrow}(V)$ has distribution function F.

Proof. For (a), recall that F is nondecreasing, $F(x) \to 0$ as $x \to -\infty$, and $F(x) \to 1$ as $x \to +\infty$. So the set $\{x : F(x) \ge y\}$ is nonempty and bounded below, and has a finite infimum.

For (b), $F(x) \ge y$ implies $x \ge F^{\leftarrow}(y)$ by definition of $F^{\leftarrow}(y)$. Conversely, as F is continuous from the right, $F(F^{\leftarrow}(y)) \ge y$, and as F is nondecreasing, $x \ge F^{\leftarrow}(y)$ implies $F(x) \ge y$.

For (c), and any x, we have by (b)

$$\Pr(F^{\leftarrow}(V) \le x) = \Pr(V \le F(x)) = U(F(x)) = F(x)$$

1

since $0 \leq F(x) \leq 1$, so (c) holds. $\square$

Recall that for any function f defined on the range of a function g, the composition $f \circ g$ is defined by $(f \circ g)(x) := f(g(x))$. We can then relate empirical distribution functions F_n for any distribution function F to those U_n for U, as follows.

Proposition 1.2 *For any distribution function F, and empirical distribution functions F_n for F and U_n for U, $U_n \circ F$ have all the properties of F_n.*

Proof. Let $V_1, \ldots, V_n$ be i.i.d. U, so that $U_n(t) = \frac{1}{n} \sum_{j=1}^{n} 1_{V_j \leq t}$ for $0 \leq t \leq 1$. Thus for any x, by Proposition 1.1(b) and (c),

$$U_n(F(x)) = \frac{1}{n} \sum_{j=1}^{n} 1_{V_j \leq F(x)}$$

$$= \frac{1}{n} \sum_{j=1}^{n} 1_{F^{\leftarrow}(V_j) \leq x}$$

$$= \frac{1}{n} \sum_{j=1}^{n} 1_{X_j \leq x}$$

where $X_j := F^{\leftarrow}(V_j)$ are i.i.d. (F). Thus $U_n(F(x))$ has all properties of $F_n(x)$. $\square$

The developments to be described in this book began (in 1933) with the Glivenko–Cantelli theorem, a uniform law of large numbers. Probability distribution functions can converge pointwise but not uniformly: for example, as $n \to \infty$, $1_{[-1/n, +\infty)}(x) \to 1_{[0, +\infty)}(x)$ for all x but not uniformly.

Theorem 1.3 (Glivenko–Cantelli) *For any distribution function F, almost surely, $\sup_x |(F_n - F)(x)| \to 0$ as $n \to \infty$.*

Proof. By Proposition 1.2, and since $U \circ F \equiv F$, it suffices to prove this for the $U[0, 1]$ distribution U. Given $\varepsilon > 0$, take a positive integer k such that $1/k < \varepsilon/2$. For each $j = 0, 1, \ldots, k$, $U_n(j/k) \to j/k$ as $n \to \infty$ with probability 1 by the ordinary strong law of large numbers. Take $n_0 = n_0(\omega)$ such that for all $n \geq n_0$ and all $j = 0, 1, \ldots, k$, $|U_n(j/k) - j/k| < \varepsilon/2$. For t outside $[0, 1]$ we have $U_n(t) \equiv U(t) = 0$ or 1. For each $t \in [0, 1]$ there is at least one $j = 1, \ldots, k$ such that $(j - 1)/k \leq t \leq j/k$. Then for $n \geq n_0$,

$$(j - 1)/k - \varepsilon/2 < U_n((j - 1)/k) \leq U_n(t) \leq U_n(j/k) < j/k + \varepsilon/2.$$

It follows that $|U_n(t) - t| < \varepsilon$, and since t was arbitrary, the theorem follows. $\square$

The next step was to consider the limiting behavior of $\alpha_n := n^{1/2}(F_n - F)$ as $n \to \infty$. For any fixed t, the central limit theorem in its most classical form, for binomial distributions, says that $\alpha_n(t)$ converges in distribution to $N(0, F(t)(1 - F(t)))$, in other words a normal (Gaussian) law, with mean 0 and variance $F(t)(1 - F(t))$.

In what follows (as mentioned in the Note after the Preface), "RAP" will mean the author's book *Real Analysis and Probability*.

For any finite set T of values of t, the multidimensional central limit theorem (RAP, Theorem 9.5.6) tells us that $\alpha_n(t)$ for t in T converges in distribution as $n \to \infty$ to a normal law $N(0, C_F)$ with mean 0 and covariance $C_F(s, t) = F(s)(1 - F(t))$ for $s \le t$.

The *Brownian bridge* (RAP, Section 12.1) is a stochastic process $y_t(\omega)$ defined for $0 \le t \le 1$ and ω in some probability space Ω, such that for any finite set $S \subset [0, 1]$, y_t for t in S have distribution $N(0, C)$, where $C = C_U$ for the uniform distribution function $U(t) = t$, $0 \le t \le 1$. So the empirical process α_n converges in distribution to the Brownian bridge composed with F, namely $t \mapsto y_{F(t)}$, at least when restricted to finite sets.

It was then natural to ask whether this convergence extends to infinite sets or the whole interval or line. Kolmogorov (1933b) showed that when F is continuous, the supremum $\sup_t \alpha_n(t)$ and the supremum of absolute value, $\sup_t |\alpha_n(t)|$, converge in distribution to the laws of the same functionals of y_F. Then, these functionals of y_F have the same distributions as for the Brownian bridge itself, since F takes $\mathbb{R}$ onto an interval including $(0, 1)$ and which may or may not contain 0 or 1; this makes no difference to the suprema since $y_0 \equiv y_1 \equiv 0$. Also, $y_t \to 0$ almost surely as $t \downarrow 0$ or $t \uparrow 1$ by sample continuity; the suprema can be restricted to a countable dense set such as the rational numbers in $(0, 1)$ and are thus measurable.

To work with the Brownian bridge process it will help to relate it to the well-known Brownian motion process x_t, defined for $t \ge 0$, also called the Wiener process. This process is such that for any any finite set $T \subset [0, +\infty)$, the joint distribution of $\{x_t\}_{t \in F}$ is $N(0, C)$ where $C(s, t) = \min(s, t)$. This process has independent increments, namely, for any $0 = t_0 < t_1 < \cdots < t_k$, the increments $x_{t_j} - x_{t_{j-1}}$ for $j = 1, \ldots, k$ are jointly independent, with $x_t - x_s$ having distribution $N(0, t - s)$ for $0 \le s < t$. Recall that for jointly Gaussian (normal) random variables, joint independence, pairwise independence, and having covariances equal to 0 are equivalent. Having independent increments with the given distributions clearly implies that $E(x_s x_t) = \min(s, t)$ and so is equivalent to the definition of Brownian motion with that covariance.

Brownian motion can be taken to be sample continuous, i.e. such that $t \mapsto x_t(\omega)$ is continuous in t for all (or almost all) ω. This theorem, proved by Norbert Wiener in the 1920s, is Theorem 12.1.5 in RAP; a proof will be indicated here.

If Z has $N(0, 1)$ distribution, then for any $c > 0$, $\Pr(Z \geq c) \leq \exp(-c^2/2)$ (RAP, Lemma 12.1.6(b)). Thus if X has $N(0, \sigma^2)$ distribution for some $\sigma > 0$, then $\Pr(X \geq c) = \Pr(X/\sigma > c/\sigma) \leq \exp(-c^2/(2\sigma^2))$. It follows that for any $n = 1, 2, \ldots$ and any $j = 1, 2, \ldots,$

$$\Pr\left(\left|x_{j/2^n} - x_{(j-1)/2^n}\right| \geq \frac{1}{n^2}\right) \leq 2\exp\left(-2^n/(2n^4)\right).$$

It follows that for any integer $K > 0$, the probability of any of the above events occurring for $j = 1, \ldots, 2^n K$ is at most $2^{n+1} K \exp(-2^n/(2n^4))$, which approaches 0 very fast as $n \to \infty$, because of the dominant factor -2^n in the exponent. Also, the series $\sum_n 1/n^2$ converges. It follows by the Borel–Cantelli Lemma (RAP, Theorem 8.3.4) that with probability 1, for all $t \in [0, K]$, for a sequence of dyadic rationals $t_n \to t$ given by the binary expansion of t, x_{t_n} will converge to some limit X_t, which equals x_t almost surely. Specifically, for $t < K$, let $t_n = (j - 1)/2^n$ for the unique $j \leq 2^n K$ such that $(j - 1)/2^n \leq t < t/2^n$. Then $t_{n+1} = t_n = 2j/2^{n+1}$ or $t_{n+1} = (2j - 1)/2^{n+1}$, so that t_{n+1} and t_n are either equal or are adjacent dyadic rationals with denominator 2^{n+1}, and the above bounds apply to the differences $x_{t_{n+1}} - x_{t_n}$.

The process X_t is sample-continuous and is itself a Brownian motion, as desired. From here on, a "Brownian motion" will always mean a sample-continuous one.

Here is a reflection principle for Brownian motion (RAP, 12.3.1). A proof will be sketched.

Theorem 1.4 *Let* $\{x_t\}_{t \geq 0}$ *be a Brownian motion,* $b > 0$ *and* $c > 0$. *Then*

$$\Pr(\sup\{x_t : t \leq b\} \geq c) = 2\Pr(x_b \geq c) = 2N(0, b)([c, +\infty)).$$

Sketch of proof: If $\sup\{x_t : t \leq b\} \geq c$, then by sample continuity there is a least time τ with $0 < \tau \leq b$ such that $x_\tau = c$. The probability that $\tau = b$ is 0, so we can assume that $\tau < b$ if it exists. Starting at time τ, x_b is equally likely to be $> c$ or $< c$. [This holds by an extension of the independent increment property or the strong Markov property (RAP, Section 12.2); or via approximation by suitably normalized simple symmetric random walks and the reflection principle for them.] Thus

$$\Pr(x_b \geq c) = \frac{1}{2}\Pr(\sup\{x_t : t \leq b\} \geq c),$$

which gives the conclusion.   □

One way to write the Brownian bridge process y_t in terms of Brownian motion is $y_t = x_t - tx_1$, $0 \leq t \leq 1$. It is easily checked that this a Gaussian process (y_t for t in any finite subset of $[0, 1]$ have a normal joint distribution, with zero means) and that the covariance $E y_s y_t = s(1 - t)$ for $0 \leq s \leq t \leq 1$, fitting the definition of Brownian bridge. It follows that the Brownian bridge process, on $[0, 1]$, is also sample continuous, i.e., we can and will take it such that $t \mapsto y_t(\omega)$ is continuous for almost all ω.

Another relation is that y_t is x_t for $0 \leq t \leq 1$ conditioned on $x_1 = 0$ in a suitable sense, namely, it has the limit of the distributions of $\{x_t\}_{0 \leq t \leq 1}$ given $|x_1| < \varepsilon$ as $\varepsilon \downarrow 0$ (RAP, Proposition 12.3.2). A proof of this will also be sketched here. Suppose we are given a Brownian bridge $\{y_t\}_{0 \leq t \leq 1}$. Let Z be a $N(0, 1)$ random variable independent of the y_t process. Define $\xi_t = y_t + tZ$ for $0 \leq t \leq 1$. Then ξ_t is a Gaussian stochastic process with mean 0 and covariance given, for $0 \leq s \leq t \leq 1$, by $E\xi_s\xi_t = s(1 - t) + 0 + 0 + st = s$, so ξ_t for $0 \leq t \leq 1$ has the distribution of Brownian motion restricted to $[0, 1]$. The conditional distribution of ξ_t given $|\xi_1| < \varepsilon$, in other words $|Z| < \varepsilon$, is that of $y_t + tZ$ given $|Z| < \varepsilon$, and since Z is independent of $\{y_t\}_{0 \leq t \leq 1}$, this conditional distribution clearly converges to that of $\{y_t\}$ as $\varepsilon \downarrow 0$, as claimed.

Kolmogorov evaluated the distributions of $\sup_t y_t$ and $\sup_t |y_t|$ explicitly. For the first (1-sided) supremum this follows from a reflection principle (RAP, Proposition 12.3.3) for y_t whose proof will be sketched:

Theorem 1.5 *For a Brownian bridge* $\{y_t\}_{0 \leq t \leq 1}$ *and any* $c > 0$,

$$\Pr(\sup_{0 \leq t \leq 1} y_t > c) = \exp(-2c^2).$$

Sketch of proof: The probability is, for a Brownian motion x_t,

$$\lim_{\varepsilon \downarrow 0} \Pr\left(\sup_{0 \leq t \leq 1} x_t > c \,\big|\, |x_1| < \varepsilon\right)$$

$$= \lim_{\varepsilon \downarrow 0} \Pr\left(\sup_{0 \leq t \leq 1} x_t > c \text{ and } |x_1| < \varepsilon\right) / \Pr(|x_1| < \varepsilon)$$

$$= \lim_{\varepsilon \downarrow 0} \Pr\left(\sup_{0 \leq t \leq 1} x_t > c \text{ and } |x_1 - 2c| < \varepsilon\right) / \Pr(|x_1| < \varepsilon)$$

where the last equality is by reflection. For ε small enough, $\varepsilon < c$, and then the last quotient becomes simply $\Pr(|x_1 - 2c| < \varepsilon)/\Pr(|x_1| < \varepsilon)$. Letting ϕ be the standard normal density function, the quotient is asymptotic as $\varepsilon \downarrow 0$ to $\phi(2c) \cdot 2\varepsilon/(\phi(0) \cdot 2\varepsilon) = \exp(-2c^2)$ as stated. $\qquad\square$

The distribution of $\sup_{0 \leq t \leq 1} |y_t|$ is given by a series (RAP, Proposition 12.3.4) as follows:

Theorem 1.6 *For any* $c > 0$, *and a Brownian bridge* y_t,

$$\Pr\left(\sup_{0 \leq t \leq 1} |y_t| > c\right) = 2\sum_{j=1}^{\infty}(-1)^{j-1}\exp(-2j^2c^2).$$

The proof is by iterated reflections: for example, a Brownian path which before time 1 reaches $+c$, then later $-c$, then returns to (near) 0 at time 1, corresponds to a path which reaches c, then $3c$, then (near) $4c$, and so on.

Doob (1949) asked whether the convergence in distribution of empirical processes to the Brownian bridges held for more general functionals (other than the supremum and that of absolute value). Donsker (1952) stated and proved (not quite correctly) a general extension. This book will present results proved over the past few decades by many researchers, first in this chapter on speed of convergence in the classical case. In the rest of the book, the collection of half-lines $(-\infty, x]$, $x \in \mathbb{R}$, will be replaced by much more general classes of sets in, and functions on, general sample spaces, for example, the class of all ellipsoids in $\mathbb{R}^3$.

To motivate and illustrate the general theory, the first section will give a revised formulation of Donsker's theorem with a statement on rate of convergence, to be proved in Section 1.4. Sections 1.2 on metric entropy and 1.3 on inequalities provide concepts and facts to be used in the rest of the book.

1.1 Empirical Processes: The Classical Case

In this section, a form of Donsker's theorem with rates of convergence will be stated for the $U[0, 1]$ distribution with distribution function U and empirical distribution functions U_n. This would imply a corresponding limit theorem for a general distribution function F via Proposition 1.2. Let $\alpha_n := n^{1/2}(U_n - U)$ on $[0, 1]$. It will be proved that as $n \to \infty$, α_n converges in law (in a sense to be made precise below) to a Brownian bridge process y_t, $0 \le t \le 1$ (RAP, before Theorem 12.1.5).

Donsker in 1952 proved that the convergence in law of α_n to the Brownian bridge holds, in a sense, with respect to uniform convergence in t on the whole interval $[0, 1]$. How to define such convergence in law correctly, however, was not clarified until much later. General definitions will be given in Chapter 3. Here, a more special approach will be taken in order to state and prove an accessible form of Donsker's theorem.

For a function f on $[0, 1]$ we have the sup norm

$$\|f\|_{\sup} := \sup\{|f(t)| : 0 \le t \le 1\}.$$

Here is a formulation of Donsker's theorem.

Theorem 1.7 *For $n = 1, 2, \ldots$, there exist probability spaces Ω_n such that:*

(a) On Ω_n, there exist n i.i.d. random variables $X_1, \ldots, X_n$ with uniform distribution in $[0, 1]$. Let α_n be the nth empirical process based on these X_i;

(b) On Ω_n a sample-continuous Brownian bridge process Y_n: $(t, \omega) \mapsto Y_n(t, \omega)$ is defined;

(c) $\|\alpha_n - Y_n\|_{\sup}$ is measurable, and for all $\varepsilon > 0$, $\Pr(\|\alpha_n - Y_n\|_{\sup} > \varepsilon) \to 0$ as $n \to \infty$.

The theorem just stated is a consequence of the following facts giving rates of convergence. Komlós, Major, and Tusnády (1975) stated a sharp rate of convergence in Donsker's theorem, namely, that on some probability space there exist X_i i.i.d. $U[0, 1]$ and Brownian bridges Y_n such that

$$P \left(\sup_{0 \le t \le 1} |(\alpha_n - Y_n)(t)| > \frac{x + c \log n}{\sqrt{n}} \right) < K e^{-\lambda x} \qquad (1.1)$$

for all $n = 1, 2, \ldots$ and $x > 0$, where c, K, and λ are positive absolute constants. If we take $x = a \log n$ for some $a > 0$, so that the numerator of the fraction remains of the order $O(\log n)$, the right side becomes $K n^{-\lambda a}$, decreasing as $n \to \infty$ as any desired negative power of n.

More specifically, Bretagnolle and Massart (1989) proved the following:

Theorem 1.8 (Bretagnolle and Massart) *The approximation (1.1) of empirical processes by Brownian bridges holds with $c = 12$, $K = 2$, and $\lambda = 1/6$ for $n \ge 2$.*

Bretagnolle and Massart's theorem is proved in Section 1.4.

1.2 Metric Entropy and Capacity

The notions in this section will be applied in later chapters, first, to Gaussian processes, then later in adapted forms, metric entropy with inclusion for sets, or with bracketing for functions, as applied to empirical processes in later chapters.

The word "entropy" is applied to several concepts in mathematics. What they have in common is apparently that they give some measure of the size or complexity of some set or transformation and that their definitions involve logarithms. Beyond this rather superficial resemblance, there are major differences. What are here called "metric entropy" and "metric capacity" are measures of the size of a metric space, which must be totally bounded (have compact completion) in order for the metric entropy or capacity to be finite. Metric entropy will provide a useful general technique for dealing with classes of sets or functions in general spaces, as opposed to Markov (or martingale) methods. The latter methods apply, as in the last section, when the sample space is $\mathbb{R}$ and the class $\mathcal{C}$ of sets is the class of half-lines $(-\infty, x]$, $x \in \mathbb{R}$, so that $\mathcal{C}$ with its ordering by inclusion is isomorphic to $\mathbb{R}$ with its usual ordering.

Let (S, d) be a metric space and A a subset of S. Let $\varepsilon > 0$. A set $F \subset S$ (not necessarily included in A) is called an *ε-net* for A if and only if for each $x \in A$, there is a $y \in F$ with $d(x, y) \leq \varepsilon$. Let $N(\varepsilon, A, S, d)$ denote the minimal number of points in an ε-net in S for A.

For any set $C \subset S$, define the *diameter* of C by

$$\text{diam}\, C := \sup\{d(x, y) : x, y \in C\}.$$

Let $N(\varepsilon, C, d)$ be the smallest n such that C is the union of n sets of diameter at most 2ε. Let $D(\varepsilon, A, d)$ denote the largest n such that there is a subset $F \subset A$ with F having n members and $d(x, y) > \varepsilon$ whenever $x \neq y$ for x and y in F.

The three quantities just defined are related by the following inequalities:

Theorem 1.9 *For any $\varepsilon > 0$ and set A in a metric space S with metric d,*

$$D(2\varepsilon, A, d) \leq N(\varepsilon, A, d) \leq N(\varepsilon, A, S, d) \leq N(\varepsilon, A, A, d) \leq D(\varepsilon, A, d).$$

Proof. The first inequality holds since a set of diameter 2ε can contain at most one of a set of points more than 2ε apart. The next holds because any ball $\overline{B}(x, \varepsilon) := \{y : d(x, y) \leq \varepsilon\}$ is a set of diameter at most 2ε. The third inequality holds since requiring centers to be in A is more restrictive. The last holds because a set F of points more than ε apart, with maximal cardinality, must be an ε-net, since otherwise there would be a point more than ε away from each point of F, which could be adjoined to F, a contradiction unless F is infinite, but then the inequality holds trivially. $\square$

It follows that as $\varepsilon \downarrow 0$, when all the functions in the Theorem go to ∞ unless S is a finite set, they have the same asymptotic behavior up to a factor of 2 in ε. So it will be convenient to choose one of the four and make statements about it, which will then yield corresponding results for the others. The choice is somewhat arbitrary. Here are some considerations that bear on the choice.

The finite set of points, whether more than ε apart or forming an ε-net, are often useful, as opposed to the sets in the definition of $N(\varepsilon, A, d)$. $N(\varepsilon, A, S, d)$ depends not only on A but also on the larger space S. Many workers, possibly for these reasons, have preferred $N(\varepsilon, A, A, d)$. But the latter may decrease when the set A increases. For example, let A be the surface of a sphere of radius ε around 0 in a Euclidean space S and let $B := A \cup \{0\}$. Then $N(\varepsilon, B, B, d) = 1 < N(\varepsilon, A, A, d)$. This was the reason, apparently, that Kolmogorov chose to use $N(\varepsilon, A, d)$.

In this book I adopt $D(\varepsilon, A, d)$ as basic. It depends only on A, not on the larger space S, and is nondecreasing in A. If $D(\varepsilon, A, d) = n$, then there are n points which are more than ε apart and at the same time form an ε-net.

Now, the *ε-entropy* of the metric space (A, d) is defined as $H(\varepsilon, A, d) := \log N(\varepsilon, A, d)$, and the *$\varepsilon$-capacity* as $\log D(\varepsilon, A, d)$. Some other authors take logarithms to the base 2, by analogy with information-theoretic entropy. In this book logarithms will be taken to the usual base e, which fits, for example, with bounds coming from moment generating functions as in the next section, and with Gaussian measures as in Chapter 2. There are a number of interesting sets of functions where $N(\varepsilon, A, d)$ is of the order of magnitude $\exp(\varepsilon^{-r})$ as $\varepsilon \downarrow 0$, for some power $r > 0$, so that the ε-entropy, and likewise the ε-capacity, have the simpler order ε^{-r}. But in other cases below, $D(\varepsilon, A, d)$ is itself of the order of a power of $1/\varepsilon$.

1.3 Inequalities

This section collects several inequalities bounding the probabilities that random variables, and specifically sums of independent random variables, are large. Many of these follow from a basic inequality of S. Bernštein and P. L. Chebyshev:

Theorem 1.10 *For any real random variable X and $t \in \mathbb{R}$,*

$$\Pr(X \geq t) \; \leq \; \inf_{u \geq 0} e^{-tu} E e^{uX}.$$

Proof. For any fixed $u \geq 0$, the indicator function of the set where $X \geq t$ satisfies $1_{\{X \geq t\}} \leq e^{u(X-t)}$, so the inequality holds for a fixed u, then take $\inf_{u \geq 0}$. $\qquad \square$

For any independent real random variables $X_1, \ldots, X_n$, let $S_n := X_1 + \cdots + X_n$.

Theorem 1.11 (Bernštein's inequality) *Let $X_1, X_2, \ldots, X_n$ be independent real random variables with mean 0. Let $0 < M < \infty$ and suppose that $|X_j| \leq M$ almost surely for $j = 1, \ldots, n$. Let $\sigma_j^2 = \text{var}(X_j)$ and $\tau_n^2 := \text{var}(S_n) = \sigma_1^2 + \cdots + \sigma_n^2$. Then for any $K > 0$,*

$$Pr\{|S_n| \geq Kn^{1/2}\} \; \leq \; 2 \cdot \exp(-nK^2/(2\tau_n^2 + 2Mn^{1/2}K/3)). \qquad (1.2)$$

Proof. We can assume that $\tau_n^2 > 0$ since otherwise $S_n = 0$ a.s. and the inequality holds. For any $u \geq 0$ and $j = 1, \ldots, n$,

$$E \exp(uX_j) \; = \; 1 + u^2 \sigma_j^2 F_j/2 \; \leq \; \exp(\sigma_j^2 F_j u^2/2) \qquad (1.3)$$

where $F_j := 2\sigma_j^{-2} \sum_{r=2}^{\infty} u^{r-2} E X_j^r/r!$, or $F_j = 0$ if $\sigma_j^2 = 0$. For $r \geq 2$, $|X_j|^r \leq X_j^2 M^{r-2}$ a.s., so $F_j \leq 2 \sum_{r=2}^{\infty} (Mu)^{r-2}/r! \leq \sum_{r=2}^{\infty} (Mu/3)^{r-2} = 1/(1 - Mu/3)$ for all $j = 1, \ldots, n$ if $0 < u < 3/M$.

Let $v := Kn^{1/2}$ and $u := v/(\tau_n^2 + Mv/3)$, so that $v = \tau_n^2 u/(1 - Mu/3)$. Then $0 < u < 3/M$. Thus, multiplying the factors on the right side of (1.3) by independence, we have

$$E \exp(u S_n) \leq \exp(\tau_n^2 u^2 / 2(1 - M u/3)) = \exp(uv/2).$$

So by Theorem 1.10, $\Pr\{S_n \geq v\} \leq e^{-uv/2}$ and

$$e^{-uv/2} = \exp(-v^2/(2\tau_n^2 + 2Mv/3)) = \exp(-nK^2/(2\tau_n^2 + 2MKn^{1/2}/3)). \quad \square$$

Here are some remarks on Bernštein's inequality. Note that for fixed K and M, if X_i are i.i.d. with variance σ^2, then as $n \to \infty$, the bound approaches the normal bound $2 \cdot \exp(-K^2/(2\sigma^2))$, as given in RAP, Lemma 12.1.6. Moreover, this is true even if $M := M_n \to \infty$ as $n \to \infty$ while K stays constant, provided that $M_n/n^{1/2} \to 0$. Sometimes, the inequality can be applied to unbounded variables X_j, replacing them by truncated ones, say replacing X_j by $f_{M_n}(X_j)$ where $f_M(x) := x 1_{\{|x| \leq M\}}$. In that case the probability

$$\Pr(|X_j| > M_n \text{ for some } j \leq n) \leq \sum_{j=1}^{n} \Pr(|X_j| > M_n)$$

needs to be small enough so that the inequality with this additional probability added to the bound is still useful.

Next, let $s_1, s_2, \ldots,$ be i.i.d. variables with $P(s_i = 1) = P(s_i = -1) = 1/2$. Such variables are called "Rademacher" variables. We have the following inequality:

Proposition 1.12 (Hoeffding) *For any $t \geq 0$, and real a_j not all 0,*

$$\Pr\left\{ \sum_{j=1}^{n} a_j s_j \geq t \right\} \leq \exp\left(-t^2 / \left(2 \sum_{j=1}^{n} a_j^2 \right) \right).$$

Proof. Since $1/(2n)! \leq 2^{-n}/n!$ for $n = 0, 1, \ldots,$ we have $\cosh x \equiv (e^x + e^{-x})/2 \leq \exp(x^2/2)$ for all x. Applying Theorem 1.10, the probability on the left is bounded above by $\inf_u \exp(-ut + \sum_{j=1}^{n} a_j^2 u^2/2)$, which by calculus is attained at $u = t/\sum_{j=1}^{n} a_j^2$, and the result follows. $\quad \square$

Proposition 1.12 can be applied as follows. Let $Y_1, Y_2, \ldots,$ be independent variables which are symmetric, in other words Y_j has the same distribution as $-Y_j$ for all j. Let s_i be Rademacher variables independent of each other and of all the Y_j. Then the sequence $\{s_j Y_j\}_{\{j \geq 1\}}$ has the same distribution as $\{Y_j\}_{\{j \geq 1\}}$. Thus to bound the probability that $\sum_{j=1}^{n} Y_j > K$, for example, we can consider the conditional probability for each $Y_1, \ldots, Y_n$,

$$\Pr\{\textstyle\sum_{j=1}^{n} s_j Y_j > K | Y_1, \ldots, Y_n\} \leq \exp(-K^2/(2 \textstyle\sum_{j=1}^{n} Y_j^2))$$

by Proposition 1.12. Then to bound the original probability, integrating over the distribution of the Y_j, one just needs to have bounds on the distribution of $\sum_{j=1}^{n} Y_j^2$, which may simplify the problem considerably.

The Bernštein inequality (Theorem 1.11) used variances as well as bounds for centered variables. The following inequalities, also due to Hoeffding, use only bounds. They are essentially the best that can be obtained, under their hypotheses, by the moment generating function technique.

Theorem 1.13 (Hoeffding) *Let* $X_1, \ldots, X_n$ *be independent random variables such that* $0 \le X_j \le 1$ *for all* j. *Let* $\overline{X} := (X_1 + \cdots + X_n)/n$ *and* $\mu := E\overline{X}$. *Then for* $0 < t < 1 - \mu$,

$$\Pr\{\overline{X} - \mu \ge t\} \le \left\{ \left(\frac{\mu}{\mu + t} \right)^{\mu + t} \left(\frac{1 - \mu}{1 - \mu - t} \right)^{1 - \mu - t} \right\}^n \le e^{-nt^2 g(\mu)},$$

where

$$g(\mu) := (1 - 2\mu)^{-1} \log((1 - \mu)/\mu) \quad \text{for } 0 < \mu < 1/2,$$

$$:= 1/(2\mu(1 - \mu)) \quad \text{for } 1/2 \le \mu \le 1.$$

For all $t > 0$,

$$\Pr(\overline{X} - \mu \ge t) \le e^{-2nt^2}. \tag{1.4}$$

Remark. For $t < 0$, the given probability would generally be of the order of $1/2$ or larger, so no small bound for it would be expected.

Proof. For any $v > 0$, the function $f(x) := e^{vx}$ is convex (its second derivative is positive), so any chord lies above its graph (RAP, Section 6.3). Specifically, if $0 < x < 1$, then $e^{vx} \le 1 - x + xe^v$. Taking expectations gives $E \exp(vX_j) \le 1 - \mu_j + \mu_j e^v$, where $\mu_j := EX_j$. (Note that the latter inequality becomes an equation for a Bernoulli variable X_j, taking only the values 0, 1.) Let $S_n := X_1 + \cdots + X_n$. Then

$$\Pr(\overline{X} - \mu \ge t) = \Pr(S_n - ES_n \ge nt) \le E \exp(v(S_n - ES_n - nt))$$

$$= e^{-vn(t+\mu)} \Pi_{j=1}^n E \exp(vX_j) \le e^{-nv(t+\mu)} \Pi_{j=1}^n (1 - \mu_j + \mu_j e^v).$$

To continue the proof, the following well-known inequality (e.g., RAP (5.1.6) and p. 276) will be useful:

Lemma 1.14 *For any nonnegative real numbers* $t_1, \ldots, t_n$, *the geometric mean is less than or equal to the arithmetic mean, in other words,*

$$(t_1 t_2 \cdots t_n)^{1/n} \le (t_1 + \cdots + t_n)/n.$$

Applying the Lemma gives

$$\Pr\{\overline{X} - \mu > t\} \leq e^{-nv(t+\mu)} \left(\frac{1}{n} \sum_{j=1}^{n} 1 - \mu_j + \mu_j e^v \right)^n$$

$$\leq e^{-nv(t+\mu)}(1 - \mu + \mu e^v)^n.$$

To find the minimum of this for $v > 0$, note that it becomes large as $v \to \infty$ since $t + \mu < 1$, while setting the derivative with respect to v equal to 0 gives a single solution, where $1 - \mu + \mu e^v = \mu e^v/(t + \mu)$ and $e^v = 1 + t/(\mu(1 - \mu - t))$. Substituting these values into the bounds gives the first, most complicated bound in the statement of the theorem. This bound can be written as $\Pr(\overline{X} - \mu \geq t) \leq \exp(-nt^2 G(t, \mu))$, where

$$G(t, \mu) := \frac{\mu + t}{t^2} \log\left(\frac{\mu + t}{\mu}\right) + \left(\frac{1 - \mu - t}{t^2}\right) \log\left(\frac{1 - \mu - t}{1 - \mu}\right).$$

The next step is to show that $\min_{0 < t < 1} G(t, \mu) = g(\mu)$ as defined in the statement of the theorem. For $0 < x < 1$ let

$$H(x) := \left(1 - \frac{2}{x}\right) \log(1 - x).$$

In $\partial G(t, \mu)/\partial t$, the terms not containing logarithms cancel, giving

$$t^2 \frac{\partial G(t, \mu)}{\partial t} = \left[1 - \frac{2}{t}(1 - \mu)\right] \log\left(1 - \frac{t}{1 - \mu}\right)$$

$$- \left[1 - \frac{2}{t}(\mu + t)\right] \log\left(1 - \frac{t}{\mu + t}\right) = H\left(\frac{t}{1 - \mu}\right) - H\left(\frac{t}{\mu + t}\right).$$

To see that H is increasing in x for $0 < x < 1$, take the Taylor series of $\log(1 - x)$ and multiply by $1 - \frac{2}{x}$ to get the Taylor series of H around 0, all whose coefficients are nonnegative. Only the first order term is 0. Thus $\partial G/\partial t > 0$ if and only if $t/(1 - \mu) > t/(\mu + t)$, or equivalently $t > 1 - 2\mu$. So if $\mu < 1/2$, then $G(t, \mu)$, for fixed μ, has a minimum with respect to $t > 0$ at $t = 1 - 2\mu$, giving $g(\mu)$ for that case as stated. Or if $\mu \geq 1/2$, then $G(t, \mu)$ is increasing in $t > 0$, with $\lim_{t \downarrow 0} G(t, \mu) = g(\mu)$ as stated for that case, using the first two terms of the Taylor series around $t = 0$ of each logarithm.

To prove (1.4) for $0 < t < 1 - \mu$, it will be shown that the minimum of $g(\mu)$ for $0 < \mu < 1$ is 2. For $\mu \geq 1/2$, g is increasing (its denominator is decreasing), and $g(1/2) = 2$. For $\mu < 1/2$, letting $w := 1 - 2\mu$ we get $g(\mu) = \frac{1}{w} \log\left(\frac{1+w}{1-w}\right)$. From the Taylor series of $\log(1 + w)$ and $\log(1 - w)$ around $w = 0$, we see that g is increasing in w, and so decreasing in μ, and converges to 2 as $\mu \to 1/2$. Thus g has a minimum at $\mu = 1/2$, which is 2.

To prove (1.4) for $0 < t = 1 - \mu$, consider $0 < s < t$ and let $s \uparrow t$. For $t > 1 - \mu$, $\Pr(\overline{X} - \mu \geq t) \leq \Pr(\overline{X} > 1) = 0 < \exp(-2nt^2)$. So (1.4) is proved for all $t > 0$ and so is the theorem. $\qquad \square$

For the empirical measure P_n, if A is a fixed measurable set, $nP_n(A)$ is a binomial random variable, and in a multinomial distribution, each n_i has a binomial distribution. So we will have need of some inequalities for binomial probabilities, defined by

$$B(k, n, p) := \sum_{0 \le j \le k} \binom{n}{j} p^j q^{n-j}, \quad 0 \le q := 1 - p \le 1,$$

$$E(k, n, p) := \sum_{k \le j \le n} \binom{n}{j} p^j q^{n-j}.$$

Here k is usually, but not necessarily, an integer. Thus, in n independent trials with probability p of success on each trial, so that q is the probability of failure, $B(k, n, p)$ is the probability of at most k successes, and $E(k, n, p)$ is the probability of at least k successes.

Theorem 1.15 (Chernoff) *We have*

$$E(k, n, p) \le \left(\frac{np}{k}\right)^k \left(\frac{nq}{n-k}\right)^{n-k} \quad if \ k \ge np, \qquad (1.5)$$

$$B(k, n, p) \le \exp(-(np - k)^2/(2npq)) \quad if \ k \le np \le n/2. \qquad (1.6)$$

Proof. These facts follow directly from the Hoeffding inequality Theorem 1.13. For (1.6), note that $B(k, n, p) = E(n - k, n, 1 - p)$ and apply the $g(\mu)$ case with $\mu = 1 - p$. $\quad\square$

If in (1.5) we set $x := nq/(n - k) \le e^{x-1}$ it follows that

$$E(k, n, p) \le (np/k)^k e^{k-np} \quad if \ k \ge np. \qquad (1.7)$$

The next inequality is for the special value $p = 1/2$:

Proposition 1.16 *If* $k \le n/2$, *then* $2^n B(k, n, 1/2) \le (ne/k)^k$.

Proof. By (1.5) and symmetry, $B(k, n, 1/2) \le (n/2)^n k^{-k} (n - k)^{k-n}$. Letting $y := n/(n - k) \le e^{y-1}$ then gives the result. $\quad\square$

A form of Stirling's formula with error bounds is:

Theorem 1.17 *For* $n = 1, 2, \ldots,$ $e^{1/(12n+1)} \le n!(e/n)^n (2\pi n)^{-1/2} \le e^{1/12n}$.

Proof. See Feller (1968), vol. 1, Section II.9, p. 54. $\quad\square$

For any real x let $x^+ := \max(x, 0)$. A *Poisson* random variable z with parameter m has the distribution given by $\Pr(z = k) = e^{-m} m^k/k!$ for each nonnegative integer k.

Lemma 1.18 *For any Poisson variable z with parameter $m \geq 1$,*

$$E(z - m)^+ \geq m^{1/2}/8.$$

Proof. We have $E(z - m)^+ = \sum_{k>m} e^{-m} m^k (k - m)/k!$. Let $j := [m]$, meaning j is the greatest integer with $j \leq m$. Then by a telescoping sum (which is absolutely convergent), $E(z - m)^+ = e^{-m} m^{j+1}/j!$. Then by Stirling's formula with error bounds (Theorem 1.17),

$$E(z - m)^+ \geq e^{-m} m^{j+1} (e/j)^j (2\pi j)^{-1/2} e^{-1/(12j)}$$

$$\geq (m^{j+1}/j^{j+\frac{1}{2}}) e^{-13/12} (2\pi)^{-1/2} \geq m^{1/2}/8. \qquad \square$$

In the following two facts, let $X_1, X_2, \dots, X_n$ be independent random variables with values in a separable normed space S with norm $\|\cdot\|$. Let $S_j := X_1 + \dots + X_j$ for $j = 1, \dots, n$.

Theorem 1.19 *(Ottaviani's inequality). If for some $\alpha > 0$ and c with $0 < c < 1$, we have $P(\|S_n - S_j\| > \alpha) \leq c$ for all $j = 1, \dots, n$, then*

$$P\{\max_{j \leq n} \|S_j\| \geq 2\alpha\} \leq P(\|S_n\| \geq \alpha)/(1 - c).$$

Proof. The proof in RAP, 9.7.2, for $S = \mathbb{R}^k$, works for any separable normed S. Here $(x, y) \mapsto \|x - y\|$ is measurable: $S \times S \mapsto \mathbb{R}$ by RAP, Proposition 4.1.7. $\qquad \square$

When the random variables X_j are symmetric, there is a simpler inequality due to P. Lévy:

Theorem 1.20 (Lévy's inequality) *Given a probability space (Ω, P), let Y be a countable set, let $X_1, X_2, \dots$, be stochastic processes defined on Ω indexed by Y; in other words, for each j and $y \in Y$, $X_j(y)(\cdot)$ is a random variable on Ω. For any bounded function f on Y let $\|f\|_Y := \sup\{|f(y)|: y \in Y\}$. Suppose that the processes X_j are independent with $\|X_j\|_Y < \infty$ a.s., and symmetric; in other words, for each j, the random variables $\{-X_j(y): y \in Y\}$ have the same joint distribution as $\{X_j(y): y \in Y\}$. Let $S_n := X_1 + \dots + X_n$. Then for each n, and $M > 0$,*

$$P\left(\max_{j \leq n} \|S_j\|_Y > M\right) \leq 2P\left(\|S_n\|_Y > M\right).$$

Note. The norm on a separable Banach space $(X, \|\cdot\|)$ can always be written in the form $\|\cdot\|_Y$ for Y countable, via the Hahn–Banach theorem: apply RAP, Corollary 6.1.5, to a countable dense set in the unit ball of X to get a countable subset Y in the dual space X' of X which is norming, i.e. on X, $\|\cdot\| = \|\cdot\|_Y$. Here X' may not be separable. On the other hand, the above Lemma applies to some nonseparable Banach spaces: the space of all bounded functions on an infinite Y with supremum norm is itself nonseparable.

Proof. Let $M_k(\omega) := \max_{j \le k} \|S_j\|_Y$. Let C_k be the disjoint events $\{M_{k-1} \le M < M_k\}$, $k = 1, 2, \ldots$, where we set $M_0 := 0$. Then for $1 \le m \le n$, $2\|S_m\|_Y \le \|S_n\|_Y + \|2S_m - S_n\|_Y$. So if $\|S_m\|_Y > M$, then $\|S_n\|_Y > M$ or $\|2S_m - S_n\|_Y > M$ or both. The transformation which interchanges X_j and $-X_j$ just for $m < j \le n$ preserves probabilities, by symmetry and independence. Then S_n is interchanged with $2S_m - S_n$, while X_j are preserved for $j \le m$. So $P(C_m \cap \{\|S_n\|_Y > M\}) = P(C_m \cap \{\|2S_m - S_n\|_Y > M\}) \ge P(C_m)/2$, and $P(M_n > M) = \sum_{m=1}^{n} P(C_m) \le 2P(\|S_n\|_Y > M)$. $\qquad\square$

1.4 *Proof of the Bretagnolle–Massart Theorem

The proof is based essentially on some lemmas, which G. Tusnády stated, on approximating symmetric binomial random variables by normal variables. Let $\mathcal{B}(n, 1/2)$ denote the symmetric binomial distribution for n trials. Thus if B_n has this distribution, B_n is the number of successes in n independent trials with probability $1/2$ of success on each trial. For any distribution function F, recall Proposition 1.1 and the *quantile function* $F^{\leftarrow}$ from $(0, 1)$ into $\mathbb{R}$ defined and used in it. If F is continuous and strictly increasing on some interval $[a, b]$, with $F(a) = 0$ and $F(b) = 1$, then $F^{\leftarrow} = F^{-1}$ from $(0, 1)$ onto (a, b). In general, whereas F is right-continuous, $F^{\leftarrow}$ is always left-continuous.

Here is one of Tusnády's lemmas (Lemma 4 of Bretagnolle and Massart 1989). Its proof will be completed in Subsection 1.4.2.

Lemma 1.21 *Let Φ be the standard normal distribution function and Y a standard normal random variable. Let Φ_n be the distribution function of $\mathcal{B}(n, 1/2)$ and set $C_n := \Phi_n^{\leftarrow}(\Phi(Y)) - n/2$. Then*

$$|C_n| \le 1 + (\sqrt{n}/2)|Y|, \tag{1.8}$$

$$|C_n - (\sqrt{n}/2)Y| \le 1 + Y^2/8. \tag{1.9}$$

By Proposition 1.1(c), if V has a $U[0, 1]$ distribution, $F^{\leftarrow}(V)$ has distribution function F. If F is continuous, we have the following in the other direction:

Proposition 1.22 *Let X be a real random variable with distribution function F. If F is continuous, then $F(X)$ has a $U[0, 1]$ distribution.*

Proof. For $0 < y < 1$, $F^{-1}(\{y\})$ is a closed interval $[a, b]$ by continuity. (It is a singleton, $a = b$, usually, but not necessarily for all y.) We have $P(F(X) =$

$y) = P(a \leq X \leq b) = F(b) - F(a-)$ where $F(a-) = \lim_{u \uparrow a} F(u) = F(a)$ because F is continuous. Now $F(b) - F(a) = y - y = 0$. Thus

$$P(F(X) \leq y) = P(F(X) < y) = 1 - P(F(X) \geq y) = 1 - P(X \geq F^{\leftarrow}(y))$$
$$= 1 - P(X \geq a) = P(X < a) = F(a-) = F(a) = y,$$

again by continuity of F, so indeed $F(X)$ has a $U[0, 1]$ distribution, completing the proof. $\qquad \square$

Thus in Lemma 1.21, $\Phi(Y)$ has a $U[0, 1]$ distribution and $\Phi_n^{\leftarrow}(\Phi(Y))$ has the symmetric binomial distribution $\mathcal{B}(n, 1/2)$. Lemma 1.21 will be shown (by a relatively short proof) to follow from:

Lemma 1.23 *Let Y be a standard normal variable and let β_n be a binomial random variable with distribution $\mathcal{B}(n, 1/2)$. Then for any integer j such that $0 \leq j \leq n$ and $n + j$ is even, we have*

$$P(\beta_n \geq (n + j)/2) \geq P(\sqrt{n}Y/2 \geq n(1 - \sqrt{1 - j/n})), \qquad (1.10)$$
$$P(\beta_n \geq (n + j)/2) \leq P(\sqrt{n}Y/2 \geq (j - 2)/2). \qquad (1.11)$$

The following form of Stirling's formula with remainder is used in the proof of Lemma 1.23.

Lemma 1.24 *Let $n! = (n/e)^n \sqrt{2\pi n} A_n$ where $A_n = 1 + \beta_n/(12n)$, which defines A_n and β_n for $n = 1, 2, \ldots$. Then $\beta_n \downarrow 1$ as $n \to \infty$.*

1.4.1 *Stirling's Formula: Proof of Lemma 1.24*

It can be checked directly that $\beta_1 > \beta_2 > \cdots > \beta_8 > 1$. So it suffices to prove the lemma for $n \geq 8$. We have $A_n = \exp((12n)^{-1} - \theta_n/(360n^3))$ where $0 < \theta_n < 1$; see Whittaker and Watson (1927, p. 252) or Nanjundiah (1959). Then by Taylor's theorem with remainder,

$$A_n = \left(1 + \frac{1}{12n} + \frac{1}{288n^2} + \frac{1}{6(12n)^3}\phi_n e^{1/12n}\right) \exp(-\theta_n/(360n^3))$$

where $0 < \phi_n < 1$. Next,

$$\beta_{n+1} \leq 12(n + 1)\left[\exp\left(\frac{1}{12(n + 1)}\right) - 1\right]$$

$$\leq 1 + \frac{1}{24(n + 1)} + \frac{1}{6(12(n + 1))^2}e^{1/(12(n+1))},$$

from which $\limsup_{n\to\infty} \beta_n \le 1$, and

$$\beta_n = 12n[A_n - 1] \ge 12n\left[\left(1 + \frac{1}{12n} + \frac{1}{288n^2}\right)\exp(-1/(360n^3)) - 1\right].$$

Using $e^{-x} \ge 1 - x$ gives

$$\beta_n \ge 12n\left[\frac{1}{12n} + \frac{1}{288n^2} - \frac{1}{360n^3}\left(1 + \frac{1}{12n} + \frac{1}{288n^2}\right)\right]$$

$$= 1 + \frac{1}{24n} - \frac{1}{30n^2}\left(1 + \frac{1}{12n} + \frac{1}{288n^2}\right).$$

Thus $\liminf_{n\to\infty} \beta_n \ge 1$ and $\beta_n \to 1$ as $n \to \infty$. To prove $\beta_n \ge \beta_{n+1}$ for $n \ge 8$ it will suffice to show that

$$1 + \frac{1}{24(n+1)} + \frac{e^{1/108}}{6 \cdot 144n^2} \le 1 + \frac{1}{24n} - \frac{1}{30n^2}\left[1 + \frac{1}{96} + \frac{1}{288 \cdot 8^2}\right]$$

or

$$\frac{e^{1/108}}{6 \cdot 144n^2} + \frac{1}{30n^2}\left[1 + \frac{1}{96} + \frac{1}{288 \cdot 64}\right] \le \frac{1}{24n(n+1)},$$

or that $0.035/n^2 \le 1/[24n(n+1)]$ or $0.84 \le 1 - 1/(n+1)$, which holds for $n \ge 8$, proving that β_n decreases with n. Since its limit is 1, Lemma 1.24 is proved. $\qquad\square$

1.4.2 *Proof of Lemma 1.23*

First, (1.10) will be proved. For any $i = 0, 1, \ldots, n$ such that $n + i$ is even, let $k := (n+i)/2$ so that k is an integer, $n/2 \le k \le n$, and $i = 2k - n$. Let $p_{ni} := P(\beta_n = (n+i)/2) = P(\beta_n = k) = \binom{n}{k}/2^n$ and $x_i := i/n$. Define $p_{ni} := 0$ for $n + i$ odd. The factorials in $\binom{n}{k}$ will be approximated via Stirling's formula with correction terms as in Lemma 1.24. To that end, let

$$CS(u, v, w, x, n) := \frac{1 + u/(12n)}{(1 + v/[6n(1-x)])(1 + w/[6n(1+x)])}.$$

By Lemma 1.24, we can write for $0 \le i < n$ and $n + i$ even

$$p_{ni} = CS(x_i, n)\sqrt{2/\pi n}\,\exp(-ng(x_i)/2 - (1/2)\log(1 - x_i^2)) \quad (1.12)$$

where $g(x) := (1+x)\log(1+x) + (1-x)\log(1-x)$ and $CS(x_i, n) := CS(\beta_n, \beta_{n-k}, \beta_k, x_i, n)$. By Lemma 1.24 and since $k \ge n/2$,

$$1^+ := 1.013251 \ge 12(e(2\pi)^{-1/2} - 1) = \beta_1 \ge \beta_{n-k} \ge \beta_k \ge \beta_n > 1.$$

Thus, for $x := x_i$, by clear or easily checked monotonicity properties,

$$CS(x, n) \leq CS(\beta_n, \beta_k, \beta_k, x, n)$$

$$= \left(1 + \frac{\beta_n}{12n}\right)\left[1 + \frac{\beta_k}{3n(1 - x^2)} + \frac{\beta_k^2}{36n^2(1 - x^2)}\right]^{-1}$$

$$\leq CS(\beta_n, \beta_k, \beta_k, 0, n) \leq CS(\beta_n, \beta_n, \beta_n, 0, n)$$

$$\leq CS(1, 1, 1, 0, n) = \left(1 + \frac{1}{12n}\right)\left[1 + \frac{1}{3n} + \frac{1}{36n^2}\right]^{-1}.$$

It will be shown next that $\log(1 + y) - 2\log(1 + 2y) \leq -3y + 7y^2/2$ for $y \geq 0$. Both sides vanish for $y = 0$. Differentiating and clearing fractions, we get a clearly true inequality. Setting $y := 1/(12n)$ then gives

$$\log CS(x_i, n) \leq -1/(4n) + 7/(288n^2). \tag{1.13}$$

To get a lower bound for $CS(x, n)$ we have by an analogous string of inequalities

$$CS(x, n) \geq \left(1 + \frac{1}{12n}\right)\left\{1 + \frac{1^+}{3n(1 - x^2)} + \frac{(1^+)^2}{36n^2(1 - x^2)}\right\}^{-1}. \tag{1.14}$$

The inequality (1.10) to be proved can be written as

$$\sum_{i=j}^{n} p_{ni} \geq 1 - \Phi(2\sqrt{n}(1 - \sqrt{1 - j/n})). \tag{1.15}$$

When $j = 0$ the result is clear. When $n \leq 4$ and $j = n$ or $n - 2$ the result can be checked from tables of the normal distribution. Thus we can assume from here on

$$n \geq 5. \tag{1.16}$$

Case 1. Let $j^2 \geq 2n$, in other words $x_j \geq \sqrt{2/n}$. Recall that for $t > 0$ we have $P(Y > t) \leq (t\sqrt{2\pi})^{-1}\exp(-t^2/2)$, e.g. Dudley (2002), Lemma 12.1.6(a). Then (1.15) follows easily when $j = n$ and $n \geq 5$. To prove it for $j = n - 2$ it is enough to show

$$n(2 - \log 2) - 4\sqrt{2n} + \log(n + 1) + 4 + \log[2\sqrt{2\pi}(\sqrt{n} - \sqrt{2})] \geq 0, \quad n \geq 5.$$

The left side is increasing in n for $n \geq 5$ and is ≥ 0 at $n = 5$.

For $5 \leq n \leq 7$ we have $(n - 4)^2 < 2n$, so we can assume in the present case that $2n \leq j^2 \leq (n - 4)^2$ and $n \geq 8$. Let $y_i := 2\sqrt{n}(1 - \sqrt{1 - i/n})$. Then it will suffice to show

$$p_{ni} \geq \int_{y_i}^{y_{i+2}} \phi(u)du, \quad i = j, j + 1, \ldots, n - 4, \tag{1.17}$$

where ϕ is the standard normal density function. Let

$$f_n(x) := \sqrt{n/2\pi(1-x)}\exp(-2n(1-\sqrt{1-x})^2). \qquad (1.18)$$

By the change of variables $u = 2\sqrt{n}(1 - \sqrt{1-x})$, (1.17) becomes

$$p_{ni} \geq \int_{x_i}^{x_{i+2}} f_n(x)dx. \qquad (1.19)$$

Clearly $f_n > 0$. To see that $f_n(x)$ is decreasing in x for $\sqrt{2/n} \leq x \leq 1 - 4/n$, note that

$$2(1-x)f_n'/f_n = 1 - 4n[\sqrt{1-x} - 1 + x],$$

so f_n is decreasing where $\sqrt{1-x} - (1-x) > 1/(4n)$. We have $\sqrt{y} - y \geq y$ for $y \leq 1/4$, so $\sqrt{y} - y > 1/(4n)$ for $1/(4n) < y \leq 1/4$. Let $y := 1 - x$. Also $\sqrt{1-x} - (1-x) > x/4$ for $x < 8/9$, so $\sqrt{1-x} - (1-x) > 1/(4n)$ for $1/n < x < 8/9$. Thus $\sqrt{1-x} - (1-x) > 1/(4n)$ for $1/n < x < 1 - 1/(4n)$, which includes the desired range. Thus to prove (1.19) it will be enough to show that

$$p_{ni} \geq (2/n)f_n(x_i), \quad i = j, j+2, \ldots, n-4. \qquad (1.20)$$

So by (1.12) it will be enough to show that for $\sqrt{2/n} \leq x \leq 1 - 4/n$ and $n \geq 8$,

$$CS(x,n)(1+x)^{-1/2}\exp[n\{4(1-\sqrt{1-x})^2 - g(x)\}/2] \geq 1. \qquad (1.21)$$

Let

$$J(x) := 4(1-\sqrt{1-x})^2 - g(x). \qquad (1.22)$$

Then J is increasing for $0 < x < 1$, since its first and second derivatives are both 0 at 0, while its third derivative is easily checked to be positive on $(0, 1)$. In light of (1.14), to prove (1.21) it suffices to show that

$$\left(1 + \frac{1}{12n}\right)e^{nJ(x)/2} \geq \sqrt{1+x}\left(1 + \frac{1^+}{3n(1-x^2)} + \frac{(1^+)^2}{36n^2(1-x^2)}\right). \qquad (1.23)$$

When $x \leq 1 - 4/n$ and $n \geq 8$ the right side is less than 1.5, using first $\sqrt{1+x} \leq \sqrt{2}$, next $x \leq 1 - 4/n$, and lastly $n \geq 8$. For $x \geq 0.55$ and $n \geq 8$ the left side is larger than 1.57, so (1.23) is proved for $x \geq 0.55$. We will next need the inequality

$$J(x) \geq x^3/2 + 7x^4/48, \quad 0 \leq x \leq 0.55. \qquad (1.24)$$

To check this one can calculate $J(0) = J'(0) = J''(0) = 0$, $J^{(3)}(0) = 3$, $J^{(4)}(0) = 7/2$, so that the right side of (1.24) is the Taylor series of J around 0 through fourth order. One then shows straightforwardly that $J^{(5)}(x) > 0$ for $0 \leq x < 1$.

It follows since $nx^2 \geq 2$ and $n \geq 8$ that $nJ(x)/2 \geq x/2 + 7/24n$. Let $K(x) := \exp(x/2)/\sqrt{1+x}$ and $\kappa(x) := (K(x)-1)/x^2$. We will next see that $\kappa(\cdot)$ is decreasing on $[0,1]$. To show $\kappa' \leq 0$ is equivalent to $e^{x/2}[4 + 4x - x^2] \geq 4(1+x)^{3/2}$, which is true at $x = 0$. Differentiating, we would like to show $e^{x/2}[6 - x^2/2] \geq 6\sqrt{1+x}$, or squaring that and multiplying by 4, $e^x(144 - 24x^2 + x^4) \geq 144(1+x)$. This is true at $x = 0$. Differentiating, we would like to prove $e^x(144 - 48x - 24x^2 + 4x^3 + x^4) \geq 144$. Using $e^x \geq 1 + x$ and algebra gives this result for $0 \leq x \leq 1$.

It follows that $K(x) \geq 1 + 0.3799/n$ when $\sqrt{2/n} \leq x \leq 0.55$. It remains to show that for $x \leq 0.55$,

$$\left(1 + \frac{1}{12n}\right)\left(1 + \frac{0.3799}{n}\right)e^{7/(24n)} \geq 1 + \frac{1^+}{3n(1-x^2)} + \frac{(1^+)^2}{36n^2(1-x^2)}.$$

At $x = 0.55$ the right side is less than $1 + 0.543/n$, so Case 1 is completed since $0.543 \leq 1/12 + 0.3799 + 7/24$.

Case 2. The remaining case is $j < \sqrt{2n}$. For any integer k, $P(\beta_n \geq k) = 1 - P(\beta_n \leq k - 1)$. For $k = (n+j)/2$ we have $k - 1 = (n + j - 2)/2$. If n is odd, then $P(\beta_n \geq n/2) = 1/2 = P(Y \geq 0)$. If n is even, then $P(\beta_n \geq n/2) - p_{n0}/2 = 1/2 = P(Y \geq 0)$. So, since $p_{n0} = 0$ for n odd, (1.10) is equivalent to

$$\frac{1}{2}p_{n0} + \sum_{0 < i \leq j-2} p_{ni} \leq P(0 \leq Y \leq 2\sqrt{n}(1 - \sqrt{1 - j/n})). \quad (1.25)$$

Given $j < \sqrt{2n}$, a family $I_0, I_1, \ldots, I_K$ of adjacent intervals will be defined such that for n odd,

$$p_{ni} \leq P(\sqrt{n}Y/2 \in I_k) \text{ with } i = 2k+1, \ 0 \leq k \leq K := (j-3)/2, \quad (1.26)$$

while for n even,

$$p_{ni} \leq P(\sqrt{n}Y/2 \in I_k) \text{ with } i = 2k, \ 1 \leq k \leq K := (j-2)/2, \quad (1.27)$$

and

$$p_{n0}/2 \leq P(\sqrt{n}Y/2 \in I_0). \quad (1.28)$$

In either case,

$$I_0 \cup I_1 \cup \cdots \cup I_K \subset [0, n(1 - \sqrt{1 - j/n})]. \quad (1.29)$$

The intervals will be defined by

$$\delta_{k+1} := (k+1)/n + k(k+1/2)(k+1)/n^{3/2}, \ k \geq 0, \quad (1.30)$$

$$\Delta_{k+1} := \delta_{k+1} + k + 1/2 = \delta_{k+1} + (i+1)/2, \ i = 2k, \ n \text{ even}, \quad (1.31)$$

$$\Delta_{k+1} := \delta_{k+1} + k + 1 = \delta_{k+1} + (i+1)/2, \ i = 2k+1, \ n \text{ odd}, \quad (1.32)$$

$$I_k := [\Delta_k, \Delta_{k+1}] \text{ with } \Delta_0 = 0. \quad (1.33)$$

It will be shown that $I_0, I_1, \ldots, I_K$ defined by (1.30) through (1.33) satisfy (1.26) through (1.29). Recall that $n \geq 5$ (1.16) and $x_i := i/n$.

Proof of (1.29). It needs to be shown that $\Delta_{K+1} \leq n(1 - \sqrt{1 - x_j})$. Since $j < \sqrt{2n}$, we have $K \leq j/2 - 1 < \sqrt{n/2} - 1$ and

$$\delta_{K+1} \leq (K + 1)/n + K(K + 1/2)/(n\sqrt{2}) \leq x_j/2 + nx_j^2/(4\sqrt{2}).$$

We have $\Delta_{K+1} = nx_j/2 - 1/2 + \delta_{K+1}$. It will be shown next that

$$1 - \sqrt{1 - x} \geq x/2 + x^2/8, \quad 0 \leq x \leq 1. \tag{1.34}$$

The functions and their first derivatives agree at 0 while the second derivative of the left side is clearly larger.

It then remains to prove that

$$1/2 + nx_j^2(1/8 - 1/4\sqrt{2}) - x_j/2 \geq 0.$$

This is true since $nx_j^2 \leq 2$ and $x_j \leq (2/8)^{1/2} = 1/2$, so (1.29) is proved.

Proof of (1.26)–(1.28). First it will be proved that

$$p_{ni} \leq \frac{\sqrt{2}}{\sqrt{\pi n}} \exp\left[-\frac{1}{4n} + \frac{7}{288n^2} - \frac{(n-1)i^2}{2n^2} + \frac{(i/n)^{2n}}{2n(1 - i^2/n^2)}\right]. \tag{1.35}$$

In light of (1.12) and (1.13), it is enough to prove, for $x := i/n$, that

$$-[ng(x) + \log(1 - x^2) - (n - 1)x^2]/2 \leq x^{2n}/2n(1 - x^2). \tag{1.36}$$

It is easy to verify that for $0 \leq t < 1$,

$$g(t) = (1 + t)\log(1 + t) + (1 - t)\log(1 - t) = \sum_{r=1}^{\infty} t^{2r}/r(2r - 1).$$

Thus the left side of (1.36) can be expanded as $\sum_{r \geq 2} x^{2r}(1 - n/(2r - 1))/2r = A + B$ where $A = \sum_{r=2}^{n-1}$ and $B = \sum_{r \geq n}$. We have

$$d^2A/dx^2 = \sum_{2 \leq r \leq (n+1)/2} (2r - n - 1)(x^{2r-2} - x^{2n-2r}),$$

which is ≤ 0 for $0 \leq x \leq 1$. Since $A = dA/dx = 0$ for $x = 0$ we have $A \leq 0$ for $0 \leq x \leq 1$. Then, $2nB \leq x^{2n}/(1 - x^2)$, so (1.35) is proved.

We have for $n \geq 5$ and $x \leq (\sqrt{2n} - 2)/n$ that $x^{2n}/(1 - x^2) < 10^{-3}$, since $n \mapsto (\sqrt{2n} - 2)/n$ is decreasing in n for $n \geq 8$ and the statement can be checked for $n = 5, 6, 7, 8$. So (1.35) yields

$$p_{ni} \leq \sqrt{2/\pi n} \exp[-0.249/n + 7/288n^2 - (n - 1)i^2/2n^2]. \tag{1.37}$$

Next we will need:

Lemma 1.25 *For any $0 \leq a < b$ and a standard normal variable Y,*

$$P(Y \in [a, b]) \geq \sqrt{1/2\pi}(b - a)\exp[-a^2/4 - b^2/4]\phi(a, b) \quad (1.38)$$

where $\phi(a, b) := [4/(b^2 - a^2)]\sinh[(b^2 - a^2)/4] \geq 1.$

Proof. Since the Taylor series of sinh around 0 has all coefficients positive, and $(\sinh u)/u$ is an even function, clearly $\sinh u/u \geq 1$ for any real u. The conclusion of the lemma is equivalent to

$$\frac{a + b}{2} \int_a^b \exp(-u^2/2)du \geq \exp(-a^2/2) - \exp(-b^2/2). \quad (1.39)$$

Letting $x := b - a$ and $v := u - a$ we need to prove

$$\left(a + \frac{x}{2}\right) \int_0^x \exp(-av - v^2/2)dv \geq 1 - \exp(-ax - x^2/2).$$

This holds for $x = 0$. Taking derivatives of both sides and simplifying, we would like to show

$$\int_0^x \exp(-av - v^2/2)dv \geq x \exp(-ax - x^2/2).$$

This also holds for $x = 0$, and differentiating both sides leads to a clearly true inequality, so Lemma 1.25 is proved. $\qquad\square$

For the intervals I_k, Lemma 1.25 yields

$$P(\sqrt{n}Y/2 \in I_k) \geq \sqrt{2/\pi n}\phi_k \exp[-(\Delta_{k+1}^2 + \Delta_k^2)/n + \log(\Delta_{k+1} - \Delta_k)] \quad (1.40)$$

where $\phi_k := \phi(2\Delta_k/\sqrt{n}, 2\Delta_{k+1}/\sqrt{n})$. The aim is to show that the ratio of the bounds (1.40) over (1.37) is at least 1.

First consider the case $k = 0$. If n is even, this means we want to prove (1.28). Using (1.37) and (1.40) and $\phi_0 \geq 1$, it suffices to show that

$$0.249/n - 7/288n^2 - 1/4n - 1/n^2 - 1/n^3 + \log(1 + 2/n) \geq 0.$$

Since $\log(1 + u) \geq u - u^2/2$ for $u \geq 0$ by taking a derivative, it will be enough to show that

$$(E)_n := 1.999/n - 3/n^2 - 7/288n^2 - 1/n^3 \geq 0,$$

and it is easily checked that $n(E)_n > 0$ since $n \geq 5$.

If n is odd, then (1.37) applies for $i = 2k + 1 = 1$, and we have $\Delta_0 = 0$, $\Delta_1 = \delta_1 + 1 = 1 + 1/n$, so (1.40) yields

$$P(\sqrt{n}Y/2 \in I_0) \geq \sqrt{2/\pi n}\exp[-(1 + 1/n)^2/n + \log(1 + 1/n)].$$

Using $\log(1 + u) \geq u - u^2/2$ again, the desired inequality can be checked since $n \geq 5$. This completes the case $k = 0$.

Now suppose $k \geq 1$. In this case, $i < \sqrt{2n} - 2$ implies $n \geq 10$ for n even and $n \geq 13$ for n odd. Let $s_k := \delta_k + \delta_{k+1}$ and $d_k := \delta_{k+1} - \delta_k$. Then for i as in the definition of Δ_{k+1},

$$\Delta_{k+1} + \Delta_k = i + s_k, \tag{1.41}$$

$$\Delta_{k+1} - \Delta_k = 1 + d_k, \tag{1.42}$$

$$s_k = \frac{2k+1}{n} + \frac{2k^3 + k}{n^{3/2}}, \tag{1.43}$$

and

$$d_k = \frac{1}{n} + \frac{3k^2}{n^{3/2}}. \tag{1.44}$$

From the Taylor series of sinh around 0 one easily sees that $(\sinh u)/u \geq 1 + u^2/6$ for all u. Letting $u := (\Delta_{k+1}^2 - \Delta_k^2)/n \geq i/n$ gives

$$\log \phi_k \geq \log(1 + i^2/6n^2). \tag{1.45}$$

We have

$$d_k \leq 3/(2\sqrt{n}) \tag{1.46}$$

since $2k \leq \sqrt{2n} - 2$ and $n \geq 10$. Next we have another lemma:

Lemma 1.26 $\log(1 + x) \geq \lambda x$ for $0 \leq x \leq \alpha$ for each of the pairs $(\alpha, \lambda) = (0.207, 0.9), (0.195, 0.913), (0.14, 0.93), (0.04, 0.98)$.

Proof. Since $x \mapsto \log(1 + x)$ is concave, or equivalently we are proving $1 + x \geq e^{\lambda x}$ where the latter function is convex, it suffices to check the inequalities at the endpoints, where they hold. $\square$

Lemma 1.26 and (1.45) then give

$$\log \phi_k \geq 0.98 i^2/6n^2 \tag{1.47}$$

since $i^2/(6n^2) \leq 1/3n \leq 0.04$, $n \geq 10$. Next,

Lemma 1.27 *We have* $\log(\Delta_{k+1} - \Delta_k) \geq \lambda d_k$ *where* $\lambda = 0.9$ *when n is even and $n \geq 20$, $\lambda = 0.93$ when n is odd and $n \geq 25$, and $\lambda = 0.913$ when $k = 1$ and $n \geq 10$. Only these cases are possible (for $k \geq 1$).*

Proof. If n is even and $k \geq 2$, then $4 \leq i = 2k < \sqrt{2n} - 2$ implies $n \geq 20$. If n is odd and $k \geq 2$, then $5 \leq i = 2k + 1 < \sqrt{2n} - 2$ implies $n \geq 25$. So only the given cases are possible.

We have $k \leq k_n := \sqrt{n/2} - 1$ for n even or $k_n := \sqrt{n/2} - 3/2$ for n odd. Let $d(n) := 1/n + 3k_n^2/n^{3/2}$ and $t := 1/\sqrt{n}$. It will be shown that $d(n)$ is decreasing in n, separately for n even and odd. For n even we would like to show that $3t/2 + (1 - 3\sqrt{2})t^2 + 3t^3$ is increasing for $0 \leq t \leq 1/\sqrt{20}$, and in fact its derivative is > 0.04. For n odd we would like to show that $3t/2 + (1 - 9/\sqrt{2})t^2 + 27t^3/4$ is increasing. We find that its derivative has no real roots and so is always positive as desired.

Since $d(\cdot)$ is decreasing for $n \geq 20$, its maximum for n even, $n \geq 20$ is at $n = 20$, and we find it is less than 0.207, so Lemma 1.26 applies to give $\lambda = 0.9$. Similarly for n odd and $n \geq 25$ we have the maximum $d(25) < 0.14$, and Lemma 1.26 applies to give $\lambda = 0.93$.

If $k = 1$, then $n \mapsto n^{-1} + 3/n^{3/2}$ is clearly decreasing. Its value at $n = 10$ is less than 0.195 and Lemma 1.26 applies with $\lambda = 0.913$. So Lemma 1.27 is proved. $\square$

It will next be shown that for $n \geq 10$

$$s_k \leq n^{-1} + k/\sqrt{n}. \tag{1.48}$$

By (1.43) this is equivalent to $2/\sqrt{n} + (2k^2 + 1)/n \leq 1$. Since $k \leq \sqrt{n/2} - 1$ one can check that (1.48) holds for $n \geq 14$. For $n = 10, 11, 12, 13$ note that k is an integer; in fact $k \leq 1$, and (1.48) holds.

After some calculations, letting $s := s_k$ and $d := d_k$ and noting that

$$\Delta_k^2 + \Delta_{k+1}^2 = \frac{1}{2}[(\Delta_{k+1} - \Delta_k)^2 + (\Delta_k + \Delta_{k+1})^2],$$

to show that the ratio of (1.40) to (1.37) is at least 1 is equivalent to showing that

$$-\frac{is}{n} - \frac{d}{n} - \frac{s^2}{2n} - \frac{d^2}{2n} - \frac{1}{2n} - \frac{7}{288n^2} - \frac{i^2}{2n^2} + \frac{0.249}{n}$$
$$+ \log(1 + d) + \log \phi_k \geq 0. \tag{1.49}$$

Proof of (1.49). First suppose that n is even and $n \geq 20$ or n is odd and $n \geq 25$. Apply the bound (1.46) for $d^2/2n$, (1.47) for $\log \phi_k$, (1.48) for s, and Lemma 1.27 for $\log(1 + d)$. Apply the exact value (1.44) of d in the d/n and λd terms. We assemble together terms with factors k^2, k, and no factor of k, getting a lower bound A for (1.49) of the form

$$A := \alpha[k^2/n^{3/2}] - 2\beta[k/n^{5/4}] + \gamma[1/n] \tag{1.50}$$

where, if n is even, so $i = 2k$ and $\lambda = 0.9$, we get

$$\alpha = 0.7 - [2.5 - 2(0.98)/3]/\sqrt{n} - 3/n,$$
$$\beta = n^{-3/4} + n^{-5/4}/2,$$
$$\gamma = 0.649 - [17/8 + 7/288]/n - 1/2n^2.$$

Note that for each fixed n, A is $1/n$ times a quadratic in $k/n^{1/4}$. Also, α and γ are increasing in n while β is decreasing. Thus for $n \geq 20$ the supremum of

$\beta^2 - \alpha\gamma$ is attained at $n = 20$ where it is < -0.06. So the quadratic has no real roots, and since $\alpha > 0$ it is always positive, thus (1.49) holds.

When n is odd, $i = 2k + 1$, $\lambda = 0.93$, and $n \geq 25$. We get a lower bound A for (1.49) of the same form (1.50) where now

$$\alpha = 0.79 - [2.5 - 2(0.98)/3]/\sqrt{n} - 3/n,$$

$$\beta = 1/2n^{1/4} + 2(1 - 0.98/6)/n^{3/4} + 1/2n^{5/4},$$

$$\gamma = 0.679 - (3.625 + 7/288 - 0.98/6)/n - 1/2n^2.$$

For the same reasons, the supremum of $\beta^2 - \alpha\gamma$ for $n \geq 25$ is now attained at $n = 25$ and is negative (less than -0.015), so the conclusion (1.49) again holds.

It remains to consider the case $k = 1$ where n is even and $n \geq 10$ or n is odd and $n \geq 13$. Here instead of bounds for s_k and d_k we use the exact values (1.43) and (1.44) for $k = 1$. We still use the bounds (1.47) for $\log \phi_k$ and Lemma 1.27 for $\log(1 + d_k)$. When n is even, $i = 2k = 2$, and we obtain a lower bound A' for (1.49) of the form $a_1/n + a_2/n^{3/2} + \cdots$. All terms n^{-2} and beyond have negative coefficients. Applying the inequality $-n^{-(3/2)-\alpha} \geq -n^{-3/2} \cdot 10^{-\alpha}$ for $n \geq 10$ and $\alpha = 1/2, 1, \ldots$, I found a lower bound $A' \geq 0.662/n - 1.115/n^{3/2} > 0$ for $n \geq 10$. The same method for n odd gave $A' \geq 0.662/n - 1.998/n^{3/2} > 0$ for $n \geq 13$. The proof of (1.10) is complete.

Proof of (1.11). For n odd, (1.11) is clear when $j = 1$, so we can assume $j \geq 3$. For n even, (1.11) is clear when $j = 2$. We next consider the case $j = 0$. By symmetry we need to prove that $p_{n0} \leq P(\sqrt{n}|Y|/2 \leq 1)$. This can be checked from a normal table for $n = 2$. For $n \geq 4$ we have $p_{n0} \leq \sqrt{2/\pi n}$ by (1.37). The integral of the standard normal density from $-2/\sqrt{n}$ to $2/\sqrt{n}$ is clearly larger than the length of the interval times the density at the endpoints, namely $2\sqrt{2/\pi n} \exp(-2/n)$. Since $\exp(-2/n) \geq 1/2$ for $n \geq 4$, the proof for n even and $j = 0$ is done.

We are left with the cases $j \geq 3$. For $j = n$, we have $p_{nn} = 2^{-n}$ and can check the conclusion for $n = 3, 4$ from a normal table. Let ϕ be the standard normal density. We have the inequality, for $t > 0$,

$$P(Y \geq t) \geq \psi(t) := \phi(t)[t^{-1} - t^{-3}] \tag{1.51}$$

(Feller 1968, p. 175). Feller does not give a proof. For completeness, here is one:

$$\psi(t) = -\int_t^\infty \psi'(x)dx = \int_t^\infty \phi(x)(1 - 3x^{-4})dx \leq P(Y \geq t).$$

To prove (1.11) via (1.51) for $j = n \geq 5$ we need to prove

$$1/2^n \leq \phi(t_n)t_n^{-1}(1 - t_n^{-2})$$

where $t_n := (n-2)/\sqrt{n}$. Clearly $n \mapsto t_n$ is increasing. For $n \geq 5$ we have $1 - t_n^{-2} \geq 4/9$ and $(2\pi)^{-1/2}e^{2-2/n} \cdot 4/9 \geq 0.878$. Thus it suffices to prove

$$n(\log 2 - 0.5) + 0.5 \log n - \log(n-2) + \log(0.878) \geq 0, \quad n \geq 5.$$

This can be checked for $n = 5, 6$ and the left side is increasing in n for $n \geq 6$, so (1.11) for $j = n \geq 5$ follows.

So it will suffice to prove $p_{ni} \leq P(\sqrt{n}Y/2 \in [(i-2)/2, i/2])$ for $j \leq i < n$. From (1.35) and Lemma 1.25, and the bound $\phi_k \geq 1$, it will suffice to prove, for $x := i/n$,

$$-\frac{1}{4n} + \frac{7}{288n^2} - \frac{(n-1)x^2}{2} + \frac{x^{2n}}{2n(1-x^2)} \leq -\frac{n[(x-2/n)^2 + x^2]}{4}$$

where $3/n \leq x \leq 1 - 2/n$. Note that $2n(1-x^2) \geq 4$. Thus it is enough to prove that

$$x - x^2/2 - x^{2n}/4 \geq 3/4n + 7/288n^2$$

for $3/n \leq x \leq 1$ and $n \geq 5$, which holds since the function on the left is concave, and the inequality holds at the endpoints. Thus (1.11) and Lemma 1.23 are proved. $\qquad\qquad\square$

1.4.3 *Proof of Lemma 1.21*

Let $G(x)$ be the distribution function of a normal random variable Z with mean $n/2$ and variance $n/4$ (the same mean and variance as for $\mathcal{B}(n, 1/2)$). Let $B(k, n, 1/2) := \sum_{0 \leq i \leq k} \binom{n}{i} 2^{-n}$. Lemma 1.23 directly implies

$$G(\sqrt{2kn} - n/2) \leq B(k, n, 1/2) \leq G(k+1) \quad \text{for } k \leq n/2. \quad (1.52)$$

Specifically, letting $k := (n-j)/2$, (1.11) implies

$$B(k, n, 1/2) \leq P(Z \geq n-k-1) = P(k+1 \geq n-Z) = G(k+1)$$

since $n - Z$ has the same distribution as Z. Then (1.10) implies

$$B(k, n, 1/2) \geq P\left(\frac{n}{2} - \frac{\sqrt{n}}{2}Y \leq -\frac{n}{2} + \sqrt{2kn}\right) = G(\sqrt{2kn} - n/2).$$

Let

$$\eta := \Phi_n^{\leftarrow}(G(Z)). \quad (1.53)$$

This definition of η from Z is called a quantile transformation. By Propositions 1.22 and 1.1(c) respectively, $G(Z)$ has a $U[0, 1]$ distribution and η a $\mathcal{B}(n, 1/2)$ distribution. It will be shown that

$$Z - 1 \leq \eta \leq Z + (Z - n/2)^2/2n + 1 \quad \text{if } Z \leq n/2 \quad (1.54)$$

and

$$Z - (Z - n/2)^2/2n - 1 \leq \eta \leq Z + 1 \text{ if } Z \geq n/2. \quad (1.55)$$

Define a sequence of extended real numbers $-\infty = c_{-1} < c_0 < c_1 < \cdots < c_n = +\infty$ by $G(c_k) = B(k, n, 1/2)$ Then one can check that $\eta = k$ on the event $A_k := \{\omega : c_{k-1} < Z(\omega) \leq c_k\}$. By (1.52), $G(c_k) = B(k, n, 1/2) \leq G(k + 1)$ for $k \leq n/2$. So, on the set A_k for $k \leq n/2$ we have $Z - 1 \leq c_k - 1 \leq k = \eta$. Note that for n even, $n/2 < c_{n/2}$, while for n odd, $n/2 = c_{(n-1)/2}$. So the left side of (1.54) is proved.

If Y is a standard normal random variable with distribution function Φ and density ϕ, then $\Phi(x) \leq \phi(x)/x$ for $x > 0$, e.g. Dudley (2002), Lemma 12.1.6(a). So we have

$$P(Z \leq -n/2) = P\left(\frac{n}{2} + \frac{\sqrt{n}}{2}Y \leq -\frac{n}{2}\right) = P\left(\frac{\sqrt{n}}{2}Y \leq -n\right)$$

$$= \Phi(-2\sqrt{n}) \leq \frac{e^{-2n}}{2\sqrt{2\pi n}} < \frac{1}{2^n}.$$

So $G(-n/2) < G(c_0) = 2^{-n}$ and $-n/2 < c_0$. Thus if $Z \leq -n/2$, then $\eta = 0$. Next note that $Z + (Z - n/2)^2/2n = (Z + n/2)^2/2n \geq 0$ always. Thus the right side of (1.54) holds when $Z \leq -n/2$ and whenever $\eta = 0$. Now assume that $Z \geq -n/2$. By (1.52), for $1 \leq k \leq n/2$

$$G((2(k - 1)n)^{1/2} - n/2) \leq B(k - 1, n, 1/2) = G(c_{k-1}),$$

from which it follows that $(2(k - 1)n)^{1/2} - n/2 \leq c_{k-1}$ and

$$k - 1 \leq (c_{k-1} + n/2)^2/2n. \quad (1.56)$$

The function $x \mapsto (x + n/2)^2$ is clearly increasing for $x \geq -n/2$ and thus for $x \geq c_0$. Applying (1.56) we get on the set A_k for $1 \leq k \leq n/2$

$$\eta = k \leq (Z + n/2)^2/2n + 1 = Z + (Z - n/2)^2/2n + 1.$$

Since $P(Z \leq n/2) = 1/2 \leq P(\eta \leq n/2)$, and η is a nondecreasing function of Z, $Z \leq n/2$ implies $\eta \leq n/2$. So (1.54) is proved.

It will be shown next that (η, Z) has the same joint distribution as $(n - \eta, n - Z)$. It is clear that η and $n - \eta$ have the same distribution and that Z and $n - Z$ do. We have for each $k = 0, 1, \ldots, n$, $n - \eta = k$ if and only if $\eta = n - k$ if and only if $c_{n-k-1} < Z \leq c_{n-k}$. We need to show that this is equivalent to $c_{k-1} \leq n - Z < c_k$, in other words $n - c_k < Z \leq n - c_{k-1}$. Thus we want to show that $c_{n-k-1} = n - c_k$ for each k. It is easy to check that $G(n - c_k) = P(Z \geq c_k) = 1 - G(c_k)$ while $G(c_k) = B(k, n, 1/2)$ and $G(c_{n-k-1}) = B(n - k - 1, n, 1/2) = 1 - B(k, n, 1/2)$. The statement about joint distributions follows. Thus (1.54) implies (1.55).

Some elementary algebra, (1.54) and (1.55) imply

$$|\eta - Z| \leq 1 + (Z - n/2)^2/2n \tag{1.57}$$

and since $Z < n/2$ implies $\eta \leq n/2$ and $Z > n/2$ implies $\eta \geq n/2$,

$$|\eta - n/2| \leq 1 + |Z - n/2|. \tag{1.58}$$

Letting $Z = (n + \sqrt{n}Y)/2$ and noting that then $G(Z) \equiv \Phi(Y)$, (1.53), (1.57), and (1.58) imply Lemma 1.21 with $C_n = \eta - n/2$.   □

1.4.4 Inequalities for the Separate Processes

We will need facts providing a modulus of continuity for the Brownian bridge and something similar for the empirical process (although it is discontinuous). Let $h(t) := +\infty$ if $t \leq -1$ and

$$h(t) := (1 + t)\log(1 + t) - t, \quad t > -1. \tag{1.59}$$

Here is a consequence of the Chernoff inequality (1.5) for binomial probabilities $E(k, n, p)$, in which k is not necessarily an integer.

Lemma 1.28 (Bennett) *For any $k \geq np$ and $q := 1 - p$ we have*

$$E(k, n, p) \leq \exp\left(-\frac{np}{q}h\left(\frac{k - np}{np}\right)\right). \tag{1.60}$$

Proof. According to Chernoff's inequality (1.5), we have

$$E(k, n, p) \leq (np/k)^k(nq/(n - k))^{n-k} \quad \text{if } k \geq np.$$

Taking logarithms, multiplying by q and simplifying it will suffice to show that for $k \geq np$

$$(n - k)q\log\left(\frac{nq}{n - k}\right) \leq kp\log\left(\frac{np}{k}\right) + k - np.$$

Taking k as a continuous variable as we may, this becomes an equality for $k = np$, so it suffices to show that the given inequality holds for $k \geq np$ if both sides are replaced by their derivatives with respect to k. Again both sides are equal when $k = np$. Differentiating again, the inequality reduces to $np \leq k$ which is true.   □

The previous lemma extends via martingales to a bound for the uniform empirical process on intervals (Lemma 2 of Bretagnolle and Massart 1989).

Lemma 1.29 *For any b with $0 < b \leq 1/2$ and $x > 0$,*

$$P(\sup_{0 \leq t \leq b} |\alpha_n(t)| > x/\sqrt{n}) \leq 2 \exp\left(-\frac{nb}{1-b}h\left(\frac{x(1-b)}{nb}\right)\right)$$

$$\leq 2\exp(-nb(1-b)h(x/(nb))). \tag{1.61}$$

Proof. From the binomial conditional distributions of multinomial variables we have for $0 \leq s \leq t < 1$

$$E(F_n(t)|F_n(u), \ u \leq s) = E(F_n(t)|F_n(s))$$
$$= F_n(s) + \frac{t-s}{1-s}(1 - F_n(s)) = \frac{t-s}{1-s} + \frac{1-t}{1-s}F_n(s),$$

from which it follows directly that

$$E\left(\frac{F_n(t)-t}{1-t}\bigg|F_n(u), \ u \leq s\right) = \frac{F_n(s)-s}{1-s};$$

in other words, the process $(F_n(t) - t)/(1 - t)$, $0 \leq t < 1$ is a martingale in t (here n is fixed). Thus, $\alpha_n(t)/(1 - t)$, $0 \leq t < 1$, is also a martingale, and for any real s the process $\exp(s\alpha_n(t)/(1 - t))$ is a submartingale, e.g. Dudley (2002, 10.3.3(b)). Then

$$P(\sup_{0 \leq t \leq b} \alpha_n(t) > x/\sqrt{n}) \leq P(\sup_{0 \leq t \leq b} \alpha_n(t)/(1 - t) > x/\sqrt{n}),$$

which for any $s > 0$ equals

$$P\left(\sup_{0 \leq t \leq b} \exp(s\alpha_n(t)/(1 - t)) > \exp(sx/\sqrt{n})\right).$$

By Doob's inequality (e.g. Dudley 2002, 10.4.2, for a finite sequence increasing up to a dense set) the latter probability is

$$\leq \inf_{s>0} \exp(-sx/\sqrt{n})E \exp(s\alpha_n(b)/(1 - b)) \leq \exp\left(-\frac{nb}{1-b}h\left(\frac{x(1-b)}{nb}\right)\right)$$

by Lemma 1.28, (1.60). In the same way, by (1.6) we get

$$P(\sup_{0 \leq t \leq b} (-\alpha_n(t)) > x/\sqrt{n}) \leq \exp(-x^2(1 - b)/(2nb)). \tag{1.62}$$

It is easy to check that $h(u) \leq u^2/2$ for $u \geq 0$, so the first inequality in Lemma 1.29 follows. It is easily shown by derivatives that $h(qy) \geq q^2 h(y)$ for $y \geq 0$ and $0 \leq q \leq 1$. For $q = 1 - b$, the bound in (1.61) then follows. □

We next have a corresponding inequality for the Brownian bridge.

Lemma 1.30 *Let $B(t)$, $0 \leq t \leq 1$, be a Brownian bridge, $0 < b < 1$ and $x >$ 0. Let Φ be the standard normal distribution function. Then*

$$P(\sup_{0 \leq t \leq b} B(t) > x) = 1 - \Phi(x/\sqrt{b(1-b)})$$

$$+ \exp(-2x^2)\left(1 - \Phi\left(\frac{(1-2b)x}{\sqrt{b(1-b)}}\right)\right). \qquad (1.63)$$

If $0 < b \leq 1/2$, then for all $x > 0$,

$$P(\sup_{0 \leq t \leq b} B(t) > x) \leq \exp(-x^2/(2b(1-b))). \qquad (1.64)$$

Proof. Let $X(t)$, $0 \leq t < \infty$ be a Wiener process. For some real α and value of $X(1)$ let $\beta := X(1) - \alpha$. It will be shown that for any real α and y

$$P\{\sup_{0 \leq t \leq 1} X(t) - \alpha t > y | X(1)\} = 1_{\{\beta > y\}} + \exp(-2y(y - \beta))1_{\{\beta \leq y\}}. \qquad (1.65)$$

Clearly, if $\beta > y$ then $\sup_{0 \leq t \leq 1} X(t) - \alpha t > y$ (let $t = 1$). Suppose $\beta \leq y$. One can apply a reflection argument as in the proof of Dudley (2002, Proposition 12.3.3), where details are given on making such an argument rigorous. Let $X(t) = B(t) + tX(1)$ for $0 \leq t \leq 1$, where $B(\cdot)$ is a Brownian bridge. We want to find $P(\sup_{0 \leq t \leq 1} B(t) + \beta t > y)$. But this is the same as $P(\sup_{0 \leq t \leq 1} Y(t) > y | Y(1) = \beta)$ for a Wiener process Y. For $\beta \leq y$, the probability that $\sup_{0 \leq t \leq 1} Y(t) > y$ and $\beta \leq Y(1) \leq \beta + dy$ is the same by reflection as $P(2y - \beta \leq Y(1) \leq 2y - \beta + dy)$. Thus the desired conditional probability, for the standard normal density ϕ, is $\phi(2y - \beta)/\phi(\beta) = \exp(-2y(y - \beta))$ as stated. So (1.65) is proved.

We can write the Brownian bridge B as $W(t) - tW(1)$, $0 \leq t \leq 1$, for a Wiener process W. Let $W_1(t) := b^{-1/2}W(bt)$, $0 \leq t < \infty$. Then W_1 is a Wiener process. Let $\eta := W(1) - W(b)$. Then η has a normal $N(0, 1 - b)$ distribution and is independent of $W_1(t)$, $0 \leq t \leq 1$. Let $\gamma := ((1 - b)W_1(1) - \sqrt{b}\eta)\sqrt{b}/x$. We have

$$P(\sup_{0 \leq t \leq b} B(t) > x | \eta, W_1(1))$$

$$= P\left(\sup_{0 \leq t \leq 1} (W_1(t) - (bW_1(1) + \sqrt{b}\eta)t) > x/\sqrt{b} | \eta, W_1(1)\right).$$

Now the process $W_1(t) - (bW_1(t) + \sqrt{b}\eta)t$, $0 \leq t \leq 1$, has the same distribution as a Wiener process $Y(t)$, $0 \leq t \leq 1$, given that $Y(1) = (1 - b)W_1(1) - \sqrt{b}\eta$. Thus by (1.65) with $\alpha = 0$,

$$P(\sup_{0 \leq t \leq b} B(t) > x | \eta, W_1(1)) = 1_{\{\gamma > 1\}} + 1_{\{\gamma \leq 1\}} \exp(-2x^2(1 - \gamma)/b). \qquad (1.66)$$

Thus, integrating gives

$$P(\sup_{0\leq t\leq b} B(t) > x) = P(\gamma > 1) + \exp(-2x^2/b)E\left(\exp(2x^2\gamma/b)1_{\{\gamma\leq1\}}\right).$$

From the definition of γ it has a $N(0, b(1 - b)/x^2)$ distribution. Since x is constant, the latter integral with respect to γ can be evaluated by completing the square in the exponent and yields (1.63).

We next need the inequality, for $x \geq 0$,

$$1 - \Phi(x) \leq \frac{1}{2}\exp(-x^2/2). \qquad (1.67)$$

This is easy to check via the first derivative for $0 \leq x \leq \sqrt{2/\pi}$. On the other hand we have the inequality $1 - \Phi(x) \leq \phi(x)/x$, $x > 0$, e.g. Dudley (2002, 12.1.6(a)), which gives the conclusion for $x \geq \sqrt{2/\pi}$.

Applying (1.67) to both terms of (1.63) gives (1.64), so the Lemma is proved.   □

1.4.5 *Proof of Theorem 1.8*

For the Brownian bridge $B(t)$, $0 \leq t \leq 1$, it is well known that for any $x > 0$

$$P(\sup_{0\leq t\leq1} |B(t)| \geq x) \leq 2\exp(-2x^2),$$

e.g. Dudley (2002, Proposition 12.3.3). It follows that

$$P(\sqrt{n}\sup_{0\leq t\leq1} |B(t)| \geq u) \leq 2\exp(-u/3)$$

for $u \geq n/6$. We also have $|\alpha_1(t)| \leq 1$ for all t and

$$P(\sup_{0\leq t\leq1} |\alpha_n(t)| \geq x) \leq D\exp(-2x^2), \qquad (1.68)$$

which is the Dvoretzky–Kiefer–Wolfowitz inequality (Dvoretzky, Kiefer, and Wolfowitz 1956) with a constant D. Massart (1990) proved (1.68) with the sharp constant $D = 2$. Massart's theorem will be proved in Section 1.5. Earlier Hu (1985) proved it with $D = 4\sqrt{2}$. $D = 6$ would suffice for present purposes. Given D, it follows that for $u \geq n/6$,

$$P(\sqrt{n}\sup_{0\leq t\leq1} |\alpha_n(t)| \geq u) \leq D\exp(-u/3).$$

For $x < 6\log 2$, we have $2e^{-x/6} > 1$ so the conclusion of Theorem 1.8 holds. holds. For $x > n/3 - 12\log n$, $u := (x + 12\log n)/2 > n/6$, so the left side of (1.1) is bounded above by $(2 + D)n^{-2}e^{-x/6}$. We have $(2 + D)n^{-2} \leq 2$ for $n \geq 2$ and $D \leq 6$.

Thus it will be enough to prove Theorem 1.8 when

$$6 \log 2 \leq x \leq n/3 - 12 \log n. \qquad (1.69)$$

The function $t \mapsto t/3 - 12 \log t$ is decreasing for $t < 36$, increasing for $t > 36$. Thus one can check that for (1.69) to be non-vacuous is equivalent to

$$n \geq 204. \qquad (1.70)$$

Let N be the largest integer such that $2^N \leq n$, so that $v := 2^N \leq n < 2v$. Let Z be a v-dimensional normal random variable with independent components, each having mean 0 and variance $\zeta := n/v$. For integers $0 \leq i < m$ let $A(i, m) := \{i + 1, \ldots, m\}$. For any two vectors $a := (a_1, \ldots, a_v)$ and $b := (b_1, \ldots, b_v)$ in $\mathbb{R}^v$, we have the usual inner product $(a, b) := \sum_{i=1}^{v} a_i b_i$. For any subset $D \subset A(0, v)$ let 1_D be its indicator function as a member of $\mathbb{R}^v$. For any integers $j = 0, 1, 2, \ldots$ and $k = 0, 1, \ldots$, let

$$I_{j,k} := A(2^j k, 2^j (k + 1)), \qquad (1.71)$$

let $e_{j,k}$ be the indicator function of $I_{j,k}$ and for $j \geq 1$, let $e'_{j,k} := e_{j-1,2k} - e_{j,k}/2$. Then one can easily check that the family $\mathcal{E} := \{e'_{j,k} : 1 \leq j \leq N, 0 \leq k < 2^{N-j}\} \cup \{e_{N,0}\}$ is an orthogonal basis of $\mathbb{R}^v$ with $(e_{N,0}, e_{N,0}) = v$ and $(e'_{j,k}, e'_{j,k}) = 2^{j-2}$ for each of the given j, k. Let $W_{j,k} := (Z, e_{j,k})$ and $W'_{j,k} := (Z, e'_{j,k})$. Then since the elements of $\mathcal{E}$ are orthogonal, it follows that the random variables $W'_{j,k}$ for $1 \leq j \leq N$, $0 \leq k < 2^{N-j}$ and $W_{N,0}$ are independent normal with

$$EW'_{j,k} = EW_{N,0} = 0, \quad \mathrm{Var}(W'_{j,k}) = \zeta 2^{j-2}, \quad \mathrm{Var}(W_{N,0}) = \zeta v. \qquad (1.72)$$

Recalling the notation of Lemma 1.21, let Φ_n be the distribution function of a binomial $\mathcal{B}(n, 1/2)$ random variable, with quantile function $\Phi_n^{\leftarrow}$. Now let $G_m(t) := \Phi_m^{\leftarrow}(\Phi(t))$.

The next theorem will give a way of linking up or "coupling" processes. Recall that a *Polish* space is a topological space metrizable by a complete separable metric.

Theorem 1.31 (Vorob'ev–Berkes–Philipp) *Let X, Y, and Z be Polish spaces with Borel σ-algebras. Let α be a law on $X \times Y$ and let β be a law on $Y \times Z$. Let $\pi_Y(x, y) := y$ and $\tau_Y(y, z) := y$ for all $(x, y, z) \in X \times Y \times Z$. Suppose the marginal distributions of α and β on Y are equal, in other words $\eta := \alpha \circ \pi_Y^{-1} = \beta \circ \tau_Y^{-1}$ on Y. Let $\pi_{12}(x, y, z) := (x, y)$ and $\pi_{23}(x, y, z) := (y, z)$. Then there exists a law γ on $X \times Y \times Z$ such that $\gamma \circ \pi_{12}^{-1} = \alpha$ and $\gamma \circ \pi_{23}^{-1} = \beta$.*

Proof. There exist conditional distributions α_y for α on X given $y \in Y$, so that for each $y \in Y$, α_y is a probability measure on X, for any Borel set $A \subset X$, the function $y \mapsto \alpha_y(A)$ is measurable, and for any integrable function f for α,

$$\int f d\alpha = \int \int f(x, y) d\alpha_y(x) d\eta(y)$$

(RAP, Section 10.2). Likewise, there exist conditional distributions β_y on Z for β. Let x and z be conditionally independent given y. In other words, define a set function γ on $X \times Y \times Z$ by

$$\gamma(C) = \int \int \int 1_C(x, y, z) d\alpha_y(x) d\beta_y(z) d\eta(y).$$

The integral is well-defined if

(a) $C = U \times V \times W$ for Borel sets U, V, and W in X, Y, and Z respectively,

(b) C is a finite union of such sets, which can be taken to be disjoint (RAP, Proposition 3.2.2 twice), or

(c) C is any Borel set in $X \times Y \times Z$, by RAP, Proposition 3.2.3, and the monotone class theorem (RAP, Theorem 4.4.2).

Also, γ is countably additive by monotone convergence (for all three integrals). So γ is a law on $X \times Y \times Z$. Clearly $\gamma \circ \pi_{12}^{-1} = \alpha$ and $\gamma \circ \pi_{23}^{-1} = \beta$. □

Now to begin the construction that will connect the empirical process with a Brownian bridge, let

$$U_{N,0} := n \qquad\qquad (1.73)$$

and then recursively as j decreases from $j = N$ to $j = 1$,

$$U_{j-1,2k} := G_{U_{j,k}}((2^{2-j}/\zeta)^{1/2} W'_{j,k}), \quad U_{j-1,2k+1} := U_{j,k} - U_{j-1,2k}, \quad (1.74)$$

$k = 0, 1, \ldots, 2^{N-j} - 1$. Note that by (1.72), $(2^{2-j}/\zeta)^{1/2} W'_{j,k}$ has a standard normal distribution, so Φ of it has a $U[0, 1]$ distribution. It is easy to verify successively for $j = N, N - 1, \ldots, 0$ that the random vector $\{U_{j,k}, 0 \le k < 2^{N-j}\}$ has a multinomial distribution with parameters $n, 2^{j-N}, \ldots, 2^{j-N}$. Let $X := (U_{0,0}, U_{0,1}, \ldots, U_{0,v-1})$. Then the random vector X has a multinomial distribution with parameters $n, 1/v, \ldots, 1/v$.

The random vector X is equal in distribution to

$$\{n(F_n((k+1)/v) - F_n(k/v)), \ 0 \le k \le v - 1\}, \qquad (1.75)$$

while for a Wiener process W, Z is equal in distribution to

$$\{\sqrt{n}(W((k+1)/v) - W(k/v)), \ 0 \le k \le v - 1\}. \qquad (1.76)$$

Without loss of generality, we can assume that the above equalities in distribution are actual equalities for some uniform empirical distribution functions F_n

and Wiener process $W = W_n$. Specifically, consider a vector of i.i.d. uniform random variables $(x_1, \ldots, x_n) \in \mathbb{R}^n$ such that

$$F_n(t) := \frac{1}{n} \sum_{j=1}^{n} 1_{\{x_j \le t\}}$$

and note that W has sample paths in $C[0, 1]$. Both $\mathbb{R}^n$ and $C[0, 1]$ are separable Banach spaces. Thus one can let $(x_1, \ldots, x_n)$ and W be conditionally independent given the vectors in (1.75) and (1.76) which have the joint distribution of X and Z, by the Vorob'ev–Berkes–Philipp Theorem 1.31. Then we define a Brownian bridge by $Y_n(t) := W_n(t) - t W_n(1)$ and the empirical process $\alpha_n(t) := \sqrt{n}(F_n(t) - t), 0 \le t \le 1$. By our choices, we then have

$$\{n(F_n(j/v) - j/v)\}_{j=0}^{v} = \left\{ \sum_{i=0}^{j-1} \left(X_i - \frac{n}{v} \right) \right\}_{j=0}^{v} \tag{1.77}$$

and

$$\{\sqrt{n} Y_n(j/v)\}_{j=0}^{v} = \left\{ \left(\sum_{i=0}^{j-1} Z_i \right) - \frac{j}{v} \sum_{r=0}^{v-1} Z_r \right\}_{j=0}^{v}. \tag{1.78}$$

Theorem 1.8 will be proved for the given Y_n and α_n. Specifically, we want to prove

$$P_0 := P\left(\sup_{0 \le t \le 1} |\alpha_n(t) - Y_n(t)| > (x + 12 \log n)/\sqrt{n} \right) \le 2 \exp(-x/6). \tag{1.79}$$

It will be shown that $\alpha_n(j/v)$ and $Y_n(j/v)$ are not too far apart for $j = 0, 1, \ldots, v$, while the increments of the processes over the intervals between the lattice points j/v are also not too large.

Let $C := 0.29$. Let M be the least integer such that

$$C(x + 6 \log n) \le \zeta 2^{M+1}. \tag{1.80}$$

Since $n \ge 204$ (1.70) and $\zeta < 2$ this implies $M \ge 2$. We have by definition of M and (1.69)

$$2^M \le \zeta 2^M \le C(x + 6 \log n) \le Cn/3 < 0.1 \cdot 2^{N+1} < 2^{N-2},$$

so $M \le N - 3$.

For each $t \in [0, 1]$, let $\pi_M(t)$ be the nearest point of the grid $\{i/2^{N-M}, 0 \le i \le 2^{N-M}\}$, or if there are two nearest points, take the smaller one. Let $D := X - Z$ and $D(m) := \sum_{i=1}^{m} D_i$. Let $C' := 0.855$ and define

$$\Theta := \{U_{j,k} \le \zeta(1 + C')2^j \text{ whenever } M + 1 < j \le N, 1 \le k < 2^{N-j}\}$$

$$\cap \{U_{j,k} \ge \zeta(1 - C')2^j \text{ whenever } M < j \le N, 1 \le k < 2^{N-j}\}.$$

Then

$$P_0 \leq P_1 + P_2 + P_3 + P(\Theta^c)$$

where

$$P_1 := P\left(\sup_{0 \leq t \leq 1} |\alpha_n(t) - \alpha_n(\pi_M(t))| > 0.28(x + 6\log n)/\sqrt{n}\right), \qquad (1.81)$$

$$P_2 := P\left(\sup_{0 \leq t \leq 1} |Y_n(t) - Y_n(\pi_M(t))| > 0.22(x + 6\log n)/\sqrt{n}\right), \qquad (1.82)$$

and, recalling (1.77) and (1.78),

$$P_3 := 2^{N-M} \max_{m \in A(M)} P\left\{\left(|D(m) - \frac{m}{v}D(v)| > 0.5x + 9\log n\right) \cap \Theta\right\}, \qquad (1.83)$$

where $A(M) := \{k2^M : k = 1, 2, \ldots\} \cap A(0, v)$.

First we bound $P(\Theta^c)$. Since by (1.74) $U_{j,k} = U_{j-1,2k} + U_{j-1,2k+1}$, we have

$$\Theta^c \subset \bigcup_{0 \leq k < 2^{N-M-2}} \{U_{M+2,k} > (1 + C')\zeta 2^{M+2}\}$$

$$\cup \bigcup_{0 \leq k < 2^{N-M-1}} \{U_{M+1,k} < (1 - C')\zeta 2^{M+1}\}.$$

Since $U_{M+2,k}$ and $U_{M+1,k}$ are binomial random variables, Lemma 1.28 gives

$$P(\Theta^c) \leq 2^{N-M-1}\left(\exp(-\zeta 2^{M+2}h(C')) + \exp(-\zeta 2^{M+1}h(-C'))\right).$$

Now $2h(C') \geq 0.5823 \geq h(-C') \geq 0.575$ (note that C' has been chosen to make $2h(C')$ and $h(-C')$ approximately equal). By definition of M (1.80), $\zeta 2^{M+1} \geq C(x + 6\log n)$, and $0.575C > 1/6$, so

$$P(\Theta^c) \leq 2^{-M}\exp(-x/6). \qquad (1.84)$$

Next, to bound P_1 *and* P_2. Let $b := 2^{M-N-1} \leq 1/2$. Since $\alpha_n(t)$ has stationary increments, we can apply Lemma 1.29. Let $u := x + 6\log n$. We have by definition of M (1.80)

$$nb = n2^{M-N-1} < Cu/2. \qquad (1.85)$$

By (1.69), $u < n/3$ so $b < C/6$. Recalling (1.59), note that $h'(t) \equiv \log(1 + t)$. Thus h is increasing. For any given $v > 0$ it is easy to check that

$$y \mapsto yh(v/y) \text{ is decreasing for } y > 0. \qquad (1.86)$$

Lemma 1.29 gives

$$P_1 \leq 2^{N-M+2} \exp\left(-nb(1-b)h\left(\frac{0.28u}{nb}\right)\right)$$

$$< 2^{N-M+2} \exp\left(-\frac{C}{2}\left[1-\frac{C}{6}\right]uh\left(0.28 \cdot \frac{2}{C}\right)\right)$$

by (1.86) and (1.85) and since $1 - b > 1 - C/6$, so one can calculate

$$P_1 \leq 2^{N-M+2}e^{-u/6} \leq 2^{2-M}\zeta^{-1}\exp(-x/6). \tag{1.87}$$

The Brownian bridge also has stationary increments, so Lemma 1.30, (1.64), and (1.85) give

$$P_2 \leq 2^{N-M+2} \exp(-(0.22u)^2/(2nb))$$

$$\leq 2^{N-M+2} \exp(-(0.22)^2 u/C) \leq 2^{2-M}\zeta^{-1}e^{-x/6} \tag{1.88}$$

since $(0.22)^2/C > 1/6$.

It remains to bound P_3. Fix $m \in A(M)$. A bound is needed for

$$P_3(m) := P\left\{\left(|D(m) - \frac{m}{\nu}D(\nu)| > 0.5x + 9\log n\right) \cap \Theta\right\}. \tag{1.89}$$

For each $j = 1, \ldots, N$ take $k(j)$ such that $m \in I_{j,k(j)}$. By the definition (1.71) of $I_{j,k}$, $k(M) = m2^{-M} - 1$ and $k(j) = [k(j-1)/2]$ for $j = 1, \ldots, N$ where $[x]$ is the largest integer $\leq x$. From here on each double subscript $j, k(j)$ will be abbreviated to the single subscript j, e.g. $e'_j := e'_{j,k(j)}$. The following orthogonal expansion holds in $\mathcal{E}$:

$$1_{A(0,m)} = \frac{m}{\nu}e_{N,0} + \sum_{M<j\leq N} c_j e'_j \tag{1.90}$$

where $0 \leq c_j \leq 1$ for $m < j \leq N$. To see this, note that $1_{A(0,m)} \perp e'_{j,k}$ for $j \leq M$ since 2^M is a divisor of m. Also, $1_{A(0,m)} \perp e'_{j,k}$ for $k \neq k(j)$ since $1_{A(0,m)}$ has all 0's or all 1's on the set where $e'_{j,k}$ has nonzero entries, half of which are $+1/2$ and the other half $-1/2$. In an orthogonal expansion $f = \sum_j c_j f_j$ we always have $c_j = (f, f_j)/\|f_j\|^2$ where $\|v\|^2 := (v, v)$. We have $\|e'_j\| = 2^{(j-2)/2}$. Now, $(1_{A(0,m)}, e'_j)$ is as large as possible when the components of e'_j equal $1/2$ only for indices $\leq m$, and then the inner product equals 2^{j-2}, so $|c_j| \leq 1$ as stated. The m/ν factor is clear.

We next have

$$e_j = 2^{j-N}e_{N,0} + \sum_{i>j}(-1)^{s(i,j,m)}2^{j+1-i}e'_i \tag{1.91}$$

where $s(i, j, m) = 0$ or 1 for each i, j, m so that the corresponding factors are ± 1, the signs being immaterial in what follows. Let $\Delta_j := (D, e'_j)$.

Then from (1.90),

$$\left| D(m) - \frac{m}{v} D(v) \right| \leq \sum_{M < j \leq N} |\Delta_j|. \tag{1.92}$$

Recall that $W'_j = (Z, e'_j)$ (see between (1.71) and (1.72)) and $D = X - Z$. Let $\xi_j := (2^{2-j}/\zeta)^{1/2} W'_j$ for $M < j \leq N$. Then by (1.72), and the preceding statement, $\xi_{M+1}, \ldots, \xi_N$ are i.i.d. standard normal random variables. We have $U_{j,k} = (X, e_{j,k})$ for all j and k from the definitions. Then $U_j = (X, e_j)$. Let $U'_j = (X, e'_j)$. By (1.74) and Lemma 1.21, (1.9),

$$|U'_j - \sqrt{U_j} \xi_j / 2| \leq 1 + \xi_j^2 / 8. \tag{1.93}$$

Let

$$L_j := |W'_j - \sqrt{U_j} \xi_j / 2| = |\xi_j| |\sqrt{U_j} - \sqrt{\zeta 2^j}| / 2$$

by definition of ξ_j. Thus

$$|\Delta_j| \leq L_j + 1 + \xi_j^2 / 8. \tag{1.94}$$

Then we have on Θ

$$|\sqrt{U_j} - \sqrt{\zeta 2^j}| = |U_j - \zeta 2^j| / (\sqrt{\zeta 2^j} + \sqrt{U_j})$$

$$\leq \frac{|U_j - \zeta 2^j|}{\sqrt{\zeta 2^j}} \cdot \frac{1}{1 + \sqrt{1 - C'}},$$

where as before $C' := 0.855$. Then by (1.74), (1.91), and (1.8) of Lemma 1.21,

$$|U_j - \zeta 2^j| \leq 2^{j-N} |U_N - n| + 2 \sum_{j < i \leq N} 2^{j-i} |U'_i|$$

$$\leq 2 + (\zeta (1 + C'))^{1/2} \sum_{j < i \leq N} 2^{j-i/2} |\xi_i|$$

on Θ, recalling that by (1.73), $U_N = U_{N,0} = n$. Let $C_2 := 1/(1 + \sqrt{1 - C'})$. It follows that

$$L_j \leq 2^{-j/2} C_2 |\xi_j| + \frac{1}{2} C_2 \sqrt{1 + C'} \sum_{j < i \leq N} 2^{(j-i)/2} |\xi_j| |\xi_i|. \tag{1.95}$$

Applying the inequality $|\xi_i| |\xi_j| \leq (\xi_i^2 + \xi_j^2)/2$, we get the bound

$$\sum_{M < j \leq N} \sum_{j < i \leq N} 2^{(j-i)/2} |\xi_i \xi_j| \leq \sum_{M < j \leq N} A_j \xi_j^2 \tag{1.96}$$

where

$$A_j := \frac{1}{2} \left(\sum_{M < r < j} 2^{(r-j)/2} + \sum_{j < i \leq N} 2^{(j-i)/2} \right).$$

Then

$$A_j \leq \frac{1}{2}\left[\frac{2^{-1/2} - 2^{(M-j)/2}}{1 - 2^{-1/2}} + \frac{2^{-1/2}}{1 - 2^{-1/2}}\right]$$

$$\leq 1 + \sqrt{2} - 2^{(M-j-2)/2}/(1 - 2^{-1/2}).$$

Let $C_3 := C_2(1 + \sqrt{2})\sqrt{1 + C'}/2 \leq 1.19067$. Then

$$\sum_{M < j \leq N} L_j \leq C_3 \sum_{M < j \leq N} \xi_j^2 \tag{1.97}$$

$$+ \sum_{M < j \leq N} 2^{-j/2}|\xi_j|C_2\left(1 - \sqrt{1 + C'}2^{(M-4)/2}|\xi_j|/(1 - 2^{-1/2})\right).$$

Let

$$C_4 := \frac{\sqrt{1 + C'}}{4(1 - 2^{-1/2})} = \frac{\sqrt{2}\sqrt{1 + C'}(\sqrt{2} + 1)}{4},$$

and for each M let $c_M := 1/(4C_42^{M/2})$. Then for any real number x, we have $x(1 - C_42^{M/2}x) \leq c_M$. It follows that

$$\sum_{M < j \leq N} L_j \leq \sum_{M < j \leq N} C_3\xi_j^2 + c_M C_2 2^{-j/2}$$

$$\leq C_2 c_M 2^{-(M+1)/2}/(1 - 2^{-1/2}) + \sum_{M < j \leq N} C_3\xi_j^2$$

$$\leq \frac{C_2 2^{-M}}{\sqrt{2}\sqrt{1 + C'}} + \sum_{M < j \leq N} C_3\xi_j^2.$$

Thus, combining (1.94) and (1.97) we get on Θ

$$\sum_{M < j \leq N} |\Delta_j| \leq N + \left(\frac{1}{8} + C_3\right)\sum_{M < j \leq N} \xi_j^2. \tag{1.98}$$

We have $E\exp(t\xi^2) = (1 - 2t)^{-1/2}$ for $t < 1/2$ and any standard normal variable ξ such as ξ_j for each j. Since $\xi_{M+1}, \ldots, \xi_N$ are independent we get

$$E\exp\left(\left(\frac{1}{3}\sum_{M < j \leq N} |\Delta_j|\right)1_\Theta\right) \leq e^{N/3}\left(1 - \frac{2}{3}\left(C_3 + \frac{1}{8}\right)\right)^{(M-N)/2}$$

$$\leq e^{N/3}2^{1.513(N-M)} \leq 2^{2N-1.5M}.$$

Markov's inequality and (1.92) then yield

$$P_3(m) \leq e^{-x/6}n^{-3}2^{2N-1.5M}.$$

Thus

$$P_3 \leq e^{-x/6} n^{-3} 2^{3N-2.5M} \leq 2^{-2.5M} e^{-x/6}. \tag{1.99}$$

Collecting (1.84), (1.87), (1.88) and (1.99) we get that $P_0 \leq (2^{3-M} \zeta^{-1} + 2^{-M} + 2^{-2.5M}) e^{-x/6}$. By (1.80) and (1.70) and since $x \geq 6 \log 2$ (1.69) and $M \geq 2$, it follows that Theorem 1.8 holds. $\qquad \square$

1.5 The DKW Inequality in Massart's Form

A. Dvoretzky, J. Kiefer, and J. Wolfowitz (1956) proved the "Dvoretzky–Kiefer–Wolfowitz" (DKW) inequality, namely that there is a constant $D < +\infty$ such that for any distribution function F on $\mathbb{R}$ and its empirical distribution functions F_n, we have for every $u > 0$,

$$\Pr(\sqrt{n} \sup_x |(F_n - F)(x)| > u) \leq D \exp(-2u^2). \tag{1.100}$$

For the case that F is the $U[0, 1]$ distribution function U, this is used in the proof of the Bretagnolle–Massart theorem (1.68) with $D = 6$.

Massart (1990) proved the following:

Theorem 1.32 (Massart) *The inequality (1.100) holds with the constant $D = 2$.*

Remark. The constant $D = 2$ is best possible because, for a Brownian bridge y and F continuous, specifically $F = U$, we have as shown by Kolmogorov (1933b) and as also follows from the Komlós–Major–Tusnády approximation, Theorem 1.8,

$$\lim_{n \to \infty} \Pr(\sqrt{n} \sup_x |(F_n - F)(x)| > u) = \Pr(\sup_t |y_t| > u)$$

$$= 2 \sum_{j=1}^{\infty} (-1)^{j-1} \exp(-2j^2 u^2),$$

where the last equality is Theorem 1.6 (or Proposition 12.3.4 of RAP). As u becomes large, the latter expression is asymptotic to $2 \exp(-2u^2)$.

Proof. It will suffice to prove Theorem 1.32 for the $U[0, 1]$ distribution U by Proposition 1.2. In fact, if F is continuous, then its range includes the open interval $(0, 1)$. Whether it contains 0 or 1 is immaterial since $(U_n - U)(0) \equiv 0 - 0 = 0$ and $(U_n - U)(1) = 1 - 1 = 0$ with probability 1. Thus, the distribution of $\sup_x |(F_n - F)(x)|$ is the same for all continuous F. For a general F, possibly discontinuous, its range is included in $[0, 1]$, so $\sup_x |(U_n - U)(F(x))| \leq \sup_{0 \leq t \leq 1} |(U_n - U)(t)|$. So Theorem 1.32 for $U[0, 1]$ implies it for arbitrary F.

Recalling $\alpha_n(t) := \sqrt{n}(U_n - U)(t)$ for $0 \le t \le 1$, let $D_n^+ := \sup_t \alpha_n(t)$, $D_n^- := \sup_t(-\alpha_n(t))$, and $D_n := \sup_t |\alpha_n(t)| = \max(D_n^+, D_n^-)$. We have the following symmetry:

Proposition 1.33 *For any* $n = 1, 2, \ldots$, D_n^+ *and* D_n^- *are equal in distribution.*

Proof. Let $X_1, \ldots, X_n$ be the i.i.d. $U[0, 1]$ variables on which U_n is based. Let $Y_j := 1 - X_j$ for $j = 1, \ldots, n$. Then $Y_1, \ldots, Y_n$ are i.i.d. $U[0, 1]$. For a function F and real x define $F(x-) := \lim_{y \uparrow x} f(y)$ if the limit (left limit) exists, as it always will for a distribution function F. It will be different from $F(x)$ if F has a jump at x, specifically, for an empirical distribution F_n and x equal to one of the X_j. Let

$$G_n(t) := \frac{1}{n} \sum_{j=1}^n 1_{Y_j \le t} = \frac{1}{n} \sum_{j=1}^n 1_{X_j \ge 1-t} = 1 - F_n((1-t)-)$$

for $0 \le t < 1$. Thus almost surely

$$
\begin{aligned}
\sup_{0 \le t < 1} G_n(t) - t &= \sup_{0 \le t < 1} [1 - t - F_n((1-t)-)] \\
&= \sup_{0 < s \le 1} (s - F_n(s-)) = \sup_{0 \le s \le 1} (s - F_n(s)),
\end{aligned}
$$

which gives the conclusion.   □

Massart (1990, Theorem 1) gives the following fact, which is interesting in itself and implies Theorem 1.32 (see the Remarks after it):

Theorem 1.34 *For any* $n = 1, 2, \ldots$ *and any* $\lambda \ge \min(\sqrt{(\log 2)/2}, \zeta n^{-1/6})$, *where* $\zeta := 1.0841$, *we have*

$$\Pr(D_n^- > \lambda) \le \exp(-2\lambda^2). \tag{1.101}$$

Remarks. If $\exp(-2\lambda^2) \le 1/2$, then $\lambda \ge \sqrt{(\log 2)/2}$, which implies the hypothesis of Theorem 1.34. Also, Proposition 1.33 implies $\Pr(D_n > \lambda) \le 2\Pr(D_n^- > \lambda)$. Further, Theorem 1.32 holds trivially if $2\exp(-2\lambda^2) > 1$, so it suffices to prove Theorem 1.34 to prove Theorem 1.32 in all cases.

Proof. If for a given $\lambda > 0$,

$$D_n^- = \sqrt{n} \sup_{0 \le t \le 1} (t - U_n(t)) > \lambda, \tag{1.102}$$

then $t - U_n(t) = \lambda/\sqrt{n}$ for some t, because between its downward jumps at the observations X_j, $t - U_n(t)$ is continuously increasing. Let $X_{(1)} < X_{(2)} < \cdots < X_{(n)}$ (almost surely) be the order statistics of $X_1, \ldots, X_n$. Thus

$$\sup_{0 \le t \le 1} t - U_n(t) = \max_{1 \le k \le n} X_{(k)} - (k-1)/n$$

(the supremum occurs just to the left of some $X_{(k)}$); the supremum is strictly positive with probability 1 because $X_{(1)} > 0$. So if (1.102) holds there is a smallest $k = 1, \ldots, n$ with $X_{(k)} - (k-1)/n > \lambda/\sqrt{n}$. Letting $X_{(0)} := 0$, we must have $t - U_n(t) = \lambda/\sqrt{n}$ for some t with $X_{(k-1)} < t < X_{(k)}$. Let $\tau_n := \tau_n(\lambda)$ be the least $t \in [0, 1]$ such that $t - U_n(t) = \lambda/\sqrt{n}$ if one exists ($D_n^- > \lambda$), otherwise let $\tau_n = 2$. If $\tau_n < 2$, then for some $j = 0, 1, \ldots, n-1$, $\tau_n - j/n = \lambda/\sqrt{n}$, i.e. $\tau_n = \frac{j}{n} + \frac{\lambda}{\sqrt{n}}$, which implies that

$$j < n - \lambda\sqrt{n}. \tag{1.103}$$

If $\lambda \geq \sqrt{n}$ then $\Pr(D_n^- > \lambda) = 0$, implying the conclusion of the theorem, so suppose $\lambda < \sqrt{n}$. Let $J \geq 0$ be the largest integer less than $n - \lambda\sqrt{n}$. $\quad\square$

The following fact, according to Massart (1990), is due to Smirnov (1944).

Proposition 1.35 (Smirnov) *For each λ with $0 < \lambda < \sqrt{n}$ and $\varepsilon := \lambda/\sqrt{n}$, and each $j = 1, \ldots, J$, $\Pr(\tau_n = \varepsilon + j/n) = p_{\lambda,n}(j)$ where*

$$p_{\lambda,n}(j) = \lambda\sqrt{n}(j + \lambda\sqrt{n})^{j-1}(n - j - \lambda\sqrt{n})^{n-j}n^{-n}\binom{n}{j}. \tag{1.104}$$

For $j = 0$, $\Pr(\tau_n = \varepsilon) = p_{\lambda,n}(0) := (1 - \varepsilon)^n$.

Proof. For each n, $\varepsilon = \lambda/\sqrt{n}$ and $i = 1, \ldots, J$ let $A_i := \{X_{(i)} \leq \varepsilon + \frac{i-1}{n}\}$. Here is a

Claim. We have $\{\tau_n = \varepsilon\} = A_1^c$ and for $j = 1, \ldots, J$, $\{\tau_n = \varepsilon + \frac{j}{n}\} = \left(\bigcap_{1 \leq i \leq j} A_i\right) \cap A_{j+1}^c$.

Proof of Claim. On A_1^c, $X_{(1)} > \varepsilon$, which is equivalent to $(U - U_n)(\varepsilon) = \varepsilon$ and so to $\tau_n = \varepsilon$. This event has probability $(1 - \varepsilon)^n$.

On A_i for $i \geq 1$, $U_n\left(\frac{i-1}{n} + \varepsilon\right) \geq i/n$ and so $(U - U_n)\left(\frac{i-1}{n} + \varepsilon\right) \leq \varepsilon - 1/n < \varepsilon$ and $\tau_n \neq \frac{i-1}{n} + \varepsilon$. Thus on $B_j := \bigcap_{1 \leq i \leq j} A_i$, $\tau_n \geq \frac{j}{n} + \varepsilon$.

On A_{j+1}^c, $X_{(j+1)} > \frac{j}{n} + \varepsilon$, and if A_j also occurs, $X_{(j)} \leq \frac{j-1}{n} + \varepsilon$, so $U_n(\varepsilon + j/n) = j/n$ and $(U - U_n)\left(\frac{j}{n} + \varepsilon\right) = \varepsilon$. Thus on $B_j \cap A_{j+1}^c$, $\tau_n = \varepsilon + j/n$. It follows that $\tau_n = \frac{j}{n} + \varepsilon$ for the unique $j \leq J$ such that $\omega \in B_j \cap A_{j+1}^c$ (these sets are disjoint), where $B_0 := \Omega$, if such a j exists, otherwise $\tau_n = 2$. This proves the Claim.

Now continuing the proof of Proposition 1.35, $X_{(1)}$ has distribution function $1 - (1 - x_1)^n$ and so density $n(1 - x_1)^{n-1}$ for $0 \leq x_1 \leq 1$. For $1 \leq i < n$, conditional on $X_{(1)}, \ldots, X_{(i)}$, $X_{(i+1)}$ is the least of $n - i$ variables i.i.d. $U[X_{(i)}, 1]$ (this conditional distribution only depends on $X_{(i)}$). Thus $\Pr(X_{(i+1)} \geq t | X_{(i)}) = ((1 - t)/(1 - X_{(i)}))^{n-i}$, and the conditional density of $X_{(i+1)}$ given $X_{(i)} = x_i$ is $(n - i)(1 - x_{i+1})^{n-i-1}/(1 - x_i)^{n-i}$. Iterating, the joint density of

$X_{(1)}, \ldots, X_{(j+1)}$ is $n!(1 - x_{j+1})^{n-j-1}/(n - j - 1)!$ for $0 \le x_1 \le x_2 \le \cdots \le x_j \le x_{j+1} \le 1$ and 0 elsewhere. Thus

$$\Pr\left(\tau_n = \varepsilon + \frac{j}{n}\right) = \frac{n!}{(n - j - 1)!} I_j J_j$$

where

$$I_j := \int_0^\varepsilon dx_1 \int_{x_1}^{\varepsilon+1/n} dx_2 \cdots \int_{x_{j-1}}^{\varepsilon+(j-1)/n} dx_j, \qquad (1.105)$$

$$J_j := \int_{\varepsilon+(j/n)}^1 (1 - x_{j+1})^{n-j+1} dx_{j+1} = \left(1 - \varepsilon - \frac{j}{n}\right)^{n-j} /(n - j), \qquad (1.106)$$

and a $(j + 1)$-fold integral equals the given product because $x_j \le \varepsilon + (j - 1)/n$ and $x_{j+1} \ge \varepsilon + j/n$ imply $x_j \le x_{j+1}$. Also, $(j/n) + \varepsilon < 1$ follows from $j \le J < n - \lambda\sqrt{n}$. So, to prove Proposition 1.35 it remains to show that

$$I_j = \varepsilon \left(\frac{j}{n} + \varepsilon\right)^{j-1} /j!. \qquad (1.107)$$

This will be proved for each η with $0 < \eta \le 1 - \frac{j-1}{n}$ in place of ε and by induction on j. Equation (1.107) holds for $j = 1$. Assume it holds for $j - 1$ for some j with $2 \le j \le J$, and for each η with $0 \le \eta \le 1 - (j - 2)/n$ in place of ε. In the integral (1.105) make the changes of variables $\xi_i = x_i - x_1$ for $i = 2, \ldots, j$ and let $\delta := \varepsilon - x_1 + \frac{1}{n}$. Then

$$I_j = \int_0^\varepsilon dx_1 \int_0^\delta d\xi_2 \int_{\xi_2}^{\delta+1/n} d\xi_3 \cdots \int_{\xi_{j-1}}^{\delta+(j-2)/n} d\xi_j. \qquad (1.108)$$

Applying the induction hypothesis to the inner $(j - 1)$-fold integral in (1.108) we get

$$I_j = \int_0^\varepsilon dx_1 \delta \left(\frac{j-1}{n} + \delta\right)^{j-2} /(j - 1)!$$

$$= \int_0^\varepsilon dx_1 \left(\varepsilon - x_1 + \frac{1}{n}\right) \left(\frac{j}{n} + \varepsilon - x_1\right)^{j-2} /(j - 1)!,$$

so setting $y := \varepsilon - x_1$ gives

$$(j - 1)! I_j = \int_0^\varepsilon \left(y + \frac{1}{n}\right) \left(y + \frac{j}{n}\right)^{j-2} dy.$$

An integration by parts then gives that $(j-1)! I_j$ equals

$$
\frac{1}{j-1}\left(y+\frac{1}{n}\right)\left(y+\frac{j}{n}\right)^{j-1}\Big|_0^\varepsilon - \frac{1}{j-1}\int_0^\varepsilon \left(y+\frac{j}{n}\right)^{j-1} dy
$$

$$
= \frac{1}{j-1}\left[\left(\varepsilon+\frac{1}{n}\right)\left(\varepsilon+\frac{j}{n}\right)^{j-1} - \frac{1}{n}\left(\frac{j}{n}\right)^{j-1} - \frac{1}{j}\left(y+\frac{j}{n}\right)^{j}\Big|_0^\varepsilon\right]
$$

$$
= \frac{1}{j-1}\left[\left(\varepsilon+\frac{1}{n}\right)\left(\varepsilon+\frac{j}{n}\right)^{j-1} - \frac{j^{j-1}}{n^j} - \frac{1}{j}\left\{\left(\varepsilon+\frac{j}{n}\right)^{j} - \left(\frac{j}{n}\right)^{j}\right\}\right]
$$

$$
= \frac{1}{j-1}\left[\left(\varepsilon+\frac{1}{n}\right)\left(\varepsilon+\frac{j}{n}\right)^{j-1} - \frac{1}{j}\left(\varepsilon+\frac{j}{n}\right)^{j}\right]
$$

$$
= \frac{1}{j-1}\left[\left(\varepsilon+\frac{j}{n}\right)^{j-1}\left\{\varepsilon+\frac{1}{n}-\frac{\varepsilon}{j}-\frac{1}{n}\right\}\right] = \varepsilon\left(\varepsilon+\frac{j}{n}\right)^{j-1}/j,
$$

which proves (1.107) and thus Proposition 1.35. □

Next, for a Brownian bridge $y=\{y_t\}_{0\le t\le 1}$, and $\lambda>0$, let $\tau_\lambda := \inf\{s>0 : y_s \ge \lambda\}$ if this is less than 1, otherwise let $\tau_\lambda = 2$. Then for $0<s<1$,

$$
\Pr(\tau_\lambda \le s) = 1 - \Phi\left(\frac{\lambda}{\sqrt{s(1-s)}}\right) + \exp(-2\lambda^2)\left(1 - \Phi\left(\frac{(1-2s)\lambda}{\sqrt{s(1-s)}}\right)\right)
$$

by Lemma 1.30. Let $f_\lambda(s) := d\Pr(\tau_\lambda \le s)/ds$. Then for the standard normal density ϕ,

$$
f_\lambda(s) = -\lambda\phi\left(\frac{\lambda}{\sqrt{s(1-s)}}\right)A_1(s) - \lambda e^{-2\lambda^2}\phi\left(\frac{(1-2s)\lambda}{\sqrt{s(1-s)}}\right)A_2(s) \quad (1.109)
$$

where

$$
A_1(s) = -\frac{-(1-s)+s}{2s^{3/2}(1-s)^{3/2}} = \frac{2s-1}{2s^{3/2}(1-s)^{3/2}},
$$

$$
A_2(s) = -\frac{2}{\sqrt{s(1-s)}} - \frac{(1-2s)^2}{2s^{3/2}(1-s)^{3/2}}
$$

$$
= \frac{-4s+4s^2-(1-2s)^2}{2s^{3/2}(1-s)^{3/2}} = -\frac{1}{2s^{3/2}(1-s)^{3/2}}.
$$

Next,

$$
-2\lambda^2 - \frac{(1-2s)^2\lambda^2}{2s(1-s)} = -\frac{\lambda^2}{2s(1-s)}.
$$

Combining terms gives a formula which Massart attributes to E. Csáki (1974),

$$f_\lambda(s) = -\frac{\lambda}{\sqrt{2\pi}} \exp\left(-\frac{\lambda^2}{2s(1-s)}\right) \cdot \frac{2s-2}{2s^{3/2}(1-s)^{3/2}}$$

$$= \frac{\lambda}{\sqrt{2\pi}} \frac{1}{s^{3/2}\sqrt{1-s}} \exp\left(-\frac{\lambda^2}{2s(1-s)}\right). \tag{1.110}$$

From the definitions we have for each $\lambda > 0$

$$\Pr(D_n^- > \lambda) = \sum_{0 \le j < n - \lambda\sqrt{n}} p_{\lambda,n}(j) \tag{1.111}$$

and using Theorem 1.5,

$$\exp(-2\lambda^2) = \Pr(\tau_\lambda \le 1) = \int_0^1 f_\lambda(s)ds. \tag{1.112}$$

The next fact is one of the main steps in the proof of Theorem 1.34.

Proposition 1.36 *Let j be a nonnegative integer with $j < n - \lambda\sqrt{n}$. Let $s = (2\varepsilon/3) + j/n$, $s' = 1 - s$, and $v_n(s) = 1/\left(s(s^2 - 1/(4n^2))\right)$. If $n\varepsilon \ge 2$, then*

$$p_{\lambda,n}(j) \le \frac{1}{n}\left(1 - \frac{\varepsilon}{3s'} + \frac{\varepsilon^2}{6s'^2}\right) E_{n,\lambda,s} \tag{1.113}$$

where

$$E_{n,\lambda,s} = \exp\left(\frac{0.4}{ns} - \frac{\varepsilon^2}{24n}(v_n(s) + v_n(s'))\right) f_\lambda(s). \tag{1.114}$$

Some lemmas and other facts will be used to prove Proposition 1.36. The first one has implications for the binomial distribution.

Lemma 1.37 *Let $0 < \varepsilon < q = 1 - p < 1$. Let*

$$h(p, \varepsilon) = (p + \varepsilon)\log\left(\frac{p + \varepsilon}{p}\right) + (q - \varepsilon)\log\left(\frac{q - \varepsilon}{q}\right).$$

For $t \ge 0$ let

$$g(t) = t - \frac{t^2}{2(1 + 2t/3)} - \log(1 + t).$$

Then

(i) g is a strictly increasing convex function with $g(t)/t \to 1/4$ as $t \to +\infty$,

(ii) For $t := \varepsilon/(q - \varepsilon)$,

$$h(p, \varepsilon) \ge \frac{\varepsilon^2}{2(p + \varepsilon/3)(q - \varepsilon/3)} + \varepsilon g(t)/t.$$

Proof. For (i), we have $g(0) = 0$, and for all $t > 0$

$$g'(t) = (t^3/9)(1 + 2t/3)^{-2}(1 + t)^{-1} > 0.$$

To see that g' is increasing, note that $(3 + 2t)^2(1 + t)/t^3$ is decreasing. So g is convex. As $t \to +\infty$, $g(t)/t \to 1 - 1/(4/3) = 1/4$ as stated.

For (ii), clearly

$$(q - \varepsilon) \log((q - \varepsilon)/q) = -(\varepsilon/t) \log(1 + t).$$

Let $x := p + \varepsilon$. Then $0 < \varepsilon \leq x$. We need to show

$$x \log x - x \log(x - \varepsilon)$$

$$\geq \frac{\varepsilon^2}{2 \left(x - \frac{2\varepsilon}{3} \right) \left(1 - x + \frac{2\varepsilon}{3} \right)} + (q - \varepsilon) \left[\frac{\varepsilon}{q - \varepsilon} - \frac{\varepsilon^2/(q - \varepsilon)^2}{2 \left(1 + 2\varepsilon/\{3(q - \varepsilon)\} \right)} \right]$$

$$= \frac{\varepsilon^2}{2 \left(x - \frac{2\varepsilon}{3} \right) \left(1 - x + \frac{2\varepsilon}{3} \right)} + \varepsilon - \frac{\varepsilon^2/(q - \varepsilon)}{2(1 + 2\varepsilon/(3(q - \varepsilon)))}.$$

The last term equals $-\varepsilon^2/(1 - x + 2\varepsilon/3)$, and so we need to show

$$x \log(x) - x \log(x - \varepsilon) - \frac{\varepsilon^2}{2(x - 2\varepsilon/3)} - \varepsilon \geq 0.$$

The left side of the last inequality is 0 when $\varepsilon = 0$. Its derivative with respect to ε is

$$(\varepsilon^3/9)(x - \varepsilon)^{-1}(x - 2\varepsilon/3)^{-2} > 0.$$

Thus (ii) holds. □

A consequence for the binomial distribution is:

Theorem 1.38 *Let S_n be a binomial (n, p) random variable and suppose that $0 < \varepsilon < q = 1 - p$. Then*

$$\Pr(S_n - np \geq n\varepsilon) \leq \exp \left(-\frac{n\varepsilon^2}{2(p + \varepsilon/3)(q - \varepsilon/3)} \right).$$

Proof. The probability is less or equal to

$$\left\{ \left(\frac{p}{p + \varepsilon} \right)^{p+\varepsilon} \left(\frac{q}{q - \varepsilon} \right)^{q-\varepsilon} \right\}^n = \exp(-nh(p, \varepsilon))$$

by Chernoff's inequality, which follows from one of Hoeffding's inequalities, Theorem 1.13. Then we can apply Lemma 1.37(ii). □

Now, to begin the proof of Proposition 1.36 for $j \geq 1$, recall Theorem 1.17 giving Stirling's formula with error bounds. From it, for $1 \leq j < n$,

$$\binom{n}{j} \leq \frac{1}{\sqrt{2\pi}} \sqrt{\frac{n}{j(n - j)}} n^n j^{-j}(n - j)^{-(n-j)} C_j$$

where $C_j := \exp(-1/(12j+1))$, using that $\frac{1}{n} - \frac{1}{n-j} < 0$. Then by (1.104),

$$p_{\lambda,n}(j) \leq \frac{\lambda}{\sqrt{2\pi}} \frac{n}{\sqrt{j(n-j)}} (j+\lambda\sqrt{n})^{-1}$$

$$\times \left(\frac{j+\lambda\sqrt{n}}{j}\right)^j \left(\frac{n-j-\lambda\sqrt{n}}{n-j}\right)^{n-j} C_j.$$

Recalling $h(\cdot, \cdot)$ as in Lemma 1.37 and that $s = 2\varepsilon/3 + j/n$ and $s' = 1 - s$, it follows that

$$p_{\lambda,n}(j) \leq \frac{\lambda}{n\sqrt{2\pi}} \left(s - \frac{2\varepsilon}{3}\right)^{-1/2} \left(s + \frac{\varepsilon}{3}\right)^{-1} \left(s' + \frac{2\varepsilon}{3}\right)^{-1/2}$$

$$\cdot \exp\left(-nh\left(s' - \frac{\varepsilon}{3}, \varepsilon\right)\right) C_j.$$

Define $\psi(t)$ for $0 \leq t < \infty$ by

$$\psi(t) = -\log(1+t) + \frac{3}{2}\log\left(1 + \frac{2t}{3}\right). \tag{1.115}$$

Setting $t = \varepsilon/(s - 2\varepsilon/3)$, which agrees with the definition of t in Lemma 1.37(ii)), that Lemma gives

$$p_{\lambda,n}(j) \leq \frac{1}{n}\left(1 + \frac{2\varepsilon}{3s'}\right)^{-1/2} \frac{s^{3/2}}{\sqrt{s - 2\varepsilon/3}(s + \varepsilon/3)} C_j \exp\left(-\frac{n\varepsilon g(t)}{t}\right) f_\lambda(s),$$

or equivalently

$$p_{\lambda,n}(j) \leq \frac{C_j}{n}\left(1 + \frac{2\varepsilon}{3s'}\right)^{-1/2} \exp\left(-\frac{n\varepsilon g(t)}{t} + \psi(t)\right) f_\lambda(s). \tag{1.116}$$

To bound the exponentiated term, we have the following.

Lemma 1.39 *Let $\theta := 0.4833$. Let g and ψ be the functions defined in Lemma 1.37 and (1.115) respectively. For $t > 0$ and $v > 0$ let*

$$T(v, t) := v^2 g(t) - vt\psi(t) + \frac{\theta t^2}{1 + 2t/3}.$$

Then $T(v, t) > 0$ for $0 < t \leq v$.

Proof. (a) First suppose $v = t > 0$. Then, straightforwardly we have

$$\frac{d}{dt}\left(\frac{T(t, t)}{t^2}\right) = \frac{-2\theta/3 - t/3 + t^2/9}{(1 + 2t/3)^2}. \tag{1.117}$$

The quadratic in the numerator has two roots, one being negative, and a positive root $t_0 = (3 + \sqrt{9 + 24\theta})/2$. The derivative in (1.117) equals $-2\theta/3 < 0$ when $t = 0$, so t_0 is a relative minimum of $T(t, t)/t^2$ and is the absolute minimum for

$t > 0$. We find $T(t_0, t_0)/t_0^2 \doteq 5.05 \cdot 10^{-6} > 5 \cdot 10^{-6} > 0$, so $T(t, t) > 0$ for all $t > 0$.

(b) Now for general $0 < t \leq v$, T is a quadratic polynomial in v for fixed t. Its derivative with respect to v is $2vg(t) - t\psi(t)$. If

$$2g(t) - \psi(t) > 0 \tag{1.118}$$

then $T(v, t)$ is increasing in v for all $v > t$ and so $T(v, t) > 0$. Or, if the discriminant $D(t)$ of the quadratic polynomial satisfies

$$D(t) = t^2 \psi^2(t) - 4g(t)\theta t^2/(1 + 2t/3) < 0,$$

or equivalently

$$\Delta(t) := (1 + 2t/3)\psi^2(t) - 4\theta g(t) < 0, \tag{1.119}$$

then $T(t, v)$ remains positive for all $v \geq t$ as it is for $v = t$ and has no roots. So to prove Lemma 1.39 it will suffice to show that

(i) $2g(t) - \psi(t) > 0$ for all $t \geq 3.37$.

(ii) $\Delta(t) < 0$ for $0 < t \leq 3.37$.

Proof of (i). We have

$$2g'(t) - \psi'(t) = \frac{t}{9}(2t^2 - 2t - 3)(1 + t)^{-1}\left(1 + \frac{2t}{3}\right)^{-2}.$$

We have $2t^2 - 2t - 3 > 0$ for $t > (1 + \sqrt{7})/2$; thus $2g - \psi$ is increasing for such t. Since $(1 + \sqrt{7})/2 \doteq 1.823 < 3.37$ we have that for $t \geq 3.37$,

$$2g(t) - \psi(t) \geq 2g(3.37) - \psi(3.37) \doteq 0.000775 > 7 \cdot 10^{-4} > 0,$$

proving (i).

For (ii), Massart states that the function $R(t) = \left(1 + \frac{2t}{3}\right)\psi^2(t)/g(t)$ is increasing for $t > 0$ (which is needed only for $t \leq 3.37$). The proof given in the first half of p. 1275 of his paper is not correct, as the function $R_0 = \psi^2/g$ is not increasing. Nevertheless it appears true that $R(t)$ is increasing by examining computed values of it on a grid, $0, 0.01, 0.02, \ldots, 3.37$. It follows that (ii) holds. □

To continue the proof of Proposition 1.36, three Claims will be stated and proved.

Claim 1. Let $\beta := 0.826$. Then for any $x \in [0, 1]$,

$$(1 + 2x)^{-1/2} \leq \left(1 - x + \frac{3x^2}{2}\right)\exp(-\beta x^3).$$

Proof of Claim 1. Let $\alpha = 135/32$ and $\rho(y) := 1 + \alpha y - \exp(2\beta y)$ for $0 \leq y \leq 1$. Then $\rho''(y) < 0$, so ρ is concave. It satisfies $\rho(0) = 0$ and $\rho(1) \geq 0.00134 >$

0. So $\rho(y) \geq 0$ for $0 \leq y \leq 1$. It follows that for $0 \leq x \leq 1$,

$$\exp(2\beta x^3) \leq 1 + \alpha x^3 \leq 1 + \alpha x^3 + \frac{1}{32}x^3(5 - 12x)^2$$

$$= \left(1 - x + 1.5x^2\right)^2 (1 + 2x).$$

Taking the square root of both sides, Claim 1 follows. $\square$

Claim 2. For $\varepsilon = \lambda/\sqrt{n}$ as usual and j, s, s', and $v_n(\cdot)$ as defined in Proposition 1.36, if $n\varepsilon \geq 2$ and $ns' \geq 1$, we have

$$\left(1 + \frac{2\varepsilon}{3s'}\right)^{-1/2} \leq \left(1 - \frac{\varepsilon}{3s'} + \frac{\varepsilon^2}{6s'^2}\right) \exp\left(-\frac{\varepsilon^2 v_n(s')}{24n}\right). \quad (1.120)$$

Proof of Claim 2. First it will be checked that $\varepsilon/(3s') \leq 1$, or equivalently $\varepsilon \leq 3s' = 3 - 2\varepsilon - 3j/n$, or $\varepsilon = \lambda/\sqrt{n} \leq 1 - j/n$, which is true since $j < n - \lambda\sqrt{n}$ by (1.103). So we can apply Claim 1 with $x = \varepsilon/(3s')$. To prove (1.120) we need to show that

$$\beta\varepsilon^3/(27s'^3) \geq \varepsilon^2 v_n(s')/(24n),$$

or equivalently

$$8\beta/9 \geq \left(n\varepsilon \left(1 - (2ns')^{-2}\right)\right)^{-1}.$$

Since $n\varepsilon \geq 2$ and $ns' \geq 1$, we have

$$\left(n\varepsilon \left(1 - (2ns')^{-2}\right)\right)^{-1} \leq \frac{2}{3} < \frac{8\beta}{9}$$

as $8\beta/9 \doteq 0.73422$. So Claim 2 is proved. $\square$

Claim 3. For v_n as defined in Proposition 1.36, and any $\varepsilon > 0$ and $s > 0$ satisfying $1/n \leq \varepsilon \leq 3s/2$, we have

$$\left(1 + 12n\left(s - \frac{2\varepsilon}{3}\right)\right)^{-1} \geq \frac{1}{12ns} + \frac{\varepsilon^2 v_n(s)}{24n}. \quad (1.121)$$

Proof of Claim 3. We have identically

$$\left(1 + 12n\left(s - \frac{2\varepsilon}{3}\right)\right)^{-1} - \frac{1}{12ns}$$

$$= (8n\varepsilon - 1)(12ns)^{-1}\left(1 + 12n\left(s - \frac{2\varepsilon}{3}\right)\right)^{-1}.$$

So (1.121) reduces to

$$\frac{8n\varepsilon - 1}{1 + 12n(s - 2\varepsilon/3)} \geq \frac{\varepsilon^2}{2} \Big/ \left(s^2 - \frac{1}{4n^2}\right),$$

or equivalently

$$2(8n\varepsilon - 1)s^2 - 12n\varepsilon^2 s + (8n\varepsilon - 1)\left(\varepsilon^2 - \frac{1}{2n^2}\right) \geq 0.$$

The derivative of the left side with respect to s is $4(8n\varepsilon - 1)s - 12n\varepsilon^2$, which is positive for $s \geq 2\varepsilon/3$, so the left side is increasing and it suffices to check the inequality for $s = 2\varepsilon/3$. So we want to show that

$$(8n\varepsilon - 1)(34n^2\varepsilon^2 - 9)/(18n^2) \geq 8n\varepsilon^3,$$

for which, since $n\varepsilon \geq 1$, it suffices to show that $(7n\varepsilon)(25n^2\varepsilon^2) \geq 144n^3\varepsilon^3$, which is true. So (1.121) holds and Claim 3 is proved. $\qquad\square$

Now to finish the proof of Proposition 1.36 for $j \geq 1$, recall (1.116), in which $t = \varepsilon/\left(s - \frac{2\varepsilon}{3}\right)$, implying $t = n\varepsilon/j \leq n\varepsilon$. Lemma 1.39 with $v = n\varepsilon$ gives

$$p_{\lambda,n}(j) \leq \frac{C_j}{n}\left(1 + \frac{2\varepsilon}{3s'}\right)^{-1/2} \exp\left(\frac{\theta}{ns}\right) f_\lambda(s).$$

Claim 2 gives an upper bound for $\left(1 + \frac{2\varepsilon}{3s'}\right)^{-1/2}$. By Claim 3, as $s - (2\varepsilon)/3 = j/n$, we have

$$\log(C_j) = -\frac{1}{12j+1} \leq -\frac{1}{12ns} - \frac{\varepsilon^2 v_n(s')}{24n}.$$

Noting that $\theta - 1/12 < 0.4$ finishes the proof for $j \geq 1$.

Proof of Proposition 1.36 for $j = 0$. In this case $s = 2\varepsilon/3$. We have $p_{\lambda,n}(0) = (1-\varepsilon)^n$ by Proposition 1.35. One can naturally define $h(p, \delta)$ when $\delta = q$ as $-\log(p)$ since for fixed q with $0 < q < 1$, $(q - \delta)\log((q - \delta)/q) \to 0$ as $\delta \uparrow q$. Then $(1 - \varepsilon)^n = \exp(-nh(1 - \varepsilon, \varepsilon))$. To apply Lemma 1.37, we will have $t = \varepsilon/(q - \varepsilon)$, which will be defined as $+\infty$ in this case with $q = \varepsilon$; for $t \to +\infty$ the limit of $g(t)/t$ is $1/4$. Then Lemma 1.37 gives

$$p_{\lambda,n}(0) = (1 - \varepsilon)^n = \exp(-nh(1 - \varepsilon, \varepsilon)) \leq \exp\left(-\frac{\lambda^2}{2ss'} - \frac{n\varepsilon}{4}\right).$$

Define $H(v)$ for $v > 0$ by

$$H(v) := \frac{3\log(3/2)}{2} - \frac{\log(2\pi)}{2} + \frac{v}{4} + \frac{0.4}{v} - \frac{\log(v)}{2}.$$

Then it is straightforward to check that

$$(1-\varepsilon)^n \le \frac{\lambda}{\sqrt{2\pi n}} s^{-3/2} \exp\left(\frac{0.4}{n\varepsilon}\right) \exp\left(-\frac{\lambda^2}{2ss'}\right) \exp(-H(n\varepsilon)).$$

We have $H'(v) = \frac{1}{4} - \frac{0.4}{v^2} - \frac{1}{2v} = 0$ if and only if $v^2 - 2v - 1.6 = 0$. The only positive root of this is at $v_0 = 1 + \sqrt{2.6}$. This is the minimum of H for $v > 0$ because $H(v) \to +\infty$ as $v \to +\infty$. Thus $H(v) \ge H(v_0) \ge 0.01534 > 0$ for all $v > 0$. It follows that

$$p_{\lambda,n}(0) \le \frac{\lambda}{\sqrt{2\pi n}} s^{-3/2} \exp\left(\frac{0.4}{n\varepsilon}\right) \exp\left(-\frac{\lambda^2}{2ss'}\right)$$

$$\le \frac{1}{n}\left(1 + \frac{2\varepsilon}{3s'}\right)^{-1/2} \exp\left(\frac{0.4}{n\varepsilon}\right) f_\lambda(s),$$

where the second equation follows on expanding $f_\lambda(s)$ by (1.110).

Since $n\varepsilon \ge 2$, it follows that

$$\frac{\varepsilon^2 v_n(s)}{24n} \le \left(16n\varepsilon\left(\frac{4}{9} - \frac{1}{16}\right)\right)^{-1} \le 9/(55n\varepsilon),$$

from which it follows that

$$\frac{0.4}{n\varepsilon} + \frac{\varepsilon^2 v_n(s)}{24n} \le \frac{0.4 + 9/55}{n\varepsilon} \le \frac{0.4}{ns}.$$

Combining gives

$$P_{\lambda,n}(0) \le \frac{1}{n}\left(1 + \frac{2\varepsilon}{3s'}\right)^{-1/2} \exp\left(\frac{0.4}{ns} - \frac{\varepsilon^2 v_n(s)}{24n}\right) f_\lambda(s).$$

Via Claim 2, (1.113) follows, and Proposition 1.36 is proved. □

In the proof of Theorem 1.34, the integral in (1.112) will be compared to Riemann sums and thereby to the sums in (1.111). The comparison will use the next lemma.

Lemma 1.40 *Let $0 < \delta \le s \le 1 - \delta$ and $s' = 1 - s$. If G is a continuous function with $G(x) > 0$ for $s - \delta \le x \le s + \delta$ and $\log(G)$ is convex, then for any $\lambda > 0$,*

$$\frac{1}{2\delta} \int_{s-\delta}^{s+\delta} G(u) \exp\left(-\frac{\lambda^2}{2u(1-u)}\right) du$$

$$\ge G(s) \exp\left(-\frac{\lambda^2}{2ss'}\right) \exp\left(-\frac{\lambda^2 \delta^2}{6}\left((s(s^2 - \delta^2))^{-1} + (s'(s'^2 - \delta^2))^{-1}\right)\right).$$

Proof. Jensen's inequality, e.g., RAP, 10.2.6, $\eta(EX) \le E\eta(X)$ for $\eta(\cdot)$ convex and X with finite mean, will be applied twice. Both times the probability measure is the uniform distribution $U[s - \delta, s + \delta]$. First, the convex function is exp, then, second, it is $\log(G)$. We get

$$\frac{1}{2\delta} \int_{s-\delta}^{s+\delta} G(u) \exp\left(-\frac{\lambda^2}{u(1-u)}\right) du$$

$$\geq \exp\left(\frac{1}{2\delta} \int_{s-\delta}^{s+\delta} \left(\log(G(u)) - \frac{\lambda^2}{2u(1-u)}\right) du\right)$$

$$\geq \exp\left(\log(G(s)) - \frac{1}{2\delta} \int_{s-\delta}^{s+\delta} \frac{\lambda^2}{2}(u^{-1} + (1-u)^{-1}) du\right).$$

The function $u \mapsto 1/u$ has a positive fourth derivative. We can apply Simpson's rule with remainder as given by Davis and Polonsky (1972, 25.4.5, p. 886): if f has a continuous fourth derivative $f^{(4)}$, $h > 0$, $x_i = x_0 + ih$ for $i = 0, 1, 2$, and $f_i = f(x_i)$, then

$$\int_{x_0}^{x_2} f(x)dx = \frac{h}{3}[f_0 + 4f_1 + f_2] - \frac{h^5}{90} f^{(4)}(\xi)$$

for some $\xi \in [x_0, x_2]$. So if $f^{(4)} \geq 0$,

$$\int_{x_0}^{x_2} f(x)dx \leq \frac{h}{3}[f_0 + 4f_1 + f_2]. \tag{1.122}$$

Thus

$$\frac{1}{2\delta} \int_{s-\delta}^{s+\delta} \frac{1}{u} du \leq \frac{1}{6}\left(\frac{1}{s+\delta} + \frac{1}{s-\delta} + \frac{4}{s}\right) = \frac{1}{s} + \frac{\delta^2}{3s(s^2 - \delta^2)}.$$

Next,

$$\frac{1}{2\delta} \int_{s-\delta}^{s+\delta} (1-u)^{-1} du = \frac{1}{2\delta} \int_{s'-\delta}^{s'+\delta} v^{-1} dv,$$

and the Lemma follows. □

Next are some identities for special integrals.

Lemma 1.41 *For any $a \geq 0$, $b \geq 0$, and $\lambda > 0$, let*

$$I_{a,b}(\lambda) = \frac{\lambda \exp(2\lambda^2)}{\sqrt{2\pi}} \int_0^1 u^{-a-1/2}(1-u)^{-b-1/2} \exp\left(-\frac{\lambda^2}{2u(1-u)}\right) du.$$

Then the following hold:

(i) $I_{1,1}(\lambda)/2 = I_{1,0}(\lambda) = 1$;

(ii) $I_{2,2}(\lambda)/2 = I_{2,1}(\lambda) = 4 + \lambda^{-2}$;

(iii) $I_{2,0}(\lambda) = 2 + \lambda^{-2}$.

Proof. Clearly $I_{a,b} \equiv I_{b,a}$. For any u with $0 < u < 1$,

$$u^{-1/2-a}(1-u)^{-1/2-b} - (1-u)^{-1/2-b}u^{1/2-a} = u^{-1/2-a}(1-u)^{1/2-b},$$

which implies for any $a \geq 1$ and $b \geq 1$ that

$$I_{a,b} = I_{a-1,b} + I_{a,b-1}. \tag{1.123}$$

For $a = b = 1$, using (1.112) without its middle term, we get (i). Next, differentiating with respect to λ gives (ii). Then, applying (1.123) with $a = 1$ and $b = 2$, (iii) follows from (i) and (ii). So Lemma 1.41 is proved. □

Now we can prove Theorem 1.34 under some conditions.

Proof of Theorem 1.34 for $n \geq 39$ and $\lambda \leq \sqrt{n}/2$. Since $n \geq 39$, the hypothesis on λ in Theorem 1.34 becomes $\lambda \geq \zeta n^{-1/6}$. Thus $n\varepsilon = \lambda\sqrt{n} \geq 3.6764$. Define the function $y(\cdot)$ by $y(x) := (e^x - 1)/x$ for $x > 0$. One can easily see (e.g. from the Taylor series) that this function is increasing. Recalling that $s \geq 2\varepsilon/3$ for $s = s_j = \frac{2\varepsilon}{3} + \frac{j}{n}$ as in Proposition 1.36 we have

$$\exp\left(\frac{0.4}{ns}\right) \leq 1 + y\left(\frac{0.6}{3.6764}\right)\frac{0.4}{ns}.$$

Let $\mu := 0.4345$. Then $\exp(0.4/(ns)) \leq 1 + \mu/(ns)$. Applying Proposition 1.36, we get another upper bound for $p_{\lambda,n}(j)$:

$$\frac{1}{n}\left(1 - \frac{\varepsilon}{3s'} + \frac{\varepsilon^2}{6s'^2}\right)\left(1 + \frac{\mu}{ns}\right)\exp\left(-\frac{\varepsilon^2(v_n(s) + v_n(s'))}{24n}\right)f_\lambda(s).$$

Preparing to apply Lemma 1.40, note that $z(\cdot)$ defined by $z(x) = \log(6x^2 - 2x + 1) - (5/2)\log(x)$ is convex for $x > 0$: calculation gives $z''(x) \cdot 2x^2(6x^2 - 2x + 1)^2 = h(x)$ where

$$h(x) = 36x^2(x - 1)^2 + 60x^2 - 20x + 5 > 0.$$

(Massart gives a different positive quartic polynomial, agreeing in the x^4, x, and constant terms but not the x^3 and x^2 terms.) Let for $0 < u < 1$

$$G(u) := \frac{\lambda}{\sqrt{2\pi}}\left(1 + \frac{\mu}{nu}\right)u^{-3/2}(1 - u)^{-1/2}\left(1 - \frac{\varepsilon}{3(1 - u)} + \frac{\varepsilon^2}{6(1 - u)^2}\right).$$

Then for $c := \log(\lambda/(6\sqrt{2\pi}\varepsilon))$, a constant with respect to u, we have

$$\log(G(u)) = c + \log\left(1 + \frac{\mu}{nu}\right) - \frac{3}{2}\log(u) + z\left(\frac{1 - u}{\varepsilon}\right),$$

in which each term is convex in u, so $\log G(\cdot)$ is convex. So Lemma 1.40 with $\delta = 1/(2n)$ gives

$$p_{\lambda,n}(j) \leq \int_{s-1/(2n)}^{s+1/(2n)}\left(1 - \frac{\varepsilon}{3(1 - u)} + \frac{\varepsilon^2}{6(1 - u)^2}\right)\left(1 + \frac{\mu}{nu}\right)f_\lambda(u)du.$$

Summing over j in (1.111) and using also (1.112) we get, in the notation of Lemma 1.41, recalling that $I_{a,b} \equiv I_{b,a}$,

$$\exp(2\lambda^2) \Pr(D_n^- > \lambda) \le I_{1,0}(\lambda) - \frac{\varepsilon}{3} I_{1,1}(\lambda) + \frac{\varepsilon^2}{6} I_{2,1}(\lambda)$$

$$+ \frac{\mu}{n} I_{2,0}(\lambda) - \frac{\varepsilon \mu}{3n} I_{2,1}(\lambda) + \frac{\varepsilon^2 \mu}{6n} I_{2,2}(\lambda).$$

By Lemma 1.41 and simple calculations we then get

$$\frac{3\sqrt{n}}{2\lambda} (\exp(2\lambda^2) \Pr(D_n^- > \lambda) - 1)$$

$$\le \eta_n(\lambda) := -1 + \left(\lambda + \frac{1}{4\lambda} + \frac{3\mu}{\lambda} + \frac{3\mu}{2\lambda^3}\right) n^{-1/2}$$

$$- \frac{\mu}{2}\left(4 + \frac{1}{\lambda^2}\right) n^{-1} + \frac{\mu}{2}\left(4\lambda + \frac{1}{\lambda}\right) n^{-3/2}. \tag{1.124}$$

Remark. Smirnov (1944), as quoted by Massart, had given the asymptotic expansion

$$\Pr(D_n^- > \lambda) \sim \exp(-2\lambda^2)\left(1 - \frac{2\lambda}{3\sqrt{n}} + O(1/n)\right) \tag{1.125}$$

if $\lambda = O(n^{1/6})$. By contrast, Massart's inequality (1.124) is one-sided, but the first term -1 on the right confirms that the $-2\lambda/(3\sqrt{n}$ term in Smirnov's expansion is valid non-asymptotically in a one-sided sense, which is what one wants.

To check that η_n is convex in λ for each n, note that

$$\mu\left(\frac{3}{\lambda} - \frac{1}{2\sqrt{n}\lambda^2} + \frac{3}{2\lambda^3}\right)$$

has positive second derivative and all other individual terms are convex. Thus, to show that $\eta_n(\lambda) < 0$ for $\zeta n^{-1/6} \le \lambda \le \sqrt{n}/2$ it will suffice to show that $a_n := \eta_n(\zeta n^{-1/6}) < 0$ and $b_n := \eta_n(\sqrt{n}/2) < 0$.

It will be shown that a_n and b_n are decreasing in n for $n \ge 39$. We have

$$a_n = \eta_n(\zeta n^{-1/6}) = -1 + \left(\zeta n^{-1/6} + \frac{n^{1/6}}{\zeta}\left(\frac{1}{4} + 3\mu\right) + \frac{3\mu}{2}\frac{n^{1/2}}{\zeta^3}\right) n^{-1/2}$$

$$- \frac{\mu}{2}\left(4 + \frac{n^{1/3}}{\zeta^2}\right) n^{-1} + \frac{\mu}{2}\left(4\zeta n^{-1/6} + \frac{n^{1/6}}{\zeta}\right) n^{-3/2}$$

$$= -1 + \frac{3\mu}{2\zeta^3} + n^{-1/3}\left(\frac{1}{\zeta}\right)\left(\frac{1}{4} + 3\mu\right) + \left(\zeta - \frac{\mu}{2\zeta^2}\right) n^{-2/3}$$

$$- 2\mu n^{-1} + \frac{\mu}{2\zeta} n^{-4/3} + 2\mu\zeta n^{-5/3}.$$

As $\zeta = 1.0841$ (Theorem 1.34) and $\mu = 0.4345$, $\zeta - \mu/(2\zeta)^2 > 0$. Terms with positive coefficients and negative powers of n are decreasing. Just one term, $-2\mu n^{-1}$, is increasing. It will suffice to show that $(3/\zeta)n^{-1/3} - 2n^{-1}$ is decreasing for $n \geq 39$, or that $3x - 2.1682x^3$ is increasing for $0 < x \leq 1/39$. Indeed its derivative is positive there. A calculation shows that b_n is a constant plus a sum of negative powers of n times positive coefficients, so it is also decreasing in n. We have $a_{39} \doteq -0.006382 < -0.006 < 0$ and $b_{39} \doteq -0.4238 < -0.4 < 0$, so both a_n and b_n are negative for all $n \geq 39$, so $\eta_n(\lambda) < 0$ for $\zeta n^{-1/6} \leq \lambda \leq \sqrt{n}/2$. So Theorem 1.34 is proved for $n \geq 39$ and $\lambda \leq \sqrt{n}/2$.

For the rest of the proof of Theorem 1.34, let $C_{\lambda,n} = \exp(2\lambda^2) \Pr(D_n^- > \lambda)$.

Proposition 1.42 Let $n \geq 2$ and let λ be such that $0 < \lambda < \sqrt{n}$. Then

(i) For $\lambda \geq \sqrt{n}/2$, $\frac{d}{d\lambda}C_{\lambda,n} \leq 0$,

(ii) $\sum_{j=1}^{n-1} j^{j-1}(n - j)^{n-j} n^{-n}\binom{n}{j} \leq 1$,

(iii) For $\lambda \geq 1/2$, we have $\frac{d}{d\lambda}C_{\lambda,n} \leq 3.61$.

Proof. (i) Let $L_{\lambda,n}(j) = \log(\exp(2\lambda^2)p_{\lambda,n}(j))$ for $0 \leq j < n - \lambda\sqrt{n}$. It will suffice by (1.111) to show that for each such j, $dL_{\lambda,n}/d\lambda < 0$ for $\lambda \geq \sqrt{n}/2$. We have by (1.35)

$$\frac{d}{d\lambda}L_{\lambda,n}(j) = 4\lambda + \frac{j}{\lambda(j + \lambda\sqrt{n})} - \frac{\lambda n^2}{(j + \lambda\sqrt{n})(n - j - \lambda\sqrt{n})}.$$

Recall that $\varepsilon = \lambda/\sqrt{n}$. Set $x = \varepsilon + j/n$, noting that $0 < x < 1$ by the restriction on j. Next,

$$\frac{d}{d\lambda}L_{\lambda,n}(j) = -\frac{n\varepsilon^2(1 - 2x)^2 - (x - \varepsilon)(1 - x)}{\lambda x(1 - x)}. \tag{1.126}$$

For $j = 0$ it follows easily that $dL_{\lambda,n}(0)/d\lambda \leq 0$, so we can assume that $j \geq 1$ and so $x \geq \varepsilon + 1/n$. To show that $n\varepsilon^2(1 - 2x)^2 - (x - \varepsilon)(1 - x) \geq 0$, or equivalently

$$\frac{j}{n} + \frac{j}{n}\varepsilon \leq 2\varepsilon\frac{j}{n} + \frac{j^2}{n^2} + n\varepsilon^2\left(2\varepsilon - 1 + \frac{2j}{n}\right)^2,$$

we can note that $4\varepsilon^2 \geq 2\varepsilon \geq 1$, which implies $n\varepsilon^2(4j^2/n^2) \geq j^2/n$. Since $j/n \leq j^2/n$, (i) holds.

(ii) Clearly $\Pr(D_n^- > \lambda)$ is nonincreasing in $\lambda \geq 0$. By (1.111) and (1.104) we have

$$0 \geq \frac{d}{d\lambda}\Pr(D_n^- > \lambda)\Big|_{\lambda=0} = \sqrt{n}\left(\sum_{j=1}^{n-1}j^{j-1}(n - j)^n j^n n^{-n}\binom{n}{j} - 1\right),$$

so (ii) follows.

(iii) First suppose $n\varepsilon \geq 2$ and $\varepsilon \leq 1/2$. We use Proposition 1.36 and the bound $\exp(0.4/(ns)) \leq \exp(0.3)$, and proceed in the same way as in the proof of Theorem 1.34 for $n \geq 39$ and $\varepsilon \leq 1/2$, with $G(u)$ replaced by $G_1(u)$ defined to have a factor $e^{0.3}$ in place of $1 + \mu/(nu)$. Then $\log G_1(u)$ is convex and we get

$$C_{\lambda,n} \leq e^{0.3} \left(I_{1,0}(\lambda) - \frac{\varepsilon}{3}I_{1,1}(\lambda) + \frac{\varepsilon^2}{6}I_{2,1}(\lambda) \right).$$

Using Lemma 1.41 and $2/\sqrt{n} \leq \lambda \leq \sqrt{n}/2$, it follows that

$$C_{\lambda,n} \leq e^{0.3} + \frac{2\lambda}{3\sqrt{n}}e^{0.3}\left(-1 + \left(\lambda + \frac{1}{4\lambda}\right) n^{-1/2} \right)$$

$$\leq e^{0.3} + \frac{2\lambda}{3\sqrt{n}}e^{0.3}\left(-\frac{3}{8} \right).$$

Combining with (i) gives

$$C_{\lambda,n} \leq \exp(\max((8/n), 0.3)) \tag{1.127}$$

for any integer $n \geq 4$ and any $\lambda > 0$.

An alternate bound for $C_{\lambda,n}$ will be helpful for small values of n. For h as defined in Lemma 1.37, we have for $j \geq 1$

$$p_{\lambda,n}(j) = \lambda\sqrt{n}\binom{n}{j}j^{j-1}(n-j)^{n-j}n^{-n}\left(\frac{j}{j + \lambda\sqrt{n}} \right)$$

$$\times \exp\left(-nh\left(1 - \varepsilon - \frac{j}{n}, \varepsilon \right) \right).$$

By Lemma 1.37 we then get

$$p_{\lambda,n}(j) \leq \lambda\sqrt{n}\binom{n}{j}j^{j-1}(n-j)^{n-j}n^{-n}\exp(-2\lambda^2).$$

Thus by (ii),

$$C_{\lambda,n} \leq \lambda\sqrt{n} + p_{\lambda,n}(0)\exp(2\lambda^2). \tag{1.128}$$

It follows from (1.126) that for $j = 0$, $dL_{\lambda,n}(0)/d\lambda < 0$, and for $1 \leq j < n - \lambda\sqrt{n}$, $dL_{\lambda,n}(j)/d\lambda < 1/\lambda$. Thus

$$\frac{d}{d\lambda}C_{\lambda,n} \leq \frac{1}{\lambda}\left(C_{\lambda,n} - p_{\lambda,n}(0)\exp(2\lambda^2) \right).$$

Combining this last inequality with (1.127) if $n \geq 14$, or (1.128) for $n \leq 13$, we get

$$\frac{d}{d\lambda}C_{\lambda,n} \leq \max(2\exp(4/7), \sqrt{13}) \leq 3.61,$$

which proves (iii) and so Proposition 1.42. $\qquad\square$

Proof of Theorem 1.34 for $n \leq 38$ or $\lambda > \sqrt{n}/2$. By Proposition 1.42 and the first part of the proof of Theorem 1.34, we can assume that $n \leq 38$. Then the assumption on λ in Theorem 1.34 reduces to $\lambda \geq \sqrt{(\log 2)/2}$, which implies $\lambda > 1/2$.

Letting $\eta := 0.01$, let $\Lambda_{\eta,n} = \{\frac{1}{2} + k\eta : k \in \mathbb{N}\} \cap [\frac{1}{2}, \sqrt{n})$. A computer calculation, reported by Massart (1990), gave

$$\max_{n \leq 38} \max_{\lambda \in \Lambda_{\eta,n}} C_{\lambda,n} \leq 0.951. \tag{1.129}$$

In a confirming computation, the maximum was found to be $\doteq 0.94955$, attained at $n = 38$ and $\lambda = 1/2$. Combining (1.129) with Proposition 1.42(iii), we get

$$\max_{n \leq 38} \sup_{1/2 \leq \lambda < \sqrt{n}} C_{\lambda,n} \leq 0.951 + 3.61\eta \leq 0.9871 < 1,$$

which finishes the proof of Theorem 1.34 for $n \leq 38$ and so completes its proof, and that of Theorem 1.32 by the Remarks after Theorem 1.34.   □

Problems

1. Find the covariance matrix on $\{0, 1/4, 1/2, 3/4, 1\}$ of:

    (a) the Brownian bridge process y_t;

    (b) $U_4 - U$. *Hint:* Recall that $n^{1/2}(U_n - U)$ has the same covariances as y_t.

2. Let $0 < t < u < 1$. Let α_n be the empirical process for the uniform distribution on $[0, 1]$.

    (a) Show that the distribution of $\alpha_n(t)$ is concentrated in some finite set A_t.
    (b) Let $f(t, y, u) := E(\alpha_n(u)|\alpha_n(t) = y)$. Show that for any y in A_t,$(u, f(t, y, u))$ is on the straight line segment joining (t, y) to $(1, 0)$.

3. Let (S, d) be a complete separable metric space. Let μ be a law on $S \times S$ and let $\delta > 0$ satisfy

$$\mu(\{(x, y) : d(x, y) > 2\delta\}) \leq 3\delta.$$

Let $\pi_2(x, y) := y$ and $P := \mu \circ \pi_2^{-1}$. Let Q be a law on S such that $\rho(P, Q) < \delta$ where ρ is Prohorov's metric. On $S \times S \times S$ let $\pi_{12}(x, y, z) := (x, y)$ and $\pi_3(x, y, z) := z$. Show that there exists a law α on $S \times S \times S$ such that $\alpha \circ \pi_{12}^{-1} = \mu$, $\alpha \circ \pi_3^{-1} = Q$, and

$$\alpha(\{(x, y, z) : d(x, z) > 3\delta\}) \leq 4\delta.$$

Hint: Use Strassen's theorem, which implies that for some law ν on $S \times S$, if $\mathcal{L}(Y, Z) = \nu$, then $\mathcal{L}(Y) = P$, $\mathcal{L}(Z) = Q$, and $\nu(\{d(Y, Z) > \delta\}) < \delta$. Then the Vorob'ev–Berkes–Philipp theorem 1.31 applies.

4. Let $A = B = C = \{0, 1\}$. On $A \times B$ let

$$\mu := \left(\delta_{(0,0)} + 2\delta_{(1,0)} + 5\delta_{(0,1)} + \delta_{(1,1)}\right)/9.$$

On $B \times C$ let $\nu := [\delta_{(0,0)} + \delta_{(1,0)} + \delta_{(0,1)} + 3\delta_{(1,1)}]/6$. Find a law γ on $A \times B \times C$ such that if $\gamma = \mathcal{L}(X, Y, Z)$, then $\mathcal{L}(X, Y) = \mu$ and $\mathcal{L}(Y, Z) = \nu$.

5. Let $I = [0, 1]$ with usual metric d. For $\varepsilon > 0$, evaluate $D(\varepsilon, I, d)$, $N(\varepsilon, I, d)$ and $N(\varepsilon, I, I, d)$. *Hint*: The ceiling function $\lceil x \rceil$ is defined as the least integer $\geq x$. Answers can be written in terms of $\lceil \cdot \rceil$.

6. For a Poisson variable X with parameter $\lambda > 0$, that is, $P(X = k) = e^{-\lambda}\lambda^k/k!$ for $k = 0, 1, 2, \ldots$, evaluate the moment generating function Ee^{tX} for all t. For $M > \lambda$, find the bound for $\Pr(X \geq M)$ given by the moment generating function inequality (Theorem 1.10).

7. The bound $Ke^{-\lambda x}$ given by the right side of (1.1) is trivial if it is larger than 1. With K and λ as given in the Bretagnolle–Massart Theorem 1.8, how large must x be for the bound to be nontrivial?

8. Based on Theorem 1.8 (Bretagnolle and Massart), for $n = 225$, find b such that the empirical process α_n satisfies

$$\sup_{0 \leq t \leq 1} |(\alpha_n - Y_n)(t)| \leq b$$

for a Brownian bridge Y_n except with probability at most 0.05.

9. Kolmogorov's test of the hypothesis H_0 that $X_1, \ldots, X_n$ are i.i.d. with a given fixed distribution function F is to form the empirical distribution function F_n based on $X_1, \ldots, X_n$ and the statistic $D_n := \sqrt{n} \sup_x |(F_n - F)(x)|$, rejecting the hypothesis at a level $\alpha > 0$ if for the observed value D_n^o of D_n if $P_0(D_n \geq D_n^o) \leq \alpha$, where P_0 is the probability assuming H_0. Suppose F is continuous. For $\alpha = 0.05, 0.01$, and 0.001, find M_α such that $2\exp(-2M_\alpha^2) = \alpha$. Evaluate the expression (twice a sum) in the Remark after Theorem 1.32, which is the asymptotic distribution under H_0 for n large, for $u = M_\alpha$ for the three αs. *Hint*: In these cases the series can be written as a power series in α. The series will converge fast, so not many terms need to be added.

Notes

Notes to Section 1.1. The Glivenko–Cantelli Theorem 1.3 also appeared as Theorem 11.4.2 in RAP. RAP's Notes to Section 11.4 say that Glivenko proved the (main) case where F is continuous. Francesco Paolo Cantelli (1875–1966) was a professor at the University of Rome from 1931 through 1951 and was chief editor of the Italian actuarial journal *Giornale dell'Istituto Italiano degli Attuari* from 1930 through 1958. Both Glivenko's and Cantelli's contributions to their theorem were published in Italian in 1933 in that journal, as was the paper of Kolmogorov (1933b), with Kolmogorov's paper immediately preceding Glivenko's, and Cantelli's some 300 pages later. I have not seen Glivenko's or Cantelli's papers in the original. Kolmogorov's paper appeared in Russian translation in his selected works in 1986.

The contributions of Doob (1949) and Donsker (1952) were mentioned in the text.

When it was realized that the formulation by Donsker (1952) was incorrect because of measurability problems, Skorokhod (1956); see also Kolmogorov (1956), defined a separable metric d on the space $D[0, 1]$ of right-continuous functions with left limits on $[0, 1]$, such that convergence for d to a continuous function is equivalent to convergence for the sup norm, and the empirical process α_n converges in law in $D[0, 1]$ to the Brownian bridge; see, for example, Billingsley (1968, Chapter 3). The formulation of Theorem 1.7 avoids the need for the Skorokhod topology and deals with measurability.

Comments on the Komlós–Major–Tusnády and Bretagnolle–Massart statements (Theorem 1.8) are given in the notes to Section 1.4.

Notes to Section 1.2. Apparently the first publication on ε-entropy was the announcement by Kolmogorov (1955). Theorem 1.9, and the definitions of all the quantities in it, are given in the longer exposition by Kolmogorov and Tikhomirov (1959, Section 1, Theorem IV).

Lorentz (1966) proposed the name "metric entropy" rather than "ε-entropy," urging that functions should not be named after their arguments, as functions of a complex variable z are not called "z-functions." The name "metric entropy" emphasizes the purely metric nature of the concept. Actually, "ε-entropy" has been used for different quantities. Posner, Rodemich, and Rumsey (1967, 1969) define an ε, δ entropy, for a metric space S with a probability measure P defined on it, in terms of a decomposition of S into sets of diameter at most ε and one set of probability at most δ. Also, Posner et al. define ε-entropy as the infimum of entropies $-\sum_i P(U_i) \log(P(U_i))$ where the U_i have diameters at most ε. So Lorentz's term "metric entropy" seems useful and has been adopted here.

Notes to Section 1.3. Sergei Bernštein (1927, pp. 159–165) published his inequality. The proof given is based on Bennett (1962, p. 34) with some incorrect, but unnecessary, steps (his (3), (4),...) removed as suggested by Giné (1974). For related and stronger inequalities under weaker conditions, such as unbounded variables, see also Bernštein (1924, 1927), Hoeffding (1963), and Uspensky (1937, p. 205).

Hoeffding (1963, Theorem 2) implies Proposition 1.12. Earlier, Chernoff (1952, (5.11)) had proved (1.5) and (1.6). Okamoto (1958, Lemma 2(b′)) rediscovered (1.6). Inequality (1.7) appeared in Dudley (1978, Lemma 2.7) and Proposition 1.16 in Dudley (1982, Lemma 3.3). On Ottaviani's inequality, Theorem 1.19, for real-valued functions see (9.7.2) and the notes to Section 9.7 in RAP. The P. Lévy inequality Theorem 1.20 is given for Banach-valued random variables in Kahane (1985, Section 2.3). For the case of real-valued random variables, it was known much earlier, see the notes to Section 12.3 in RAP.

Notes to Section 1.4. The proof in this section for Theorem 1.8, due to Bretagnolle and Massart (1989), is an expanded version of their proof. It was included in the lecture notes of a MaPhySto course given in Aarhus, Denmark, in August 1999, and has been revised slightly for inclusion here.

Komlós, Major, and Tusnády (1975) formulated a construction giving a joint distribution of α_n and Y_n, and this construction has been accepted by later workers. But Komlós, Major, and Tusnády gave hardly any proof for (1.1). Csörgő and Révész (1981) sketched a method of proof of (1.1) based on lemmas of G. Tusnády, Lemmas 1.21 and 1.23. The implication from Lemma 1.23 to 1.21 is not difficult, but Csörgő and Révész did not include a proof of Lemma 1.23. Bretagnolle and Massart (1989) gave a proof of the lemmas and of the inequality (1.1) with specific constants, Theorem 1.8. Bretagnolle and Massart's proof was rather compressed and some readers have had difficulty following it. Csörgő and Horváth (1993, pp. 116–139) expanded the proof while making it more elementary and gave a proof of Lemma 1.23 for $n \geq n_0$ where n_0 is at least 100. Section 1.4 gives a detailed and in some minor details corrected version of the original Bretagnolle and Massart proof of the lemmas for all n, overlapping in part with the Csörgő and Horváth proof, and then it proves (1.1) for some constants, as given by Bretagnolle and Massart and largely following their proof.

Sharpness of the Komlós–Major–Tusnády rate, up to constants, follows from Theorem 4.4.2 of Csörgő and Révész (1981), whose proof is outlined.

Mason and van Zwet (1987) gave another proof of the inequality (1.1) and an extended form of it for subintervals $0 \leq t \leq d/n$ with $1 \leq d \leq n$ and $\log n$ replaced by $\log d$, without Tusnády's inequalities and without specifying the constants c, K, λ. Some parts of the proof sketched by Mason and van Zwet are given in more detail by Mason (1998).

Bennett (1962) proved Lemma 1.28 with the inequality (1.60); see also Shorack and Wellner 1986, p. 440, (3). Bennett showed also that the Chernoff inequality for $p \leq 1/2$ (1.6) follows from (1.60).

Vorob'ev (1962) proved Theorem 1.31 for finite sets. Then Berkes and Philipp (1977, Lemma A1) proved it for separable Banach spaces. Their proof carries over to the present case. Vorob'ev (1962) treated more complicated families of joint distributions on finite sets, as did Shortt (1984) for more general measurable spaces.

In Theorem 1.31, the assumption that X, Y, and Z are Polish can be weakened: they could instead be any Borel sets in Polish spaces (RAP, Section 13.1). Still more generally, since the proof of Theorem 10.2.2 in RAP depends just on tightness, it is enough to assume that X, Y, and Z are universally measurable subsets of their completions, in other words, measurable for the completion of any probability measure on the Borel sets (RAP, Section 11.5). Shortt (1983) treats universally measurable spaces and considers just what hypotheses on X, Y, and Z are necessary.

Notes to Section 1.5. This section is based on the paper of Massart (1990), with some details inserted or revised. Massart states Proposition 1.35 without proof, referring to Smirnov (1944), which I have not seen, for a proof. Birnbaum and Tingey (1951) state and prove one Theorem, given in their formula (3.0), which is very close to Proposition 1.35. One difference is that they treat $\Pr(D_n^- \leq \lambda) = 1 - \Pr(D_n^- > \lambda)$ and expand the latter as a sum. Birnbaum and Tingey cite a paper by Smirnov from 1939, not 1944. Their sketched proof is very parallel to but different from the detailed one given above.

2

Gaussian Processes; Sample Continuity

2.1 General Empirical and Gaussian Processes

In Chapter 1 we needed essentially just two Gaussian processes, the Brownian motion $\{x_t\}_{t \geq 0}$ and the Brownian bridge $\{y_t\}_{0 \leq t \leq 1}$, to get a limit for a classical empirical process $\{\sqrt{n}(F_n - F)(x)\}_{x \in \mathbb{R}}$ as $\{y_{F(x)}\}_{x \in \mathbb{R}}$. It will be seen how that limit can be extended.

Let $(S, \mathcal{S}, P)$ be a probability space, where S might often be a Euclidean space $\mathbb{R}^d$ but now with $d \geq 2$. Let $X_1, X_2, \ldots$, be independent, identically distributed variables with values in S and distribution P. Let P_n be the empirical probability measure $P_n := \frac{1}{n} \sum_{j=1}^n \delta_{X_j}$ on S, recalling that $\delta_x(A) = 1_A(x) = 1$ for $x \in A$ and 0 otherwise. For any given set $A \in \mathcal{S}$, $\sqrt{n}(P_n - P)(A)$ will converge in distribution by the original central limit theorem (for binomial probabilities, de Moivre 1733) to a variable $G_P(A)$ with distribution $N(0, P(A)(1 - P(A)))$. If B is another set in $\mathcal{S}$, then the random vector $\sqrt{n}((P_n - P)(A), (P_n - P)(B))$ will converge in distribution as $n \to \infty$ to $(G_P(A), G_P(B))$, by the multivariate central limit theorem (e.g. RAP, Theorem 9.5.6), having a normal bivariate distribution with mean 0, and the covariance

$$E(G_P(A)G_P(B)) = P(A \cap B) - P(A)P(B) = \mathrm{Cov}_P(1_A, 1_B). \quad (2.1)$$

Similarly, for any finite collection $\{A_j\}_{j=1}^k$ of sets in $\mathcal{S}$, $\sqrt{n}\{(P_n - P)(A_j)\}_{j=1}^k$ will converge in distribution to $\{G_P(A_j)\}_{j=1}^k$ which has a k-variate normal distribution with mean 0 and covariances given by (2.1) for $A = A_i$, $B = A_j$.

For any function f and probability measure Q on S such that $E_Q f = \int f \, dQ$ is defined and $\int f^2 dQ < \infty$, let $Qf := E_Q(f)$. This will always hold for $Q = P_n$. If it also holds for $Q = P$, then $\sqrt{n}(P_n - P)(f)$ will converge in distribution to a random variable $G_P(f)$ with distribution $N(0, \mathrm{Var}_P(f))$ where $\mathrm{Var}_P(f)$ is the variance of f for P, by the usual central limit theorem. If the hypotheses hold for functions $f_1, \ldots, f_k$ with respect to P, then the random vector $\sqrt{n}\{(P_n - P)(f_j)\}_{j=1}^k$ will converge in distribution as $n \to \infty$ to $\{G_P(f_j)\}_{j=1}^k$, which have a k-variate normal distribution with mean 0 and covariances given by

$$E(G_P(f)G_P(g)) = (f, g)_{0,P} := \text{Cov}_P(f, g) = E_P(fg) - E_P f E_P g. \quad (2.2)$$

The main question this book treats is to find conditions under which the convergence in distribution extends to infinite classes $\mathcal{C}$ of sets or $\mathcal{F}$ of functions, with respect to uniform convergence over such a class. In Chapter 1 this was done in $\mathbb{R}$ for the family of half-lines $\mathcal{C} = \{(-\infty, x] : x \in \mathbb{R}\}$. This chapter will consider Gaussian processes with general index sets T, such as G_P when T is a class $\mathcal{C}$ of sets or a class $\mathcal{F}$ of functions. For a central limit theorem to hold uniformly over a class $\mathcal{C}$ or $\mathcal{F}$, we will need the limit process G_P to be reasonably well behaved on the class, as, for example, the Brownian bridge sample paths $t \mapsto y_t(\omega)$ can be and are taken to be continuous in t for (almost) all ω.

2.2 Some Definitions

A real-valued *stochastic process* consists of a set T, a probability space $(\Omega, \mathcal{A}, P)$, and a map $(t, \omega) \mapsto X_t(\omega)$ from $T \times \Omega$ into $\mathbb{R}$, such that for each $t \in T$, $X_t(\cdot)$ is measurable from Ω into $\mathbb{R}$. The process is called *Gaussian* iff for every finite subset F of T, the law $\mathcal{L}(\{X_t\}_{t \in F})$ is a normal distribution on $\mathbb{R}^F$, where the *law* $\mathcal{L}(X)$ of a random variable or vector X means the probability measure which is its distribution.

For any set S let $\mathbb{R}^S$ be the set of all real-valued functions on S. Whenever $A \subset B$, there is a natural mapping (projection) $\pi_{B,A}$ of $\mathbb{R}^B$ onto $\mathbb{R}^A$, namely, restriction of functions to A. The Kolmogorov existence theorem for stochastic processes, applied to real-valued processes, is as follows: let T be any set, and suppose for any finite subset $F \subset T$, we are given a probability measure P_F on $\mathbb{R}^F$, a finite-dimensional real vector space with its usual Borel σ-algebra. Suppose that the P_F are consistent in the sense that for each nonempty set $E \subset F$ finite, the image measure of P_F under $\pi_{F,E}$ is P_E. Then by Kolmogorov's theorem (e.g. RAP, Theorem 12.1.2) there exists a probability measure P_T on $\mathbb{R}^T$, defined on the smallest σ-algebra making all $\pi_{T,F}$ measurable for F finite, such that the image measure of P_T by $\pi_{T,F}$ is P_F for all finite F. This P_T gives a stochastic process where $\Omega = \mathbb{R}^T$, and for each $t \in T$ and $\omega \in \Omega$ we take $X_t(\omega) := \omega(t)$.

Kolmogorov's theorem may not be surprising if one realizes that $\mathbb{R}^T$ is the class of *all* real-valued functions on T, which have a real value at each t, but may vary wildly as t varies.

Kolmogorov's theorem specializes to Gaussian processes as follows. If a finite set F has cardinality $d \geq 1$, let $\mu \in \mathbb{R}^F$ be any vector, and let

$C: F \times F \to \mathbb{R}$, regarded as a $d \times d$ matrix, be symmetric and nonnegative definite. Then we know that there exists a uniquely determined normal (Gaussian) probability measure $N(\mu, C)$ with mean μ and covariance matrix C. Let T again be any set and suppose for each nonempty finite $F \subset T$ we have $\mu_F \in \mathbb{R}^F$ and $C_F : F \times F \to \mathbb{R}$, which is symmetric and nonnegative definite. Suppose that these are mutually consistent in the sense that whenever $E \subset F$, the restriction of μ_F to E is μ_E, and the restriction of C_F to $E \times E$ is C_E. Then the probability measures $P_F = N(\mu_F, C_F)$ are consistent, so Kolmogorov's theorem applies, and the resulting process is Gaussian with the given means and covariances.

Let X be a real vector space. Recall that a *seminorm* is a function $\| \cdot \|$ from X into the nonnegative real numbers such that $\|x + y\| \le \|x\| + \|y\|$ for all x and y in X and $\|cx\| = |c| \|x\|$ for all real c and $x \in X$. The seminorm $\| \cdot \|$ is called a *norm* if $\|x\| = 0$ only for $x = 0$ in X, and then $(X, \| \cdot \|)$ is called a *normed linear space*. A norm defines a metric by $d(x, y) := \|x - y\|$. A normed linear space complete for this metric is called a *Banach space*. As with any metric space, it is called *separable* if it has a countable dense subset. A probability distribution P defined on a separable Banach space will be assumed to be defined on the Borel σ-algebra generated by the open sets, unless another σ-algebra is specified. Then P will be called a *law*.

Some Banach spaces with especially pleasant properties are Hilbert spaces, defined as follows. For a real vector space X, an *inner product* is a function $(\cdot, \cdot)$ from $X \times X$ into $\mathbb{R}$ which is symmetric, i.e. $(x, y) = (y, x)$ for all x and y in X; bilinear, meaning that for each fixed x, $(x, \cdot)$ is linear from X into $\mathbb{R}$, thus for each y, $(\cdot, y)$ is linear from X into $\mathbb{R}$; and positive definite, meaning that $(x, x) > 0$ for all $x \ne 0$ in S.

Example: For X any finite-dimensional Euclidean space, the usual dot product of vectors is an inner product.

Any inner product $(\cdot, \cdot)$ defines a norm by $\|x\| = (x, x)^{1/2}$: one can easily see that for any constant c,

$$\|cx\| = (cx, cx)^{1/2} = [c^2(x, x)]^{1/2} = |c| \|x\|.$$

The subadditivity $\|x + y\| \le \|x\| + \|y\|$ follows from the Cauchy–Schwarz inequality for inner products, $|(x, y)| \le \|x\| \|y\|$, which is proved in the same way as the usual Cauchy–Schwarz inequality, by expanding $(x + ty, x + ty) \ge 0$ for all real t as a quadratic in t.

If a norm is defined by an inner product, then the inner product is uniquely determined, as $(x, y) \equiv (\|x + y\|^2 - \|x - y\|^2)/4$. Those norms definable from inner products are characterized by the fact called the *parallelogram law*: $\|x + y\|^2 + \|x - y\|^2 = 2\|x\|^2 + 2\|y\|^2$.

If X is complete for the norm $\|x\|$ defined by an inner product, then it is called a *Hilbert space*. Among the most-encountered Hilbert spaces are the L^2 spaces defined as follows. Let $(S, \mathcal{S}, \mu)$ be a σ-finite measure space. Let $\mathcal{L}^2(S, \mathcal{S}, \mu)$ be the set of all measurable real functions f on S such that $\int_S f^2 d\mu < +\infty$. Let $L^2(X, \mathcal{S}, \mu)$ be the set of equivalence classes of functions in $\mathcal{L}^2$ for equality μ-almost everywhere. On L^2 we have the inner product $(f, g) = \int_S f(x)g(x)d\mu(x)$, which does not depend on the choice of f and g from their equivalence classes. A proof that this is in fact an inner product is easy.

Let $(X, \|\cdot\|)$ be a separable Banach space. A law P on X will be called *Gaussian* or *normal* iff for every continuous linear form $f \in X'$, $P \circ f^{-1}$ is a normal law on $\mathbb{R}$. Recall that a law on a finite-dimensional real vector space is normal if and only if every real linear form is normally distributed (RAP, Theorem 9.5.13).

2.2.1 *The Isonormal Process*

Let H be a Hilbert space with inner product $(\cdot, \cdot)$. There exists a Gaussian process L indexed by H with mean $EL(f) = 0$ for all $f \in H$ and covariance $E(L(f)L(g)) = (f, g)$, where the covariances are nonnegative definite by the assumptions on an inner product. This process exists by the general existence theorem for Gaussian processes mentioned earlier in Section 2.2. It is called the *isonormal* process on H. For any $x \in H$, $L(x)$ is a Gaussian random variable with distribution $N(0, \|x\|^2)$, and for any $x_1, \ldots, x_n \in H$, $(L(x_1), \ldots, L(x_n))$ have a jointly normal distribution with covariance given by the inner products (x_i, x_j).

The isonormal process has the following linearity property (also in RAP, Theorem 12.1.4):

Theorem 2.1 *For an isonormal process L on a Hilbert space H, and any $x, y \in H$ and real constant c, with probability 1,*

$$L(cx + y) = cL(x) + L(y). \tag{2.3}$$

Proof. Expanding $E((L(cx + y) - cL(x) - L(y))^2)$ and using the given covariances for L, we find that the given expectation is 0, and the conclusion follows. □

Remark. The set of probability 0 on which (2.3) fails can depend on c, x, and y. The equation can be taken to hold with probability 1 for all real c and all x and y in a suitable subset, in Theorem 3.2 below.

Two stochastic processes $\{X_t, \ t \in T\}$ and $\{Y_t, \ t \in T\}$ defined on the same index set, but possibly on different probability spaces, will be said to have *the same laws*, or one will be said to be a *version* of the other, if for each finite subset F of T, $\{X_t, \ t \in F\}$ and $\{Y_t, \ t \in F\}$ have the same law. If X_t and Y_t are defined on the same probability space, then one is said to be a *modification* of the other if for each $t \in T$, we have $P(X_t = Y_t) = 1$. On the relationship of versions and modifications, especially for the isonormal process restricted to a set, see Appendix I.

Any Gaussian process with mean 0 can be factored through L in the following sense:

Theorem 2.2 *Let $\{X_t\}_{t \in T}$ be any Gaussian process with mean 0 defined on a probability space $(\Omega, \mathcal{A}, Q)$. Let H be the Hilbert space $L^2(\Omega, \mathcal{A}, Q)$ and L the isonormal process on H. For each $t \in T$ let $Y_t := L(X_t(\cdot))$. Then the process $\{Y_t\}_{t \in T}$ is a version of $\{X_t\}_{t \in T}$.*

Proof. Since L is a Gaussian process with mean 0, so is $\{Y_t\}_{t \in T}$. For any $s, t \in T$, the covariance $E(Y_s Y_t)$ equals the inner product of $X_s(\cdot)$ and $X_t(\cdot)$ in $L^2(\Omega, \mathcal{A}, Q)$, which is $E(X_s X_t)$, so the covariances of the two processes are the same, and being Gaussian, they are versions of each other. $\square$

One can get the Gaussian process G_P with mean 0 and covariance (2.2), which is the limit of empirical processes $\sqrt{n}(P_n - P)$, from an isonormal process as follows. Given $(S, \mathcal{S}, P)$, let H be the Hilbert space $L^2(S, \mathcal{S}, P)$ and let W_P be the isonormal process on H. Set

$$G_P(f) := W_P(f - E_P f),$$

which equals $W_P(f) - E_P f W_P(1)$ almost surely for any given f. Clearly, this gives a Gaussian process with mean 0 and covariance (2.2). From this representation, one can see by Theorem 2.1 that for any f and g in $\mathcal{L}^2$ and real c, with probability 1

$$G_P(cf + g) = cG_P(f) + G_P(g). \tag{2.4}$$

Suppose a Gaussian process $\{X_t\}_{t \in T}$ with mean 0 is defined on a space T, such as a Euclidean space or a subset of one, which already has a usual metric e. Many authors on Gaussian processes, instead of factorizing through L as in Theorem 2.2, adopt the idea in an alternate form by keeping T but using on it the "natural metric" (more precisely a pseudometric) defined by the process, namely, for s and t in T,

$$e_X(s, t) := (E((X_s - X_t)^2))^{1/2}.$$

A pseudometric on T is a function ρ on $T \times T$ which is a metric except possibly that $\rho(x, y) = 0$ for $x \neq y$. Then $t \mapsto X_t$ is an isometry (it preserves distances) from T with e_X into the Hilbert space $H = L^2(\Omega, \mathcal{A}, Q)$ with its usual metric. For two examples, for the Brownian motion process $\{x_t\}_{t \geq 0}$, the natural metric would be $|s - t|^{1/2}$, the square root of the usual metric. For the Brownian bridge $\{y_t\}_{0 \leq t \leq 1}$ the natural "metric" is $(|t - s| - (t - s)^2)^{1/2}$, a pseudometric since $e_y(0, 1) = 0$. For other processes the natural metric may have no simple relation to the usual metric.

For the isonormal process L, often the Hilbert space H will be taken to be separable and infinite-dimensional. In the finite-dimensional case we have a representation as follows. For any real Hilbert space H, the map i_H from H into its own dual Banach space, defined by $i_H(x)(y) := (y, x)$, is linear, one-to-one, and onto. The dual space is itself a Hilbert space, and i_H preserves the Hilbert structure. If H is finite-dimensional, then we have the standard normal probability measure $N(0, I)$ defined on it, where I is the identity matrix or operator. The next fact is immediate:

Proposition 2.3 *For any finite-dimensional Hilbert space H, the mapping i_H, with the probability measure $N(0, I)$ on H, is a version of the isonormal process L.*

The last proposition does not extend to H infinite-dimensional because a probability measure "$N(0, I)$" does not exist on H: let $\{e_j\}_{j=1}^\infty$ be an infinite orthonormal set. Then $Z_j := L(e_j)$ are i.i.d. $N(0, 1)$ and $\sum_{j=1}^\infty Z_j^2 = +\infty$ almost surely.

For any subset $A \subset H$, $L(A)^*$ will be defined as ess.sup$_{x \in A} L(x)$, the smallest random variable Y such that $Y \geq L(x)$ a.s. for all $x \in A$, where Y is determined up to a.s. equality. Here "ess.sup" stands for "essential supremum." It will be shown in the next lemma that $L(A)^*$ is well-defined for A separable. It will be seen later (proof of Theorem 2.19) that for A nonseparable it is also well-defined and equal to $+\infty$ a.s. Similarly let $|L(A)|^* := $ ess.sup$_{x \in A} |L(x)|$.

Lemma 2.4 *For any separable subset $A \subset H$, $L(A)^*$ and $|L(A)|^*$ are well-defined up to almost sure equality.*

Proof. If B is countable, then $L(B)^*$ exists and equals sup$_{x \in B} L(x)$, clearly. Let B be any countable dense subset of A. Let $Y := L(B)^*$. Then Y is measurable. If U is any random variable such that $U \geq L(x)$ a.s. for all $x \in A$, then clearly $U \geq Y$ a.s. For each $x \in A$, there is a sequence $y_n \in B$ such that $\|y_n - x\|^2 < 1/n^2$ for all $n = 1, 2, \ldots$, and then by Chebyshev's inequality and the Borel–Cantelli lemma, $L(y_n) \to L(x)$ a.s. Thus $L(x) \leq Y$ a.s., and as seen above Y is the smallest random variable with this property, so $L(A)^*$ is well-defined and equals Y a.s. The proof with absolute values is the same. $\square$

Definitions A set C in H is called a *GB-set* iff $|L(C)|^* < \infty$ a.s. Also, C will be called a *GC-set* iff it is totally bounded and the restriction of L to C can be chosen so that each of its sample functions $x \mapsto L(x)(\omega)$, $x \in C$, is uniformly continuous on C.

2.3 Bounds for Gaussian Vectors

First, some bounds for one-dimensional Gaussian variables will be given. Let Φ be the standard normal distribution function and ϕ its density function, so $\phi(x) = (2\pi)^{-1/2} \exp(-x^2/2)$ for all real x, and $\Phi(x) = \int_{-\infty}^{x} \phi(u) du$.

Proposition 2.5 *Let X be a real-valued random variable with a normal distribution $N(0, \sigma^2)$. Then*

(a) for any $M > 0$, $\Pr(|X| > M) \leq \exp\left(-M^2/\left(2\sigma^2\right)\right)$;

(b) if $M/\sigma \geq 1$, then

$$\frac{\sigma}{M}\phi\left(\frac{M}{\sigma}\right) \leq \Pr(|X| > M) \leq \frac{2\sigma}{M}\phi\left(\frac{M}{\sigma}\right).$$

Proof. Replacing X by X/σ, we can assume $\sigma = 1$. For (a) we want to prove $2\Phi(-c) \leq \exp(-c^2/2)$ for any $c \geq 0$. This holds for $c = 0$ and follows by differentiating both sides for $0 \leq c \leq (2/\pi)^{1/2}$. For larger c it follows from $\Phi(-c) \leq \phi(c)/c$ (RAP, Lemma 12.1.6(a)), as does the right side of (b). For the left side of (b), note that ϕ is a convex function for $x \geq 1$ since there $\phi''(x) = (x^2 - 1)\phi(x) \geq 0$. Thus, the region between the graph of ϕ and the x axis for $x \geq c$ includes a right triangle with right vertex at $(c, 0)$, a vertex at $(c, \phi(c))$, and whose hypotenuse is along the tangent line to the graph of ϕ at c. This triangle is easily seen to have area $\phi(c)/(2c)$, which finishes the proof. $\square$

This section will prove an extension of inequality (a) to infinite-dimensional Gaussian variables such as those taking values in separable Banach spaces. It will be said that a law P on a separable Banach space $(X, \| \cdot \|)$ has mean 0 if $\int \|x\| dP(x) < \infty$ and $\int f(x) dP(x) = 0$ for each $f \in X'$. Recall the dual norm $\|f\|' := \sup\{|f(x)| : \|x\| \leq 1\}$. Here is one of the main results.

Theorem 2.6 (Landau–Shepp–Marcus–Fernique) *Let P be a normal law with mean 0 on a separable Banach space X. For $f \in X'$ let $\sigma^2(f) := \int f^2 dP$. Then $\tau^2 := \sup\{\sigma^2(f) : \|f\|' \leq 1\} < \infty$ and*

$$\int \exp(\alpha\|x\|^2) dP(x) < \infty \quad \text{for any} \quad \alpha < 1/(2\tau^2).$$

By Proposition 2.5(a), the theorem holds in the one-dimensional case, and by the left side of part (b), the condition $\alpha < 1/(2\tau^2)$ is best possible. Before the theorem is proved in general, some other facts will be brought in.

Definition Let X be a real vector space and $\mathcal{B}$ a σ-algebra of subsets of X. Then $(X, \mathcal{B})$ is called a *measurable vector space* if both

(a) Addition is jointly measurable from $X \times X$ to X, and

(b) Scalar multiplication is jointly measurable from $\mathbb{R} \times X$ to X (for the usual Borel σ-algebra on $\mathbb{R}$).

Example. Let X be a topological vector space, namely, a vector space with a topology for which (a) and (b) hold with "measurable" replaced by "continuous." Suppose the topology of X is metrizable and separable. For a Cartesian product of two separable metric spaces, since their topologies have countable bases, the Borel σ-algebra in the product equals the product σ-algebra of the Borel σ-algebras in the two spaces (RAP, Proposition 4.1.7). Thus X with its Borel σ-algebra is a measurable vector space.

The notion of normal law cannot be defined for general measurable vector spaces by way of linear forms, as it was for Banach spaces in the last section, since there exist measurable vector spaces, such as spaces $L^p[0, 1]$ for $0 < p < 1$, which have nontrivial normal measures but turn out to have no nontrivial measurable linear forms (Appendix F). Fernique (1970) proposed the following ingenious definition:

Definition A probability measure P on a measurable vector space $(X, \mathcal{B})$ will be called *centered Gaussian* iff for variables U and V independent with law P (say, coordinates on the product $X \times X$ for the product law $P \times P$) and any θ with $0 < \theta < 2\pi$, $U \cos \theta + V \sin \theta$ and $-U \sin \theta + V \cos \theta$ are also independent with distribution P.

If $X = \mathbb{R}$, the transformation of $(U, V) \in \mathbb{R}^2$ in the last definition is a rotation through an angle θ. Normal laws with mean 0 on finite-dimensional real vector spaces are centered Gaussian in this sense, as can be seen from covariances. Conversely, a law on $X = \mathbb{R}$ satisfying the above definition of "centered Gaussian," even for one value of θ with $\sin(2\theta) \neq 0$, must be normal according to the "Darmois–Skitovič" theorem; see the notes for this section. We will not need the full strength of the latter theorem below, but the following facts will be proved.

Proposition 2.7 *Let $(X, \mathcal{A})$ and $(Y, \mathcal{B})$ be two measurable vector spaces. Let P be a centered Gaussian measure on $(X, \mathcal{A})$ and T a measurable linear map from X into Y. Then the image measure $Q := P \circ T^{-1}$ is centered Gaussian on $(Y, \mathcal{B})$,*

Proof. Let $(x, \xi) \in X \times X$ with distribution $P \times P$. Then (Tx, Ty) has distribution $Q \times Q$ on $Y \times Y$. Also for each θ, $((\cos \theta)x + (\sin \theta)\xi, (-\sin \theta)x + (\cos \theta)\xi)$ has distribution $P \times P$ on $X \times X$. Now

$$((\cos \theta)Tx + (\sin \theta)T\xi, (-\sin \theta)Tx + (\cos \theta)Ty)$$

$$\equiv (T((\cos \theta)x + (\sin \theta)\xi), T((-\sin \theta)x + (\cos \theta)\xi))$$

has distribution $Q \times Q$ on $Y \times Y$, so Q is centered Gaussian. $\qquad \square$

Proposition 2.8 *For any finite dimension* d, *a centered Gaussian law* P *on* $\mathbb{R}^d$ *with Borel* σ-*algebra is a centered normal law in the usual sense,* $N(0, \Sigma)$ *for some nonnegative definite symmetric matrix* Σ.

Proof. First suppose $d = 1$. Then $P \times P$ on $\mathbb{R}^2$ is invariant under all rotations. A rotation through $\theta = \pi$ shows that P is symmetric, $dP(x) \equiv dP(-x)$. Let f be the characteristic function of P, $f(t) := \int_{-\infty}^{\infty} e^{itx} dP(x)$. Then f is real-valued, $f(0) = 1$, and $f(t) \equiv f(-t)$. Any point $(t, u) \in \mathbb{R}^2$ can be rotated to a point on a coordinate axis, so $f(t)f(u) \equiv f((t^2 + u^2)^{1/2})$. Let $h(t) := \log f(|t|^{1/2})$ where it is defined and finite, i.e. where $f > 0$, as is true at least in a neighborhood of 0. Then $h(t + u) \equiv h(t) + h(u)$ for $t, u \geq 0$ and, perhaps, small enough. Where both sides are defined and finite we have $h(qu) \equiv qh(u)$ first when q is an integer, then when it is rational, and then for general real q by continuity. Since h thus is bounded on finite intervals where it is defined, it is defined and continuous on the whole line, with $h(t) \equiv ct$ for some constant $c = h(1)$, so $f(t) = \exp(ct^2)$ for all t, and $c \leq 0$ since $|f(t)| \leq 1$. Thus $P = N(0, \sigma^2)$ where $\sigma^2 = -2c$ (RAP, Proposition 9.4.2, Theorem 9.5.1).

Now for general d, for any linear form f from $\mathbb{R}^d$ into $\mathbb{R}$, $P \circ f^{-1}$ is centered Gaussian by Proposition 2.7 and thus is a law $N(0, \sigma^2)$ by the $d = 1$ case. It follows that P is a law $N(0, \Sigma)$ by RAP, Theorem 9.5.13. $\qquad\square$

Given a normal measure $P = N(0, C)$ on a finite-dimensional space X and a vector subspace Y of X, it follows from the structure of normal measures (RAP, Theorem 9.5.7) that $P(Y) = 0$ or 1. This fact extends to general measurable vector spaces:

Theorem 2.9 *(0-1 law) Let* $(X, \mathcal{B})$ *be a measurable vector space and* Y *a vector subspace with* $Y \in \mathcal{B}$. *Then for any centered Gaussian law* P *on* X, $P(Y) = 0$ *or 1.*

Proof. Let U and V be independent in X with law P. For $0 \leq \theta \leq \pi/2$, let $A(\theta)$ be the event

$$A(\theta) := \{U \cos \theta + V \sin \theta \in Y, \ -U \sin \theta + V \cos \theta \notin Y\}.$$

If $\theta \neq \phi$, with $0 \leq \phi \leq \pi/2$, then

$$\cos \theta \sin \phi - \sin \theta \cos \phi = \sin(\phi - \theta) \neq 0,$$

so if $y_1 := u \cos \theta + v \sin \theta \in Y$ and $y_2 := u \cos \phi + v \sin \phi \in Y$, then u and v can be solved for as linear combinations of y_1, y_2, so they are in Y and $-u \sin \theta + v \cos \theta \in Y$, $-u \sin \phi + v \cos \phi \in Y$. So the sets $A(\theta)$ are disjoint for different values of $\theta \in [0, \pi/2]$. By definition of centered Gaussian, these sets all have the same probability, which thus must be 0. Taking $\theta = 0$ gives $0 = \Pr(U \in Y) \Pr(V \notin Y) = P(Y)P(X \backslash Y)$, so $P(Y) = 0$ or 1. $\qquad\square$

A measurable function $\| \cdot \|$ from a measurable vector space X into $[0, \infty]$ will be called a *pseudo-seminorm* iff $Y := \{x \in X : \|x\| < \infty\}$ is a vector subspace of X and $\| \cdot \|$ is a seminorm on Y, that is, $\|cx\| = |c| \|x\|$ for each real c and $x \in Y$, and so for all $x \in X$, with $0 \cdot \infty := 0$, and $\|x + y\| \leq \|x\| + \|y\|$ for all $x, y \in Y$, and so for all $x, y \in X$.

By the 0–1 law (Theorem 2.9), for any pseudo-seminorm $\| \cdot \|$ and centered Gaussian P on X, $P(\| \cdot \| < \infty) = 0$ or 1. Likewise, $P(\| \cdot \| = 0) = 0$ or 1.

If S is a countable set, then $\mathbb{R}^S$, the set of all real-valued functions on S, with product topology, is a separable metric topological linear space, hence a measurable vector space. If P is the law of a Gaussian stochastic process $\{x_t, \ t \in S\}$ on $\mathbb{R}^S$, with $Ex_t = 0$ for all $t \in S$, then P is centered Gaussian on $\mathbb{R}^S$. The supremum "norm" $\|\{y_t, \ t \in S\}\| := \sup_t |y_t|$ is clearly a pseudo-seminorm on $\mathbb{R}^S$.

Here is a step toward proving Theorem 2.6:

Lemma 2.10 (Landau–Shepp–Fernique) *Let $(X, \mathcal{B})$ be a measurable vector space, P a centered Gaussian measure, and $\| \cdot \|$ a pseudo-seminorm on X with $P(\| \cdot \| < \infty) > 0$. Then for some $\varepsilon > 0$,*

$$\int \exp\left(\alpha \|x\|^2\right) dP(x) < \infty \quad \text{for} \quad 0 < \alpha < \varepsilon.$$

Proof. As noted above, $P(\| \cdot \| < \infty) = 1$. Let U and V be independent with distribution P in X. The definition of centered Gaussian for $\theta = -\pi/4$ yields, for any real s and t,

$$P(\| \cdot \| \leq s)P(\| \cdot \| > t)$$
$$= \Pr\left\{ \|(U - V)/2^{1/2}\| \leq s, \ \|(U + V)/2^{1/2}\| > t \right\}. \qquad (2.5)$$

Note that

$$2^{1/2} \min(\|U\|, \|V\|) \geq \|(U + V)/2^{1/2}\| - \|(U - V)/2^{1/2}\|,$$

where the event that the right side is undefined, equaling $\infty - \infty$, has zero probability and so can be neglected. Thus on the event on the right in (2.5), we have $\|U\| > (t - s)/2^{1/2}$ and $\|V\| > (t - s)/2^{1/2}$. So

$$P(\| \cdot \| \leq s)P(\| \cdot \| > t) \leq P\left(\| \cdot \| > (t - s)/2^{1/2}\right)^2. \qquad (2.6)$$

Choose s with $0 < s < \infty$ large enough so that $q := P(\| \cdot \| \leq s) > 1/2$. Define a sequence t_n recursively by $t_0 := s$, $t_{n+1} := s + 2^{1/2} t_n$, $n = 0, 1, \ldots$. Then we have by induction

$$t_n = \left(2^{1/2} + 1\right)\left(2^{(n+1)/2} - 1\right) s.$$

So t_n increases up to $+\infty$ with n. Let $x_n := P(\| \cdot \| > t_n)/q$, so $x_n \le x_{n-1}^2$. By induction, we then have

$$P(\| \cdot \| > t_n) \le q \left((1 - q)/q\right)^{2^n} .$$

It follows that

$$E \exp \left(\alpha \| \cdot \|^2\right) \le q e^{\alpha s^2} + \sum_{n=0}^{\infty} P(t_n < \| \cdot \| \le t_{n+1}) \exp \left(\alpha t_{n+1}^2\right)$$

where the nth term of the latter sum is bounded above by

$$q \left(q^{-1} - 1\right)^{2^n} \exp \left[\alpha \left(2^{1/2} + 1\right)^2 \left(2^{(n+2)/2} - 1\right)^2 s^2\right].$$

The sum will be finite if

$$\sum_{n=0}^{\infty} \exp \left\{2^n \left[\log \frac{1-q}{q} + 4 \left(2^{1/2} + 1\right)^2 \alpha s^2\right]\right\} < \infty,$$

which holds for $\alpha \le (\log \frac{q}{1-q})/(24s^2)$, proving the Lemma. $\qquad \square$

Theorem 2.6 will be a corollary of the following fact:

Theorem 2.11 *Let* $(X, \mathcal{B})$ *be a measurable vector space and* P *a centered Gaussian law on* X. *Let* $\{y_n\}_{n \ge 1}$ *be a sequence of measurable linear forms:* $X \to \mathbb{R}$. *Let* $\|x\| := \sup_n |y_n(x)|$. *Suppose* $P(\|x\| < \infty) > 0$. *Then* $\tau := (\sup_n \int y_n^2 dP)^{1/2} < \infty$, *and* $E \exp(\alpha \|x\|^2) < \infty$ *if and only if* $\alpha < 1/(2\tau^2)$.

Proof. For each n, $\| \cdot \| \ge |y_n|$. It is easily checked that $P \circ y_n^{-1}$ is centered Gaussian and thus by Proposition 2.8 is a law $N(0, \sigma_n^2)$, $\sigma_n^2 \le \tau^2$. Now $P(|y_n| \ge \sigma_n) > c > 0$ for all n ($c > 0.3$). Thus if σ_n are unbounded, $P(\| \cdot \| \ge \sigma_n) > c$ for all n gives a contradiction. Then to prove "only if," we have $E \exp \left(|y_n|^2/(2\tau^2)\right) = \tau/(\tau^2 - \sigma_n^2)^{1/2}$, or $= +\infty$ if $\sigma_n^2 = \tau^2$. Taking the supremum over n gives $E \exp(\|x\|^2/(2\tau^2)) = +\infty$.

Now to prove "if," recall the space ℓ^∞ of all bounded sequences of real numbers with supremum norm. This is a nonseparable Banach space. With the smallest σ-algebra making the coordinates measurable, it is a measurable vector space (by the way, this σ-algebra is smaller than the Borel σ-algebra for the supremum norm). Let $Y(x) := \{y_n(x)\}_{n \ge 1}$. Let S be the vector subspace of X where $\| \cdot \|$ is finite. Then $P(S) = 1$ by Theorem 2.9 and Y is linear, measurable, and preserves norms from S into ℓ^∞. So it will be enough to prove the theorem in ℓ^∞ with coordinates $\{y_n\}$.

For any finite k, for $T_k : x \mapsto \{y_n(x)\}_{n=1}^k$, $P \circ T_k^{-1}$ is centered Gaussian on $\mathbb{R}^k$ by Proposition 2.7, thus it is some normal law $N(0, \Sigma_k)$ by Proposition 2.8. By Gram–Schmidt orthonormalization in $L^2(P)$ (RAP, 5.4.6) we can write $y_n = \sum_{j=1}^{m(n)} a_{nj} g_j$ for all n, where g_j are linear functions on ℓ^∞ (finite linear combinations of coordinates), are orthonormal in $L^2(P)$ and are normally distributed, so they are i.i.d. $N(0, 1)$. If the y_n are linearly independent in $L^2(P)$, then $m(n) \equiv n$. Otherwise, $m(n + 1) = m(n) + 1$ or $m(n)$ according as y_{n+1} is or is not linearly independent of the y_j for $j \le n$.

Let $a_{nj} = 0$ for $j > m(n)$. Each g_i is in turn a linear combination of $y_1, \ldots, y_n$ for the least n such that $m(n) \ge i$.

For each n, $\int y_n^2 dP = \sum_j a_{nj}^2 \le \tau^2$. For $k = 1, 2, \ldots$, $n = 1, 2, \ldots$, let $V_{kn} := \sum_{j>k} a_{nj} g_j$. Since $a_{nj} = 0$ for $j > n$, the sum defining V_{kn} runs over $k < j \le n$, and there is no problem of convergence. Let $\mathcal{B}_{-k}$ be the smallest σ-algebra for which g_j are measurable for all $j > k$. Then for any $0 \le j \le k$ and n, we have $V_{kn} = E(V_{jn}|\mathcal{B}_{-k})$. Let $\|V_k\| := \sup_n |V_{kn}| \le +\infty$. Then for $\alpha > 0$ we have the inequalities

$$\exp\left(\alpha\|V_k\|^2\right) = \exp\left(\alpha \sup_n |V_{kn}|^2\right) \exp\left(\alpha \sup_n |E(V_{jn}|\mathcal{B}_{-k})|^2\right)$$

$$\le \exp\left(\alpha\{E(\sup_n |V_{jn}||\mathcal{B}_{-k})\}^2\right) = \exp\left(\alpha\{E(\|V_j\||\mathcal{B}_{-k})\}^2\right)$$

$$\le E\left(\exp(\alpha\|V_j\|^2)|\mathcal{B}_{-k}\right)$$

by conditional Jensen's inequality (RAP, 10.2.7) if the expectations are finite. First, taking $j = 0$, $E \exp(\alpha\|V_0\|^2) < \infty$ for some $\alpha > 0$ by Lemma 2.10. Then for $j = 0 \le k$, the inequalities hold and give for that α, $E \exp\left(\alpha\|V_k\|^2\right) \le E \exp\left(\alpha\|V_0\|^2\right) < \infty$ for all k. So for almost all $y \in \ell^\infty$, $V_k(y) := \{V_{kn}(y)\}_{n\ge 1} \in \ell^\infty$. Let $W_{-k} := \exp(\alpha\|V_k\|^2)$, $k = 0, 1, \ldots$. Then by the inequalities for general $0 \le j \le k$, $\{(W_j, \mathcal{B}_j) : j = \ldots, -2, -1, 0\}$ is a submartingale (RAP, Section 10.3) and in view of its index set, a reversed submartingale. For any $s > 0$ and finite k, by the Doob maximal inequality (RAP, 10.4.2),

$$\Pr\{\max_{0\le j\le k} \|V_j\| \ge s\} = \Pr\left\{\max_{0\le j\le k} W_{-j} \ge \exp\left(\alpha s^2\right)\right\}$$

$$\le E W_0 / \exp\left(\alpha s^2\right) < \infty$$

by choice of α. Choose s large enough so that $E W_0 \exp(-\alpha s^2) < 1/2$. Then letting $k \to \infty$,

$$\Pr(\limsup_{j\to\infty} \|V_j\| \ge s) \le \Pr(\sup_j \|V_j\| \ge s) < 1/2.$$

Now $\limsup_{j\to\infty} \|V_j\|$ is measurable for the "tail σ-algebra" $\bigcap_j \mathcal{B}_{-j}$. Thus $\Pr(\limsup_{j\to\infty} \|V_j\| \ge s) = 0$ by the Kolmogorov 0-1 law (RAP, 8.4.4). Then for $0 < \varepsilon < 1/2$, there is a $k(\varepsilon) < \infty$ such that $\Pr(\|V_k\| \ge s) \le \varepsilon$ for $k \ge k(\varepsilon)$ since if $\Pr(\|V_k\| \ge s) > \varepsilon$ for infinitely many values of k, then $\Pr(\limsup_{k\to\infty} \|V_k\| \ge s) \ge \varepsilon$. Then by the last line of the proof of

Lemma 2.10,

$$E\left(\exp(\beta\|V_k\|^2)\right) < \infty \quad \text{for } \beta \le \frac{1}{24s^2}\log\left(\frac{1-\varepsilon}{\varepsilon}\right). \tag{2.7}$$

Now take any α with $0 < \alpha < 1/(2\tau^2)$. Choose γ with $\alpha < \gamma < 1/(2\tau^2)$, then $\varepsilon > 0$ small enough so that

$$\zeta := \frac{\alpha\gamma}{(\gamma^{1/2} - \alpha^{1/2})^2} \le \frac{\log((1-\varepsilon)/\varepsilon)}{24s^2}. \tag{2.8}$$

Let $k = k(\varepsilon)$ and $U_k(y) := y - V_k(y)$. Then

$$\alpha^{1/2}\|y\| \le \alpha^{1/2}\|U_k(y)\| + \alpha^{1/2}\|V_k(y)\|$$
$$= (\alpha/\gamma)^{1/2}\gamma^{1/2}\|U_k(y)\| + \left(1 - (\alpha/\gamma)^{1/2}\right)\zeta^{1/2}\|V_k(y)\|.$$

Since $t \mapsto \exp(t^2)$ is a convex function (RAP, Section 6.3, see Problem 1),

$$\exp\left(\alpha\|y\|^2\right) \le (\alpha/\gamma)^{1/2}\exp\left(\gamma\|U_k\|^2\right)$$
$$+ \left(1 - (\alpha/\gamma)^{1/2}\right)\exp\left(\zeta\|V_k\|^2\right).$$

By (2.7) and (2.8), $E\exp(\zeta\|V_k\|^2) < \infty$. Now, by the Cauchy inequality, for each n,

$$U_{kn}^2 = \left(\sum_{i=1}^k a_{ni}g_i\right)^2 \le \left(\sum_{i=1}^k g_i^2\right)\left(\sum_{j=1}^k a_{nj}^2\right),$$

so

$$E\left(\exp\left(\gamma\|U_k\|^2\right)\right) \le E\left\{\exp\left(\gamma\left(\sum_{i=1}^k g_i^2\right)\sup_n\sum_{j=1}^k a_{nj}^2\right)\right\}$$
$$\le E\exp\left\{\left(\sum_{i=1}^k g_i^2\right)\gamma\tau^2\right\} = (1 - 2\gamma\tau^2)^{-k/2} < \infty.$$

So Theorem 2.11 is proved. $\qquad\qquad\square$

Proof of Theorem 2.6. Let $\{x_n\}_{n=1}^\infty$ be dense in X. For each n, by the Hahn–Banach theorem and a corollary (RAP, 6.1.5) there is a $y_n \in X'$ with $\|y_n\|' = 1$ and $|y_n(x_n)| = \|x_n\|$. Then for all $x \in X$, we have $\sup_n |y_n(x)| = \|x\|$, so the Theorem follows from Theorem 2.11. $\qquad\qquad\square$

2.4 Inequalities and Comparisons for Gaussian Distributions

The main result of this section will show that if a set of Gaussian random variables is large enough in the sense defined in Section 1.2, meaning that the number of variables more than ε apart grows rather fast as $\varepsilon \downarrow 0$, then it is almost surely unbounded. The main steps in the proof will be some inequalities, one due to Slepian and another to Sudakov and Chevet.

In the next proof, the following well-known relations will be used. For any random variable $Y \geq 0$ with distribution function G we have

$$\int_0^\infty \Pr(Y > r)\,dr = \int_0^\infty 1 - G(r)\,dr = \int_0^\infty \int_r^\infty dG(y)\,dr \qquad (2.9)$$

$$= \int_0^\infty \int_0^y dr\,dG(y) = \int_0^\infty y\,dG(y) = EY.$$

If Y is any random variable with a finite expectation, let $Y^+ := \max(Y, 0) \geq 0$ and $Y^- := -\min(Y, 0) \geq 0$, so that $Y = Y^+ - Y^-$ and we get

$$EY = \int_0^{+\infty} \Pr(Y > t)\,dt - \int_0^{+\infty} \Pr(Y < -t)\,dt. \qquad (2.10)$$

Theorem 2.12 (Slepian's inequality) *Let $X_1, \ldots, X_n$ be real random variables with a normal joint distribution $N(0, r)$ on $\mathbb{R}^n$. For any real $\lambda_1, \ldots, \lambda_n$, let $P_n(r) := P_n(r, \{\lambda_j\}_{j=1}^n) := \Pr\{X_j \geq \lambda_j \text{ for all } j = 1, \ldots, n\}$. Let q be another covariance matrix, with $r_{ii} = q_{ii}$ for all $i = 1, \ldots, n$ and $r_{ij} \geq q_{ij}$ for all i and j. Then $P_n(r) \geq P_n(q)$.*

If $(X_1, \ldots, X_n)$ have distribution $N(0, r)$ and $(Y_1, \ldots, Y_n)$ have distribution $N(0, q)$ then $E \max(X_1, \ldots, X_n) \leq E \max(Y_1, \ldots, Y_n)$.

Proof. If for some i, $r_{ii} = q_{ii} = 0$, then $P_n(r) = P_n(q) = 0$ if $\lambda_i > 0$, whereas if $\lambda_i \leq 0$, then $X_i \geq \lambda_i$ holds with probability 1 for both distributions, and we can eliminate X_i and reduce n until $r_{ii} = q_{ii} > 0$ for all i.

Suppose next that r is nonsingular, so that it is strictly positive definite and $N(0, r)$ has a density g_n given by Fourier inversion of its characteristic function as

$$g_n(x_1, \ldots, x_n) := g_n(x_1, \ldots, x_n; r) = (2\pi)^{-n} \int_{-\infty}^\infty \cdots \int_{-\infty}^\infty$$

$$\cdot \exp\left(-i \sum_{j=1}^n x_j t_j - \frac{1}{2} \sum_{k,m=1}^n r_{km} t_k t_m\right) dt_1 \cdots dt_n$$

(RAP, Theorem 9.5.4). Since r is symmetric, it is given by the $n(n+1)/2$ variables r_{km}, $1 \leq k \leq m \leq n$. The partial derivatives $\partial^2 g_n/\partial x_k \partial x_m$ can be evaluated by differentiating under the integral signs, applying Corollary A.7 (in Appendix A below) twice, thus multiplying the integrand by $-t_k t_m$ (RAP, Theorem 9.4.4). The same integral results from taking $\partial g_n/\partial r_{km}$ for $k < m$, where $\partial/\partial r_{km}$ can be taken under the integral sign by Proposition A.16 (since

r is strictly positive definite). So

$$\partial^2 g_n/\partial x_k \partial x_m \equiv \partial g_n/\partial r_{km}, \quad k \neq m. \tag{2.11}$$

Now

$$P_n(r) = \int_{\lambda_n}^{\infty} \cdots \int_{\lambda_1}^{\infty} g_n(x_1, \ldots, x_n)\, dx_1 \ldots dx_n \tag{2.12}$$

where $g_n(x) = (2\pi)^{-n/2}(\det r)^{-1/2} \exp((-r^{-1}x, x)/2)$ (RAP, 9.5.8). For a positive definite symmetric matrix s, by Proposition A.16, applied to $t = s_{ij}$ and $\psi(x) = x_i x_j$ (or $x_i^2/2$ if $i = j$), the integral of $\exp(-(sx, x)/2)$ over any measurable region in $\mathbb{R}^n$, specifically the orthant $\{\lambda_i \leq x_i < \infty, \ i = 1, \ldots, n\}$, can be differentiated under the integral sign with respect to any component of s. Then, since the functions $r \mapsto r^{-1}$ and $r \mapsto (\det r)^{-1/2}$ are smooth for r in the set of symmetric, (strictly) positive definite matrices, the integral (2.12) can be differentiated under the integral sign with respect to any r_{km}, $k < m$. Then by (2.11), we need to evaluate $\int_{\lambda_n}^{\infty} \cdots \int_{\lambda_1}^{\infty} \partial^2 g_n/\partial x_k \partial x_m\, dx_1 \ldots dx_n$. Since the integrand is absolutely integrable, we can do the integrations in any order and replace $\int_{\lambda_m}^{\infty} \int_{\lambda_k}^{\infty} dx_k dx_m$ by $\lim_{M \to \infty} \int_{\lambda_m}^{M} \int_{\lambda_k}^{M} dx_k dx_m$. Here we may as well assume that $k = 1$, $m = 2$. Now since g_n is a smooth function, for $M \geq \max(\lambda_1, \lambda_2)$,

$$\int_{\lambda_2}^{M} \int_{\lambda_1}^{M} \partial^2 g_n/\partial x_1 \partial x_2\, dx_1 dx_2$$

$$= g_n(M, M, x_3, \ldots, x_n) - g_n(M, \lambda)2, x_3, \ldots, x_n)$$

$$- g_n(\lambda_1, M, x_3, \ldots, x_n) + g_n(\lambda_1, \lambda_2, x_3, \ldots, x_n)$$

$\to g_n(\lambda_1, \lambda_2, x_3, \ldots, x_n)$ as $M \to \infty$. Thus

$$\partial P_n(r)/\partial r_{12} = \int_{\lambda_n}^{\infty} \cdots \int_{\lambda_3}^{\infty} g_n(\lambda_1, \lambda_2, x_3, \ldots, x_n)\, dx_3 \ldots dx_n \geq 0.$$

Likewise, $\partial P_n(r)/\partial r_{km} \geq 0$ for all $k < m$.

In the general case, let $\varepsilon > 0$ and $0 < \lambda < 1$. Let I be the identity matrix and $\rho := \lambda r + (1 - \lambda)q + \varepsilon I$, which is positive definite. Then

$$\frac{dP_n(\rho)}{d\lambda} = \sum_{k<m} \frac{\partial P_n(\rho)}{\partial \rho_{km}} \frac{d\rho_{km}}{d\lambda} = \sum_{k<m} \frac{\partial P_n(\rho)}{\partial \rho_{km}} (r_{km} - q_{km}) \geq 0.$$

Integrating from 0 to 1 with respect to λ gives $P_n(r + \varepsilon I) \geq P_n(q + \varepsilon I)$. The laws $N(0, r + \varepsilon I)$ converge to $N(0, r)$ as $\varepsilon \downarrow 0$. (To prove this, let X and Y be independent with $\mathcal{L}(X) = N(0, r)$ and $\mathcal{L}(Y) = N(0, I)$. Then $\mathcal{L}(X + \sqrt{\varepsilon}Y) = N(0, r + \varepsilon I)$ and as $\varepsilon \downarrow 0$, $X + \sqrt{\varepsilon}Y \to X$ a.s., hence in law.) Likewise, $N(0, q + \varepsilon I)$ converges to $N(0, q)$. Since $r_{kk} = q_{kk} > 0$ for all k, for $P = N(0, r)$ or $N(0, q)$, $P(X_k = \lambda_k) = 0$, and the boundary of the orthant $\{X_k \geq \lambda_k$

for $k = 1, \ldots, n\}$ has zero probability. Then, using the portmanteau theorem (RAP, 11.1.1(d)) it follows that $P_n(r) \geq P_n(q)$.

Taking all $\lambda_i = -\lambda$ and applying the conclusion to $(-X_1, \ldots, -X_n)$, which does not change the distribution, we get

$$N(0, r)(\max_{1 \leq j \leq n} X_j \leq \lambda) \geq N(0, q)(\max_{1 \leq j \leq n} X_j \leq \lambda).$$

Then applying (2.10) gives the last conclusion. □

Example Let $r_{ij} = 1$, i, $j = 1, 2, 3$, and let q be the 3×3 identity matrix. Let $s = r$ except that $s_{13} = s_{31} = 0$. Then s is not nonnegative definite, in other words it is not a covariance matrix, because $s_{ij} = 1$ for $\{i, j\} \neq \{1, 3\}$ implies $X_1 = X_2 = X_3$ a.s. with a $N(0, 1)$ distribution while $s_{13} = s_{31} = 0$ implies X_1 and X_3 are independent.

Thus, in the above proof, if we change r_{km} to q_{km} for one pair of indices k, m at a time, it is possible to pass through values of the matrix which are not positive definite. Then, the "normal density" $g_n(x_1, \ldots, x_n, r)$ is not well-defined, and integrals of its "characteristic function" diverge.

Recall that the *correlation (coefficient)* $r(X, Y)$ of two nonconstant variables X and Y with finite second moments is defined by

$$r(X, Y) := E((X - EX)(Y - EY))/(\sigma_X \sigma_Y)$$

where $\sigma_X := \sigma(X)$ is the standard deviation $(E(X - EX)^2)^{1/2}$.

Corollary 2.13 *Let* $X_1, \ldots, X_n$ *and* $Y_1, \ldots, Y_n$ *be two sets of jointly normally distributed variables with mean 0,* $\sigma(X_i) > 0$ *and* $\sigma(Y_i) > 0$ *for all* i, *and* $r(X_i, X_j) \geq r(Y_i, Y_j)$ *for all* $i \neq j = 1, \ldots, n$. *Then*

$$\Pr\{X_i \geq 0, \ i = 1, \ldots, n\} \geq \Pr\{Y_i \geq 0, \ i = 1, \ldots, n\}.$$

Proof. Replacing each X_i by $X_i/\sigma(X_i)$ and Y_i by $Y_i/\sigma(Y_i)$ does not change the events being considered or the correlations, and gives covariances to which Slepian's inequality applies. □

Let C be a jointly Gaussian set of random variables with mean 0, and with the $\mathcal{L}^2$ metric $d(X, Y) := (E(X - Y)^2)^{1/2}$. Recall that for $\varepsilon > 0$, $D(\varepsilon, C) := D(\varepsilon, C, d) := \sup\{n: \text{ for some } X_1, \ldots, X_n \in C, \ d(X_i, X_j) > \varepsilon, \ 1 \leq i < j \leq n\}$. In the following, by Theorem 1.9, $D(\varepsilon, C)$ could be replaced equivalently by $N(\varepsilon, C, d)$. An inequality called the Sudakov minoration (Theorem 2.22 below) is closely related. V. N. Sudakov around 1965 first discovered a fact close to the following. S. Chevet in 1970 first published a proof.

Theorem 2.14 (Sudakov–Chevet) *If* $\limsup_{\varepsilon \downarrow 0} \varepsilon^2 \log D(\varepsilon, C) = +\infty$, *then* $\sup\{|X|: X \in C\} = +\infty$ *almost surely.*

Proof. Let Φ be the standard normal distribution function and $G(x) := 1 - \Phi(x)$. Most of the proof is in the next inequality. $\square$

Lemma 2.15 (Chevet) *Let* $X_1, \ldots, X_n$ *be jointly normally distributed with mean 0 and such that for some* $M < \infty$ *and some* ε *with* $0 < \varepsilon \le 1$, *we have* $d(0, X_j) \le M$ *for all* j *and* $d(X_i, X_j) > \varepsilon$ *for all* $i \ne j$. *Let* $K := 2^{1/2}(M^2 + 1)$. *Then*

$$G(1) \Pr\{X_j \le 1, \ j = 1, \ldots, n\}$$
$$< 2^{-n-1} + (2\pi)^{-1/2} \int_0^\infty \exp(-t^2/2) \Phi(Kt/\varepsilon)^n dt.$$

Proof. Let $e_1, \ldots, e_n$ be orthonormal variables such that $X_1, \ldots, X_n$ are in their linear span, by the Gram–Schmidt process, RAP 5.4.6. Then $e_1, \ldots, e_n$ are i.i.d. $N(0, 1)$. Let e_{n+1} be another $N(0, 1)$ variable independent of $e_1, \ldots, e_n$. Let $Y_i := X_i - e_{n+1}$, $i = 1, \ldots, n$. Then

$$G(1) \Pr\{X_j \le 1, \ j = 1, \ldots, n\}$$
$$= \Pr\{e_{n+1} \ge 1 \ \text{and} \ X_j \le 1, \ 1 \le j \le n\} \le \Pr\{Y_i \le 0, \ i = 1, \ldots, n\}.$$

Let $b_{ij} := r(Y_i, Y_j)$ and $\|Y\| := d(0, Y)$. Let θ_{ij} be the angle between Y_i and Y_j at 0, so that $b_{ij} = \cos(\theta_{ij})$. Let $U := X_i$ and $V := X_j$ with $i \ne j$. Then $u := \|U\| \le M$, $v := \|V\| \le M$, and $\|U - V\| > \varepsilon$. So

$$\varepsilon^2 \le \|U - V\|^2 = u^2 - 2(U, V) + v^2,$$

and $2(U, V) \le u^2 + v^2 - \varepsilon^2$. Thus

$$[(U, V) + 1]^2 = (U, V)^2 + 2(U, V) + 1$$
$$\le u^2 v^2 + u^2 + v^2 + 1 - \varepsilon^2,$$

so

$$b_{ij} = \frac{(U, V) + 1}{[(u^2 + 1)(v^2 + 1)]^{1/2}} \le \left[1 - \frac{\varepsilon^2}{(u^2 + 1)(v^2 + 1)}\right]^{1/2}$$
$$\le 1 - \frac{\varepsilon^2}{2(M^2 + 1)^2} = 1 - \varepsilon^2/K^2 \le 1/(1 + \varepsilon^2/K^2).$$

Let $f_i := \varepsilon e_i/K - e_{n+1}$, $f_{ij} := r(f_i, f_j) = 1/(1 + \varepsilon^2/K^2)$ for $i \neq j$, $b_{ii} = f_{ii} := 1$. Then Slepian's inequality (above), specifically Corollary 2.13, gives

$$\Pr\{Y_i \leq 0, \; i = 1, \ldots, n\} \; \leq \; \Pr\{f_i \leq 0, \; i = 1, \ldots, n\}$$

$$= \Pr\{\varepsilon e_i/K \leq e_{n+1}, \; i = 1, \ldots, n\}$$

$$= (2\pi)^{-1/2} \int_{-\infty}^{\infty} \Pr\{e_i \leq tK/\varepsilon, \; i = 1, \ldots, n\} \exp(-t^2/2)dt$$

$$= (2\pi)^{-1/2} \int_{-\infty}^{\infty} \Phi(tK/\varepsilon)^n \exp(-t^2/2)dt$$

$$= (2\pi)^{-1/2} \int_{0}^{\infty} \exp(-t^2/2)[\Phi^n(Kt/\varepsilon) + \Phi^n(-Kt/\varepsilon)]dt$$

$$\leq 2^{-n-1} + (2\pi)^{-1/2} \int_{0}^{\infty} \exp(-t^2/2)\Phi^n(Kt/\varepsilon)dt,$$

proving the Lemma. □

Now to prove Theorem 2.14, take $\varepsilon_k \downarrow 0$ such that $\varepsilon_k^2 \log D(\varepsilon_k, C) \geq k$, $k = 1, 2, \ldots$. In Lemma 2.15 let $\varepsilon = \varepsilon_k$, $n = D(\varepsilon_k, C)$. If the variances of the variables in C are unbounded, let $X(n) \in C$ with $\sigma(X(n)) \geq n$. Then clearly $|X(n)|$ are unbounded a.s. So we can assume that for some $M < \infty$, $\sigma(X) \leq M$ for all $X \in C$.

For $1 \leq B < \infty$, the probability that $|X_j| \leq B$ for $j = 1, \ldots, n$ is the probability that $|X_j/B| \leq 1$ for each j. To apply Lemma 2.15, we can then replace ε_k by ε_k/B, giving the same bound except for replacing Kt/ε_k by s/ε_k where $s := KtB$.

Since $\Phi^n = (1 - G)^n \leq e^{-nG}$, it will be enough using the dominated convergence theorem to show that $D(\varepsilon_k, C)G(s/\varepsilon_k) \to +\infty$ as $k \to \infty$ for every $s > 0$. As $x \to +\infty$, $G(x) \sim (2\pi)^{-1/2}x^{-1} \exp(-x^2/2)$ (RAP, Lemma 12.1.6), so for k large, $G(s/\varepsilon_k) \geq \varepsilon_k(3s)^{-1} \exp(-(s/\varepsilon_k)^2/2)$. So it is now enough to prove that $\log D(\varepsilon_k, C) + \log \varepsilon_k - c\varepsilon_k^{-2} \to +\infty$ as $k \to \infty$ for any $c < \infty$. Now $\log \varepsilon > -\varepsilon^{-2}$ for small $\varepsilon > 0$, so the $\log \varepsilon_k$ term can be removed. By assumption, $\varepsilon_k^2 \log D(\varepsilon_k, C) - c \to +\infty$, and Theorem 2.14 follows on multiplying by $\varepsilon_k^{-2} \to +\infty$ also. □

Example Let G_n be i.i.d. $N(0, 1)$ variables and $X_n := G_n/(\log n)^{1/2}$, $n \geq 2$. Then $d(X_j, X_k) > (\log j)^{-1/2}$ for $j < k$, so $D(\varepsilon, \{X_j\}_{2 \leq j \leq n}) \geq n - 1$ if $\varepsilon \leq (\log(n - 1))^{-1/2}$. So for $0 < \varepsilon < 1$, $D(\varepsilon, \{X_j\}_{j \geq 2}) \geq [\exp(\varepsilon^{-2})] \geq \exp(\varepsilon^{-2})/2$ where $[x]$ denotes the greatest integer $\leq x$. So $\log D(\varepsilon, \{X_j\}_{j \geq 2}) \geq$

$\varepsilon^{-2} - \log 2 \geq \varepsilon^{-2}/2$ for $0 < \varepsilon < 1/2$. On the other hand for each n, Proposition 2.5 gives $\Pr\{|X_n| \geq 2\} \leq \exp(-2\log n) = 1/n^2$, and so by the Borel–Cantelli Lemma, $\limsup_{n\to\infty} |X_n| \leq 2$ and $\sup_n |X_n| < +\infty$ a.s., so Theorem 2.14 is sharp.

Here is a further fact related to Slepian's inequality that avoids the rather restrictive assumption $r_{ii} = q_{ii}$:

Theorem 2.16 *Let $N(0, C)$ and $N(0, D)$ be two normal measures with mean 0 on $\mathbb{R}^n$. Let $X = \{X_i\}_{i=1}^n$ have law $N(0, C)$, and let $Y = \{Y_i\}_{i=1}^n$ have law $N(0, D)$. Suppose that for all $i, j = 1, \ldots, n$, we have $E((Y_i - Y_j)^2) \leq E((X_i - X_j)^2)$, in other words for each i, j,*

$$D_{ii} + D_{jj} - 2D_{ij} \leq C_{ii} + C_{jj} - 2C_{ij}. \tag{2.13}$$

Then

(a) $E\{\max_{1\leq i,j\leq n}(Y_i - Y_j)\} \leq E\{\max_{1\leq i,j\leq n}(X_i - X_j)\}$ and

(b) $E \max_i Y_i \leq E \max_i X_i$.

Proof. First, it will be shown that we can assume C and D are nonsingular. Suppose the Theorem holds in that case, and let C, D be possibly singular. Then for any $t > 0$, $C + t^2 I$ and $D + t^2 I$ are nonsingular and the hypotheses hold for them, thus the conclusion. Now, $N(0, C + t^2 I)$ is the law of $X + tZ$ where Z has law $N(0, I)$ and is independent of X, and likewise for D and Y. Then, letting $t \downarrow 0$, the Theorem follows for C, D.

Suppose then that C and D are nonsingular. For $0 \leq \lambda \leq 1$ let $C_\lambda := \lambda C + (1 - \lambda)D$ and let $g(\lambda) := E \max_{1\leq i,j\leq n} U_i - U_j$ where $U := \{U_i\}_{i=1}^n$ has law $N(0, C_\lambda)$. It will be enough to show that $dg(\lambda)/d\lambda \geq 0$ for $0 \leq \lambda \leq 1$, taking a right derivative at 0 and a left derivative at 1. We have $g(\lambda) = \int \max_{1\leq i,j\leq n}(x_i - x_j)f_\lambda(x)dx$ where f_λ is the density of $N(0, C_\lambda)$, namely,

$$f_\lambda(x) = (2\pi)^{-n/2}(\det C_\lambda)^{-1/2} \exp(-(C_\lambda^{-1}x, x)/2).$$

Since C and D are nonsingular, they are strictly positive definite. Thus for some $\gamma > 0$, $(C_\lambda y, y) \geq \gamma|y|^2$ for all y and $0 \leq \lambda \leq 1$. Thus $\det C_\lambda$ is bounded below and $(\det C_\lambda)^{-1/2}$ is bounded above for $0 \leq \lambda \leq 1$. Here γ can also be chosen so that $(C_\lambda^{-1}x, x) \geq \gamma|x|^2$ for all $x \in \mathbb{R}^n$ and $0 \leq \lambda \leq 1$. The mappings $\lambda \mapsto C_\lambda$ and $\lambda \to C_\lambda^{-1}$ are both smooth (C^∞), and the map $(C, x) \mapsto (\det C)^{-1/2} \exp(-(Cx, x))$ is a smooth map on the open set of strictly positive definite symmetric matrices C and all $x \in \mathbb{R}^n$. It follows that the functions $f_\lambda(x)$ and their difference-quotients and partial derivatives with respect to λ and x are uniformly integrable on $\mathbb{R}^n$ with respect to x and remain so if multiplied by any function bounded above in absolute value by a polynomial. Thus by Theorem A.2 of Appendix A, we can differentiate under the integral sign:

$$\frac{dg(\lambda)}{d\lambda} = \int \max_{1 \le r,s \le n} (x_r - x_s)(\partial f_\lambda(x)/\partial\lambda)dx. \qquad (2.14)$$

As in (2.11) we have $\partial^2 f_\lambda/\partial x_k^2 = 2\partial f_\lambda/\partial r_{kk}$ for $k = 1,\ldots,n$. By this and (2.11) itself we have

$$\frac{\partial f_\lambda(x)}{\partial\lambda} = \frac{1}{2}\sum_{i,j=1}^{n} \frac{d(C_\lambda)_{ij}}{d\lambda} \frac{\partial^2 f_\lambda}{\partial x_i \partial x_j} \qquad (2.15)$$

where $d(C_\lambda)_{ij}/d\lambda = C_{ij} - D_{ij}$. Let $\mathcal{S} := \mathcal{S}(\mathbb{R}^n)$ be the space of all C^∞ functions f from $\mathbb{R}^n$ into $\mathbb{R}$ such that for every polynomial $P(\cdot)$ and every n-tuple $p := (p_1,\ldots,p_n)$ of nonnegative integers with $|p| := p_1 + \ldots p_n$, $\sup_x |P(x)D^p f(x)| < \infty$ where $D^p f(x) := \partial^{|p|} f(x)/\partial x_1^{p_1} \ldots \partial x_n^{p_n}$. After multiplying $P(x)$ by some power of $1 + |x|^2$ where $|x|^2 := x_1^2 + \cdots + x_n^2$, we see that each $P(x)D^p f(x)$ is integrable on $\mathbb{R}^n$. Here $\mathcal{S}(\mathbb{R}^n)$ is Laurent Schwartz's space of rapidly decreasing functions. It is easily seen that any normal density such as f_λ is in $\mathcal{S}(\mathbb{R}^n)$.

Let $u(x) := \max_i x_i$ and for each $i = 1,\ldots,n$ let $v_i := v_i(x) := \max_{j \ne i} x_j$. Let $dx/dx_i := \Pi_{j \ne i} dx_j$. For any $i \ne j$ define a function g_{ij} by $g_{ij}(\{x_r\}_{r \ne i}) := \{y_r\}_{r=1}^n$ where $y_r := x_r$ for $r \ne i$ and $y_i := y_j$. Let V_{ij} be the measure defined for suitable functions ϕ on $\mathbb{R}^n$ by $\int \phi dV_{ij} := \int_{A(j)} \phi(g_{ij}(\{x_r\}_{r \ne i})dx/dx_i$ where $A(j)$ is the set of $\{x_r\}_{r \ne i}$ such that $x_r \le x_j$ for all $r \ne i$, or equivalently $v_i = x_j$. In other words, V_{ij} is a measure on the subset where $x_r \le x_i$ for all r of the $(n-1)$-dimensional hyperplane $\{x_i = x_j\}$, given by dx/dx_i or equivalently by dx/dx_j. On $A(j)$, as an image by a linear map ϕ_{ij} of Lebesgue measure on $\mathbb{R}^{n-1}$, V_{ij} is a multiple (by $2^{-1/2}$) of Lebesgue measure on the hyperplane $\{x_i = x_j\}$. Thus, $V_{ij} = V_{ji}$ for $1 \le i < j \le n$.

Lemma 2.17 *For any $\phi \in \mathcal{S}(\mathbb{R}^n)$,*

(a) For any $i = 1,\ldots,n$, $\int u(x)(\partial\phi/\partial x_i)dx = -\int_{x_i \ge v_i} \phi(x)dx.$

(b) For any $i \ne j$ with $1 \le i, j \le n$,

$$\int u(x)(\partial^2\phi/\partial x_i\partial x_j)dx = -\int \phi dV_{ij}. \qquad (2.16)$$

(c) For each $i = 1,\ldots,n$,

$$\int u(x)(\partial^2\phi/\partial x_i^2)dx = \int \phi(x)|_{x_i=v_i} dx/dx_i. \qquad (2.17)$$

Proof. For (a), we have $\int u(x)(\partial\phi/\partial x_i)dx = \int(u - v_i)(x)(\partial\phi/\partial x_i)dx$ since v_i does not depend on x_i. Integrating with respect to x_i first we get

$$\int\int_{v_i(x)}^{\infty} (x_i - v_i)\frac{\partial\phi}{\partial x_i}dx_i\frac{dx}{dx_i}.$$

Integrating by parts in the inner integral, the boundary terms are 0 both at $x_i = v_i$ and at $x_i = \infty$ since $\phi \in \mathcal{S}$, so (a) follows.

Then to prove (b), apply (a) to $\partial \phi / \partial x_j$. Integrating first with respect to x_j, (b) follows.

For (c), applying (a) to $\partial \phi / \partial x_i$ and integrating first with respect to x_i gives (c). □

Now continuing the proof of Theorem 2.16, note that we have $\max_{i,j}(X_i - X_j) = \max_i X_i - \min_j X_j$ and $\min_j X_j = -\max(-X_j)$. Normal laws with mean 0 are symmetric, so $\{-X_j\}_{j=1}^n$ and $\{X_j\}_{j=1}^n$ have the same law. Thus,

$$E \max_{i,j}(X_i - X_j) = 2E \max_i X_i. \tag{2.18}$$

Then, from equations (2.14), (2.15), (2.18), (2.16), and (2.17), we get

$$\frac{dg}{d\lambda} = 2 \sum_{1 \le i < j \le n} (D_{ij} - C_{ij}) \int f_\lambda dV_{ij} + \sum_{i=1}^{n}(C_{ii} - D_{ii}) \int f_\lambda(x)|_{x_i = v_i} \frac{dx}{dx_i}.$$

Now, for Lebesgue measure dx/dx_i on $\{x_j\}_{j \ne i}$, each set $\{x_j = x_r\}$ for $j \ne i \ne r \ne j$ has measure 0, and thus $\sum_{j \ne i} 1_{v_i = x_j} = 1$ almost everywhere. So

$$\int f_\lambda(x)|_{x_i = v_i} dx/dx_i = \sum_{j \ne i} \int f_\lambda dV_{ij},$$

and

$$\frac{dg}{d\lambda} = \sum_{i=1}^{n}\sum_{j \ne i}(D_{ij} - C_{ij} + C_{ii} - D_{ii}) \int f_\lambda dV_{ij}.$$

Symmetrizing the sum, interchanging j and i, gives

$$\frac{dg}{d\lambda} = \frac{1}{2} \sum_{i=1}^{n}\sum_{j \ne i}(C_{ii} + C_{jj} + 2D_{ij} - 2C_{ij} - D_{ii} - D_{jj}) \int f_\lambda dV_{ij} \ge 0$$

by (2.13). Thus (a) of Theorem 2.16 is proved. Then since (2.18) also holds for Y, (b) follows and Theorem 2.16 is proved. □

Here is a another fact related to Slepian's inequality.

Theorem 2.18 *Let T be a set and $\{X_t\}_{t \in T}$ and $\{Y_t\}_{t \in T}$ two Gaussian processes with mean 0. Let $d_X(s, t) := E((X_s - X_t)^2)^{1/2}$ and likewise define d_Y. Suppose that $d_X(s, t) \le d_Y(s, t)$ for all $s, t \in T$, and that T has a countable subset S dense for d_Y, thus also for d_X. Suppose $E \sup_{t \in S} Y_t < \infty$. Then* ess. $\sup_{t \in T} Y_t := \{Y_t\}_{t \in T}^* := \sup_{t \in U} Y_t$ *is defined up to almost sure equality for all countable d_Y-dense subsets U of T, and*

$$E\{X_t\}_{t \in T}^* \le E\{Y_t\}_{t \in T}^* < \infty. \tag{2.19}$$

We also have

$$E(\text{ess. sup}_{s,t \in T} X_s - X_t) \le E(\text{ess. sup}_{s,t \in T} Y_s - Y_t). \qquad (2.20)$$

Also, let $\{|X_t|\}_{t \in T}^* := \text{ess. sup}_{t \in T} |X_t|$ *and likewise for* Y_t. *If* $\inf_{t \in S} |X_t| = 0$ *almost surely, then*

$$E\{|X_t|\}_{t \in T}^* \le 2E\{|Y_t|\}_{t \in T}^* < \infty. \qquad (2.21)$$

Proof. For (2.19) one can apply Theorem 2.16(b) for finite sets increasing up to S (or U) and use the proof of Lemma 2.4.

For (2.20), since $\{-X_t\}$ is equal in distribution to $\{X_t\}$, we have $E \text{ ess. sup}_{t \in T} X_t = E \text{ ess. sup}_{s \in T} -X_s$ and the left side of (2.20) equals $2E \text{ ess. sup}_{t \in T} X_t$. Doing the same with Y_t, we get that (2.20) follows from (2.19).

To prove (2.21), for each ω and each $s \in T$, we have $X_s \le \sup_{t \in S} X_s - X_t$, taking $t \in S$ with $X_t(\omega) \to 0$. Likewise, we have $-X_t \le \sup_{s \in S} X_s - X_t$ for $s \in S$ such that $X_s(\omega) \to 0$. It follows that ess. $\sup_{t \in T} |X_t| \le$ ess. $\sup_{s,t \in T} X_s - X_t$. Since $Y_s - Y_t \le |Y_s| + |Y_t|$, we get (2.21). $\qquad \square$

2.5 Sample Boundedness

First is a nice characterization of the GB property, due to Sudakov (1971, 1973).

Theorem 2.19 *Let C be a subset of a Hilbert space H. Then C is a GB-set if and only if $EL(C)^* < +\infty$.*

Proof. If C is nonseparable then by Theorem 2.14, it has a countable subset B which is not a GB-set, so $L(C)^* \ge L(B)^* = +\infty$ almost surely, and the equivalence holds. So suppose C is separable. Let B be a countable dense subset of C. Then $L(B)^* = L(C)^*$ a.s. by the proof of Lemma 2.4. Let $B = \{x_j\}_{j \ge 1}$. If $E \sup_j L(x_j) < +\infty$, then $\sup_j L(x_j) < +\infty$ almost surely, and since $-L$ is a version of L, also $\inf_j L(x_j) > -\infty$ almost surely, so B and hence C is a GB-set. Conversely suppose B is a GB-set. Define a probability measure P_B on the measurable linear space ℓ^∞ of all bounded sequences $\{y_j\}_{j \ge 1}$ of real numbers with the smallest σ-algebra making all the coordinates measurable, where $\{y_j\}_{j \ge 1}$ have the joint distribution of $\{L(x_j)\}_{j \ge 1}$. By Theorem 2.11, $E \sup_j |L(x_j)| < +\infty$. $\qquad \square$

One can make a stronger statement. Let g be a convex, increasing function from $[0, \infty)$ onto itself. If Y is a random variable such that $Eg(\delta Y) < \infty$ for some $\delta > 0$, let $\|Y\|_g := \inf\{c > 0 : Eg(|Y|/c) \le 1\}$. Then $\|\cdot\|_g$ is a seminorm on such random variables (Appendix H). If there is no such $\delta > 0$, let $\|Y\|_g := +\infty$.

Theorem 2.20 *There is an absolute constant $M < \infty$ such that for any subset C of a Hilbert space H, and $g(x) := \exp(x^2) - 1$,*

$$\|L(C)^*\|_g \leq \| |L(C)|^*\|_g \leq ME(|L(C)|^*).$$

Proof. If C is not a GB-set, all three expressions will be infinite, so suppose C is a GB-set, which we can then take to be countable by Theorem 2.14. Then by Theorem 2.11 as in the previous proof, $\| |L(C)|^*\|_g < \infty$.

The first inequality in the theorem is clear. For the second, suppose there is no such $M < \infty$. Then there are countable GB-sets $C_j \subset H$ with $\| |L(C_j)|^*\|_g \geq j^3 E|L(C_j)|^*$ for $j = 1, 2, \ldots$. By homogeneity, we can assume $E|L(C_j)|^* = 1$ for each j. Let $H_1, H_2, \ldots$ be infinite-dimensional Hilbert spaces and form the direct sum $\mathcal{H} := \oplus_j H_j$, so that H_j are taken as orthogonal subspaces of $\mathcal{H}$. We can take $C_j \subset H_j$ for each j. Let $D_j := C_j / j^2$ for each j. Let $D := \cup_j D_j \subset \mathcal{H}$. Then

$$E|L(D)|^* = E \max_j |L(D_j)|^* \leq \sum_j E|L(D_j)|^* = \sum_j j^{-2} < \infty,$$

so D is a GB-set. Thus by Theorem 2.14 again, $\| |L(D)|^*\|_g < \infty$. But for each j, $\| |L(D)|^*\|_g \geq \| |L(D_j)|\|_g \geq j$, a contradiction, so Theorem 2.20 is proved. $\square$

Next, bounds for expected suprema will be developed, closely related to the Sudakov–Chevet theorem 2.14. Let $M_n := \max(Z_1, \ldots, Z_n)$ be the maximum of n i.i.d. standard normal variables Z_i. Then EM_n is bounded below as follows.

Lemma 2.21 *For all $n \geq 1$, $EM_n \geq (\log n)^{1/2}/12$.*

Remark. The constant $1/12$ can be improved to $(\pi \log 2)^{-1/2}$, by a less elementary proof (Fernique 1997, (1.7.1) and references given there).

Proof. For $n = 1$ we have $0 \geq 0$. For $n = 2$, $EM_2 > (\log 2)^{1/2}/12$ can be found by a direct calculation (Problem 1). The following proof is for $n \geq 3$. Let $\alpha := (8\pi)^{-1/2}$. By Proposition 2.5, we have

$$P(Z \geq (\log n)^{1/2}) \geq \frac{1}{2(2\pi \log n)^{1/2}} e^{-\log n/2} = \alpha(n \log n)^{-1/2} \geq \frac{\alpha}{n}$$

where $\log n \leq n$ for all $n \geq 1$ since $x \leq e^x$ for all x. Now, by its Taylor series, $\log(1 - x) \leq -x$ for $0 < x < 1$, so

$$\left(1 - \frac{\alpha}{n}\right)^n = \exp\left(n \log\left(1 - \frac{\alpha}{n}\right)\right) \leq \exp\left(n\left(-\frac{\alpha}{n}\right)\right) = e^{-\alpha}.$$

Then $P(M_n \geq (\log n)^{1/2}) \geq 1 - e^{-\alpha} \geq 0.18$. Recall that for any real-valued function f, we let $f^+ := \max(f, 0)$ and $f^- := -\min(f, 0)$. Thus $E(M_n^+) \geq 0.18(\log n)^{1/2}$.

Next, $-M_n^-$ is nondecreasing as n increases, so for $n \geq 3$, $E(-M_n)^- \geq E(-M_3^-)$. Also, $M_3^- = 0$ unless Z_1, Z_2 and Z_3 are all negative, which has probability $1/8$. We have $E(-M_3^-|M_3 < 0) \geq E(Z_1|Z_1 < 0) = -(2/\pi)^{1/2}$. Thus $E(-M_3^-) \geq (2/\pi)^{1/2}/8 \geq -0.1$. It follows that $EM_n \geq (\log n)^{1/2}/12$ for all $n \geq 3$ and so for all $n \geq 1$. □

Here is a bound in terms of expectation and packing numbers, extending the more special preceding lemma. Part (b) is the form most often applied.

Theorem 2.22 (Sudakov minoration) *(a) For any countable subset S of a Hilbert space H with its usual metric, and any $\varepsilon > 0$,*

$$E \sup_{x \in S} L(x) \geq \frac{1}{17}\varepsilon(\log D(\varepsilon, S, d))^{1/2}.$$

(b) Let X_t, $t \in T$, be a Gaussian process with mean 0 defined on a countable parameter space T. On T take the pseudometric d_X defined by the process, $d_X(s, t) := [E((X_s - X_t)^2)]^{1/2}$. Then for any $\varepsilon > 0$,

$$E \sup_{t \in T} X_t \geq \frac{1}{17}\varepsilon(\log D(\varepsilon, T, d_X))^{1/2}.$$

Remark. The constant $1/17$ can be improved to $(2\pi \log 2)^{-1/2}$ (Fernique 1997, Theorem 4.1.4).

Proof. (a) Fix any $\varepsilon > 0$ and let $m := D(\varepsilon, S, d)$. Take m points $x_1, \ldots, x_m \in S$ with $\|x_i - x_j\|^2 > \varepsilon^2$ for $i \neq j$ and i.i.d. $N(0, 1)$ variables Z_i. Let $V_i := \varepsilon Z_i/2^{1/2}$. Then $E\left((V_i - V_j)^2\right) = \varepsilon^2 \leq E\left((L(x_i) - L(x_j))^2\right)$ for all $i, j = 1, \ldots, m$. It follows then from Lemma 2.21 and the comparison theorem 2.16 that

$$E \sup_{x \in S} L(x) \geq E \max_{1 \leq i \leq m} L(x_i) \geq E \max_{1 \leq i \leq m} V_i \geq (\varepsilon/2^{1/2})EM_m$$

$$\geq (\varepsilon/2^{1/2})(\log m)^{1/2}/12 \geq \varepsilon(\log D(\varepsilon, S, d))^{1/2}/17,$$

which finishes the proof of part (a).

(b): The map $t \mapsto X_t$ takes T into a Hilbert space $H = L^2(\Omega, P)$, with $\{L(X_t(\cdot))\}_{t \in T}$ equal in distribution to $\{X_t\}_{t \in T}$. So the result follows from part (a). □

2.6 Gaussian Measures and Convexity

There are several useful inequalities about normal measures and convex sets. Convex sets were treated in RAP, Sections 6.2 and 6.6.

A set C in a vector space is called *symmetric* if $-C := \{-x \colon x \in C\} = C$. A function f is called *even* if $f(-x) = f(x)$ for all x. Thus, the indicator function of a set is even if and only if the set is symmetric.

For sets A, B in a vector space and a constant c let $cA := \{ca \colon a \in A\}$ and $A + B := \{a + b \colon a \in A, b \in B\}$.

Theorem 2.23 *Let C be a convex, symmetric set in $\mathbb{R}^k$. Let f be a nonnegative, even function in $\mathcal{L}^1(\mathbb{R}^k, \mathcal{B}, V)$ where V is Lebesgue measure λ^k and $\mathcal{B}$ is the Borel σ-algebra. Suppose that for every $t > 0$, $K_t := \{x \colon f(x) > t\}$ is convex. Then for $0 \le \alpha \le 1$, any $y \in \mathbb{R}^k$, and $dx := dV(x)$,*

$$\int_C f(x + \alpha y) dx \ge \int_C f(x + y) dx.$$

Proof Since f is integrable, both sides of the stated inequality are finite. First suppose that f is the indicator function of a set K. Then K is convex and symmetric. We need to prove

$$V(C \cap (K - \alpha y)) \ge V(C \cap (K - y)). \tag{2.22}$$

For these measures even to make sense, we need to take care of a measurability problem. Not all convex sets are Borel sets: for example, in $\mathbb{R}^2$, the unit disk together with an arbitrary subset of the boundary unit circle is always convex. But we do have:

Lemma 2.24 *Any convex set D in $\mathbb{R}^k$ is Lebesgue measurable, and its boundary ∂D has measure 0, $V(\partial D) = 0$.*

Proof. Either D is included in some $(k - 1)$-dimensional hyperplane, in which case $V(D) = 0$ and we are done, or D has a nonempty interior U (RAP, Theorem 6.2.6). Then D and U have the same closure and boundary (RAP, Proposition 6.2.10; also, any point of D is a vertex of a k-dimensional simplex included in D). Let $p \in U$. By translation we can assume $p = 0$. A straight line L through 0 can intersect ∂U in at most two points: suppose $0, q, r$ are on L in that order, $q \ne r$, $q, r \in \partial U$. Now D includes a neighborhood W of 0, $W = \{x \colon |x| < \delta\}$ for some $\delta > 0$. Let $\lambda := |q|/|r|$. Take points $r_n \in D$ with $r_n \to r$. Then the convex combinations $\lambda r_n + (1 - \lambda)w$, $w \in W$, yield all points in a neighborhood of q, so $q \in U$, a contradiction. Then by spherical coordinates (RAP, Section 4.4, problems 8,9), $V(\partial D) = 0$ so D is Lebesgue measurable. $\square$

Now continuing the proof of Theorem 2.23, let $\lambda := (1 + \alpha)/2$, so that $\lambda(-y) + (1 - \lambda)y = -\alpha y$, and $1/2 \le \lambda \le 1$. Then $K - \alpha y = \lambda(K - y) +$

$(1 - \lambda)(K + y)$ since K is convex, so

$$C \cap (K - \alpha y) \supset \lambda(C \cap (K - y)) + (1 - \lambda)(C \cap (K + y))$$

because C is also convex. Thus

$$V(C \cap (K - \alpha y)) \ge V\{\lambda(C \cap (K - y)) + (1 - \lambda)(C \cap (K + y))\}.$$

Since both C and K are symmetric, $C \cap (K - y) = -\{C \cap (K + y)\}$ and $V(C \cap (K - y)) = V(C \cap (K + y))$. If C and K are compact, then the Brunn–Minkowski inequality (RAP, 6.6.1(b)), with $A = C \cap (K - y)$, $B = C \cap (K + y)$ and still $\lambda = (1 + \alpha)/2$ gives (2.22) as desired.

Recall that if U is any open set in a metric space and $r > 0$, with $B(x, r) := \{y: d(x, y) < r\}$, then $U_r := \{x \in U : B(x, r) \subset U\}$ is always a closed subset of U. It is easily seen that if U is convex and/or symmetric, so is U_r for any $r > 0$. Thus any open (symmetric) convex set is the union of an increasing sequence of compact (symmetric) convex sets $K_n := \{x \in U_{1/n} : |x| \le n\}$. Thus for any convex set L in $\mathbb{R}^k$, by Lemma 2.24 there exist compact convex sets $L(n)$ such that $1_{L(n)} \uparrow 1_L$ almost everywhere for Lebesgue measure. Applying this to the different convex sets $L = A, B$ shows that (2.22) holds for any convex symmetric sets C and K with $V(K) < \infty$, so that $V(K) = 0$ or K is bounded.

Next suppose f is bounded. Then we can assume $0 \le f \le 1$. Take simple functions approaching f, specifically

$$f_n := \sum_{k=0}^{n-1} \frac{k}{n} 1_{\{k/n < f \le (k+1)/n\}} = \frac{1}{n} \sum_{j=1}^{n} 1_{\{f > j/n\}},$$

since $k/n < f \le (k + 1)/n$ if and only if $f > j/n$ for exactly k values of $j \ge 1$. So by linearity and (2.22), the conclusion holds for f_n for each n. By dominated convergence, it then holds for any bounded f. Then if f is unbounded, let $g_n := \min(f, n)$. Then the result holds for g_n for each n and $g_n \uparrow f$, so by monotone convergence it holds for f. $\qquad\square$

Theorem 2.25 *Let X be a random variable with values in $\mathbb{R}^k$ whose law has a density f satisfying the hypotheses of Theorem 2.23. Let Y be any other $\mathbb{R}^k$-valued random variable independent of X. Then for any convex, symmetric set C and $0 \le \alpha \le 1$,*

$$\Pr\{X + \alpha Y \in C\} \ge \Pr\{X + Y \in C\}.$$

Proof. For $P := \mathcal{L}(Y)$, by Theorem 2.23, with $-y$ in place of y,

$$\Pr\{X + \alpha Y \in C\} = \int\!\!\int_{C - \alpha y} f(x) dx \, dP(y) = \int\!\!\int_C f(z - \alpha y) dz \, dP(y)$$

$$\ge \int\!\!\int_C f(z - y) dz \, dP(y) = \Pr(X + Y \in C). \qquad\square$$

Corollary 2.26 *Let Z and X be random variables with values in $\mathbb{R}^k$, $\mathcal{L}(Z) = N(0, A)$ and $\mathcal{L}(X) = N(0, D)$, where D and $A - D$ are nonnegative definite and symmetric. Let C be a convex symmetric set in $\mathbb{R}^k$. Then*

$$\Pr(X \in C) \geq \Pr(Z \in C).$$

Proof. Let Y be independent of X with $\mathcal{L}(Y) = N(0, A - D)$. Then $\mathcal{L}(X + Y) = \mathcal{L}(Z)$, and Theorem 2.25 for $\alpha = 0$ gives the result. $\square$

Corollary 2.27 *Let $X_1, \ldots, X_n$ and $Y_1, \ldots, Y_n$ both be jointly Gaussian with mean 0 and such that $\{EY_iY_j - EX_iX_j\}_{1 \leq i, j \leq n}$ is nonnegative definite. Then for any M,*

$$\Pr\{ \max_{1 \leq j \leq n} |X_j| > M\} \leq \Pr\{ \max_{1 \leq j \leq n} |Y_j| > M\}.$$

Proof. Corollary 2.26 applies, taking $C := \{t \in \mathbb{R}^n : \max_j |t_j| \leq M\}$, a convex, symmetric set, then taking complements. $\square$

For any set C in a vector space V, and a vector space W of real linear forms on V, the *polar* of C is defined by $C^{*1} := \{w \in W : w(x) \leq 1 \text{ for all } x \in C\}$. If C is symmetric in V then C^{*1} is also symmetric in W and equals $\{w \in W : |w(x)| \leq 1 \text{ for all } x \in C\}$.

For $V = \mathbb{R}^k$, W will be understood to be $\mathbb{R}^k$ also, defining linear forms via the usual inner product. Recall the standard normal law $N(0, I)$ on $\mathbb{R}^k$. For a set C in a vector space V and a linear transformation A defined on V, AC will mean $\{Ax : x \in C\}$.

Corollary 2.28 *Let A be a linear transformation from $\mathbb{R}^k$ into itself with norm $\|A\| := \sup\{\|Ax\| : \|x\| \leq 1\} \leq 1$, for the usual Euclidean norm $\|\cdot\|$ on $\mathbb{R}^k$. Then for any symmetric subset C of $\mathbb{R}^k$, $N(0, I)((AC)^{*1}) \geq N(0, I)(C^{*1})$.*

Proof. If $\mathcal{L}(G) = N(0, I)$, then $G \in (AC)^{*1}$ means $(G, Ax) \leq 1$ for all $x \in C$, for the usual inner product, or equivalently $(A^tG, x) \leq 1$ for all $x \in C$. Now $\mathcal{L}(A^tG) = N(0, A^tA)$, where $I - A^tA$ is nonnegative definite, and $N(0, I)((AC)^{*1}) = N(0, A^tA)(C^{*1})$. Since C^{*1} is a convex, symmetric set, Corollary 2.26 applies and gives the result as stated. $\square$

The next fact is closely related:

Lemma 2.29 *Let B be a linear operator from H into itself with $\|B\| \leq 1$ and C a (nonempty) subset of H. Then for any $t \geq 0$,*

$$\Pr\{|L(BC)|^* \leq t\} \geq \Pr\{|L(C)|^* \leq t\}.$$

Proof. If $t = 0$, then the right side is 0 unless $C = \{0\}$, in which case both sides of the inequality equal 1 and it holds. So assume $t > 0$. First suppose C is finite. We have $|L(C)|^* \leq t$ if and only if $|L(C/t)|^* \leq 1$. Let V be the linear span of

C, a finite-dimensional Hilbert space. A version of L restricted to V is given by $L(v) = (v, w)$ for the given inner product where w has distribution $N(0, I)$ on V. Since the joint distribution of $L(x)$ for $x \in C$ is uniquely determined, Corollary 2.28 for the set $\{\pm x/t : x \in C\}$ gives the result. In general, let finite sets F_n increase up to a countable dense set in C. Then $|L(C)|^* \le t$ if and only if $|L(F_n)|^* \le t$ for all n (except possibly on a set of 0 probability) so

$$\Pr\{|L(C)|^* \le t\} = \lim_{n \to \infty} \Pr\{|L(F_n)|^* \le t\}$$

$$\le \lim_{n \to \infty} \Pr\{|L(BF_n)|^* \le t\} = \Pr\{|L(BC)|^* \le t\}. \qquad \square$$

2.7 Sample Continuity

A function f from a set C in a vector space V into $\mathbb{R}$ will be called *prelinear* iff for any $c_1, \ldots, c_n \in C$ and $a_1, \ldots, a_n \in \mathbb{R}$ such that $a_1 c_1 + \cdots + a_n c_n = 0$, we have $a_1 f(c_1) + \cdots + a_n f(c_n) = 0$.

A GB-set must be totally bounded (by the Sudakov–Chevet theorem 2.14). Since a uniformly continuous function on a totally bounded set must be bounded, every GC-set is a GB-set.

Lemma 2.30 *For any prelinear function f on a set C in a real vector space V into $\mathbb{R}$, let*

$$g(x_1 c_1 + \cdots + x_n c_n) := x_1 f(c_1) + \cdots + x_n f(c_n)$$

for any $x_1, \ldots, x_n \in \mathbb{R}$ and $c_1, \ldots, c_n \in C$. Then g is a well-defined linear function from the linear span of C into $\mathbb{R}$ which extends f. Such an extension exists if and only if f is prelinear.

Proof. To show that g is well-defined, suppose $x_1 c_1 + \cdots + x_n c_n = y_1 d_1 + \cdots + y_m d_m$ for some $c_1, \ldots, c_n, d_1, \ldots, d_m \in C$ and $x_1, \ldots, x_n, y_1, \ldots, y_m \in \mathbb{R}$. Then $x_1 c_1 + \cdots + x_n c_n - y_1 d_1 - \cdots - y_m d_m = 0$, so since f is prelinear, $x_1 f(x_1) + \cdots + x_n f(c_n) - y_1 f(d_1) - \cdots - y_m f(d_m) = 0$, so the apparent candidates to be defined as $g(x_1 c_1 + \cdots + x_n c_n)$ are equal, and g is well-defined. Then g is clearly linear and extends f. On the other hand if f is not prelinear, suppose $x_1 c_1 + \cdots + x_n c_n = 0$ but $x_1 f(c_1) + \cdots + x_n f(c_n) \ne 0$. Then $x_1 c_1 = -x_2 c_2 - \cdots - x_n c_n$ but $x_1 f(c_1) \ne -x_2 f(c_2) - \cdots - x_n f(x_n)$, so f has no linear extension to the linear span of C. $\qquad \square$

Now, a *finite-dimensional projection* (fdp) will be an orthogonal projection (RAP, end of Section 5.3) of H onto a finite-dimensional subspace of H. For a sequence $\{\pi_n\}$ of such projections, $\pi_n \uparrow I$ will mean that the range of π_n is included in that of π_{n+1} for all n and that the union of all the ranges is dense in H. Since $\pi_n x$ is the nearest point to x in the range of π_n (RAP, Theorems 5.3.6 and 5.3.8), it follows that $\|\pi_n x - x\| \to 0$ as $n \to \infty$ for all $x \in H$. For any orthogonal projection π, let $\pi^\perp := I - \pi$, the orthogonal projection onto the orthogonal complement of the range of π (RAP, 5.3.8).

Lemma 2.31 *Whenever fdp's $\pi_n \uparrow I$, there is an orthonormal basis of H which includes an orthonormal basis of the range of π_n for each n.*

Proof. Begin with an orthonormal basis of the range of π_1, then recursively given an orthonormal basis of the range of π_n, adjoin an orthonormal basis of the range of $\pi_{n+1} - \pi_n$. For each n, $\pi_{n+1} \circ \pi_n \equiv \pi_n \equiv \pi_n \circ \pi_{n+1}$, so $\pi_n^\perp \pi_{n+1} = \pi_{n+1} \pi_n^\perp = \pi_{n+1} - \pi_n$. Also, $\pi_{n+1} = \pi_{n+1} \circ (I - \pi_n + \pi_n) = (\pi_{n+1} - \pi_n) + \pi_n$, so the range of π_{n+1} is the direct sum of two orthogonal subspaces, $\operatorname{ran}(\pi_{n+1} - \pi_n) \perp \operatorname{ran}(\pi_n)$. Taking the union of the bases over n gives an orthonormal set whose linear span is dense in H, and thus is a basis (RAP, Theorem 5.4.9). $\qquad\square$

If C is a totally bounded set in a metric space, then the set V_1 of all uniformly continuous real-valued functions on C is a vector space. For any real function f on C let $\|f\|_C := \sup\{|f(x)| : x \in C\}$. Then V_1 with norm $\|\cdot\|_C$ is naturally isometric to the space $C(K)$ of all continuous functions on the completion K of C, where K is compact. Since $C(K)$ is separable for the supremum norm (RAP, Corollary 11.2.5), V_1 is separable.

Let $C \subset H$ and let V_2 be the set of prelinear elements of V_1. Each element h of H defines a function on C by $x \mapsto (x, h)$, $x \in C$. Let H_C be the completion of H for $\|\cdot\|_C$. Note that each element of H_C naturally defines a uniformly continuous, prelinear function on C as a uniform limit of uniformly continuous, prelinear functions. Let V_3 be the set of functions on C so defined. Then $V_3 \subset V_2$. (Often, $V_3 = V_2$, but whether $V_3 = V_2$ in all cases will not be settled here.)

Let V be a set of functions on C. Say that L on C can be *realized* on V if there is a probability measure μ on V such that the process $(v, x) \mapsto v(x)$, $v \in V$, $x \in C$, has the joint distributions of L restricted to C: for any $x_1, \ldots, x_n$ in C, $v \mapsto v(x_i)$ are jointly Gaussian with mean 0 and covariances (x_i, x_j), $i, j = 1, \ldots, n$.

From the definition, μ would be defined on the smallest σ-algebra $\mathcal{B}_C$ making all evaluations $v \mapsto v(x)$ measurable for $x \in C$. If D is a countable dense set in C, V is a set of continuous functions on C and $v, w \in V$, then

$$\|v - w\|_C := \sup\{|(v - w)(y)| : y \in C\} = \sup\{|(v - w)(y)| : y \in D\},$$

so $v \mapsto \|v - w\|_C$ is $\mathcal{B}_C$ measurable for w fixed. If also V is a set of bounded functions on C, separable for $\|\cdot\|_C$, as V_1, V_2 and V_3 are, then all open sets for the $\|\cdot\|_C$ topology are in $\mathcal{B}_C$ (RAP, Proposition 2.1.4 and its proof), so $\mathcal{B}_C$ equals the Borel σ-algebra.

Given a set A in a vector space, the *symmetric convex hull* of A is the smallest convex set including A and $-A = \{-x: \ x \in A\}$, and is the set of all finite convex combinations $\sum_{i=1}^{n} \lambda_i a_i$, $a_i \in A \cup -A$, with $\lambda_i \geq 0$ and $\sum_{i=1}^{n} \lambda_i = 1$, for all positive integers n. The *closed symmetric convex hull* of A for some topology (in this case the Hilbert norm) is the closure of the symmetric convex hull. Here is a set of characterizations of GC-sets:

Theorem 2.32 *The following are equivalent for a totally bounded set C in H:*

(a) *C is a GC-set;*

(a′) *The closed, symmetric convex hull of C is a GC-set;*

(b) *For any $\varepsilon > 0$, $\Pr(|L(C)|^* < \varepsilon) > 0$;*

(c) *There exist fdp's $\pi_n \uparrow I$ such that $\liminf_{n\to\infty} |L(\pi_n^{\perp}C)|^* = 0$ a.s.;*

(d) *For some sequence $\pi_n \uparrow I$ of fdp's, $|L(\pi_n^{\perp}C)|^* \to 0$ in probability;*

(d′) *For some sequence $\pi_n \uparrow I$ of fdp's, $|L(\pi_n^{\perp}C)|^* \to 0$ almost surely;*

(e) *For every sequence $\pi_n \uparrow I$ of fdp's, $|L(\pi_n^{\perp}C)|^* \to 0$ in probability;*

(e′) *For every sequence $\pi_n \uparrow I$ of fdp's, $|L(\pi_n^{\perp}C)|^* \to 0$ almost surely;*

(f) *L can be realized on the completion V_3 of H for $\|\cdot\|_C$;*

(g) *L on C can be realized on the space V_2 of uniformly continuous, prelinear functions;*

(h) *L on C can be realized on the space V_1 of uniformly continuous functions.*

Proof. For any f, $g \in H$ and constant c, $L(cf + g) = cL(f) + L(g)$ a.s., since $L(cf + g) - cL(f) - L(g)$ has mean and variance both 0.

For any fdp π, the processes $L \circ \pi$ and $L \circ \pi^{\perp}$, when restricted to a countable set, are independent and satisfy $L = L \circ \pi + L \circ \pi^{\perp}$ a.s.

Let us first consider the properties (d), (d′), (e), (e′). Clearly (e′) implies (e) which implies (d).

To show (d) implies (d′), let Y be a countable dense subset of C. For $k = 1, 2, \ldots$, let $Y^k = \{(\pi_j - \pi_k)y : y \in Y, \ j \geq k\}$. Then Y^k is countable and dense in $\pi_k^{\perp}C$. By Lemma 2.4, $|L(\pi_k^{\perp}C)|^* = \sup_{x\in Y^k} |L(x)|$ for all k a.s. For $\varepsilon > 0$, choose $k = k(\varepsilon)$ sufficiently large so that

$$\Pr\left\{\sup_{\substack{j\geq k \ y\in Y}} |L((\pi_j - \pi_k)y)| > \varepsilon\right\}$$

$$= \Pr\left\{\sup_{x\in Y^k} |L(x)| > \varepsilon\right\} = \Pr\left\{|L(\pi_k^{\perp}C)|^* > \varepsilon\right\} \leq \varepsilon.$$

Then

$$\Pr\left\{\sup_{\substack{i,j\geq k \ y\in Y}} |L((\pi_i - \pi_j)y)| > 2\varepsilon\right\} \leq 2\varepsilon.$$

So $\Pr\{\sup_{i,j \geq k} \sup_{y \in Y} |L((\pi_i - \pi_j)y)| > 2\varepsilon\} \downarrow 0$ as $k \to \infty$. Thus,

$$\sup_{i,j \geq k} \sup_{y \in Y} |L((\pi_i - \pi_j)y)| \to 0$$

a.s. as $k \to \infty$. Since

$$|L(\pi_k^\perp C)|^* \leq \sup_{i,j \geq k} \sup_{y \in Y} |L((\pi_i - \pi_j)y)|$$

a.s., it follows that $|L(\pi_k^\perp C)|^* \to 0$ a.s., i.e. (d') holds.

To show (d') implies (e'), let Q_m be another sequence of fdp's with $Q_m \uparrow I$. Then for k fixed as above, let $e_1, \ldots, e_r$ be a basis for the range of π_k. Then $Q_m^\perp e_j \to 0$ in H for each j. Since r is fixed and C is bounded, $|L(Q_m^\perp \pi_k C)|^* \to 0$ in probability. We also have $\Pr\{|L(Q_m^\perp \pi_k^\perp C)|^* > \varepsilon\} \leq \Pr\{|L(\pi_k^\perp C)|^* > \varepsilon\}$ by Lemma 2.29. The latter is $\leq \varepsilon$ by choice of k. Since $|L(Q_m^\perp C)|^* \leq |L(Q_m^\perp \pi_k C)|^* + |L(Q_m^\perp \pi_k^\perp C)|^*$ a.s., we have $|L(Q_m^\perp C)|^* \to 0$ in probability as $m \to \infty$. By the last paragraph, the convergence is almost sure. So the properties (d), (d'), (e), and (e') are equivalent.

These properties clearly imply (c). To see that they imply (f), note that in the above proof, each $M_n(\cdot)(\omega)$ is the inner product with some element of H, and M_n almost surely converges uniformly on C to M. Each M_n defines a measurable function from Ω into H, thus into V_3. Hence M defines a random variable with values in V_3 (RAP, Theorem 4.2.2). So (f) follows.

Clearly (f) implies (g) implies (h) implies (a).

On the other hand, (d) implies that the M_n converge uniformly also on the closed, symmetric convex hull $\overline{\mathrm{sco}}(C)$ of C. In fact, for any fdp π, $|L(\pi^\perp \overline{\mathrm{sco}}(C))|^* = |L(\pi^\perp C)|^*$ a.s., since $|L(-x)| = |L(x)|$ a.s. for any x and finite convex combinations with rational coefficients of elements of $Y \cup -Y$ give a countable dense set in $\overline{\mathrm{sco}}(C)$. The limit of the M_n is again M, a version of L, now on $\overline{\mathrm{sco}}(C)$. So (d) implies (a'), which implies (a) clearly.

Next, to see that (d) implies (b), given $\varepsilon > 0$, take a fdp π such that $\Pr\{|L(\pi^\perp C)|^* > \varepsilon/2\} < 1/2$. Also $\Pr\{|L(\pi C)|^* < \varepsilon/2\} > 0$, since πC is a bounded set in a finite-dimensional space, and since $L \circ \pi$ and $L \circ \pi^\perp$ can be taken to be independent (on Y), we get $\Pr\{|L(C)|^* < \epsilon\} > 0$, proving (b).

Next it will be shown that (b) implies (c). For $\varepsilon > 0$, and fdp's $\pi_n \uparrow I$, Lemma 2.29 implies $\Pr\{|L(\pi_n^\perp C)|^* < \varepsilon\} \geq \Pr\{|L(C)|^* < \varepsilon\} \geq \delta$ for some $\delta > 0$ for all n. The event D that $|L(\pi_n^\perp C)|^* < \varepsilon$ for infinitely many n, that is, $\bigcap_{m \geq 1} \bigcup_{n \geq m} \{|L(\pi_n^\perp C)|^* < \varepsilon\}$, thus has probability at least δ. But D is a "tail event," since it depends on the sequence of independent random variables $L(e_j)$ only for $j \geq k$ for k arbitrarily large. It follows that D has probability 0 or 1 (Kolmogorov's zero-one law, RAP, 8.4.4), thus probability 1. This yields (c).

Next (c) implies (a): for any $\varepsilon > 0$, suppose that $|L(\pi_n^\perp C)|^* < \varepsilon/2$ for some ω and n. Then M_n, being linear on the finite-dimensional bounded set $\pi_n C$, is uniformly continuous there, so for some $\gamma > 0$, $\|x - y\| < \gamma$ implies

$|M_n(x) - M_n(y)| < \varepsilon/2$, for x, $y \in \pi_n(Y)$ and thus for x and y in Y, and then since $M_n + L \circ \pi_n^\perp = L$ almost surely on Y, $|L(x) - L(y)| < \varepsilon$. Thus $L(\cdot)(\omega)$ is almost surely uniformly continuous on Y, hence again extends to a uniformly continuous function on C which is a version of L, giving (a).

It will now be enough to prove that (a) implies (d). Given $\varepsilon > 0$, take a version of L and $\delta > 0$ such that

$$\Pr\{\sup\{|L(x) - L(y)| : x, y \in Y, \|x - y\| < \delta\} \geq \varepsilon\} < \varepsilon.$$

Take a finite-dimensional linear subspace F of H such that $F \cap C$ is within $\delta/2$ of every point of C. We can assume that $Y \cap F$ is dense in $F \cap C$. Let π be the orthogonal projection onto F. Since Y is countable, we have $L(x - y) = L(x) - L(y)$ and $L(\pi^\perp(x - y)) = L(\pi^\perp x) - L(\pi^\perp y)$ almost surely for all x, $y \in Y$. Then by Lemma 2.29,

$$\varepsilon > \Pr\{\sup\{|L(\pi^\perp x) - L(\pi^\perp y)| : x, y \in Y, \|x - y\| < \delta\} \geq \varepsilon\}$$

$$\geq \Pr\{\sup\{|L(\pi^\perp x)| : x \in Y\} \geq \varepsilon\}$$

since for any $x \in Y$ there is a y in $F \cap Y$ with $\|x - y\| < \delta$ and $\pi^\perp y = 0$. Letting $\varepsilon = 1/n \downarrow 0$, $n \to \infty$, (d) holds. $\qquad\square$

Recall that a Borel probability measure on a separable Banach space B is called *Gaussian* if every continuous linear form in B' has a Gaussian distribution. It follows that the norm $\| \cdot \|$ on B satisfies some inequalities on the upper tail of its distribution for μ (Landau–Shepp–Marcus–Fernique bounds, Theorem 2.6). In particular, $\int \|x\|^2 d\mu(x) < \infty$.

Theorem 2.33 *Let* $(B, \| \cdot \|)$ *be a separable Banach space. Let* μ *be a Gaussian probability measure with mean* 0 *on the Borel sets of* B. *Then the unit ball* $B_1' := \{f : \|f\|' \leq 1\}$ *in the dual Banach space* B' *is a compact GC-set in* $L^2(B, \mu)$.

Proof. A Cauchy sequence $\{y_n\}$ in B_1' for the $L^2(\mu)$ norm converges in $L^2(\mu)$. Consider the weak-star topology on B_1', in other words the topology of pointwise convergence on B. The functions in B_1' are uniformly equicontinuous, indeed Lipschitz with the uniform bound $|f(x) - f(y)| \leq \|x - y\|$, $f \in B_1'$, x, $y \in B$. Thus, in B_1', pointwise convergence on B is equivalent to convergence on a countable dense set, so the weak-star topology on B_1' is metrizable (cf. RAP, Theorem 2.4.4).

Any linear function on B is given by its values on $B_1 := \{x \in B : \|x\| \leq 1\}$, and pointwise convergence on B is equivalent to pointwise convergence on B_1. The set of all functions from B_1 into $[-1, 1]$, with the topology of pointwise convergence, is compact by Tychonoff's theorem (RAP, 2.11). Clearly B_1' is a closed subset of this compact space, so it is also compact. (Compactness of

B'_1 in the weak* topology for a general Banach space B is known as Alaoglu's theorem, e.g. Dunford and Schwartz 1958, pp. 424–426.)

So $\{y_n\}$ has a subsequence converging pointwise on B to some element y of B'_1. For jointly Gaussian variables, pointwise convergence (convergence in probability) implies L^2 convergence, so y is the L^2 limit of $\{y_n\}$ and B'_1 is compact in $L^2(\mu)$.

The natural mapping T of B' into $L^2(\mu)$ has an adjoint T^* taking $L^2(\mu)$ into B'', the dual space of $(B', \|\cdot\|')$. There is a natural map of B into B'' given by $x \mapsto (h \mapsto h(x))$ for $x \in B$ and $h \in B'$. The map is an isometry (RAP, Corollary 6.1.5, of the Hahn–Banach theorem 6.1.4). So B can be viewed as a linear subspace of B''. If it is all of B'', then B is called reflexive. In the present case, whether or not B is reflexive, T^* actually has values in B :

Lemma 2.34 *Let B be a separable Banach space and μ a measure on B such that $\int \|x\|^2 d\mu(x) < \infty$. Then for the natural mapping T of B' into $H :=$ $L^2(\mu)$, the adjoint T^* on H' has values in B.*

Proof. For any $h \in H$ and $y \in B'$,

$$(T^*h)(y) = (h, Ty) = \int_B h(x)y(x)\,d\mu(x) = y(u)$$

where $u \in B$ is defined by the Bochner integral $u = \int_B h(x)x\,d\mu(x)$ (Appendix E, Theorem E.7). The linear form y can be taken under the integral sign since the Bochner integral, when it exists, equals the Pettis integral (Appendix E). □

So let J be the range of T^*, a linear subspace of B, and S its closure, a Banach subspace of B. Note that each element of S is uniformly continuous on B'_1 for the $H = L^2(\mu)$ norm topology, since it is a limit in the norm $\|\cdot\|$ on B, and thus uniformly on B'_1, of such functions. It will be shown that $\mu(S) = 1$. If $S \neq B$, take a countable dense subset $\{x_m\}$ of $B\backslash S$. By the Hahn–Banach theorem (RAP, 6.1.4) for each $m = 1, 2, \ldots$, there is a $u_m \in B'_1$ such that $u_m = 0$ on S and $u_m(x_m) = d(x_m, S) := \inf\{\|x_m - y\|: y \in S\}$.

For any $x \in B\backslash S$, let $\varepsilon := d(x, S) > 0$ and take m with $\|x - x_m\| < \varepsilon/2$. Then $|u_m(x)| > d(x_m, S) - \varepsilon/2 > \varepsilon - \varepsilon/2 - \varepsilon/2 = 0$, so $u_m(x) \neq 0$. So $S = \bigcap_m u_m^{-1}\{0\}$. For each m, to show that $\mu(u_m = 0) = 1$ is equivalent to showing that $T(u_m) = 0$. If not, let $T(u_m) = v \neq 0$. Then $0 < (v, v) = (T(u_m), v) = u_m(T^*v) = 0$ since $T^*v \in S$, a contradiction. So $\mu(u_m = 0) = 1$ for each of the countably many values of m. It follows that $\mu(S) = 1$.

Let K be the closure of the range of T in H. Then K is a Hilbert space. A limit of Gaussian random variables with mean 0 in $L^2(\mu)$ is also such a random variable, so K consists of such random variables, and any finite set of them has a joint normal distribution. Thus the identity from K to itself is an isonormal process L. For this L, we can apply Theorem 2.32, where S is the space V_3 of Theorem 2.32(f). So B'_1 is a GC-set, proving Theorem 2.33. □

The next fact is a direct consequence of Theorem 2.32:

Corollary 2.35 *For any two GC-sets C, D, their union C ∪ D is also a GC-set.*

Proof. Condition (e) or (e′) in Theorem 2.32 holds on C and on D and so, clearly, on $C \cup D$.   □

2.8 A Metric Entropy Condition Implying Sample Continuity

Recall that a stochastic process $X_t(\omega)$, $t \in T$, is said to be *sample-bounded* on T if $\sup_{t \in T} X_t$ is finite for almost all ω. If T is a topological space, then the process is said to be *sample-continuous* if for almost all ω, $t \mapsto X_t(\omega)$ is continuous. The isonormal process is not sample-continuous on the Hilbert space H: let $\{e_n\}$ be an orthonormal sequence. Then $L(e_n)$ are i.i.d. $N(0, 1)$ variables. Thus if $a_n \to 0$ slowly enough, specifically if $a_n(\log n)^{1/2} \to \infty$ as $n \to \infty$, $L(a_n e_n)$ are almost surely unbounded (by Theorem 2.14). So not all bounded sets or even compact sets are GB-sets or GC-sets. Such sets must be small enough in a metric entropy sense. This section will prove a sufficient condition based on metric entropy (defined in Section 1.2), while Section 2.10 will give a characterization based on what is called generic chaining.

A metric entropy sufficient condition for sample continuity of L will actually give a quantitative bound for the continuity. Let (T, d) be a metric space, or, more generally, let d be a pseudometric on T. A function J will be called a *sample modulus* for a real stochastic process $\{X_t,\ t \in T\}$ iff there is a process Y_t with the same laws as X_t and such that for almost all ω there is an $M(\omega) < \infty$ such that for all $s,\ t \in T$, $|Y_s - Y_t|(\omega) \le M(\omega)J(d(s, t))$.

Whenever J is a sample modulus for L on $C \subset H$, and $\{X_t,\ t \in T\}$ is a Gaussian process with mean 0 and $\{X_t(\cdot):\ t \in T\} = C$, then J is also a sample modulus for the process X_t, with the intrinsic pseudo-metric $d_X(s, t) :=$ $(E(X_s - X_t)^2)^{1/2}$ on T.

Recall from Section 1.2 the definition of $D(\varepsilon, C) := D(\varepsilon, C, d)$, the maximum number of points of C more than ε apart for d. Similarly let $N(\varepsilon, C) :=$ $N(\varepsilon, C, d)$. In Hilbert space d will be the usual metric. In the following, when it is said that a stochastic process X_t "can be chosen" with some properties, it means there is a process V_t on the same parameter space T and probability space Ω such that for each $t \in T$, $\Pr(V_t = X_t) = 1$ (V is a modification of X) such that V has the given properties. Now the main theorem of this section can be stated. Forms of it with expectations are given in Theorem 2.37(a) and 2.38.

Theorem 2.36 *For any $C \subset H$, if $\int_0^\infty (\log N(t, C))^{1/2} dt < \infty$, then C is a GC-set, and if*

$$f(x) := f_C(x) := \int_0^x (\log N(t, C))^{1/2} dt, \qquad x > 0, \qquad (2.23)$$

then f is a sample modulus for L on C.

Note. If C is bounded, then $N(t, C) = 1$ and $\log N(t, C) = 0$ for t large enough, and $N(\cdot, C)$ is a nonincreasing function, so integrability of $(\log N(t, C))^{1/2}$ is only an issue near $t = 0$. If $f(x) = +\infty$ for some $x > 0$, then $f(x) = +\infty$ for all $x > 0$, so it still provides a sample modulus but only a trivial one. By Theorem 1.9, $N(t, C)$ could be replaced equivalently by $D(t, C)$.

Proof. We have $f(1) < \infty$, and we can assume C is infinite. Then $H(\varepsilon) := H(\varepsilon, C) \to +\infty$ as $\varepsilon \downarrow 0$. Sequences $\delta_n \downarrow 0$ and $\varepsilon(n) := \varepsilon_n \downarrow 0$ will be defined recursively as follows. Let $\varepsilon_1 := 1$. Given $\varepsilon_1, \ldots, \varepsilon_n$, let

$$\delta_n := 2 \inf \{\varepsilon : H(\varepsilon) \leq 2H(\varepsilon_n)\},$$

$$\varepsilon_{n+1} := \min(\varepsilon_n/3, \delta_n).$$

Then $\varepsilon_n \leq 3(\varepsilon_n - \varepsilon_{n+1})/2$. Also, if $\varepsilon_{n+1} = \delta_n$, then

$$\int_{\varepsilon(n+1)}^{\varepsilon(n)} H(x)^{1/2} dx \leq 2H(\varepsilon_n)^{1/2} \varepsilon_n,$$

while otherwise $\varepsilon_{n+1} = \varepsilon_n/3$ and

$$\int_{\varepsilon(n+1)}^{\varepsilon(n)} H(x)^{1/2} dx \leq 2H(\varepsilon_{n+1})^{1/2} \varepsilon_{n+1}.$$

It follows that for any $n = 1, 2, \ldots$,

$$\frac{2}{3} \sum_{m=n}^{\infty} H(\varepsilon_m)^{1/2} \varepsilon_m \leq \sum_{m=n}^{\infty} (\varepsilon_m - \varepsilon_{m+1}) H(\varepsilon_m)^{1/2}$$

$$\leq f(\varepsilon_n) \leq 4 \sum_{m=n}^{\infty} \varepsilon_m H(\varepsilon_m)^{1/2}.$$

So the convergence of the above sums is equivalent to that of the integral defining $f(1)$, and they all converge.

For each n, there is a set $A_n \subset C$ such that for any $x \in C$ we have $\|x - y\| \leq 2\delta_n$ for some $y \in A_n$, and the number of elements of A_n is bounded by $\mathrm{card}(A_n) \leq \exp(2H(\varepsilon_n))$. Let $G_n := \{x - y : x, y \in A_{n-1} \cup A_n\}$. Then $\mathrm{card}(G_n) \leq 4 \exp(4H(\varepsilon_n))$. Let

$$P_n := \Pr\{\max\{|L(z)|/\|z\| : 0 \neq z \in G_n\} \geq 3H(\varepsilon_n)^{1/2}\}.$$

If Φ is the standard normal distribution function, then for $T > 0$, $1 - \Phi(T) \leq \frac{1}{2} \exp(-T^2/2)$ by Proposition 2.5(a). Then

$$P_n \leq 4 \exp\{4H(\varepsilon_n) - 9H(\varepsilon_n)/2\} = 4 \exp(-H(\varepsilon_n)/2).$$

Since $H(\varepsilon_{n+2}) \geq H(\delta_n/3) \geq 2H(\varepsilon_n)$, $\sum_n P_n$ is dominated for $n \geq 2$ by a sum of two geometric series, one for n even and one for n odd, and so converges. Then for almost all ω there is an $n_0(\omega)$ such that for all $n \geq n_0(\omega)$, $|L(z)| < 3\|z\| H(\varepsilon_n)^{1/2}$ for all $z \in G_n$.

Either $\delta_m = \varepsilon_{m+1}$ or $\delta_m \leq 2\varepsilon_m = 6\varepsilon_{m+1}$, so $\delta_m \leq 6\varepsilon_{m+1}$ for all $m \geq 1$, and $\sum_{n \geq 2} \delta_{n-1} H(\varepsilon_n)^{1/2} < \infty$. For each $x \in C$ choose $A_n(x) \in A_n$ with $\|x - A_n(x)\| \leq 2\delta_n$. Then $\|A_{n-1}(x) - A_n(x)\| \leq 2\delta_{n-1} + 2\delta_n$ and for almost all ω, $L(A_n(x))(\omega)$ is a Cauchy sequence for all $x \in C$. Define a process M by $M(x)(\omega) := \lim_{n \to \infty} L(A_n(x))(\omega)$, $x \in C$, or $M(x)(\omega) := 0$ if the sequence does not converge. Then since $L(\cdot)$ is continuous in probability, for each $x \in C$, $M(x) = L(x)$ a.s., and M and L have the same laws (on C). From the definition of sample modulus we can then assume that $L \equiv M$. So except for ω in the set of 0 probability where $n_0(\omega)$ is undefined, $L(A_n(x))(\omega) \to L(x)(\omega)$ as $n \to \infty$ for all $x \in C$.

If s, $t \in C$ and $\varepsilon_{n+1} < \|s - t\| \leq \varepsilon_n$, then $\|A_n(s) - A_n(t)\| \leq \|s - t\| + 4\delta_n$, and if $n \geq n_0(\omega)$, then

$$|L(s) - L(t)|(\omega) \leq |L(A_n(s)) - L(A_n(t))|(\omega)$$

$$+ \sum_{m=n}^{\infty} |L(A_m(s)) - L(A_{m+1}(s))|(\omega) + |L(A_m(t)) - L(A_{m+1}(t))|(\omega)$$

$$\leq (\|s - t\| + 4\delta_n)3H(\varepsilon_n)^{1/2} + \sum_{m=n}^{\infty} 8\delta_m \cdot 3H(\varepsilon_{m+1})^{1/2}.$$

Thus

$$|L(s) - L(t)|(\omega) \leq 75\|s - t\|H(\|s - t\|)^{1/2} + 144 \sum_{m>n} \varepsilon_m H(\varepsilon_m)^{1/2}$$

$$\leq 291 f(\|s - t\|).$$

This is valid for $\|s - t\| \leq \varepsilon_{n_0(\omega)}$. The hypothesis implies that C is totally bounded, so whenever $n_0(\omega) < \infty$, $|L(\cdot)(\omega)|$ is bounded on C, say by $K(\omega)$. So the conclusion of Theorem 2.36 holds with $M(\omega)$ defined as $\max\left(291, 2K(\omega)/f\left(\varepsilon_{n_0(\omega)}\right)\right)$. $\square$

It follows from Theorem 2.36 that C is a GC-set if as $\varepsilon \downarrow 0$, $N(\varepsilon, C) = O(\exp(\varepsilon^{-p}))$ for some $p < 2$, or if $N(\varepsilon, C) = O(\exp(\varepsilon^{-2}|\log \varepsilon|^{-r}))$ for some $r > 2$. On the other hand, Theorem 2.14 implies that C is not a GB-set if as $\varepsilon \downarrow 0$, eventually $N(\varepsilon, C) \geq \exp(\varepsilon^{-p})$ for some $p > 2$ or $N(\varepsilon, C) \geq \exp(\varepsilon^{-2}|\log \varepsilon|^{s})$ for some $s > 0$. It turns out that the gap cannot be closed further: if $N(\varepsilon, C)$ is of the order of $\exp(\varepsilon^{-2}|\log \varepsilon|^{-r})$ for $0 \leq r \leq 2$, there are examples showing that C may or may not be a GB-set (see Problems 14 and 15). So a characterization of the GB-property cannot be given in terms of metric entropy, although it comes rather close. For a characterization in other terms, see the next section.

Remark. If C is a GC-set, then $L(\cdot)(\cdot)$ can be chosen such that for all ω, $x \mapsto L(x)(\omega)$ is continuous for $x \in C$. Then for any countable dense subset A of C, $L(C)^* = \sup_{x \in A} L(x)$ a.s.

Next, the same integral as in Theorem 2.36 yields a bound for expectations of certain suprema.

Theorem 2.37 *Let $C \subset H$ be nonempty and let $D := \operatorname{diam} C = \sup_{x,y \in C} \|x - y\|$. Let $B := B(C) := \{x - y : x, y \in C\}$. Then for f_C as in Theorem 2.36,*

(a) $E|L(B)|^* \leq 81 f_C(D/4)$ and

(b) $EL(C)^* \leq 81 f_C(D/4)$.

Remarks. All three quantities in (a), (b) are invariant under translation, replacing C by $\{c + u : c \in C\}$ for any fixed u. But $E|L(C)|^*$ does not have such invariance, and becomes unbounded as $\|u\| \to \infty$, so for it we cannot have an upper bound $Kf(D)$, $K < \infty$.

If the constant 81 is replaced by a larger one, one can have, instead of the quantities on the left in (a) and (b), Young–Orlicz norms (Appendix H) $\| \cdot \|_g$ where $g(x) := \exp(x^2) - 1$; see Theorem 2.20.

Proof. Note that $\log N(t, C) = 0$ for $t \geq D/2$, so $f(D/2) = f(+\infty)$. If $f(x) = +\infty$ for some (and hence all) $x > 0$, then (a) and (b) hold trivially (under the given definitions). If $f(D) < \infty$, then we can take L sample-continuous on C by Theorem 2.36.   □

Before proving Theorem 2.37, here is a consequence:

Theorem 2.38 *Let $C \subset H$ with, for f_C as defined in Theorem 2.36, $f_C(x) < +\infty$ for some, or equivalently all, $x > 0$. Then for any $\delta > 0$,*

$$E \left(\sup\{|L(x) - L(y)| : x, y \in C, \|x - y\| \leq \delta\}^* \right) \leq Kf_C(\delta/4) \quad (2.24)$$

for $K = 162\sqrt{2}$. Thus if $\{X_t\}_{t \in T}$ is a Gaussian process with mean 0, it can be chosen so that for any $\delta > 0$,

$$E(\sup\{|X_s - X_t| : s, t \in T, d_X(s,t) \leq \delta\} \leq K \int_0^{\delta/4} \sqrt{\log N(u, T, d_X)} du.$$
$$(2.25)$$

Proof. As in Theorem 2.37 let $B := B(C) := \{x - y : x, y \in C\}$. For any $\varepsilon > 0$ we have $N(\varepsilon, B_\delta) \leq N(\varepsilon, B) \leq N(\varepsilon/2, C)^2$. We also have $\operatorname{diam}(B_\delta) \leq 2\delta$. Let $D_\delta := B(B_\delta) := \{x - y : x, y \in B_\delta\}$. Then $N(\varepsilon, D_\delta) \leq N(\varepsilon/4, C)^4$, so by Theorem 2.36, D_δ is a GC-set, and by Theorem 2.32, we can take L to be prelinear on it. Applying Theorem 2.37(a) we get

$$E \sup\{|L(x) - L(y)| : x, y \in C, \|x - y\| \leq \delta\} = E|L(B_\delta)|^* \leq E|L(D_\delta)|^*$$

(because $0 \in B_\delta$, so $B_\delta \subset D_\delta$)

$$\leq 81 f_{B_\delta}(2\delta/4) \leq 81 f_B(\delta/2)$$

$$\leq 81 \int_0^{\delta/2} \sqrt{2 \log N(\varepsilon/2, C)} d\varepsilon = 162\sqrt{2} \int_0^{\delta/4} \sqrt{\log N(u, C)} du,$$

proving (2.24). For (2.25), if the right side is $+\infty$ it holds trivially. If it is finite, apply the equality in distribution of $\{X_t\}_{t \in T}$ and $\{L(X_t(\cdot))\}_{t \in T}$ for $X_t(\cdot) \in H = L^2(P)$, with $d_X(s, t) = \|X_s(\cdot) - X_t(\cdot)\|$ as usual.   □

To prove Theorem 2.37 here is, first:

Lemma 2.39 *Let* $g_0(u) := 2u\phi(\Phi^{-1}(1/(2u)))$ *for* $u > 1/2$, *where* ϕ *and* Φ *are the standard normal density and distribution function, respectively, and* $\Phi^{-1}(y) := x$ *such that* $\Phi(x) = y$, $0 < y < 1$. *Then* g_0 *is concave. For any random variable* Z *with distribution* $N(0, \sigma^2)$ *and any event* A *with* $P(A) > 0$,

$$\int_A |Z| dP \leq \sigma P(A) g_0(1/P(A)). \tag{2.26}$$

Proof. Let $h(v) := g_0(v/2) = v\phi(\Phi^{-1}(1/v))$ for $v > 1$. Then $h'(v) =$

$$\phi(\Phi^{-1}(1/v)) + v\phi(\Phi^{-1}(1/v))(-\Phi^{-1}(1/v))\frac{1}{\phi(\Phi^{-1}(1/v))} \cdot \left(-\frac{1}{v^2}\right)$$

$$= \phi(\Phi^{-1}(1/v)) + \frac{1}{v}\Phi^{-1}(1/v),$$

$$h''(v) = \Phi^{-1}(1/v) \cdot \left(\frac{1}{v^2}\right) - \frac{1}{v^2}\Phi^{-1}(1/v)$$

$$+ \frac{1}{v} \cdot \frac{1}{\phi(\Phi^{-1}(1/v))} \cdot \left(-\frac{1}{v^2}\right) = -\frac{1}{v^3\phi(\Phi^{-1}(1/v))} < 0,$$

so h is concave for $v > 1$ and g_0 is concave for $u > 1/2$.

Next, the left side of (2.26) is maximized for fixed $P(A) > 0$ when A is a set $\{|Z| \geq r\}$ for some $r \geq 0$, by the Neyman–Pearson lemma (e.g. Lehmann 1986, p. 74). Then $P(A) = 2\Phi(-r/\sigma)$ and

$$\int_A |Z| dP = (2/\pi)^{1/2}\sigma \cdot \exp(-r^2/(2\sigma^2)).$$

Letting $x = r/\sigma$ we need to prove, for $x \geq 0$,

$$2\phi(x) \leq 2\Phi(-x)g_0(1/(2\Phi(-x))). \tag{2.27}$$

Setting $u := 1/(2\Phi(-x))$, so that $x = -\Phi^{-1}(1/(2u))$, shows that (2.27) holds, with equality.   □

Lemma 2.40 *If* $Z_1, \ldots, Z_N$ *are each normally distributed with mean 0 and variance* $\leq \sigma^2$, *then* $E \max_{1 \leq j \leq N} |Z_j| \leq \sigma g_0(N)$.

Proof. The probability space Ω is a union of disjoint events A_i such that $|Z_i| = \max_{1 \le j \le N} |Z_j|$ on A_i. Thus

$$E \max_{1 \le j \le N} |Z_j| = \sum_{j=1}^{N} \int_{A_j} |Z_j| dP \le \sum_{j=1}^{N} \sigma P(A_j) g_0(1/P(A_j)) \le \sigma g_0(N)$$

by Lemma 2.39, from which g_0 is concave. $\qquad\square$

Lemma 2.41 *Let* $g_1(u) := K(\log(1 + u))^{1/2}$ *for* $u > 0$ *where*

$$K := (2 + [4 + \log 4]/(\log(3/2)))^{1/2}.$$

Then $g_1(u) \ge g_0(u)$ *for all* $u \ge 1$.

Proof. "Mills' ratio" satisfies, for $x \ge 0$, Komatsu's inequality

$$M(x) := \Phi(-x)/\phi(x) \ge 2/(x + (x^2 + 4)^{1/2})$$

(RAP, Section 12.1, Problem 7). Thus $M(x) \ge 1/(x^2 + 4)^{1/2}$. Let $x := -\Phi^{-1}(1/(2u)) \ge 0$, so $u = 1/(2\Phi(-x))$. Then we need to show that

$$g_1(1/(2\Phi(-x))) \ge \phi(x)/\Phi(-x) = 1/M(x),$$

so it will be enough to prove $g_1(1/(2\Phi(-x))) \ge (x^2 + 4)^{1/2}$. Since $\Phi(-x) \le \exp(-x^2/2)$ for $x \ge 0$ (RAP, Lemma 12.1.6(b)), and g_1 is nondecreasing, it will be enough to show

$$g_1(\exp(x^2/2)/2) \ge (x^2 + 4)^{1/2}, \quad x \ge 0.$$

Letting $y := \exp(x^2/2)/2$, we need to show

$$g_1(y) = K(\log(1 + y))^{1/2} \ge (4 + 2\log(2y))^{1/2}, \quad y \ge 1/2,$$

or

$$K^2 \log(1 + y) \ge 4 + 2\log 2 + 2\log y, \quad y \ge 1/2,$$

which follows from the definition of K. $\qquad\square$

Now to prove Theorem 2.37, $D = 0$ if and only if C consists of a single point. Then both sides of (a) and (b) are 0 and they hold. So assume $D > 0$. Let $\varepsilon_k := D/2^k$ and $N_k := N(\varepsilon_k/2, C)$, $k = 0, 1, \dots$. Then for each $k = 0, 1, 2, \dots$, there is a set C_k of N_k points x_{kj}, $j = 1, \dots, N_k$, such that for all $x \in C$, $\|x - x_{kj}\| \le \varepsilon_k$ for some j. Then $N_0 = 1$, so $C_0 = \{x_{01}\}$ for some x_{01}. For each $k = 1, 2, \dots$, and $j = 1, \dots, N_k$, choose and fix a point $y_{kj} = x_{k-1,i}$ for some i such that $\|x_{kj} - y_{kj}\| \le \varepsilon_{k-1}$. Let W_k be the set of all variables $L(x_{kj}) - L(y_{kj})$, $j = 1, \dots, N_k$. Then by Lemma 2.40,

$$E \max\{|z| : z \in W_k\} \le \varepsilon_{k-1} g_0(N_k).$$

For any $u_k \in C_k$ there is a sequence of points $u_j \in C_j$, $j = 0, \ldots, k$, such that $L(u_j) - L(u_{j-1}) \in W_j$, $j = 1, \ldots, k$. Thus

$$E \sup\{|L(x) - L(y)| : x, y \in \bigcup_{i=1}^{k} C_i\} \leq 2 \sum_{j=1}^{k} \varepsilon_{j-1} g_0(N_j).$$

The union of all the C_i is dense in C, so by sample continuity and monotone convergence,

$$E_C := E \sup\{|L(x) - L(y)| : x, y \in C\} \leq 2 \sum_{j=1}^{\infty} \varepsilon_{j-1} g_0(N_j)$$

$$= 4D \sum_{j=1}^{\infty} g_0(N_j)/2^j.$$

By Lemma 2.41, where $K < 4$, we get

$$E_C \leq 16D \sum_{j=1}^{\infty} (\log(1 + N_j))^{1/2}/2^j.$$

For all $j \geq 1$, $N_j \geq 2$, so $[\log(1 + N_j)/\log N_j]^{1/2} \leq (\log 3/\log 2)^{1/2} < 1.26$. Thus

$$E_C \leq 20.2D \sum_{j=1}^{\infty} (\log N_j)^{1/2}/2^j \leq 81 \sum_{j=1}^{\infty} \int_{\varepsilon_{j+2}}^{\varepsilon_{j+1}} (\log N(t, C))^{1/2} dt = 81 f(D/4),$$ proving (a). Then for any fixed $y \in C$,

$$\sup_{x \in C} L(x) \leq L(y) + \sup_{x \in C} |L(x) - L(y)|,$$

so (b) follows and Theorem 2.37 is proved. $\qquad\square$

2.9 Gaussian Concentration Inequalities

This section is based on excerpts from the book of Ledoux (2001). Let (S, d) be a metric space and μ a probability measure on the Borel sets of S. For a set $A \subset S$ and $r > 0$, let $A^r := \{y \in S : d(x, y) < r \text{ for some } x \in S\}$. The *concentration function* for μ (and the given metric d) is defined by

$$\alpha_\mu(r) := \alpha_{\mu,d}(r) := \sup\{1 - \mu(A^r) : A \subset S, \mu(A) \geq 1/2\}. \quad (2.28)$$

Here A ranges over Borel sets. Each set A^r is open, for any set A. In a simple example let $S = \mathbb{R}$ with usual metric and $\mu = N(0, 1)$. For $A = (-\infty, 0]$ we have $1 - \mu(A^r) = 1 - \Phi(r) \leq \exp(-r^2/2)$ where Φ is the standard normal distribution function. The choice of A turns out to be an extremal one among all sets with $\mu(A) \geq 1/2$, so that in fact $\alpha_\mu(r) \leq \exp(-r^2/2)$ (Ledoux 2001, (1.4)) although that will not be proved here. Inequalities of the same form, possibly with right-hand side of the form $b \exp(-cr^2)$ for some $b > 0$ and $c > 0$, hold in multidimensional and even infinite-dimensional spaces. Such an inequality, although not with best constants, will be proved.

There is a wide choice of measurable sets A with $\mu(A) \geq 1/2$. Among the most useful are the following. Let F be a real-valued measurable function on S, which thus is a random variable on the probability space (S, μ). Let $m(F)$ be a median of F (which is not necessarily unique). Then the sets $A_1 := \{x : F(x) \leq m(F)\}$ and $A_2 := \{x : F(x) \geq m(F)\}$ satisfy $\mu(A_i) \geq 1/2$, $i = 1, 2$. First, let $A = A_1$. In order that sets A^t should be meaningful in relation to F it is useful that F be a Lipschitz function, meaning that the Lipschitz seminorm $\|F\|_L := \sup_{x \neq y} |F(x) - F(y)|/d(x, y)$ is finite. Let $K := \|F\|_L$. Then for any $r > 0$ we have

$$\Pr(F - m(F) \geq Kr) \leq \alpha_\mu(r) \tag{2.29}$$

since the given event is included in the complement of A_1^r. Using also $A = A_2$ we would get

$$\Pr(|F - m(F)| \geq Kr) \leq 2\alpha_\mu(r). \tag{2.30}$$

Thus if we have good Gaussian-type bounds on $\alpha_\mu(r)$, the values of Lipschitz functions with bounded Lipschitz seminorms will be concentrated close to their medians (and, it will be seen in Proposition 2.46, also close to their means).

One class of examples of Lipschitz functions is bounded linear functionals on normed linear spaces, whose Lipschitz seminorms equal their dual norms. One can take the maximum, or supremum, of two or more Lipschitz functions with bounded $\| \cdot \|_L$ and get a Lipschitz function, as follows:

Proposition 2.42 *Let (S, d) be a metric space. Let $\mathcal{F}$ be a collection of real-valued Lipschitz functions on S for d such that $M := \sup_{f \in \mathcal{F}} \|f\|_L < +\infty$. Let $F(x) := \sup_{f \in \mathcal{F}} f(x)$ and suppose that $F(x) < +\infty$ for all x. Then F is a Lipschitz function with $\|F\|_L \leq M$.*

Proof. Suppose that $|F(x) - F(y)| > Md(x, y)$ for some $x, y \in S$. By symmetry we can assume that $F(x) > F(y) + Md(x, y)$. Then for some $f \in \mathcal{F}$, $f(x) > F(y) + Md(x, y) \geq f(y) + Md(x, y)$, but this contradicts $\|f\|_L \leq M$, finishing the proof. $\qquad\square$

The hypotheses of the preceding proposition clearly hold whenever $\mathcal{F}$ is a finite set of Lipschitz functions. By taking suprema, a rather large class of Lipschitz functions can be generated starting with some basic ones.

Let $\{X_t\}_{t \in T}$ be any Gaussian process with mean 0 and G a finite subset of T. Let $F := \max_{t \in G} X_t$. By way of Theorem 2.2 and Proposition 2.3, we can view each X_t as a linear function f_t on a finite-dimensional Hilbert space having the probability measure $\gamma = N(0, I)$. Let $\sigma_t := (E(X_t^2))^{1/2}$. Then f_t

is Lipschitz with $\|f_t\|_L = \sigma_t$ and by Proposition 2.42, F is represented as a Lipschitz function with

$$\|F\|_L \leq \sigma_G := \max_{t \in G} \sigma_t. \tag{2.31}$$

Here is a concentration inequality (one part of Ledoux 2001, Theorem 2.15). Ledoux (2001, Corollary 2.6) gives a Gaussian concentration inequality with $\exp(-r^2/2)$ in place of $2\exp(-r^2/4)$ and $F - EF$ in place of $F - m(F)$, but with a harder proof.

Theorem 2.43 *Let $d = 1, 2, \ldots$ and let γ be the standard normal law $N(0, I)$ on $\mathbb{R}^d$. Then for all $r > 0$,*

$$\alpha_\gamma(r) \leq 2\exp(-r^2/4).$$

An inequality to be used in the proof is one that Ledoux (2001, p. 33) calls a functional version of a multiplicative Brunn–Minkowski inequality.

Theorem 2.44 *Let f, g, and h be nonnegative, measurable functions on $\mathbb{R}^d$ with g and h integrable (for Lebesgue measure) and let $0 < \theta < 1$. Suppose that for all x and $y \in \mathbb{R}^d$,*

$$f(\theta x + (1 - \theta)y) \geq g(x)^\theta h(y)^{1-\theta}. \tag{2.32}$$

Then

$$\int f(z)\,dz \geq \left(\int g(x)\,dx \right)^\theta \left(\int h(y)\,dy \right)^{1-\theta}. \tag{2.33}$$

Remarks. If $\theta = 0$ or 1, the conclusion holds trivially. If $\int g\,dx$ or $\int h\,dx$ is $+\infty$ and the other is positive, the conclusion still follows via approximation by increasing sequences of integrable functions. If one of the two integrals is $+\infty$ and the other is 0, the best available lower bound for $\int f\,dx$ is 0, as is seen by setting $f \equiv h \equiv 0, g \equiv 1$.

Proof. The proof will be by induction on d. First, let $d = 1$. Ledoux (2001, p. 34) says in half a sentence "we may assume ... by approximation that" g and h "are continuous with strictly positive values." In the following proof, more than a page is spent on proving one can take them continuous and then another half-page on taking them strictly positive. After that the remaining page of the proof proceeds almost exactly as in Ledoux's book.

For any M with $0 < M < +\infty$, if we replace g by $g_M := \min(g, M)$ and h by $h_M := \min(h, M)$, the assumption (2.32) still holds, and if the conclusion does, we can let $M \uparrow +\infty$ and use monotone convergence to get the conclusion in general, so we can assume that for some M with $1 \leq M < +\infty, 0 \leq g(x) \leq M$ and $0 \leq h(x) \leq M$ for all x. Similarly, we can assume that for some $L < +\infty, g(x) = h(x) = 0$ for $|x| \geq L$.

Let λ be Lebesgue measure on $\mathbb{R}$. Given any $\varepsilon > 0$, by dominated convergence there is a $\delta > 0$ with $\delta \leq \varepsilon/M$ such that if $\lambda(B) < \delta$, then $\int_B (g + h) \, dx < \varepsilon$. To approximate g and h by continuous functions, first, by Lusin's theorem (RAP, Theorem 7.5.2) there is a compact set $K \subset [-L, L]$ such that (g, h) from K to $\mathbb{R}^2$, thus g and h from K into $\mathbb{R}$, are continuous, and $\lambda([-L, L] \setminus K) < \delta$. We have $0 \leq g 1_K \leq g$ and $0 \leq h 1_K \leq h$. Thus (2.32) holds for $g 1_K$ and $h 1_K$ in place of g and h, with $\int (h - h 1_K) \, dx < \varepsilon$ and $\int (g - g 1_K) \, dx < \varepsilon$.

There exist continuous functions g_n with $g_n \equiv g$ on K, $0 \leq g_n \leq M$, $g_n = 0$ outside $[-2L, 2L]$, and $g_n \downarrow g 1_K$ as $n \to +\infty$. Likewise there exist h_n for h. Recall that the *support* of a function ϕ on a topological space is the closure of the set on which $\phi \neq 0$.

I claim the following: if ϕ is a continuous function from $\mathbb{R}^2$ into $\mathbb{R}$ having compact support J, then the function $\phi_1(x) := \sup_y \phi(x, y)$ is continuous on $\mathbb{R}$ and has compact support. To see this, we have $J \subset A \times B$ for some compact subsets A and B of $\mathbb{R}$. Let $x_n \to x \in A$. If $\phi_1(x_n) \not\to \phi_1(x)$, taking a subsequence we can assume that $\phi_1(x_n) = \phi(x_n, y_n)$ where $y_n \to y$ for some $y \in B$, so $\phi(x_n, y_n) \to \phi(x, y) \leq \phi_1(x)$. If $\phi(x, y) < \phi_1(x)$, let $\phi_1(x) = \phi(x, u)$ for some u. Then for n large enough, $\phi(x_n, u) > \phi(x_n, y_n)$, contradicting the choice of y_n and proving the claim.

If g and h are continuous on $\mathbb{R}$ with compact supports A and B respectively, then the function $F(z, y) := g(z - y) h(y)$ is continuous on $\mathbb{R}^2$, and it has compact support since it is 0 if $y \notin B$ or if $z - y \notin A$, and so if $z \notin A + B := \{a + b : a \in A, b \in B\}$, thus F is 0 outside the compact set $(A + B) \times B$.

For each n, the function $G_n(z, y) := g_n([z - (1 - \theta)y]/\theta)^\theta h_n(y)^{1-\theta}$ is similarly continuous with compact support, thus by the claim, so is $f_n(z) := \sup_y G_n(z, y)$. It follows that $f_n(\theta x + (1 - \theta)y) \geq g_n(x)^\theta h_n(y)^{1-\theta}$ for all x and y. Since g_n and h_n are nonincreasing sequences of functions with respect to n, so is G_n and thus f_n. We have

$$G_n(z, y) \downarrow G_0(z, y) := (g 1_K)([z - (1 - \theta)y]/\theta)^\theta (h 1_K)(y)^{1-\theta}$$

for all z and y. I claim that $f_n(z) \downarrow f_0(z) := \sup_y G_0(z, y)$ for all z. Clearly $f_n(z) \geq f_0(z)$ for all n and z. Suppose that for some z, $f_n(z) \downarrow c > f_0(z)$. Then for some y_n, $G_n(z, y_n) \geq c > 0$ for all n. Taking a subsequence, we can assume that $y_n \to y$ for some y. Let $x := [z - (1 - \theta)y]/\theta$, so that $\theta x + (1 - \theta)y = z$. If $y \notin K$, then by Dini's theorem (RAP, Theorem 2.4.10), $h_n \to 0$ uniformly in some neighborhood of y, so $h_n(y_n) \to 0$ and since $0 \leq g_n \leq M$, $G_n(z, y_n) \to 0$, a contradiction. So $y \in K$. Likewise, $x \in K$. Thus for all $n = 0, 1, \ldots$, $G_n(z, y) = g(x)^\theta h(y)^{1-\theta}$, and $f_0(z) \geq g(x)^\theta h(y)^{1-\theta}$.

Let $x_n := [z - (1 - \theta)y_n]/\theta \to x$. We have

$$\limsup_{n \to \infty} h_n(y_n) \leq \limsup_{n \to \infty} h_1(y_n) = h_1(y) = h(y),$$

and likewise $\limsup_{n\to\infty} g_n(x_n) \le g(x)$, so

$$\limsup_{n\to\infty} G_n(z, y_n) \le g(x)^\theta h(y)^{1-\theta} \le f_0(z),$$

contradicting the choice of z. Thus $f_n \downarrow f_0$ as claimed.

Since f_n are uniformly bounded by M^2 and all have supports included in $[-2L, 2L]$, by dominated convergence we have $\int f_n(z)\,dz \downarrow \int f_0(z)\,dz$. By the definitions, the hypothesis (2.32) holds for the continuous functions f_n, g_n, h_n in place of f, g, h, noting also that $f_0 \le f$, and if we can prove the conclusion for f_n, g_n, and h_n, it will follow for f, $g1_K$, and $h1_K$. Then letting $\varepsilon \downarrow 0$ it will follow for f, g, and h. So we can assume f, g, and h are each continuous with supports included in $[-2L, 2L]$.

Let ψ be a continuous real function on $\mathbb{R}$, with $0 < \psi(x) \le 1$ and $\psi(-x) = \psi(x)$ for all x, $\psi(x) = 1$ for $|x| \le 2L$, $\psi(x)$ decreasing as $|x|$ increases for $|x| > 2L$, and with ψ going to 0 at ∞ fast enough so that ψ^θ and $\psi^{1-\theta}$ are both integrable. Suppose for some constant ζ with $0 < \zeta \le 1$, where eventually $\zeta \to 0$, we replace g by $g + \zeta\psi$ and h by $h + \zeta\psi$. It will be shown for all y and z, setting $x := [z - (1 - \theta)y]/\theta$ so that $z = \theta x + (1 - \theta)y$, we have

$$(g + \zeta\psi)(x)^\theta (h + \zeta\psi)(y)^{1-\theta}$$

$$\le f(z) + (M + 1)\left[(\zeta\psi(z))^\theta + (\zeta\psi(z))^{1-\theta} + \psi(z)\{\zeta^\theta\theta^{-1} \right.$$

$$\left. + \zeta^{1-\theta}(1 - \theta)^{-1}\}\right]. \tag{2.34}$$

To prove this first suppose $|z| \ge 2L$. Clearly $|z| \le \theta|x| + (1-\theta)|y| \le \max(|x|, |y|)$, so either $|x| \ge 2L$ or $|y| \ge 2L$. If, for example, $\max(|x|, |y|) = |y| \ge |z|$, then $\psi(y) \le \psi(z)$. For any u with $|u| \ge 2L$ we have $g(u) = h(u) = 0$. Thus the left side of (2.34) is bounded above by $(M + 1)[(\zeta\psi(z))^\theta + (\zeta\psi(z))^{1-\theta}]$, which implies (2.34).

So, suppose $|z| < 2L$. Then $\psi(z) = 1$. The derivative with respect to ζ of the left side of (2.34) is

$$\theta(g + \zeta\psi)(x)^{\theta-1}\psi(x)(h + \zeta\psi)(y)^{1-\theta} + (g + \zeta\psi)(x)^\theta(1 - \theta)(h + \zeta\psi)(y)^{-\theta}$$

$$\times \psi(y) \le \theta(\zeta\psi)(x)^{\theta-1}\psi(x)(M + 1)^{1-\theta} + (M + 1)^\theta(1 - \theta)(\zeta\psi(y))^{-\theta}$$

$$\times \psi(y) \le (M + 1)[\zeta^{\theta-1} + \zeta^{-\theta}].$$

The indefinite integral of this bound from 0 to ζ gives that the left side of (2.34) is bounded above by $f(z) + (M + 1)[\zeta^\theta\theta^{-1} + \zeta^{1-\theta}(1 - \theta)^{-1}]$, so (2.34) is proved in both cases.

Thus if we replace g, h, and f by $g + \zeta\psi$, $h + \zeta\psi$, and the right side of (2.34) respectively, hypothesis (2.32) holds, and each function is integrable, with integrals approaching those of g, h, and f respectively as $\zeta \downarrow 0$. So we can assume g and h are continuous and everywhere strictly positive. Clearly we can assume $\int g\,dx = 1$ and $\int h\,dy = 1$.

Now, the function $x \mapsto \int_{-\infty}^{x} g(u)\,du$ is C^1, with a strictly positive derivative g, from $\mathbb{R}$ onto $(0, 1)$. So this function has an inverse $t \mapsto x(t)$ from $(0, 1)$ onto $\mathbb{R}$, with a continuous derivative $x'(t) = 1/(g(x(t)))$ for all $t \in (0, 1)$. Likewise, $y \mapsto \int_{-\infty}^{y} h(v)\,dv$ is C^1 with a positive derivative h and has an inverse $t \mapsto y(t)$ from $(0, 1)$ onto $\mathbb{R}$ with a continuous derivative $y'(t) \equiv 1/h(y(t))$. Let $z(t) := \theta x(t) + (1 - \theta)y(t)$ for $0 < t < 1$. For any $a > 0$ and $b > 0$ we have $\theta a + (1 - \theta)b \geq a^\theta b^{1-\theta}$. (Setting $u := a/b > 0$ this is equivalent to $\theta u + 1 - \theta \geq u^\theta$, which holds with equality for $u = 1$ and follows for all u by taking derivatives with respect to u.) It follows that for $0 < t < 1$

$$z'(t) = \theta x'(t) + (1 - \theta)y'(t) \geq (x'(t))^\theta (y'(t))^{1-\theta}. \qquad (2.35)$$

Since $z' > 0$ is continuous it follows from the hypothesis (2.32) and (2.35) that

$$
\begin{aligned}
\int f(x)\,dx = \int f(z)\,dz &= \int_0^1 f(z(t))z'(t)\,dt \\
&\geq \int_0^1 g(x(t))^\theta h(y(t))^{1-\theta}(x'(t))^\theta (y'(t))^{1-\theta}\,dt \\
&= \int_0^1 [g(x(t))x'(t)]^\theta [h(y(t))y'(t)]^{1-\theta}\,dt \\
&= 1.
\end{aligned}
$$

Thus the case of dimension $d = 1$ is proved. Now suppose $d > 1$ and that the conclusion holds in dimension $d - 1$. Let f, g, and h on $\mathbb{R}^d$ satisfy the hypotheses. For each $x \in \mathbb{R}^{d-1}$ and $u \in \mathbb{R}$ let $f_u(x) := f(x, u)$ and likewise define g_u and h_u. If $u = \theta u_0 + (1 - \theta)u_1$ for some real u_j, then

$$f_u(\theta x + (1 - \theta)y) \geq g_{u_0}(x)^\theta h_{u_1}(y)^{1-\theta}$$

for any x, $y \in \mathbb{R}^{d-1}$. It follows by the induction hypothesis, where the integrability of f_u, g_{u_0}, and h_{u_1} holds for Lebesgue almost all u, u_0, and u_1 (which will suffice for the conclusion), that

$$\int_{\mathbb{R}^{d-1}} f_u(z)\,dz \geq \left(\int_{\mathbb{R}^{d-1}} g_{u_0}(x)\,dx\right)^\theta \left(\int_{\mathbb{R}^{d-1}} h_{u_1}(y)\,dy\right)^{1-\theta}.$$

This gives the hypothesis (2.32) for the one-dimensional case with functions of u, where in the hypothesis $u = \theta u_0 + (1 - \theta)u_1$, and of u_0 and u_1, so by the conclusion in that case

$$\int f(z)\,dz = \int \left(\int_{\mathbb{R}^{d-1}} f_u(x)\,dx\right)du \geq \left(\int g(v)\,dv\right)^\theta \left(\int h(w)\,dw\right)^{1-\theta},$$

which finishes the proof of Theorem 2.44. $\qquad\qquad\square$

Now to continue the proof of Theorem 2.43, for a bounded measurable function j on $\mathbb{R}^d$ and $c > 0$, define the "infimum-convolution" $Q_c j$ by

$$(Q_c j)(x) := \inf_{v \in \mathbb{R}^d} \left[j(v) + \frac{c}{2} |x - v|^2 \right]$$

for all $x \in \mathbb{R}^d$. It's easily seen that $Q_c j$ is a well-defined real-valued function. It will be shown that

$$\int \exp(Q_{1/2} j(x)) d\gamma(x) \int \exp(-j(y)) d\gamma(y) \leq 1. \tag{2.36}$$

For this, one will apply Theorem 2.44 for $\theta = 1/2$, then square both sides of the conclusion. Let ϕ_d be the density of $\gamma = N(0, I)$. Let $f := \phi_d$, $g(x) := \exp(Q_{1/2} j(x)) \phi_d(x)$, and $h(y) := \exp(-j(y)) \phi_d(y)$. Hypothesis (2.32) then requires that $f((x + y)/2) \geq \sqrt{g(x) h(y)}$ for all x and y. To check this, note that the normalizing constants of the ϕ_d's cancel. Taking logarithms of both sides, we get an inequality that holds by definition of $Q_{1/2} j$. The integrability of f and h is immediate, and that of g is easy (let $v = x$). Thus Theorem 2.44 does apply and gives the conclusion (2.36) after squaring.

Now the bound (2.36) will be extended to any measurable function j such that $0 \leq j(x) \leq +\infty$. Let j_k be nonnegative bounded, measurable functions increasing up to j. Then (2.36) holds for each j_k. By monotone convergence, $\int \exp(-j_k) d\gamma$ decreases down to $I_0 := \int e^{-j} d\gamma \geq 0$ and $\int \exp(Q_{1/2} j_k(x)) d\gamma(x)$ increases up to $I_1 := \int (\exp(Q_{1/2} j(x)) d\gamma(x) \leq +\infty$. If $I_0 > 0$, then $I_1 < +\infty$ by Fatou's lemma and (2.36) holds for j. If $I_0 = 0$, then possibly $I_1 = +\infty$, but if we make the convention that $(+\infty) \cdot 0 \leq 1$ in this case, then (2.36) still holds.

Next the following will be proved:

Proposition 2.45 *For any measurable set $A \subset \mathbb{R}^d$ and $\rho > 0$ we have*

$$\gamma(\{x : \inf_{v \in A} |v - x|^2/4 \geq \rho\}) \leq e^{-\rho}/\gamma(A).$$

Proof. If $\gamma(A) = 0$, in other words $\lambda(A) = 0$, then one may say that the right side is $+\infty$ and the inequality holds trivially. So suppose $\gamma(A) > 0$.

Apply (2.36) to the function $j(x) = 0$ on A and $j = +\infty$ outside it. Then $\int e^{-j} d\gamma = \gamma(A) > 0$, so both factors on the left in (2.36) are finite, and it gives $\int \exp(Q_{1/2} j(x)) d\gamma(x) \leq 1/\gamma(A)$. It follows from the Markov–Chebyshev inequality that $\gamma(\{x : Q_{1/2} j(x) \geq \rho\}) \leq e^{-\rho}/\gamma(A)$. For the given j we have

$$Q_{1/2} j(x) = \inf_v \{j(v) + |x - v|^2/4\} = \inf_{v \in A} |x - v|^2/4,$$

and the proposition is proved. □

Now to finish the proof of Theorem 2.43, let $\rho := r^2/4$, so $r = 2\sqrt{\rho}$. Then for any x, $x \in A^r$ if and only if for some $v \in A$, $|x - v| < r$, or $|x - v|^2/4 < r^2/4 = \rho$, so the theorem follows from Proposition 2.45 in light of the definition (2.28) of α_μ. □

The bound for $\alpha_\gamma(r)$ given by Theorem 2.43 can be used to bound the probabilities of deviations of a Lipschitz function F from its median $m(F)$ in (2.29) and (2.30) for $\mu = \gamma$. For any Lipschitz function F on $\mathbb{R}^d$ we have $|F(x)| \leq |F(0)| + \|F\|_L|x|$ for all x and consequently that $E_\gamma F = \int F \, d\gamma$ exists and is finite. Also note here that with respect to γ, any continuous function F (in particular any Lipschitz F) has a unique median: if it had a nonunique median there would be an interval $[a, b]$ of medians with $a < b$ such that $\gamma(F^{-1}((a, b))) = 0$. But F must take values $\geq b$ and $\leq a$, thus by the intermediate value theorem, it takes all values in (a, b), so $F^{-1}((a, b))$ is a nonempty open set and must have probability > 0 for γ.

To compare the mean and median we have:

Proposition 2.46 *There exists an absolute constant C such that for any d, any Lipschitz function F on $\mathbb{R}^d$ with $K = \|F\|_L$ and its unique median $m(F)$ for γ, we have $|EF - m(F)| \leq CK$. Thus for any $r > C$,*

$$\Pr(F - EF \geq Kr) \leq \alpha_\gamma(r - C) \tag{2.37}$$

and

$$\Pr(|F - EF| \geq Kr) \leq 2\alpha_\gamma(r - C). \tag{2.38}$$

Proof. For any random variable U let $U^+ := \max(U, 0)$. Letting $U := (F - m(F))/\|F\|_L$ we get $EU \leq E(U^+)$ bounded by (2.9) for $Y = U^+$, then using (2.29) and, for example, Theorem 2.43 to get a bound by an absolute constant C, not depending on F or the dimension d. The other conclusions then follow. □

Next, the essential supremum, or essential supremum of absolute values, of L on a GB-set satisfies concentration inequalities. Recall $L(A)^*$ and $|L(A)|^*$ from Lemma 2.4.

Theorem 2.47 *Let H be a Hilbert space and A a GB-set in H. Let $\sigma := \sup_{x \in A} \|x\|$. Let $Y = L(A)^*$ or $|L(A)|^*$. Then*

(a) For any $r > 0$, $\Pr(Y - m(Y) > \sigma r) \leq 2\exp(-r^2/4)$.

(b) For any $r > C$ from Proposition 2.46,

$$\Pr(Y - EY > \sigma r) \leq 2\exp(-(r - C)^2/4)$$

and

$$\Pr(|Y - EY| > \sigma r) \le 4\exp(-(r - C)^2/4).$$

Proof. A GB-set, being totally bounded by Theorem 2.14, is bounded, so $\sigma < +\infty$. By the proof of Lemma 2.4, we can assume that A is countable (replacing A by a countable dense subset B). Consider finite subsets $B_k \uparrow A$. For each k and each $x \in B_k$, we can view $L(x)$ or $|L(x)|$ as a Lipschitz function on $\mathbb{R}^d$ where d is the cardinality of B_k, with Lipschitz seminorm $\le \sigma$, by Proposition 2.3 and (2.31). Then, applying Proposition 2.42, we get that $L(B_k)^*$ or $|L(B_k)|^*$ has the same property. Thus (2.29) applies with $F = Y$ and $K = \sigma$. Then applying Theorem 2.43 gives (a) for $A = B_k$ for each k. Since the bound does not depend on k, we can let $k \to \infty$ and get the bound as stated for A.

For part (b), we can then just apply Proposition 2.46. This finishes the proof. $\qquad\square$

2.10 Generic Chaining

There are some differing definitions of subgaussian random variable or process. Here is a condition used by Talagrand (2005, (0.4)).

Definition. Let (S, d) be a metric space. A real-valued stochastic process $\{X_t\}_{t \in S}$ will be called *T-subgaussian* iff it is centered, i.e. $EX_t = 0$ for all $t \in S$, and for every $u > 0$ and $s \ne t$ in S we have

$$\Pr(|X_s - X_t| \ge u) \le 2\exp\left(-\frac{u^2}{2d(s, t)^2}\right). \tag{2.39}$$

Note that although (2.39) puts strong restrictions on the distributions of differences of values of the process, and the distributions of X_t must satisfy $E|X_t| < \infty$ for all t since $EX_t = 0$, nothing in the definition of T-subgaussian requires that $E(X_t^2) < \infty$ or that $E|X_t|^p < \infty$ for any $p > 1$.

A centered Gaussian process is always T-subgaussian for the L^2 metric $d(s, t) = (E[(X_s - X_t)^2])^{1/2}$ if it is a metric, by Proposition 2.5(a); otherwise, for any $s \ne t$ such that $X_s = X_t$ a.s., the left side of (2.39) is 0, and we may say the inequality holds with $0 \le 2\exp(-\infty) = 0$. For any Hilbert space H the isonormal process L is T-subgaussian on H, or any subset, for the usual metric on H.

A *decomposition* or *partition* of a set S will mean a collection of finitely many disjoint nonempty sets whose union is S. If (S, d) is a metric space and $A \subset S$, the *diameter* diam A is defined as $\sup_{s,t \in A} d(s, t)$. A sequence $\{\mathcal{A}_k\}_{k \ge 0}$ of partitions of S will be called *nested* if for each $k = 1, 2, \ldots$, each set in

A_k is a subset of some set in A_{k-1}. A nested sequence of partitions A_k will be called a *Talagrand sequence* if A_0 consists of the one set S, and for each $k = 1, 2, \ldots, A_k$ contains $M(k)$ sets where

$$1 \leq M(k) \leq N_k := 2^{2^k}. \tag{2.40}$$

If X_t and Y_t are stochastic processes with the same index set S and on the same probability space, the two processes are said to be *modifications* of each other if for each t, $X_t = Y_t$ with probability 1. Thus $\{X_t\}_{t \in S}$ will be said to have a *sample-bounded modification* if it has a modification $\{Y_t\}_{t \in S}$ which is sample-bounded. Talagrand's generic chaining method gives a characterization of sample boundedness for centered Gaussian processes. It can be compared to the facts based on covering or packing numbers (metric entropy) as in Theorems 2.14 and 2.36, which do not give characterizations but are perhaps more accessible.

Given a sequence $\{A_k\}_{k \geq 0}$ of partitions, a $t \in S$ and a k, let $A_k(t)$ be the set $A \in A_k$ such that $t \in A$.

For any (totally bounded) metric space (S, d), define $\gamma_2(S, d)$ as the infimum over all Talagrand sequences $\{A_k\}_{k \geq 0}$ of $\sup_{t \in S} \sum_{k=0}^{\infty} 2^{k/2} \operatorname{diam} A_k(t)$.

Theorem 2.48 *If (S, d) is a metric space such that $\gamma_2(S, d) < \infty$ and $\{X_t\}_{t \in S}$ is a T-subgaussian stochastic process, then it has a sample-bounded modification $\{Y_t\}_{t \in S}$. Moreover, Y_t can be chosen such that*

$$E \sup_t Y_t < 7\gamma_2(S, d). \tag{2.41}$$

Proof. Suppose $\gamma_2(S, d) < +\infty$. Let $\{A_k\}_{k \geq 0}$ be a Talagrand sequence of partitions of S such that

$$G := \sup_{t \in S} \sum_{k=0}^{\infty} 2^{k/2} \operatorname{diam} A_k(t) < \infty.$$

We have

$$\Delta_k := \max_{A \in A_k} \operatorname{diam} A \to 0 \tag{2.42}$$

as $k \to \infty$, since if not, for some $\varepsilon > 0$, $\Delta_k \geq \varepsilon$, implying $G \geq 2^{k/2}\varepsilon \to +\infty$, for infinitely many k, in fact for all k since the partitions are nested and so Δ_k are nonincreasing in k. Then $G = +\infty$ is a contradiction. Let $M(0) := 1$. For each $k = 0, 1, \ldots,$ form a set $T_k \subset S$ having $M(k)$ points by choosing just one point from each of the $M(k)$ sets in A_k. Since the partitions are nested, we can and do choose T_k such that $T_k \subset T_{k+1}$ for all k. For any $t \in S$, let $\pi_k(t)$ be the one point in $T_k \cap A_k(t)$. For any $t \in S$, $u > 0$, and $k = 1, 2, \ldots,$ we have by (2.39)

$$\Pr\left\{\left|X_{\pi_k(t)} - X_{\pi_{k-1}(t)}\right| > u 2^{k/2} d(\pi_k(t), \pi_{k-1}(t))\right\} \leq 2\exp\left(-u^2 2^{k-1}\right).$$

Let

$$\Omega_u := \bigcap_{k \geq 1} \bigcap_{t \in S} \left\{ \left| X_{\pi_k(t)} - X_{\pi_{k-1}(t)} \right| \leq u 2^{k/2} d(\pi_k(t), \pi_{k-1}(t)) \right\}.$$

The number of possible pairs $(\pi_k(t), \pi_{k-1}(t))$ is at most 2^{2^k} by (2.40) since $\pi_k(t)$ uniquely determines $\pi_{k-1}(t)$. Thus

$$\Pr\left(\Omega_u^c\right) \leq q(u) := \sum_{k=1}^{\infty} 2^{2^{k+1}} \exp\left(-u^2 2^{k-1}\right).$$

Now $u^2 2^{k-1} \geq \frac{1}{2}u^2 + u^2 2^{k-2} \geq \frac{1}{2}u^2 + 2^{k+1}$ for $u \geq 3$, and then

$$q(u) \leq e^{-u^2/2} \sum_{k=1}^{\infty} 2^{2^{k+1}} e^{-2^{k+1}} < e^{-u^2/2}. \tag{2.43}$$

For any $t \in S$ and $k \geq 1$, both $\pi_{k-1}(t) \in A_{k-1}(t)$ and $\pi_k(t) \in A_k(t) \subset A_{k-1}(t)$ since the partitions are nested, so $d(\pi_k(t), \pi_{k-1}(t)) \leq \operatorname{diam} A_{k-1}(t)$. We have for any t that

$$\sum_{k=1}^{\infty} 2^{k/2} \operatorname{diam} A_{k-1}(t) = \sum_{k=0}^{\infty} 2^{(k+1)/2} \operatorname{diam} A_k(t) \leq \sqrt{2}G < 2G.$$

Thus for any $u > 0$ and $\omega \in \Omega_u$, for all $t \in S$,

$$\sum_{k=1}^{\infty} \left| X_{\pi_k(t)} - X_{\pi_{k-1}(t)} \right| \leq 2uG. \tag{2.44}$$

As $u \to +\infty$, $\Pr(\Omega_u^c) \to 0$ by (2.43), so almost every ω is in some Ω_u, and the telescoping series

$$Y_t := X_{t_0} + \sum_{k=1}^{\infty} X_{\pi_k(t)} - X_{\pi_{k-1}(t)} = \lim_{k \to \infty} X_{\pi_k(t)}$$

converges almost surely. More precisely, define Y_t as the limit if it exists and 0 otherwise, so it is a measurable function of the countable set of random variables X_s for $s \in T_k$ for all k. If $t \in T_k$ for some k, then since the sets T_j are nested we have $\pi_j(t) = t$ for all $j \geq k$ and simply $Y_t = X_t$. Condition (2.39) implies that the process $t \mapsto X_t(\cdot)$ is continuous in probability, and thus by (2.42) for each t, $Y_t = X_t$ a.s., so Y_t is a modification of X_t. We have $\sup_t |Y_t - X_{t_0}| \leq 2uG$ for all $\omega \in \Omega_u$, so $\sup_t |Y_t| \leq |X_{t_0}| + 2uG < \infty$ a.s., proving that $\{Y_t\}_{t \in S}$ is sample-bounded and $\{X_t\}_{t \in S}$ has a sample-bounded modification as claimed.

Next, let $\eta(\omega) := \sup_{t \in T} Y_t$, which is measurable because the sup can be restricted to the countable set $\bigcup_k T_k$. We have $E\eta = E(\eta - X_{t_0})$ since $EX_{t_0} = 0$ (the process is centered by definition of T-subgaussian) and

$$0 \leq \eta - X_{t_0} = \sup_{t \in S}(Y_t - X_{t_0}) \leq 2uG$$

for each $\omega \in \Omega_u$ by (2.44). Let $Z := (\eta - X_{t_0})/(2G)$. Then by (2.43), since $Z \geq 0$,

$$EZ = \int_0^\infty \Pr(Z > u)du \leq 3 + \int_3^\infty \exp(-u^2/2)du < 3.1,$$

so $E\eta < 6.2G$. Varying partitions and letting $G \downarrow \gamma_2(S, d)$, (2.41) follows, proving Theorem 2.48. $\qquad\square$

For the isonormal process, a converse holds:

Theorem 2.49 *A set $C \subset H$ for a Hilbert space H is a GB-set in H if and only if $\gamma_2(C, d) < \infty$ for the usual metric d on H.*

Proof. For "if," the isonormal process L is T-subgaussian, so Theorem 2.48 applies. For "only if," suppose C is a GB-set in H. The first step is the following, related to the Sudakov minoration. As usual for $x \in H$ and $r > 0$, $B(x, r) = \{y : \|y - x\| < r\}$.

Proposition 2.50 *There are finite, absolute constants L_1 and L_2 satisfying the following. Let $a > 0$ and let t_i for $i = 1, \ldots, m$ be points of H. Assume that $\|t_i - t_{i'}\| \geq a$ for $i \neq i'$. Let $\sigma > 0$ and for each $i = 1, \ldots, m$ let H_i be a nonempty set included in $B(t_i, \sigma)$ and $J := \bigcup_{i=1}^m H_i$. Then*

$$EL(J)^* \geq \frac{a}{L_1}\sqrt{\log m} - L_2\sigma\sqrt{\log m} + \min_{1 \leq i \leq m} EL(H_i)^*. \qquad (2.45)$$

Remark. If $\sigma \leq a/(2L_1L_2)$ then (2.45) implies

$$EL(J)^* \geq \frac{a}{2L_1}\sqrt{\log m} + \min_{1 \leq i \leq m} EL(H_i)^*. \qquad (2.46)$$

Proof. We can assume that $m \geq 2$. For each $i \leq m$ let

$$Y_i := L(H_i)^* - L(t_i).$$

For each $t \in H_i$, $\|t - t_i\| \leq 2\sigma$, and so for each $u \geq C$, by Theorem 2.47(b),

$$\Pr(|Y_i - EY_i| \geq 2\sigma u) \leq 4\exp(-(u - C)^2/4).$$

Letting $V := \max_{1 \leq i \leq m} |Y_i - EY_i|$ we then have for each $u \geq C$

$$\Pr(V \geq 2\sigma u) \leq 4m\exp(-(u - C)^2/4). \qquad (2.47)$$

By the identity (2.9) we then have

$$EV/(2\sigma) \leq C + \int_C^\infty \min(1, 4m\exp(-(u - C)^2/4)\,du.$$

Letting $u_0 := C + 2\sqrt{\ln(4m)}$, the minimum is < 1 if and only if $u > u_0$, and the integral from u_0 to $+\infty$ is < 1, so for some absolute constant L_2 and all

$m \geq 2$,

$$EV \leq \sigma(4C + 2 + 4\sqrt{\ln(4m)}) \leq L_2\sigma\sqrt{\ln m}. \tag{2.48}$$

For each $i = 1, \ldots, m$, $Y_i \geq EY_i - V \geq \min_{1 \leq j \leq m} EY_j - V$. So

$$\max_{t \in H_i} L(t) = Y_i + L(t_i) \geq L(t_i) + \min_{1 \leq j \leq m} EY_j - V$$

and

$$\max_{t \in H} L(t) \geq \max_{1 \leq i \leq m} L(t_i) + \min_{1 \leq j \leq m} EY_j - V.$$

Then, take expectations of both sides, apply the Sudakov minoration in expectation form, Theorem 2.22, to the first term on the right, use $EL(t_i) = 0$ for each i in the second, and apply (2.48) to the third, proving the proposition. □

To continue the proof of "only if" in Theorem 2.49, recall that $N_n \equiv 2^{2^n}$ and that $\Delta(A)$ is the diameter of a set A. For any set $A \subset H$ let $F_0(A) := EL(A)^*$, which for A finite or countable is the same as $E \sup_{t \in A} L(t)$. For L_1 and L_2 as in Proposition 2.50 let $r := \max(4, 2L_1L_2)$. We have the following fact, which produces partitions nearly as good as those desired, and combined with a short argument following it will lead to the proof.

Theorem 2.51 *For each $n = 0, 1, 2, \ldots$, let $m := N_{n+1}$ and $\theta(n) := 2^{n/2}$. Let T be a GB-set in H and assume that for any $s \in T$, any a with $0 < a \leq r\Delta(T)$, and any $t_1, \ldots, t_m$ in $B(s, ar)$ such that $\|t_i - t_{i'}\| \geq a$ whenever $i \neq i'$, and any nonempty sets H_i such that $H_i \subset B(t_i, a/r)$, we have*

$$F_0\left(\bigcup_{1 \leq i \leq m} H_i\right) \geq a\theta(n+1) + \min_{1 \leq i \leq m} F_0(H_i). \tag{2.49}$$

Then there exists a nested sequence $\{A_k\}_{k \geq 0}$ of partitions of T with $\mathrm{card}(A_n) \leq N_{n+1}$ for each n such that for a universal constant K,

$$\sup_{t \in T} \sum_{k=0}^{\infty} 2^{k/2} \Delta(A_k(t)) \leq Kr\left(\frac{F_0(T)}{\sqrt{2} - 1} + \Delta(T)\right). \tag{2.50}$$

Proof. The hypothesis will be applied only to a of the form r^{-j-1} where j in the set $\mathbb{Z}$ of integers (positive, negative, or 0), so that in (2.49), $a/r = r^{-j-2}$.

The sequence $\{A_n\}$ of partitions will be defined recursively. For each n and each $C \in A_n$, a point $t_C \in C$, an integer $j(C) \in \mathbb{Z}$ and numbers $b_\ell(C)$ for $\ell = 0, 1$, and 2 will be defined, and it will be shown that the following properties hold:

$$C \subset B(t_C, r^{-j(C)}), \tag{2.51}$$

so that $\Delta(C) \leq 2r^{-j(C)}$,

$$F_0(C) \leq b_0(C), \tag{2.52}$$

and for all $t \in C$,

$$F_0(C \cap B(t, r^{-j(C)-1})) \leq b_1(C) \tag{2.53}$$

and

$$F_0(C \cap B(t, r^{-j(C)-2})) \leq b_2(C). \tag{2.54}$$

It will also be shown in all cases that

$$b_1(C) \leq b_0(C) \tag{2.55}$$

and for each $C \in \mathcal{A}_n$, letting $\varepsilon_n := F_0(T)/2^n$, that

$$b_0(C) - r^{-j(C)-1}\theta(n) \leq b_2(C) \leq b_0(C) + \varepsilon_n. \tag{2.56}$$

Last but not least it will be shown for each $n \geq 0$, each $C \in \mathcal{A}_n$, and each $A \in \mathcal{A}_{n+1}$ with $A \subset C$ that

$$\sum_{\ell=0}^{2} b_\ell(A) + (1 - 2^{-1/2})r^{-j(A)-1}\theta(n + 1)$$

$$\leq \sum_{\ell=0}^{2} b_\ell(C) + \frac{1}{2}(1 - 2^{-1/2})r^{-j(C)-1}\theta(n) + \varepsilon_{n+1}. \tag{2.57}$$

For $n = 0$, define $\mathcal{A}_0 : \{T\}, b_0(T) := b_1(T) := b_2(T) := F_0(T)$ and choose any $t_T \in T$. Let $j(T)$ be the largest integer such that $T \subset B(t_T, r^{-j(T)})$. It follows that

$$r^{-j(T)-1} \leq \Delta(T) \leq 2r^{-j(T)}. \tag{2.58}$$

For $n = 0$, (2.51) through (2.56) all clearly hold and (2.57) does not yet apply. The cardinality of $\mathcal{A}_0$ is $1 < N_1 = 4$.

Now for the recursion-induction step, suppose that for a given $n = 0, 1, \ldots$, a partition $\mathcal{A}_n$ with cardinality at most N_{n+1} and the points t_A and numbers $j(A)$ and $b_\ell(A)$ have been defined with all the given properties (2.51) through (2.57) holding. We want to define a partition $\mathcal{A}_{n+1}$ so that all properties continue to hold and the sequence of partitions is nested. Each set $C \in \mathcal{A}_n$ will be decomposed into at most $m := N_{n+1}$ sets in $\mathcal{A}_{n+1}$, so since $N_{n+1}^2 = N_{n+2}$, the desired bound on the cardinality of $\mathcal{A}_{n+1}$ will hold.

Let $D_0 := C$ and $j := j(C)$. Choose $t_1 \in C$ such that

$$F_0(C \cap B(t_1, r^{-j-2})) \geq \sup_{t \in C} F_0(C \cap B(t, r^{-j-2})) - \varepsilon_{n+1}. \tag{2.59}$$

Set $A_1 := C \cap B(t_1, r^{-j-1})$.

For $1 \leq i < m - 1$, if t_j and A_j for $j = 1, \ldots, i$ have been defined, let $D_i := C \setminus \bigcup_{1 \leq q \leq i} A_i$. If D_i is empty, the decomposition of C is finished. If not, choose $t_{i+1} \in D_i$ such that

$$F_0(D_i \cap B(t_{i+1}, r^{-j-2}) \geq \sup_{t \in D_i} F_0(D_i \cap B(t, r^{-j-2})) - \varepsilon_{n+1}. \quad (2.60)$$

Let $A_{i+1} := D_i \cap B(t_{i+1}, r^{-j-1})$. If eventually $A_{m-1} \neq \emptyset$ is defined, then let $D_{m-1} := C \setminus \bigcup_{i<m} A_i$. If D_{m-1} is empty the decomposition of C is finished, otherwise let $A_m := D_{m-1}$. Thus C is decomposed into at most $m = N_{n+1}$ sets A_i as desired.

Let A be a set in this decomposition. If $A = A_m$ then define $j(A) := j = j(C)$, $t_A := t_C$, $b_0(A) := b_0(C)$, $b_1(A) := b_1(C)$, and

$$b_2(A) := b_0(C) - r^{-j-1}\theta(n+1) + \varepsilon_{n+1}.$$

Then (2.51), (2.52), (2.53), and (2.55) hold for A since they did for C by induction hypothesis. Writing (2.56) for A in place of C and $n+1$ in place of n, it holds by definition of $b_2(A)$.

To prove (2.54) for $A = A_m$, take any point $t \in A$ and call it t_m. For $1 \leq i \leq m$, we have $t_i \in D_{i-1}$ by the definitions. It follows that if $i' < i$ then $d(t_i, t_{i'}) \geq r^{-j-1}$. Then by (2.49) for $a = r^{-j-1}, s = t_C$ using (2.51), and $H_i := D_i \cap B(t_{i+1}, r^{-j-2})$ it follows that

$$F_0(C) \geq r^{-j-1}\theta(n+1) + \min_{0 \leq l \leq m-1} \left(F_0(D_i \cap B(t_{i+1}, r^{-j-2})) \right). \quad (2.61)$$

By (2.60), since $t_m \in D_i$ for $1 \leq i \leq m-1$ and then since $A \subset D_i$, we have

$$F_0(D_i \cap B(t_{i+1}, r^{-j-2})) \geq F_0(D_i \cap B(t_m, r^{-j-2})) - \varepsilon_{n+1}$$
$$\geq F_0(A \cap B(t_m, r^{-j-2})) - \varepsilon_{n+1}.$$

Since $F_0(C) \leq b_0(C)$ by induction hypothesis (2.52), it follows from the preceding display and (2.61) that

$$b_0(C) \geq r^{-j-1}\theta(n+1) - \varepsilon_{n+1} + F_0(A \cap B(t_m, r^{-j-2})).$$

Recalling that t_m is an arbitrary point of $A = A_m$, and the definition of $b_2(A)$, (2.54) is proved in this case.

To prove (2.57), by the definitions including $\theta(n) := 2^{n/2}$ we have

$$\sum_{\ell=0}^{2} b_\ell(A) + (1 - 2^{-1/2})r^{-j(A)-1}\theta(n+1)$$

$$= 2b_0(C) + b_1(C) - 2^{-1/2}r^{-j-1}\theta(n+1) + \varepsilon_{n+1} \quad (2.62)$$
$$\leq 2b_0(C) + b_1(C) - r^{-j-1}\theta(n) + \varepsilon_{n+1}.$$

By (2.56) we have $b_0(C) \leq b_2(C) + r^{-j-1}\theta(n)$, and so (2.57) follows.

The case $A = A_m$ is now finished. Let $A = A_i$ for some $i < m$. Define $j(A) := j + 1$ and $t_A := t_i$, so that

$$A = A_i \subset B(t_i, r^{-j-1}) = B(t_A, r^{-j(A)})$$

and (2.51) holds for A. Define

$$b_0(A) := b_2(A) := b_1(C), \quad b_1(A) := \min(b_1(C), b_2(C)).$$

The conditions (2.55) and (2.56) for A are immediate. To prove (2.52) for A, we have in view of (2.53) for C

$$F_0(A) \leq F_0(C \cap B(t_i, r^{-j-1})) \leq b_1(C) = b_0(A).$$

To prove (2.53) for A, using (2.54) for C, for any $t \in A$,

$$F_0(A \cap B(t, r^{-j(A)-1})) \leq F_0(C \cap B(t, r^{-j-2}))$$

$$\leq \min(b_1(C), b_2(C)) = b_1(A).$$

Relation (2.54) for A follows from (2.52) for A because $b_2(A) = b_0(A)$.

To prove (2.57) for A, we have

$$\sum_{\ell=0}^{2} b_\ell(A) \leq 2b_1(C) + b_2(C) \leq \sum_{\ell=0}^{2} b_\ell(C) \tag{2.63}$$

since $b_1(C) \leq b_0(C)$ by (2.55) for it. We have $j(A) = j(C) + 1$, and $r^{-1}\theta(n + 1) \leq \theta(n)/2$ since $r \geq 4$, we have

$$r^{-j(A)-1}\theta(n + 1) \leq \frac{1}{2} r^{-j(C)-1}\theta(n),$$

which together with (2.63) proves (2.57) for A.

The inductive definition and proof of the given properties of the partitions $\mathcal{A}_n$ and associated points and numbers is now complete, and it remains to prove the conclusion (2.50). By (2.57) for each $t \in T$ and any $n = 0, 1, \ldots$, setting $j_n(t) := j(A_n(t))$,

$$\sum_{\ell=0}^{2} b_\ell(A_{n+1}(t)) + \left(1 - 2^{-1/2}\right) r^{-j_{n+1}(t)-1}\theta(n + 1)$$

$$\leq \sum_{\ell=0}^{2} b_\ell(A_n(t)) + \frac{1}{2}\left(1 - 2^{-1/2}\right) r^{-j_n(t)-1}\theta(n) + \varepsilon_{n+1}.$$

We have $b_\ell(T) = F_0(T)$ for each ℓ, and since each $b_\ell(A) \geq 0$ by (2.52) through (2.54), summing the previous relations from $n = 0$ to q gives

$$(1 - 2^{-1/2})\sum_{n=0}^{q} r^{-j_{n+1}(t)-1}\theta(n + 1) \leq 4F_0(T) + \frac{1}{2}(1 - 2^{-1/2})\sum_{n=0}^{q} r^{-j_n(t)-1}\theta(n),$$

$$\tag{2.64}$$

and so, since the term for $n + 1 = q + 1$ is ≥ 0,

$$\frac{1}{2}(1 - 2^{-1/2}) \sum_{n=0}^{q} r^{-j_n(t)-1} \theta(n) \leq 4 F_0(T) + (1 - 2^{-1/2}) r^{-j(T)-1} \theta(0).$$

By (2.51), $\Delta(A_n(t)) \leq 2r^{-j_n(t)}$ and by (2.58), $r^{-j(T)-1} \leq \Delta(T)$. Thus (2.50) follows, and Theorem 2.51 is proved. $\qquad\square$

To finish the proof of "only if" in Theorem 2.49, note that Proposition 2.50 implies that the hypotheses of Theorem 2.51 hold. The cardinalities of the A_n given are bounded by N_{n+1} rather than N_n as one could like. To remedy this, define a sequence $\{B_n\}$ of partitions by $B_n := \{T\}$ for $n = 0$ and $B_n := A_{n-1}$ for $n \geq 1$ (so that $B_1 = \{T\}$ also). Then B_n has cardinality $\leq N_n$ for each n and $\{B_n\}$ is a Talagrand sequence. We have

$$\sum_{n=1}^{\infty} 2^{n/2} \Delta(B_n(t)) = \sqrt{2} \sum_{k=0}^{\infty} 2^{k/2} \Delta(A_k(t)),$$

and the supremum of the latter over all t is finite by Theorem 2.51, while $\Delta(B_0(t)) = \Delta(T)$ is bounded uniformly in t, so $\gamma_2(T, d) < \infty$ and the proof of Theorem 2.49 is complete. $\qquad\square$

Next, (2.41) implies the following relation:

$$E \sup_{s,t \in S} |Y_t - Y_s| = E[\sup_{t \in S} Y_t - \inf_{s \in S} Y_s] = 2E \sup_{t \in S} Y_t < 14\gamma_2(S, d). \quad (2.65)$$

Now to relate the above chaining method to metric entropy, recall that for $\varepsilon > 0$ and $A \subset S$, $N(\varepsilon, A) := N(\varepsilon, A, d)$ is the minimum number of sets of diameter at most 2ε that cover A. Theorem 2.36 states that if $C \subset H$, a Hilbert space, with

$$\int_0^{\infty} \sqrt{\log N(t, C)}\, dt < \infty, \quad (2.66)$$

then C is a GC-set (and hence a GB-set). Suppose that

$$\int_0^{\infty} \sqrt{\log N(t, S)}\, dt < \infty.$$

We can assume that S is infinite since otherwise there is no problem about sample boundedness or uniform continuity. Then $N(\varepsilon, C) \uparrow +\infty$ as $\varepsilon \downarrow 0$. Also, (2.66) implies that (S, d) is totally bounded, since otherwise $N(t, C) = +\infty$ for all small enough $t > 0$, so $N(t, C) < +\infty$ for all $t > 0$. For $k = 1, 2, \ldots$, let

$$\varepsilon_k := \inf\{\varepsilon > 0 : N(\varepsilon, S) \leq 2^{2^{k-1}}\}. \quad (2.67)$$

Thus there is a collection of at most $2^{2^{k-1}}$ sets B_j of diameters at most $3\varepsilon_k$ which cover S. Define a partition C_k consisting of all those sets $B_j \setminus \bigcup_{i<j} B_i$

which are nonempty, which also contains at most $2^{2^{k-1}}$ sets, each with diameter at most $3\varepsilon_k$.

Now recursively we define a nested sequence of partitions as follows. Let $\mathcal{A}_0 = \{S\}$. Given a partition $\mathcal{A}_k$ for $k \geq 0$, containing at most 2^{2^k} sets, let $\mathcal{A}_{k+1}$ consist of all nonempty intersections of a set in $\mathcal{A}_k$ and a set in $\mathcal{C}_{k+1}$. Then $\mathcal{A}_{k+1}$ is a partition, consists of sets with diameter at most $3\varepsilon_{k+1}$, and has in it at most

$$2^{2^k + 2^{k+1-1}} = 2^{2^{k+1}}$$

sets, so the recursion can continue and $\{\mathcal{A}_k\}_{k \geq 0}$ is a nested, Talagrand sequence of partitions of S.

We have diam $A_k(t) \leq 3\varepsilon_k$ for all $k \geq 1$ and all t. Clearly ε_k is a nonincreasing sequence as k increases. The integral in (2.67) is bounded below by

$$\sqrt{\log 2} \sum_{k=1}^{\infty} (\varepsilon_k - \varepsilon_{k+1}) 2^{(k-1)/2}$$

$$\geq \sqrt{\log 2} \sum_{k=2}^{\infty} \varepsilon_k \left[2^{(k-1)/2} - 2^{(k-2)/2} \right] = \sqrt{\log 2} \left(\frac{1}{\sqrt{2}} - \frac{1}{2} \right) \frac{1}{3} \sum_{k=2}^{\infty} 2^{k/2} \cdot 3\varepsilon_k,$$

which implies that $\gamma_2(S, d) < \infty$, and so, gives that a set $C \subset H$ satisfying (2.66) is a GB-set.

In the above argument, the diameters of all sets in $\mathcal{A}_k$ were bounded by the same bound, in this case $3\varepsilon_k$, as is required by methods based on covering or packing numbers (metric entropy). The generic chaining method is more flexible in that it allows the diameters of different sets in the partition $\mathcal{A}_k$ to be quite different.

2.11 Homogeneous and Quasi-homogeneous Sets in H

We know that there are GB-sets in a Hilbert space that are not GC-sets, for example, if $\{e_n\}_{n \geq 1}$ is an orthonormal basis, the set $\{e_n / \sqrt{\log n}\}_{n \geq 2}$, or, to get a compact set, that sequence together with 0. The non-GC property of the set is local around 0, as outside of any neighborhood of 0, the set is finite. Thus in a sense the set is highly inhomogeneous. For more homogeneous sets in a sense to be defined, the GB- and GC-properties are equivalent:

Theorem 2.52 Let $T \subset H$. Suppose there exists a law (Borel probability measure) μ on T such that for some $M < \infty$ $f(r) := \sup_{x \in T} \mu(B(x, r)) < M \inf_{y \in T} \mu(B(y, r))$ for all $r > 0$. Then the following are equivalent:

  (i) T is a GC-set;

  (ii) T is a GB-set;

  (iii) $\int_0^1 (\log D(\varepsilon, T))^{1/2} d\varepsilon < \infty$.

Proof. From the definitions, (i) always implies (ii). We know that (iii) always implies (i) by Theorem 2.36. It remains to show that (ii) implies (iii). It follows from (ii) that T is totally bounded, by the Sudakov minoration Theorem 2.14.

We can assume that T is infinite. It follows that as $r \downarrow 0$, we have $f(r)/M \downarrow 0$ and so $f(r) \downarrow 0$. We can assume that L_1 and L_2 in Proposition 2.50 are both ≥ 2. Let $\rho := 1/(3L_1 L_2)$.

For any $r > 0$ there are $D(r, T)$ points x_i in T more than r apart. Thus the balls $B(x_i, r/2)$ are disjoint and $D(r, T) f(r/2)/M \leq 1$.

The integral in (iii) can be restricted to the range $(0, \text{diam}(T)]$. It will suffice to show that

$$I := \int_0^\rho \sqrt{\log(M/f(r/2))} \, dr < \infty. \tag{2.68}$$

We have $I = \sum_{j=1}^\infty I_j$ where

$$I_j := \int_{\rho^{j+1}}^{\rho^j} \sqrt{\log(M/f(r/2))} \, dr < u_j := \rho^j \sqrt{\log(1/f(\rho^{j+1}/2))}. \tag{2.69}$$

Let $j_1 := 1$ and recursively, given j_k for some $k \geq 1$, let j_{k+1} be the least j such that $f(\rho^j/2) < f(\rho^{j_k})^2/M^2$. Let

$$v_k := \sum_{j_k \leq j < j_{k+1}} u_j.$$

If $j_{k+1} = j_k + 1$, then simply $v_k = u_{j_k}$. If $j_{k+1} \geq j_k + 2$ then

$$v_k = u_{j_{k+1}-1} + \sum_{j=j_k}^{j_{k+1}-2} u_j \leq v_{k1} + v_{k2}$$

where

$$v_{k1} := \rho^{j_{k+1}-1} \sqrt{\log\left(1/\rho^{j_{k+1}}/2\right)} \quad \text{and} \quad v_{k2} := \rho^{j_k} \sqrt{2\log\left(1/\rho^{j_{k+1}}/2\right)}. \tag{2.70}$$

To apply Proposition 2.50, consider a ball $A := B(x, \rho^{j_k})$ for any $x \in T$ and $k = 1, 2, \ldots$. For the metric space (A, d) (where d is the usual metric on H, restricted to A) and $\varepsilon > 0$ let

$$D_\geq(\varepsilon, A) := \sup\{m : \text{ for some } t_1, \ldots, t_m \in A, \ d(t_i, t_j) \geq \varepsilon \text{ for } i \neq j\}.$$

Let $a := \rho^{j_{k+1}+1}/2$ and $m := D_\geq(a, A)$. Take t_i from the definition. Then the open balls $B(t_i, a)$ cover A, so

$$f\left(\rho^{j_k}\right)/M \leq \mu(A) \leq mf\left(\rho^{j_{k+1}+1}/2\right).$$

Thus by definition of j_{k+1},

$$m \geq f\left(\rho^{j_k}\right)/\left(Mf\left(\rho^{j_{k+1}+1}/2\right)\right) > 1/\sqrt{f\left(\rho^{j_{k+1}+1}/2\right)}.$$

Let $\sigma := \rho^{j_{k+4}} < \rho a < a/(2L_1 L_2)$ and $H_i := B(t_i, \sigma)$ for $i = 1, \ldots, m$. Then Proposition 2.50 and (2.46) apply and give

$$EL\left(B\left(x, \rho^{j_k}\right)\right)^* \geq \frac{\rho^{j_{k+1}+1}}{4L_1} \sqrt{\frac{1}{2}\log\left(\frac{1}{f\left(\rho^{j_{k+1}+1}/2\right)}\right)} + \min_i EL\left(B\left(t_i, \rho^{j_{k+4}}\right)\right)^*.$$

Iterating this with k replaced by $k+4$ and x by any t_i gives

$$+\infty > 2^{5/2}L_1 EL(T)^* \geq \sum_{s=0}^{\infty} \rho^{j_{u+4s}} \sqrt{\log\left(\frac{1}{f\left(\rho^{j_{u+4s}+1}/2\right)}\right)}$$

for $u = 2, 3, 4, 5$, and therefore

$$\sum_{w=2}^{\infty} \rho^{j_w} \sqrt{\log\left(\frac{1}{f\left(\rho^{j_w+1}/2\right)}\right)} < +\infty.$$

It follows that $\sum_{k=1}^{\infty} v_{k1} + v_{k2} < \infty$ for v_{ki} defined by (2.70), and so (iii) holds, completing the proof. $\qquad\square$

Corollary 2.53 (Fernique) *Let (T, d) be a metric space such that there is a group G of 1–1 transformations g of T onto itself for which*

(a) *d is G-invariant: for all s, $t \in T$ and $g \in G$,*

$$d(g(s), g(t)) = d(s, t);$$

(b) *There is a law (Borel probability measure) μ on T which is G-invariant, i.e. $\mu \circ g^{-1} = \mu$ for all $g \in G$;*

(c) *G acts transitively on T : for all s, $t \in T$ there is a $g \in G$ with $g(s) = t$.*

Then the hypotheses and thus the conclusion of Theorem 2.52 hold.

Proof. The hypotheses imply that for each $r > 0$, $\mu(B(x, r))$ is the same for all $x \in T$. Thus Theorem 2.52 applies for any $M \geq 1$. $\qquad\square$

For examples of the situation in Corollary 2.53 consider the following. A *topological group* is a group G with a topology under which the group operation $(g, h) \mapsto gh$ is jointly continuous from $G \times G$ into G and the inverse $g \mapsto g^{-1}$ is continuous from G into G. Only Hausdorff topologies will be considered. A reference for the following is Nachbin (1965). If G is locally compact, there exist so-called left and right *Haar measures* μ_l and μ_r, which are strictly positive on all nonempty open sets, finite on all compact sets, and such that for every Borel set $A \subset G$ and $g \in G$, $\mu_l(gA) = \mu_l(A)$ and $\mu_r(Ag) = \mu_r(A)$. Each of μ_l and μ_r is unique up to a positive multiplicative constant. G is called *unimodular* if the left and right Haar measures coincide. All compact groups are unimodular (Nachbin 1965, Chapter 2, Proposition 13, p. 81). Thus, for every compact Hausdorff topological group G, there is a probability measure μ on the Borel sets which is both left and right invariant and is unique with either property. It will be called *the* Haar measure on G.

Let E be a Hausdorff topological space and G a topological group. Then E is said to be a *homogeneous space* under G if we have a jointly continuous map $(g, x) \mapsto gx$ from $G \times E$ onto E such that $(gh)x = g(hx)$ for all $g, h \in G$ and $x \in E$, $ex = x$ for all $x \in E$ where e is the identity element of G, and such that for every $x, x' \in E$ there is some $g \in G$ such that $gx = x'$.

If G is compact then E is necessarily also compact. In that case there is a unique probability measure m on the Borel sets of E such that m is G-invariant, meaning that for each $g \in G$, the map $x \mapsto gx$ of E to itself preserves m (Nachbin 1965, Chapter 3, Theorem 1 p. 138, Corollary 4, p. 140).

If G is a compact, metrizable group, then there exists a metric d on G which is two-sided invariant, meaning that for any $g, h, j \in G$ we have $d(g, h) = d(jg, jh) = d(gj, hj)$ (Hewitt and Ross 1979, Theorem 8.6, p. 71). If E is a homogeneous space under G, then for any $x, y \in E$ we know that $G_{x,y} := \{g \in G : gx = y\}$ is nonempty. Let $\rho_d(x, y) := \inf\{d(g, e) : g \in G_{x,y}\}$. The infimum is actually attained, by compactness and joint continuity. Clearly ρ_d is nonnegative. By the joint continuity, $G_{x,y}$ is closed, and for $x \neq y$ it does not contain e. It follows that $\rho_d(x, y) > 0$. We have $\rho_d(x, y) \equiv \rho_d(y, x)$ because by the two-sided invariance of d, $d(g, e) \equiv d(g^{-1}, e)$. For the triangle inequality, given any x, y, and $z \in E$, if $gx = y$ and $hy = z$ with $\rho_d(x, y) = d(g, e)$ and $\rho_d(y, z) = d(h, e)$, then $hgx = z$, so

$$\rho_d(x, z) \leq d(hg, e) \leq d(hg, g) + d(g, e) = d(h, e) + d(g, e)$$

$$= \rho_d(x, y) + \rho_d(y, z).$$

So ρ_d is a metric on E. It clearly satisfies $\rho_d(x, y) = \rho_d(gx, gy)$ for any $x, y \in E$ and $g \in G$, i.e., ρ_d is *invariant* under the action of G.

Thus all the hypotheses of Corollary 2.53 hold whenever T is a homogeneous space under the action of a compact metrizable group G. There are many examples of such T and G. One class of them is as follows. In $\mathbb{R}^d$ for any $d \geq 2$, let S^{d-1} be the unit sphere $\{x \in \mathbb{R}^d : |x| = 1\}$ and let G be the group $O(d)$ of all orthogonal transformations U of $\mathbb{R}^d$ onto itself, in other words linear transformations U such that $Ux \cdot Uy \equiv x \cdot y$ for the usual inner product. One can also take $SO(d)$, the special orthogonal group of orthogonal transformations (given by matrices) with determinant 1. For $d = 2$, we get $T = S^1$, the unit circle $x^2 + y^2 = 1$ in $\mathbb{R}^2$, and $G = SO(2)$ is the group of rotations, The unique invariant probability measure is $dm(\theta) = d\theta/(2\pi)$. Here d may be any rotationally invariant metric (or pseudometric) on the circle. (Unlike m, d is not at all unique; for example, if ρ is a G-invariant metric, so is ρ^α for any α with $0 < \alpha \leq 1$.) (See Problems 11 and 12.)

To see how Theorem 2.52 applies beyond Corollary 2.53, suppose one wants to prove sample-continuity of a Gaussian process on a locally compact but not compact metric space, such as a Euclidean space or a noncompact manifold. Then it suffices to prove sample continuity on each of a family of compact sets whose interiors form a base for the topology, such as balls or cubes in Euclidean spaces. Then one can often define a measure, such as Lebesgue measure in a Euclidean space, restrict it to a compact set C, and normalize it to have mass 1 to get a law μ. Here $\mu(B(x, r))$ may not depend on x while $B(x, r)$ is included in the interior of C, but become smaller as x approaches the boundary of C, yet the hypothesis of Theorem 2.52 still holds. See for example Problem 13.

2.12 Sample Continuity and Compactness

This section will show that for a Gaussian process X_t indexed by a compact metric space, or other suitable parameter space such as an open or closed set in a Euclidean space, sample continuity reduces to that of the isonormal process on some subsets, and continuity of the nonrandom function $t \mapsto EX_t$ (Corollary 2.56).

Let $(T, \mathcal{T})$ and $(W, \mathcal{U})$ be two topological spaces. Let $\{X_t, \ t \in T\}$ be a stochastic process defined over a probability space $(\Omega, \mathcal{B}, P)$ with values in W, meaning that for each $t \in T$ and Borel set $B \subset W$, $X_t^{-1}(B) \in \mathcal{B}$. (Recall that the σ-algebra of Borel sets is generated by the open sets and that it is equivalent to assume $X_t^{-1}(U) \in \mathcal{B}$ for each $U \in \mathcal{U}$.) Let $\{Y_t, \ t \in T\}$ be another process with values in W, possibly defined over a different probability space $(\Omega', \mathcal{B}', P')$. Recall that the processes $\{X_t\}$ and $\{Y_t\}$ have the *same laws* iff for every $n = 1, 2, \ldots$, and $t_1, \ldots, t_n \in T$, the law of $\{X_{t_j}\}_{j=1}^n$ on the product σ-algebra in W^n is the same as that of $\{Y_{t_j}\}_{j=1}^n$. Two processes with the same laws are said to be *versions* of each other. A process $\{X_t\}_{t \in T}$ will be called *version-continuous* iff there is a process Y with the same laws such that for all $\omega' \in \Omega'$, $t \mapsto Y_t(\omega')$ is continuous from T into W. (Equivalently, continuity need only hold for almost all ω' since then without changing the laws, for the set of measure 0 of values of ω' for which $Y_t(\omega')$ is not continuous, one can replace it by a fixed continuous function, say having a constant value in W.)

Now let $W = \mathbb{R}$ with usual topology and suppose $\{X_t, \ t \in T\}$ is a Gaussian process. Suppose also that (T, e) is a metric space with the metric topology on T. We have

Theorem 2.54 *A Gaussian process $\{X_t\}$ indexed by a metric space T, defined on a probability space (Ω, P), is version-continuous if and only if both*

(a) *the nonrandom function $t \mapsto EX_t$ is continuous, and*

(b) *the process $\{X_t - EX_t\}$ is version-continuous.*

Then, $t \mapsto X_t(\cdot)$ is continuous into $L^2(P)$.

Proof. "If" is clear. To prove "only if," suppose X_t is sample-continuous. For any sequence $t_n \to t \in T$, sample continuity implies that $X_{t_n} \to X_t$ almost surely, and therefore in probability. For jointly Gaussian random variables, convergence in probability is equivalent to convergence in $L^2(P)$, since for the Gaussian variables $Y_n := X_{t_n} - X_t$ to converge to 0 in probability, the means EY_n must converge to 0, and so must the variances. Thus $EX_{t_n} \to EX_t$. Since T is a metric space, (a) follows. Then by subtracting the continuous function EX_t, (b) follows. $\square$

So in studying sample continuity or version-continuity of Gaussian processes we may as well restrict ourselves to processes with mean 0. Let X_t be such a process, $t \in T$. Each $X_t(\cdot)$ is an element of a Hilbert space H, namely $L^2(P)$. Consider the isonormal process L on this H. Then since L is Gaussian, has mean 0, and preserves covariances, we see that $L(X_t)$ has the same laws as X_t.

If $h(\cdot)$ is a continuous function from T into a Hilbert space H, with range $C := \{h(t): t \in T\}$, and if L restricted to C is version-continuous, then the process $L \circ h$ is clearly version-continuous. Conversely, if (T, e) is compact and h is 1–1, then h is a a homeomorphism (RAP, Theorem 2.2.11). Then, version continuity of L on C and $L \circ h$ on T are equivalent. So, for (T, e) compact and $t \mapsto X_t(\cdot)$ one-to-one, version continuity of the Gaussian process X_t reduces to that of L on a subset C of H. (Theorem 2.55 and Corollary 2.56 below will show that the 1–1 assumption is not actually necessary.) If T is locally compact, for example an open or closed subset of some $\mathbb{R}^k$, then continuity is equivalent to continuity on each compact subset.

Let T be a set and d a pseudo-metric on T: for all $x, y, z \in T$, $d(x, y) = d(y, x)$ and $d(x, z) \le d(x, y) + d(y, z)$, $d(x, x) = 0$, but possibly $d(x, y) = 0$ for some $x \ne y$. Recall that for a set $S \subset T$, the diameter (with respect to d) is defined by diam $S := \text{diam}_d S := \sup\{d(x, y): x, y \in S\}$.

The next fact holds for general, not necessarily Gaussian processes.

Theorem 2.55 *If (T, e) is a compact metric space, h is a continuous function from T onto a metric space K and $Y(x, \omega)$, $x \in K$, $\omega \in \Omega$, is a stochastic process on K with values in a complete separable metric space S, then $Y \circ h$ is version-continuous on T if and only if Y is on K.*

Remark If $Y(x, \omega) \equiv Y(x)$, a nonrandom function, then the result is a known fact in general topology (RAP, Theorem 2.2.11). The difficulty in the proof here is that if $Y \circ h$ is version-continuous, it is not clear that the corresponding sample-continuous process X can be written as $Y' \circ h$ for a process Y' on K.

Proof. "If" is obvious. Conversely let $Y \circ h$ be version-continuous and take a process X on T with the same laws as $Y \circ h$ and $t \mapsto X(t, \omega) := X_t(\omega)$

continuous on T for all ω. Let ρ and ζ be the metrics on K and S respectively. Let $d(s, t) := \rho(h(s), h(t))$, a pseudo-metric on T. Let A be a countable dense subset of T, and $B := h(A) := \{h(a): a \in A\}$, so B is a countable dense subset of K. For $\gamma, \delta > 0$ and any countable set $F \subset T$ define a random variable by

$$D(F, \delta, \gamma) := \sup \{\zeta (X_s, X_t): s, t \in F, \ d(s, t) < \delta, \ e(s, t) < \gamma\}.$$

This is measurable since F is countable and S is separable (RAP, Proposition 4.1.7). Let $D(F, \delta) := D(F, \delta, 1 + \mathrm{diam}_e T)$, and

$$UC := \bigcap_{n=1}^{\infty} \bigcup_{m=1}^{\infty} \{D(A, 1/m) \le 1/n\},$$

so that UC is measurable. If $P(UC) = 1$, then the sample functions $t \mapsto X_t(\omega)$ are almost surely uniformly continuous with respect to d on A. Since A is countable and $Y \circ h$ has the same laws as X, $Y \circ h$ also has sample functions almost surely uniformly continuous for d on A. Equivalently, Y has sample functions almost surely uniformly continuous for ρ on B. For any $x \in K$ let $Y'(x) := \lim\{Y(u): u \to x, \ u \in B\}$. Almost surely, all these limits exist (by uniform continuity and since S is complete), and $x \mapsto Y'(x)$ is continuous. Since X and $Y' \circ h$ both have continuous sample functions on T and have the same law on A, they have the same laws on T, and so does $Y \circ h$ by choice of X. If follows that Y' has the same laws as Y on K, so Y is version-continuous as desired.

Otherwise, $P(UC) < 1$. Then for some $\varepsilon > 0$,

$$\inf_{\delta>0} P(D(A, \delta) > 3\varepsilon) > 3\varepsilon.$$

Then by inclusion (monotone convergence, with $\delta = 1/m$),

$$P \left(\bigcap_{\delta>0} \{D(A, \delta) > 3\varepsilon\} \right) > 3\varepsilon. \tag{2.71}$$

On the other hand, continuity of $t \mapsto X_t$ and compactness of T imply that for some $\gamma > 0$, and any countable set $C \subset T$,

$$P \{D (C, 1 + \mathrm{diam}_d T, \gamma) > \varepsilon\} < \varepsilon. \tag{2.72}$$

(Otherwise, take a countable union of countable sets for $\gamma = 1/n$, $n = 1, 2, \ldots$, to get a contradiction.)

T is a finite union of e-open sets T_j with $\mathrm{diam}_e T_j < \gamma$. For each $i \ne j$, if there are $s \in T_i$ and $t \in T_j$ with $h(s) = h(t)$, then we say $\langle i, j \rangle \in \mathcal{L}$, and let us choose and fix such $s = s(i, j)$ and $t = t(i, j)$. Let C be the union of A and the set of all $s(i, j)$ and $t(i, j)$. Since $\mathcal{L}$ is finite, we can assume that

$X_{s(i,j)}(\omega) = X_{t(i,j)}(\omega)$ for all ω and $\langle i, j \rangle \in \mathcal{L}$. Let

$$J := \bigcap_{n=1}^{\infty} \{D(C, 1/n) > 3\varepsilon\} \cap \{D(C, 1 + \operatorname{diam}_d T, \gamma) \le \varepsilon\}.$$

Then (2.71) and (2.72) give $P(J) > 3\varepsilon - \varepsilon > \varepsilon$. Fix an $\omega \in J$ and choose $s_n \in C$ and $t_n \in C$ such that $d(s_n, t_n) < 1/n$ and $\zeta(X_{s_n}, X_{t_n})(\omega) > 3\varepsilon$. By compactness, we can assume that the sequences s_n and t_n both converge for e and hence also for d. Let $s_n \to s$ and $t_n \to t$. Then $d(s, t) = 0$, $X_{s_n}(\omega) \to X_s(\omega)$ and $X_{t_n}(\omega) \to X_t(\omega)$ as $n \to \infty$. Let $s \in T_i$ and $t \in T_j$. If $i = j$ we have, since $\omega \in J$, $3\varepsilon \le \zeta(X_s, X_t)(\omega) \le \varepsilon$, a contradiction. If $i \ne j$, then $\langle i, j \rangle \in \mathcal{L}$. For n large enough, $s_n \in T_i$ and $t_n \in T_j$, so

$$3\varepsilon < \zeta\left(X_{s_n}, X_{t_n}\right)(\omega) \le \zeta\left(X_{s_n}, X_{s(i,j)}\right)(\omega) + \zeta\left(X_{t(i,j)}, X_{t_n}\right)(\omega)$$
$$\le \varepsilon + \varepsilon = 2\varepsilon < 3\varepsilon,$$

again a contradiction. □

Now recall that a totally bounded set C in a Hilbert space H is called a GC-set iff L restricted to C has a version with uniformly continuous sample functions.

Corollary 2.56 *A Gaussian process $\{X_t, t \in T\}$ with mean 0 on a compact metric space (T, e) is version-continuous if and only if both $t \mapsto X_t(\cdot) \in H := L^2(P)$ is continuous and its range K is a GC-set.*

Proof. Apply Theorem 2.55 with $h(t) := X_t(\cdot)$, $K := h(T)$, ρ the usual metric in H, and $S = \mathbb{R}$ with its usual metric; again $L(X_t(\cdot))$ has the same laws as X_t. If X_t is version-continuous, then $t \mapsto X_t(\cdot)$ is continuous into H by Theorem 2.54, and the rest follows. □

Example. If X_t is a Gaussian process defined for $t \in \mathbb{R}$, suppose X_t is periodic of period 2π, $X_t \equiv X_{t+2\pi}$ for all t. Suppose that $E((X_t - X_s)^2) > 0$ for $|s - t| < 2\pi$. Then we can write the process as $X_t = Y(e^{it})$ where Y is a process indexed by the unit circle $T^1 := \{z: |z| = 1\}$ in the complex plane, which is compact. Version continuity for X and Y are equivalent, and $z \mapsto Y(z)(\cdot)$ is 1–1 from T^1 into $H := L^2(P)$, so version continuity is equivalent to that of L on the range of Y in H (without needing Theorem 2.55 and Corollary 2.56). On the other hand, any process indexed by $\mathbb{R}$ is version continuous if and only if it is so on each compact interval $[-N, N]$, where in this example for $N \ge \pi$, the process is not 1–1 into H.

Recall that a sample function of a stochastic process X_t is a function $t \mapsto X_t(\omega)$ for a fixed ω. The usual metric on Hilbert space is the natural one for an isonormal process, but the GC-property holds for other metrics in the following sense:

Theorem 2.57 *Let C be a subset of Hilbert space H. Then C is a GC-set if and only if there exists a metric ρ on C such that (C, ρ) is totally bounded, and the sample functions of the isonormal process L on C can be chosen to be ρ-uniformly continuous a.s.*

Proof. "Only if" is immediate where ρ is the usual metric. To prove "if," take a version of L such that on a set of probability one, the sample functions of L are ρ-uniformly continuous on C. Then L extends to a Gaussian process $t \mapsto X_t$ on the compact completion M of C for ρ. Here X_t is version-continuous, and so by Corollary 2.56, C is included in, and thus is, a GC-set. $\qquad\square$

2.13 Two-Series and One-Series Theorems

The following material was not needed so far in the text, but it can be helpful in some of the problems.

For independent real random variables X_n, Lévy's equivalence theorem says that three ways for the series $\sum_{n=1}^{\infty} X_n$ to converge are equivalent: almost surely, in probability or in distribution (e.g. RAP, Theorem 9.7.1). Let the variables, truncated to have absolute values ≤ 1, be $X_n^1 = X_n$ if $|X_n| \leq 1$ and 0 otherwise. The three-series theorem (e.g. RAP, Theorem 9.7.3) says that the almost sure convergence of $\sum_n X_n$ is equivalent to convergence of all three of three series of numbers: $\sum_{n=1}^{\infty} P(|X_n| > 1)$, $\sum_n EX_n^1$ (which need not converge absolutely), and $\sum_n \text{Var}(X_n^1)$.

If the variables satisfy further conditions, the conditions can simplify. Specifically, we have the following "two-series" theorem.

Theorem 2.58 *Let X_n be independent, nonnegative real random variables. Then for $\sum_{n=1}^{\infty} X_n$ to converge almost surely (to a finite limit random variable), it suffices that $\sum_{n=1}^{\infty} EX_n < +\infty$. It is equivalent that the following two series should both converge:*

(a) $\sum_{n=1}^{\infty} P(X_n > 1)$,

(b) $\sum_{n=1}^{\infty} E(X_n^1)$.

Proof. The sequence of partial sums $S_n = \sum_{j=1}^{n} X_j$ is nondecreasing up to some limit $S_\infty \leq +\infty$. If $\sum_{n=1}^{\infty} EX_n < +\infty$ then by Fatou's lemma or the monotone convergence theorem, $ES_\infty < +\infty$, so S_∞ is finite almost surely and the series converges almost surely.

Now suppose both series (a) and (b) converge. Then by series (a) and the Borel–Cantelli lemma, almost surely $X_n = X_n^1$ for all $n \geq n_0(\omega)$ large enough. We have $\sum_{n=1}^{\infty} X_n^1$ converging almost surely by series (b) and the first part of the proof, and so $\sum_{n=1}^{\infty} X_n$ converges almost surely.

Conversely, if $\sum_{n=1}^{\infty} X_n$ converges almost surely, then both series (a) and (b) must converge by the three-series theorem. This completes the proof. □

So, for nonnegative variables one need not consider the variances of the truncated variables X_n^1. It is actually easy to see why this is, namely,

$$\mathrm{Var}(X_n^1) \le E((X_n^1)^2) \le E(X_n^1)$$

for all n, so convergence of series (b), for nonnegative variables, implies that of series (c) in the three-series theorem.

For series of independent normal variables X_n with $EX_n = 0$, convergence will reduce to that of one series of numbers, the variances. There are also other aspects of convergence that can be included. A series is said to converge *unconditionally* if it can be rearranged in any order and converges to the same limit. For a given series $\sum_n a_n$ of real numbers, it is known that unconditional convergence is equivalent to absolute convergence, i.e. $\sum_n |a_n| < \infty$. If $(S, \| \cdot \|)$ is a Banach space, then a series $\sum_n s_n$ of elements of S is said to converge *absolutely* iff $\sum_n \|s_n\| < \infty$. It is known that in infinite-dimensional Banach spaces, unconditional convergence of series is not equivalent to absolute convergence, and we will see that in a Hilbert space H. Here is an equivalence, where the one series of real numbers is the one in part (e):

Theorem 2.59 Let G_i be independent $N(0, \sigma_i^2)$ random variables defined on a probability space (Ω, P). Then the following are equivalent:

(a) $\sum_{i=1}^{\infty} G_i$ converges almost surely;

(b) $\sum_{i=1}^{\infty} G_i^2 < \infty$ almost surely;

(c) $\sum_{i=1}^{\infty} G_i$ converges in the Hilbert space $H = L^2(\Omega, P)$;

(d) The series in (c) converges unconditionally in H;

(e) $\sum_{i=1}^{\infty} \sigma_i^2 < \infty$.

Proof. By the Lévy equivalence theorem, (a) is equivalent to convergence in probability. For jointly Gaussian random variables, convergence in probability is equivalent to convergence in L^2, as was noted in the proof of Theorem 2.33 (for a Gaussian variable to be close to 0 in probability, its mean, in this case 0, and its variance must be small). Thus (a) is equivalent to (c). A series $\sum_{i=1}^{\infty} s_i$ of orthogonal elements of H converges in H if and only if $\sum_{i=1}^{\infty} \|s_i\|^2 < \infty$. Thus (e) is equivalent to (c) as well as (a). By the two-series Theorem 2.58, (e) implies (b). Conversely, if (b) holds, it follows from the Borel–Cantelli Lemma that $\sum_i P(|G_i| > 1) < \infty$, and so $\sigma_i^2 \to 0$ as $i \to \infty$. By the two-series theorem, $\sum_i E(G_i^2 1_{|G_i| \le 1}) < \infty$, and the terms are asymptotic to σ_i^2, so (e) holds and (b) is equivalent to (e).

Convergence of the series in (e), if it holds, is clearly absolute and unconditional, thus its convergence is equivalent to (d). So the theorem is proved. □

Example. Suppose $\sigma_i^2 = 1/i^2$ for all i. Then all the equivalent conditions in Theorem 2.59 hold. We have $\|G_i\| = 1/i$ in H, so the series $\sum_{i=1}^{\infty} G_i$ does not converge absolutely in H, giving an example where unconditional convergence is not equivalent to absolute convergence in H, an infinite-dimensional Hilbert space. Moreover, the series in (a) does not converge absolutely: with probability 1, by the two-series theorem, $\sum_{i=1}^{\infty} |G_i| = +\infty$, and so for almost all ω, $\sum_i G_i(\omega)$ does not converge unconditionally.

Problems

1. If X and Y are i.i.d. $N(0, 1)$, evaluate $E \max(X, Y)$.

2. Evaluate $E \exp(\alpha \|X\|^2))$ (finite or infinite) as a function of $\alpha > 0$ if

(a) $\mathcal{L}(X) = N(0, 1)$ in $\mathbb{R}$, $\|X\| = |X|$;

(b) $\mathcal{L}(X) = N(0, C)$ in $\mathbb{R}^2$, $C = \left(\begin{smallmatrix} 1 & 0 \\ 0 & 2 \end{smallmatrix}\right)$, and $\|(x_1, x_2)\| = (x_1^2 + x_2^2)^{1/2}$.

3. Let H be the Hilbert space $L^2([0, +\infty), \lambda)$ where λ is Lebesgue measure. As usual let 1_A be the indicator function of a set A, i.e. $1_A(x) = 1$ for $x \in A$ and 0 otherwise.

(a) Show that for the isonormal process L on H, $x_t = L(1_{[0,t]})$ for $t \geq 0$ gives a Brownian motion (is a Gaussian process with correct mean and covariance).

(b) For $0 \leq t \leq 1$ find functions g_t such that $y_t = L(g_t)$ gives a Brownian bridge.

4. Let H be a Hilbert space with orthonormal basis $\{e_j\}_{j \geq 1}$. Let G_n be independent with laws $N(0, \sigma_n^2)$.

(a) Under what conditions on σ_n^2 does $\sum G_n e_n$ converge almost surely in the norm of H? *Hint:* Apply the two-series Theorem 2.58 to suitable real-valued random variables.

(b) If $G = \sum_n G_n e_n$ in H as in (a), where the sum converges almost surely, find for what $\alpha > 0$ we have $E \exp(\alpha \|G\|^2) < \infty$. After doing this directly, compare with the results of Section 2.3.

5. Let G_n be i.i.d. $N(0, 1)$ variables and $a_n > 0$ for each n. Under what conditions on a_n is $\sum_n a_n |G_n| < \infty$ a.s.? *Hint:* Use the two-series theorem.

6. (a) Show that for any set A in a real vector space V, and any vector space W of linear forms on V, the polar A^{*1} of A, defined by $A^{*1} := \{w \in W : w(v) \leq 1$ for all $v \in A\}$, is convex in W.

(b) Let $C \subset V$ be a set and D its convex hull, the smallest convex set including C. Show that $D^{*1} = C^{*1}$.

(c) In $V = \mathbb{R}^2$ let C be the unit square $\{0 \le x \le 1, \ 0 \le y \le 1\}$. Evaluate the polar C^{*1}, where $W = \mathbb{R}^2$ and $w(v) := w_1 v_1 + w_2 v_2$.

7. Let H be a Hilbert space with orthonormal basis $\{e_n\}_{n\ge1}$. For $c_n > 0$ let $E(\{c_n\}_{n\ge1}) := \{\sum_{n\ge1} x_n e_n : \sum_{n\ge1} x_n^2/c_n^2 \le 1\}$, an infinite-dimensional ellipsoid. Show that E is a GB-set if and only if $\sum_n c_n^2 < \infty$.

8. With notation as in the previous problem, let $C := \{e_n/(\log n)^{1/2} : n \ge 2\}$. Show that C is not a GC-set (it is a GB-set as shown in the example before Theorem 2.16).

9. Let ψ be a characteristic function on $\mathbb{R}$, so that $\psi(t) = \int_{-\infty}^{\infty} e^{ixt} dP(x)$ for some law P on $\mathbb{R}$, which we assume is symmetric, $P(A) = P(-A)$ for all Borel sets A. (Then ψ must be real-valued.) Show that there exists a Gaussian process X_t, $t \in \mathbb{R}$ with mean 0 and covariance $E X_s X_t = \psi(s - t)$ for all real s, t.

10. (a) For each $t > 0$ let ψ_t be a "triangle function" on $\mathbb{R}$ with $\psi_t(0) = 1$, and for some $t > 0$, $\psi_t(s) = 0$ whenever $|s| \ge t$, while ψ_t is linear on each interval $[-t, 0]$ and $[0, t]$. Show that ψ_t satisfies the conditions of the previous problem. *Hint:* Find its (inverse) Fourier transform and show that it is a probability density.

(b) Let ψ be a continuous real-valued function on $\mathbb{R}$ which is even, $\psi(-x) \equiv \psi(x)$, $\psi(0) = 1$, and on $[0, \infty)$, ψ is nonincreasing, nonnegative, and convex. Show that ψ is a characteristic function. *Hint:* Use a mixture of triangle functions. First consider the case that ψ is piecewise linear and is 0 outside some finite interval. Take limits of such piecewise linear functions.

11. For $\alpha > 0$ let $\psi(x) = 1 - (\log(1/|x|))^{-\alpha}$ for x in some neighborhood of 0 (piecewise linear elsewhere).

(a) Show that there exists such a ψ satisfying the conditions of Problem 9, assuming Problem 10.

(b) What can be said about sample-continuity of the Gaussian process X_t for different values of α?

12. Let $X_t(\omega) := \sum_{n=-\infty}^{\infty} G_n(\omega) \cos(nt)$ where G_n are independent random variables with laws $N(0, \sigma_n^2)$ and $\sum_n \sigma_n^2 < \infty$. Show that the process $t \mapsto X_t$ is version-continuous if and only if $\{X_t(\cdot) : 0 \le t \le 2\pi\}$ is a GC-set in $L^2(\Omega)$.

13. Let $X_t(\omega) := \sum_{n=1}^{\infty} G_n(\omega) \cos(nt) + H_n \sin(nt)$ where G_n and H_n for all n are independent random variables (also independent for different n) where for each n, both G_n and H_n have laws $N(0, \sigma_n^2)$ and $\sum_n \sigma_n^2 < \infty$. Show that the

process $t \mapsto X_t$ is sample-continuous if and only if $T := \{X_t(\cdot) : 0 \le t \le 2\pi\}$ is a GC-set in $L^2(\Omega)$, and if so, T satisfies the hypothesis of Theorem 2.36. *Hint*: Show that Corollary 2.53 applies, in that for points $p_t := (\cos t, \sin t)$ of the unit circle, $d(s, t) := [E(X_s - X_t)^2]^{1/2}$ is rotationally invariant, although it is in general not a usual metric such as distance in $\mathbb{R}^2$ or arc length distance.

14. Let $\{X_t\}_{t\in\mathbb{R}}$ be a stationary Gaussian process with mean 0, where stationarity means that for any $n = 1, 2, \ldots$, any $t_1, \ldots, t_n \in \mathbb{R}$, and any $h \in \mathbb{R}$, the joint distribution of $\{X_{t_j}\}_{j=1}^n$ is the same as that of $\{X_{t_j+h}\}_{j=1}^n$. Suppose that $t \mapsto X_t(\cdot)$ is continuous in probability, or equivalently into $L^2(\Omega)$. Show that:

(a) The process is version-continuous if and only if it is when restricted to the interval $[-1, 1]$.

(b) Show further that for the set $C := \{X_t(\cdot)\}_{-1 \le t \le 1}$ in $H = L^2(\Omega, P)$, version-continuity holds if and only if C satisfies the hypothesis of Theorem 2.36. *Hint*: Show that the hypothesis of Theorem 2.52 holds with $M = 2$ and μ equal to Lebesgue measure over 2, even though the (pseudo)metric on $[-1, 1]$ induced by the process is not in general the usual one.

15. Let e_n be orthonormal in H and $C := \{a_n(\log n)^{-1/2} e_n\}_{n \ge 2}$ where $a_n \to 0$ as $n \to \infty$.

(a) Show that every such set is a GC-set.

(b) By taking $a_n \to 0$ slowly enough, Show that for any $r > 0$ there exist GC-sets C with $D(\varepsilon, C) \ge \exp[1/(\varepsilon^2 |\log \varepsilon|^r)]$ for $\varepsilon > 0$ small enough.

16. (A further extension of problem 10). For $c > 0$ let $\psi(x) := 1 - (\log(1/|x|))^{-1}(\log\log(1/|x|))^{-c}$ for x in some neighborhood of 0 (piecewise linear elsewhere).

(a) Show that there exists such a ψ satisfying the conditions of Problem 9, assuming Problem 10.

(b) What can be said about sample-continuity of the Gaussian process X_t for different values of c?

(c) Show that for any $r < 2$ there exist non-GB-sets C such that for $\varepsilon > 0$ small enough, with $D(\varepsilon, C) \le \exp(1/(\varepsilon^2 |\log \varepsilon|^r)$. Compare with Problem 15, part (b), to see that for $0 < r < 2$, one cannot tell from $D(\varepsilon, C)$ whether C is a GC-set or not.

17. Let v_k be the Lebesgue volume of the unit ball in $\mathbb{R}^k$. Then it is known that $v_k = \pi^{k/2}/\Gamma(1 + (k/2))$ for $k = 1, 2, \ldots$. For any $c_i > 0$, the ellipsoid $\mathcal{E}_k := \mathcal{E}(\{c_i\}_{i=1}^k) := \{x : \sum_{i=1}^k x_i^2/c_i^2 \le 1\} \subset \mathbb{R}^k$ has volume $v_k c_1 \ldots c_k$ for any $c_i > 0$, $i = 1, \ldots, k$. For $\varepsilon > 0$ let $m := D(\varepsilon, \mathcal{E}_k)$.

(a) Show that $m \ge c_1 c_2 \cdots c_k/\varepsilon^k$.

(b) If $c_j \geq \varepsilon$ for $j = 1, \ldots, k$, show that $m(\varepsilon/2)^k \leq 2^k c_1 c_2 \cdots c_k$.

(c) If $c_j = j^{-\alpha}$ for $j = 1, 2, \ldots,$ for the infinite-dimensional ellipsoid $\mathcal{E}$ equal to $E(\{c_j\}_{j \geq 1}$ as in Problem 7, give upper and lower bounds for $D(\varepsilon, \mathcal{E})$ as $\varepsilon \downarrow 0$. *Hint*: Choose k, depending on ε, to give as good bounds as possible. Recall Stirling's formula $k!/[(k/e)^k (2\pi k)^{1/2}] \to 1$ as $k \to \infty$ (Theorem 1.17). Are your bounds consistent with Theorems 2.14 and 2.36 and the result of Problem 7?

18. If g is a Young–Orlicz modulus, $x^2 = o(\log g(x))$ as $x \to +\infty$, and Y is a $N(0, 1)$ random variable, show that $\|Y\|_g = +\infty$.

19. If g is a Young–Orlicz modulus, $\log g(x)) = O(x^2)$ as $x \to +\infty$, and Y is a $N(0, 1)$ random variable, show that $\|Y\|_g < \infty$.

20. Let $f(x) = \pi - x$ for $0 < x < \pi$ and $f(x) = -x - \pi$ for $-\pi < x < 0$. Find the Fourier series

$$f \sim c + \sum_{n=1}^{\infty} a_n \sin(nx) + b_n \cos(nx)$$

in $L^2((-\pi, \pi))$ and use it to prove $\sum_{n=1}^{\infty} n^{-2} = \pi^2/6$.

21. (a) Find a numerical value of the absolute constant C in Proposition 2.46 used in bounding the difference of the mean EF and median $m(F)$ of a Lipschitz function F with respect to $N(0, I)$, $|EF - m(F)| \leq C\|F\|_L$. Use the method of proof for that Proposition and the bound $\alpha_y(r) \leq 2\exp(-r^2/4)$ given in Theorem 2.43. Also use that for any $r > 0$, $P(F - m(F) > r) \leq 1/2$ by definition of median, since $P(F \leq m(F)) \geq 1/2$.

(b) Find a value of C by another method using the inequality not proved in this text (but in Ledoux's 2001 book), $P(F - EF \geq r\|F\|_L) \leq \exp(-r^2/2)$, so if the right side is $\leq 1/2$, then $m(F) - EF \leq r\|F\|_L$, and considering $-F$, $|m(F) - EF| \leq r\|F\|_L$.

Notes

Notes to Section 2.3. Proposition 2.5(a) improves on RAP, Lemma 12.1.6(b) by a factor of 2. Lemma 2.10 was first proved by Landau and Shepp (1971), then by Fernique (1970) whose note giving a much shorter proof appeared in print earlier. Fernique's statement about an error in Landau and Shepp's paper apparently had to do with an earlier, unpublished draft of the paper. The main theorems 2.6 and 2.11, giving the best possible upper bound for α, were then proved independently by Marcus and Shepp (1972) and Fernique (1971). The current exposition is based mainly, though not entirely, on Fernique (1975).

The Darmois–Skitovič theorem says that if $X_1, \ldots, X_n$ are independent real random variables where for some constants $a_1, \ldots, a_n$ and $b_1, \ldots, b_n$, $a_1 X_1 + \cdots + a_n X_n$ is independent of $b_1 X_1 + \cdots + b_n X_n$, with $a_i b_i \neq 0$, then for that i, X_i has a normal distribution: see Darmois (1951) and Skitovič (1954). C. R. Rao (1973, pp. 158–163 and 218) lists various characterizations of normal distributions.

Notes to Section 2.4. Slepian (1962, Lemma 1) proved his inequality. Sudakov (1969) announced a result somewhat weaker than Theorem 2.14. Sudakov (1971, Theorem 5) announced a stronger result, corrected in Sudakov (1973) to be what is here Theorem 2.14. Lemma 2.15 is essentially from Chevet (1970). Fernique (1975, Théorème 2.1.2, Corollaire 2.1.3) proved Theorem 2.16. The proof given is as in Fernique (1997, pp. 59, 63–67), who mentions an idea of Kahane (cf. Kahane 1986). Fernique actually proved the more general (2.19) and (2.20) in Theorem 2.18. Inequality (2.21) is given in Giné and Zinn (1986) assuming that T is countable. To allow T uncountable (but separable), a different hypothesis is used here. Sudakov (1973, Proposition 7) proved the inequality in Theorem 2.22, assuming $D(\varepsilon, S, d) \geq 10$, and with the constant $(1 - e^{-1})/2$ in place of $1/17$ (for $D(\varepsilon, S, d) < 10$ it is easy to check that the inequality must hold, possibly with a smaller constant).

Notes to Section 2.5. Sudakov (1971) stated that a set K in Hilbert space is GB if and only if its mixed volume $h_1(K)$ homogeneous of degree 1 is finite. In my review of that paper in *Mathematical Reviews* I wrote that "Up to the present, no such geometric criterion for the GB-property was known, so that the theorem is of great interest." Then Sudakov (1973) pointed out that finiteness of $h_1(K)$ is equivalent to that of $EL(K)^*$. My unduly brief review of the 1973 paper in *Math. Revs.* did not even mention $EL(K)^*$ although it turned out to appear much more often in later literature.

Notes to Section 2.6. Theorems 2.23 and 2.25 and Corollaries 2.26 and 2.27, and essentially Corollary 2.28, are due to T. W. Anderson (1955). These facts were to some extent rediscovered by L. Gross (1962). Both used the Brunn–Minkowski inequality in their proofs. See also Borell (1974, 1975a,b), Gordon (1985), and Kahane (1986).

Notes to Section 2.7. Lemma 2.29 is essentially due to T. W. Anderson (1955). Lemmas 2.30 and 2.31 are straightforward. In Dudley (1967a, Theorem 4.6), parts of which are due to Gross (1962), most of the conditions in Theorem 2.32 were proved equivalent for convex, symmetric sets. Feldman (1972) proved part (b'), about the symmetric convex hull of C.

Notes to Section 2.8. In apparently the first public use of a metric entropy hypothesis to obtain sample continuity of Gaussian processes, V. N. Sudakov, in a talk at the International Congress of Mathematicians in Moscow in 1965, announced that if $\log N(\varepsilon, C) = O(\varepsilon^{-r})$ as $\varepsilon \downarrow 0$ for some $r < 2$, then C is a GC-set. Sudakov (1969) was his first publication on the topic. Specific bounds on metric entropy implying sample continuity were suggested by previous work of Fernique (1964) for processes indexed by a real parameter. The fact that $\int_0^1 (\log N(\varepsilon, C))^{1/2} d\varepsilon < \infty$ implies C is a GC-set was given (with an equivalent series instead of the integral) in Dudley (1967a). Theorem 2.36 and its proof were given in Dudley (1973). Theorem 2.37 has often been attributed to me, but I did not prove it. I only gave the integral on the right side. Sudakov (1973) first gave the expectation on the left, but as mentioned in the Notes to Section 2.5, I did not for some time take note of it. One source for the formulation and a proof (with an extension) of Theorem 2.37 is Pisier (1983). The proof given above is based (with some changes) on Ledoux and Talagrand (1991, Section 11.1) where the Theorem is extended, with different functions g_0, to a large class of non-Gaussian stochastic processes. Komatsu's inequality, used in the proof of Lemma 2.41, is quoted, with hints for the proof, in Itô and McKean (1974, p. 17), citing as original source Komatsu (1955).

Notes to Section 2.9. As mentioned at the beginning of the section and at several points in it, the section is based on the book Ledoux (2001). Some details not given in the book are filled in.

Notes to Section 2.10. This section is based on early parts of the book of Talagrand (2005).

Notes to Section 2.11. Fernique (1975) proved that for Gaussian processes satisfying a homogeneity condition like that in Corollary 2.53, the metric entropy integral condition (or the corresponding condition on $N(\varepsilon, T)$, equivalent by Theorem 1.9 above) is necessary and sufficient for sample continuity.

Notes to Section 2.12. I have no reference for Theorem 2.55. The facts in this section on Gaussian processes were to some extent stated, but not proved, in Dudley (1967a, 1973). Andersen and Dobrić (1988, Lemma 8, (4.6); preprint, 1985) first published proofs of Corollary 2.56 and Theorem 2.57 according to Giné and Zinn (1986, p. 58).

3

Foundations of Uniform Central Limit Theorems; Donsker Classes

3.1 Definitions: Convergence in Law

Empirical processes $\sqrt{n}(P_n - P)$ as mentioned in Section 2.1 will be defined here with more detail and precision than there. Let $(S, \mathcal{B}, P)$ be a probability space, to be called the *sample space*. Examples to have in mind for S are Euclidean spaces such as the plane $\mathbb{R}^2$. To form empirical measures, one would like to take variables $X_1, X_2, \ldots$, i.i.d. with law P. To do this, take a countable product S^∞ of copies of $(S, \mathcal{B}, P)$ (RAP, Theorem 8.2.2) and let X_i be the coordinates on the product. A product may be taken with another probability space. *Throughout the rest of this book, X_i will be defined as such coordinates* unless something is said to the contrary. An example showing that use of coordinates on product spaces, rather than just having $X_1, X_2, \ldots$ i.i.d. (P), makes a difference will be given at the end of Section 5.3. Recall that the *product σ-algebra* is the smallest for which all the coordinates are measurable.

Then, we can form the *empirical measures* $P_n := \frac{1}{n} \sum_{i=1}^n \delta_{X_i}$, and the *empirical process* $\nu_n := n^{1/2}(P_n - P)$. So P_n is a probability measure on S, defined on $\mathcal{B}$ and actually on all subsets of S, for any values of $X_1, \ldots, X_n$. Each ν_n is a finite signed measure of total charge 0.

Recall (Section 2.1) the Gaussian process $G_P(f)$ defined for $f \in \mathcal{L}^2(S, \mathcal{B}, P)$: G_P has mean 0 and covariance equal to the covariance for P (2.2). Given $f \in \mathcal{L}^2(P)$, let $\pi_0(f) := f - \int f \, dP$. Then $\pi_0(f) \in \mathcal{L}_0^2(P)$.

A *semi-inner product* $(\cdot, \cdot)$ on $V \times V$ for a vector space V satisfies the definition of inner product except that possibly $(u, u) = 0$ for $u \neq 0$.

For $f, g \in \mathcal{L}^2(P)$, again recalling (2.2),

$$E(G_P(f)G_P(g)) = (f, g)_{0, P} = (f - \pi_0(f), g - \pi_0(g))$$

133

where $(\cdot, \cdot)$ (without subscripts) is the usual semi-inner product on $\mathcal{L}^2(P)$, $(f, g) := (f, g)_P := \int fg \, dP$, and $(\cdot, \cdot)_{0,P}$ is another semi-inner product.

Let $\mathcal{L}_0^2(P)$ be the set of all functions $f \in \mathcal{L}^2(P)$ such that $\int f \, dP = 0$. Then π_0 is linear from $\mathcal{L}^2(P)$ onto $\mathcal{L}_0^2(P)$. Let $L_0^2(P)$ be the set of all equivalence classes of elements of $\mathcal{L}_0^2(P)$ for equality a.s. (P). On $\mathcal{L}_0^2(P)$, $(\cdot, \cdot)_{0,P} \equiv (\cdot, \cdot)_P$.

Thus, restricted to $L_0^2(P)$, G_P is an isonormal process. Let C be the one-dimensional space of constant functions c as a subspace of $\mathcal{L}^2(P)$. Then G_P is 0 on C, while the spaces C and $\mathcal{L}_0^2(P)$ are orthogonal complements of each other (RAP, Theorem 5.3.8) in $\mathcal{L}^2(P)$. For any f and g in $\mathcal{L}^2(P)$ and $c \in \mathbb{R}$, $G_P(cf + g) = cG_P(f) + G_P(g)$ a.s. (2.4).

Recall the notion of pseudometric as defined after (2.4). The covariance for P defines a pseudometric on $\mathcal{L}^2(P)$ by $\rho_P(f, g) := (E(G_P(f) - G_P(g))^2)^{1/2}$. Since $f = g$ a.s. implies $G_P(f) = G_P(g)$ a.s., ρ_P defines a pseudometric on $L^2(P)$, which is the usual Hilbert metric on $L_0^2(P)$. Thus, *the results of Chapter 2 for the isonormal process L apply to G_P on $L_0^2(P)$, with inner product $(\cdot, \cdot)_P = (\cdot, \cdot)_{0,P}$ there, and the Hilbert metric is ρ_P*.

The Brownian bridge process y_t is a special case of the G_P process where P is Lebesgue measure on $[0, 1]$ and $y_t = G_P(1_{[0,t]})$; it is easily checked that this has the right covariance.

The Brownian bridge process can be taken to have continuous sample paths, in other words, to be continuous as a function of t for each ω (RAP, Theorem 12.1.5). But except in special cases, such as that the sample space S is a finite set, the spaces $\mathcal{L}^2(P)$ and $\mathcal{L}_0^2(P)$ are infinite-dimensional, in the sense that they contain infinite orthonormal sets. We saw in Section 2.8 that an isonormal process on an infinite-dimensional Hilbert space H is not sample-continuous. Then G_P is not sample-continuous on the whole space $\mathcal{L}^2(P)$. We will be concerned then with suitable subsets of $\mathcal{L}^2(P)$. A class $\mathcal{F} \subset \mathcal{L}^2$ will be called *pregaussian* if a G_P process $(f, \omega) \mapsto G_P(f)(\omega)$ can be defined on some probability space such that for each ω, $f \mapsto G_P(f)(\omega)$ is bounded and uniformly continuous for ρ_P from $\mathcal{F}$ into $\mathbb{R}$.

Given $\mathcal{F} \subset \mathcal{L}^2(P)$, let $\pi_0(\mathcal{F})$ be the set of all functions $\pi_0(f)$, $f \in \mathcal{F}$. For any $f \in \mathcal{L}^2(P)$, $\rho_P(f, \pi_0(f)) = 0$, and $G_P(f) = G_P(\pi_0(f))$ a.s. I claim that $\mathcal{F}$ is pregaussian if and only if $\pi_0(\mathcal{F})$ is: if $\pi_0(\mathcal{F})$ is pregaussian, then $f \mapsto G_P(\pi_0(f))$ has the desired properties on $\mathcal{F}$. Conversely if $\mathcal{F}$ is pregaussian, take a probability space $(\Omega, \mathcal{A}, Q)$ and a G_P process on $\mathcal{F}$ over this probability space such that $G_P(\cdot)(\omega)$ is bounded and ρ_P-uniformly continuous on $\mathcal{F}$ for all ω. For each $g \in \pi_0(\mathcal{F})$ there is a nonempty set $C_{\mathcal{F}}(g)$ of constants c such that $g + c \in \mathcal{F}$. For any $c, d \in C_{\mathcal{F}}(g)$ and any ω, $G_P(g + c)(\omega) = G_P(g + d)(\omega)$ since $\rho_P(g + c, g + d) = 0$. Define $H_P(g)(\omega)$ as any such $G_P(g + c)(\omega)$. Then H_P has the desired properties of G_P on $\pi_0(\mathcal{F})$, as claimed.

Now recall the definition of GC-set from Subsection 2.2.1 above. We have the following:

Theorem 3.1 *Let $\mathcal{F} \subset \mathcal{L}^2(P)$. Let $\mathcal{F}'$ be the set of all equivalence classes in $L_0^2(P)$ of functions in $\pi_0(\mathcal{F})$. Then $\mathcal{F}$ is pregaussian if and only if $\mathcal{F}'$ is a GC-set in $L_0^2(P)$.*

Proof. Let L be the isonormal process on $L^2(P)$. The three stochastic processes indexed by $\mathcal{F}$, $\{G_P(f): f \in \mathcal{F}\}$, $\{G_P(\pi_0(f)): f \in \mathcal{F}\}$, and $\{L(\pi_0(f)): f \in \mathcal{F}\}$ are equal in distribution. By definition a GC-set is totally bounded and L on it can be chosen with uniformly continuous, thus bounded sample functions. So the "if" part follows. Conversely if $\mathcal{F}$ is pregaussian, $\mathcal{F}'$ must be a GC-set. □

Recall the notion of prelinear function (Lemma 2.30). A G_P process Y on a class $\mathcal{F} \subset \mathcal{L}^2(S, \mathcal{B}, P)$ for a probability space $(S, \mathcal{B}, P)$ will be called *coherent* if for each $\omega \in \Omega$, the function $f \mapsto Y(f)(\omega)$ on $\mathcal{F}$ is bounded, ρ_P-uniformly continuous, and prelinear.

Theorem 3.2 *Given a probability space $(S, \mathcal{B}, P)$ and $\mathcal{F} \subset \mathcal{L}^2(S, \mathcal{B}, P)$, $\mathcal{F}$ is pregaussian if and only if there exists a coherent G_P process on $\mathcal{F}$.*

Proof. "If" follows from the definition of pregaussian. For the converse, apply Theorems 3.1 and 2.32. □

Now, if a class $\mathcal{F} \subset \mathcal{L}^2(P)$ is pregaussian we can ask whether v_n converges in distribution, or in law, to G_P with respect to uniform convergence over $\mathcal{F}$. Recall that for random variables Y_n with values in a separable metric space S, convergence in law of Y_n to Y_0 is defined to mean that $Eg(Y_n) \to Eg(Y_0)$ as $n \to \infty$ for every bounded continuous function g on S. But empirical processes take values in nonseparable metric spaces in general. Consider the following:

Example. Let U be the $U[0, 1]$ distribution function as in Section 1.1 above. Let X_1 have this distribution function and let U_1 be the empirical distribution function, so that $U_1(x) = 1$ if $x \geq X_1$ and 0 otherwise. The empirical process $\sqrt{n}(U_n - U)$ for $n = 1$ is just $U_1 - U$. For each possible value y of X_1 we get a function $G_y(x) = (U_1 - U)(x) = -x$ for $0 \leq x < y$ and $1 - x$ for $y \leq x \leq 1$ (equal to 0 outside $[0, 1]$). As in Section 1.1, consider the supremum norm and distance, for a bounded real-valued function g on $[0, 1]$, $\|g\|_{\sup} := \sup_{0 \leq x \leq 1} |g(x)|$ and for two such functions g and h, $d_{\sup}(g, h) := \|g - h\|_{\sup}$. Let $\mathcal{G}_1$ be the set of all G_y for $0 \leq y \leq 1$. For $y' \neq y$ in $[0, 1]$ and the corresponding U_1' we have

$$d_{\sup}(G_{y'}, G_y) = \|(U_1' - U) - (U_1 - U)\|_{\sup} = \|U_1' - U_1\|_{\sup} = 1 \quad (3.1)$$

since if, for example, $y' < y$, then for each x with $y' \leq x < y$ we have $U_1'(x) = 1$ and $U_1(x) = 0$. So, in the metric $d_{\sup}$, $\mathcal{G}_1$ is a discrete, complete (and thus closed in any set including it) set with any two points in it at distance 1 apart. Thus every subset F of $\mathcal{G}_1$ is also complete and closed. There exists a bounded continuous function H on the space $\ell^\infty[0, 1]$ of all bounded real functions on $[0, 1]$ which equals 1 on F and 0 on $\mathcal{G}_1 \setminus F$, by the Tietze extension theorem (such an H can be defined explicitly as $H(g) := \max(0, 1 - d(F, g))$ where $d(F, g) := \inf_{G \in F} \|G - g\|_{\sup}$). If A is a non-Lebesgue measurable subset of $[0, 1]$, which exists assuming the axiom of choice (RAP, Theorem 3.4.4) and $F := \{G_y : y \in A\}$, then for X_1 having a $U[0, 1]$ distribution, $EH(G_y) = \int_0^1 H(G_y) dy = \lambda(A)$, which is undefined, where λ is Lebesgue measure.

So, the definition of convergence in distribution, or weak convergence, of probability measures on the Borel sets of separable metric spaces, in terms of integrals of bounded continuous functions converging, does not apply to empirical processes, because the integrals of some bounded continuous functions do not exist. In a nonseparable metric space, we cannot necessarily expect to have distributions defined on all Borel sets. A new definition of convergence in law is needed to take care of this nonmeasurability. The definition will involve the notion of upper integral. Let g be a real-valued, not necessarily measurable function defined on a space X where $(X, \mathcal{S}, \mu)$ is a measure space. Let $\overline{\mathbb{R}}$ be the set $[-\infty, \infty]$ of extended real numbers. Then the *upper integral* is defined by

$$\textstyle\int^* g d\mu := \inf\{\int h d\mu : h \geq g, \ h \text{ measurable and } \overline{\mathbb{R}}\text{-valued}\},$$

which will be undefined if there exists a measurable $h \geq g$ with $\int h d\mu = \infty - \infty$ undefined, unless there is also a measurable $\psi \geq g$ with $\int \psi d\mu = -\infty$, in which case $\int^* g d\mu$ will also be defined as $-\infty$. There always exists at least one measurable $h \geq g$, namely $h \equiv +\infty$.

We will be dealing often with compositions of functions. If f is a function whose domain includes the range of g, then either $f(g)$ or $f \circ g$ will denote the function such that $(f \circ g)(x) \equiv f(g(x))$.

A function, which may not be measurable, from a probability space into a metric space will be called a *random element*. Now here is a definition of convergence *in* law where only the limit variable Y_0 necessarily *has* a law:

Definition. Let (S, d) be any metric space. Let $(\Omega_n, \mathcal{A}_n, Q_n)$ be probability spaces for $n = 0, 1, 2, \ldots$, and Y_n, $n \geq 0$, functions from Ω_n into S. Suppose that Y_0 takes values in some separable subset of S and is measurable for the Borel sets on its range. Then Y_n will be said to converge to Y_0 *in law* as $n \to \infty$, in symbols $Y_n \Rightarrow Y_0$, if for every bounded continuous real-valued function g on S,

$$\int^* g(Y_n) dQ_n \ \to \ \int g(Y_0) dQ_0 \text{ as } n \to \infty.$$

Note. If $\int G(Y_0)dQ$ is defined for all bounded continuous real G (as it must be if $Y_n \Rightarrow Y_0$, by definition), then the image measure $Q \circ Y_0^{-1}$ is defined on all Borel subsets of S (RAP, Theorem 7.1.1). Such a law does have a separable support except perhaps in some set-theoretically pathological cases (Appendix C).

For g bounded, $\int^* g(Y_n)dQ_n$ is always defined and finite. Then, here is a general definition of when the central limit theorem for empirical measures holds with respect to uniform convergence over a class $\mathcal{F}$ of functions. The metric space S will be the space $\ell^\infty(\mathcal{F})$ of all bounded real-valued functions on $\mathcal{F}$, with metric given by the supremum norm $\|H\|_{\mathcal{F}} := \sup\{|H(f)| : f \in \mathcal{F}\}$.

Definition. Let $(\Omega, \mathcal{A}, P)$ be a probability space and $\mathcal{F} \subset \mathcal{L}^2(P)$. Then $\mathcal{F}$ will be called a *Donsker class* for P, or *P-Donsker class*, or be said to satisfy the central limit theorem (for empirical measures) for P, if $\mathcal{F}$ is pregaussian for P, G_P is coherent, and $\nu_n \Rightarrow G_P$ in $\ell^\infty(\mathcal{F})$.

Later, a number of rather large classes $\mathcal{F}$ of functions will be shown to be Donsker classes for various laws P. The next few sections develop some of the needed theory.

3.2 Measurable Cover Functions

In the last section, convergence in law was defined in terms of upper integrals. The notion of upper integral is related to that of measurable cover. Let $(\Omega, \mathcal{A}, Q)$ be a probability space. Then for a possibly nonmeasurable set $A \subset \Omega$, a set B is called a *measurable cover* of A if $A \subset B$, $B \in \mathcal{A}$, and $P(B) = \inf\{P(C) : A \subset C, C \text{ measurable}\}$. If B and C are measurable covers of the same set A, then clearly so is $B \cap C$. It follows that $B = C$ up to a set of measure 0, in other words $P(B \triangle C) = 0$ where $\triangle$ denotes the symmetric difference, or equivalently $P(1_B = 1_C) = 1$.

For any set $A \subset \Omega$ let $P^*(A) := \inf\{P(B) : A \subset B, B \text{ measurable }\}$. Then for any measurable cover B of A, clearly $P^*(A) = P(B)$.

Let $\mathcal{L}^0 := \mathcal{L}^0(\Omega, \mathcal{A}, P, \overline{\mathbb{R}})$ denote the set of all measurable functions from Ω into $\overline{\mathbb{R}}$. Then $\mathcal{L}^0$ is a lattice: for any $f, g \in \mathcal{L}^0$, $f \vee g := \max(f, g)$ and $f \wedge g := \min(f, g)$ are in $\mathcal{L}^0$. But this $\mathcal{L}^0$ is not a vector space since we could have, for example, $f = +\infty$ and $g = -\infty$, so $f + g$ would be undefined.

The map $y \mapsto \tan^{-1} y$ is one-to-one from $\overline{\mathbb{R}}$ onto $[-\pi/2, \pi/2]$. Then a metric on $\overline{\mathbb{R}}$ is defined from the usual metric on $[-\pi/2, \pi/2]$ by $\bar{d}(x, y) := |\tan^{-1} x - \tan^{-1} y|$. On $\mathcal{L}^0$ we have the Ky Fan metric (RAP, Theorem 9.2.2) defined by

$$d(f, g) := \inf\{\varepsilon > 0 : P(\bar{d}(f(x), g(x)) > \varepsilon) \le \varepsilon\}.$$

Then $d(f, g) = 0$ if and only if $P(f = g) = 1$.

For any set $\mathcal{J} \subset \mathcal{L}^0(\Omega, \mathcal{A}, P, \overline{\mathbb{R}})$, a function $f \in \mathcal{L}^0$ is called an *essential infimum* of $\mathcal{J}$, or $f := \text{ess. inf } \mathcal{J}$, iff for all $j \in \mathcal{J}$, $f \leq j$ a.s. and for any $g \in \mathcal{L}^0$ such that $g \leq j$ a.s. for all $j \in \mathcal{J}$, we have $g \leq f$ a.s. If f and g are two essential infima of the same set $\mathcal{J}$, then clearly $f = g$ a.s. A set $\mathcal{J}$ of functions will be called a *lower semilattice* if for any $f, g \in \mathcal{J}$, we have $\min(f, g) \in \mathcal{J}$.

Theorem 3.3 *For any probability space* $(\Omega, \mathcal{A}, P)$ *and set* $\mathcal{J} \subset \mathcal{L}^0(\Omega, \mathcal{A}, P, \overline{\mathbb{R}})$, *an essential infimum of* $\mathcal{J}$ *exists. If for some function* $f : \Omega \mapsto \mathbb{R}$ *we have* $\mathcal{J} = \{j \in \mathcal{L}^0 : j \geq f \text{ everywhere}\}$, *then* $f^* := \text{ess. inf } \mathcal{J}$ *can be chosen so that* $f^* \geq f$ *everywhere. Also,* $\int f^* dP$ *and* $E^* f := \int^* f dP$ *are both defined and equal if either of them is well-defined (possibly infinite), for example, if* f^* *is bounded below.*

Proof. Let $\mathcal{J}_1$ be the class of all functions $\min(f_1, \ldots, f_m)$ for $f_1, \ldots, f_m \in \mathcal{J}$ and $m = 1, 2, \ldots$. Then $\mathcal{J}_1$ is a lower semilattice. For $f \in \mathcal{L}^0$, $f = \text{ess.inf } \mathcal{J}_1$ if and only if $f = \text{ess.inf } \mathcal{J}$. So we can assume $\mathcal{J}$ is a lower semilattice. For each $j \in \mathcal{J}$, $\tan^{-1} j$ is a measurable function with values in $[-\pi/2, \pi/2]$. Take $j_m \in \mathcal{J}$ such that $\int \tan^{-1} j_m dP \downarrow \inf_{j \in \mathcal{J}} \int \tan^{-1} j dP$. Then $\min(j_1, \ldots, j_m)$ is in $\mathcal{J}$ and decreases to g as $m \to \infty$ for some $g \in \mathcal{L}^0(\Omega, \mathcal{A}, P, \overline{\mathbb{R}})$. For any $h \in \mathcal{J}, \min(h, j_1, \ldots, j_m) \downarrow \min(h, g)$ so $\int \tan^{-1} \min(h, g) dP = \int \tan^{-1}(g) dP$ and $g \leq h$ a.s., so g satisfies the definition of ess.inf $\mathcal{J}$. If $\mathcal{J} = \{h \in \mathcal{L}^0 : h \geq f\}$, then $\mathcal{J}$ is a lower semilattice and for g constructed as above, $g \geq f$ everywhere.

By the definitions, $\int^* f dP \leq \int f^* dP$ if either side is well-defined, and the inequality is an equation by the definition of essential infimum. $\square$

Definition. For any f as in Theorem 3.3, f^* will mean a function as shown to exist in the theorem with $f^* \geq f$ everywhere and will be called a *measurable cover function* of f.

Recall that in Chapter 2, $L(A)^*$ was the essential supremum of $L(x)$ for $x \in A$, and so, the essential infimum of random variables Y such that for each $x \in A$, $Y \geq L(x)$ a.s. — a different, although related, notion.

If f is real valued and bounded above by some finite valued measurable function, then f^* is a measurable real-valued function. But whenever there exist nonmeasurable sets $A_n \downarrow \emptyset$ with $P^*(A_n) \equiv 1$, as for Lebesgue measure (e.g. RAP, Section 3.4, Problem 2), let $f := n$ on $A_n \setminus A_{n+1}$. Then f is real valued but $f^* = +\infty$ a.s.

The next two lemmas on measurable cover functions are basic.

Lemma 3.4 *For any two functions* $f, g : \Omega \mapsto (-\infty, \infty]$, *we have*

(a) $(f + g)^* \leq f^* + g^*$ *a.s., and*

(b) $(f - g)^* \geq f^* - g^*$ *whenever both sides are defined a.s.*

Proof. (a) We have $-\infty < f^* \leq +\infty$ and $-\infty < g^* \leq +\infty$ everywhere, so $f^* + g^*$ is an everywhere defined, measurable function $\geq f + g$, and (a) follows. For part (b), on the measurable set where $g^* = +\infty$, on which by assumption f^* is finite a.s., the right side is $-\infty$ and the inequality holds. Where g^* is finite, g is also finite and $f = (f - g) + g$, so $f^* \leq (f - g)^* + g^*$ by (a), so $f^* - g^* \leq (f - g)^*$, since this holds where $(f - g)^* < \infty$ and where $(f - g)^* = \infty$. $\qquad\square$

Lemma 3.5 *Let S be a vector space with a seminorm $\|\cdot\|$. Then for any two functions X, Y from Ω into S, $\|X + Y\|^* \leq (\|X\| + \|Y\|)^* \leq \|X\|^* + \|Y\|^*$ a.s. and $\|cX\|^* = |c| \|X\|^*$ a.s. for any real c.*

Proof. The first inequality is clear, the second follows from Lemma 3.4, and the equation is clear (for $c = 0$ and $c \neq 0$). $\qquad\square$

Next, in some cases of independence, the upper-star operation can be distributed over products or sums.

Lemma 3.6 *Let $(\Omega_j, \mathcal{A}_j, P_j)$, $j = 1, \ldots, n$, be any n probability spaces. Let f_j be functions from Ω_j into $\overline{\mathbb{R}}$. Suppose either*

(a) $f_j \geq 0$, $j = 1, \ldots, n$, or

(b) $f_1 \equiv 1$ and $n = 2$.

Then on the Cartesian product $\prod_{j=1}^{n}(\Omega_j, \mathcal{A}_j, P_j)$ with $x := (x_1, \ldots, x_n)$, if $f(x) := \prod_{j=1}^{n} f_j(x_j)$, we have $f^*(x) = \prod_{j=1}^{n} f_j^*(x_j)$ a.s., where $0 \cdot \infty$ is set equal to 0.

(c) Or, if $f_j(x_j) > -\infty$ for all x_j, $j = 1, \ldots, n$, and $g(x_1, \ldots, x_n) := f_1(x_1) + \cdots + f_n(x_n)$, then $g^*(x_1, \ldots, x_n) = f_1^*(x_1) + \cdots + f_n^*(x_n)$ a.s.

Proof. First, for (c), by induction we can assume $n = 2$. We have

$$g^*(u, v) \geq g(u, v) = f_1(u) + f_1(v) \qquad (3.2)$$

for all u, v by the definitions. We have $g^*(u, v) \leq f_1^*(u) + f_2^*(v)$ a.s. by Lemma 3.4(a), and if equality does not hold a.s., there is a rational t such that on a measurable set C of positive probability in the product space, $g^*(u, v) < t < f_1^*(u) + f_2^*(v)$, and there exist rational q, r with $q + r > t$ such that C can be chosen with $f_1^*(u) > q$ and $f_2^*(v) > r$ for $(u, v) \in C$. Let $C_u := \{v : (u, v) \in C\}$. By the Tonelli–Fubini theorem there is a measurable set $D \subset \Omega_1$ with $P_1(D) > 0$ such that $P_2(C_u) > 0$ for all $u \in D$. If $f_1 \leq q$ on D, then $f_1^* \leq q$ a.s. on D, but for any $u \in D$ and $v \in C_u \neq \emptyset$ we have $f_1^*(u) > q$, a contradiction. So choose and fix a $u \in D$ with $f_1(u) > q$. Then for any $v \in C_u$, $q + f_2(v) < f_1(u) + f_2(v) \leq g^*(u, v)$, so $f_2(v) < g^*(u, v) - q$ and $f_2^*(v) \leq g^*(u, v) - q$ for almost all $v \in C_u$. For any such v, $q + f_2^*(v) < q + r$ and $f_2^*(v) < r$, a contradiction. So (c) is proved.

Now for products, in case (a) or (b), clearly $f^*(x) \leq \Pi_{j=1}^n f_j^*(x_j)$ a.s., with $1^* \equiv 1$. For the converse inequality we can assume $n = 2$, by induction in case (a). Suppose $f^*(x) < f_1^*(x_1)f_2^*(x_2)$ with positive probability. Then for some rational r, $f^*(x) < r < f_1^*(x_1)f_2^*(x_2)$ with positive probability. If $f_1 \equiv 1$ this gives $f(x) \leq f^*(x) < r < f_2^*(x_2)$ on a set of positive probability. Then by the Tonelli–Fubini theorem, for some x_1, $f_2(x_2) \leq f^*(x_1, x_2) < r < f_2^*(x_2)$ on a set of x_2 with $P_2 > 0$, contradicting the choice of f_2^*.

So assume $f_1 \geq 0$ and $f_2 \geq 0$. Then as in case (c), using the analogue of (3.2) for products of nonnegative functions rather than sums, there are rationals a, b with $ab > r$, $a > 0$, $b > 0$, such that on a set C in the product with positive probability, $f_1^*(x_1) > a$, $f_2^*(x_2) > b$, and $f^*(x_1, x_2) < r$. Again by the Tonelli–Fubini theorem, there is a measurable set $D \subset \Omega_1$ with $P_1(D) > 0$ and $P_2(C_u) > 0$, $u \in D$, and there is a point u of D where $f_1(u) > a$. Then for any $v \in C_u$, $f_2(v) \leq f^*(u, v)/a$, so $f_2^*(v) \leq f^*(u, v)/a$ for almost all $v \in C_u$. For such a v we have $af_2^*(v) < ab$ and $f_2^*(v) < b$, a contradiction, finishing the proof. $\square$

For the next fact here is some notation: given two functions f, g and a σ-algebra $\mathcal{S}$ on the range of f, let $(f, g)(x) := (f(x), g(x))$ and $f^{-1}(\mathcal{S}) := \{f^{-1}(A): A \in \mathcal{S}\}$.

Lemma 3.7 *Let* $(\Omega, \mathcal{A}, P) = \Pi_{i=1}^3 (\Omega_i, \mathcal{S}_i, P_i)$ *with coordinate projections* Π_i: $\Pi_i(x_1, x_2, x_3) := x_i$, $i = 1, 2, 3$. *Let* $\mathcal{S}_1 \otimes \mathcal{S}_2$ *denote the product σ-algebra on* $\Omega_1 \times \Omega_2$. *Then for any bounded real function* f *on* $\Omega_1 \times \Omega_3$ *and* $g(x_1, x_2, x_3) := f(x_1, x_3)$, *conditional expectations of* g^* *satisfy*

$$E(g^*|(\Pi_1, \Pi_2)^{-1}(\mathcal{S}_1 \otimes \mathcal{S}_2)) = E(g^*|\Pi^{-1}(\mathcal{S}_1)) \ \text{ a.s. for } P.$$

Proof. By Lemma 3.6(b), for $\Omega_2 \times (\Omega_1 \times \Omega_3)$, g^* equals P-a.s. a measurable function not depending on x_2, thus independent of $\Pi_2^{-1}(\mathcal{S}_2)$. Let $\mathcal{S}$ be the collection of all sets $A \in (\Pi_1, \Pi_2)^{-1}(\mathcal{S}_1 \otimes \mathcal{S}_2)$ such that g^* and $h := E(g^*|\Pi_1^{-1}(\mathcal{S}_1))$ have the same integral over A. Then $\mathcal{S}$ contains all finite disjoint unions of sets $(\Pi_1, \Pi_2)^{-1}(B_1 \times B_2) = \Pi_1^{-1}(B_1) \cap \Pi_2^{-1}(B_2)$, $B_i \in \mathcal{S}_i$, $i = 1, 2$, since both g^* and h are independent of $\Pi_2^{-1}(\mathcal{S}_2)$. Now $\mathcal{S}$ is easily seen to be a monotone class, so it equals all of $(\Pi_1, \Pi_2)^{-1}(\mathcal{S}_1 \otimes \mathcal{S}_2)$ (RAP, Theorem 4.4.2). $\square$

Lemma 3.8 *Let X be a real-valued function on a probability space* $(\Omega, \mathcal{A}, P)$. *Then for any* $t \in \mathbb{R}$,

(a) $P^*(X > t) = P(X^* > t)$.

(b) For any $\varepsilon > 0$, $P^*(X \geq t) \leq P(X^* \geq t) \leq P^*(X \geq t - \varepsilon)$.

Proof. Clearly $\{X > t\} \subset \{X^* > t\}$ and $\{X \geq t\} \subset \{X^* \geq t\}$, so we have "≤" in (a) and the first inequality in (b).

Take a measurable cover $A \supset \{X > t\}$, so $P^*(X > t) = P(A)$. Then $X^* \leq t$ a.s. outside A, so $P(X^* > t) \leq P(A)$, proving (a). Thus for $0 < \delta \leq \varepsilon$, $P(X^* > t - \delta) \leq P^*(X > t - \varepsilon)$. Letting $\delta \downarrow 0$ proves (b). $\qquad \square$

Let $(\Omega, \mathcal{A}, P)$ be a probability space. For a function f from Ω into $[-\infty, \infty]$ let $\int_* f dP := \sup\{\int g dP : g \text{ measurable}, g \leq f\}$. Let f_* be the essential supremum of all measurable functions $g \leq f$. Then just as for f^*, f_* is well-defined up to a.s equality and $\int_* f dP = \int f_* dP$ whenever either side is defined, as in Theorem 3.3.

It is easy to check that $f_* = -((-f)^*)$ and that $\int_* f dP = -(\int^* -f dP)$. So the convergence in law $Y_n \Rightarrow Y_0$ for functions into a metric space S as defined in Section 3.1 implies that $\int_* g(Y_n) dQ_n \to \int g(Y_0) dQ_0$ for every bounded continuous real-valued function g on S.

For the next fact, recall that if $(S, \mathcal{S}, \mu)$ is a measure space, an *atom* of μ is a point $x \in S$ or the singleton $\{x\}$ if $\{x\} \in \mathcal{S}$ and $\mu(\{x\}) > 0$. The measure μ will be called *purely atomic* if there is a countable set $A \in \mathcal{S}$ such that each $x \in A$ is an atom and $\mu(S \setminus A) = 0$. Next, here is a one-sided Tonelli–Fubini theorem for starred functions:

Theorem 3.9 *Let* $(X, \mathcal{A}, P) \times (Y, \mathcal{B}, Q)$ *be a product of two probability spaces. For a real-valued function* $f \geq 0$ *on* $X \times Y$, *define* f^* *with respect to* $P \times Q$. *For each* $x \in X$ *let* $(E_2^* f)(x) := \int^* f(x, y) dQ(y)$. *Then*

$$E_1^* E_2^* f(x, y) \leq \int f^*(x, y) d(P \times Q)(x, y), \tag{3.3}$$

where $E_1^* = E^*$ *with respect to* P. *Also, if* Q *is purely atomic, with* $\sum_j Q(\{y_j\})$ $= 1$ *for some* $y_j \in Y$, *and* $E_2(\cdot) := \int \cdot dQ$, *then*

$$E_1^* E_2 f(X, Y) \leq E^* f(X, Y) = E_2 E_1^* f(X, Y). \tag{3.4}$$

Proof. By the usual Tonelli–Fubini theorem, we have

$$\int f^*(x, y) d(P \times Q)(x, y) = \int \int f^*(x, y) dQ(y) dP(x), \tag{3.5}$$

and $\int f^*(x, y) dQ(y)$ is measurable in x. For each $x \in X$ let $f_x(y) := f(x, y)$. Then $f(x, y) \leq f^*(x, y)$, which is measurable in y, so $f_x^*(y) \leq f^*(x, y)$ for Q-almost all y. Thus since $E^* f = Ef^*$ by Theorem 3.3,

$$(E_2^* f)(x) = \int f_x^*(y) dQ(y) \leq \int f^*(x, y) dQ(y),$$

and (3.3) follows. Next, to prove (3.4), note that on Y, since Q is purely atomic, all real-valued functions are measurable (for the completion of Q), so $E_2^* \equiv E_2$, the left inequality follows from (3.3), and the equality from (3.5), so (3.4) is proved. $\qquad \square$

Remark. Under usual set-theoretic hypotheses, there exists an "ordinal triangle" set $A \subset I \times I, I := [0, 1]$ such that for all $x \in I, \{y : (x, y) \in A\}$ is countable, and for all $y \in I$, $\{x : (x, y) \in A\}$ has countable complement in I. (The set is described at the beginning of Chapter 5.) Clearly $\int_0^1 \int_0^1 1_A(x, y) dy dx = 0 < 1 = \int_0^1 \int_0^1 1_A(x, y) dx dy$. The use of stars does not change this pathology: for $f = 1_A$ in (3.3), the left side is 0 and the right side is 1.

We also have a one-sided monotone convergence theorem with stars:

Theorem 3.10 *Let $(\Omega, \mathcal{A}, P)$ be a probability space and let f_j be real-valued functions on Ω such that $f_j \uparrow f$, i.e. $f_j(x) \uparrow f(x)$ for all $x \in \Omega$. If $E^* f_1 > -\infty$, then $E^* f_j \uparrow E^* f$ as $j \to \infty$.*

Proof. By Theorem 3.3, $E^* f_j = E f_j^*$ for each j, and f_j^* increase a.s. up to some function g where clearly $g \leq f^*$ almost surely. If $g < f^*$ on some set A with $P(A) > 0$, then $f_j \leq g$ on A implies $f \leq g$ and so $f^* \leq g$ on A, a contradiction. The theorem follows. $\square$

Note that there exist subsets $A_j := A(j)$ of $[0, 1]$ with outer measure $\lambda^*(A_j) = 1$ for all j and $A_j \downarrow \emptyset$ (RAP, Problem 3.4.2). Letting $f_j := 1_{A(j)}$, we have $f_j \downarrow 0$, and $E^* f_j = 1$ for all j, so that the monotone convergence theorem fails for E^* for decreasing sequences. Next is a Fatou lemma with stars.

Theorem 3.11 *Let $(\Omega, \mathcal{A}, P)$ be a probability space and f_j any nonnegative real-valued functions on Ω. Then $E^* \liminf_{j \to \infty} f_j \leq \liminf_{j \to \infty} E^* f_j$.*

Proof. For $1 \leq k \leq m$ we have $\inf_{n \geq k} f_n \leq f_m$ and so $E^* \inf_{n \geq k} f_n \leq E^* f_m$, thus $E^* \inf_{n \geq k} f_n \leq \inf_{m \geq k} E^* f_m$. Taking the supremum over k on both sides and using the previous monotone convergence theorem gives the result. $\square$

We need a notion of independence for random elements. Let $(A_j, \mathcal{A}_j, P_j)$, $j = 1, 2, \ldots$, be probability spaces, and form a product $\prod_{j=1}^n (A_j, \mathcal{A}_j, P_j) = (B, \mathcal{B}, P)$ with points $x := \{x_j\}_{j=1}^n$. If X_j are functions on B of the form $X_j = h_j(x_j)$, $j = 1, \ldots, n$, where each h_j is a function on A_j (not necessarily measurable), then we call X_j *independent random elements*. If the h_j are measurable, this implies independence in the usual sense.

Suppose given a vector space S with a seminorm $\| \cdot \|$. For a class $\mathcal{F}$ of measurable functions on a sample space $(X, \mathcal{B})$, S may be the class of all bounded functions on $\mathcal{F}$ with the supremum norm $\| \cdot \|_{\mathcal{F}}$, or its subspace of prelinear functions. Random elements X_j with values in S will be called *symmetric* if we can write $x_j = \langle y_j, z_j \rangle$ where y_j and z_j are independent and have the same distribution in some space D_j ($A_j = D_j \times D_j$ with $P_j = Q_j \times Q_j$ for some Q_j) and $X_j = \psi(y_j) - \psi(z_j)$ for some function ψ from D_j into S.

With these definitions, the Lévy inequality (Theorem 1.20) can be extended to starred norms, with about the same proof:

Lemma 3.12 *If X_j are independent symmetric random elements and $S_j = X_1 + \cdots + X_j$, then for any $r \geq 0$, we have*

(a) $\Pr\left(\max_{j \leq n} \|S_j\|^* > r\right) \leq 2\Pr\left(\|S_n\|^* > r\right)$, *and*

(b) $\Pr\left(\max_{j \leq n} \|X_j\|^* > r\right) \leq 2\Pr\left(\|S_n\|^* > r\right)$.

Proof. For (a), let $M_k(\omega) := \max_{j \leq k} \|S_j\|^*$. Let C_k be the disjoint events $\{M_{k-1} \leq r < \|S_k\|^*\}$, $k = 1, 2, \ldots$, where $M_0 := 0$. By Lemma 3.5,

$$2\|S_m\|^* \leq \|S_n\|^* + \|2S_m - S_n\|^*, \quad 1 \leq m \leq n,$$

so if $\|S_m\|^* > r$, then either $\|S_n\|^* > r$ or $\|2S_m - S_n\|^* > r$ or both. The transformation which interchanges y_j and z_j for $j > m$ preserves probabilities and interchanges S_n and $2S_m - S_n$, so interchanges $\|S_n\|^*$ and $\|2S_m - S_n\|^*$, while preserving all X_j for $j \leq m$. Thus

$$\Pr\left(C_m \cap \{\|S_n\|^* > r\}\right) = \Pr\left(C_m \cap \{\|2S_m - S_n\|^* > r\}\right)$$
$$\geq \frac{1}{2}\Pr\left(C_m\right).$$

Thus $\Pr(M_n > r) = \sum_{m=1}^{n} \Pr(C_m) \leq 2\Pr\left(\|S_n\|^* > r\right)$. For (b), the proof is similar: replace S_i for $i = j$, k or m by X_i, and use the transformation interchanging y_j and z_j for $j \neq m$, which does not change any $\|X_j\|^*$ or $\|S_n\|^*$. $\qquad\square$

3.3 Convergence Almost Uniformly and in Outer Probability

In Section 3.1, the definition of convergence of laws was adapted to define convergence in law for random elements which may not have laws defined. In this section the same will be done for convergence in probability and almost sure convergence.

Let $(\Omega, \mathcal{A}, Q)$ be a probability space, (S, d) a metric space, and f_n functions from Ω into S. Then f_n will be said to converge to f_0 in *outer probability* if $d(f_n, f_0)^* \to 0$ in probability as $n \to \infty$, or equivalently, by Lemma 3.8, for every $\varepsilon > 0$, $Q^*\{d(f_n, f_0) > \varepsilon\} \to 0$ as $n \to \infty$.

Also, f_n is said to converge to f_0 *almost uniformly* if as $n \to \infty$, $d(f_n, f_0)^* \to 0$ almost surely.

The following is immediate:

Proposition 3.13 *Almost uniform convergence always implies convergence in outer probability.*

If f_n are all measurable functions, then clearly convergence in outer probability is equivalent to the usual convergence in probability, and almost uniform convergence to almost sure convergence.

Now some definitions will be given for Glivenko–Cantelli properties, which are laws of large numbers for empirical measures.

Definition. If $(X, \mathcal{A}, P)$ is a probability space and $\mathcal{F}$ is a class of integrable real-valued functions, $\mathcal{F} \subset \mathcal{L}^1(X, A, P)$, then $\mathcal{F}$ will be called a *strong* (resp. *weak) Glivenko–Cantelli class for P* iff as $n \to \infty$, $\|P_n - P\|_{\mathcal{F}} \to 0$ almost uniformly (resp. in outer probability).

In the following Proposition, part (C) is what is usually called "Egorov's theorem" for almost surely convergent sequences of measurable functions (RAP, Theorem 7.5.1).

Proposition 3.14 *Let $(\Omega, \mathcal{A}, Q)$ be a probability space, (S, d) a metric space, and f_n any functions from Ω into S for $n = 0, 1, \dots$. Then the following are equivalent:*

(A) $f_n \to f_0$ almost uniformly.

(B) For any $\varepsilon > 0$, $Q^*\{\sup_{n \geq m} d(f_n, f_0) > \varepsilon\} \downarrow 0$ as $m \to \infty$.

(C) For any $\delta > 0$ there is some $B \in \mathcal{A}$ with $Q(B) > 1 - \delta$ such that $f_n \to f_0$ uniformly on B.

(D) There exist measurable $h_n \geq d(f_n, f_0)$ with $h_n \to 0$ a.s.

Proof. (A) implies that

$$(\sup_{n \geq m} d(f_n, f_0))^* \leq \sup_{n \geq m}(d(f_n, f_0)^*) \downarrow 0 \text{ a.s. as } m \to \infty,$$

which implies (B).

Assuming (B), for $k = 1, 2, \dots$, let $C_k := \{\sup_{n \geq m(k)} d(f_n, f_0) > 1/k\}$, where $m(k)$ is large enough so that $Q^*(C_k) < 2^{-k}$. Take measurable covers B_k for C_k, so $C_k \subset B_k$, $B_k \in \mathcal{A}$ and $Q(B_k) < 2^{-k}$. For $r = 1, 2, \dots$, let $A_r := \Omega \setminus \bigcup_{k > r} B_k$. Then $Q(A_r) > 1 - 2^{-r}$ and $f_n \to f_0$ uniformly on A_r, so (C) holds.

Now assume (C). Take $C_k \in \mathcal{A}$, $k = 1, 2, \dots$, such that $Q(C_k) \uparrow 1$ and $f_n \to f_0$ uniformly on C_k. We can take $C_1 \subset C_2 \subset \cdots$. Take m_k such that $d(f_n, f_0) < 1/k$ on C_k for all $n \geq m_k$. Then $d(f_n, f_0)^* \leq 1/k$ on C_k, so $d(f_n, f_0)^* \to 0$ a.s., proving (A). Clearly, (A) and (D) are equivalent. □

Example. In $[0, 1]$ with Lebesgue measure P let $A_1 \supset A_2 \supset \cdots$ be sets with $P^*(A_n) = 1$ and $\bigcap_{n=1}^{\infty} A_n = \emptyset$ (e.g. RAP, Section 3.4, Problem 2; Cohn, 1980, p. 35). Then $1_{A_n} \to 0$ everywhere and, in that sense, almost surely, but not almost uniformly. Note also that 1_{A_n} does not converge to 0 in law as defined in Section 3.1. To avoid such pathology, almost uniform convergence is helpful.

Proposition 3.15 *Let* (S, d) *and* (Y, e) *be two metric spaces and* $(\Omega, \mathcal{A}, Q)$ *a probability space. Let* f_n *be functions from* Ω *into* S *for* $n = 1, 2, \ldots$, *such that* $f_n \to f_0$ *in outer probability as* $n \to \infty$. *Assume that* f_0 *has separable range and is measurable (for the Borel* σ-*algebra on* S). *Let* g *be a continuous function from* S *into* Y. *Then* $g(f_n) \to g(f_0)$ *in outer probability.*

Proof. Given $\varepsilon > 0$, $k = 1, 2, \ldots$, let $B_k := \{x \in S : d(x, y) < 1/k$ implies $e(g(x), g(y)) \leq \varepsilon$, $y \in S\}$. Then each B_k is closed and $B_k \uparrow S$ as $k \to \infty$. Fix k large enough so that $Q(f_0^{-1}(B_k)) > 1 - \varepsilon$. Then

$$\{e(g(f_n), g(f_0)) > \varepsilon\} \cap f_0^{-1}(B_k) \subset \{d(f_n, f_0) \geq 1/k\}.$$

Thus

$$Q^*\{e(g(f_n), g(f_0)) > \varepsilon\} < \varepsilon + Q^*\{d(f_n, f_0) \geq 1/k\} < 2\varepsilon$$

for n large enough. $\qquad\square$

Lemma 3.16 *Let* $(\Omega, \mathcal{A}, Q)$ *be a probability space and* $\{g_n\}_{n=0}^{\infty}$ *a uniformly bounded sequence of real-valued functions on* Ω *such that* g_0 *is measurable. If* $g_n \to g_0$ *in outer probability, then* $\limsup_{n \to \infty} \int^* g_n dQ \leq \int g_0 dQ$.

Proof. Let $|g_n(x)| \leq M < \infty$ for all n and all $x \in \Omega$. We can assume $M = 1$. Given $\varepsilon > 0$, for n large enough $Q^*(|g_n - g_0| > \varepsilon) < \varepsilon$. Let A_n be a measurable set on which $|g_n - g_0| \leq \varepsilon$ with $Q(\Omega \setminus A_n) < \varepsilon$. Then

$$\int g_n^* dQ \leq \varepsilon + \int_{A_n}^* g_n dQ \leq 2\varepsilon + \int_{A_n} g_0 dQ \leq 3\varepsilon + \int g_0 dQ.$$

Letting $\varepsilon \downarrow 0$ completes the proof. $\qquad\square$

On any metric space, the σ-algebra will be the Borel σ-algebra unless something is said to the contrary.

Corollary 3.17 *If* f_n *are functions from a probability space into a metric space,* $f_n \to f_0$ *in outer probability and* f_0 *is measurable with separable range, then* $f_n \Rightarrow f_0$.

Proof. Apply Proposition 3.15 to $g = G$ for any bounded continuous G and Lemma 3.16 to $g_n := G \circ f_n$ and to $g_n = -G \circ f_n$. $\qquad\square$

3.4 Perfect Functions

For a function g defined on a set A let $g[A] := \{g(x) : x \in A\}$. It will be useful that under some conditions on a measurable function g and general real-valued f, $(f \circ g)^* = f^* \circ g$. Here are some equivalent conditions:

Theorem 3.18 *Let* $(X, \mathcal{A}, P)$ *be a probability space,* $(Y, \mathcal{B})$ *any measurable space, and g a measurable function from X to Y. Let Q be the restriction of* $P \circ g^{-1}$ *to* $\mathcal{B}$*. For any real-valued function f on Y, define* f^* *for Q. Then the following are equivalent:*

(a) For any $A \in \mathcal{A}$ *there is a* $B \in \mathcal{B}$ *with* $B \subset g[A]$ *and* $Q(B) \geq P(A);$

(b) For any $A \in \mathcal{A}$ *with* $P(A) > 0$ *there is a* $B \in \mathcal{B}$ *with* $B \subset g[A]$ *and* $Q(B) > 0;$

(c) For every real function f on Y, $(f \circ g)^* = f^* \circ g$ *a.s.;*

(d) For any $D \subset Y$*,* $(1_D \circ g)^* = 1_D^* \circ g$ *a.s.*

Proof. Clearly (a) implies (b). To show (b) implies (c), note that always $(f \circ g)^* \leq f^* \circ g$. Suppose $(f \circ g)^* < f^* \circ g$ on a set of positive probability. Then for some rational r, $(f \circ g)^* < r < f^* \circ g$ on a set $A \in \mathcal{A}$ with $P(A) > 0$. Let $g[A] \supset B \in \mathcal{B}$ with $Q(B) > 0$. Then $f \circ g < r$ on A implies $f < r$ on B, so $f^* \leq r$ on B a.s., contradicting $f^* \circ g > r$ on A.

Clearly (c) implies (d).

Now, to show (d) implies (a), given $A \in \mathcal{A}$, let $D := Y \setminus g[A]$. Then we can take $1_D^* = 1_C$ for some $C \in \mathcal{B}$: let C be the set where $1_D^* \geq 1$. Then $D \subset C$ and $1_D \circ g = (1_D \circ g)^* = 0$ a.s on A. Let $B := Y \setminus C$. Then $B \subset g[A]$, and

$$Q(B) = 1 - Q(C) = 1 - \int 1_D^* d(P \circ g^{-1}),$$

which by the image measure change of variables theorem (e.g. RAP, Theorem 4.1.11) equals

$$1 - \int 1_D^* \circ g \, dP = 1 - \int (1_D \circ g)^* dP \geq P(A). \qquad \square$$

Note. In (a) or (b), if the direct image $g[A] \in \mathcal{B}$, we could just set $B := g[A]$. But, for any uncountable complete separable metric space Y, there exists a complete separable metric space S (for example, a countable product $\mathbb{N}^\infty$ of copies of $\mathbb{N}$) and a continuous function f from S into Y such that $f[B]$ is not a Borel set in Y (RAP, Theorem 13.2.1, Proposition 13.2.5). If f is only required to be Borel measurable, then S can also be any uncountable complete metric space (RAP, Theorem 13.1.1).

A function g satisfying any of the four conditions in Theorem 3.18 will be called *perfect* or *P-perfect*. Coordinate projections on a product space are, as one would hope, perfect:

Proposition 3.19 *Suppose* $A = X \times Y$*, P is a product probability* $v \times m$*, and g is the natural projection of A onto Y. Then g is P-perfect.*

Proof. Here $P \circ g^{-1} = m$. For any $B \subset A$ let $B_y := \{x : (x, y) \in B\}$, $y \in Y$. If B is measurable, then by the Tonelli–Fubini theorem, for $C := \{y : v(B_y) >$

0}, C is measurable, $C \subset g[B]$, and $P(B) \leq m(C)$, so condition (a) of Theorem 3.18 holds. □

Theorem 3.20 *Let* $(\Omega, \mathcal{A}, P)$ *be a probability space and* (S, d) *a metric space. Suppose that for* $n = 0, 1, \ldots,$ *$(Y_n, \mathcal{B}_n)$ is a measurable space, g_n a perfect measurable function from Ω into Y_n, and f_n a function from Y_n into S, where f_0 has separable range and is measurable. Let $Q_n := P \circ g_n^{-1}$ on $\mathcal{B}_n$ and suppose $f_n \circ g_n \to f_0 \circ g_0$ in outer probability as $n \to \infty$. Then $f_n \Rightarrow f_0$ as $n \to \infty$ for f_n on $(Y_n, \mathcal{B}_n, Q_n)$.*

Before proving this, here is an example:

Proposition 3.21 *Theorem 3.20 can fail without the hypothesis that g_n be perfect.*

Proof. Let $C \subset I := [0, 1]$ satisfy $0 = \lambda_*(C) < \lambda^*(C) = 1$ for Lebesgue measure λ (RAP, Theorem 3.4.4). Let $P = \lambda^*$, giving a probability measure on the Borel sets of C (RAP, Theorem 3.3.6). Let $\Omega = C$, $f_0 \equiv 0$, $Y_n = I$, $f_n := 1_{I \backslash C}$ for $n \geq 1$, and let g_n be the identity from C into Y_n for all n. Then $f_n \circ g_n \equiv 0$ for all n, so $f_n \circ g_n \to f_0 \circ g_0$ in outer probability (and in any other sense). Let $\mathcal{B}_n$ be the Borel σ-algebra on $Y_n = I$ for each n. Let G be the identity from I into $\mathbb{R}$. Then $\int^* G(f_n) dQ_n = \int^* f_n d\lambda = 1$ for $n \geq 1$, while $\int G(f_0) dQ_0 = 0$, so f_n does not converge to f_0 in law. □

After Theorem 3.20 is proved, it will follow that the g_n in the last proof are not perfect, as can also be seen directly, from condition (c) or (d) in Theorem 3.18.

Proof of Theorem 3.20. By Corollary 3.17, $f_n \circ g_n \Rightarrow f_0 \circ g_0$. Let H be any bounded, continuous, real-valued function on S. Then by an image measure change of variables (RAP, Theorem 4.1.11),

$$\int^* H(f_n(g_n)) dP \;\to\; \int H(f_0(g_0)) dP \;=\; \int H(f_0) dQ_0.$$

Also,

$$\int^* H(f_n(g_n)) dP = \int H(f_n(g_n))^* dP \quad \text{by Theorem 3.3}$$

$$= \int (H \circ f_n)^*(g_n) dP \quad \text{by Theorem 3.18}$$

$$= \int (H \circ f_n)^* dQ_n \quad \text{(RAP, Theorem 4.1.11)}$$

$$= \int^* H(f_n) dQ_n \quad \text{by Theorem 3.3,}$$

and the Theorem follows. □

In Proposition 3.19, $X \times Y$ could be an arbitrary product probability space, but projection is a rather special function. The following fact will show that all measurable functions on reasonable domain spaces are perfect.

Recall that a law P is called *tight* if $\sup\{P(K) : K \text{ compact}\} = 1$. A set $\mathcal{P}$ of laws is called *uniformly tight* if for every $\varepsilon > 0$ there is a compact K such that $P(K) > 1 - \varepsilon$ for all $P \in \mathcal{P}$. Also, a metric space (S, d) is called *universally measurable (u.m.)* if for every law P on the completion of S, S is measurable for the completion of P (RAP, Section 11.5). So any metric space which is a Borel subset of its completion is u.m.

Theorem 3.22 *Let (S, d) be a u.m. separable metric space. Let P be a probability measure on the Borel σ-algebra of S. Then any Borel measurable function g from S into a separable metric space Y is perfect for P.*

Note. In view of Appendix C, the hypothesis that Y be separable is not very restrictive.

Proof. Let A be any Borel set in S with $P(A) > 0$. Let $0 < \varepsilon < P(A)$. By the extended Lusin theorem (Theorem D.1, Appendix D) there is a closed set F with $P(F) > 1 - \varepsilon/2$ such that g restricted to F is continuous. Since P is tight (RAP, Theorems 11.5.1 and 7.1.3), there is a compact set $K \subset A$ with $P(K) > \varepsilon$. Then $C := F \cap K$ is compact, $C \subset A$, $P(C) > 0$, and g is continuous on C, so $g[C]$ is compact, $g[C] \subset g[A]$, and $(P \circ g^{-1})(g[C]) \geq P(C) > 0$, so the conclusion follows from Theorem 3.18. $\square$

Let $(\Omega, \mathcal{A}, P)$ be a probability space and g a measurable function from Ω into Y where $(Y, \mathcal{B})$ is a measurable space. Let $Q := P \circ g^{-1}$ on $\mathcal{B}$. Call g *quasiperfect* for P or *P-quasiperfect* if for every $C \subset Y$ with $g^{-1}C \in \mathcal{A}$, C is measurable for the completion of Q. Then the probability space $(\Omega, \mathcal{A}, P)$ is called *perfect* if every real-valued function G on Ω, measurable for the usual Borel σ-algebra on $\mathbb{R}$, is quasiperfect.

Example. A measurable, quasiperfect function g on a finite set need not be perfect: let $X = \{a_1, a_2, a_3, a_4, a_5, a_6\}$, $U := \{a_1, a_2\}$, $V := \{a_3, a_4\}$, $W := \{a_5, a_6\}$, $\mathcal{A} := \{\emptyset, U, V, W, U \cup V, U \cup W, V \cup W, X\}$, $P(U) = P(V) = P(W) = 1/3$, $Y := \{0, 1, 2\}$, $g(a_1) := g(a_3) := 0$, $g(a_2) := g(a_5) := 1$, $g(a_4) := g(a_6) := 2$. Let $\mathcal{B} := \{\emptyset, Y\}$. For $C \subset Y$, $g^{-1}(C) \in \mathcal{A}$ if and only if $C \in \mathcal{B}$, so g is quasiperfect. But, $P(U) > 0$ and $g[U]$ does not include any nonempty set in $\mathcal{B}$, so g is not perfect.

Proposition 3.23 *Any perfect function is quasiperfect.*

Proof. Let $C \subset Y$, $A := g^{-1}(C) \in \mathcal{A}$. By Theorem 3.18 take $B \subset g[A]$ with $B \in \mathcal{B}$, $Q(B) \geq P(A)$. Then $B \subset C$, so $Q(B) = P(g^{-1}(B)) \leq P(g^{-1}(C)) =$

$P(A)$, and $Q(B) = P(A)$. Thus the inner measure $Q_*(C) = P \circ g^{-1}(C)$. Likewise, $Q_*(Y \setminus C) = (P \circ g^{-1})(Y \setminus C)$, so $Q^*(C) = (P \circ g^{-1})(C)$ and C is Q-completion measurable. $\qquad \square$

3.5 Almost Surely Convergent Realizations

First recall a theorem of Skorohod (RAP, Theorem 11.7.2): if (S, d) is a complete separable metric space, and P_n are laws on S converging to a law P_0, then on some probability space there exist S-valued measurable functions X_n such that $\mathcal{L}(X_n) = P_n$ for all n and $X_n \to X_0$ almost surely. This section will prove an extension of Skorohod's theorem to our current setup.

Having almost uniformly convergent realizations shows that the definition of convergence in law for random elements is reasonable and will be useful in some later proofs on convergence in law.

Suppose $f_n \Rightarrow f_0$ where f_n are random elements, in other words functions not necessarily measurable, except for $n = 0$, defined on some probability spaces (Ω_n, Q_n) into a possibly nonseparable metric space S. We want to find random elements Y_n "having the same laws" as f_n for each n such that $Y_n \to Y_0$ almost surely or better, almost uniformly. At first look it is not clear what "having the same laws" should mean for random elements f_n, $n \geq 1$, not having laws defined on any nontrivial σ-algebra . A way that turns out to work is to define $Y_n = f_n \circ g_n$ where g_n are functions from some other probability space Ω with probability measure Q into Ω_n such that each g_n is measurable and $Q \circ g_n^{-1} = Q_n$ for each n. Thus the argument of f_n will have the same law Q_n as before. It turns out moreover that the g_n should be not only measurable but perfect.

Before stating the theorem, here is an example to show that there may really be no way to define a σ-algebra on S on which laws could be defined and yield an equivalence as in the next theorem, even if S is a finite set.

Example. Let $(X_n, \mathcal{A}_n, Q_n) = ([0, 1], \mathcal{B}, \lambda)$ for all n (λ = Lebesgue measure, $\mathcal{B}$ = Borel σ-algebra). Take sets $C(n) \subset [0, 1]$ with $0 = \lambda_*(C(n)) < \lambda^*(C(n)) = 1/n^2$ (RAP, Theorem 3.4.4). Let S be the two-point space $\{0, 1\}$ with usual metric. Then $f_n := 1_{C(n)} \to 0$ in law and almost uniformly, but each "law" $\beta_n := Q_n \circ f_n^{-1}$ is only defined on the trivial σ-algebra $\{\emptyset, S\}$. The only larger σ-algebra on S is 2^S, but no β_n for $n \geq 1$ is defined on 2^S.

Theorem 3.24 *Let (S, d) be any metric space, $(X_n, \mathcal{A}_n, Q_n)$ any probability spaces, and f_n a function from X_n into S for each $n = 0, 1, \ldots$. Suppose f_0 has separable range S_0 and is measurable (for the Borel σ-algebra on S_0). Then $f_n \Rightarrow f_0$ if and only if there exists a probability space $(\Omega, \mathcal{S}, Q)$ and perfect measurable functions g_n from $(\Omega, \mathcal{S})$ to $(X_n, \mathcal{A}_n)$ for each $n = 0, 1, \ldots$, such* *that* $Q \circ g_n^{-1} = Q_n$ *on* $\mathcal{A}_n$ *for each* n *and* $f_n \circ g_n \to f_0 \circ g_0$ *almost uniformly as* $n \to \infty$.

Notes. Proposition 3.21 and the "if and only if" in Theorem 3.24 show that the hypothesis that g_n be perfect cannot just be dropped from the Theorem.

Proof. "If" follows from Proposition 3.13 and Theorem 3.20. "Only if" will be proved very much as in RAP (Theorem 11.7.2).

Let Ω be the Cartesian product $\Pi_{n=0}^{\infty} X_n \times I_n$ where each I_n is a copy of $[0, 1]$. Here g_n will be the natural projection of Ω onto X_n for each n.

Let $P := Q_0 \circ f_0^{-1}$ on the Borel σ-algebra of S, concentrated in the separable subset S_0. A set $B \subset S$ will be called a *continuity set* (RAP, Section 11.1) for P if $P(\partial B) = 0$ where ∂B is the boundary of B. Then,

Lemma 3.25 *For any* $\varepsilon > 0$ *there are disjoint open continuity sets* U_j, $j = 1, \ldots, J$, *for some* $J < \infty$, *where for each* j, *diam* $U_j := \sup\{d(x, y) : x, y \in U_j\} < \varepsilon$, *and with* $\sum_{j=1}^{J} P(U_j) > 1 - \varepsilon$.

Proof. Let $\{x_j\}_{j=1}^{\infty}$ be dense in S_0. Let $B(x, r) := \{y \in S : d(x, y) < r\}$ for $0 < r < \infty$ and $x \in S_0$. Then $B(x_j, r)$ is a continuity set of P for all but at most countably many values of r. Choose r_j with $\varepsilon/3 < r_j < \varepsilon/2$ such that $B(x_j, r_j)$ is a continuity set of P for each j. The continuity sets form an algebra (RAP, Proposition 11.1.4). Let

$$U_j := B(x_j, r_j) \setminus \bigcup_{i < j} \{y : d(x_i, y) \le r_i\}.$$

Then U_j are disjoint open continuity sets of diameters $< \varepsilon$ with $\sum_{j=1}^{\infty} P(U_j) = 1$, so there is a $J < \infty$ with $\sum_{j=1}^{J} P(U_j) > 1 - \varepsilon$.   □

Now to continue the proof of Theorem 3.24, for each $k = 1, 2, \ldots$, by the last Lemma take disjoint open continuity sets $U_{kj} := U(k, j)$ of P for $j = 1, 2, \ldots, J_k := J(k) < \infty$, with $\text{diam}(U_{kj}) < 1/k$, $P(U_{kj}) > 0$, and

$$\sum_{j=1}^{J(k)} P(U_{kj}) > 1 - 2^{-k}. \tag{3.6}$$

For any open set U in S with complement F, let $d(x, F) := \inf\{d(x, y) : y \in F\}$. For $r = 1, 2, \ldots$, let $F_r := \{x : d(x, F) \ge 1/r\}$. Then F_r is closed and $F_r \uparrow U$ as $r \to \infty$. There is a continuous h_r on S with $0 \le h_r \le 1$, $h_r = 1$ on F_r and $h_r = 0$ outside F_{2r}: let $h_r(x) := \min(1, \max(0, 2rd(x, F) - 1))$.

For each j and k, let $F(k, j) := S \setminus U_{kj}$. Take $r := r(k, j)$ large enough so that $P(F(k, j)_r) > (1 - 2^{-k})P(U_{kj})$. Let h_{kj} be the h_r as defined above for such an r and H_{kj} the h_{2r}.

For n large enough, say for $n \geq n_k$, we have

$$\int_* h_{kj}(f_n)dQ_n > (1 - 2^{-k})P(U_{kj}) \quad \text{and} \quad \int^* H_{kj}(f_n)dQ_n < (1 + 2^{-k})P(U_{kj})$$

for all $j = 1, \ldots, J_k$. We may assume $n_1 \leq n_2 \leq \cdots$.

For every $n = 0, 1, \ldots$, let $f_{kjn} := (h_{kj} \circ f_n)_*$ for Q_n, so that by Theorem 3.3, since $h_{kj} \geq 0$, we can assume that $f_{kjn} \geq 0$ everywhere,

$$\int_* h_{kj}(f_n)dQ_n = \int f_{kjn}dQ_n, \quad 0 \leq f_{kjn} \leq h_{kj}(f_n),$$

and f_{kjn} is $\mathcal{A}_n$-measurable. For $n \geq 1$ let $B_{kjn} := \{f_{kjn} > 0\} \in \mathcal{A}_n$. Let $B_{kj0} := f_0^{-1}(U_{kj}) \in \mathcal{A}_0$. For each k and n, the $B_{kjn} \subset f_n^{-1}(U_{kj})$ are disjoint for $j = 1, \ldots, J_k$, and $H_{kj}(f_n) = 1$ on B_{kjn}, so for $n \geq n_k$,

$$(1 - 2^{-k})P(U_{kj}) < Q_n(B_{kjn}) < (1 + 2^{-k})P(U_{kj}), \quad \text{and} \quad Q_0(B_{kj0}) = P(U_{kj}). \tag{3.7}$$

Let $T_n := X_n \times I_n$. Let μ_n be the product law $Q_n \times \lambda$ on T_n where λ is Lebesgue measure on the Borel σ-algebra $\mathcal{B}$ in I_n. For each $k \geq 1$, $n \geq n_k$, and $j = 1, \ldots, J_k$, let

$$C_{kjn} := B_{kjn} \times [0, F(k, j, n)] \subset T_n,$$

$$D_{kjn} := B_{kj0} \times [0, G(k, j, n)] \subset T_0,$$

where F and G are defined so that

$$\mu_n(C_{kjn}) = \mu_0(D_{kjn}) = \min(Q_n(B_{kjn}), Q_0(B_{kj0})).$$

Then for each k, j, and $n \geq n_k$, we have by (3.7), since $1/(1 + 2^{-k}) > 1 - 2^{-k}$,

$$1 - 2^{-k} < \min(F, G)(k, j, n) < \max(F, G)(k, j, n) = 1. \tag{3.8}$$

Let

$$C_{k0n} := T_n \setminus \bigcup_{j=1}^{J(k)} C_{kjn}, \quad D_{k0n} := T_0 \setminus \bigcup_{j=1}^{J(k)} D_{kjn}.$$

For $k = 0$ let $J_0 := J(0) := 0$, $C_{00n} := T_n$, $D_{00n} := T_0$, and $n_0 := 0$.

For each $n = 1, 2, \ldots$, let $k(n)$ be the unique k such that $n_k \leq n < n_{k+1}$. Then for $n \geq 1$, T_n is the disjoint union of sets $W_{nj} := C_{k(n)jn}$, $j = 0, 1, \ldots, J_{k(n)}$. We also have

If $j \geq 1$ and $(v, s) \in W_{nj}$, then $v \in B_{k(n)jn}$ so $f_n(v) \in U_{k(n)j}$. $\tag{3.9}$

Next, for each n, T_0 is the disjoint union of sets $E_{nj} := D_{k(n)jn}$ for $j = 0, 1, \ldots, J_{k(n)}$. Then $\mu_n(W_{nj}) = \mu_0(E_{nj})$ for each n and j, and if $j \geq 1$ or $k(n) = 0$, then $\mu_0(E_{nj}) > 0$.

For x in T_0 and each n,

$$\text{let } j(n, x) \text{ be the } j \text{ such that } x \in E_{nj}. \qquad (3.10)$$

Let $L := \{x \in T_0 : \mu_0(E_{nj(n,x)}) > 0 \text{ for all } n\}$. Then $T_0 \setminus L \subset \bigcup_i E_{n(i)0}$ for some (possibly empty or finite) sequence $n(i)$ such that $\mu_0(E_{n(i)0}) = 0$ for all i. Thus $\mu_0(L) = 1$.

For $x \in L$ and any measurable set $B \subset T_n$, in other words B is in the product σ-algebra $\mathcal{A}_n \otimes \mathcal{B}$, recall that $\mu_n(W_{nj}) = \mu_0(E_{nj})$ and let

$$P_{nj}(B) := \mu_n(B \cap W_{nj})/\mu_0(E_{nj}), \quad P_{nx} := P_{nj(n,x)}. \qquad (3.11)$$

Then P_{nx} is a probability measure on $\mathcal{A}_n \otimes \mathcal{B}$. Let ρ_x be the product measure $\Pi_{n=1}^{\infty} P_{nx}$ on $T := \Pi_{n=1}^{\infty} T_n$ (RAP, Theorem 8.2.2).

Lemma 3.26 *For any measurable set $H \subset T$ (for the infinite product σ-algebra with $\mathcal{A}_n \otimes \mathcal{B}$ on each factor), $x \mapsto \rho_x(H)$ is measurable on $(T_0, \mathcal{A}_0 \otimes \mathcal{B})$.*

Proof. Let $\mathcal{H}$ be the collection of all H for which the assertion holds. Given n, P_{nx} is one of finitely many laws, each obtained for x in a measurable subset E_{nj}. Thus if Y_n is the natural projection of T onto T_n and $H = Y_n^{-1}(B)$ for some $B \in \mathcal{A}_n \otimes \mathcal{B}$ then $H \in \mathcal{H}$.

If $H = \bigcap_{i \in M} Y_{m(i)}^{-1}(B_i)$, where $B_i \in \mathcal{A}_{m(i)} \otimes \mathcal{B}$ and M is finite, we may assume the $m(i)$ are distinct. Then

$$\rho_x(H) = \Pi_{i \in M} \rho_x(Y_{m(i)}^{-1}(B_i)),$$

so $H \in \mathcal{H}$. Then, any finite, disjoint union of such intersections is in $\mathcal{H}$. Such unions form an algebra. If $H_n \in \mathcal{H}$ and $H_n \uparrow H$ or $H_n \downarrow H$, then $H \in \mathcal{H}$. As the smallest monotone class containing an algebra is a σ-algebra (RAP, Theorem 4.4.2), the Lemma follows. $\qquad \square$

Now returning to the proof of Theorem 3.24, $\Omega = T_0 \times T$. For any product measurable set $C \subset \Omega$ and $x \in T_0$, let $C_x := \{y \in T : (x, y) \in C\}$, and $Q(C) := \int \rho_x(C_x) d\mu_0(x)$. Here $x \mapsto \rho_x(C_x)$ is measurable if C is a finite union of products $A_i \times F_i$ where $A_i \in \mathcal{A}_0 \otimes \mathcal{B}$ and F_i is product measurable in T. Such a union equals a disjoint union (RAP, Proposition 3.2.2). Thus by monotone classes again, $x \mapsto \rho_x(C_x)$ is measurable on T_0 for any product measurable set $C \subset \Omega$. Thus Q is defined. It is then clearly a countably additive probability measure.

Let p be the natural projection of T_n onto X_n. Recall that $P_{nx} = P_{nj}$ for all $x \in E_{nj}$, by (3.10) and (3.11). The marginal of Q on X_n, in other words

$Q \circ g_n^{-1}$, is by (3.11) again

$$\sum_{j=0}^{J(k(n))} \mu_0(E_{nj})P_{nj} \circ p^{-1} \; = \; \mu_n \circ p^{-1} \; = \; Q_n.$$

Thus Q has marginal Q_n on X_n for each n as desired.

By (3.6), $\sum_{k=1}^{\infty} Q_0(X_0 \setminus \bigcup_{j=1}^{J(k)} f_0^{-1}(U_{kj})) < \sum_k 2^{-k} < \infty$. So Q_0-almost every $y \in X_0$ belongs to $\bigcup_{j=1}^{J(k)} f_0^{-1}(U_{kj})$ for all large enough k. Also if $t \in I_0$ and $t < 1$, then by (3.8), $t < G(k, j, n)$ for all $j \geq 1$ as soon as $1 - 2^{-k} > t$ and $n \geq n_k$. Thus for μ_0-almost all (y, t), there is an m such that $(y, t) \in \bigcup_{j=1}^{J(k(n))} E_{nj}$ for all $n \geq m$. If $x := (y, t) \in E_{nj}$ for $j \geq 1$, then $y \in B_{k(n)j0}$, so $f_0(y) \in U_{k(n)j}$. Also, by (3.11), $P_{nx} = P_{nj}$ is concentrated in W_{nj}. For $(v, s) \in W_{nj}$, $f_n(v) \in U_{k(n)j}$ by (3.9). Since $\text{diam}(U_{kj}) < 1/k$ for each $j \geq 1$,

$$Q^*(d(f_n(g_n), f_0(g_0)) > 1/k(n) \text{ for some } n \geq m)$$

$$\leq \mu_0(\{(y, t) \in E_{n0} \text{ for some } n \geq m\}) \to 0$$

as $m \to \infty$, so $f_n(g_n) \to f_0(g_0)$ almost uniformly.

Lastly, it will be shown that the g_n are perfect. Suppose $Q(A) > 0$ for some A. Now $Q(A) = \int \rho_x(A_x)d\mu_0(x)$. First let $n \geq 1$. Then for some x, $\rho_x(A_x) > 0$. If $\mu_0(E_{n0}) = 0$, we take $x \notin E_{n0}$. Now $T = T_n \times T^{(n)}$ where $T^{(n)} := \Pi_{1 \leq i \neq n}T_i$. Then, on T, $\rho_x = P_{nx} \times Q_{nx}$ for a law $Q_{nx} = \Pi_{m \neq n}P_{mx}$ on $T^{(n)}$. Let $A(x) := A_x$. By the Tonelli–Fubini theorem,

$$\rho_x(A_x) \; = \; \int \int 1_{A(x)}(u, v)dP_{nx}(u)dQ_{nx}(v).$$

Thus for some v, $\int 1_{A(x)}(u, v)dP_{nx}(u) > 0$. Choose and fix such a v as well as x. Now $P_{nx} = P_{nj}$ for $j = j(n, x)$ with $\mu_0(E_{nj}) > 0$. Let $u = (s, t)$, $s \in X_n$, and $t \in I_n$. Then since $P_{nj} = Q_n \times \lambda$ restricted to a set of positive measure and normalized,

$$0 \; < \; \int \int 1_{A(x)}(s, t, v)dQ_n(s)dt.$$

Choose and fix a t with

$$0 \; < \; \int 1_{A(x)}(s, t, v)dQ_n(s).$$

Let $C := \{s \in X_n : (s, t, v) \in A_x\}$. Then $Q_n(C) > 0$. Clearly $C \subset g_n[A]$, so g_n is perfect for $n \geq 1$ by Theorem 3.18.

To show g_0 is perfect, we have $\mu_0 = Q_0 \times \lambda$, and $\rho_x(A_x) > 0$ for $x = (y, t)$ in a set X with $\mu_0(X) > 0$. There is a $t \in I_0$ such that $Q_0(C) > 0$ where $C := \{y : (y, t) \in X\}$. Then $C \subset g_0[A]$, so g_0 is perfect, finishing the proof of Theorem 3.24. $\qquad \square$

3.6 Conditions Equivalent to Convergence in Law

Conditions equivalent to convergence of laws on separable metric spaces are given in the portmanteau theorem (RAP, Theorem 11.1.1) and metrization theorem (RAP, Theorem 11.3.3). Here, the conditions will be extended to general random elements for the theory being developed in this chapter.

For any probability space $(\Omega, \mathcal{A}, Q)$ and real-valued function f on Ω let $E^*f := \int^* f\, dP$, $E_* f := \int_* f\, dP$. If (S, d) is a metric space and f is a real-valued function on S, recall (RAP, Section 11.2) that the Lipschitz seminorm of f is defined by

$$\|f\|_L := \sup\{|f(x) - f(y)|/d(x, y) : x \neq y\},$$

and f is called a *Lipschitz* function if $\|f\|_L < \infty$. The bounded Lipschitz norm is defined by $\|f\|_{BL} := \|f\|_L + \|f\|_{\sup}$ where $\|f\|\sup := \sup_x |f(x)|$. Then f is called a bounded Lipschitz function if $\|f\|_{BL} < \infty$, and $\|\cdot\|_{BL}$ is a norm on the space of all such functions.

The extended portmanteau theorem about to be proved is an adaptation of RAP, Theorem 11.1.1 and some further facts based on the last section (Theorem 3.24). The proof to be given includes relatively easy implications, some of which consist of putting in stars at appropriate places in the proofs in RAP.

Theorem 3.27 *For any metric space (S, d) and $n = 0, 1, \ldots$, let $(X_n, \mathcal{A}_n, Q_n)$ be a probability space and f_n a function from X_n into S. Suppose f_0 has separable range S_0 and is measurable. Let $P := Q_0 \circ f_0^{-1}$ on S. Then the following are equivalent:*

(a) $f_n \Rightarrow f_0$;

(a') $\limsup_{n \to \infty} E^* G(f_n) \leq E G(f_0)$ for all bounded continuous real-valued G on S;

(b) $E^* G(f_n) \to E G(f_0)$ as $n \to \infty$ for every bounded Lipschitz function G on S;

(b') (a') holds for all bounded Lipschitz G on S;

(c) $\sup\{|E^* G(f_n) - E G(f_0)| : \|G\|_{BL} \leq 1\} \to 0$ as $n \to \infty$;

(d) For any closed $F \subset S$, $P(F) \geq \limsup_{n \to \infty} Q_n^*(f_n \in F)$;

(e) For any open $U \subset S$, $P(U) \leq \liminf_{n \to \infty}(Q_n)_*(f_n \in U)$;

(f) For any continuity set A of P in S, $Q_n^*(f_n \in A) \to P(A)$ and $(Q_n)_*(f_n \in A) \to P(A)$ as $n \to \infty$;

(g) There exist a probability space $(\Omega, \mathcal{S}, Q)$ and measurable functions g_n from Ω into X_n and h_n from Ω into S such that the g_n are perfect, $Q \circ g_n^{-1} = Q_n$ and $Q \circ h_n^{-1} = P$ for all n, and $d(f_n \circ g_n, h_n) \to 0$ almost uniformly.

Moreover, (g) remains equivalent if any of the following changes are made in it: "almost uniformly" can be replaced by "in outer probability"; we can take $h_n = f_0 \circ \gamma_n$ for some measurable functions γ_n from Ω into X_0, which can be assumed to be perfect; and we can take γ_n to be all the same, $\gamma_n \equiv \gamma_1$ for all n.

Proof. Clearly (a) implies (a′). Conversely, interchanging G and $-G$, (a′) implies

$$\liminf_{n \to \infty} E^* G(f_n) \geq \liminf_{n \to \infty} E_* G(f_n) \geq EG(f_0),$$

and (a) follows, so (a) and (a′) are equivalent.

Clearly (a) implies (b), which is equivalent to (b′) just as (a) is to (a′). To show (b) implies (c), let T be the completion of S. Then all the f_n take values in T. Each bounded Lipschitz function G on S extends uniquely to such a function on T, and the functions $G \circ f_n$ on X_n are exactly the same. So we can assume in this step that S and S_0 are complete.

Let $\varepsilon > 0$. By Ulam's theorem (RAP, Theorem 7.1.4), take a compact $K \subset S_0$ with $P(K) = Q_0(f_0 \in K) > 1 - \varepsilon$. Recall that $d(x, K) := \inf\{d(x, y) : y \in K\}$ and $K^\varepsilon := \{x : d(x, K) < \varepsilon\}$. Let $g(x) := \max(0, 1 - d(x, K)/\varepsilon)$ for $x \in S$. Then g is a bounded Lipschitz function with $\|g\|_{BL} \leq 1 + 1/\varepsilon < \infty$. Clearly $1_K \leq g \leq 1_{K^\varepsilon}$. Since $E_* g(f_n) \to Eg(f_0)$ as $n \to \infty$, we have for n large enough

$$(Q_n)_*(f_n \in K^\varepsilon) \geq E_* g(f_n) > Eg(f_0) - \varepsilon > 1 - 2\varepsilon. \tag{3.12}$$

Let B be the set of all G on S with $\|G\|_{BL} \leq 1$. Then the functions in B are uniformly equicontinuous, so the set of restrictions of functions in B to K is totally bounded for the supremum distance over K by the Arzelà–Ascoli theorem (RAP, Theorem 2.4.7). Let $G_1, \ldots, G_J$ for some finite J be functions in B such that for each $G \in B$, $\sup_{x \in K} |(G - G_j)(x)| < \varepsilon$ for some $j = 1, \ldots, J$. Next, for any $G \in B$, choose such a j. Then

$$|E^* G(f_n) - EG(f_0)| \leq |E^* G(f_n) - E^* G_j(f_n)|$$
$$+ |E^*(G_j(f_n)) - E(G_j(f_0))| + |E(G_j(f_0) - G(f_0))|,$$

$$\tag{3.13}$$

a sum of three terms. For the last term, splitting the integral into two parts according as $f_0 \in K$ or not, the first part is bounded by ε and the second by 2ε since $|G_j - G| \leq 2$ everywhere and $Q_0(f_0 \notin K) < \varepsilon$.

The middle term on the right side of (3.13) is bounded above by

$$\max_{j \leq J} |E^* G_j(f_n) - EG_j(f_0)|,$$

which is less than ε for n large enough by (b).

The first term on the right in (3.13) is a sum of two parts, one over a measurable subset A_n of X_n where $f_n \in K^\varepsilon$ with $Q_n(A_n) > 1 - 2\varepsilon$ by (3.12), and the other over the complement of A_n. Since G, $G_j \in B$ and $|G - G_j| < \varepsilon$ on K, we have $|G - G_j| < 3\varepsilon$ on K^ε. It follows by Lemma 3.4(a) that on A_n, $(G \circ f_n)^* (\leq G_j \circ f_n)^* + 3\varepsilon$, and likewise $(G_j \circ f_n)^* (\leq G \circ f_n)^* + 3\varepsilon$, so $|(G \circ f_n)^* - (G_j \circ f_n)^*| \leq 3\varepsilon$ on A_n, and the first part is bounded above by 3ε. The second part is bounded by $2(1 - Q_n(A_n)) < 4\varepsilon$. So for n large enough the left side of (3.13) is less than 11ε uniformly for $G \in B$ and (c) follows.

Clearly (c) implies (b), so (b) and (c) are equivalent. Next, (b) implies (d) by taking bounded Lipschitz functions g_k decreasing down to 1_F as $k \to \infty$, specifically the functions g in the proof of (b) implies (c) with F in place of K and $\varepsilon = 1/k$.

Next, (d) and (e) are equivalent by taking complements; (d) and (e) together easily imply (f).

For any bounded continuous function G, all but countably many of the sets $\{G < t\}$ are continuity sets of P. It follows that G can be approximated uniformly within ε by a simple function $\sum t_i 1_{A_i}$ where A_i are continuity sets $\{t_{i-1} \leq G < t_i\}$. So (f) implies (a), and (a) through (f) are all equivalent.

Theorem 3.24 says that (a) implies (g), in the strongest form where $h_n \equiv h_1 \equiv f_0 \circ \gamma_1$ and γ_1 is perfect.

The weakest form of (g), with convergence in outer probability and h_n depending on n and not necessarily a composition of f_0 with any (perfect) function, will be shown to imply (b'). Given $\varepsilon > 0$, take n large enough so that

$$Q_n^* \{ d(f_n \circ g_n, h_n) > \varepsilon \} < \varepsilon.$$

So on a measurable set A_n with $Q(A_n) > 1 - \varepsilon$, we have $d(f_n \circ g_n, h_n) \leq \varepsilon$, and so for any $G \in B$ (in other words $\|G\|_{BL} \leq 1$) we have

$$|G(f_n(g_n)) - G(h_n)| \leq \varepsilon 1_{A_n} + 2 \cdot 1_{A_n^c},$$

so

$$E^* G(f_n(g_n)) \leq EG(h_n) + 3\varepsilon = EG(f_0) + 3\varepsilon$$

(since h_n has the same distribution as f_0) and

$$E^* G(f_n(g_n)) = E(((G \circ f_n) \circ g_n)^*) = E((G \circ f_n)^* \circ g_n)$$

by Theorem 3.3 and since g_n is perfect, and where all the expectations are with respect to Q. The latter integral by the image measure theorem (RAP, Theorem 4.1.11) equals $E^*(G(f_n))$ (for Q_n). Letting $\varepsilon \downarrow 0$, (b') follows, so all of the forms of (g) are equivalent to any of (a) through (f). $\square$

Theorem 1.7, in a special case, proved a form of convergence which in that case is now easily seen to imply the conditions in Theorem 3.27: the one in (c), for example.

It will be shown that convergence in law is also equivalent to convergence in some analogues of the Prohorov and dual-bounded-Lipschitz metrics which metrize convergence of laws on separable metric spaces as shown in RAP, Theorem 11.3.3. We will have analogues of metrics, rather than actual metrics, because of the non-symmetry between the nonmeasurable random elements f_n and limiting measurable random variable f_0.

Definitions. Let $(X_m, \mathcal{A}_m, Q_m)$ be probability spaces, $m = 0, 1$, and (S, d) a metric space. Let f_m be functions from X_m into S, $m = 0, 1$, such that f_0 is measurable and has separable range. Let $P := Q_0 \circ f_0^{-1}$. Then let

$$\beta(f_1, f_0) := \sup\{|E^* G(f_1) - E G(f_0)| : \|G\|_{BL} \le 1\}.$$

Define the extended Prohorov distance $\rho(f_1, f_0)$ as the infimum of all $\varepsilon > 0$ such that $P(F) \le (Q_1)_*(f_1 \in F^\varepsilon) + \varepsilon$ for every nonempty closed set $F \subset S$.

In the following theorem, $\beta(f_m, f_0)$ and $\rho(f_m, f_0)$ are defined for any $m = 1, 2, \ldots$, just as for $m = 1$.

Theorem 3.28 *For any metric space (S, d), probability spaces $(X_m, \mathcal{A}_m, Q_m)$, $m = 0, 1, 2, \ldots$, and functions f_m from X_m into S, where f_0 has separable range and is measurable, the following are equivalent:*

(i) $f_m \Rightarrow f_0$;

(ii) $\beta(f_m, f_0) \to 0$ as $m \to \infty$;

(iii) $\rho(f_m, f_0) \to 0$ as $m \to \infty$.

Proof. (i) and (ii) are equivalent as (a) and (c) in Theorem 3.27. To show that (ii) implies (iii), we have:

Lemma 3.29 *For any f_1 and f_0 as in the statement of Theorem 3.28,*

$$\rho(f_1, f_0) \le \min(1, (2\beta(f_1, f_0))^{1/2}).$$

Proof. Let F be any closed set in S and $P = Q_0 \circ f_0^{-1}$. Given $0 < \varepsilon \le 1$, let $g(x) := \max(0, 1 - d(x, F)/\varepsilon)$ as before. Then $\|g\|_{BL} \le 1 + 1/\varepsilon$, and $(Q_1)_*(f_1 \in F^\varepsilon) \ge \int_* g(f_1) dQ_1 \ge \int g(f_0) dQ_0 - \beta(f_1, f_0)(1 + \frac{1}{\varepsilon}) \ge P(F) - \varepsilon$ if $\beta(f_1, f_0) \le \varepsilon^2/2$. Clearly $\rho(f_1, f_0) \le 1$ in all cases, so for $\varepsilon := (2\beta(f_1, f_0))^{1/2}$ when that is < 1, the Lemma follows. $\square$

Now to continue the proof of Theorem 3.28, applying the Lemma to f_m in place of f_1 for each m shows that (ii) implies (iii). To show that (iii) implies (i), let U be an open set in S. For $k = 1, 2, \ldots$, let $F_k := \{x : d(x, y) < 1/k \text{ implies } y \in U\}$. Then F_k are closed sets and $F_k \uparrow U$ as $k \to \infty$. For n large enough, $\rho(f_n, f_0) < 1/k$ and then $P(F_k) \le (Q_n)_*(f_n \in F_k^{1/k}) + \frac{1}{k} \le$

$(Q_n)_*(f_n \in U) + \frac{1}{k}$, so

$$P(F_k) \le \liminf_{n \to \infty}(Q_n)_*(f_n \in U) + \frac{1}{k}.$$

Then letting $k \to \infty$ gives

$$P(U) \le \liminf_{n \to \infty}(Q_n)_*(f_n \in U).$$

This is (e) of Theorem 3.27, so the proof of Theorem 3.28 is complete.   □

Next is a version of Theorem 3.28 where we have two indices.

Theorem 3.30 *Let (S, d) be a metric space and $(\Omega, \mathcal{S}, Q)$ a probability space. Suppose that for each $m, n = 1, 2, \ldots$, f_{mn} is a function from Ω into S, and f_0 is a measurable function from Ω into S with separable range. Then the following are equivalent:*

(i) $f_{m,n} \Rightarrow f_0$, i.e. for every bounded continuous real function G on S,

$$\int^* G(f_{m,n})dQ \to \int G(f_0)dQ \quad as \ m, n \to \infty;$$

(ii) $\beta(f_{mn}, f_0) \to 0$ as $m, n \to \infty$;

(iii) $\rho(f_{mn}, f_0) \to 0$ as $m, n \to \infty$.

Proof. For a double sequence a_{mn} of real numbers, let

$$\limsup_{m,n \to \infty} a_{mn} := \inf_k \sup_{m \ge k, n \ge k} a_{mn}, \quad \liminf_{m,n \to \infty} a_{mn} := \sup_k \inf_{m \ge k, n \ge k} a_{mn}.$$

Then the steps in the proof of Theorem 3.27 showing that (a) through (f) are equivalent extend directly to convergence as $m \to \infty$ and $n \to \infty$, if $\limsup_{n \to \infty}$ is replaced by $\limsup_{m,n \to \infty}$ (not by $\limsup_{m \to \infty} \limsup_{n \to \infty}$ or the iterated lim sup in the reverse order, which may be different), and likewise for lim inf. Thus, the proof of Theorem 3.28 also extends.   □

We have the following;

Theorem 3.31 (Continuous mapping theorem) *Under the conditions of Theorem 3.28, if (T, e) is another metric space, G is a continuous function from S into T, and $f_m \Rightarrow f_0$, then $G(f_m) \Rightarrow G(f_0)$.*

Proof. This follows directly from the definition of convergence in law $\Rightarrow$.   □

The next fact is straightforward:

Proposition 3.32 *Suppose f_m, $m \ge 0$, are measurable random variables taking values in a separable metric space S, so that laws $\mathcal{L}(f_m)$ exist on the Borel σ-algebra of S. Then convergence $f_m \Rightarrow f_0$ is equivalent to convergence of the laws $\mathcal{L}(f_m) \to \mathcal{L}(f_0)$ in the usual sense.*

3.7 Asymptotic Equicontinuity

Recall from Section 3.1 the definitions of the empirical measures P_n, empirical process ν_n, pseudometric ρ_P, and the Gaussian process G_P. Recall that a set $\mathcal{F} \subset \mathcal{L}^2(P)$ is called *pregaussian* if a G_P process restricted to $\mathcal{F}$ exists whose sample functions $f \mapsto G_P(f)(\omega)$ are (almost) all bounded and uniformly continuous for ρ_P on $\mathcal{F}$. Such a G_P process is called *coherent* if, in addition, its sample functions are prelinear on $\mathcal{F}$ as in Lemma 2.30. By Theorem 3.2 we can and will assume that on a pregaussian $\mathcal{F}$, G_P is coherent.

Proposition 3.33 *For any probability measure P and pregaussian set $\mathcal{F} \subset \mathcal{L}^2(P)$, the symmetric convex hull $sco(\mathcal{F})$ of $\mathcal{F}$ is also pregaussian and there exists a G_P process defined on the linear span of $\mathcal{F}$ and constant functions which is 0 on the constant functions and coherent on $sco(\mathcal{F})$.*

Proof. Apply Theorems 3.1 and 2.32. □

For a signed measure ν and measurable function f such that $\int f \, d\nu$ is defined, let $\nu(f) := \int f \, d\nu$.

A class $\mathcal{F}$ of functions will be said to satisfy the *asymptotic equicontinuity condition* for P and a pseudometric τ on $\mathcal{F}$, or $\mathcal{F} \in AEC(P, \tau)$ for short, if for every $\varepsilon > 0$ there is a $\delta > 0$ and an n_0 large enough such that for $n \geq n_0$,

$$\Pr^*\{\sup\{|\nu_n(f - g)| : f, g \in \mathcal{F}, \tau(f, g) < \delta\} > \varepsilon\} < \varepsilon. \tag{3.14}$$

Then $\mathcal{F} \in AEC(P)$ will mean $\mathcal{F} \in AEC(P, \rho_P)$.

Theorem 3.34 *Let $\mathcal{F} \subset \mathcal{L}^2(X, \mathcal{A}, P)$. Then the following are equivalent:*

(I) $\mathcal{F}$ is a Donsker class for P, in other words $\mathcal{F}$ is P-pregaussian and $\nu_n \Rightarrow G_P$ in $\ell^\infty(\mathcal{F})$;

(II) (a) $\mathcal{F}$ is totally bounded for ρ_P and (b) $\mathcal{F}$ satisfies the asymptotic equicontinuity condition for P, $\mathcal{F} \in AEC(P)$;

(III) There is a pseudometric τ on $\mathcal{F}$ such that $\mathcal{F}$ is totally bounded for τ and $\mathcal{F} \in AEC(P, \tau)$.

Proof. (I) implies (II): if $\mathcal{F}$ is a Donsker class and $\varepsilon > 0$, then since $\mathcal{F}$ is pregaussian, it is totally bounded for ρ_P by Theorem 3.1 and the definition of GC-sets, so (a) holds. Take $\delta > 0$ small enough so that for any coherent G_P process,

$$\Pr\{\sup\{|G_P(f) - G_P(g)| : \rho_P(f, g) < \delta\} > \varepsilon/3\} < \varepsilon/2.$$

By almost uniformly convergent realizations (Theorem 3.24) defined on some probability space, for each $n \geq n_0$ large enough we can assume $\Pr^*\{\|\nu_n - G_P\|_{\mathcal{F}} \geq \varepsilon/3\} < \varepsilon/2$. If $\|\nu_n - G_P\|_{\mathcal{F}} < \varepsilon/3$ and $|G_P(f) - G_P(g)| \leq \varepsilon/3$, then $|\nu_n(f) - \nu_n(g)| < \varepsilon$. So the asymptotic equicontinuity holds with the δ

and n_0 chosen, and (b) holds. (More precisely, it holds on the original probability space because the functions g_n in Theorem 3.24 are perfect.) Thus (I) implies (II). (II) implies (III) directly, with $\tau = \rho_P$.

To show (III) implies (I), suppose $\mathcal{F}$ is τ-totally bounded and belongs to $AEC(P, \tau)$. Let UC $:=$ UC$(\mathcal{F})$ denote the set of all real-valued functions on $\mathcal{F}$ uniformly continuous for τ. Then UC is a separable subspace of $\ell^\infty(\mathcal{F})$ for $\| \cdot \|_{\mathcal{F}}$ since $\mathcal{F}$ is totally bounded, and we in effect have the space of continuous functions on the compact completion (RAP, Corollary 11.2.5).

For any finite subset $\mathcal{G}$ of $\mathcal{F}$, by the finite-dimensional central limit theorem (RAP, Theorem 9.5.6) we can let $n \to \infty$ in (3.14) and get

$$\Pr^*\{\sup\{|G_P(f) - G_P(g)| : f, g \in \mathcal{G}, \tau(f, g) < \delta\} > \varepsilon\} \le \varepsilon.$$

Letting $\mathcal{G}$ increase up to a countable τ-dense set $\mathcal{H} \subset \mathcal{F}$, we get by monotone convergence

$$\Pr^*\{\sup\{|G_P(f) - G_P(g)| : f, g \in \mathcal{H}, \tau(f, g) < \delta\} > \varepsilon\} \le \varepsilon.$$

We can let $\varepsilon \downarrow 0$ and $\delta = \delta(\varepsilon) \downarrow 0$ and conclude that G_P is almost surely uniformly continuous for τ on $\mathcal{H}$. For each $f \in \mathcal{F}$, considering also the countable set $\{f\} \cup \mathcal{H}$, on the set of probability 1 where G_P is τ-uniformly continuous, the limit $G_P^*(f) = \lim\{G_P(h) : h \in \mathcal{H}, \tau(h, f) \to 0\}$ exists, and equals $G_P(f)$ almost surely. It follows that we can take G_P to be uniformly continuous for τ on all of $\mathcal{F}$.

Recall (Section 3.1) that $\pi_0(f) = f - \int f \, dP$ for any $f \in \mathcal{L}^2(P)$. We have $\nu_n(f) \equiv \nu_n(\pi_0(f))$ and a.s. $G_P(f) = G_P(\pi_0(f))$. For the pseudometric τ, if $\tau(f, g) = 0$ for some $f, g \in \mathcal{F}$, then $\pi_0(f) = \pi_0(g)$; in other words $f - g$ must be a constant. It suffices to consider the family $\pi_0(\mathcal{F}) := \{\pi_0(f) : f \in \mathcal{F}\}$ in place of $\mathcal{F}$. On $\pi_0(\mathcal{F})$, ρ is a metric. So by Theorem 2.57, since G_P is isonormal for $(\cdot, \cdot)_{0,P}$ it follows that $\mathcal{F}$ is pregaussian. So G_P has a law μ_3 defined on the Borel sets of the separable Banach space UC, in view of Theorems 3.1 and 2.32 (a) implies (h).

Given $\varepsilon > 0$, take $\delta > 0$ from (3.14) and a finite set $\mathcal{G} \subset \mathcal{F}$ such that for each $f \in \mathcal{F}$ there is a $g \in \mathcal{G}$ such that $\tau(f, g) < \delta$. Then $\mathbb{R}^{\mathcal{G}}$ is the set of all real-valued functions on $\mathcal{G}$. Let μ_2 be the law of G_P on $\mathbb{R}^{\mathcal{G}}$ and let μ_{23} be the law on $\mathbb{R}^{\mathcal{G}} \times$ UC where G_P on $\mathcal{G}$ in $\mathbb{R}^{\mathcal{G}}$ is just the restriction of G_P on UC. So μ_{23} has marginals μ_2 and μ_3.

Let $\mu_{1,n}$ be the law of ν_n on $\mathcal{G}$, so $\mu_{1,n}$ is also defined on $\mathbb{R}^{\mathcal{G}}$. Then by the finite-dimensional central limit theorem again, the laws $\mu_{1,n}$ converge to μ_2 on $\mathbb{R}^{\mathcal{G}}$. So for the Prohorov metric ρ, since it metrizes convergence of laws (RAP, Theorem 11.3.3), $\rho(\mu_{1,n}, \mu_2) < \varepsilon$ for n large enough. Take $n \ge n_0$ also, then fix n. By Strassen's theorem (RAP, Corollary 11.6.4), there is a law μ_{12} on $\mathbb{R}^{\mathcal{G}} \times \mathbb{R}^{\mathcal{G}}$ with marginals $\mu_{1,n}$ and μ_2 such that $\mu_{1,2}\{(x, y) : |x - y| > \varepsilon\} \le \varepsilon$. By the Vorob'ev–Berkes–Philipp theorem (1.31), there is a Borel measure

μ_{123} on $\mathbb{R}^\mathcal{G} \times \mathbb{R}^\mathcal{G} \times$ UC having marginals μ_{12} and μ_{23} on the appropriate spaces.

The next step is to link up ν_n with its restriction to the finite set $\mathcal{G}$. The Vorob'ev–Berkes–Philipp theorem may not apply here since ν_n on $\mathcal{F}$ may not be in a Polish space, at least not one that seems apparent. (About nonmeasurability on nonseparable spaces see the remarks at the end of Section 1.1.) Here we can use instead:

Lemma 3.35 *Let S and T be Polish spaces and* $(\Omega, \mathcal{A}, P)$ *a probability space. Let Q be a law on* $S \times T$ *with marginal q on S. Let V be a random variable on* Ω *with values in S and law* $\mathcal{L}(V) = q$. *Suppose there is a real random variable U on* Ω *independent of V with continuous distribution function* F_U. *Then there is a random variable* $W : \Omega \mapsto T$ *such that the joint law* $\mathcal{L}(V, W)$ *of* (V, W) *is Q.*

Proof. Every Polish space is Borel-isomorphic to some compact subset of $[0, 1]$, either the whole interval, a finite set, or a convergent sequence and its limit (RAP, Theorem 13.1.1). Since the lemma involves only measurability and not topological properties of the Polish spaces we can assume $S = T = [0, 1]$. Recall that for any real-valued random variable X with continuous distribution function F, $F(X)$ has a uniform distribution in $[0, 1]$: Proposition 1.22.

So taking $F_U(U)$, we can assume U is uniformly distributed in $[0, 1]$. By way of regular conditional probabilities (RAP, Section 10.2; Bauer, 1981) we can write $Q = \int Q_x dq(x)$ where for each x, Q_x is a probability measure on T, so that for any measurable set A in (the square) $S \times T$, $Q(A) = \int \int 1_A(x, y) dQ_x(y) dq(x)$ (RAP, Theorems 10.2.1 and 10.2.2). Let F_x be the distribution function of Q_x and

$$F_x^{-1}(t) := \inf\{u : F_x(u) \ge t\}, \quad 0 < t < 1.$$

Then for any real z and $0 < t < 1$, $F_x^{-1}(t) \le z$ if and only if $F_x(z) \ge t$. Now $x \mapsto F_x(z)$ is measurable for any fixed z. It follows that $x \mapsto F_x^{-1}(t)$ is measurable for each t, $0 < t < 1$. For each x, F_x^{-1} is left continuous and nondecreasing in t. It follows that $(x, t) \mapsto F_x^{-1}(t)$ is jointly measurable. Thus for $W(\omega) := F_{V(\omega)}^{-1}(U(\omega))$, $\omega \mapsto W(\omega)$ is measurable. For each x we have the image measure $\lambda \circ (F_x^{-1})^{-1} = Q_x$ (RAP, Proposition 9.1.2). So for any bounded Borel function g,

$$\int g\, dQ = \int_0^1 \int_0^1 g(x, y) dQ_x(y) dq(x) \quad \text{by Theorem 10.2.1 of RAP}$$

$$= \int_0^1 \int_0^1 g(x, F_x^{-1}(y)) dy\, dq(x) \quad \text{by the image measure theorem}$$

$$= \int_0^1 \int_0^1 g(x, F_x^{-1}(y)) d(q \times \lambda)(x, y) \quad \text{(Tonelli–Fubini theorem)}$$

$$= E(g(V, F_V^{-1}(U))) = Eg(V, W),$$

since U is independent of V and $\mathcal{L}(V) = q$ and by the image measure theorem again. So $\mathcal{L}(V, W) = Q$, proving Lemma 3.35. $\qquad\square$

Now given $\varepsilon > 0$, let $(\Omega, \mathcal{S}, Q)$ be a probability space on which all the empirical processes ν_n and an independent U are defined, specifically a countable product of copies of the probability space $(X, \mathcal{A}, P)$ times one copy of $[0, 1]$ with Lebesgue measure λ. Then Lemma 3.35 applies to Ω with $S = \mathbb{R}^{\mathcal{G}}$ and $T = \mathbb{R}^{\mathcal{G}} \times \mathrm{UC}$, where V is ν_n restricted to $\mathcal{G}$, and $Q = \mu_{123}$ on $S \times T$. On Ω we then have processes ν_n and G_P defined on $\mathcal{F}$, which by construction and the τ-asymptotic equicontinuity condition (3.14) are within 3ε of each other uniformly on $\mathcal{F}$ except with a probability at most 3ε.

Let $\varepsilon \downarrow 0$ through the sequence $\varepsilon = 1/k, k = 1, 2, \ldots$. Let the approximation just shown hold for $n \geq n_k$ on a probability space $(\Omega_k, \mathcal{S}_k, Q_k)$. We can assume n_k is nondecreasing in k. Let A_{kn} be the ν_n process defined on Ω_k and G_{kn} the corresponding G_P process on Ω_k. Let $n_0 := 1$ and let $(\Omega_0, \mathcal{S}_0, Q_0)$ be a probability space on which ν_n processes A_{0n} and G_P processes $G_{0n} := G_0$ are defined, with G_0 independent of the A_{0n} processes. Let $A_n := A_{kn}$ and $G_n := G_{kn}$ if and only if $n_k \leq n < n_{k+1}$ for $k = 0, 1, \ldots$. Then for all n, A_n is a ν_n process and G_n is a G_P process. On the probability space $(\Omega', Q') := (\prod_{k \geq 0} \Omega_k, \prod_{k \geq 0} Q_k)$, all A_n and G_n are defined and $\|A_n - G_n\|_{\mathcal{F}} \to 0$ in outer probability, so by Theorem 3.27, $\nu_n \Rightarrow G_P$ on $\mathcal{F}$ and (III) implies (I), proving Theorem 3.34. $\qquad\square$

3.8 Unions of Donsker Classes

It will be shown in this section that the union of any two Donsker classes $\mathcal{F}$ and $\mathcal{G}$ is a Donsker class. This is not surprising: one might think it was enough, given the asymptotic equicontinuity conditions for the separate classes, for a given $\varepsilon > 0$, to take the larger of the two n_0's and the smaller of the two δ's. But it is not so easy as that. For example, $\mathcal{F}$ and $\mathcal{G}$ could both be finite sets, with distinct elements of $\mathcal{F}$ at distance, say, more than 0.2 apart for ρ_P, and likewise for $\mathcal{G}$, but there may be some element of $\mathcal{F}$ very close to an element of $\mathcal{G}$. So the equicontinuity condition on the union will not just follow from the conditions on the separate families.

Given a probability measure P, $\mathcal{F} \subset \mathcal{L}^2(P)$, $\varepsilon > 0$, $\delta > 0$, and a positive integer n_0, it will be said that $\mathrm{AE}(\mathcal{F}, n_0, \varepsilon, \delta)$ holds if for all $n \geq n_0$,

$$\mathrm{Pr}^*\{\sup\{|\nu_n(f) - \nu_n(g)| : f, g \in \mathcal{F}, \rho_P(f, g) < \delta\} > \varepsilon\} < \varepsilon.$$

Then the *asymptotic equicontinuity condition*, as in the previous section, holds for $\mathcal{F}$ and P if and only if for every $\varepsilon > 0$ there is a $\delta > 0$ and an n_0 such that $\mathrm{AE}(\mathcal{F}, n_0, \varepsilon, \delta)$ holds. The asymptotic equicontinuity condition, together with total boundedness of $\mathcal{F}$ for ρ_P, is equivalent to the Donsker property of $\mathcal{F}$ for P (Theorem 3.34).

Theorem 3.36 (K. S. Alexander) *Let $(\Omega, \mathcal{A}, P)$ be a probability space and let $\mathcal{F}_1$ and $\mathcal{F}_2$ be two Donsker classes for P. Then $\mathcal{F} := \mathcal{F}_1 \cup \mathcal{F}_2$ is also a Donsker class for P.*

Proof. (M. Arcones). Given $\varepsilon > 0$, take $\delta_i > 0$ and $n_i < \infty$ such that for $i = 1, 2$, $AE(\mathcal{F}_i, n_i, \varepsilon/3, \delta_i)$ holds. $\mathcal{F}_1$ and $\mathcal{F}_2$, being Donsker classes, are pregaussian, and G_P is an isonormal process for $(\cdot, \cdot)_{0,P}$, so $\mathcal{F}$ is pregaussian by Corollary 2.35. So there is an $\alpha > 0$ such that for a suitable version of G_P,

$$\Pr\{\sup\{|G_P(f) - G_P(g)| : \rho_P(f, g) < \alpha, \ f, g \in \mathcal{F}\} \geq \varepsilon/3\} < \varepsilon/3.$$

Let $\delta := \min(\delta_1, \delta_2, \alpha/3)$. Take finite sets $\mathcal{H}_i \subset \mathcal{F}_i$, $i = 1, 2$, such that for each i and $f \in \mathcal{F}_i$ there is an $h := \tau_i f \in \mathcal{H}_i$ with $\rho_P(f, h) < \delta$. Since $\mathcal{H} := \mathcal{H}_1 \cup \mathcal{H}_2$ is finite, by the finite-dimensional central limit theorem (RAP, Theorem 9.5.6) ν_n restricted to $\mathcal{H}$ converges in law to G_P restricted to $\mathcal{H}$. Let $F(\mathcal{H}, \alpha, \varepsilon/3)$ be the set of all $y \in \mathbb{R}^{\mathcal{H}}$ such that $|y(f) - y(g)| \geq \varepsilon/3$ for some $f, g \in \mathcal{H}$ with $\rho_P(f, g) < \alpha$. Then $F(\mathcal{H}, \alpha, \varepsilon/3)$ is closed and has probability less than $\varepsilon/3$ for the law of G_P on $\mathbb{R}^{\mathcal{H}}$. Thus by the portmanteau theorem (RAP, Theorem 11.1.1), there is an m such that

$$AE(\mathcal{H}, m, \varepsilon/3, \alpha) \text{ holds.} \tag{3.15}$$

Let $n_0 := \max(n_1, n_2, m)$. It will be shown that $AE(\mathcal{F}, n_0, \varepsilon, \delta)$ holds. By the asymptotic equicontinuity conditions in each $\mathcal{F}_i$ and since $n_0 \geq \max(n_1, n_2)$ and $\delta \leq \min(\delta_1, \delta_2)$, there is a set of probability less than $\varepsilon/3$ for each $i = 1, 2$ such that outside these sets, $|\nu_n(f) - \nu_n(g)| \leq \varepsilon/3$ for any f, g in the same $\mathcal{F}_i$ with $\rho_P(f, g) < \delta$. For pairs f, g with $\rho_P(f, g) < \delta$, with $f \in \mathcal{F}_1$ and $g \in \mathcal{F}_2$, we have $\rho_P(\tau_1 f, \tau_2 g) < \alpha$ since $\rho_P(f, \tau_1 f) < \delta$, $\rho_P(g, \tau_2 g) < \delta$, and $3\delta \leq \alpha$. Thus by (3.15), $|\nu_n(\tau_1 f) - \nu_n(\tau_2 g)| \leq \varepsilon/3$ for all such f, g, outside of another set of probability at most $\varepsilon/3$. Thus

$$|\nu_n(f) - \nu_n(g)| \leq |\nu_n(f) - \nu_n(\tau_1 f)| + |\nu_n(\tau_1 f) - \nu_n(\tau_2 g)|$$
$$+ |\nu_n(\tau_2 g) - \nu_n(g)|$$
$$\leq \varepsilon/3 + \varepsilon/3 + \varepsilon/3 = \varepsilon$$

for all $f \in \mathcal{F}_1$ and $g \in \mathcal{F}_2$ except on a set of probability at most $\varepsilon/3 + \varepsilon/3 + \varepsilon/3 = \varepsilon$. $\square$

3.9 Sequences of Sets and Functions

This section will show how the asymptotic equicontinuity condition in Theorem 3.34 can be applied to prove that some sequences of sets and functions are

Donsker classes. In Chapters 6 and 7, other sufficient conditions for the Donsker property will be given that will apply to uncountable families of sets and functions. For two measurable sets A and B, let $\rho_P(A, B) := \rho_P(1_A, 1_B)$.

Theorem 3.37 *Let $(X, \mathcal{A}, P)$ be a probability space and $\{C_m\}_{m \geq 1}$ a sequence of measurable sets. If*

$$\sum_{m=1}^{\infty} (P(C_m)(1 - P(C_m)))^r < \infty \text{ for some } r < \infty, \tag{3.16}$$

then the sequence $\{C_m\}_{m \geq 1}$ is a Donsker class for P. Conversely, if the sets C_m are independent for P, then the sequence is a Donsker class only if (3.16) holds.

Proof. Suppose (3.16) holds. Then the positive integers can be decomposed into two subsequences, over one of which $P(C_m) \to 0$ and over the other $P(C_m) \to 1$. It is enough to prove the Donsker property separately for each subsequence by Theorem 3.36. For any measurable set A with complement A^c, $\nu_n(A^c) \equiv -\nu_n(A)$ and $G_P(A^c) \equiv -G_P(A)$. The transformation of these processes into their negatives preserves convergence in law if it holds. So we can assume $P(C_m) \downarrow 0$ as $m \to \infty$ and then $\sum_m P(C_m)^r < \infty$. Also, $\{C_m\}_{m \geq 1}$ is totally bounded for ρ_P. By Theorem 3.36 we can assume $p_m := P(C_m) \leq 1/2$ for all m.

For any i and m such that $P(C_i \triangle C_m) = 0$, we will have almost surely for any n that $P_n(C_i) = P_n(C_m)$. So we can assume that $P(C_i \triangle C_m) > 0$ for all $i \neq m$.

For any m such that $P(C_m) = 0$ we will have $P_n(C_m) = 0$ almost surely for any n and then $\nu_n(C_m) = 0$, so we can assume $P(C_m) > 0$ for all m.

Let $0 < \varepsilon < 1$. Suppose we can find M and N such that for all $n \geq N$

$$\Pr\left\{\sup_{m \geq M} |\nu_n(C_m)| > \varepsilon\right\} < \varepsilon. \tag{3.17}$$

Then for J large enough, $p_m < p_M/2$ for $m \geq J$. Let $\gamma := \min\{P(C_i \triangle C_j) : 1 \leq i < j \leq J\}$, $\alpha := \min(\gamma, p_M)/2$. Then for $n \geq N$

$$\sup\{|\nu_n(C_i) - \nu_n(C_j)| : P(C_i \triangle C_j) < \alpha\}$$

$$\leq \sup\{|\nu_n(C_i) - \nu_n(C_j)| : i, j \geq M\} \leq 2\sup\{|\nu_n(C_j)| : j \geq M\} \leq 2\varepsilon$$

with probability at least $1 - \varepsilon$, proving the asymptotic equicontinuity condition.

So it will be enough to prove (3.17). For that, recalling the binomial probabilities defined in Section 1.3, it will suffice to find M and N such that for $n \geq N$

$$\sum_{m=M}^{\infty} E(np_m + \varepsilon n^{1/2}, n, p_m) < \varepsilon/2 \tag{3.18}$$

and

$$\sum_{m=M}^{\infty} B(np_m - \varepsilon n^{1/2}, n, p_m) < \varepsilon/2. \tag{3.19}$$

Let $q_m := 1 - p_m$. For (3.19), the Chernoff inequality (1.6) gives

$$B(np_m - \varepsilon n^{1/2}, n, p_m) \le \exp(-\varepsilon^2/(2p_m q_m)).$$

For some K, $1 \le K < \infty$, $p_m \le Km^{-1/r}$ for all m. Choose M large enough so that

$$\sum_{m=M}^{\infty} \exp(-m^{1/r}\varepsilon^2/(2K)) < \varepsilon/2$$

to give (3.19) for all n.

The other side, (3.18), is harder. Bernstein's inequality Theorem 1.11 gives, if $2pn^{1/2} \ge \varepsilon$ and $0 < p \le 1/2$, with $q := 1 - p$, that

$$E(np + \varepsilon n^{1/2}, n, p) \le \exp(-\varepsilon^2/(2pq + \varepsilon n^{-1/2})) \le \exp(-\varepsilon^2/(6pq)).$$

Then

$$\sum_{m=M}^{\infty} \{E(np_m + \varepsilon n^{1/2}, n, p_m) : 2p_m n^{1/2} \ge \varepsilon\}$$

$$\le \sum_{m=M}^{\infty} \{\exp(-\varepsilon^2/(6p_m)) : 2p_m n^{1/2} \ge \varepsilon\} \tag{3.20}$$

$$\le \sum_{m=M}^{\infty} \exp(-\varepsilon^2 m^{1/r}/(6K)) < \varepsilon/4$$

for all n if M is large enough.

It remains to treat the sum, which will be called S_2, of (3.18) restricted to values of m with

$$2p_m n^{1/2} < \varepsilon. \tag{3.21}$$

For $p := p_m$, inequality (1.7) implies

$$E(np + \varepsilon n^{1/2}, n, p) \le (np/(np + \varepsilon n^{1/2}))^{np + \varepsilon n^{1/2}} e^{\varepsilon n^{1/2}}.$$

Let $y := y(n, m, \varepsilon) := n^{1/2} p_m/\varepsilon$, so $y > 0$. Let $f(x) \equiv (1 + x)\log(1 + x^{-1})$. Then

$$\varepsilon n^{1/2} - (np + \varepsilon n^{1/2})\log(1 + \varepsilon/(pn^{1/2})) = \varepsilon n^{1/2}(1 - f(y)).$$

For $x > 0$, $f'(x) < 0$. By (3.21), $y < 1/2$, so $f(y) > f(1/2) > 3/2$. Thus $1 < 2f(y)/3$, $\varepsilon n^{1/2}(1 - f(y)) \leq -\varepsilon n^{1/2} f(y)/3$, and

$$S_2 \leq \sum_{m=1}^{\infty} \{\exp(-(\varepsilon n^{1/2} + np_m)[\log(1 + \varepsilon/(n^{1/2}p_m))]/3) : 2p_m n^{1/2} < \varepsilon\}$$

$$\leq \sum_{m=1}^{\infty} \{(p_m n^{1/2}/\varepsilon)^{\varepsilon n^{1/2}/3} : 2p_m n^{1/2} < \varepsilon\}$$

since $1/y \leq (1 + \frac{1}{y})^{1+y}$, so $(1 + \frac{1}{y})^{-(1+y)} \leq y$. Now since $p_m \leq Km^{-1/r}$ we have $S_2 \leq S_3 + S_4$ where

$$S_3 := \sum_{m=1}^{\infty} \{(p_m n^{1/2}/\varepsilon)^{\varepsilon n^{1/2}/3} : 2Km^{-1/r} n^{1/2} < \varepsilon\}$$

$$\leq (Kn^{1/2}/\varepsilon)^{\varepsilon n^{1/2}/3} \sum_{m>G} m^{-\varepsilon n^{1/2}/(3r)}$$

where $G := (2Kn^{1/2}/\varepsilon)^r \geq 2$. Then for any $n_1 > (3r/\varepsilon)^2$ and $n \geq n_1$,

$$S_3 \leq (Kn^{1/2}/\varepsilon)^{\varepsilon n^{1/2}/3} \int_{G-1}^{\infty} x^{-\varepsilon n^{1/2}/(3r)} dx$$

$$\leq (Kn^{1/2}/\varepsilon)^{\varepsilon n^{1/2}/3} (\varepsilon n^{1/2}(3r)^{-1} - 1)^{-1}(G - 1)^{1-\varepsilon n^{1/2}/(3r)}.$$

For K, ε and r fixed and $n \to \infty$, the logarithm of the latter expression is asymptotic to $-\varepsilon n^{1/2}(\log 2)/3 \to -\infty$, so $S_3 \to 0$. Thus $S_3 \leq \varepsilon/8$ for $n \geq n_2$ for some n_2. Finally,

$$S_4 := \sum_{m=1}^{\infty} \{(p_m n^{1/2}/\varepsilon)^{\varepsilon n^{1/2}/3} : 2p_m n^{1/2} < \varepsilon \leq 2Km^{-1/r} n^{1/2}\}$$

$$\leq (2Kn^{1/2}/\varepsilon)^r 2^{-\varepsilon n^{1/2}/3} \to 0$$

as $n \to \infty$, so $S_4 < \varepsilon/8$ for $n \geq n_3$ for some n_3. Thus for $n \geq \max(n_1, n_2, n_3)$, $S_2 < \varepsilon/4$. This and (3.20) give (3.18). So (3.16) implies that $\{C_m\}_{m \geq 1}$ is a Donsker class.

For the converse, if the measurable sets $\{C_m\}_{m \geq 1}$ are independent for P and form a Donsker class for P, it will be shown that (3.16) holds. Note that for each n, $P_n(C_m)$ are independent random variables for $m = 1, 2, \ldots$. Let $A := \{m : p_m \leq 1/2\}$. Suppose that for all n, $\sum_{m \in A} P_m^n = +\infty$. Then A is infinite. For each n, $\Pr\{P_n(C_m) = 1$ for infinitely many $m\} = 1$ by the Borel–Cantelli lemma. Now $P_n(C_m) = 1$ implies $v_n(C_m) = n^{1/2}(1 - p_m) \geq n^{1/2}/2$ for $m \in A$. Similarly, if for all n, $\sum_{m \notin A}(1 - p_m)^n = +\infty$, then

$$\Pr(P_n(C_m) = 0 \text{ for infinitely many } m \notin A) = 1,$$

and $P_n(C_m) = 0$, $m \notin A$, implies $\sqrt{n}(P_n - P)(C_m) \leq -\sqrt{n}/2$.

Let $\mathcal{F} := \{1_{C_m}\}_{m \geq 1}$. As $\mathcal{F}$ is countable, $\|\nu\|_{\mathcal{F}}$ and $\|G_P\|_{\mathcal{F}}$ are measurable random variables. It has been shown that if either $\sum_{m \in A} p_m^n = +\infty$ or $\sum_{m \notin A} (1 - p_m)^n = +\infty$ for all n, then $\|\nu_n\|_{\mathcal{F}} \geq \sqrt{n}/2$ almost surely for all n. As $\mathcal{F}$ is Donsker and so pregaussian, there is an $M < \infty$ such that $\Pr(\|G_P\|_{\mathcal{F}} > M/2) < 1/4$. For $\phi \in \ell^\infty(\mathcal{F})$ let $H_M(\phi) := \min(M, \|\phi\|_{\mathcal{F}})$. Then H_M is a bounded Lipschitz, thus continuous function on $\ell^\infty(\mathcal{F})$. We have $E H_M(G_P) \leq 3M/4$. For $n \geq 4M^2$, $E H_M(\nu_n) \geq M$, a contradiction since $\nu_n \Rightarrow G_P$ by definition of Donsker class. It follows that for some n large enough,

$$\sum_m ((1 - p_m)p_m)^n \leq \sum_{m \in A} p_m^n + \sum_{m \notin A} (1 - p_m)^n < +\infty,$$

which gives (3.16). $\qquad\square$

Next, we consider sequences of functions. For a probability space $(A, \mathcal{A}, P)$ and $f \in \mathcal{L}^2(A, \mathcal{A}, P)$ let $\sigma_P^2(f) := \int f^2 dP - (\int f \, dP)^2$ (the variance of f). Here is a sufficient condition for the Donsker property of a sequence $\{f_m\}$ which is easy to prove, yet turns out to be optimal of its kind:

Theorem 3.38 *If* $\{f_m\}_{m \geq 1} \subset \mathcal{L}^2(P)$ *and* $\sum_{m=1}^\infty \sigma_P^2(f_m) < \infty$, *then* $\{f_m\}_{m \geq 1}$ *is a Donsker class for* P.

Proof. We can assume that $\sigma_P^2(f_m) > 0$ for all m since the set of f_m with 0 variance is clearly P-Donsker, and we can apply Theorem 3.36. Since ν_n and G_P are the same on $f_m - Ef_m$ as on f_m a.s. for all m, we can assume $Ef_m = 0$ for all m. Then $f_m \to 0$ in $\mathcal{L}_0^2$, so the sequence $\{f_m\}$ is totally bounded for ρ_P.

For any $0 < \varepsilon < 1$, $n \geq 1$ and $m \geq 1$, by Chebyshev's inequality

$$\sum_{j \geq m} \Pr(|\nu_n(f_j)| \geq \varepsilon/2) \leq 4 \sum_{j \geq m} \sigma_P^2(f_j)/\varepsilon^2 < \varepsilon$$

for $m \geq m_0$ for some $m_0 < \infty$. We have a.s. for all n, $\nu_n(f_j) = \nu_n(f_k)$ for all j and k with $\sigma_P^2(f_j - f_k) = 0$. Let

$$\alpha := \inf\{\sigma_P^2(f_j - f_k) : \sigma_P^2(f_j - f_k) > 0, \; j < m_0\}.$$

Then $\alpha > 0$ since in some ρ_P neighborhood of f_j there are only finitely many f_k. Let $\delta := \min(\alpha, 1)$. Then

$$\Pr(|\nu_n(f_j) - \nu_n(f_k)| > \varepsilon \text{ for some } j, k \text{ with } \sigma_P^2(f_j - f_k) < \delta\} < \varepsilon,$$

implying the asymptotic equicontinuity condition and so finishing the proof by Theorem 3.34. $\qquad\square$

The following shows that Theorem 3.38, although it does not imply the first half of Theorem 3.37, is sharp in one sense:

Proposition 3.39 *Let* $A := [0, 1]$ *and* $P := U[0, 1] :=$ *Lebesgue measure on* A. *Let* $a_m > 0$ *satisfy* $\sum_{m=1}^{\infty} a_m = +\infty$. *Then there is a sequence* $\{f_m\} \subset \mathcal{L}^2(A, \mathcal{A}, P)$ *with* $\sigma_P^2(f_m) \le a_m$ *for all* m *where* $\{f_m\}$ *is not a Donsker class.*

Proof. We can assume $a_m \downarrow 0$. There exist $c_m \downarrow 0$ such that $\sum_m a_m c_m = +\infty$ (see Problem 12). In A let C_m be independent sets with $P(C_m) = a_m c_m$ for each m (see Problem 13). Let $f_m := c_m^{-1/2} 1_{C_m}$. Then $\sigma_P^2(f_m) \le \int f_m^2 dP = a_m$. For each n, almost surely $P_n(C_m) \ge 1/n$ for infinitely many m. Then $\sup_m \nu_n(f_m) = +\infty$ a.s., so the asymptotic equicontinuity condition fails and $\{f_m\}$ is not a Donsker class for P.   □

3.10 Closure of Donsker Classes under Sequential Limits

If $(X, \mathcal{A})$ is a measurable space and $\mathcal{F} \subset \mathcal{L}^0(X, \mathcal{A})$, a function $F \in \mathcal{L}^0(X, \mathcal{A})$ will be called an *envelope function* for $\mathcal{F}$ iff for all $x \in X$ and $f \in \mathcal{F}, |f(x)| \le F(x)$. We have the following stability of the Donsker property under pointwise sequential limits:

Theorem 3.40 *Let* $(X, \mathcal{A}, P)$ *be a probability space, and* $\mathcal{F} \subset \mathcal{L}^2(X, \mathcal{A}, P)$ *a Donsker class for* P. *Suppose* $\mathcal{F}$ *has an envelope* $F \in \mathcal{L}^2(X, \mathcal{A}, P)$. *Let* $\mathcal{G}$ *be the class of all functions* $g : X \to \mathbb{R}$ *such that for some* $g_m \in \mathcal{F}, g_m(x) \to g(x)$ *for all* $x \in X$. *Then* $\mathcal{G}$ *is also* P-*Donsker.*

Proof. Clearly, $\mathcal{G} \subset \mathcal{L}^0(X, \mathcal{A}, P)$. By dominated convergence using $F \in \mathcal{L}^2$, we have $\mathcal{G} \subset \mathcal{L}^2(X, \mathcal{A}, P)$. By Theorem 3.34, (a) implies (b), we have the asymptotic equicontinuity condition for $\mathcal{F}$ and ρ_P. Let it hold for a given $\varepsilon > 0$, $\delta > 0$, and n_0, and take any $n \ge n_0$. Let $f, g \in \mathcal{G}$ satisfy $\rho_P(f, g) < \delta$. Take f_m and g_m in $\mathcal{F}$ with $f_m(x) \to f(x)$ and $g_m(x) \to g(x)$ for all x. Then $f_m \to f$ and $g_m \to g$ in $\mathcal{L}^2(X, \mathcal{A}, P)$. It follows that $\rho_P(f_m, g_m) < \delta$ for m large enough. Thus the supremum in the definition of the asymptotic equicontinuity condition for the given δ is actually the same for $\mathcal{G}$ as it is for $\mathcal{F}$, so the condition holds for $\mathcal{G}$ with the same δ and n_0 for the given ε. Likewise, $\mathcal{G}$ is totally bounded for ρ_P. By Theorem 3.34, (b) implies (a), $\mathcal{G}$ is Donsker for P.   □

3.11 Convex Hulls of Donsker Classes

Let $\mathcal{F}$ be a class of real-valued functions on a set X. Recall that the symmetric convex hull of $\mathcal{F}$ is the set of all functions $\sum_{i=1}^{m} c_i f_i$ for $f_i \in \mathcal{F}, c_i \in \mathbb{R}$, any finite m, and $\sum_{i=1}^{m} |c_i| \le 1$. If $0 < M < \infty$, let $H(\mathcal{F}, M)$ denote M times the symmetric convex hull of $\mathcal{F}$.

Let $\bar{H}_s(\mathcal{F}, M)$ be the smallest class $\mathcal{G}$ of functions including $H(\mathcal{F}, M)$ such that whenever $g_n \in \mathcal{G}$ for all n and $g_n(x) \to g(x)$ as $n \to \infty$ for all x, we have $g \in \mathcal{G}$.

We have the following:

Theorem 3.41 *Let $(X, \mathcal{A}, P)$ be a probability space and $\mathcal{F} \subset \mathcal{L}^2(P)$ a class of functions. Suppose $\mathcal{F}$ has an envelope function F in $\mathcal{L}^2(P)$. If $\mathcal{F}$ is a Donsker class for P, then so is $\bar{H}_s(\mathcal{F}, M)$ for any M.*

Proof. The process G_P can and will be chosen to have a distribution μ concentrated on a separable subspace of $\ell^\infty(\mathcal{F})$. By Theorem 3.27, (a) implies (g), there is some probability space $(\Omega, \mathcal{S}, Q)$ and measurable functions g_n from Ω into X^n and h_n from Ω into $\ell^\infty(\mathcal{F})$ and such that the g_n are perfect, $Q \circ g_n^{-1} = P^n$, $Q \circ h_n^{-1} = \mu$ for all n, and $\|v_n \circ g_n - h_n\|_{\mathcal{F}} \to 0$ almost uniformly, where v_n is the function $v_n(x_1, \ldots, x_n) = \sqrt{n}(P_n - P)$ with $P_n = \frac{1}{n}\sum_{j=1}^{n}\delta_{x_j}$ from X^n into $\ell^\infty(\mathcal{F})$. Let $v_n' := v_n \circ g_n$. Let $G_P^{(n)} := h_n$ on Ω, which is a version of G_P for each n since it has distribution μ. So we have

$$\|v_n' - G_P^{(n)}\|_{\mathcal{F}} \to 0 \tag{3.22}$$

almost uniformly. $H(\mathcal{F}, 1)$ is pregaussian by Proposition 3.33, so each $G_P^{(n)}$ extends, using prelinearity (Theorem 2.32) to be uniformly continuous almost surely on $H(\mathcal{F}, 1)$. It is easily seen that in (3.22), the norm over $\mathcal{F}$ can be replaced by the norm over $H(\mathcal{F}, 1)$ without increasing it. We can apply Theorem 3.27 (g) implies (a) since the perfect functions g_n are unchanged. So $H(\mathcal{F}, 1)$ is a Donsker class. We can multiply by M, multiplying the norms all by M so that we still have almost uniform convergence. We can take sequential pointwise limits by Theorem 3.40. This finishes the proof. $\qquad\square$

Problems

1. Let $(\Omega, \mathcal{A}, P)$ be a probability space, (S, d) a (possibly nonseparable) metric space and let x_n, $n = 0, 1, 2, \ldots$, be points of S. Let $f_n(\omega) \equiv x_n$ for all ω. Show that $f_n \Rightarrow f_0$ if and only if $x_n \to x_0$. *Hint:* Define $H(x) = d(x, x_0)$. Show that H is continuous, in fact, $|H(x) - H(y)| \le d(x, y)$, so H is Lipschitz. Let $G(x) = \min(H(x), 1)$. Then show that G is bounded and continuous, with $G(x) \to 0$ if and only if $x \to x_0$. Use G for one direction.

2. In a general (possibly nonseparable) metric space show that if $X_0 \equiv p \in S$ is a constant random variable then random elements $X_n \Rightarrow X_0$ if and only if $X_n \to X_0$ in outer probability. *Hint:* Use a function G as in the last problem with p in place of x_0.

3. Let (T, d) be any metric space and $(\Omega, \mathcal{A}, Q)$ a probability space. Let f_n for $n \geq 0$ be functions from Ω into T such that $f_n \Rightarrow f_0$. Let g_n for $n \geq 1$ be functions from Ω into T such that $d(g_n, f_n) \to 0$ in outer probability. Show that $g_n \Rightarrow f_0$. Hint: Apply Theorem 3.27 (a) $\Leftrightarrow$ (b'), using (a) in the hypothesis and (b') in the conclusion.

4. Let T be the set of all bounded real-valued functions on $[0, 1]$ and $C[0, 1]$ the space of all continuous functions on $[0, 1]$. Let the norm on T and its subspace $C[0, 1]$ be $\|f\|_{\sup} := \sup_{0 \leq t \leq 1} |f(t)|$. Suppose on some probability space $(\Omega, \mathcal{A}, P)$, Brownian bridge processes Y_n are defined for $n = 0, 1, \ldots$, having an arbitrary joint distribution, but such that for each, $t \mapsto Y_n(t, \omega)$ is continuous as a function of $t \in [0, 1]$.

(a) Show that $Y_n \Rightarrow Y_0$. Hint: As all Y_n take values in the separable Banach space $C[0, 1]$ and have the same distribution, it should not be hard to show that they converge in distribution (law).

(b) Suppose that on $(\Omega, \mathcal{A}, P)$ we also have empirical processes $\alpha_n(t)$, $0 \leq t \leq 1$, for the $U[0, 1]$ distribution, defined with their usual joint distribution. Show that $\mathcal{C} := \{[0, t] : 0 \leq t \leq 1\}$ is a Donsker class for P; i.e., show that $\alpha_n \Rightarrow Y_0$. Hint: Never mind what Y_n might have been in part (a). Instead choose Y_n based on the Komlós–Major–Tusnády (–Bretagnolle–Massart) Theorem 1.8 and use the result of Problem 3.

(c) Show that for any probability measure P on the Borel sets of $S = \mathbb{R}$, $\mathcal{C} = \{(-\infty, t] : t \in \mathbb{R}\}$ is a Donsker class for P. Hint: Let F be the distribution function of P. One needs to compose the processes in part (b) with F. (This gives essentially "Donsker's theorem" as M. D. Donsker stated it in 1954.)

5. Let A_k be independent sets in a probability space $(\Omega, \mathcal{A}, P)$ such that

$$\sum_k [P(A_k)(1 - P(A_k))]^n = +\infty$$

for all $n = 1, 2, \ldots$. Show that $\{A_k\}_{k \geq 1}$ is not a Glivenko–Cantelli class, i.e. $\sup_k |(P_n - P)(A_k)|$ does not converge to 0 in probability as $n \to \infty$. Hint: See the last part of the proof of Theorem 3.37.

6. Let $(A, \mathcal{A}, P)$ be a probability space and let $\mathcal{F} \subset \mathcal{L}^2(A, \mathcal{A}, P)$ be a Donsker class for P.

(a) Show that the convex hull of $\mathcal{F}$, namely $co(\mathcal{F}) :=$

$$\left\{ \sum_{j=1}^k \lambda_j f_j : f_j \in \mathcal{F}, \ j = 1, \ldots, k, \ \lambda_j \geq 0, \ \sum_{j=1}^k \lambda_j = 1, \ k = 1, 2, \ldots \right\}$$

is a Donsker class. *Hint*: Use Theorem 2.32 to get that $co(\mathcal{F})$ is pregaussian and G_P can be taken to be prelinear on $co(\mathcal{F})$. Then use almost surely convergent realizations (Theorem 3.24).

(b) For any fixed $k < \infty$, show that $\{\sum_{j=1}^k f_j : f_j \in \mathcal{F}$ for $j = 1, \ldots, k\}$ is also a Donsker class. *Hints*: Use induction and Theorem 3.36 on unions. Thus take $k = 2$. It is easy to show that $2\mathcal{F} := \{2f : f \in \mathcal{F}\}$ is Donsker. Then apply part (a).

7. Let $c > 0$. For Lebesgue measure λ on $[0, 1]$, the *Poisson process* with intensity measure $c\lambda$ is defined by first choosing n at random having a Poisson distribution with parameter c, so that $P(n = k) = e^{-c}c^k/k!$ for $k = 0, 1, \ldots$, then setting $Y_c := \sum_{j=1}^n \delta_{X_j}$ where X_j are i.i.d. with law λ on $[0, 1]$. In the Banach space of bounded functions on $[0, 1]$ with supremum norm, prove that as $c \to \infty$ (along any sequence) the random functions $t \mapsto (Y_c - c\lambda)c^{-1/2}([0, t])$, $0 \le t \le 1$, converge in law to the Brownian motion process x_t, $0 \le t \le 1$. (Recall that $x_t = L(1_{[0,t]})$.) *Hints*: For $c = c_k \to \infty$ let $n = n_k$ be Poisson (c_k). For $F(t) \equiv t$, $0 \le t \le 1$, write $Y_c(t) := Y_c([0, t]) = nF_n(t)$, so

$$c^{-1/2}(Y_c(t) - ct) = (n/c)^{1/2}[n^{1/2}(F_n - F)(t)] + c^{-1/2}(n - c)t.$$

By Donsker's theorem take Brownian bridges $y^{(n)}$ such that $n^{1/2}(F_n - F)$ is close to $y^{(n)}$ for n large. Also, $c^{-1/2}(n - c)$ is close in distribution by the central limit theorem to a random variable Z_c with distribution $N(0, 1)$. Show one can take $n(\omega)$ independent of $X_1, X_2, \ldots$, then $y_t(\omega) := y_t^{(n(\omega))}(\omega)$ is a Brownian bridge. Show one can get Z_c to be independent of $\{y_t\}_{0 \le t \le 1}$. Then $y_t + Z_c t$ has the distribution of Brownian motion on $[0, 1]$. Apply the method of Problem 3 as in Problem 4.

8. Let P be a law on a separable, infinite-dimensional Hilbert space H such that $\int \|x\|^2 dP(x) < \infty$ and with mean 0, so that $\int (x, h)dP(x) = 0$ for all $h \in H$. Let $X_1, X_2, \ldots$ be i.i.d. in H with law P and $S_n := X_1 + \cdots + X_n$.

(a) Show that the central limit theorem holds in H, i.e. $S_n/n^{1/2}$ converges in law to some normal measure on H. *Hint*: Prove using variances that the laws of $S_n/n^{1/2}$ are tight.

(b) Show that the class of functions $x \mapsto (x, h)$ for $h \in H$ with $\|h\| \le 1$ (the unit ball of the dual space of H) is a Donsker class of functions on H for P.

Hints: Part (a): Let $\{e_n\}$ be an orthonormal basis so that for any $x \in H$, $x = \sum_n x_n e_n$ with $\|x\|^2 = \sum_n x_n^2$. Thus $\sum_n Ex_n^2$ is finite. Show that for some $c_n \to 0$ slowly enough, $\sum Ex_n^2/c_n^2$ is still finite. For any K let C_K be the set of x such that $\sum_n (x_n/c_n)^2 \le K^2$ (an infinite-dimensional ellipsoid). Show that each C_K is compact and that the laws of $S_n/n^{1/2}$ are uniformly tight using the sets

C_K. Thus subsequences of these laws converge. Show that they all converge to the same Gaussian limit law, since by the Stone-Weierstrass theorem, functions depending on only finitely many x_j are dense in the continuous functions on each C_K.

9. *The two-sample empirical process.* Let $X_1, \ldots, X_m, Y_1, \ldots, Y_n$ be i.i.d. with the uniform distribution $U[0, 1]$ on $[0, 1]$. Let F_m be the empirical distribution function based on $X_1, \ldots, X_m$ and G_n likewise based on $Y_1, \ldots, Y_n$. Show that in the space of all bounded functions on $[0, 1]$ with supremum norm,

$$\left(\frac{mn}{m+n} \right)^{1/2} (F_m - G_n) \Rightarrow y$$

as $m, n \to \infty$ where $t \mapsto y_t$, $0 \le t \le 1$ is a Brownian bridge process. *Hints:* Extend the definition of convergence in law to two indices m, n both going to $+\infty$ and otherwise unrestricted. For m, n large, $m^{1/2}(F_m - F)$ is close to a Brownian bridge $Y(m)$ and $n^{1/2}(G_n - F)$ is close to an independent Brownian bridge $Z(n)$ by the Komlós–Major–Tusnády–Bretagnolle–Massart theorem. It follows as in Problems 3 and 4 that $[(mn)/(m + n)]^{1/2}(F_m - G_n)$ is close to $(n/(m + n))^{1/2}Y(m) - (m/(m + n))^{1/2}Z(n)$, which is a Brownian bridge.

10. Let $f_n(x) = \cos(2\pi nx)$ for $0 \le x \le 1$ with law $U[0, 1]$ on $[0, 1]$. For real c let $\mathcal{F}_c$ be the sequence of functions $n^{-c} f_n$ for all $n = 1, 2, \ldots$. For what values of c is $\mathcal{F}_c$ pregaussian? A Donsker class? *Hints:* If $c \le 0$, show easily that the class is not pregaussian and so not Donsker. If $c > 0$ then it is pregaussian, by metric entropy. If $c > 1/2$, it is Donsker by Theorem 3.38. Show it is Donsker for $0 < c \le 1/2$ by the Bernstein inequality and the asymptotic equicontinuity condition.

11. In $\mathbb{R}$, for any law P on $\mathbb{R}$, show that for any fixed $k < \infty$, $\mathcal{C} := \{\bigcup_{j=1}^k (a_j, b_j] : a_j \le b_j \text{ for all } j\}$ is a Donsker class, i.e. $\mathcal{F} := \{1_C : C \in \mathcal{C}\}$ is a Donsker class. *Hint:* Apply Donsker's theorem as in Problem 4 and take differences (and sums). For $k = 2$ reduce to the case of disjoint intervals and apply Problem 6. Do induction to get general k.)

12. Show that as stated in the proof of Proposition 3.39, for any $a_m > 0$ with $\sum_m a_m = +\infty$ there are $c_m \downarrow 0$ with $\sum_m c_m a_m = +\infty$. *Hint:* Take a sequence m_k such that $\sum\{a_m : m_k \le m < m_{k+1}\} > k$ for each $k = 1, 2, \ldots$. Let c_m have the same value for $m_k \le m < m_{k+1}$.

13. Show that as also stated in the proof of Proposition 3.39, in $[0, 1]$ for $P = U[0, 1]$ there exist independent sets C_m with any given probabilities. *Hint:* Use binary expansions. By decomposing the set of positive integers into a countable union of countably infinite sets, show that $([0, 1], P)$ is isomorphic as a probability space to a countable Cartesian product of copies of itself.

14. Suppose that $\{C_m\}_{m\geq 1}$ are independent for P, $\sum_{m=1}^{\infty} P(C_m)(1 - P(C_m)) = +\infty$ and $c_m \to \infty$. Show that $\{c_m 1_{C_m}\}_{m\geq 1}$ is not a Donsker class.

15. Do f_n in the proof of Proposition 3.21 converge in outer probability: to f_0, to something else, or not to any function? Based on your answer, explain whether there is a contradiction with Corollary 3.17, and if not, why not.

Notes

Notes to Section 3.1. For any metric space (S, d), let $\mathcal{B}_b(S, d)$ be the σ-algebra generated by all balls $B(x, r) := \{y : d(x, y) < r\}$, $x \in S$, $r > 0$. Then $\mathcal{B}_b(S, d)$ is always included in the Borel σ-algebra $\mathcal{B}(S, d)$ generated by all the open sets, with $\mathcal{B}_b(S, d) = \mathcal{B}(S, d)$ if (S, d) is separable.

Suppose Y_n are functions from a probability space (Ω, P) into S, measurable for $\mathcal{B}_b(S, d)$. Then each Y_n has a law $\mu_n = P \circ Y_n^{-1}$ on $\mathcal{B}_b(S, d)$.

Dudley (1966, 1967b) defined convergence in law of Y_n to Y_0 to mean that $\int^* H d\mu_n \to \int H d\mu_0$ for every bounded continuous real-valued function H on S. Hoffmann-Jørgensen (1984) gave the newer definition adopted generally and here, where the upper integrals and integral are taken over Ω, not S, so that the laws μ_n are not necessarily defined on any particular σ-algebra in S. Hoffmann-Jørgensen's monograph was published in 1991, apparently without major revision (its latest reference is from 1981).

Andersen (1985a,b), Andersen and Dobrić (1987, 1988), and Dudley (1985a) developed further the theory based on Hoffmann-Jørgensen's definition.

Notes to Section 3.2. Blumberg (1935) defined the measurable cover function f^*, see also Goffman and Zink (1960). (I thank the late Rae M. Shortt for pointing out Blumberg's paper.) Later, Eames and May (1967) also defined f^*. Lemmas 3.4 through 3.8 are more or less as in Dudley and Philipp (1983, Section 2), except that Lemma 3.6(c) is new here (in the first, 1999 edition). Theorem 3.3 and its proof are as in Vulikh (1961/1967, pp. 78–79) for Lebesgue measure on an interval (the proof needs no change). Luxemburg and Zaanen (1983, Lemma 94.4 p. 222) also prove existence of essential suprema and infima of families of measurable (extended) real functions.

Notes to Section 3.3. This section was based on parts of Dudley (1985a).

Notes to Section 3.4. Perfect probability spaces were apparently first defined by Gnedenko and Kolmogorov (1949), Section 3, and their theory was carried on among others by Ryll-Nardzewski (1953), Sazonov (1962), and Pachl (1979).

Perfect functions are defined and treated in Hoffmann-Jørgensen (1984,1985) and Andersen (1985a,b); see also Dudley (1985a).

Notes to Section 3.5. The existence of almost surely convergent random variables with a given converging sequence of laws was first proved by Skorohod (1956) for complete separable metric spaces, then Dudley (1968) for any separable metric space, with a re-exposition in RAP, Section 11.7, and by Wichura (1970) for laws on the σ-algebra generated by balls in an arbitrary metric space as mentioned in the notes to Section 3.1. The current version was given in Dudley (1985a).

Notes to Section 3.6. Hoffmann-Jørgensen (1984), who defined convergence in law in the sense adopted in this chapter, also developed the theory of it as in this section, and partly in a more general form (with nets instead of sequences, and other classes of functions in place of the bounded Lipschitz functions).

Andersen and Dobrić (1987, Remark 2.13) pointed out that the portmanteau theorem (as in Topsøe, 1970, Theorem 8.1) "can be extended to the nonmeasurable case. The proof of this extension is the same as the ordinary proof." Much the same might be said of other equivalences in this section. Dudley (1990, Theorem A) gave a form of the portmanteau theorem and (Theorem B) of the metrization theorem 3.28.

But not all facts or proofs from the separable case extend so easily: for example, in the separable case, there is an inequality for the two metrics, $\beta \leq 2\rho$, in the opposite direction to Lemma 3.29, which follows from Strassen's theorem on nearby variables with nearby laws (RAP, Corollary 11.6.5), but Strassen's theorem seems not to extend well to the nonmeasurable case (Dudley 1994).

Notes to Section 3.7. An early form of the asymptotic equicontinuity condition appeared in Dudley (1966, Proposition 2) and a later form in Dudley (1978). The equivalence with a different pseudometric τ in Theorem 3.34 is due to Giné and Zinn (1986, p. 58).

Lemma 3.35 is essentially contained in the proof of Skorohod (1976, Theorem 1), as Erich Berger kindly pointed out. See also Eršov (1975).

Notes to Section 3.8. Alexander (1987, Corollary 2.7) stated Theorem 3.36 but did not publish a proof of it, although he had written out an unpublished proof several years earlier. He says that the result is "an extension of a slightly weaker result of Dudley (1981)," where $\mathcal{F}_2$ is finite, but this author himself doesn't think his 1981 result was only "slightly weaker"! The proof presented was suggested by Miguel Arcones in Berkeley during the fall of 1991, but I take responsibility for any possible errors in it. Apparently van der Vaart (1996, Theorem A.3) first published a proof.

Notes to Section 3.9. Theorem 3.37 first appeared in Dudley (1978, Section 2), Theorem 3.38 in Dudley (1981), and Proposition 3.39 in Dudley (1984).

4
Vapnik–Červonenkis Combinatorics

This chapter will treat some classes of sets satisfying a combinatorial condition. In Chapter 6 it will be shown that under a mild measurability condition to be treated in Chapter 5, these classes have the Donsker property, for all probability measures P on the sample space, and satisfy a law of large numbers (Glivenko–Cantelli property) uniformly in P. Moreover, for either of these limit-theorem properties of a class of sets (without assuming any measurability), the Vapnik–Červonenkis property is necessary (Section 6.4).

The name Červonenkis is sometimes transliterated into English as Chervonenkis. The present chapter will be self-contained, not depending on anything earlier in this book, except in some examples.

4.1 Vapnik–Červonenkis Classes of Sets

Let X be any set and $\mathcal{C}$ a collection of subsets of X. For $A \subset X$ let $\mathcal{C}_A := \mathcal{C} \sqcap A := A \sqcap \mathcal{C} := \{C \cap A : C \in \mathcal{C}\}$. Let $\mathrm{card}(A) := |A|$ denote the cardinality (number of elements) of A and $2^A := \{B : B \subset A\}$. Let $\Delta^{\mathcal{C}}(A) := |\mathcal{C}_A|$. If $A \sqcap \mathcal{C} = 2^A$, then $\mathcal{C}$ is said to *shatter* A. If A is finite, then $\mathcal{C}$ shatters A if and only if $\Delta^{\mathcal{C}}(A) = 2^{|A|}$.

Let $m^{\mathcal{C}}(n) := \max\{\Delta^{\mathcal{C}}(F) : F \subset X, |F| = n\}$ for $n = 0, 1, \ldots,$ or if $|X| < n$ let $m^{\mathcal{C}}(n) := m^{\mathcal{C}}(|X|)$. Then $m^{\mathcal{C}}(n) \le 2^n$ for all n. Let

$$V(\mathcal{C}) := \begin{cases} \inf\{n : m^{\mathcal{C}}(n) < 2^n\}, & \text{if this is finite,} \\ +\infty, & \text{if } m^{\mathcal{C}}(n) = 2^n \text{ for all } n, \end{cases}$$

$$S(\mathcal{C}) := \begin{cases} \sup\{n : m^{\mathcal{C}}(n) = 2^n\}, & \mathcal{C} \ne \emptyset, \\ -1, & \text{if } \mathcal{C} \text{ is empty.} \end{cases}$$

175

Then $S(\mathcal{C}) \equiv V(\mathcal{C}) - 1$, and $S(\mathcal{C})$ is the largest cardinality of a set shattered by $\mathcal{C}$, or $+\infty$ if arbitrarily large finite sets are shattered. So, $V(\mathcal{C})$ is the smallest n, if one exists, such that no set of cardinality n is shattered by $\mathcal{C}$. If $V(\mathcal{C}) < \infty$, or equivalently if $S(\mathcal{C}) < \infty$, $\mathcal{C}$ will be called a *Vapnik–Červonenkis class* or *VC class*. In the (very large) machine learning literature relating to VC classes, $S(\mathcal{C})$ is called the *VC dimension* of $\mathcal{C}$.

If X is finite, with n elements, then clearly 2^X is a VC class, with $S(2^X) = n$. Let $_N C_{\le k} := \sum_{j=0}^k \binom{N}{j}$, where

$$\binom{N}{j} := \begin{cases} N!/(j!(N-j)!), & j = 0, 1, \ldots, N, \\ 0, & j > N. \end{cases}$$

Then $_N C_{\le k}$ is the number of combinations of N things, at most k at a time. (In an older notation $_N C_k := \binom{N}{k}$.) "Pascal's triangle" of identities for binomial coefficients extends to the $_N C_{\le k}$:

Proposition 4.1 $_N C_{\le k} = {}_{N-1}C_{\le k} + {}_{N-1}C_{\le k-1}$ *for* $k = 1, 2, \ldots,$ *and* $N = 1, 2, \ldots$.

Proof. For each $j = 1, 2, \ldots, N$, we have by the classical Pascal's triangle

$$\binom{N}{j} = \binom{N-1}{j} + \binom{N-1}{j-1}, \text{ and } \binom{N}{0} = \binom{N-1}{0} = 1.$$

Summing over j, the conclusion follows. If $N \le k$, we get $2^N = 2^{N-1} + 2^{N-1}$. $\qquad\square$

For a non-VC class $\mathcal{C}$ we have $m^{\mathcal{C}}(n) = 2^n$ for all n. For a VC class, the next fact, which is fundamental in the Vapnik–Červonenkis theory, will imply that $m^{\mathcal{C}}(n)$ grows only as a polynomial rather than exponentially in n.

Theorem 4.2 (Sauer's Lemma) *If* $m^{\mathcal{C}}(n) > {}_n C_{\le k-1}$, *where* $k \ge 1$, *then* $m^{\mathcal{C}}(k) = 2^k$. *Hence if* $S(\mathcal{C}) < \infty$, *then* $m^{\mathcal{C}}(n) \le {}_n C_{\le S(\mathcal{C})}$ *for all* n.

Proof. The proof is by induction on k and n. For $k = 1$, $_n C_{\le 0} = 1 < m^{\mathcal{C}}(n)$ implies that $\mathcal{C}$ contains at least two elements, so for some singleton $G = \{x\}$, $\Delta^{\mathcal{C}}(G) = 2$ as desired. If $k > n$, then $_n C_{\le k-1} = 2^n \ge m^{\mathcal{C}}(n)$, so the assumption implies $k \le n$.

Now assume that the theorem holds whenever $k \le K$ and $n \ge k$, for all $\mathcal{C}$. Fix $k := K + 1$. For $n < k$, as noted, it holds vacuously. For $n = k$, the hypothesis $m^{\mathcal{C}}(n) > {}_n C_{\le n-1} = 2^n - 1$ implies $m^{\mathcal{C}}(n) = 2^n$ as desired. Then to continue the proof by induction on n for k fixed, supposing the statement holds for $n \le N$, it will be proved for $n = N + 1$. Let $H_n := \{x_1, \ldots, x_n\}$ be a set with n elements. Suppose $\Delta^{\mathcal{C}}(H_n) > {}_n C_{\le K}$. Let $H_N := \{x_1, \ldots, x_N\}$. If

$\Delta^{\mathcal{C}}(H_N) > {}_N C_{\leq K}$, then by induction assumption, $m^{\mathcal{C}}(k) = 2^k$ as desired. So assume

$$\Delta^{\mathcal{C}}(H_N) \leq {}_N C_{\leq K}. \tag{4.1}$$

Let $\mathcal{C}_n := H_n \sqcap \mathcal{C} := \{A \cap H_n : A \in \mathcal{C}\}$. Call a set $E \subset H_N$ *full* iff both E and $E \cup \{x_n\}$ belong to $\mathcal{C}_n$. Let f be the number of full sets. Then the map $A \mapsto A \cap H_N$ takes $\mathcal{C} \sqcap H_n$ onto $\mathcal{C} \sqcap H_N$ and is two-to-one onto full sets and one-to-one onto non-full sets in $\mathcal{C} \sqcap H_N$. Thus

$$\Delta^{\mathcal{C}}(H_n) = \Delta^{\mathcal{C}}(H_N) + f. \tag{4.2}$$

Let $\mathcal{F}$ be the collection of all full sets. Suppose $f = \Delta^{\mathcal{F}}(H_N) > {}_N C_{\leq K-1}$. Then by induction assumption there is a $G \subset H_N$ with $\text{card}(G) = K$ and $\Delta^{\mathcal{F}}(G) = 2^K$. For $J := G \cup \{x_n\}$ we then have $\text{card}(J) = k$ and $\Delta^{\mathcal{C}}(J) = 2^k$ as desired.

In the remaining case, $f \leq {}_N C_{\leq K-1}$. Then by (4.1) and (4.2),

$$\Delta^{\mathcal{C}}(H_n) \leq {}_N C_{\leq K} + {}_N C_{\leq K-1} = {}_n C_{\leq K}$$

by Proposition 4.1, a contradiction. So the first sentence of the Theorem is proved. The second follows from the definition of $S(\mathcal{C})$. $\qquad\square$

For fixed k, ${}_n C_{\leq k}$ is easily seen to be a polynomial in n of degree k, with leading term $n^k/k!$. Thus, the next fact is not far from optimal:

Proposition 4.3 (Vapnik and Červonenkis) *For any nonnegative integers n and k with $n \geq k+2$, ${}_n C_{\leq k} \leq 1.5 n^k/k!$. For $k \geq 1$ we have $1.5 n^k/k! \leq (ne/k)^k$.*

Proof. The latter inequality holds by the simplest form of the Stirling formula with error bounds, Theorem 1.17, $k! \geq (k/e)^k \sqrt{2\pi k}$, and since $1.5 < \sqrt{2\pi}$.

The first inequality clearly holds for $k = 0$, so assume $k \geq 1$. By the binomial theorem, $n^{k-1}(k + n) \leq (n + 1)^k$, so

$$n^{k-1}/(k - 1)! + n^k/k! \leq (n + 1)^k/k!. \tag{4.3}$$

The proof will be done by induction on n and k. For $k = 1$ the inequality is $n + 1 \leq 1.5n$, which holds for all $n \geq 2$, and so for $n \geq k + 2 = 3$.

For $n = k + 2$, the desired inequality is

$$2^n - n - 1 \leq 1.5 n^{n-2}/(n - 2)! = 1.5(n - 1)n^{n-1}/n!.$$

This can be checked directly for $n = 3, 4, 5$, and 6. Stirling's formula with an error bound (Theorem 1.17) gives

$$n! \leq \left(\frac{n}{e}\right)^n (2\pi n)^{1/2} e^{1/(12n)},$$

so it will be enough to prove

$$2^n \left(\frac{n}{e}\right)^n (2\pi n)^{1/2} e^{1/(12n)} \leq 1.5(n-1)n^{n-1}, \quad n \geq 7,$$

which follows from $(e/2)^n \geq 2n^{1/2}$, $n \geq 7$; for $f(x) := (e/2)^x$ and $g(x) := 2x^{1/2}$ it is straightforward to check that $f(7) \geq g(7)$, $f'(7) \geq g'(7)$, $f'' > 0$ and $g'' < 0$, so $f(x) \geq g(x)$ for all $x \geq 7$.

Now suppose the Proposition has been proved for $n = k + i$, $i = 2, \ldots, j$, and for $n = k + J$, $J := j + 1$, for $k = 1, \ldots, K$, as we have done for $j = 2$ and for $K = 1$. We need to prove (4.3) for $n = k + J$ and $k = K + 1$. We have $k + j = K + J$ and

$$
\begin{aligned}
{}_nC_{\leq k} &= {}_{k+J}C_{\leq k} \\
&\leq {}_{k+j}C_{\leq k} + {}_{K+J}C_{\leq K} && \text{by Proposition 4.1} \\
&\leq 1.5(k+j)^k/k! + 1.5(K+J)^K/K! && \text{by induction hypotheses} \\
&\leq 1.5n^k/k! && \text{by (4.3),}
\end{aligned}
$$

completing the proof.   □

Combining Theorem 4.2 and Proposition 4.3 gives $m^{\mathcal{C}}(n) \leq 1.5n^k/k!$ for $n \geq k + 2$ where $k := S(\mathcal{C})$. To see that Theorem 4.2 is sharp, let X be an infinite set and $\mathcal{C}$ the collection of all subsets of X with cardinality k. Then $S(\mathcal{C}) = k$ and the inequality in the second sentence of the theorem becomes an equality for all n.

Let

$$\text{dens}(\mathcal{C}) := \inf\{r > 0 : \text{ for some } K < \infty, \ m^{\mathcal{C}}(n) \leq Kn^r \text{ for all } n \geq 1\}.$$

Then we have

Corollary 4.4 *For any set X and $\mathcal{C} \subset 2^X$, $\text{dens}(\mathcal{C}) \leq S(\mathcal{C})$. Conversely if $\text{dens}(\mathcal{C}) < \infty$ then $S(\mathcal{C}) < \infty$.*

Proof. By Theorem 4.2 and Proposition 4.3, there is a K such that $m^{\mathcal{C}}(n) \leq Kn^{S(\mathcal{C})}$ for all $n \geq S(\mathcal{C}) + 2$. The same holds for all $n \geq 1$, possibly with a larger K, so $\text{dens}(\mathcal{C}) \leq S(\mathcal{C})$. Conversely if $\text{dens}(\mathcal{C}) < \infty$, then since $m^{\mathcal{C}}(n) \leq Kn^r < 2^n$ for n large we have $S(\mathcal{C}) < \infty$.   □

Note that $S(\mathcal{C})$ can be determined by one large shattered set while $\text{dens}(\mathcal{C})$ has to do with the behavior of $\mathcal{C}$ on arbitrarily large finite sets. For example, if X is a set with $\text{card}(X) = n$ and $\mathcal{C} = 2^X$, then $S(\mathcal{C}) = n$ while $\text{dens}(\mathcal{C}) = 0$.

For any set X, it is immediate that if $\mathcal{C} \subset \mathcal{D} \subset 2^X$, then $S(\mathcal{C}) \leq S(\mathcal{D})$ and $\text{dens}(\mathcal{C}) \leq \text{dens}(\mathcal{D})$.

The following is straightforward since for any set X, the map $A \mapsto X \setminus A$ is one-to-one from 2^X onto itself, and for any $A, B, C \subset X$, $A \cap B \neq C \cap B$ if and only if $(X \setminus A) \cap B \neq (X \setminus C) \cap B$:

Proposition 4.5 *If X is any set, $\mathcal{C} \subset 2^X$ and $\mathcal{D} := \{X \setminus A : A \in \mathcal{C}\}$, then for all $B \subset X$, $\Delta^{\mathcal{C}}(B) = \Delta^{\mathcal{D}}(B)$, so $m^{\mathcal{C}}(n) = m^{\mathcal{D}}(n)$ for all n, $S(\mathcal{D}) = S(\mathcal{C})$ and $dens(\mathcal{D}) = dens(\mathcal{C})$.*

4.2 Generating Vapnik–Červonenkis Classes

First, here are some examples of *non*-VC classes for which some uniform limit theorems for empirical measures fail.

First, let $X = [0, 1]$ and let $\mathcal{C}$ be the class of all finite subsets of X. Let P be the uniform (Lebesgue) law on $[0, 1]$. Clearly, $S(\mathcal{C}) = +\infty$, and $\mathcal{C}$ is not a VC class. Also, for any possible value of P_n, we will have $P_n(A) = 1$ for some $A = \{X_1, \ldots, X_n\} \in \mathcal{C}$ while $P(A) = 0$. Thus $\sup_{A \in \mathcal{C}}(P_n - P)(A) = 1$ for all n, so $\mathcal{C}$ is not a Glivenko–Cantelli class for P, in other words,

$$\|P_n - P\|_{\mathcal{C}} := \sup_{A \in \mathcal{C}} |(P_n - P)(A)|$$

does not approach 0 as $n \to \infty$ in any sense, e.g. in outer probability, since it is identically 1. It follows that $\mathcal{C}$ is also not a Donsker class for P.

Note that all functions 1_A for $A \in \mathcal{C}$ equal 0 almost surely for P. Thus, the whole class $\mathcal{F} := \{1_A : A \in \mathcal{C}\}$ reduces to the one point 0 in the space $L^2(P)$ of equivalence classes for equality almost everywhere of functions in $\mathcal{L}^2(P)$, that is, measurable, square-integrable functions. Thus for purposes of empirical processes, functions equal a.s. P are not the same, and we need to deal with classes $\mathcal{F} \subset \mathcal{L}^2(P)$ of actual real-valued functions, not equivalence classes. Then, the integral $\int f d(P_n - P)$ will be well-defined for any $f \in \mathcal{L}^2(P)$. This integral is linear in f and thus prelinear for $f \in \mathcal{F}$ for any set $\mathcal{F} \subset \mathcal{L}^2(P)$. For the empirical process $\nu_n = n^{1/2}(P_n - P)$ we will not be taking versions or modifications as was done for Gaussian processes (Appendix I).

Next, let $\mathcal{C}_2$ be the collection of all closed, convex subsets of $\mathbb{R}^2$. Let S^1 be the unit circle $\{(x, y) : x^2 + y^2 = 1\}$. For any finite subset F of S^1, the convex polygon with vertices in F (a singleton if $|F| = 1$, or a line segment if $|F| = 2$) is in $\mathcal{C}_2$ and its intersection with S^1 is F. Thus $S(\mathcal{C}) = +\infty$ and $\mathcal{C}$ is not a VC class. Let P be the uniform law $dP(\theta) = 2\theta/(2\pi)$ on S^1. Then the Glivenko–Cantelli and Donsker properties fail for P just as in the previous example.

Classes with $S(\mathcal{C})$ finite, in other words Vapnik–Červonenkis classes, can be formed in various ways. Here is one. Let G be a collection of real-valued functions on a set X. Let

$$\text{pos}(g) := \{x : g(x) > 0\}, \ \text{nn}(g) := \{x : g(x) \geq 0\}, \ g \in G,$$

$$\text{pos}(G) := \{\text{pos}(g) : g \in G\}, \ \text{nn}(G) := \{\text{nn}(g) : g \in G\},$$

$$U(G) := \text{pos}(G) \cup \text{nn}(G).$$

Theorem 4.6 *Let H be an m-dimensional real vector space of functions on a set X, f any real function on X, and $H_1 := \{f + h : h \in H\}$. Then $S(\text{pos}(H_1)) = S(\text{nn}(H_1)) = m$. If H contains the constants, then also $S(U(H_1)) = m$.*

Proof. First it will be shown that $S(\text{pos}(H_1)) = m$. Clearly $\text{card}(X) \geq m$. If $\text{card}(X) = m$, then $H = H_1$ is the set $\mathbb{R}^X$ of all real-valued functions on X, so the result holds.

Otherwise, let $A \subset X$ with $\text{card}(A) = m + 1$. Let G be the vector space $\{af + h : a \in \mathbb{R}, h \in H\}$. Let $r_A : G \mapsto \mathbb{R}^A$ be the restriction of functions in G to A. If r_A is not onto, take $0 \neq v \in \mathbb{R}^A$ where v is orthogonal to $r_A(G)$ for the usual inner product $(\cdot, \cdot)_A$.

Let $A_+ := \{x \in A : v(x) > 0\}$. We can assume A_+ is nonempty, replacing v by $-v$ if necessary. If $A_+ = A \cap \text{pos}(g)$ for some $g \in G$, then $(r_A(g), v)_A > 0$, a contradiction. So $\text{pos}(G)$ doesn't shatter A.

Suppose instead that $r_A(G) = \mathbb{R}^A$. Then r_A is 1–1 on G, $f \notin H$, and $r_A(H_1)$ is a hyperplane in $\mathbb{R}^A$ not containing 0, so that for some $v \in \mathbb{R}^A$, $(j, v)_A = -1$ for all $j \in r_A(H_1)$. Again let $A_+ := \{x \in A : v(x) > 0\}$. If $A_+ = A \cap \text{pos}(f + h)$ for some $h \in H$, then $(f + h, v)_A \geq 0$, a contradiction (here A_+ may be empty). Thus $\text{pos}(H_1)$ never shatters A, so $S(\text{pos}(H_1)) \leq m$.

For each $x \in X$, a linear form δ_x is defined on H by $\delta_x(h) := h(x)$, $h \in H$. Let H' be the vector space of all real linear forms on H. Then H' is m-dimensional. Let H'_X be the linear span in H' of the set of all δ_x, $x \in X$,

$$H'_X := \left\{ \sum_{j=1}^{r} a_j \delta_{x_j} : x_j \in X, \ a_j \in \mathbb{R}, \ r = 1, 2, \ldots \right\}. \qquad (4.4)$$

The map $h \mapsto (\psi \mapsto \psi(h)) : h \in H, \ \psi \in H'_X$, is 1–1 and linear from H into $(H'_X)'$, so H'_X is m-dimensional. Take $B = \{x_1, \ldots, x_m\} \subset X$ such that the δ_{x_i} are linearly independent in H'. So $r_B(H) = \mathbb{R}^B$, $r_B(H_1) = \mathbb{R}^B$, and $\text{pos}(r_B(H_1)) = 2^B$, so $S(\text{pos}(H_1)) = m$.

Then $S(\text{nn}(H_1)) = m$ by taking complements (Proposition 4.5). If H contains the constant functions, then the sets $\text{nn}(f)$, $f \in H_1$, are the same as the sets $\{f \geq t\}$, $f \in H_1$, $t \in \mathbb{R}$, and the sets $\text{pos}(f)$, $f \in H_1$, are the same as the sets $\{f > t\}$, $f \in H_1$, $t \in \mathbb{R}$. Now for any finite subset A of X, $f \in H_1$ and $t \in \mathbb{R}$, since f takes only finitely many values on A, there exist s and u such that $A \cap \{f \geq t\} = A \cap \{f > s\}$ and $A \cap \{f > t\} = A \cap \{f \geq u\}$. So in this case $S(U(H_1)) = m$. $\qquad \square$

Examples. (I) Let $H := \mathcal{P}_{d,k}$ be the space of all polynomials of degree at most k on $\mathbb{R}^d$. Then for each d and k, H is a finite-dimensional vector space of functions, so $\text{pos}(H)$ is a Vapnik–Červonenkis class. For $k = 2$, it follows specifically that the set of all ellipsoids in $\mathbb{R}^d$ is included in a Vapnik–Červonenkis class and thus is one.

(II) Let $X = \mathbb{R}$. Let H be the 1-dimensional space of linear functions $f(x) = cx$, $x \in \mathbb{R}$, $c \in \mathbb{R}$. Then $S(\text{pos}(H)) = S(\text{nn}(H)) = 1$ by Theorem 4.6, but $U(H)$ shatters $\{0, 1\}$. Since sets in $U(H)$ are convex (half-lines), it follows that $S(U(H)) = 2$. So the condition that H contains the constants cannot just be dropped from Theorem 4.6 for $U(H)$.

Let X be a real vector space of dimension m. Let H be the space of all real *affine* functions on X, in other words, functions of the form $h + c$ where h is real linear and c is any real constant. Then H has dimension $m + 1$, and $\text{pos}(H)$ is the set of all open half-spaces of X. Letting $f = 0$ in Theorem 4.6 for this H gives a special case known as Radon's Theorem. On the other hand, Theorem 4.6 for $f = 0$ with general X and H follows from Radon's Theorem via the following stability fact:

Theorem 4.7 *If X and Y are sets, F is a function from X into Y, $\mathcal{C} \subset 2^Y$, and $F^{-1}(\mathcal{C}) := \{F^{-1}(A) : A \in \mathcal{C}\}$, then $S(F^{-1}(\mathcal{C})) \leq S(\mathcal{C})$. If F is onto Y, then $S(F^{-1}(\mathcal{C})) = S(\mathcal{C})$.*

Proof. Let $F^{-1}(\mathcal{C})$ shatter $\{x_1, \ldots, x_m\}$ where $x_i \neq x_j$ for $i \neq j$. Then $F(x_i) \neq F(x_j)$ for $i \neq j$ and $\mathcal{C}$ shatters $\{F(x_1), \ldots, F(x_m)\}$. So $S(F^{-1}(\mathcal{C})) \leq S(\mathcal{C})$. If F is onto Y and $H \subset Y$ with $\text{card}(H) = m$, choose $G \subset X$ such that F takes G 1–1 onto H. Then if $\mathcal{C}$ shatters H, $F^{-1}(\mathcal{C})$ shatters G, so $S(F^{-1}(\mathcal{C})) = S(\mathcal{C})$. $\qquad\square$

Now let X be any set and G a finite-dimensional real vector space of real functions on X. Then there is a natural map $F : x \mapsto \delta_x$ from X into the space of linear functions on G. Then by Theorem 4.7 one could deduce Theorem 4.6 from its special case where X is an m- or $(m + 1)$-dimensional real vector space and f and all functions in H are affine, so that sets in $\text{pos}(H_1)$ are open half-spaces.

Next it will be seen how a bounded number of Boolean operations preserves the Vapnik–Červonenkis property.

Theorem 4.8 *Let X be a set, $\mathcal{C} \subset 2^X$, and for $k = 1, 2, \ldots,$ let $\mathcal{C}^{(k)}$ be the union of all (Boolean) algebras generated by k or fewer elements of $\mathcal{C}$. Then $\text{dens}(\mathcal{C}^{(k)}) \leq k \cdot \text{dens}(\mathcal{C})$, so if $S(\mathcal{C}) < \infty$, then $S(\mathcal{C}^{(k)}) < \infty$.*

Proof. Let $\text{dens}(\mathcal{C}) = r$, so that for any $\varepsilon > 0$ there is some $M < \infty$ such that $m^{\mathcal{C}}(n) \leq Mn^{r+\varepsilon}$ for all n.

For any $A \subset X$ we have $A \sqcap \mathcal{C}^{(k)} = A \sqcap (A \sqcap \mathcal{C})^{(k)}$. An algebra $\mathcal{A}$ with k generators $A_1, \ldots, A_k$ has at most 2^k atoms, which are those nonempty sets that are intersections of some of the A_i and the complements of the rest. Sets in $\mathcal{A}$ are unions of atoms, so $|\mathcal{A}| \leq 2^{2^k}$. Thus $|A \sqcap \mathcal{C}^{(k)}| \leq 2^{2^k} |A \sqcap \mathcal{C}|^k \leq$

$2^{2^k} M^k |A|^{k(r+\varepsilon)}$. Letting $\varepsilon \downarrow 0$ gives $\operatorname{dens}(\mathcal{C}^{(k)}) \leq k \cdot \operatorname{dens}(\mathcal{C})$. If $S(\mathcal{C}) < \infty$, then by Corollary 4.4, $S(\mathcal{C}^{(k)}) < \infty$.   □

The constant 2^{2^k} is very large if k is at all large. Let $\mathcal{C}^{(\cap k)}$ be the class of all intersections of at most k sets in $\mathcal{C}$. Then $\mathcal{C}^{(\cap k)} \subset \mathcal{C}^{(k)}$. For $\mathcal{C}^{(\cap k)}$, bounds in the preceding proof can be replaced by $|A \cap \mathcal{C}^{(\cap k)}| \leq |A \cap \mathcal{C}|^k \leq M^k |A|^{k(r+\varepsilon)}$, so that the constant 2^{2^k} is not needed.

Theorems 4.6 and 4.8 can be combined to generate Vapnik–Červonenkis classes. For example, half-spaces in $\mathbb{R}^d$ form a VC class. Intersections of at most k half-spaces give convex polytopes with at most k faces, so these form a VC class.

Remarks. Let X be an infinite set, $r = 1, 2, \ldots$, and $\mathcal{C}_r$ the collection of all subsets of X with at most r elements. Then clearly $\operatorname{dens}(\mathcal{C}_r) = S(\mathcal{C}_r) = r$. It is easy to check that $\mathcal{D} := \mathcal{C}_r^{(k)}$ consists of all sets B such that either B or $X \setminus B$ has at most kr elements. Thus $m^{\mathcal{D}}(n) \leq 2(_n C_{\leq kr})$, with $m^{\mathcal{D}}(n) = 2(_n C_{\leq kr})$ for $n \geq 2kr + 1$. So $\operatorname{dens}(\mathcal{D}) = kr$ since $_n C_{\leq j}$ is a polynomial in n of degree j. Thus the inequality $\operatorname{dens}(\mathcal{C}^{(k)}) \leq k \cdot \operatorname{dens}(\mathcal{C})$ is sharp. But it does not always hold for $S(\cdot)$ in place of $\operatorname{dens}(\cdot)$: if $\mathcal{C}$ is the collection of open half-spaces in $\mathbb{R}^d$, $d \geq 1$, then $S(\mathcal{C}) = d + 1$ by Radon's theorem. For example, taking $\mathcal{C}$ as the set of d half-spaces $\{x_j > 0\}$ for $j = 1, \ldots, d$, we see that $\mathcal{C}^{(d)}$ shatters a set of 2^d points, one in each coordinate orthant, so $S(\mathcal{C}^{(d)}) \geq 2^d > d(d+1)$ for $d \geq 5$.

Classes with $V(\mathcal{C}) = 0$ or 1 are easily characterized:

Proposition 4.9 *A class $\mathcal{C}$ of subsets of a set X has $V(\mathcal{C}) = 0$, or equivalently $S(\mathcal{C}) = -1$, if and only if $\mathcal{C}$ is empty. Also, $V(\mathcal{C}) = 1$, or equivalently $S(\mathcal{C}) = 0$, if and only if $\mathcal{C}$ contains exactly one set. Thus $S(\mathcal{C}) \geq 1$ if and only if $\mathcal{C}$ contains at least two sets.*

Proof. Clearly $\mathcal{C}$ shatters the empty set if and only if $\mathcal{C}$ contains at least one set. If $\mathcal{C}$ contains at least two sets, then for some $A, B \in \mathcal{C}$ and $x \in X$, $x \in A \setminus B$. Then $\mathcal{C}$ shatters $\{x\}$, so $S(\mathcal{C}) \geq 1$. Conversely if $S(\mathcal{C}) \geq 1$, then clearly $\mathcal{C}$ contains at least two sets.   □

A collection $\mathcal{C}$ of sets is said to be *linearly ordered by inclusion* if for any $A, B \in \mathcal{C}$, either $A \subset B$ or $B \subset A$. Here are two sufficient conditions for $S(\mathcal{C}) = 1$:

Theorem 4.10 *If $\mathcal{C}$ is a collection of at least two subsets of a set X, then $S(\mathcal{C}) = 1$ if either*

(a) $\mathcal{C}$ is linearly ordered by inclusion, or

(b) Any two sets in $\mathcal{C}$ are disjoint.

Proof. In any case $S(\mathcal{C}) \geq 1$. If $\mathcal{C}$ is linearly ordered by inclusion, suppose it shatters $\{x, y\}$ for some $x \neq y$. Let $A, B \in \mathcal{C}$, $A \cap \{x, y\} = \{x\}$, $B \cap \{x, y\} = \{y\}$. But $A \subset B$ or $B \subset A$, giving a contradiction.

If the sets in $\mathcal{C}$ are disjoint, then we can argue as in part (a), and now take $C \in \mathcal{C}$ with $\{x, y\} \subset C$, but C cannot be disjoint from A or B, a contradiction. $\qquad\square$

Example. Let $X = \mathbb{R}$ and let $\mathcal{C}$ be the collection of half-lines $(-\infty, x]$ for all $x \in \mathbb{R}$. Then $\mathcal{C}$ is linearly ordered by inclusion, so $S(\mathcal{C}) = 1$. Applying empirical processes $\sqrt{n}(P_n - P)$ to this class of sets gives the classical empirical processes $\sqrt{n}(F_n - F)$ of Chapter 1.

Section 4.4 will go more into detail about classes of index 1.

4.3 *Maximal Classes

Starred sections are referred to later, if at all, only in other starred sections.

Let $\mathcal{C} \subset \mathcal{A}$ be classes of subsets of a set X. Then $\mathcal{C}$ will be called $(\mathcal{A}, n)$-*maximal* if $S(\mathcal{C}) = n$ and if $\mathcal{C} \subset \mathcal{D}$ strictly and $\mathcal{D} \subset \mathcal{A}$, then $S(\mathcal{D}) > n$. If $\mathcal{A}$ is the class 2^X of all subsets of X, then $\mathcal{C}$ will be called n-*maximal*. If $\mathcal{C}$ is n-maximal, then clearly $\mathcal{C}$ is $(\mathcal{A}, n)$-maximal for any $\mathcal{A}$ such that $\mathcal{C} \subset \mathcal{A} \subset 2^X$.

In view of Proposition 4.9, classes $\mathcal{C}$ with $S(\mathcal{C}) = i$, $i = -1$ or 0, are empty or contain just one set respectively, and so are always i-maximal. Thus n-maximality is interesting only for $n \geq 1$.

Examples. 1. For any set X, let $\mathcal{C}$ consist of $\emptyset$ (the empty set) and all singletons $\{x\}$ for $x \in X$. Then $\mathcal{C}$ is clearly 1-maximal.

2. Let $X = \mathbb{R}$. Let $\mathcal{LH}$ consist of $\emptyset$, $\mathbb{R}$, and all left half-lines, closed $(-\infty, x]$ or open $(-\infty, x)$, for $x \in \mathbb{R}$. In other words, $\mathcal{LH}$ is the collection of all subsets $A \subset \mathbb{R}$ such that whenever $x < y \in A$ then also $x \in A$. Then clearly $S(\mathcal{LH}) = 1$ since for $x < y$ and $A \in \mathcal{LH}$, $A \cap \{x, y\} \neq \{y\}$. But if any subset of $\mathbb{R}$ not in $\mathcal{LH}$ is adjoined, then some 2-element set is shattered, so $\mathcal{LH}$ is 1-maximal.

3. Let $X = \mathbb{R}$ and let Co consist of all subintervals of $\mathbb{R}$, namely $\emptyset$, $\mathbb{R}$, any closed or open, left or right half-line, and any bounded interval, open or closed at either end. In other words, Co is the class of all convex subsets of $\mathbb{R}$. Then $S(Co) = 2$; in fact, Co shatters every 2-element subset of $\mathbb{R}$, while if $x < y < z$ and $A \in Co$, then $A \cap \{x, y, z\} \neq \{x, z\}$. On the other hand if any set not in Co is adjoined to it, its index becomes 3, so Co is 2-maximal.

Here is an existence theorem for maximal classes:

Theorem 4.11 *Let X be a set and $\mathcal{D} \subset 2^X$. Suppose that $\mathcal{C} \subset \mathcal{D}$ and $S(\mathcal{C}) = n$. Then there exists a $(\mathcal{D}, n)$-maximal class $\mathcal{B}$ with $\mathcal{C} \subset \mathcal{B}$.*

Proof. Zorn's Lemma (RAP, Section 1.5) will be applied. Let $(\mathcal{B}_\alpha)_{\alpha \in I}$ be such that $\mathcal{C} \subset \mathcal{B}_\alpha \subset \mathcal{D}$ for all α in the index set I and such that the $\mathcal{B}_\alpha$ are linearly ordered (form a *chain*) by inclusion, with $S(\mathcal{B}_\alpha) = n$ for all α. Let $\mathcal{A} := \bigcup_\alpha \mathcal{B}_\alpha$. To show that $S(\mathcal{A}) = n$, suppose $\mathcal{A}$ shatters a set F with $|F| = n + 1$. Each of the 2^{n+1} subsets of F is induced by a set in some $\mathcal{B}_\alpha$. Since there are only finitely many of these sets and the $\mathcal{B}_\alpha$ are linearly ordered by inclusion, there is some α such that $\mathcal{B}_\alpha$ shatters F, a contradiction. So the chain $\{\mathcal{B}_\alpha\}_{\alpha \in I}$ has an upper bound. So by Zorn's Lemma the collection of all $\mathcal{B}$ with $\mathcal{C} \subset \mathcal{B} \subset \mathcal{D}$ and $S(\mathcal{B}) = n$ has a maximal element (for inclusion). $\qquad \square$

The following fact is straightforward:

Proposition 4.12 *For any set* X, $Y \subset X$, $\mathcal{C} \subset 2^X$, *and* $\mathcal{C}_Y := \mathcal{C} \sqcap Y$, *we have* $S(\mathcal{C}_Y) \le S(\mathcal{C})$.

Recall that $\mathbb{Z}_2 := \{0, 1\}$ with addition mod 2, in other words the usual addition except that $1 + 1 = 0$. For any set X, the group $\mathbb{Z}_2^X$ of all functions from X into $\mathbb{Z}_2$, with the natural addition $(f + g)(x) := f(x) + g(x)$ in $\mathbb{Z}_2$, provides a group structure for the collection 2^X of all subsets of X. Addition of indicator functions mod 2 corresponds to the symmetric difference $A \triangle B := (A \setminus B) \cup (B \setminus A)$, so that $1_A + 1_B = 1_{A \triangle B}$ mod 2. For any fixed set $A \subset X$, the translation $1_B \mapsto 1_A + 1_B$ takes $\mathbb{Z}_2^X$ one-to-one and onto itself. If the functions are restricted to a subset $Y \subset X$, translation still takes $\mathbb{Z}_2^Y$ one-to-one and onto itself. For any $A \subset X$ and $\mathcal{C} \subset 2^X$, let $A \triangle \mathcal{C} := \{A \triangle C : C \in \mathcal{C}\}$. Then for any finite $F \subset X$, $\mathcal{C}$ shatters F if and only if $A \triangle \mathcal{C}$ does. It follows that:

Proposition 4.13 *For any fixed set* $A \subset X$ *and class* $\mathcal{C} \subset 2^X$, *we have* $S(\mathcal{C}) = S(A \triangle \mathcal{C})$, *and* $\mathcal{C}$ *is* n-*maximal if and only if* $A \triangle \mathcal{C}$ *is.*

Next, we have:

Proposition 4.14 *If* $\mathcal{C}$ *is an* n-*maximal class of subsets of a set* X, *and* $n \ge 1$, *then* $\bigcup_{A \in \mathcal{C}} A = X$ *and* $\bigcap_{A \in \mathcal{C}} A = \emptyset$.

Proof. Suppose there is an x such that $x \notin A$ for all $A \in \mathcal{C}$. Take any $B \in \mathcal{C}$ and let $\mathcal{D} := \mathcal{C} \cup (B \cup \{x\})$. Then by n-maximality, $\mathcal{D}$ must shatter some set F of cardinality $n + 1 \ge 2$, and evidently $x \in F$. Thus $\mathcal{D}$ contains at least $2^n \ge 2$ sets which contain x, but it contains only one, a contradiction. This proves the first conclusion. The second follows on taking complements, setting $A = X$ in Proposition 4.13. $\qquad \square$

On $\mathbb{Z}_2^X = 2^X$ there is a product topology coming from the discrete topology on $\mathbb{Z}_2$. The product topology is compact by Tychonoff's theorem (RAP, Theorem 2.2.8).

Proposition 4.15 *For any set X, any n-maximal class $C \subset 2^X$ is closed and so compact in 2^X.*

Proof. Suppose $C_\alpha \to C$ is a convergent net in 2^X with $C_\alpha \in C$ for all α. Then for any finite set $F \subset X$, there is some α with $C_\alpha \cap F = C \cap F$, so $S(C \cup \{C\}) = S(C)$ and $C \in C$. So C is closed (RAP, Theorem 2.1.3). $\square$

A class C of subsets of a set X will be called *complemented* if $X \setminus A \in C$ for every $A \in C$.

Theorem 4.16 *If $S(C) = n$, $C \subset A$ strictly, and C is complemented, then C is not (A, n)-maximal.*

Proof. For any finite set $F \subset X$ and $G \subset F$, $G \in C \sqcap F$ if and only if $F \setminus G \in C \sqcap F$. So if $|F| = n + 1$, then $|C \sqcap F| \leq 2^{n+1} - 2$. So, for any $A \in A \setminus C$, $S(C \cup \{A\}) = n$. $\square$

If $\mathcal{F}$ is a k-dimensional real vector space of real-valued functions on a set X containing the constants and C is the collection $U(\mathcal{F})$ of all sets $\{x : f(x) > 0\}$ or $\{x : f(x) \geq 0\}$ for all $f \in \mathcal{F}$ and real t, then $S(C) = k$ by Theorem 4.6. Since C is complemented, it is never k-maximal.

Let X be any set and $C = C_k$ the collection of all subsets of X with at most k elements. Then clearly $S(C) = k$. Also, C is k-maximal since if $A \notin C$, $A \subset X$, then $|A| > k$, and if B is any subset of A with $|B| = k + 1$, then B is shattered by $C \cup \{A\}$. For $C = C_k$ we have $m^C(n) \equiv {}_nC_{\leq k}$, which is the maximum possible value of $m^C(n)$ by Sauer's Lemma (Theorem 4.2). The following example shows that not all k-maximal classes have these values of $m^C(n)$:

Example. Let $X = \{1, 2, 3, 4\}$,

$$G = \{\{4\}, \{1, 3\}, \{2, 3\}, \{3, 4\}, \{1, 2, 3\}, \{1, 2, 3, 4\}\}.$$

Let C be the complement of G in 2^X. Then it can be checked that C is 2-maximal but $|C| = 10 < {}_4C_{\leq 2} = 11$.

4.4 *Classes of Index 1

In this section the structure of classes C with $S(C) = 1$ will be treated. Recall that for classes of two or more sets, disjoint classes and classes linearly ordered by inclusion have $S(C) = 1$ (Theorem 4.10). A common extension of these two kinds of classes is given by treelike partial orderings, defined as follows.

A binary relation $\leq$ on a set X will be called a *quasi-order* if it is transitive: $x \leq y$ and $y \leq z$ imply $x \leq z$, and reflexive: $x \leq x$ for all $x \in X$. The quasi-order is called a *partial order* if also $x \leq y$ and $y \leq x$ imply $x = y$. For any set S, inclusion ($\subset$) is a partial order on 2^S or any subset of 2^S.

Let $\leq$ be a quasi-order on a set X. Then two elements x and y of X are called *comparable* if at least one of $x \leq y$ and $y \leq x$ holds, or *incomparable* if neither holds. A quasi-order $\leq$ on X will be called *fully comparable* if any two elements of X are comparable. A quasi-order $\leq$ will be called *sub-fully comparable* if for any $y \in X$ and $L_y := \{x : x \leq y\}$, the restriction of $\leq$ to L_y is fully comparable. A fully comparable partial order is called *linear*. A sub-fully comparable partial order will be called *treelike*.

Theorem 4.17 *Let $C \subset 2^X$ contain at least two sets and satisfy, for any $x \neq y$ in X,*

$$A \cap \{x, y\} = \emptyset \text{ for some } A \in C. \tag{4.5}$$

In particular it suffices that $\emptyset \in C$. Then the following are equivalent:

(a) $S(C) = 1$;

(b) For every $Y \subset X$, the inclusion partial ordering of $C_Y := Y \sqcap C$ is treelike;

(c) For every $Y \subset X$ with $|Y| = 2$, the partial ordering of $C_Y := Y \sqcap C$ by inclusion is treelike.

Proof. In proving (a) implies (b), since $S(C_Y) \leq S(C)$ and classes with $S(C) < 1$ contain at most one set and trivially have a treelike ordering by inclusion, we can assume $Y = X$. If C does not have a treelike ordering by inclusion, there is a set $D \in C$ and $B \subset D$, $C \subset D$ such that B and C are not comparable, so there exist some $x \in B \setminus C$ and $y \in C \setminus B$. Take A from assumption (4.5). But then the sets $A, B, C,$ and D shatter $\{x, y\}$ and $S(C) \geq 2$, a contradiction. So (b) holds.

Now (b) implies (c) directly. If (c) holds and $|Y| = 2$, since 2^Y does not have a treelike ordering by inclusion, C must not shatter Y, so (a) follows. $\square$

Proposition 4.18 *Let X be a set and $A \subset 2^X$ where $\emptyset \in A$ and for any B and C in A, $B \cap C \in A$. If C is $(A, 1)$-maximal and satisfies (4.5) for any $x \neq y$ in X, then $\emptyset \in C$ and $B \cap C \in C$ for any B and C in C.*

Proof. If $x \neq y$ and (4.5) holds for A, then $A \cap \{x, y\} = \emptyset = \emptyset \cap \{x, y\}$, so adjoining $\emptyset$ to C does not induce any additional subsets of sets with two elements, and $S(C \cup \{\emptyset\}) = 1$ and by maximality $\emptyset \in C$.

Suppose $B, C \in C$ and $S(C \cup \{B \cap C\}) > 1$. Then for some $x \neq y$ in X, $B \cap C \cap \{x, y\} \neq D \cap \{x, y\}$ for all $D \in C$. Then by (4.5), we can assume $x \in B \cap C$. If $\{x, y\} \subset B \cap C \subset B$, then taking $D = B$ would give a contradiction, so $y \notin B \cap C$. Now $B \cap C \cap \{x, y\} = \{x\} \neq D \cap \{x, y\}$ for $D = B$ or C implies $y \in B$ and $y \in C$, again a contradiction. So $B \cap C \in C$. $\square$

Proposition 4.19 *Let X be a set and C a finite class of subsets of X with $S(C) = 1$ such that for any $x \neq y$ in X, (4.5) holds. Let $D := D(C)$ consist* of $\emptyset$ and all intersections of nonempty subclasses of $\mathcal{C}$. Then $S(\mathcal{D}) = 1$. For each nonempty set $D \in \mathcal{D}$ there is a $C := C(D) \in \mathcal{D}$ such that $C \subset D$ strictly $(C \neq D)$ and if B is any set in $\mathcal{D}$ with $B \subset D$ strictly, then $B \subset C$.

Proof. By Theorem 4.11, let $\mathcal{C} \subset \mathcal{E}$ with $\mathcal{E}$ 1-maximal. Then by Proposition 4.18 for $\mathcal{A} = 2^X$ and induction, $\mathcal{C} \subset \mathcal{D} \subset \mathcal{E}$, so $S(\mathcal{D}) = 1$. Clearly, $\mathcal{D}$ is finite. For each nonempty $D \in \mathcal{D}$, by Theorem 4.17, $\{B \in \mathcal{D} : B \subset D\}$ is linearly ordered by inclusion and contains $\emptyset$, so it has a largest element $C(D)$ other than D itself. $\qquad\square$

Proposition 4.20 *Under the hypotheses of Proposition 4.19, the sets $D \setminus C(D)$ for distinct nonempty $D \in \mathcal{D}$ are all disjoint and are nonempty.*

Proof. Let $A \neq B$ in $\mathcal{D}$. If $B \subset A$, then $B \subset C(A)$, so B and a fortiori $B \setminus C(B)$ are disjoint from $A \setminus C(A)$. Otherwise, $A \cap B \subset B$ strictly, and then $A \cap B \subset C(B)$, so again $A \setminus C(A)$ is disjoint from $B \setminus C(B)$. That $A \setminus C(A) \neq \emptyset$ for $A \neq \emptyset$ follows from the definitions. $\qquad\square$

A *graph* is a nonempty set S together with a set E of unordered pairs $\{x, y\}$ for some $x \neq y$ in S. Then S will be called the set of *nodes* and E the set of *edges* of the graph. The graph (S, E) is called a *tree* if

(a) It is connected, in other words, for any x and y in S there is a finite n and $x_i \in S$, $i = 0, 1, \ldots, n$, such that $x_0 = x$, $x_n = y$, and $\{x_{k-1}, x_k\} \in E$ for $k = 1, \ldots, n$.

(b) The graph is *acyclic*, which means that there is no cycle, where a *cycle* is a set of distinct $x_1, \ldots, x_n \in S$ such that $n \geq 3$, and letting $x_0 := x_n$, $\{x_{k-1}, x_k\} \in E$ for $k = 1, \ldots, n$.

Theorem 4.21 *(a) For m nodes, for any positive integer m, there exist connected graphs with $m - 1$ edges.*

(b) A connected graph with m nodes cannot have fewer than $m - 1$ edges.

(c) A connected graph with m nodes has exactly $m - 1$ edges if and only if it is a tree.

Proof. (a) is clear. (b) will be proved by induction. It is clearly true for $m = 1, 2$. Suppose (S, E) is a connected graph with $|S| = m$, $|E| \leq m - 2$, and $m \geq 3$. The edges in E contain at most $2m - 4$ nodes, counted with multiplicity, so at least 4 nodes appear in only one edge each, or some node is in no edge, a contradiction. Select a node in only one edge and delete it and the edge that contains it. The remaining graph must be connected, but is not by induction assumption, a contradiction, so (b) holds.

For (c), let (S, E) be a connected graph with $|S| = m$ and $|E| = m - 1$. If the graph contains a cycle, we can delete any one edge in the cycle and the graph remains connected, contradicting (b). So (S, E) is a tree.

Conversely, let (S, E) be a tree with $|S| = m$. It will be proved by induction that $|E| \leq m - 1$. This is clearly true for $m = 1, 2$. Suppose $|E| \geq m \geq 3$. Take a maximal set $C := \{x_1, \dots, x_k\} \subset S$ such that the x_i are distinct and $\{x_{j-1}, x_j\} \in E$ for $j = 2, \dots, k$. Then there is no $y \neq x_2$ with $\{x_1, y\} \in E$: y cannot be any x_j, $j \geq 3$, or there would be a cycle, and there is no such $y \notin C$ since C and k are maximal. So we can delete the node x_1 and the edge $\{x_1, x_2\}$ from the graph, leaving a graph which is still a tree with $m - 1$ nodes and at least $m - 1$ edges, contradicting the induction hypothesis and so proving (c).   □

Let the class $\mathcal{D}$ in Propositions 4.19 and 4.20 form the nodes of a graph G whose edges are the pairs $\{C(D), D\}$ for $D \in \mathcal{D}$, $D \neq \emptyset$.

Proposition 4.22 *The graph G is a tree.*

Proof. If $\mathcal{D}$ has m elements, then there are exactly $m - 1$ pairs $\{C(D), D\}$, for $D \in \mathcal{D}$, $D \neq \emptyset$. Starting with any $D \in \mathcal{D}$, we have a decreasing sequence of sets $D \supset C(D) \supset C(C(D)) \supset \cdots$ which must end with the empty set, so all sets in $\mathcal{D}$ are connected in G via the empty set and G is connected. Then by Theorem 4.21 it is a tree.   □

Proposition 4.23 *Let X be a finite set. Let $\mathcal{C}$ be 1-maximal in X and suppose (4.5) holds for all $x \neq y$ in X. Then $\mathcal{C} = \mathcal{D}(\mathcal{C})$ as defined in Proposition 4.19. The sets $D \setminus C(D)$ for nonempty $D \in \mathcal{C}$ are all the singletons $\{x\}$, $x \in X$. If $|X| = m$ then $|\mathcal{C}| = m + 1$.*

Proof. $\mathcal{C} = \mathcal{D}(\mathcal{C})$ by Proposition 4.19. Suppose that for some $D \in \mathcal{C}$, $D \setminus C(D)$ has two or more elements. Then for some $B, C := C(D) \subset B \subset D$ where both inclusions are strict. It will be shown that $S(\mathcal{C} \cup \{B\}) = 1$. If not, then for some $x \neq y$, $\mathcal{C} \cup \{B\}$ shatters $\{x, y\}$, so $B \cap \{x, y\} \neq F \cap \{x, y\}$ for all $F \in \mathcal{C}$. Letting $F = C$ shows that $B \cap \{x, y\} \neq \emptyset$. Likewise, letting $F = D$ shows that $B \cap \{x, y\} \neq \{x, y\}$. So we can assume $B \cap \{x, y\} = \{x\}$. Taking $F = D$ shows that $y \in D$. If $G \cap \{x, y\} = \{y\}$ for some $G \in \mathcal{C}$, then $y \in G \cap D \in \mathcal{C}$, and $G \cap D \subset D$ strictly, so $G \cap D \subset C$ and $y \in C \subset B$, giving a contradiction. So $\mathcal{C} \cup \{B\}$ does not shatter $\{x, y\}$, and $S(\mathcal{C} \cup \{B\}) = 1$, contradicting 1-maximality of $\mathcal{C}$.

So, each set $D \setminus C(D)$ for $\emptyset \neq D \in \mathcal{C}$ is a singleton. Each singleton $\{x\}$ equals $D \setminus C(D)$ for at most one $D \in \mathcal{C}$ by Proposition 4.20. By Proposition 4.14, $X = \bigcup_{D \in \mathcal{C}} D$. For any $x \in X$ take $D_1 \in \mathcal{C}$ with $x \in D_1$. Let $D_{n+1} := C(D_n)$ for $n = 1, 2, \dots$. For some m, $D_m = \emptyset$, and $\{x\} = D_j \setminus C(D_j)$ for some $j < m$. So all singletons are of the form $D \setminus C(D)$, $D \in \mathcal{C}$. This gives a 1–1 correspondence between singletons and nonempty sets in $\mathcal{D}$, so there are exactly m such sets and $|\mathcal{D}| = m + 1$.   □

Suppose in this paragraph (only) that C is a class of two or more sets such that (4.5) holds with $\emptyset$ replaced by $\{x, y\}$. Then the class of complements, $\mathcal{N} := \{X \setminus C : C \in \mathcal{C}\}$, satisfies the original hypotheses of Proposition 4.17. If $\mathcal{C}$ is 1-maximal, so is $\mathcal{N}$ by Proposition 4.13. So Theorem 4.17 and Propositions 4.18 through 4.22 apply to $\mathcal{N}$, and so does Proposition 4.23 if X is finite. Then, $\mathcal{C}$ itself has a "cotreelike" ordering, where for each $C \in \mathcal{C}$, $\{D \in \mathcal{C} : C \subset D\}$ is linearly ordered by inclusion. Propositions 4.18 and 4.19 apply to $\mathcal{C}$ if $\emptyset$ is replaced by X and intersections by unions; in Proposition 4.19, we will have an immediate successor $D(C) \supset C$ instead of a predecessor. We take sets $D(C) \setminus C$ instead of $D \setminus C(D)$ in Propositions 4.20 and 4.23. The resulting tree (Proposition 4.22) then branches out as sets become smaller rather than larger.

Next will be several facts in the general case, i.e. without the hypothesis (4.5).

Theorem 4.24 *Let X be any set and $\mathcal{C}$ any collection of subsets with $S(\mathcal{C}) = 1$. Then for any $C \in \mathcal{C}$, the collection $\mathcal{C}_{X \setminus C} := \{B \setminus C : B \in \mathcal{C}\}$ satisfies (4.5) for any $x \neq y$ as a collection of subsets of $X \setminus C$. Likewise, $\mathcal{C}_{C \setminus} := \{C \setminus B : B \in \mathcal{C}\}$ satisfies (4.5) for any $x \neq y$ as a collection of subsets of C, $S(\mathcal{C}_{C \setminus}) \leq 1$, and $S(\mathcal{C}_{X \setminus C}) \leq 1$.*

Proof. Letting $B = C$ shows that both classes $\mathcal{C}_{C \setminus}$ and $\mathcal{C}_{X \setminus C}$ contain $\emptyset$, so (4.5) holds for them. Both have index $S \leq 1$ by Propositions 4.12 and 4.13. $\square$

So, for an arbitrary class $\mathcal{C}$ with $S(\mathcal{C}) = 1$, we have by Theorem 4.17 a treelike inclusion partial ordering in one part $X \setminus C$ of X and a cotreelike ordering in the complementary part C, for any $C \in \mathcal{C}$. If also $X \setminus C$ happens to be in $\mathcal{C}$, both orderings are linear. To see how the two orderings fit together in general, Proposition 4.13 gives:

Corollary 4.25 *Let $\mathcal{C}$ be any class of sets with $S(\mathcal{C}) = 1$ and $A \in \mathcal{C}$. Let $\mathcal{D} := A \triangle \Delta \mathcal{C}$. Then $S(\mathcal{D}) = 1$ and $\emptyset \in \mathcal{D}$. If $\mathcal{C}$ is 1-maximal, so is $\mathcal{D}$. Then Theorem 4.17, Proposition 4.18, and if $\mathcal{C}$ is finite, Propositions 4.19, 4.20, 4.22, and if X is finite, 4.23, apply to $\mathcal{D}$.*

The last sentence in Proposition 4.23 has a converse and extension:

Proposition 4.26 *Let X be finite with m elements and $\mathcal{C} \subset 2^X$ with $S(\mathcal{C}) = 1$. Then $\mathcal{C}$ is 1-maximal if and only if $|\mathcal{C}| = m + 1$.*

Proof. For any fixed $C \in \mathcal{C}$, we can replace $\mathcal{C}$ by $C \triangle \Delta \mathcal{C}$ without loss of generality by Theorem 4.13. So we can assume $\emptyset \in \mathcal{C}$, and then (4.5) holds for all x, y. Then Proposition 4.23 implies "only if."

Conversely, let $S(\mathcal{C}) = 1$ and $|\mathcal{C}| = m + 1$. Let $\mathcal{C} \subset \mathcal{D}$ strictly. Then $|\mathcal{D}| > m + 1$, so by Sauer's Lemma (Theorem 4.2), $S(\mathcal{D}) \geq 2$. So $\mathcal{C}$ is 1-maximal. $\square$

Now, $m + 1 = {}_m C_{\leq 1}$, which is the maximum value of $m^C(m)$ for $S(C) = 1$ by Sauer's Lemma (Theorem 4.2). The example at the end of Section 4.3 shows that Proposition 4.26 in the form $|C| = {}_m C_{\leq k}$, $k = 1$, does not extend to k-maximality for $k > 1$.

Next it will be shown that 1-maximality can be relativized to subsets. For a set X, a subset $Y \subset X$, and a class $C \subset 2^X$, recall that $C_Y := C \sqcap Y := \{A \cap Y : A \in C\}$.

Theorem 4.27 *If C is 1-maximal and $\emptyset \neq Y \subset X$, then C_Y is a 1-maximal class of subsets of Y.*

Proof. Let $A \in C$. Without changing 1-maximality, C can be replaced by $A \triangle \triangle C$ (Proposition 4.13). So we can assume $\emptyset \in C$. We can also assume that $|X| \geq 2$.

Case 1. Suppose X is finite, $|X| = m < \infty$. Then by Proposition 4.23 or 4.26, $|C| = m + 1$. Let $x \in X$ and $Y := X \setminus \{x\}$. Let $B := \{y \in Y : \{x, y\} \in C_{\{x,y\}}$ and $\{y\} \in C_{\{x,y\}}\}$. A set $C \subset Y$ will be called *full* if $C \in C$ and $C \cup \{x\} \in C$. To continue the proof of the theorem, we have the following:

Lemma 4.28 *If $A \subset Y$ and A is full, then $A = B$.*

Proof. Suppose A is full. First suppose $y \in A \setminus B$. Then $A \cup \{x\} \in C$ implies $\{x, y\} \in C_{\{x,y\}}$ and $A \in C$ implies $\{y\} \in C_{\{x,y\}}$, contradicting $y \notin B$.

Next, suppose $y \in B \setminus A$. Then $\{x\} = (A \cup \{x\}) \cap \{x, y\}$, $\emptyset = \emptyset \cap \{x, y\}$, and $y \in B$ imply that C shatters $\{x, y\}$, a contradiction. So $A = B$, proving the Lemma. $\quad\square$

Continuing the proof of Theorem 4.27, any two distinct sets in C have different intersections with Y, except possibly for B and $B \cup \{x\}$. Now $|C| = m + 1$ by Proposition 4.26, so $|C_Y| \geq m$ and $|C_Y| = m$ since $S(C_Y) \leq 1$ and $|Y| = m - 1$. So if $m \geq 2$, C_Y is a 1-maximal class of subsets of Y by Proposition 4.26 again. Then by induction downward, C_Y is 1-maximal in Y for any $Y \subset X$, $Y \neq \emptyset$, proving the Theorem if X is finite.

Case 2. Let X be general and $\emptyset \neq Y$ finite. Suppose C_Y is not 1-maximal in Y. Then by Case 1, for any finite $Z \supset Y$, C_Z is not 1-maximal in Z. Let $\mathcal{E}$ be the class of all $B \subset Y$ such that $B \notin C_Y$ and $S(\{B\} \cup C_Y) \leq 1$. Suppose for each $B \in \mathcal{E}$ there is some finite $Z(B) \supset Y$ such that there is no $A \subset Z(B)$ with $S(\{A\} \cup C_{Z(B)}) \leq 1$ and $A \cap Y = B$. Let $Z := \bigcup_{B \in \mathcal{E}} Z(B)$, a finite set including Y. So C_Z is strictly included in some class $\mathcal{D}$ which is 1-maximal in Z, by Theorem 4.11 (with $\mathcal{A} = 2^Z$), and $\mathcal{D}_Y$ is 1-maximal by Case 1. So $B \in \mathcal{D}_Y$ for some $B \in \mathcal{E}$, say $D \cap Y = B$ for some $D \in \mathcal{D}$. Then $A := D \cap Z(B)$ gives a contradiction.

So, for some $B \in \mathcal{E}$ and any finite $Z \supset Y$, there is an $A \subset Z$ such that $A \cap Y = B$ and $S(C_Z \cup \{A\}) \leq 1$. Let $\mathcal{D}_Z$ be the class of all subsets D of X for which $D \cap Z$ is such a set A. Then $\mathcal{D}_Z$ is compact in the product topology of

2^X (which was treated in Proposition 4.15). For any finite $U \supset Y$ and $V \supset Y$, $\mathcal{D}_U \cap \mathcal{D}_V \supset \mathcal{D}_{U \cup V} \neq \emptyset$. So the intersection of all $\mathcal{D}_U$ is nonempty. Let $D \in \mathcal{D}_U$ for all finite $U \supset Y$. Then $S(\mathcal{C} \cup \{D\}) = 1$, so $D \in \mathcal{C}$, but $D \cap Y = B \notin \mathcal{C}_Y$, a contradiction. This finishes the proof in Case 2.

Case 3. Let X be infinite and U any infinite subset. Let $\mathcal{F}(U) := \{D \subset U : D \cap Y \in \mathcal{C}_Y$ for all finite $Y \subset U\}$. Then clearly $\mathcal{C}_U \subset \mathcal{F}(U)$. Conversely, if $D \in \mathcal{F}(U)$, then for each finite $Z \subset X$, let $Y := Z \cap U$. Let $\mathcal{G}(Z) := \{A \in \mathcal{C} : A \cap Y = D \cap Y\}$. Then $\mathcal{G}(Z)$ is compact by Proposition 4.15, nonempty since $D \in \mathcal{F}(U)$, and decreases as Z increases. Thus there is some $C \in \bigcap_Z \mathcal{G}_Z$. Then for any finite Z, $C \cap Z \cap U = D \cap Z \cap U$, so $C \cap U = D \cap U = D$. So $D \in \mathcal{C}_U$ and $\mathcal{C}_U = \mathcal{F}(U)$.

For each finite, nonempty $Y \subset U$, $\mathcal{C}_Y$ is 1-maximal in Y by Case 2. Suppose $\mathcal{F}(U)$ is not 1-maximal in U. Let $A \subset U$, $A \notin \mathcal{F}(U)$, $S(\mathcal{E}) \leq 1$ where $\mathcal{E} = \{A\} \cup \mathcal{F}(U)$. For each finite $Y \subset U$, $\mathcal{C}_Y \subset \mathcal{E}_Y$ so $\mathcal{C}_Y = \mathcal{E}_Y$. Let $\mathcal{H}_Y := \{B \in \mathcal{C} : B \cap Y = A \cap Y\}$. Then $\mathcal{H}_Y \neq \emptyset$, and as shown above, $\bigcap_Y \mathcal{H}_Y \neq \emptyset$. Let $B \in \bigcap_Y \mathcal{H}_Y$. Then $A = B \cap U \in \mathcal{C}_U = \mathcal{F}(U)$, a contradiction. So $\mathcal{C}_U$ is 1-maximal in U, and Theorem 4.27 is proved. $\qquad \square$

For any set X and $\mathcal{C} \subset 2^X$, let $x \leq_{\mathcal{C}} y$ iff $x = y$ or $y \in \bigcup_{B \in \mathcal{C}} B$ and for all $A \in \mathcal{C}$, $y \in A$ implies $x \in A$. Then $\leq_{\mathcal{C}}$ is a quasi-order (as defined early in this section) but in general not a partial order. The treelike partial orderings as in Theorem 4.17 were on collections of sets. Now orderings will be defined on X.

Theorem 4.29 *If $S(\mathcal{C}) = 1$, $\bigcup_{B \in \mathcal{C}} B = X$, and $\mathcal{C}$ satisfies (4.5) for all $x \neq y$, then $\leq_{\mathcal{C}}$ is a sub-fully comparable quasi-order on X. If $\mathcal{C}$ is also 1-maximal, then $\leq_{\mathcal{C}}$ is a partial order. Conversely, for any quasi-order $\leq$ on a set X, let $\mathcal{C} := \mathcal{C}_{\leq} := \{A \subset X : A$ is fully comparably quasi-ordered by $\leq$ and $x \in A$ whenever $x \leq y \in A\}$. Then $S(\mathcal{C}) \leq 1$. If $\leq$ is a treelike partial order and $X \neq \emptyset$, then $\mathcal{C}$ is 1-maximal.*

Proof. Suppose $S(\mathcal{C}) = 1$ and (4.5) holds for all $x \neq y$. Let $x \leq_{\mathcal{C}} z$ and $y \leq_{\mathcal{C}} z$. Suppose $z \in B \in \mathcal{C}$. If x and y are not comparable for $\leq_{\mathcal{C}}$, take $C, D \in \mathcal{C}$ with $x \in C \setminus D$, $y \in D \setminus C$. Then $\mathcal{C}$ shatters $\{x, y\}$, a contradiction. So $\{x : x \leq_{\mathcal{C}} z\}$ is fully comparably quasi-ordered by $\leq_{\mathcal{C}}$ and $\leq_{\mathcal{C}}$ is sub-fully comparable.

If $\mathcal{C}$ is 1-maximal, suppose $x \leq_{\mathcal{C}} y$ and $y \leq_{\mathcal{C}} x$. If $x \neq y$, the only subsets of $\{x, y\}$ induced by $\mathcal{C}$ are $\emptyset$ and $\{x, y\}$, contradicting Theorem 4.27. So $x = y$ and $\leq_{\mathcal{C}}$ is a partial order.

Next, let $\leq$ be a quasi-order on X and $\mathcal{C} := \mathcal{C}_{\leq}$. Take any $x \neq y$. If $x \leq y$, then $\{y\} \notin \mathcal{C}_{\{x,y\}}$, or if $y \leq x$, then $\{x\} \notin \mathcal{C}_{\{x,y\}}$. If x and y are not comparable for $\leq$, then $\{x, y\} \notin \mathcal{C}_{\{x,y\}}$. So $S(\mathcal{C}) \leq 1$.

If $\leq$ is a treelike partial order and $X \neq \emptyset$, then $\emptyset \in \mathcal{C}$. For any $x \in X$, $L_x := \{y : y \leq x\} \in \mathcal{C}$, so $S(\mathcal{C}) = 1$.

Let $A \subset X$ and suppose A is not fully comparably ordered by $\leq$. Take $x, y \in A$ which are not comparable for $\leq$. Since $\emptyset$, L_x, and L_y are all in C, $C \cup \{A\}$ shatters $\{x, y\}$, and $S(C \cup \{A\}) \geq 2$.

Let $B \subset X$ and suppose there exist $x \leq y \in B$ with $x \notin B$. Then $L_x \in C$, $L_y \in C$, and $B \cap \{x, y\} = \{y\}$ imply that $S(C \cup \{B\}) \geq 2$. This has now been shown for any $B \notin C$, so C is 1-maximal. $\square$

Example. Let $X = \{1, 2, 3, 4, 5\}$ and

$$\mathcal{G} = \{\emptyset, \{1\}, \{5\}, \{2, 5\}, \{1, 2, 4\}, \{1, 2, 3, 4\}, \{1, 2, 3, 5\}, \{1, 2, 3, 4, 5\}\}.$$

Let C be the complement of $\mathcal{G}$ in 2^X. Then it can be checked that C is 3-maximal but for $Y = \{1, 2, 3, 4\}$, C_Y is not 3-maximal in Y. So in Theorem 4.27, "1-maximal" and $Y \neq \emptyset$ cannot be replaced by "3-maximal" and "$|Y| \geq 3$" respectively.

Recall that a linearly ordered subset of a partially ordered set is called a *chain*.

Theorem 4.30 *Let C be a 1-maximal class of subsets of a set X containing $\emptyset$.*

(I) Then $B \in C$ if and only if both (a) B is a chain for $\leq_C$ and (b) if $x \leq_C y \in B$, then $x \in B$.

(II) If X is finite, $B \in C$ if and only if $B = \emptyset$ or for some $z \in X$, $B = \{x : x \leq_C z\}$.

Proof. To prove "only if" in (I), (b) holds by definition of $\leq_C$. To prove (a), suppose $x, y \in B$ are not comparable for $\leq_C$. By Theorem 4.27 applied to singletons Y, we have $X = \cup_{C \in C} C$. Thus for some $D \in C$, $y \in D$ and $x \notin D$, and for some $E \in C$, $x \in E$ and $y \notin E$. So $C \supset \{\emptyset, D, E, B\}$ shatters $\{x, y\}$, a contradiction. Thus (a) holds.

Conversely, suppose (a) and (b) hold. Suppose $C \cup \{B\}$ shatters some $\{x, y\}$. If $x \leq_C y$, then $C \cap \{x, y\} \neq \{y\}$ for $C \in C$ or $C = B$, a contradiction. So x and y are not comparable for $\leq_C$. Then $C \cap \{x, y\}$ contains $\emptyset$, $\{x\}$ and $\{y\}$, and so not $\{x, y\}$. Also $B \cap \{x, y\} \neq \{x, y\}$, giving another contradiction. So $S(C \cup \{B\}) = 1$ and since C is 1-maximal, $B \in C$, proving "if."

For (II), a B of the given form satisfies (b) clearly, and (a) holds because $\leq_C$ is treelike by Theorem 4.29, so $B \in C$ by part (I). Conversely, if $B \in C$ it is a chain for $\leq_C$ by (I), so if it is nonempty, it has a largest element z, and then $B = \{x : x \leq_C z\}$ by (a) and (b). $\square$

4.5 *Combining VC Classes

Recalling the density as in Corollary 4.4, the following is clear:

Theorem 4.31 *For any set X, if $\mathcal{A} \subset 2^X$ and $C \subset 2^X$, then*

$$dens(\mathcal{A} \cup C) = \max(dens(\mathcal{A}), dens(C)).$$

For the Vapnik–Červonenkis index we have instead:

Proposition 4.32 *For any set* X, $\mathcal{A} \subset 2^X$ *and* $\mathcal{C} \subset 2^X$, $S(\mathcal{A} \cup \mathcal{C}) \leq S(\mathcal{A}) + S(\mathcal{C}) + 1$. *This bound is best possible: for any nonnegative integers* k *and* m *there exist* X, $\mathcal{A}$ *and* $\mathcal{C} \subset 2^X$ *with* $S(\mathcal{A}) = k$, $S(\mathcal{C}) = m$ *and* $S(\mathcal{A} \cup \mathcal{C}) = k + m + 1$.

Proof. By Theorem 4.2, if $k := S(\mathcal{A})$, $m := S(\mathcal{C})$, and $n \geq k + m + 2$,

$$m^{\mathcal{A} \cup \mathcal{C}}(n) \leq m^{\mathcal{A}}(n) + m^{\mathcal{C}}(n) \leq {}_nC_{\leq k} + {}_nC_{\leq m}$$

$$= \sum_{j=0}^{k} \binom{n}{j} + \sum_{j=n-m}^{n} \binom{n}{j} < 2^n,$$

so $S(\mathcal{A} \cup \mathcal{C}) < n$. It follows that $S(\mathcal{A} \cup \mathcal{C}) \leq k + m + 1$ as claimed. Conversely, given k and m, let $n = k + m + 1$ and let X be a set with n members. Let $\mathcal{A}$ be the set of all subsets of X with at most k members and $\mathcal{C}$ the set of subsets of X with at least $n - m = k + 1$ members. Then $S(\mathcal{A}) = k$, $S(\mathcal{C}) = m$ and $\mathcal{A} \cup \mathcal{C} = 2^X$, so $S(\mathcal{A} \cup \mathcal{C}) = n$. $\qquad\square$

Let X be a set and $\mathcal{C}$, $\mathcal{D}$ any two collections of subsets of X. Let

$$\mathcal{C} \sqcap \mathcal{D} := \{C \cap D : C \in \mathcal{C}, D \in \mathcal{D}\}, \quad \mathcal{C} \sqcup \mathcal{D} := \{C \cup D : C \in \mathcal{C}, D \in \mathcal{D}\}.$$

If $\mathcal{A}$ is a class of subsets of another set Y, let

$$\mathcal{C} \boxtimes \mathcal{A} := \{C \times A : C \in \mathcal{C}, A \in \mathcal{A}\}.$$

For such classes, we have:

Theorem 4.33 *For any* $\mathcal{C} \subset 2^X$ *and* $\mathcal{D} \subset 2^X$ *or* 2^Y *let* $k := dens(\mathcal{C})$ *and* $m := dens(\mathcal{D})$. *Then we have:* $dens(\mathcal{C} \,\square\, \mathcal{D}) \leq k + m$ *for* $\square = \sqcap$, $\sqcup$ *or* $\boxtimes$.

Proof. For any of the three operations we have $m^{\mathcal{C} \square \mathcal{D}}(n) \leq m^{\mathcal{C}}(n)m^{\mathcal{D}}(n)$, and the conclusions follow. $\qquad\square$

For the Vapnik–Červonenkis index the behavior of the $\square$ operations is not so simple. For $k, m = 0, 1, 2, \ldots$, and $\square = \sqcup, \sqcap$ or $\boxtimes$ let $\square(k, m) := \max\{S(\mathcal{C} \,\square\, \mathcal{D}) : S(\mathcal{C}) = k, S(\mathcal{D}) = m\}$. Here the maximum is taken where X and Y may be infinite sets. Then we have:

Theorem 4.34 *For any* $k = 0, 1, 2, \ldots$ *and* $m = 0, 1, 2, \ldots$, *and* $\square = \sqcup, \sqcap$ *or* $\boxtimes$, *we have* $\square(k, m) < \infty$.

Proof. By Theorem 4.2 and Proposition 4.3, if $S(\mathcal{C}) = k$ and $S(\mathcal{D}) = m$, then

$$m^{\mathcal{C} \square \mathcal{D}}(n) \leq m^{\mathcal{C}}(n)m^{\mathcal{D}}(n) \leq 9n^{k+m}/(4k!m!) < 2^n$$

for $n \geq n_0(k, m)$ large enough. $\qquad\square$

Theorem 4.35 *For any* $k, m = 0, 1, 2, \ldots$, $\sqcap(k, m) = \sqcup(k, m) = \boxtimes(k, m)$.

Proof. The first equation follows from taking complements (Proposition 4.5). For the second, for given k and m by Theorem 4.34 it is enough to consider large enough finite sets in place of X and Y, and then we can assume $X = Y$. We have $\sqcap(k, m) \le \boxtimes(k, m)$ by restricting to the diagonal in $X \times X$ and applying Proposition 4.12.

In the other direction let Π_X and Π_Y be the projections of $X \times Y$ onto X and Y respectively. Let $\mathcal{F} := \{\Pi_X^{-1}(C) : C \in \mathcal{C}\}$, $\mathcal{B} = \{\Pi_Y^{-1}(A) : A \in \mathcal{A}\}$. By Theorem 4.7, $S(\mathcal{F}) = S(\mathcal{C})$ and $S(\mathcal{B}) = S(\mathcal{A})$. Since $\Pi_X^{-1}(C) \cap \Pi_Y^{-1}(A) \equiv C \times A$ it follows that $S(\mathcal{F} \sqcap \mathcal{B}) \ge S(\mathcal{C} \boxtimes \mathcal{A})$, and the Theorem is proved.   □

Let $S(k, m)$ be the common value of the quantities in Theorem 4.35. Theorem 4.34 can be improved as follows. For any nonnegative integers j, k let $\theta(j, k) := \sup\{r \in \mathbb{N} : ({}_rC_{\le j})({}_rC_{\le k}) \ge 2^r\}$. Then $\theta(j, k) < \infty$ for each j, k by Proposition 4.3 and we have:

Proposition 4.36 $S(j, k) \le \theta(j, k)$ *for any* $j, k \in \mathbb{N}$.

Proof. Let $S(\mathcal{C}) = j$ and $S(\mathcal{D}) = k$. Then for any $n > \theta(j, k)$, by Sauer's Lemma (Theorem 4.2), and Proposition 4.3 again,

$$m^{\mathcal{C} \sqcup \mathcal{D}}(n) \le m^{\mathcal{C}}(n) m^{\mathcal{D}}(n) \le ({}_nC_{\le j})({}_nC_{\le k}) < 2^n.$$

This finishes the proof.   □

Can the values $S(k, m)$ be computed? The next two theorems and proposition will give some information.

Theorem 4.37 *Let X be a set, $\mathcal{C}, \mathcal{D} \subset 2^X$, and $\mathcal{C} \sqcup \mathcal{D} = 2^X$. Let $A \subset X$ and suppose for all $B \in \mathcal{C}$, either $B \subset A$ or $B \subset A^c$. Then $\mathcal{D}$ shatters either A or A^c.*

Proof. Suppose $\mathcal{D}$ does not shatter A. Then take $H \subset A$ such that $D \cap A \ne H$ for all $D \in \mathcal{D}$. Take any $E \subset A^c$. Then $E \cup H = C \cup D$ for some $C \in \mathcal{C}$ and $D \in \mathcal{D}$. If $C \subset A^c$ then $D \cap A = H$, a contradiction. So $C \subset A$ and $D \cap A^c = E$. Thus $\mathcal{D}$ shatters A^c.   □

For any set Y recall that $|Y|$ denotes the number of elements of Y. Here is an upper bound for $S(1, k)$ that will be shown to be exact for $k = 1, 2, 3$ in Proposition 4.39.

Theorem 4.38 *For any* $k = 1, 2, \ldots,$ $S(1, k) \le 2k + 1$.

Proof. Suppose $|X| = 2k + 2$, $\mathcal{C} \sqcup \mathcal{D} = 2^X$, $S(\mathcal{C}) = 1$, and $S(\mathcal{D}) = k$. We can assume by Theorem 4.11 that $\mathcal{C}$ is 1-maximal. Then $\cup_{B \in \mathcal{C}} B = X$ by Theorem 4.27 applied to singletons Y. We have $\emptyset \in \mathcal{C} \cap \mathcal{D}$. Thus by Theorem 4.17, $\mathcal{C}$ has a treelike partial ordering by inclusion, which induces such a treelike partial ordering on X by Theorem 4.29. Let Y be the set of elements of X having at least one predecessor for this ordering. Each $y \in Y$ has a smallest predecessor $f(y) \notin Y$. For each $B \subset Y$, we have $B = C \cup D, C \in \mathcal{C}, D \in \mathcal{D}$, where $C = \emptyset$, $B = D$ since if $y \in C \cap Y$, $f(y) \in C \setminus Y$. So $\mathcal{D}$ shatters Y and Y has at most k elements.

Let r be the number of values of f, say $t_1, \ldots, t_r$. Then Y is decomposed into disjoint subsets $Y_1, \ldots, Y_r$ such that $f = t_j$ on Y_j for each j. Let $C := (Y \cup \operatorname{ran} f)^c$. Then $|C| \geq 2$. Let $n_j := |Y_j|, j = 1, \ldots, r$. Then

$$2k + 2 = |C| + \sum_{j=1}^{r}(n_j + 1). \tag{4.6}$$

It will be shown that there exist subsets $E \subset C$ and $I \subset \{1, \ldots, r\}$ such that

$$|E| + \sum_{j \in I}(n_j + 1) = k + 1. \tag{4.7}$$

Let K be the largest possible value $\leq k + 1$ of the left side of (4.7). Suppose $K \leq k$. Then

$$K = |C| + \sum_{j \in J}(n_j + 1) \tag{4.8}$$

for some $J \subset \{1, \ldots, r\}$ since elements of C could be put into E one at a time. We then have by (4.6)

$$\sum_{j \notin J}(n_j + 1) = 2k + 2 - K \geq k + 2. \tag{4.9}$$

Let n_0 be the smallest value of n_j for $j \notin J$. Then $n_0 \geq |C| + 1$, or another j could be put in I for a suitable E on the left side of (4.7), giving a larger K. Since each $n_j \leq k$, there must be at least two $j \notin J$ by (4.9). Thus $r - 2 + 2n_0 \leq |Y| \leq k, r \leq k - 2|C|$, and by (4.6),

$$2k + 2 - |C| = \sum_{j=1}^{r}(n_j + 1) \leq 2k - 2|C|$$

and $|C| \leq -2$, a contradiction, so $K = k + 1$ and (4.7) is proved.

Thus there is a set $A \subset X$ with $|A| = k + 1$, $A := E \cup \bigcup_{j \in I} Y_j \cup \{t_j\}$, with E and I from (4.7). Let $B \in \mathcal{C}$. Then by Theorem 4.30(I)(a), either $B \subset Y_j \cup \{t_j\}$ for some j or B is a singleton. Thus either $B \subset A$ or $B \subset A^c$. So Theorem 4.37 applies and $S(\mathcal{D}) \geq k + 1$, a contradiction. $\qquad\square$

For $k = 1, 2, 3$ we have, where the lower bound for $k = 2$ is due to L. Birgé,

Proposition 4.39 $S(1, k) = 2k + 1$ for $k = 1, 2, 3$.

Proof. By Theorem 4.38 we need to show $S(1, k) \geq 2k + 1$ for $k = 1, 2, 3$. Sets $\{a, b, \ldots, d\}$ will be denoted $ab \cdots d$, e.g. 1246 $:= \{1, 2, 4, 6\}$. For $k = 1$ let $X := 123, \mathcal{C} := \{\emptyset, 1, 2, 3\}$, and $\mathcal{D} := \{\emptyset, 1, 2, 23\}$. Then clearly $S(\mathcal{C}) = 1$, $S(\mathcal{D}) = 1$, and $S(\mathcal{C} \sqcup \mathcal{D}) = 3$, so $S(1, 1) \geq 3$.

For $k = 2$ let $X := 12345, \mathcal{C} := \{\emptyset, 1, 2, 3, 4, 45\}$,

$$\mathcal{D} := \{\emptyset, 1, 2, 3, 5, 12, 13, 15, 23, 25, 234, 235, 2345\}.$$

Then one can check that $S(\mathcal{C}) = 1$, $S(\mathcal{D}) = 2$, and $\mathcal{C} \sqcup \mathcal{D} = 2^X$. So $S(1, 2) \geq 5$.

To show that $S(1, 3) = 7$, take the set $X := \{0, 1, 2, 3, 4, 5, 6\}$. We will find classes $\mathcal{C} \subset 2^X$ and $\mathcal{E} \subset 2^X$ with $S(\mathcal{C}) = 1$, $S(\mathcal{E}) = 3$, and $\mathcal{C} \sqcup \mathcal{E} = 2^X$.

Let $\mathcal{C} := \{\emptyset, 0, 1, 12, 3, 34, 5, 56\}$. Then $\mathcal{C}$ has a treelike partial ordering by inclusion and $S(\mathcal{C}) = 1$.

A set with k elements is called a k-set. $\mathcal{E}$ will contain the following subsets of X: the 0-set $\emptyset$; all 1-sets; all 2-sets except 12 and 34; all 3-sets not including 12 or 34; all 4-sets included in 01234; and the 5-sets 01234 and 12346. Then $\mathcal{E}$ shatters some 3-sets, e.g., 246. To show $S(\mathcal{E}) = 3$ we need to show $\mathcal{E}$ shatters no 4-set. $\mathcal{E}$ shatters no 4-set containing 5 since there is no set A in $\mathcal{E}$ with cardinality $|A| \geq 4$ containing 5.

A 4-set $B \subset 01234$ includes at least one of the pairs 12 or 34. By symmetry, suppose $12 \subset B$. Each set C in $\mathcal{E}$ including 12 contains at least two of 0, 3 and 4, so $|C \cap B| \geq 3$ and $C \cap B \neq 12$. Thus $\mathcal{E}$ does not shatter B. It remains to consider 4-sets containing 6 and not 5. There is no $A \in \mathcal{E}$ including 06 with $|A| \geq 4$. Thus $\mathcal{E}$ does not shatter any 4-set including 06. The sets 1236 and 1246 are not shattered by $\mathcal{E}$ because the subset 126 is not cut from them. Likewise the sets 1346 and 2346 are not shattered because 346 is not cut from them. Thus $S(\mathcal{E}) = 3$.

To show $\mathcal{C} \sqcup \mathcal{E} = 2^X$, clearly $\mathcal{C} \sqcup \mathcal{E}$ contains all 0- and 1-sets, and it is easy to check that it contains all 2-sets and all 3-sets A not including 12 or 34. If $A \supset 12$, then $A = 12 \cup c$ where $12 \in \mathcal{C}$ and $c \in \mathcal{E}$, and likewise for 34.

$\mathcal{C} \sqcup \mathcal{E}$ contains $X = 56 \cup 01234$, $012345 = 5 \cup 01234$, and $012346 = 0 \cup 12346$. Each other 6-set is the union of $56 \in \mathcal{C}$ and a 4-set in $\mathcal{E}$ included in 01234.

A 5-set containing 5 and not 6 is the union of $5 \in \mathcal{C}$ and a 4-set $\subset 01234$. A 5-set F containing 6 and not 5 includes at least one of the pairs $P_1 = 12$ or $P_2 = 34$. If it includes both pairs, it is in $\mathcal{E}$. If it includes just one pair P_j, we have

(*) $\qquad\qquad F = A \cup (F \setminus A), \ A \in \mathcal{C}, \ F \setminus A \in \mathcal{E},$

for $A = P_j$.

If a 5-set $F \supset 56$ includes a pair P_j, then (*) follows likewise. Otherwise it holds for $A = 56$. The remaining 5-set, 01234, is in $\mathcal{E}$.

A 4-set $\subset 01234$ is in $\mathcal{E}$. A 4-set F containing 5 or 6 or both includes at most one pair P_j. If it includes P_j, (*) holds for $A = P_j$. So suppose F includes neither pair P_j. If $56 \subset F$, then (*) holds for $A = 56$. If $5 \in F$ and $6 \notin F$, then (*) holds for $A = 5$. The remaining case is $6 \in F$ and $5 \notin F$. At least one of $a = 0, 1$, or 3 is in F, and (*) holds for $A = a$. The proof of the case $k = 3$ of Proposition 4.39 is complete. $\qquad\square$

For classes satisfying stronger conditions, more is true:

Theorem 4.40 *Let X and Y be sets, $\mathcal{C} \subset 2^X$ and $\mathcal{D} \subset 2^Y$. If $\mathcal{C}$ is linearly ordered by inclusion, then $S(\mathcal{C} \boxtimes \mathcal{D}) \leq S(\mathcal{D}) + 1$.*

Proof. Let $m = S(\mathcal{D})$. We can assume that $m < \infty$ and $X \in \mathcal{C}$. Let $F \subset X \times Y$ with $|F| = m + 2$. Suppose $\mathcal{C} \boxtimes \mathcal{D}$ shatters F. The subsets of $\Pi_X F \subset X$ induced by $\mathcal{C}$ are linearly ordered by inclusion. Let G be the next largest other than $\Pi_X F$ itself and $p \in \Pi_X F \setminus G$. Take y such that $(p, y) \in F$. Then all subsets of F containing (p, y) are induced by sets of the form $\Pi_X F \times D$, $D \in \mathcal{D}$. Thus $H := F \setminus \{(p, y)\}$ is shattered by such sets. Now $|H| = m + 1$, and Π_Y must be one-to-one on H or it could not be shattered by sets of the given form. But then $\mathcal{D}$ shatters $\Pi_Y H$, giving a contradiction. $\qquad\square$

Theorem 4.41 *For any set X and $\mathcal{C}, \mathcal{D} \subset 2^X$, if $\mathcal{C}$ is linearly ordered by inclusion, then $S(\mathcal{C} \boxdot \mathcal{D}) \leq S(\mathcal{D}) + 1$ for $\boxdot = \sqcap$ or $\sqcup$.*

Proof. First, consider $\boxdot = \sqcap$. Again let $m = S(\mathcal{D})$, and we can assume $m < \infty$. Let $F \subset X$ and $|F| = m + 2$. Suppose $\mathcal{C} \sqcap \mathcal{D}$ shatters F. Now $\mathcal{C}_F$ is linearly ordered by inclusion. We can assume that $X \in \mathcal{C}$, so $F \in \mathcal{C}_F$. Let G be the next largest element of $\mathcal{C}_F$ and $p \in F \setminus G$. Each set $A \subset F$ containing p is of the form $C \cap D \cap F$, $C \in \mathcal{C}$, $D \in \mathcal{D}$, so we must have $C \cap F = F$, and $A = D \cap F$. So $\mathcal{D}$ shatters $F \setminus \{p\}$, a contradiction.

Since the complements of a class linearly ordered by inclusion are again linearly ordered by inclusion, the case of unions follows by taking complements (Proposition 4.5). $\qquad\square$

Then by Theorem 4.10 and induction we have:

Corollary 4.42 *Let $\mathcal{C}_i$ be classes of subsets of a set X and $\mathcal{C} := \{\bigcap_{i=1}^n C_i : C_i \in \mathcal{C}_i, i = 1, \ldots, n\}$, where each $\mathcal{C}_i$ is linearly ordered by inclusion. Then $S(\mathcal{C}) \leq n$.*

Definition. For any set X and Vapnik–Červonenkis class $\mathcal{C} \subset 2^X$, $\mathcal{C}$ will be called *bordered* if for some $F \subset X$, with $|F| = S(\mathcal{C})$, and $x \in X \setminus F$, F is shattered by sets in $\mathcal{C}$ all containing x.

Theorem 4.43 *Let $\mathcal{C}_i \subset 2^{X(i)}$ be bordered Vapnik–Červonenkis classes for $i = 1, 2$. Then $S(\mathcal{C}_1 \boxtimes \mathcal{C}_2) \geq S(\mathcal{C}_1) + S(\mathcal{C}_2)$.*

Proof. Take $F_i \subset X(i)$ and x_i as in the definition of "bordered." Let $H :=$ $(\{x_1\} \times F_2) \cup (F_1 \times \{x_2\})$. Then $|H| = S(\mathcal{C}_1) + S(\mathcal{C}_2)$. For any $V_i \subset F_i$, $i =$ $1, 2$, take $C_i \in \mathcal{C}_i$ with $C_i \cap F_i = V_i$ and $x_i \in C_i$. Then $(C_1 \times C_2) \cap H =$ $(\{x_1\} \times V_2) \cup (V_1 \times \{x_2\})$, so $\mathcal{C}_1 \boxtimes \mathcal{C}_2$ shatters H. $\qquad\square$

Theorem 4.43 extends by induction to any number of factors. One consequence is:

Corollary 4.44 *In $\mathbb{R}$ let $\mathcal{J}$ be the set of all intervals, which may be open or closed, bounded or unbounded on each side. In other words $\mathcal{J}$ is the set of all convex subsets of $\mathbb{R}$. In $\mathbb{R}^m$ let $\mathcal{C}$ be the collection of all rectangles parallel to the axes, $\mathcal{C} := \{\Pi_{i=1}^m J_i : J_i \in \mathcal{J}, i = 1, \ldots, m\}$. Then $S(\mathcal{C}) = 2m$. Let $\mathcal{D}$ be the set of all left half-lines $(-\infty, x]$ or $(-\infty, x)$ for $x \in \mathbb{R}$. Let $\mathcal{T} := \{\Pi_{i=1}^m H_i : H_i \in \mathcal{D}, i = 1, \ldots, m\}$, so $\mathcal{T}$ is the class of lower orthants parallel to the given axes. Then $S(\mathcal{C}) = m$.*

Proof. The class $\mathcal{D}$ is linearly ordered by inclusion and is bordered with $S(\mathcal{D}) = 1$. So is the class of half-lines $[a, \infty)$ or (a, ∞), $a \in \mathbb{R}$. The class $\mathcal{J}$ of all intervals in $\mathbb{R}$ is bordered with $S(\mathcal{J}) = 2$. The results now follow from Corollary 4.42 with $n = 2m$ and $n = m$, and Theorem 4.43 and induction. $\quad\square$

Proposition 4.45 *Let $\mathcal{I}$ be the set of all intervals in $\mathbb{R}$. Let Y be any set and $\mathcal{C} \subset 2^Y$ with $Y \in \mathcal{C}$. Then in $\mathbb{R} \times Y$, $S(\mathcal{J} \boxtimes \mathcal{C}) \leq 2 + S(\mathcal{C})$.*

Proof. If $S(\mathcal{C}) = +\infty$ the result is clear, so suppose $m := S(\mathcal{C}) < \infty$. Let $F \subset \mathbb{R} \times Y$ with $|F| = 3 + m$ and suppose $\mathcal{J} \boxtimes \mathcal{C}$ shatters F. Let (x_i, y_i), $i = 1, \ldots, m + 3$, be the points of F. Let $u := \min_i x_i$ and $v := \max_j x_j$. Let $p := (u, y_i) \in F$ and $q := (v, y_j) \in F$. All subsets of F which include $\{p, q\}$ must be induced by sets of the form $\mathbb{R} \times C$, $C \in \mathcal{C}$. So Π_Y must be one-to-one on $F \setminus \{p, q\}$, and $\mathcal{C}$ shatters $\Pi_Y(F \setminus \{p, q\})$ of cardinality $m + 1$, a contradiction. $\qquad\square$

Next is a necessary condition for a class $\mathcal{C}$ to be of index 1. Recall that a *chain* of sets is a class of sets linearly ordered by inclusion. For any class $\mathcal{D}$ of sets let $\mathcal{D}' := \{A^c : A \in \mathcal{D}\}$.

Theorem 4.46 (A. Smoktunowicz) *In a set X, let $\mathcal{C} \subset 2^X$ and $S(\mathcal{C}) = 1$.*

(i) If $\emptyset \in \mathcal{C}$ then for some chains $\mathcal{A}$ and $\mathcal{B}$, $\mathcal{C} \subset \mathcal{A} \sqcap \mathcal{B}$.

(ii) In general, for some chains $\mathcal{A}_i$, $i = 1, 2, 3, 4$,

$$\mathcal{C} \subset (\mathcal{A}_1 \sqcap \mathcal{A}_2) \sqcup (\mathcal{A}_3 \sqcap \mathcal{A}_4)'.$$

Proof. For (ii), for any $A \in \mathcal{C}$, $\mathcal{C} \sqcap A^c$ and $\mathcal{C}' \sqcap A$ are VC classes of index 1, containing $\emptyset$, and

$$\mathcal{C} \subset (\mathcal{C} \sqcap A^c) \sqcup ((\mathcal{C}' \sqcap A)' \sqcap A).$$

Assuming (i), $C' \sqcap A \subset \mathcal{B}_3 \sqcap \mathcal{B}_4$ for some chains $\mathcal{B}_3, \mathcal{B}_4$ of subsets of A. Letting $\mathcal{A}_j := \mathcal{B}_j \sqcup A^c$, $j = 3, 4$, we have $(C' \sqcap A)' \sqcap A \subset (\mathcal{A}_3 \sqcap \mathcal{A}_4)'$. So (i) implies (ii).

To prove (i), by Theorem 4.11 we can assume $\mathcal{C}$ is 1-maximal. Thus since $\emptyset \in \mathcal{C}$, $\mathcal{C}$ has a treelike partial ordering by inclusion by Theorem 4.17. First suppose X is finite. By Theorem 4.29, take the treelike partial ordering of X induced by $\mathcal{C}$.

Any chain is included in a maximal chain (for inclusion), and in a finite set of n elements, a maximal chain is of the form

$$\{\emptyset, \{a_1\}, \{a_1, a_2\}, \dots, \{a_1, \dots, a_n\}\}$$

and thus is equivalent to defining a linear ordering of the set, $a_1 < a_2 < \cdots < a_n$. To define our two chains $\mathcal{A}$, $\mathcal{B}$ we will thus define two linear orderings $<_A$, $<_B$ of X. This will be done recursively as follows. Take the elements of X having no predecessors (there must be at least one) and call them $a_1, \dots, a_k$ for some choice of indices. Let $a_1 <_A a_2 <_A \cdots <_A a_k$, $a_k <_B a_{k-1} <_B \cdots <_B a_1$.

Next, suppose a_j has immediate successors $a_{j1}, \dots, a_{jr}$. Let $a_j <_A a_{j1} <_A \cdots <_A a_{jr} <_A a_{j+1}$ where "$<_A a_{j+1}$" is omitted if $j = k$. Also let $a_j <_B a_{jr} <_B a_{j,r-1} <_B \cdots <_B a_{j1} <_B a_{j-1}$ where "$<_B a_{j-1}$" is omitted if $j = 1$. Iterating such definitions we get two linear orderings of X, each defining a chain of sets as above, so we get chains $\mathcal{A}, \mathcal{B}$.

By Theorem 4.30, every element of $\mathcal{C}$ is a set of the form

$$C := \{a_{j_1}, a_{j_1 j_2}, \dots, a^{(m)} := a_{j_1 j_2 \cdots j_m}\}$$

where $a_{j_1} \leq_C a_{j_1 j_2} \leq_C \cdots \leq_C a^{(m)}$ and "$\leq_C$" can be replaced by either $<_A$ or $<_B$. It can be checked easily that

$$C = \{x \in X : x \leq_A a^{(m)}\} \cap \{x \in X : x \leq_B a^{(m)}\}.$$

Thus C is the intersection of a set in $\mathcal{A}$ and a set in $\mathcal{B}$, and the finite case is done.

Now suppose X is infinite, $\emptyset \in \mathcal{C} \subset 2^X$ and $S(\mathcal{C}) = 1$. Let $\mathcal{H} := \{J \subset X : J \neq \emptyset, |J| < \infty\}$. For each $J \in \mathcal{H}$ take chains $\mathcal{P}_{Ji}$, $i = 1, 2$, such that $\mathcal{C} \sqcap J \subset \mathcal{P}_{J1} \sqcap \mathcal{P}_{J2}$. Let h be a non-point ultrafilter of subsets of $\mathcal{H}$, in other words:

(1) h is a nonempty collection of nonempty subsets of $\mathcal{H}$.
(2) If $\mathcal{A}, \mathcal{B} \in h$ then $\mathcal{A} \cap \mathcal{B} \in h$.
(3) If $\mathcal{A} \in h$ and $\mathcal{A} \subset \mathcal{B} \subset \mathcal{H}$, then $\mathcal{B} \in h$.
(4) For all $\mathcal{A} \subset \mathcal{H}$, either $\mathcal{A} \in h$ or $\mathcal{A}^c \in h$.
(5) For each $J \in \mathcal{H}$, $\{J\} \notin h$.

The first three conditions make h a filter, the fourth an ultrafilter, and the fifth a non-point ultrafilter. Non-point ultrafilters exist by the axiom of choice (RAP, Theorem 2.2.4 and the statement after its proof).

Given any indexed family of nonempty sets $\{A_J\}_{J \in \mathcal{H}}$, $\Pi_{J \in \mathcal{H}} A_J$ is the set of all $\{a_J\}_{J \in \mathcal{H}}$ such that $a_J \in A_J$ for all $J \in \mathcal{H}$. For $\{a_J\}_{J \in \mathcal{H}}$, $\{b_J\}_{J \in \mathcal{H}}$ in $\Pi_{J \in \mathcal{H}} A_J$, let $\{a_J\}_{J \in \mathcal{H}} \equiv_h \{b_J\}_{J \in \mathcal{H}}$ if and only if for some $\mathcal{A} \in h$, $a_J = b_J$ for all $J \in \mathcal{A}$. Let $\lim_h A_J$ be the set of equivalence classes of members of $\Pi_{J \in \mathcal{H}} A_J$ for the relation $\equiv_h$. Let $\{a_J\}_{J \in \mathcal{H}}^{(\equiv)}$ be the equivalence class to which $\{a_J\}_{J \in \mathcal{H}}$ belongs.

If $\mathcal{B}_J$ is a class of subsets of A_J for each $J \in \mathcal{H}$, then for each element $\mathcal{Z}$ of $\lim_h \mathcal{B}_J$, where $\mathcal{Z} = \{B_j\}_{J \in \mathcal{H}}^{(\equiv)}$ for some $B_J \in \mathcal{B}_J$ for each J, define a set $\mathcal{E}_{\mathcal{Z}} \subset \lim_h A_J$ by $\{a_J\}_{J \in \mathcal{H}}^{(\equiv)} \in \mathcal{E}_{\mathcal{Z}}$ if and only if for some $\mathcal{A} \in h$, $a_J \in B_J$ for all $J \in \mathcal{A}$. Let $(\lim)_h \mathcal{B}_J := \{\mathcal{E}_{\mathcal{Z}} : \mathcal{Z} \in \lim_h \mathcal{B}_J\}$.

Let $\overline{X} := \lim_h J$, $\overline{\mathcal{C}} := (\lim)_h \mathcal{C} \sqcap J$ and for $i = 1, 2$ let $\overline{\mathcal{P}}_i := (\lim)_h \mathcal{P}_{Ji}$. To see that each $\overline{\mathcal{P}}_i$ is a chain of subsets of $\overline{X}$, we can take $i = 1$. Let $\mathcal{W}$, $\mathcal{Z} \in \lim_h \mathcal{P}_{J1}, \{B_J\}_{J \in \mathcal{H}} \in \mathcal{W}, \{C_J\}_{J \in \mathcal{H}} \in \mathcal{Z}$. Let $\mathcal{J} := \{J \in \mathcal{H} : B_J \subset C_J\}$. Then either $\mathcal{J} \in h$ or $\mathcal{J}^c \in h$. If $\mathcal{J} \in h$ then clearly $\mathcal{E}_{\mathcal{W}} \subset \mathcal{E}_{\mathcal{Z}}$. Otherwise, $\mathcal{J}^c \in h$, and since $\mathcal{P}_{J1}$ is a chain for each J, $C_J \subset B_J$ for all $J \in \mathcal{J}^c$ and $\mathcal{E}_{\mathcal{Z}} \subset \mathcal{E}_{\mathcal{W}}$.

It is easy to check that $\overline{\mathcal{C}} \subset \overline{\mathcal{P}}_1 \sqcap \overline{\mathcal{P}}_2$. There is a natural 1–1 map i of X into $\overline{X}$ by $\{a_J\}_{J \in \mathcal{H}}^{(\equiv)} \in i(x)$ if and only if for some $\mathcal{A} \in h$, $a_J = x$ for all $J \in \mathcal{A}$. So we can view X as a subset of $\overline{X}$. Each $\overline{\mathcal{P}}_j \sqcap X$ is a chain of subsets of X, and

$$\mathcal{C} \subset \overline{\mathcal{C}} \sqcap X \subset (\overline{\mathcal{P}}_1 \sqcap X) \sqcap (\overline{\mathcal{P}}_2 \sqcap X),$$

completing the proof.   □

Section 4.4 describes the structure of classes $\mathcal{C}$ with $S(\mathcal{C}) = 1$, but the structure of VC classes with $S(\mathcal{C}) = k$ for $k > 1$ apparently is not known in general. Smoktunowicz (1997) showed that the class $\mathcal{L}$ of lines in the plane, a VC class with $S(\mathcal{L}) = 2$, cannot be obtained from finitely many VC classes of index 1 and finitely many applications of the operations $\sqcap$, $\sqcup$, and taking complements. Theorem 4.46 reduces the proof from "VC classes of index 1" to "chains."

4.6 Probability Laws and Independence

Let $(X, \mathcal{A}, P)$ be a probability space. Recall the pseudo-metric $d_P(A, B) := P(A \triangle B)$ on $\mathcal{A}$, where $A \triangle B := (A \setminus B) \cup (B \setminus A)$. Recall also that for any (pseudo-) metric space (S, d) and $\varepsilon > 0$, $D(\varepsilon, S, d)$ denotes the maximum number of points more than ε apart (Section 1.2).

Definition. For a measurable space $(X, \mathcal{A})$ (a set X and a σ-algebra $\mathcal{A}$ of subsets of X) and $\mathcal{C} \subset \mathcal{A}$ let $s(\mathcal{C}) := \inf\{w : \text{there is a } K = K(w, \mathcal{C}) < \infty \text{ such that for every law } P \text{ on } \mathcal{A} \text{ and } 0 < \varepsilon \leq 1, D(\varepsilon, \mathcal{C}, d_P) \leq K\varepsilon^{-w}\}$.

This index $s(\mathcal{C})$ turns out to equal the density:

Theorem 4.47 *For any measurable space* $(X, \mathcal{A})$ *and* $\mathcal{C} \subset \mathcal{A}$, $dens(\mathcal{C}) = s(\mathcal{C})$.

Proof. Let P be a probability measure on $\mathcal{A}$. Suppose $A_1, \ldots, A_m \in \mathcal{C}$ and $d_P(A_i, A_j) > \varepsilon > 0$ for $1 \leq i < j \leq m$, with $m \geq 2$. Let $X_1, X_2, \ldots$ be i.i.d. (P), specifically coordinates on a countable product of copies of $(X, \mathcal{A}, P)$. Then for $n = 1, 2, \ldots$,

$$\Pr\{\text{for some } i \neq j, \ X_k \notin A_i \triangle A_j \text{ for all } k \leq n\} \leq m(m-1)(1-\varepsilon)^n/2 < 1$$

for n large enough, $n > -\log(m(m-1)/2)/\log(1-\varepsilon)$. Let $P_n :=$ $\frac{1}{n}\sum_{r=1}^{n} \delta_{X_r}$ (as usual) be empirical measures for P. For such n, there is positive probability that $P_n(A_i \triangle A_j) > 0$ for all $i \neq j$, and so A_i and A_j induce different subsets of $\{X_1, \ldots, X_n\}$ and $m^{\mathcal{C}}(n) \geq m$. For any $r > dens(\mathcal{C})$ there is an $M = M(r, \mathcal{C}) < \infty$, where we can take $M \geq 2$, such that $m^{\mathcal{C}}(n) \leq Mn^r$ for all n. Note that $-\log(1-\varepsilon) \geq \varepsilon$. Thus for $m \geq 2$, $m \leq M(2\log(m^2))^r \varepsilon^{-r}$, or $m(\log m)^{-r} \leq M_1 \varepsilon^{-r}$ for some $M_1 = M_1(r, \mathcal{C}) = 4^r M$. For any $\delta > 0$ and C large enough $(\log m)^r \leq Cm^\delta$ for all $m \geq 1$, so for all $m \geq 0$, $m^{1-\delta} \leq M_2 \varepsilon^{-r}$ for $0 < \varepsilon \leq 1$ for some $M_2 = M_2(r, \mathcal{C}, \delta)$. Thus $m \leq (M_2 \varepsilon^{-r})^{1/(1-\delta)}$. Letting $r \downarrow dens(\mathcal{C})$ and $\delta \downarrow 0$ gives $dens(\mathcal{C}) \geq s(\mathcal{C})$.

In the converse direction, let $|A| := \text{card}(A)$. Since it is not the case that $m^{\mathcal{C}}(n) \leq kn^t$ for all $n \geq 1$, for $r < t < dens(\mathcal{C})$ and $k = 1, 2, \ldots$, let $A_k \subset X$ with $A_k \neq \emptyset$ and $|A_k \sqcap \mathcal{C}| > k|A_k|^t$. Then $|A_k| \to \infty$ as $k \to \infty$. Let $B_0 := A_1$. Other sets B_j will be defined recursively. Let $B_0, \ldots, B_{n-1}$ be disjoint subsets of X and let $C(n) := \bigcup_{0 \leq j < n} B_j$. Let

$$k(n) := 2^{|C(n)|+2^n}.$$

Let $B_n := A_{k(n)} \setminus C(n)$. So all the B_n are defined and disjoint. Each set in $B_n \sqcap \mathcal{C}$ is induced by at most $2^{|C(n)|}$ different sets in $A_{k(n)} \sqcap \mathcal{C}$. Thus

$$|B_n \sqcap \mathcal{C}| \geq |A_{k(n)} \sqcap \mathcal{C}|/2^{|C(n)|} \geq 2^{2^n}|A_{k(n)}|^t \geq 2^{2^n}|B_n|^t.$$

Since B_n is nonempty, it follows that

$$|B_n \sqcap \mathcal{C}| \geq 2^{2^n}, \quad \text{and hence } |B_n| \geq 2^n.$$

Let

$$\alpha_n := |B_n|^{-t/r}, \quad S := \sum_{n=0}^{\infty} |B_n|^{1-t/r} < \infty.$$

Let P be the probability measure on $\bigcup_{n=1}^{\infty} B_n \subset X$ giving mass α_n/S to each point of B_n for each n. The distinct sets in $B_n \sqcap \mathcal{C}$ are at d_P-distance at least α_n/S apart, and so are a set of elements of $\mathcal{C}$ which induce the subsets of B_n.

So for all n,

$$D(\alpha_n/(2S), \mathcal{C}, d_P) \geq |B_n \sqcap \mathcal{C}| \geq 2^{2^n}|B_n|^t = 2^{2^n}\alpha_n^{-r} = 2^{2^n}S^{-r}(\alpha_n/S)^{-r}.$$

For $\varepsilon := \alpha_n/(2S) \to 0$ as $n \to \infty$, this implies

$$D(\varepsilon, \mathcal{C}, d_P) \geq 2^{2^n}(2S)^{-r}\varepsilon^{-r}.$$

Since $2^n \to \infty$ as $\alpha_n \downarrow 0$ this implies $r \leq s(\mathcal{C})$. Then letting $r \uparrow \mathrm{dens}(\mathcal{C})$, the proof is complete. $\square$

Corollary 4.48 *For any probability space $(X, \mathcal{A}, P)$ and VC class $\mathcal{C} \subset \mathcal{A}$, the class $\mathcal{F} = \{1_A : A \in \mathcal{C}\}$ is pregaussian for P.*

Proof. Take any $r > S(\mathcal{C})$. By Corollary 4.4 and Theorem 4.47, there is a constant $K < +\infty$ such that $D(\varepsilon, \mathcal{C}, d_P) \leq K\varepsilon^{-r}$ for $0 < \varepsilon \leq 1$. For f and g in $\mathcal{L}^2(P)$ and ρ_P as defined in Section 3.1 we have $\rho_P(f, g) \leq e_P(f, g) := \left(\int(f - g)^2 dP\right)^{1/2}$. For $A, B \in \mathcal{A}$, we have $e_P(1_A, 1_B) = d_P(A, B)^{1/2}$. It follows that for $0 < \varepsilon < 1$,

$$D(\varepsilon, \mathcal{F}, \rho_P) \leq D(\varepsilon, \mathcal{F}, e_P) \leq D(\varepsilon^2, \mathcal{C}, d_P) \leq K^2\varepsilon^{-2r}. \qquad (4.10)$$

As the set $\mathcal{F}$ is bounded with respect to e_P, there is an $M < \infty$ such that for $\varepsilon \geq M$, $D(\varepsilon, \mathcal{F}, \rho_P) = D(\varepsilon, \mathcal{F}, e_P) = 1$. Thus the integral in the metric entropy sufficient condition for the GC property, Theorem 2.36, is finite, using the last sentence over $M \leq \varepsilon < \infty$ where $\log D(\varepsilon) = 0$, and (4.10) for $0 < \varepsilon \leq M$ where $\sqrt{\log(1/\varepsilon)}$ is integrable. So $\mathcal{F}$ is pregaussian. $\square$

To get the Donsker property, however, will require additional measurability assumptions (Chapters 5 and 6).

There is a notion of independence for sets without probability. To define it, for any set X and subset $A \subset X$ let $A^1 := A$ and $A^{-1} := X \setminus A$. Sets $A_1, \ldots, A_m$ are called *independent*, or *independent as sets*, if for every function $s(\cdot)$ from $\{1, \ldots, m\}$ into $\{-1, +1\}$, $\bigcap_{j=1}^m A_j^{s(j)} \neq \emptyset$. Such intersections, when they are nonempty, are called *atoms* of the Boolean algebra generated by $A_1, \ldots, A_m$. Thus for $A_1, \ldots, A_m$ to be independent as sets means that the Boolean algebra they generate has the maximum possible number, 2^m, of atoms.

If $A_1, \ldots, A_m$ are independent as sets, then one can define a probability law on the algebra they generate for which they are jointly independent in the usual probability sense and for which $P(A_i) = 1/2$, $i = 1, \ldots, m$. For example, choose a point in each atom and put mass $1/2^m$ at each point chosen. Or, if desired, given any q_i, $0 \leq q_i \leq 1$, one can define a probability measure Q for which the A_i are jointly independent and have $Q(A_i) = q_i$, $i = 1, \ldots, n$.

For a set X and $\mathcal{C} \subset 2^X$ let

$$I(\mathcal{C}) := \sup\{m : A_1, \ldots, A_m \text{ are independent as sets for some } A_i \in \mathcal{C}\}.$$

Theorem 4.49 *For any set* X, $C \subset 2^X$, *and* $n = 1, 2, \ldots$, *if* $S(C) \geq 2^n$, *then* $I(C) \geq n$. *Conversely if* $I(C) \geq 2^n$, *then* $S(C) \geq n$. *So* $I(C) < \infty$ *if and only if* $S(C) < \infty$. *In both cases,* 2^n *cannot be replaced by* $2^n - 1$.

Proof. Clearly, if a set Y has n or more independent subsets, then $|Y| \geq 2^n$. Conversely if $|Y| = 2^n$, we can assume that Y is the set of all strings of n digits each equal to 0 or 1. Let A_j be the set of strings for which the jth digit is 1. Then the A_j are independent. It follows that if $S(C) \geq 2^n$, then $I(C) \geq n$, while if $|Y| = 2^n - 1$ and $C = 2^Y$, then $S(C) = 2^n - 1$, while $I(C) < n$ as stated.

Conversely if B_j are independent as sets for $j = 1, \ldots, 2^n$, $B_j \in C$, let $A(i) := A_i$ be independent subsets of $\{1, \ldots, 2^n\}$ for $i = 1, \ldots, n$. Choose $x_i \in \bigcap_{j \in A(i)} B_j \cap \bigcap_{j \notin A(i)} (X \setminus B_j)$. Then $x_i \in B_j$ if and only if $j \in A(i)$. For each set $S \subset \{1, \ldots, n\}$,

$$\bigcap_{i \in S} A_i \cap \bigcap_{i \notin S} (\{1, \ldots, 2^n\} \setminus A_i) = \{j\}$$

for some $j := j_S \in \{1, \ldots, 2^n\}$. Then $j \in A_i$ if and only if $i \in S$, and

$$B_j \cap \{x_1, \ldots, x_n\} = \{x_i : i \in S\}.$$

So C shatters $\{x_1, \ldots, x_n\}$ and $S(C) \geq n$, as stated. If C consists of $2^n - 1$ (independent) sets, then clearly $S(C) < n$. $\square$

For any set X, $C \subset 2^X$ and $Y \subset X$, recall that $C_Y := Y \cap C := \{Y \cap C : C \in C\}$. Let $\mathrm{At}(C|Y)$ be the set of atoms of the algebra of subsets of Y generated by C_Y, where in the cases to be considered, C_Y will be finite because C or Y is. Let $\Delta_C(Y) := |\mathrm{At}(C|Y)|$ be the number of such atoms. Let $m_C^Y(n) := \sup\{\Delta_A(Y) : A \subset C, |A| \leq n\} \leq 2^n$. Let

$$\mathrm{dens}^*(C) := \inf\{s \geq 0 : \text{ for some } C < \infty, \; m_C^X(n) \leq Cn^s \text{ for all } n\}.$$

For any $x \in X$ let $C_x := \{A \in C : x \in A\}$. Let $C'_Y := \{C_y : y \in Y\}$.

Theorem 4.50 *For any set* X *and* $A \subset C \subset 2^X$, *with* A *finite,*

(a) $\Delta_A(X) = \Delta^{C'_X}(A)$.

(b) For $n = 1, 2, \ldots$, $m_C^X(n) = m^{C'_X}(n)$.

(c) $S(C'_X) = I(C)$.

(d) $\mathrm{dens}^*(C) = \mathrm{dens}(C'_X) \leq I(C)$.

Proof. For any $B \subset A$ let

$$\alpha(B) := \bigcap_{B \in B} B \cap \bigcap_{A \in A \setminus B} (X \setminus A).$$

Then $\alpha(B)$ is an atom, in $\mathrm{At}(A|X)$, if and only if it is nonempty. Now $y \in \alpha(B)$ if and only if $A \cap C_y = B$, so (a) follows. Then taking the maximum over A

with $|\mathcal{A}| = n$ on both sides of (a) gives (b), which then implies (c) and (d). The last inequality follows from Corollary 4.4.   □

4.7 Vapnik–Červonenkis Properties of Classes of Functions

The notion of VC class of sets has several extensions to classes of functions.

Definitions. Let X be a set and $\mathcal{F}$ a class of real-valued functions on X. Let $\mathcal{C} \subset 2^X$. If f is any real-valued function, each set $\{f > t\}$ for $t \in \mathbb{R}$ will be called a *major set* of f. The class $\mathcal{F}$ will be called a *major class* for $\mathcal{C}$ if all the major sets of each $f \in \mathcal{F}$ are in $\mathcal{C}$. If $\mathcal{C}$ is a Vapnik–Červonenkis class, then $\mathcal{F}$ will be called a *VC major* class (for $\mathcal{C}$).

The *subgraph* of a real-valued function f will be the set $\{(x, t) \in X \times \mathbb{R} : 0 \le t \le f(x) \text{ or } f(x) \le t \le 0\}$. If $\mathcal{D}$ is a class of subsets of $X \times \mathbb{R}$, and for each $f \in \mathcal{F}$, the subgraph of f is in $\mathcal{D}$, then $\mathcal{F}$ will be called a *subgraph class* for $\mathcal{D}$. If $\mathcal{D}$ is a VC class in $X \times \mathbb{R}$, then $\mathcal{F}$ will be called a *VC subgraph class*.

Recall from Section 3.11 the definitions of symmetric convex hull $H(\mathcal{F}, M)$ (times M) and its sequential pointwise closure $H_s(\mathcal{F}, M)$. A class $\mathcal{F}$ of functions such that $\mathcal{F} \subset \bar{H}_s(\mathcal{G}, M)$ for some $M < \infty$ and a given $\mathcal{G}$ will be called a *VC subgraph hull class* if $\mathcal{G}$ is a VC subgraph class, and a *VC hull class* if $\mathcal{G} = \{1_C : C \in \mathcal{C}\}$ where $\mathcal{C}$ is a VC class of sets.

So there are at least four possible ways to extend the notion of VC class to classes of functions. Some implications hold between these different conditions, but no two of them are equivalent. The next theorem deals with some of the easier cases of implication or non-implication.

Theorem 4.51 *Let $\mathcal{F}$ be a uniformly bounded class of nonnegative real-valued functions on a set X. Then*

(a) If $\mathcal{F}$ is the set of indicators of members of a VC class of sets, then $\mathcal{F}$ is also a VC major class, a VC subgraph class, and a VC hull class.

(b) If $\mathcal{F}$ is a VC major class then it is a VC hull class.

(c) There exist VC hull classes $\mathcal{F}$ which are not VC major.

(d) There exist VC subgraph classes $\mathcal{F}$ which are not VC major.

Remark. There are VC major classes which are not VC subgraph, as seen in Problem 10 and a remark at the end of Section 4.8.

Proof. (a): The indicators of a VC class $\mathcal{C}$ of sets clearly form a VC major class (for $\mathcal{C}$) and a VC hull class. Also, the class of sets $A \times C$ for $C \in \mathcal{C}$, for a fixed set A, here $[0, 1]$, form a VC class, for example by Theorem 4.34. So (a) holds.

(b): Let $\mathcal{F}$ be a VC major class for a VC class $\mathcal{C}$ of subsets of X. Let $|f(x)| \leq K < \infty$ for all $f \in \mathcal{F}$ and $x \in X$, with $K > 0$. Then for any $f \in \mathcal{F}$, $\{x : f(x) > -2K\} = X$, so $X \in \mathcal{C}$.

For any $f \in \mathcal{F}$, let $g := (f + K)/(2K)$. The class of all such g is clearly VC major for the same $\mathcal{C}$. We have $0 \leq g \leq 1$. Let

$$g_n := \frac{1}{n} \sum_{j=1}^{n-1} 1_{\{g > j/n\}} = \sum_{j=0}^{n-1} \frac{j}{n} 1_{\{j/n < g \leq (j+1)/n\}}.$$

Then $g_n(x) \to g(x)$ as $n \to \infty$ for all x. [In fact, $g(x) - 1/n \leq g_n(x) \leq g(x)$ for all x, so $g_n \to g$ uniformly.] For each n let $f_n = 2K g_n - K 1_X$. Then as $n \to \infty$, $f_n(x) \to f(x)$ for all x, and $f_n \in H(\{1_A : A \in \mathcal{C}\}, 2K)$. So $\mathcal{F}$ is VC hull (for the same VC class $\mathcal{C}$).

(c) Let $X = \mathbb{R}^2$. Let $\mathcal{C}$ be the set of open lower left quadrants $\{(x, y) : x < a, y < b\}$ for all $a, b \in \mathbb{R}$. Then $S(\mathcal{C}) = 2$ by Corollary 4.44. Let $\mathcal{F}$ be the set of all sums $\sum_{k=1}^{\infty} 1_{C(k)}/2^k$ where $C(k) \in \mathcal{C}$. Then clearly $\mathcal{F}$ is a VC hull class. The sets where $f > 0$ for f in $\mathcal{F}$ are exactly the countable unions of sets in $\mathcal{C}$. But such unions are not a VC class; for example, they shatter any finite subset of the line $x + y = 1$. So $\mathcal{F}$ is not a VC major class, and (c) is proved.

(d): Let $f_n := n^{-1} + n^{-2} 1_{B_n}$ for $n = 1, 2, \ldots$, for any measurable sets B_n. Then $f_n \downarrow 0$, so the subgraphs of the functions f_n are linearly ordered by inclusion and form a VC class of index 1 (Theorem 4.10(a)). Now $\{f_n > 1/n\} = B_n$ for each n, and the sequence $\{B_n\}$ need not form a VC class; for example, let B_n be a sequence of independent sets (see Section 4.6). Then $\{f_n\}$ is not a VC major class, proving (d). $\qquad \square$

4.8 Classes of Functions and Dual Density

For a metric space (S, d) and $\varepsilon > 0$ recall $D(\varepsilon, S, d)$, the maximum number of points more than ε apart. For a probability measure Q and $1 \leq p < \infty$ we have the L^p metric $d_{p,Q}(f, g) := (\int |f - g|^p dQ)^{1/p}$. For a class $\mathcal{F} \subset \mathcal{L}^p(Q)$ let $D^{(p)}(\varepsilon, \mathcal{F}, Q) := D(\varepsilon, \mathcal{F}, d_{p,Q})$. Let $D^{(p)}(\varepsilon, \mathcal{F})$ be the supremum of $D(\varepsilon, \mathcal{F}, d_{p,Q})$ over all laws Q concentrated in finite sets.

If $\mathcal{F}$ is a class of measurable real-valued functions on a measurable space $(X, \mathcal{A})$, let $F_{\mathcal{F}}(x) := \sup_{f \in \mathcal{F}} |f(x)|$. Then a measurable function F will be called an *envelope function* for $\mathcal{F}$ if and only if $F_{\mathcal{F}} \leq F$. If $F_{\mathcal{F}}$ is measurable it will be called *the* envelope function of $\mathcal{F}$. For any law P on $(X, \mathcal{A})$, $F_{\mathcal{F}}^*$ is an envelope function for $\mathcal{F}$, which in general depends on P.

Given $\mathcal{F}$, an envelope function F for it, $\varepsilon > 0$ and $1 \leq p < \infty$, let $D_F^{(p)}(\varepsilon, \mathcal{F}, Q)$ be the supremum of m such that there exist $f_1, \ldots, f_m \in \mathcal{F}$ for which $\int |f_i - f_j|^p dQ > \varepsilon^p \int F^p dQ$ for all $i \neq j$.

The next fact extends part of Theorem 4.47 to families of functions. The proof is also similar.

Theorem 4.52 *Let* $1 \le p < \infty$. *Let* $(X, \mathcal{A}, Q)$ *be a probability space and* $\mathcal{F}$ *be a VC subgraph class of measurable real-valued functions on* X. *Let* $\mathcal{F}$ *have an envelope* $F \in \mathcal{L}^p(X, \mathcal{A}, Q)$ *with* $0 < \int F dQ$. *Let* $\mathcal{C}$ *be the collection of subgraphs in* $X \times \mathbb{R}$ *of functions in* $\mathcal{F}$. *Then for any* $W > S(\mathcal{C})$ *there is an* $A < \infty$ *depending only on* W *and* $S(\mathcal{C})$ *such that*

$$D_F^{(p)}(\varepsilon, \mathcal{F}, Q) \le A(2^{p-1}/\varepsilon^p)^W \quad \text{for } 0 < \varepsilon \le 1. \tag{4.11}$$

Proof. We can assume that $F \ge 1$ everywhere on X. Given $0 < \varepsilon < 1$, take a maximal m and $f_1, \ldots, f_m$ as in the definition of $D_F^{(p)}$. First, suppose $p = 1$. For any measurable set B let $(FQ)(B) := \int_B F dQ$. Let $Q_F := FQ/QF$ where $QF := \int F dQ$, so Q_F is a probability measure. Let $k = k(m, \varepsilon)$ be the smallest integer such that $e^{k\varepsilon/2} > \binom{m}{2}$. Then $k \le 1 + (4\log m)/\varepsilon$. Let $X_1, \ldots, X_k$ be i.i.d. (Q_F). Given X_i, let Y_i be uniformly distributed on the interval $[-F(X_i), F(X_i)]$, and such that the vectors $(X_i, Y_i) \in \mathbb{R}^2$ are independent for $i = 1, \ldots, k$. Let C_j be the subgraph of f_j, $j = 1, \ldots, m$. Then for all i and $j \ne s$

$$\Pr((X_i, Y_i) \in C_j \Delta C_s) = \int [|f_j(X_i) - f_s(X_i)|/(2F(X_i))] dQ_F(X_i)$$

$$= \int |f_j - f_s| dQ/(2 \int F dQ) > \varepsilon/2.$$

Thus by independence

$$\Pr((X_i, Y_i) \notin C_j \Delta C_s \text{ for all } i = 1, \ldots, k) \le \left(1 - \frac{\varepsilon}{2}\right)^k \le e^{-k\varepsilon/2}, \text{ and}$$

$$\Pr\{(X_i, Y_i) \notin C_j \Delta C_s \text{ for all } i = 1, \ldots, k \text{ and some } j \ne s\} \le \binom{m}{2} e^{-k\varepsilon/2} < 1.$$

Thus with positive probability, for all $j \ne s$ there is some $(X_i, Y_i) \in C_j \Delta C_s$. Fix X_i and Y_i, $i = 1, \ldots, k$, for which this happens. Then the sets $C_j \cap \{(X_i, Y_i)\}_{i=1}^k$ are distinct for all j, so $m^{\mathcal{C}}(k) \ge m$.

Let $S := S(\mathcal{C})$. By the Sauer and Vapnik–Červonenkis lemmas (Theorem 4.2 and Proposition 4.3), $m^{\mathcal{C}}(k) \le 1.5 k^S/S!$ for $k \ge S + 2$. It follows that for some constant C depending only on S, $m^{\mathcal{C}}(k) \le Ck^S$ for all $k \ge 1$, where $C \le 2^{S+1} - 1$. We can assume that $C \ge 1$. So

$$m \le Ck^S \le C((1 + 4\log m)/\varepsilon)^S.$$

For any $\alpha > 0$ there is an m_0 such that $1 + 4\log m \le m^\alpha$ for $m \ge m_0$ and then $m^{1-\alpha S} \le C/\varepsilon^S$. Choosing α small enough so that $\alpha S < 1$ we have $m \le C/\varepsilon^{S/(1-\alpha S)}$ for $m \ge m_0$. For any $W > S$ we can solve $W = \frac{S}{1-\alpha S}$ by $\alpha = \frac{W-S}{WS}$. Then $m \le A\varepsilon^{-W}$ for $A := \max(m_0, C)$, and then m_0 and A are functions of W and S, finishing the proof for $p = 1$.

Now suppose $p > 1$. Let $Q_{F,p} := F^{p-1}Q/Q(F^{p-1})$. Then for $i \neq j$,

$$\varepsilon^p \int F^p dQ \;<\; \int |f_i - f_j|^p dQ \;\leq\; \int |f_i - f_j|(2F)^{p-1}dQ$$
$$= \int |f_i - f_j| dQ_{2F,p} \cdot Q((2F)^{p-1}),$$

and $Q_{2F,p} = Q_{F,p}$. Thus by the $p = 1$ case,

$$D_F^{(p)}(\varepsilon, \mathcal{F}, Q) \;\leq\; D_F^{(1)}(\delta, \mathcal{F}, Q_{F,p}) \;\leq\; A\delta^{-W}$$

where $\delta := \varepsilon^p Q(F^p)/[Q(F)Q((2F)^{p-1})$. Now by Hölder's inequality, $QF = Q(F \cdot 1) \leq (Q(F^p))^{1/p}$ and $Q(F^{p-1}) = Q(F^{p-1} \cdot 1) \leq (Q(F^p))^{(p-1)/p}$. So $\delta \geq \varepsilon^p/2^{p-1}$ and the conclusion follows. □

The following fact is a continuation of Theorem 4.51.

Theorem 4.53 *Let $\mathcal{F}$ be a uniformly bounded class of functions on a set X.*

(a) If $\mathcal{F}$ is a VC subgraph class, then

For some $r < \infty$ and $M < \infty$, $D^{(2)}(\varepsilon, \mathcal{F}) \leq M\varepsilon^{-r}$ for $0 < \varepsilon < 1$. (4.12)

(b) There exist classes $\mathcal{F}$ satisfying (4.12) which are not VC hull.

(c) There exist VC subgraph classes which are not VC hull.

Proof. (a) There is a finite constant envelope function K for $\mathcal{F}$, so for any $f, g \in \mathcal{F}$ and law γ, $\int (f - g)^2 d\gamma \leq 2K \int |f - g| d\gamma$. Thus (a) follows from Theorem 4.52 with $r = 2W$.

(b) It will be shown that there exist sequences $\mathcal{F} = \{b_n 1_{A_n}\}_{n \geq 1}$, where we can take $b_n = 1/n^v$ for any positive integer v, such that $\mathcal{F}$ is not VC hull. Clearly such a sequence will satisfy (4.12). For this two lemmas will be helpful:

Lemma 4.54 *Let $A_1, \ldots, A_n$ be jointly independent events in a probability space (X, P) with $P(A_i) = 1/2$, $i = 1, \ldots, n$. Let $B_1, \ldots, B_n$ be any events. Let $D := \bigcup_{1 \leq j \leq n} A_j \Delta B_j$. Then the algebra $\mathcal{B}$ generated by $B_1, \ldots, B_n$ has at least $2^n(1 - P(D))$ atoms.*

Proof. For any $F \subset \{1, \ldots, n\}$ let

$$A_F := \bigcap_{j \in F} A_j \cap \bigcap_{j \notin F, \; j \leq n} X \setminus A_j,$$

and define B_F likewise. The atoms of $\mathcal{B}$ are those B_F which are nonempty. For each F, $P(A_F) = 1/2^n$. If a point of A_F is not in D, then it is also in B_F, which then is nonempty. Since D can include at most $2^n P(D)$ of the 2^n events A_F, the Lemma follows. □

Lemma 4.55 *Suppose A_j are independent events with $P(A_j) = 1/2$ for all j, and $\mathcal{C}$ is a class of events such that for some $K < \infty$ and $u < \infty$, for each* *j* there is an event D_j such that $P(A_j \triangle D_j) < \eta_j$, where D_j is in an algebra generated by at most Kj^u elements of $\mathcal{C}$ and $\sum_j \eta_j < 1$. Then $\mathcal{C}$ is not a VC class.

Proof. Let $\alpha := 1 - \sum_{j=1}^{\infty} \eta_j > 0$. By Lemma 4.54, for each $m = 1, 2, \ldots$, the algebra $\mathcal{D}_m$ generated by $D_1, \ldots, D_m$ has at least $2^m \alpha$ atoms. On the other hand, $\mathcal{D}_m$ is generated by at most $\sum_{j=1}^{m} Kj^u \leq K \int_1^{m+1} x^u dx \leq K(m + 1)^{u+1}/(u + 1)$ sets in $\mathcal{C}$. By Theorems 4.49 and 4.50(c), $\mathcal{C}'_X$ is a VC class and then by Theorem 4.50(b) and Corollary 4.4, there are $t < \infty$ and $C < \infty$ such that the number of atoms of the algebra generated by k elements of $\mathcal{C}$ is at most Ck^t. Then $2^m \alpha$ is bounded above by a polynomial in m of degree at most $(u + 1)t$, a contradiction, so Lemma 4.55 is proved. $\qquad\square$

Now to prove Theorem 4.53(b), let $X := [0, 1]$, let $P := U[0, 1]$ be the uniform (Lebesgue) law and let A_m be independent sets with $P(A_m) = 1/2$. Let v be a positive integer and $\mathcal{F} := \{1_{A_m}/m^v\}_{m \geq 1}$. Then (4.12) holds for $\mathcal{F}$. Suppose $\mathcal{F}$ is a VC hull class for some $\mathcal{C}$.

Suppose $P(A) = 1/2$ and that $a + b1_A \in \bar{H}_s(\mathcal{C}, M)$ for some $a \geq 0$, $b > 0$ and $0 < M < \infty$. We can take $M = 1$, replacing a by a/M and b by b/M. Let $r := S(\mathcal{C}) < \infty$. Let $d_P(C, D) := P(C \triangle D)$ for measurable sets C, D. Then for any $w > r$, specifically for $w := r + 1$, there is some $C < \infty$ with $D(\varepsilon, \mathcal{C}, d_P) < C\varepsilon^{-w}$ for $0 < \varepsilon < 1$ by Corollary 4.4 and Theorem 4.47. Given $0 < \beta < 1$, take $C_j \in \mathcal{C}$ and t_j with $\sum_j |t_j| \leq 1$ such that $P(|a + b1_A - \sum_j t_j 1_{C_j}|) < \beta$. Choose $D_i \in \mathcal{C}$, $i = 1, \ldots, m$, where $m \leq C\beta^{-r-1}$, such that for each j there is an $i := i(j)$ with $P(C_j \triangle D_i) \leq \beta$. Let $f := \sum_j t_j 1_{D_{i(j)}}$. Then $P(|a + b1_A - f|) < 2\beta$. Let $B := \{f > a + b/2\}$. Then B is in the algebra generated by $D_1, \ldots, D_m$, and $P(A \triangle B) < 4\beta/b$.

Apply what has just been shown to $A = A_k$ for $k = 1, \ldots, m$, with $a = a_k = 0$, $b = b_k = 1/k^v$, and $\beta = \beta_k = 1/(8k^{2+v})$. Then there are sets $B = B_k$ and $m = m_k \leq Nk^u$ for some $N < \infty$ where $u := (r + 1)(2 + v) < \infty$. To apply Lemma 4.55, let $\eta_k = 1/(2k^2)$ to get a contradiction. So (b) is proved.

For (c), let $a = a_k = 1/k$ and $b = b_k = 1/k^2$ to get a VC subgraph class as in the proof of Theorem 4.51(d). The same argument as for (b), now with $v = 2$, again gives a contradiction, so (c) holds. $\qquad\square$

It will be shown in Proposition 10.12 below that there are VC major (thus VC hull) classes which do not satisfy (4.12) (and so, in particular, are not VC subgraph classes).

Problems

1. Let $\mathcal{C}$ be the class of all unions of two intervals in $\mathbb{R}$. Evaluate $S(\mathcal{C})$. *Hint:* Try it first directly; if you like, look at the more general Problem 11.

2. If $S(\mathcal{C}) = 3$ find the upper bounds for $m^{\mathcal{C}}(n)$ given by Theorem 4.2 and by Proposition 4.3.

3. Show that for $\operatorname{dens}(\mathcal{C}) = 0$, $S(\mathcal{C})$, which is finite by Corollary 4.4, can be arbitrarily large. *Hint*: Let $\mathcal{C}$ be finite.

4. Find the smallest n such that there is a set X with $|X| = n$ and $\mathcal{C} \subset 2^X$ with $S(\mathcal{C}) = 1$ where neither (a) nor (b) in Theorem 4.10 holds.

Hints on Problems 5–7: If $\mathcal{C}$ is a collection of convex sets in $\mathbb{R}^d$ and shatters a set F, then no point in F is in the convex hull of the other points. Then, the convex hull of F is a polyhedron of which each point of F is a vertex. In the plane, it is a polygon. To get a lower bound $S(\mathcal{C}) \geq k$ it is enough to find one set of k elements that is shattered. Try the vertices of a regular k-gon. To get upper bounds, use facts such as Theorem 4.6 and Proposition 4.36.

5. Let $\mathcal{C}$ be the set of all interiors of ellipses in $\mathbb{R}^2$, with arbitrary centers and semiaxes in any two perpendicular directions. Give upper and lower bounds for $S(\mathcal{C})$.

6. A half-plane in $\mathbb{R}^2$ is a set of the form $\{(x, y) : ax + by \geq c\}$ for real a, b, c with a and b not both 0. Define a wedge as an intersection of two half-planes. Let $\mathcal{C}$ be the collection of all wedges in $\mathbb{R}^2$. Show that $S(\mathcal{C}) \geq 5$. Also find an upper bound for $S(\mathcal{C})$.

7. Let $\mathcal{C}$ be the set of all interiors of triangles in $\mathbb{R}^2$. Show that $S(\mathcal{C}) \geq 7$. Also give an upper bound for $S(\mathcal{C})$.

8. Show that the lower bounds for $S(\mathcal{C})$ in Problems 6 and 7 are the values of $S(\mathcal{C})$. *Hint*: For a convex polygon, the set F of vertices can be arranged in cyclic order, say clockwise around the boundary of the polygon, $v_1, v_2, \ldots, v_n, v_1$. Show that if a half-plane J contains v_i and v_j with $i < j$, then it includes either $\{v_i, v_{i+1}, \ldots, v_j\}$ or $\{v_j, v_{j+1}, \ldots, v_n, v_1, \ldots, v_i\}$. Thus find what kind of set the intersection of J and F must be. From that, find what occurs if two or three half-planes are intersected (or unioned, via complements).

9. In the example at the end of Section 4.3, for each set $A \subset X$ with 3 elements, find a specific subset of A not in $A \sqcap \mathcal{C}$.

10. Let $\mathcal{F}$ be the class of all probability distribution functions on $\mathbb{R}$. Show that $\mathcal{F}$ is a VC major class but not a VC subgraph class. *Hint*: Show that the subgraphs of functions in $\mathcal{F}$ shatter all sets $\{(x_j, y_j)\}_{j=1}^n$ with $x_1 < \cdots < x_n$ and $0 < y_1 < \cdots < y_n < 1$.

11. Let $\mathcal{C}(j)$ be the class of all unions of j intervals in $\mathbb{R}$ for $j = 1, 2, \ldots$. Show that $S(\mathcal{C}(j)) = 2j$ for all j and that for any finite set $F \subset \mathbb{R}$ with $|F| = n$ we have $\Delta^{\mathcal{C}(j)}(F) = {}_nC_{\leq 2j}$ (the largest possible value by Sauer's Lemma). *Hints*: One can take $F = \{1, 2, \ldots, n\}$. For $A \subset F$ and $x, y \in F$ let $x =_A y$ mean

that $A \cap \{x, y\} = \emptyset$ or $\{x, y\}$, otherwise $x \neq_A y$. If $A \neq \emptyset$ let $j_1 := j_1(A)$ be the least element of A. If $j_1(A), \ldots, j_k(A)$ are defined, let $j_{k+1}(A)$ be the least $j > j_k(A)$ such that $j \neq_A j_k(A)$ and $j \leq n$, if there is such a j. Show that there is a 1–1 correspondence between subsets $A \subset F$ and finite sequences $(j_1, j_2, \ldots, j_r)$ for $r = r(A) = 1, \ldots, n$ where $r(\emptyset) := 0$. Show that $A \in \mathcal{C}(j) \sqcap F$ if and only if $r(A) \leq 2j$.

12. For each $d = 1, 2, \ldots$, let $\mathcal{P}_d$ be the vector space of all polynomials of degree at most d on $\mathbb{R}$.

(a) Show that pos($\mathcal{P}_d$) shatters *all* sets F of $d + 1$ points in $\mathbb{R}$. (It shatters no set of $d + 2$ points, by what fact?) *Hints*: If for a given finite set F, a polynomial $f > 0$ on a set $A \subset F$ and $f \leq 0$ on $F \setminus A$, then for some $\varepsilon > 0$, $f - \varepsilon$ is still > 0 on A, is a polynomial of the same degree, and is < 0 on $F \setminus A$. Let $F = \{x_1, \ldots, x_{d+1}\}$ where $x_1 < x_2 < \cdots < x_{d+1}$. If for some consecutive points $x_j < x_{j+1}$ with $j = 1, \ldots, d$, one of x_j and x_{j+1} is in A but not the other, the polynomial f should have a simple zero somewhere in the interval (x_j, x_{j+1}). Find a polynomial f with just such zeroes, show that its degree is at most d, and take $-f$ if necessary, then show that $f(x) > 0$ for $x \in F$ if and only if $x \in A$.

(b) Prove directly, without using an earlier fact, that positivity sets of functions in $\mathcal{P}_d$ cannot shatter a set F of $d + 2$ points. *Hint*: Let A be the set of all $x_j \in F$ with j odd, and show that the argument in part (a) can be reversed.

A set of n distinct points of $\mathbb{R}^d$ is said to be in *general position* if no $k + 2$ of them are in any k-dimensional hyperplane for $k = 1, \ldots, d - 1$.

13. (a) Recall the definition of half-planes in the plane from Problem 6. Show that the collection $\mathcal{H}(2)$ of all half-planes in the plane shatters every set of three points in general position, but not every set of three points.

(b) An open *half-space* in $\mathbb{R}^d$ will be a set of the form $\{x := \{x_j\}_{j=1}^d : c_0 + \sum_{j=1}^d c_j x_j > 0\}$ where not all c_j for $1 \leq j \leq d$ are 0. In $\mathbb{R}^3$, we know (why?) that the set $\mathcal{H}(3)$ of all half-spaces shatters some set of 4 points. Show that every set of 4 points in general position is shattered.

(c) Give an example of a set of 4 points in $\mathbb{R}^3$, no three of which are in any line, but that is not shattered.

14. It is a known fact, although not proved in this book, that for the class $\mathcal{H}(d)$ of all half-spaces in $\mathbb{R}^d$, $m^{\mathcal{H}(d)}(n) = P_d(n) := 2_{n-1}C_{\leq d}$. Let $Q_d(n) := {}_nC_{\leq d+1}$, which is the largest possible value of $m^{\mathcal{C}}(n)$ for VC classes $\mathcal{C}$ with $S(\mathcal{C}) = d + 1$. In this case $S(\mathcal{H}(d)) = d + 1$.

(a) Evaluate the two polynomials $P_2(n)$ and $Q_2(n)$ explicitly. Show that they both equal 2^n for $n = 1, 2$, and 3 but not for $n = 4$.

(b) Find an example of a set F of four points in $\mathbb{R}^2$ such that the number $\Delta^{\mathcal{H}(2)}(F)$ of sets $A \cap F$ for $A \in \mathcal{H}(2)$ achieves its maximum $P_2(4)$. *Hint*: Take the vertices of a square.

(c) If the four points in F are the vertices of a triangle and a point in the interior of the triangle, then what is the maximum?

Notes

Notes to Section 4.1. The definitions of $\Delta^{\mathcal{C}}$, $m^{\mathcal{C}}$ and (in effect) $V(\mathcal{C})$ appeared in the announcement by Vapnik and Červonenkis (1968). In their 1971 paper they had a weaker form of Theorem 4.2 with "$\geq {}_nC_{\leq k}$" instead of "$> {}_nC_{\leq k-1}$." The theorem as stated appears in Sauer (1972, Theorem 1). The quantities ${}_nC_{\leq k}$ first appeared in mathematics, to my knowledge, for general $n \geq k$, in work of Schläfli (1901), who showed that this is the maximum number of open regions into which $\mathbb{R}^k$ is decomposed by n hyperplanes $\{x : (v_j, x) = c_j\}$ where $v_j \neq 0$ in $\mathbb{R}^k$, $c_j \in \mathbb{R}$, and $(v, x) := v \cdot x$ is the usual inner product. Schläfli showed that the maximum is attained when the hyperplanes are in general position, meaning in this case that any k or fewer of the v_j are linearly independent. Steiner (1826) had proved these facts for $k \leq 3$. Cover (1965), Harding (1967), and Watson (1969) considered $m^{\mathcal{H}(k)}(n)$ and showed that it is $\leq 2_{n-1}C_{\leq k}$. Schläfli (1901, p. 211), for the case $c_j \equiv 0$ of $(k-1)$-dimensional linear subspaces, and the class $\mathcal{H}_0(k)$ of half-spaces bounded by them, had shown that $m^{\mathcal{H}_0(k)}(n) \leq 2_{n-1}C_{\leq k-1}$, also attained when the subspaces are in general position in the same sense.

Steele (1978b, and earlier in his 1975 Ph.D. thesis) seems to have coined the term "shatter."

"Pascal's triangle," published by Pascal in 1653, had been known since much earlier times. Some sources given are the Persian mathematicians Al-Karaji (953–1029) and the later but more famous Omar Khayyam (1048–1131); the Chinese mathematicians Yang Hui (whose name is said to be used in China) and Jia Xian (1010-1070); and in India, Halayudha (about 975) who referred to a book, apparently now lost, by an author some 1,000 years earlier.

Sauer (1972) says he is answering a question Erdös asked him in 1970. The question as posed could have been answered by the not yet sharp, although path-breaking, combinatorial results of Vapnik and Červonenkis (1968, 1971), not cited by Sauer (1972) nor Shelah (1972).

Shelah (1972) is mainly about infinite sets. Other, published questions of Erdös and co-authors are answered. Shelah (p. 54) cites Sauer (1972) and states Sauer's Lemma, saying "Perles and I [had] prove[d]" it earlier, but I could not find documentation of their proof.

Vapnik and Červonenkis in their 1974 book also gave Theorem 4.2, then proved Proposition 4.3. Assouad (1981, (3.2); 1983, Proposition 2.4) defined "dens" and noted Proposition 4.5.

Notes to Section 4.2. Theorem 4.6, when H is the space of linear functions on $\mathbb{R}^m$ and $f = 0$, is a classical fact known as Radon's Theorem, proved by Radon (1921, p. 114) and reviewed by Danzer, Grünbaum, and Klee (1963, p. 103). Theorem 4.6 for general H and $f = 0$ appeared in Dudley (1978, Theorem 7.2); Wenocur and Dudley (1981) proved it for any f. Assouad (1983) noted Theorem 4.7.

Notes to Section 4.3. This section is based on the paper Dudley (1985b).

Notes to Section 4.4. This section is also largely based on Dudley (1985b), except that Theorem 4.30 was added, just after the first edition of this book was published (1999), in Dudley (2000). Theorem 4.21 and other characterizations of trees are given in Harary (1969), pp. 32–33.

Notes to Section 4.5. Some of the facts in this section are from Dudley (1984, Section 9.2). Those on the density, Theorems 4.31 and 4.33, are due to Assouad (1983). Assouad also told me Proposition 4.36 in a letter. Theorem 4.41 and Corollary 4.44 are from Wenocur and Dudley (1981). Theorems 4.30 and 4.38, and the case $k = 3$ of Proposition 4.39, appeared in Dudley (2000). Theorem 4.46 appeared in Smoktunowicz (1997).

Notes to Section 4.6. Theorem 4.47 is partly in Dudley (1978, Section 7) and was stated in the current form by Assouad (1981, 1983). Haussler (1995) gave a sharper form. Assouad (1983) also proved Theorem 4.49, first defined dens*($\cdot$), and proved Theorem 4.50.

Notes to Section 4.7. VC subgraph classes have been called "VC graph" classes (Alexander 1984, 1987) or "polynomial classes" (Pollard 1984, pp. 17, 34; 1985). This section is based on a small part of Dudley (1987).

Notes to Section 4.8. Theorem 4.52 is essentially Lemma 25 of Pollard (1984, p. 27) and had appeared earlier in Pollard (1982). Theorem 4.53(b) and (c) are based on Section 3 of Dudley (1987).

5

Measurability

The example after Theorem 3.2 showed that for a continuous distribution function F such as for $U[0, 1]$, the set of all possible functions $\sqrt{n}(F_n - F)$, even for $n = 1$, is nonseparable in the sup norm, and all its subsets are closed, including those corresponding to nonmeasurable sets of possible values of the observation X_1. Therefore, the classical definition of convergence in law, or weak convergence, which works in separable metric spaces, does not work in this case, So, in Chapter 3, functions f^* and upper expectations E^* were used to get around measurability problems.

But, in the classical Glivenko–Cantelli theorem, saying that $\sup_x |(F_n - F)(x)| \to 0$ almost surely as $n \to \infty$ for any distribution function F on $\mathbb{R}$ and its empirical distribution functions F_n (RAP, Theorem 11.4.2), there is no measurability problem. The supremum is measurable, as it can be restricted to rational x by right-continuity of F_n and F. The collection $\mathcal{C}$ of left half-lines $(-\infty, x]$ is linearly ordered by inclusion and so has $S(\mathcal{C}) = 1$, and for it, not only the Glivenko–Cantelli theorem but, after suitable formulations (Theorem 1.8 or, less specifically, Chapter 3), the uniform central limit theorem (Donsker property) holds for any probability measure P on the Borel sets of $\mathbb{R}$.

Another example of a class $\mathcal{C}$ linearly ordered by inclusion will be given, where the Glivenko–Cantelli property actually fails, although $\sup_{A \in \mathcal{C}} |(P_n - P)(A)|$ is measurable (it is identically 1). The notion of linear ordering was defined before Theorem 4.17. A linearly ordered set $(X, \preceq)$ is said to be *well-ordered* iff every nonempty subset contains a least element. Thus, for example, with usual orderings, the set $\mathbb{N}$ of nonnegative integers is well-ordered, but the set $\mathbb{Z}$ of all integers is not. It follows from the axiom of choice that every set can be well-ordered (Zermelo's theorem, RAP, Theorem 1.5.1), and in particular, there exist well-ordered sets $(X, \preceq)$ such that X is uncountable.

213

Example. This is related to the "ordinal triangle" counterexample in integration theory, showing why measurability is needed in the Tonelli–Fubini Theorem on Cartesian product integrals. Let $(\Omega, \preceq)$ be an uncountable well-ordered set such that for each $x \in \Omega$, the initial segment $I_x := \{y : y \preceq x\}$ is countable. (In terms of ordinals, Ω is, or is order-isomorphic to, the least uncountable ordinal.) Let S be the σ-algebra of subsets of Ω consisting of sets that are countable or have countable complement. Let P be the probability measure on S which is 0 on countable sets and 1 on sets with countable complement. Then

$$\int \int 1_{y \preceq x} dP(y) dP(x) = 0 < \int \int 1_{y \preceq x} dP(x) dP(y) = 1.$$

Since all other hypotheses of the Tonelli–Fubini theorem hold, the function $(x, y) \mapsto 1_{y \preceq x}$ must not be measurable for the product σ-algebra, even if S is replaced by any larger σ-algebra of subsets of Ω to which P can be extended. For example, according to the continuum hypothesis, we could take Ω to be $[0, 1]$ (where the well-ordering is unrelated to the usual ordering), and P to be Lebesgue measure or any other nonatomic law on $[0, 1]$.

Now, consider the class C of sets I_x for each $x \in \Omega$. Each of these sets is countable by assumption. The sets are linearly ordered by inclusion since $\preceq$ is a linear ordering. Thus $S(C) = 1$ by Theorem 4.10. But, C is not a weak or strong Glivenko–Cantelli class as defined in Section 3.3 (still less a Donsker class), since for any possible $X_1, \ldots, X_n$, a maximum $x := \max(X_1, \ldots, X_n)$ for the well-ordering exists, so $P_n(I_x) = 1$ while $P(I_x) = 0$, so $\sup_{A \in C} |(P_n - P)(A)| \equiv 1$.

To find hypotheses that will avoid problems as in the example just given, suppose we have a class $\mathcal{F}$ of functions on X, where $(X, \mathcal{A})$ is a measurable space, for example $\mathcal{F} = \{1_A : A \in C\}$ for a class $C \subset \mathcal{A}$. The mapping $(f, x) \mapsto f(x)$ seems very natural, but is it jointly measurable in a useful sense? Sometimes it is useful to consider a kind of parametrization, via a measurable space $(Y, \mathcal{B})$, with a mapping $y \mapsto f_y$ from Y onto $\mathcal{F}$. We will want the mapping $(x, y) \mapsto f_y(x)$ to be jointly measurable on $X \times Y$. This property, called (image) *admissibility*, will be the topic of Section 5.2. In the above examples, for the usual ordering $\leq$ of $\mathbb{R}$, the set $\{(x, y) : x \leq y\}$ is jointly measurable (Borel, in fact closed in $\mathbb{R}^2$, and any open set in $\mathbb{R}^2$ is a countable union of open rectangles), but we saw that in an uncountable well-ordered set $(X, \preceq)$, $\{(x, y) : x \preceq y\}$ was not jointly measurable in $X \times X$. A convenient condition to assume on the parametrizing set Y is that it is a Suslin measurable space, a property to be defined and treated, in combination with admissibility, in Section 5.3. The admissibility and Suslin properties will hold for classes of sets (or functions) encountered in practice. It will be shown in Chapter 6 that these conditions, together with the Vapnik–Červonenkis or related properties, are enough to imply Glivenko–Cantelli and Donsker properties.

5.1 Sufficiency

This section could nearly be starred, as it is referred to later only once in this chapter and once in Chapter 6.

Suppose that a probability measure P is known to be in a certain family $\mathcal{P}$ of laws and we have observed $X_1, \ldots, X_n$ i.i.d. (P), but nothing else is known about P. A *statistic*, T, which is a measurable function of $X_1, \ldots, X_n$, will roughly speaking be said to be sufficient for $\mathcal{P}$ if, given T, no further information about $X_1, \ldots, X_n$ is useful in making decisions or inferences about $P \in \mathcal{P}$. A precise definition is given below. This section will show that the empirical measure P_n is sufficient even when $\mathcal{P}$ is the family of all probability measures on a measurable space.

Note that P_n is a symmetric function of the X_i in the sense that it is preserved by any permutation of the indices $1, \ldots, n$. Once P_n is given, knowing that the X_i were observed in a certain order will not help in making inferences about P.

Here is the formal definition of sufficiency: let $(S, \mathcal{B})$ be a measurable space (a set S and a σ-algebra $\mathcal{B}$ of subsets of S). Let $\mathcal{Q}$ be a set of probability laws on $(S, \mathcal{B})$. A sub-σ-algebra $\mathcal{D}$ of $\mathcal{B}$ is called *sufficient* for $\mathcal{Q}$ iff for every $C \in \mathcal{B}$ there is some $\mathcal{D}$-measurable function g_C such that for every $Q \in \mathcal{Q}$, the conditional probability

$$Q(C|\mathcal{D}) = g_C \text{ almost surely for } Q. \tag{5.1}$$

The essential point is that g_C does not depend on Q in $\mathcal{Q}$.

Most often, there will be some $n > 1$, a measurable space $(X, \mathcal{A})$ and a family $\mathcal{P}$ of laws on $(X, \mathcal{A})$ such that S is the n-fold Cartesian product X^n with the product σ-algebra $\mathcal{B} = \mathcal{A}^n$ and $\mathcal{Q} = \mathcal{P}^n := \{P^n : P \in \mathcal{P}\}$, where P^n is the n-fold Cartesian product $P \times P \times \cdots \times P$ (RAP, Theorem 4.4.6).

The meaning of sufficiency is clarified by the factorization theorem, to be proved next. A family $\mathcal{P}$ of probability measures on a measurable space $(S, \mathcal{B})$ is said to be *dominated* by a measure μ if every $P \in \mathcal{P}$ is absolutely continuous with respect to μ. Then we have the density (Radon–Nikodym derivative) $dP/d\mu$ (RAP, Section 5.5).

If there is a nonatomic law on $(S, \mathcal{B})$, the family $\mathcal{P}$ of all laws on $(S, \mathcal{B})$ is not dominated. Factorization is still useful in that case, in the proof of Theorem 5.6 below.

Theorem 5.1 (Factorization theorem) *Let $(S, \mathcal{B})$ be a measurable space, $\mathcal{D}$ a sub-σ-algebra of $\mathcal{B}$, and $\mathcal{P}$ a family of probability measures on $\mathcal{B}$, dominated by a σ-finite measure μ. Then $\mathcal{D}$ is sufficient for $\mathcal{P}$ if and only if there is a $\mathcal{B}$-measurable function $h \geq 0$ such that for all $P \in \mathcal{P}$, there is a $\mathcal{D}$-measurable*

function f_P with $dP/d\mu = f_P h$ almost everywhere for μ. We can take $h \in$
$\mathcal{L}^1(S, \mathcal{B}, \mu)$.

Proof. Two measures are called *equivalent* if each is absolutely continuous with respect to the other. Two families $\mathcal{P}$ and $\mathcal{Q}$ of probability measures on a σ-algebra $\mathcal{B}$ are called *equivalent* if and only if for all $B \in \mathcal{B}$, $(P(B) = 0$ for all $P \in \mathcal{P})$ is equivalent to $(Q(B) = 0$ for all $Q \in \mathcal{Q})$.

Lemma 5.2 *If a family $\mathcal{P}$ of probability measures on a measurable space $(S, \mathcal{B})$ is dominated by a σ-finite measure μ, there is a countable subfamily of $\mathcal{P}$ equivalent to $\mathcal{P}$.*

Proof. For each $P \in \mathcal{P}$, let f_P be (a specific version of) the Radon–Nikodym derivative (density) $dP/d\mu$ and let $K_P := \{x : f_P(x) > 0\}$. A set $K \in \mathcal{B}$ such that for some $P \in \mathcal{P}$, $K \subset K_P$ and $\mu(K) > 0$, will be called a *kernel*. A *chain* is any union of disjoint kernels (necessarily a countable union).

If μ is not finite let $S = \bigcup_{n \geq 1} A_n$ where A_n are disjoint measurable sets with $0 < \mu(A_n) < \infty$. Let $\nu(A) := \sum_{n=1}^{\infty} \mu(A \cap A_n)/[2^n \mu(A_n)]$. Then ν and μ are equivalent (mutually absolutely continuous), so replacing μ by ν, we can assume μ is finite.

Let $C_1, C_2, \ldots$ be chains such that $\sup_n \mu(C_n) = \sup\{\mu(C) : C$ a chain$\}$. Let $D_1 := C_1$. Given a chain D_{n-1}, let $C_n = \bigcup_{j \geq 1} K_{nj}$ where K_{nj} are disjoint kernels (for n fixed). Let $D_{nj} := K_{nj} \setminus D_{n-1}$ if it is a kernel, i.e., if $\mu(K_{nj} \setminus D_{n-1}) > 0$. Otherwise, let D_{nj} be empty. Let $D_n := D_{n-1} \cup \bigcup_{j \geq 1} D_{nj}$, which is a chain. Let $D := \bigcup_n D_n$, a chain with maximal μ. Then $D = \bigcup_n K_n$ for some disjoint kernels K_n. Choose $P := P_n := P(n)$ in $\mathcal{P}$ such that $K_n \subset K_{P(n)}$ and $P_n(K_n) > 0$. To show that $\{P_m\}$ is equivalent to $\mathcal{P}$, suppose not. Then for some $B \in \mathcal{B}$, $P_n(B) = 0$ for all n and $P(B) > 0$ for some $P \in \mathcal{P}$. If $P(B \setminus D) > 0$, then some kernel is included in $B \setminus D$, contradicting the choice of D. So we can assume $B \subset D$ and for some n, $P(B \cap K_n) > 0$. Then $P_n(B \cap K_n) > 0$ since $\mu(B \cap K_n) > 0$ and K_n is a kernel for P_n. This contradiction completes the proof. $\square$

Lemma 5.3 *Any countable family of σ-finite measures μ_n is equivalent to one finite measure μ.*

Proof. First, as in the last proof, we can assume $\mu_n(S) \leq 1$ for all n. Then, let $\mu := \sum_{n \geq 1} \mu_n/2^n$, which is finite and equivalent to $\{\mu_n\}_{n \geq 1}$. $\square$

The next step in the proof of Theorem 5.1 is:

Theorem 5.4 *Under the hypotheses of Theorem 5.1, the following are equivalent:*

(a) $\mathcal{D}$ is sufficient for $\mathcal{P}$;

(b) There is a probability measure λ such that $\{\lambda\}$ is equivalent to $\mathcal{P}$ and $dP/d\lambda$ can be chosen to be $\mathcal{D}$-measurable for all $P \in \mathcal{P}$;

(c) There is a probability measure λ which dominates $\mathcal{P}$ such that $dP/d\lambda$ can be chosen to be $\mathcal{D}$-measurable for all $P \in \mathcal{P}$.

Proof. First it will be shown that (c) implies (a). For any $B \in \mathcal{B}$ let $f_B :=$ $E_\lambda(1_B|\mathcal{D})$. Then for any $P \in \mathcal{P}$ and $A \in \mathcal{D}$, by RAP, Theorem 10.1.9,

$$\int_A f_B dP = \int_A E_\lambda(1_B|\mathcal{D})\frac{dP}{d\lambda}d\lambda = \int_A 1_B \frac{dP}{d\lambda}d\lambda = P(A \cap B),$$

so $f_B = E_P(1_B|\mathcal{D})$ and $\mathcal{D}$ is sufficient.

(a) implies (b): if $\mathcal{D}$ is sufficient, take a sequence $\{P_n\} \subset \mathcal{P}$, equivalent to $\mathcal{P}$, by Lemma 5.2. Let $\lambda := \sum_{n \geq 1} P_n/2^n$. For each $B \in \mathcal{B}$ take $f_B = E_P(1_B|\mathcal{D})$ P-a.s. for all $P \in \mathcal{P}$. Since $0 \leq f_B \leq 1$ P-almost surely for all P in $\mathcal{P}$, also $0 \leq f_B \leq 1$ almost surely for λ. Since $P_n(B \cap A) = \int_A f_B dP_n$ for all n and all $A \in \mathcal{D}$, it follows that $\lambda(B \cap A) = \int_A f_B d\lambda$ so $f_B = E_\lambda(1_B|\mathcal{D})$.

To show that for all $P \in \mathcal{P}$, $dP/d\lambda$ is $\mathcal{D}$-measurable, note that for each B in $\mathcal{B}$,

$$\int_B \frac{dP}{d\lambda}d\lambda = P(B) = \int f_B \frac{dP}{d\lambda}d\lambda = \int E_\lambda\left(f_B\frac{dP}{d\lambda}\Big|\mathcal{D}\right)d\lambda,$$

which by RAP, Theorem 10.1.9 again, equals

$$\int E_\lambda(1_B|\mathcal{D})E_\lambda\left(\frac{dP}{d\lambda}\Big|\mathcal{D}\right)d\lambda = \int E_\lambda\left(1_B E_\lambda\left(\frac{dP}{d\lambda}\Big|\mathcal{D}\right)\Big|\mathcal{D}\right)d\lambda$$

$$= \int 1_B E_\lambda\left(\frac{dP}{d\lambda}\Big|\mathcal{D}\right)d\lambda.$$

Since $dP/d\lambda$ and $E_\lambda(dP/d\lambda|\mathcal{D})$ have the same integral over all sets in $\mathcal{B}$, they must be equal P-a.s., so that $dP/d\lambda$ is equal P-a.s. to a $\mathcal{D}$-measurable function. So (a) implies (b). Clearly (b) implies (c), so Theorem 5.4 is proved. $\square$

Proof of the factorization theorem 5.1. If $\mathcal{D}$ is sufficient, take λ from the last theorem and let $h := d\lambda/d\mu$. Letting $f_P = dP/d\lambda$ we have the desired factorization, with $h \in \mathcal{L}^1(S, \mathcal{B}, \mu)$ (in fact $\int h \, d\mu = 1$).

Conversely, if factorization holds (with h not necessarily integrable), then by Lemma 5.2, take $\{P_n\}_{n \geq 1}$ equivalent to $\mathcal{P}$, and $\lambda := \sum_{n \geq 1} P_n/2^n$, so that $\{\lambda\}$ is equivalent to P by Lemma 5.3. Then λ is absolutely continuous with respect to μ and $d\lambda/d\mu = hk$ where $k(x) := \sum_{n \geq 1} f_{P_n}(x)/2^n$ for all x. Also, k is $\mathcal{D}$-measurable. For each $P \in \mathcal{P}$ let $g_P(x) := f_P(x)/k(x)$ if $k(x) > 0$ and $g_P(x) := 0$ otherwise. Then $\lambda(k = 0) = 0$, and for each $P \in \mathcal{P}$, g_P is $\mathcal{D}$-measurable, P is absolutely continuous with respect to λ, and $dP/d\lambda = g_P$. So $\mathcal{D}$ is sufficient for $\mathcal{P}$ by Theorem 5.4. $\square$

Given a statistic T, i.e. a measurable function, from S into Y for measurable spaces $(S, \mathcal{B})$ and $(Y, \mathcal{F})$, let $\mathcal{D} := T^{-1}(\mathcal{F}) := \{T^{-1}(A) : A \in \mathcal{F}\}$, a σ-algebra. For a family $\mathcal{Q}$ of laws on $(S, \mathcal{B})$, T is called a *sufficient statistic* for $\mathcal{Q}$ iff $\mathcal{D}$ is sufficient for $\mathcal{Q}$. If T is sufficient we can write $f_P = g_P \circ T$ for some $\mathcal{F}$-measurable function g_P by RAP, Theorem 4.2.8. Sufficiency, defined in terms of conditional probabilities of measurable sets, can be extended to suitable conditional expectations:

Theorem 5.5 *Let $\mathcal{D}$ be sufficient for a family $\mathcal{P}$ of laws on a measurable space $(S, \mathcal{B})$. Then for any measurable real-valued function f on $(S, \mathcal{B})$ which is integrable for each $P \in \mathcal{P}$, there is a $\mathcal{D}$-measurable function g such that $g = E_P(f|\mathcal{D})$ a.s. for all $P \in \mathcal{P}$.*

Proof. When f is the indicator function of a set in $\mathcal{B}$, the assertion is the definition of sufficiency. It then follows for any simple function, which is a finite linear combination of such indicators. If f is nonnegative, there is a sequence of nonnegative simple functions increasing up to f and the conclusion holds (RAP, Proposition 4.1.5 and Theorem 10.1.7). Then any f satisfying the hypothesis can be written as $f = f^+ - f^-$ for f^+ and f^- nonnegative and the result follows. $\square$

Let μ and ν be two probability measures on the same measurable space $(V, \mathcal{U})$. Take the Lebesgue decomposition (RAP, Theorem 5.5.3) $\nu = \nu_{ac} + \nu_s$ where ν_{ac} is absolutely continuous, and ν_s is singular, with respect to μ. Let $A \in \mathcal{U}$ with $\nu_s(A) = \mu(V \setminus A) = 0$, so $\nu_{ac}(V \setminus A) = 0$. Then the *likelihood ratio* $R_{\nu/\mu}$ is defined as the Radon–Nikodym derivative $d\nu_{ac}/d\mu$ on A and $+\infty$ on $V \setminus A$. By uniqueness of the Hahn decomposition of V for $\nu_s - \mu$ (RAP, Theorem 5.6.1), $R_{\nu/\mu}$ is defined up to equality $(\mu + \nu_s)$- and so $(\mu + \nu)$-almost everywhere.

Theorem 5.6 *For any family $\mathcal{P}$ of laws on a measurable space $(S, \mathcal{B})$ and sub-σ-algebra $\mathcal{D} \subset \mathcal{B}$, if $\mathcal{D}$ is sufficient for $\mathcal{P}$, then for all $P, Q \in \mathcal{P}$, $R_{Q/P}$ can be taken to be $\mathcal{D}$-measurable, i.e., is equal $(P + Q)$-almost everywhere to a $\mathcal{D}$-measurable function.*

Proof. Suppose $\mathcal{D}$ is sufficient for $\mathcal{P}$. Then it is also sufficient for $\{P, Q\}$, which is dominated by $\mu := P + Q$. So by factorization (Theorem 5.1) there are $\mathcal{D}$-measurable functions f_P and f_Q and a $\mathcal{B}$-measurable function h such that $dP/d\mu = f_P h$ and $dQ/d\mu = f_Q h$. Then $R_{Q/P} = f_Q h/(f_P h) = f_Q/f_P$, where $y/0 := +\infty$ if $y > 0$ and 0 if $y = 0$, is $\mathcal{D}$-measurable (note that $R_{Q/P}$ does not depend on the choice of dominating measure). $\square$

Suppose we observe $X_1, \ldots, X_n$ i.i.d. with law P or Q but we do not know which and want to decide. Suppose we have no *a priori* reason to favor a choice of P or Q, only the data. Then it is natural to evaluate the likelihood ratio R_{Q^n/P^n}

and choose Q if $R_{Q^n/P^n} > 1$ and P if $R_{Q^n/P^n} < 1$, while if $R_{Q^n/P^n} = 1$ we still have no basis to prefer P or Q. More generally, decisions between P and Q can be made optimally in terms of minimizing error probabilities or expected losses by way of the likelihood ratio $R_{Q/P}$ or R_{Q^n/P^n} as appropriate (the Neyman–Pearson Lemma, Lehmann 1991, pp. 74, 125; Bickel and Doksum 2001, Theorem 4.2.1). By Theorem 5.6, if $\mathcal{B}$ is sufficient for $\mathcal{P}^n$ for some $\mathcal{P} \supset \{P, Q\}$, then R_{Q^n/P^n} is $\mathcal{B}$-measurable. Specifically, if T is a sufficient statistic, then by Theorem 5.6 and RAP, Theorem 4.2.8, R_{Q^n/P^n} is a measurable function of T. Thus, no information in $(X_1, \ldots, X_n)$ beyond T is helpful in choosing P or Q. In this sense, the definition of sufficiency fits with the informal notion of sufficiency given at the beginning of the section.

It will be shown that empirical measures are sufficient in a sense to be defined. Let $\mathcal{S}_n$ be the sub-σ-algebra of $\mathcal{A}^n$ consisting of sets invariant under all permutations of the coordinates.

Theorem 5.7 $\mathcal{S}_n$ *is sufficient for* $\mathcal{P}^n := \{P^n : P \in \mathcal{P}\}$ *where $\mathcal{P}$ is the set of all laws on* $(X, \mathcal{A})$.

Proof. Let S_n be the symmetric group of all permutations of $\{1, 2, \ldots, n\}$. For each $\Pi \in S_n$ and $x := (x_1, \ldots, x_n) \in X^n$, set $f_\Pi(x) := (x_{\Pi(1)}, \ldots, x_{\Pi(n)})$. Then f_Π is a 1–1 measurable transformation of X^n onto itself with measurable inverse and preserves the product law P^n for each law P on $(X, \mathcal{A})$. For any $C \in \mathcal{A}^n$, we have

$$P^n(C|\mathcal{S}_n) = \frac{1}{n!} \sum_{\Pi \in S_n} 1_{f_\Pi(C)}(\cdot)$$

almost surely for P^n since for any $B \in \mathcal{S}_n$,

$$P^n(B \cap C) = \frac{1}{n!} \sum_{\Pi \in S_n} P^n(C \cap f_\Pi^{-1}(B)) = \frac{1}{n!} \sum_{\Pi \in S_n} P^n(f_\Pi(C) \cap B).$$

The conclusion follows. $\qquad\square$

For example, if X has just two points, say $X = \{0, 1\}$, and $S := \sum_{i=1}^n x_i$, then $\mathcal{S}_n$ is the smallest σ-algebra for which S is measurable. In this case no σ-algebra strictly smaller than $\mathcal{S}_n$ is sufficient ($\mathcal{S}_n$ is "minimal sufficient"). For each $B \in \mathcal{A}$ and $x = (x_1, \ldots, x_n) \in X^n$, let

$$P_n(B)(x) := \frac{1}{n} \sum_{j=1}^n 1_B(x_j).$$

So P_n is the usual empirical measure, except that in this section, $x \mapsto P_n(B)(x)$ is a measurable function, or statistic, on a measurable space, rather than a probability space, since no particular law P or P^n has been specified as yet. Here, $P_n(B)(x)$ is just a function of B and x.

For a collection $\mathcal{F}$ of measurable functions on $(X^n, \mathcal{A}^n)$, let $\mathcal{S}_{\mathcal{F}}$ be the smallest σ-algebra making all functions in $\mathcal{F}$ measurable. Then $\mathcal{F}$ will be called *sufficient* if and only if $\mathcal{S}_{\mathcal{F}}$ is sufficient.

Theorem 5.8 *For any measurable space $(X, \mathcal{A})$ and for each $n = 1, 2, \ldots$, the empirical measure P_n is sufficient for $\mathcal{P}^n$ where $\mathcal{P}$ is the set of all laws on $(X, \mathcal{A})$. In other words the set $\mathcal{F}$ of functions $x \mapsto P_n(B)(x)$, for all $B \in \mathcal{A}$, is sufficient. In fact the σ-algebra $\mathcal{S}_{\mathcal{F}}$ is exactly $\mathcal{S}_n$.*

Proof. Clearly $\mathcal{S}_{\mathcal{F}} \subset \mathcal{S}_n$. To prove the converse inclusion, for each set $B \in \mathcal{A}^n$ let $S(B) := \bigcup_{\Pi \in S_n} f_\Pi(B) \in \mathcal{S}_n$. Then if $B \in \mathcal{S}_n$, $S(B) = B$. Let $\mathcal{E} := \{C \in \mathcal{A}^n : S(C) \in \mathcal{S}_{\mathcal{F}}\}$. We want to prove $\mathcal{E} = \mathcal{A}^n$.

Now $\mathcal{E}$ is a monotone class: if $C_n \in \mathcal{E}$ and $C_n \uparrow C$ or $C_n \downarrow C$, then $C \in \mathcal{E}$. Also, since $S(\cdot)$ commutes with finite unions, any finite union of sets in $\mathcal{E}$ is in $\mathcal{E}$. So it will be enough to prove

$$A_1 \times \cdots \times A_n \in \mathcal{E} \text{ for any } A_j \in \mathcal{A}, \ j = 1, \ldots, n, \tag{5.2}$$

since the collection $\mathcal{C}$ of finite unions of such sets, which can be taken to be disjoint, is an algebra and the smallest monotone class including $\mathcal{C}$ is $\mathcal{A}^n$ (RAP, Propositions 3.2.2 and 3.2.3 and Theorem 4.4.2). Here by another finite union the A_i can be replaced by $B_{j(i)}$ where $B_1, \ldots, B_r$ are atoms of the algebra generated by $A_1, \ldots, A_n$, so $r \leq 2^n$. So we just need to show that for all $j(1), \ldots, j(n)$ with $1 \leq j(i) \leq r$, $i = 1, \ldots, n$, we have $S(B_{j(1)} \times \cdots \times B_{j(n)}) \in \mathcal{S}_{\mathcal{F}}$. Now, $x \in S(B_{j(1)} \times \cdots \times B_{j(n)})$ if and only if for each $i = 1, \ldots, r$, $P_n(B_i) = k_i/n$ where k_i is the number of values of s such that $j(s) = i$, $s = 1, \ldots, n$. So $\mathcal{S}_{\mathcal{F}} = \mathcal{S}_n$, and by Theorem 5.7, the conclusion follows. $\square$

For some subclasses $\mathcal{C} \subset \mathcal{A}$, the restriction of P_n to $\mathcal{C}$ may be sufficient, and handier than the values of P_n on the whole σ-algebra $\mathcal{A}$. Recall that a class $\mathcal{C}$ included in a σ-algebra $\mathcal{A}$ is called a *determining class* if any two measures on $\mathcal{A}$, equal and finite on $\mathcal{C}$, are equal on all of $\mathcal{A}$. If $\mathcal{C}$ generates the σ-algebra $\mathcal{A}$, $\mathcal{C}$ is not necessarily a determining class unless, for example, it is an algebra (RAP, Theorem 3.2.7 and the example after it).

Sufficiency of $P_n(A)$ for $A \in \mathcal{C}$ can depend on n. Let $X = \{1, 2, 3, 4, 5\}$, $\mathcal{A} = 2^X$, and $\mathcal{C} = \{\{1, 2, 3\}, \{2, 3, 4\}, \{3, 4, 5\}\}$. Then $\mathcal{C}$ is sufficient for $n = 1$, but not for $n = 2$ since, for example, $(\delta_1 + \delta_4)/2 \equiv (\delta_2 + \delta_5)/2$ on $\mathcal{C}$. This is a case where $\mathcal{C}$ generates $\mathcal{A}$ but is not a determining class.

Theorem 5.9 *Let (X, d) be a separable metric space which is a Borel subset of its completion, with Borel σ-algebra $\mathcal{A}$. Suppose $\mathcal{C} = \{C_k\}_{k=1}^{\infty}$ is a countable determining class for $\mathcal{A}$. Then for each $n = 1, 2, \ldots$, the sequence $\{P_n(C_k)\}_{k=1}^{\infty}$ is sufficient for the class $\mathcal{P}^n$ of all laws P^n on $(X^n, \mathcal{A}^n)$ where $P \in \mathcal{P}$, the class of all laws on $(X, \mathcal{A})$.*

Proof. For $n = 1, 2, \ldots$, let I_n be the finite set $\{j/n : j = 0, 1, \ldots, n\}$. Let I_n^∞ be a countable product of copies of I_n with the product σ-algebra defined by the σ-algebra of all subsets of I_n. We have:

Lemma 5.10 *Under the hypotheses of Theorem 5.9, for each $n = 1, 2, \ldots$, and $A \in \mathcal{A}$, there is a Borel measurable function f_A on I_n^∞ such that for any $x_1, \ldots, x_n \in X$ and $P_n := n^{-1} \sum_{j=1}^n \delta_{x_j}$, we have $P_n(A) = f_A(\{P_n(C_k)\}_{k=1}^\infty)$.*

Proof. Since $\mathcal{C}$ is a determining class, the function f_A exists. We need to show it is measurable.

If X is uncountable, then by the Borel isomorphism theorem (RAP, Theorem 13.1.1) we can assume $X = [0, 1]$. Or, if X is countable, then $\mathcal{A}$ is the σ-algebra of all its subsets, and we can assume $X = \{0\} \cup Y$ where $Y \subset \{1/k\}_{k\geq 1}$. Then X is always complete and included in $[0, 1]$. Let

$$X^{(n)} := \{x := \{x_j\}_{j=1}^n \in X^n : x_1 \leq x_2 \leq \cdots \leq x_n\}.$$

The map $x \mapsto P_n^{(x)}$ is 1–1 from $X^{(n)}$ into the set of all possible empirical measures P_n, since if $Q := P_n^{(x)} = P_n^{(y)}$, with $x, y \in X^{(n)}$, then $x_1 = y_1 =$ the smallest u such that $Q(\{u\}) > 0$. Next, $x_2 = y_2 = x_1$ if and only if $Q(\{x_1\}) \geq 2/n$, while otherwise $x_2 = y_2 =$ the next-smallest v such that $Q(\{v\}) > 0$, and so on.

Now, $X^{(n)}$ is a Borel subset of X^n. The completion of X^n for any of the usual product metrics is isometric to S^n where S is the completion of X. Clearly X^n is a Borel subset of S^n. It follows that $X^{(n)}$ is a Borel subset of its completion. Here the following will be useful:

Lemma 5.11 *On I_n^∞, the product σ-algebra $\mathcal{B}^\infty$, the smallest σ-algebra for which all the coordinates are measurable, equals the Borel σ-algebra $\mathcal{B}_\infty$ of the product topology $\mathcal{T}$.*

Proof. Clearly $\mathcal{B}^\infty \subset \mathcal{B}_\infty$ since the coordinates are continuous. Conversely, $\mathcal{T}$ has a base $\mathcal{R}$ consisting of all sets $\Pi_{j=1}^\infty A_j$ where $A_j = I_n$ for all but finitely many j; $\mathcal{R}$ is countable (since I_n is finite) and consists of sets in $\mathcal{B}^\infty$, and every $U \in \mathcal{T}$ is a countable union of sets in $\mathcal{R}$, so $U \in \mathcal{B}^\infty$ and $\mathcal{B}_\infty = \mathcal{B}^\infty$ so Lemma 5.11 is proved. $\square$

Now to continue the proof of Lemma 5.10, the map $f : x \mapsto \{P_n^{(x)}(C_k)\}_{k=1}^\infty$ is Borel measurable from $X^{(n)}$ into I_n^∞. Since $\{C_k\}_{k=1}^\infty$ are a determining class, f is one-to-one. Thus by Appendix G, Theorem G.6, f has Borel image $f[X^{(n)}]$ in I_n^∞ and f^{-1} is Borel measurable from $f[X^{(n)}]$ onto $X^{(n)}$. Then f^{-1} extends to a Borel measurable function h on all of I_n^∞ into $X^{(n)}$, since $X^{(n)}$ is Borel-isomorphic to $\mathbb{R}$, or a countable subset, with Borel σ-algebra, by the Borel isomorphism theorem (RAP, Theorem 13.1.1 again), and thus the extension works as for real-valued functions (RAP, Theorem 4.2.5). For any

$A \in \mathcal{A}$, $g_A : x \mapsto P_n^{(x)}(A)$ is Borel measurable. Thus $f_A \equiv g_A \circ h$ is Borel measurable, and Lemma 5.10 is proved. $\qquad\square$

Now to prove Theorem 5.9, we know from Theorem 5.8 that the smallest σ-algebra $\mathcal{S}_\mathcal{F}$ making all functions $x \mapsto P_n(B)(x)$ measurable for $B \in \mathcal{A}$ is sufficient. By Lemma 5.10, $\mathcal{S}_\mathcal{F}$ is the same as the smallest σ-algebra making $x \mapsto P_n(C_k)$ measurable for all k, which finishes the proof. $\qquad\square$

In the real line $\mathbb{R}$, the closed half-lines $(-\infty, x]$ form a determining class. In other words, as is well known, a probability measure P on the Borel σ-algebra of $\mathbb{R}$ is uniquely determined by its distribution function F (RAP, Theorem 3.2.6). It follows that the half-lines $(-\infty, q]$ for q rational are a determining class: for any real x, take rational $q_k \downarrow x$, then $F(q_k) \downarrow F(x)$. Thus we have:

Corollary 5.12 *In $\mathbb{R}$, the empirical distribution functions defined by $F_n(x) := P_n((-\infty, x])$ for all x are sufficient for the family $\mathcal{P}^n$ of all laws P^n on $\mathbb{R}^n$ where P varies over all laws on the Borel σ-algebra in $\mathbb{R}$.*

5.2 Admissibility

Let $\mathcal{F}$ be a family of real-valued functions on a set S, measurable for a σ-algebra $\mathcal{B}$ on S. Then there is a natural function, called here the *evaluation map*, $\mathcal{F} \times S \mapsto \mathbb{R}$ given by $(f, x) \mapsto f(x)$. It turns out that for general $\mathcal{F}$ there may not exist any σ-algebra of subsets of $\mathcal{F}$ for which the evaluation map is jointly measurable. This section is about the possible existence of such a σ-algebra and its uses.

Let $(S, \mathcal{B})$ be a measurable space. Then $(S, \mathcal{B})$ will be called *separable* if $\mathcal{B}$ is generated by some countable subclass $\mathcal{C} \subset \mathcal{B}$ and $\mathcal{B}$ contains all singletons $\{x\}$, $x \in S$. In this section $(S, \mathcal{B})$ will be assumed to be such a space. Let $\mathcal{F}$ be a collection of real-valued functions on S. (The following definition is unrelated to the usage of "admissible" for estimators in statistics.)

Definition. $\mathcal{F}$ is called *admissible* iff there is a σ-algebra $\mathcal{T}$ of subsets of $\mathcal{F}$ such that the evaluation map $(f, x) \mapsto f(x)$ is jointly measurable from $(\mathcal{F}, \mathcal{T}) \times (S, \mathcal{B})$ (with product σ-algebra) to $\mathbb{R}$ with Borel sets. Then $\mathcal{T}$ will be called an *admissible structure* for $\mathcal{F}$.

$\mathcal{F}$ will be called *image admissible via* $(Y, \mathcal{S}, T)$ if $(Y, \mathcal{S})$ is a measurable space and T is a function from Y onto $\mathcal{F}$ such that the map $(y, x) \mapsto T(y)(x)$ is jointly measurable from $(Y, \mathcal{S}) \times (S, \mathcal{B})$ with product σ-algebra to $\mathbb{R}$ with Borel sets.

To apply these definitions to a family $\mathcal{C}$ of sets let $\mathcal{F} = \{1_A : A \in \mathcal{C}\}$.

Remarks. There is no assumption of separability of measurable spaces in the definition just given. In the next three theorems, $(S, \mathcal{B})$ is assumed separable, but still there is no restriction on the σ-algebras $\mathcal{T}$ on $\mathcal{F}$ or $\mathcal{S}$ on Y.

If $\mathcal{G} \subset \mathcal{F}$ with $\mathcal{F}$ admissible, then clearly so is $\mathcal{G}$, with σ-algebra $\mathcal{G} \sqcap \mathcal{T}$.

For an example, let (K, d) be a compact metric space and let $\mathcal{F}$ be a set of continuous real-valued functions on K, compact for the supremum norm. Then the functions in $\mathcal{F}$ are uniformly equicontinuous on K by the Arzelà–Ascoli theorem (RAP, Theorem 2.4.7). It follows that $(f, x) \mapsto f(x)$ is jointly continuous for the supremum norm on $f \in \mathcal{F}$ and d on K. Since both spaces are separable metric spaces, the map is also jointly measurable, so that $\mathcal{F}$ is admissible.

If a family $\mathcal{F}$ is admissible, then it is image admissible, taking T to be the identity. In regard to the converse direction here is an example. Let $S = [0, 1]$ with usual Borel σ-algebra $\mathcal{B}$. Let $(Y, \mathcal{S})$ be a countable product of copies of $(S, \mathcal{B})$. For $y = \{y_n\}_{n=1}^{\infty} \in Y$ let $T(y)(x) := 1_J(x, y)$ where $J := \{(x, y) : x = y_n \text{ for some } n\}$. Let $\mathcal{C}$ be the class of all countable subsets of S and $\mathcal{F}$ the class of indicator functions of sets in $\mathcal{C}$. Then it is easy to check that $\mathcal{F}$ is image admissible via $(Y, \mathcal{S}, T)$. If a σ-algebra $\mathcal{T}$ is defined on $\mathcal{F}$ by setting $\mathcal{T} := \{F \subset \mathcal{F} : T^{-1}(F) \in \mathcal{S}\}$, then $\mathcal{T}$ is not countably generated (see Problem 5(b)) although $\mathcal{S}$ is. This example shows how sometimes image admissibility may work better than admissibility.

Theorem 5.13 *For any separable measurable space $(S, \mathcal{B})$, there is a subset Y of $[0, 1]$ and a 1–1 function M from S onto Y which is a measurable isomorphism (is measurable and has measurable inverse) for the Borel σ-algebra on Y.*

Remarks. Note that Y is not necessarily a measurable subset of $[0, 1]$. On the other hand if $(S, \mathcal{B})$ is given as a separable metric space which is a Borel subset of its completion, with Borel σ-algebra, then $(S, \mathcal{B})$ is measurably isomorphic either to a countable set, with the σ-algebra of all its subsets, or to all of $[0, 1]$ by the Borel isomorphism theorem (RAP, Theorem 13.1.1).

Proof. Recall that the Borel subsets of Y as a metric space with usual metric are the same as the intersections with Y of Borel sets in $[0, 1]$, since the same is true for open sets.

Let $\mathcal{C} := \{C_j\}_{j \geq 1}$ be a countable set of generators of $\mathcal{B}$. Consider the map $f : x \mapsto \{1_{C_j}(x)\}_{j \geq 1}$ from S into a countable product 2^{∞} of copies of $\{0, 1\}$ with product σ-algebra. Then f is 1–1 and onto its range Z. Thus it preserves all set operations, specifically countable unions and complements. So it is easily seen that f is a measurable isomorphism of S onto Z.

Next consider the map $g : \{z_j\} \mapsto \sum_j 2z_j/3^j$ from 2^{∞} into $[0, 1]$, actually onto the Cantor set C (RAP, proof of Proposition 3.4.1). Then g is continuous from the compact space 2^{∞} with product topology onto C. It is easily seen that g is 1–1. Thus g is a homeomorphism (RAP, Theorem 2.2.11) and a measurable isomorphism for Borel σ-algebras. The Borel σ-algebra on 2^{∞}

equals the product σ-algebra (Lemma 5.11). So the restriction of g to Z is a measurable isomorphism onto its range Y. It follows that the composition $g \circ f$, called the *Marczewski function*

$$M(x) := \sum_{n=1}^{\infty} 2 \cdot 1_{C_n}(x)/3^n$$

is 1–1 from S onto $Y \subset I := [0, 1]$ and is a measurable isomorphism onto Y. $\square$

Let $(S, \mathcal{B})$ be a separable measurable space where $\mathcal{B}$ is generated by a sequence $\{C_i\}$. By taking the union of the finite algebras generated by $C_1, \ldots, C_n$ for each n, we can and do take $\mathcal{C} := \{C_i\}_{i \geq 1}$ to be an algebra.

Let $\mathcal{F}_0$ be the class of all finite sums $\sum_{i=1}^n c_i 1_{C_i}$ for rational $c_i \in \mathbb{R}$, and $n = 1, 2, \ldots$. Then "Borel classes" or "Banach classes" are defined as follows by transfinite recursion (RAP, 1.3.2). Let $(\Omega, <)$ be an uncountable well-ordered set such that for each $\beta \in \Omega$, $\{\alpha \in \Omega : \alpha < \beta\}$ is countable. (Specifically, one can take Ω to be the set of all countable ordinals with their usual ordering.) For any countable set $A \subset \Omega$, $\{y : y \leq x$ for some $x \in A\}$ is countable, so there is a $z \in \Omega$ with $x < z$ for all $x \in A$. For each $\alpha \in \Omega$ there is a next larger element called $\alpha + 1$. Let 0 be the smallest element of Ω. For each $\alpha \in \Omega$, given $\mathcal{F}_\alpha$, let $\mathcal{F}_{\alpha+1}$ be the set of all limits of everywhere pointwise convergent sequences of functions in $\mathcal{F}_\alpha$. If $\beta \in \Omega$ is not of the form $\alpha + 1$ (β is a "limit ordinal"), $\beta > 0$ and $\mathcal{F}_\alpha$ is defined for all $\alpha < \beta$ let $\mathcal{F}_\beta$ be the union of all $\mathcal{F}_\alpha$ for $\alpha < \beta$. Note that $\mathcal{F}_\alpha \subset \mathcal{F}_\beta$ whenever $\alpha < \beta$. Let $U := \bigcup_{\alpha \in \Omega} \mathcal{F}_\alpha$.

Theorem 5.14 *For $(S, \mathcal{B})$ separable, U is the set of all measurable real functions on S.*

Proof. Clearly, each function in U is measurable. Conversely, the class of all sets B such that $1_B \in U$ is a monotone class and includes the generating algebra $\mathcal{A}$, so it is the σ-algebra $\mathcal{B}$ of all measurable sets (RAP, Theorem 4.4.2). Likewise, for a fixed $A \in \mathcal{A}$ and constants c, d, the collection of all sets B such that $c1_A + d1_B \in U$ is $\mathcal{B}$. Then for a fixed $B \in \mathcal{B}$, the set of all $C \in \mathcal{B}$ such that $c1_C + d1_B \in U$ is all of $\mathcal{B}$. By a similar proof for a sum of n terms, we get that all simple functions $\sum_{i=1}^n c_i 1_{B(i)}$ are in U for any $c_i \in \mathbb{R}$ and $B(i) \in \mathcal{B}$. Since any measurable real function f is the limit of a sequence of simple functions $f_n + g_n$ where $f_n \to \max(f, 0)$ and $g_n \to \min(f, 0)$ (RAP, Proposition 4.1.5), every measurable real function on S is in U. $\square$

On admissibility there is the following main theorem:

Theorem 5.15 (Aumann) *Let $I := [0, 1]$ with usual Borel σ-algebra. Given a separable measurable space $(S, \mathcal{B})$ and a class $\mathcal{F}$ of measurable real-valued functions on S, the following are equivalent:*

(i) $\mathcal{F} \subset \mathcal{F}_\alpha$ for some $\alpha \in \Omega$;

(ii) There is a jointly measurable function $G : I \times S \mapsto \mathbb{R}$ such that for each $f \in \mathcal{F}$, $f = G(t, \cdot)$ for some $t \in I$;

(iii) There is a separable admissible structure for $\mathcal{F}$;

(iv) $\mathcal{F}$ is admissible;

(v) $2^{\mathcal{F}}$ is an admissible structure for $\mathcal{F}$;

(vi) $\mathcal{F}$ is image admissible via some $(Y, \mathcal{S}, T)$.

Remarks. The specific classes $\mathcal{F}_\alpha$ depend on the choice of the countable family $\mathcal{A}$ of generators, but condition (i) does not: if $\mathcal{C}$ is another countable set of generators of $\mathcal{B}$ with corresponding classes $\mathcal{G}_\alpha$, then for any $\alpha \in \Omega$ there are $\beta \in \Omega$ and $\gamma \in \Omega$ with $\mathcal{F}_\alpha \subset \mathcal{G}_\beta$ and $\mathcal{G}_\alpha \subset \mathcal{F}_\gamma$.

Proof. (ii) implies (iii): for each $f \in \mathcal{F}$ choose a unique $t \in I$ and restrict the Borel σ-algebra to the set of t's chosen. Then (iii) follows.

Clearly (iii) implies (iv), which is equivalent to (v).

(iv) implies (iii): note that any real-valued measurable function G (for the Borel σ-algebra on the range $\mathbb{R}$ as usual) is always measurable for some countably generated sub-σ-algebra, for example, generated by $\{G > q\}$, q rational. Let G be the evaluation map $G(f, t) := f(t)$. The product σ-algebra $\mathcal{T} \otimes \mathcal{B}$ is the union of the σ-algebras generated by countable sets of rectangles $A_i \times B_i$ for $A_i \in \mathcal{T}$ and $B_i \in \mathcal{B}$. The σ-algebra generated by countably many σ-algebras on $\mathcal{F} \times S$ generated in this way is also generated in the same way. So the evaluation map G is measurable for such a sub-σ-algebra. Let $\mathcal{D}$ be the σ-algebra of subsets of $\mathcal{F}$ generated by the A_i. For any two distinct functions f, g in $\mathcal{F}$, $f(x) \neq g(x)$ for some x. The map $h \mapsto h(x)$ is $\mathcal{D}$ measurable. So $\{f\}$ is the intersection of those A_i that contain f and the complements of the others, and (iii) follows. So (iii) through (v) are equivalent.

(iii) implies (ii): by Theorem 5.13, there is a subset $Y \subset I := [0, 1]$ with Borel σ-algebra and a measurable isomorphism $t \mapsto G(t, \cdot)$ from Y onto $\mathcal{F}$. The assumed admissibility implies that $(t, x) \mapsto G(t, x)$ is jointly measurable. By the general extension theorem for real-valued measurable functions (RAP, Theorem 4.2.5), although Y is not necessarily measurable, we can assume G is jointly measurable on $I \times S$, proving (ii). So (ii) through (v) are equivalent.

(ii) implies (i): On $[0, 1] \times S$, take generators of the form $A_i \times B_i$, where A_i are Borel, $B_i \in \mathcal{B}$, and $\{A_i\}_{i \geq 1}$ and $\{B_i\}_{i \geq 1}$ are algebras. Then by Theorem 5.14, G belongs to some $\mathcal{F}_\alpha$ on $S \times T$. It will be shown by transfinite induction on α that the sections $G(t, \cdot)$ on S all belong to $\mathcal{F}_\alpha$ for the generators $\{B_i\}$.

If $G \in \mathcal{F}_0$ on $I \times S$, then $G = \sum_{i=1}^n c_i 1_{A_i \times B_i}$ for some c_i, n, Borel A_i, and $B_i \in \mathcal{B}$. For each $t \in I$, $G(t, s) = \sum_{i=1}^n c_i 1_{A_i}(t) 1_{B_i}(s)$, so $G(t, \cdot) \in \mathcal{F}_0$ on S. Suppose the statement holds for a given α. Let $G \in \mathcal{F}_{\alpha+1}$ on $I \times S$. Then for some $G_k \in \mathcal{F}_\alpha$ on $I \times S$, $G_k(t, s) \to G(t, s)$ as $k \to \infty$ for all $t \in I$ and $s \in S$. For each t and k, $G_k(t, \cdot) \in \mathcal{F}_\alpha$ on S by induction assumption. Thus $G(t, \cdot) \in \mathcal{F}_{\alpha+1}$ on S as desired. Let β be a limit ordinal (not a successor $\alpha + 1$ of any α), and suppose the statement holds for each $\alpha < \beta$. Then by definition of $\mathcal{F}_\beta$, it also holds for $\alpha = \beta$. This completes the proof by induction, and so (ii) implies (i).

(i) implies (ii): for this we need universal functions, defined as follows. A jointly measurable function $G : I \times S \mapsto \mathbb{R}$ will be called a *universal class α function* if every function $f \in \mathcal{F}_\alpha$ on S is of the form $G(t, \cdot)$ for some $t \in I$. (G itself will not necessarily be of class α on $I \times S$.) Recall by the way that an open, universal open set U in $\mathbb{N}^\infty \times S$ exists for any separable metric space S (RAP, Proposition 13.2.3), where "universal open" means that for every open set $V \subset S$, there is an $x \in \mathbb{N}^\infty$ such that $V = \{y : (x, y) \in U\}$.

Theorem 5.16 (Lebesgue) *For any $\alpha \in \Omega$ there exists a universal class α function $G : I \times S \mapsto \mathbb{R}$.*

Proof. For $\alpha = 0$, $\mathcal{F}_\alpha$ is a countable sequence $\{f_k\}_{k \geq 1}$ of functions. Let $G(1/k, x) := f_k(x)$ and $G(t, x) := 0$ if $t \neq 1/k$ for all k. Then G is jointly measurable and a universal class 0 function.

For general $\alpha > 0$, by transfinite induction (RAP, Section 1.3) suppose there is a universal class β function for all $\beta < \alpha$. First suppose α is a successor, $\alpha = \beta + 1$ for some β. Let H be a universal class β function on $I \times S$.

Let I^∞ be the countable product of copies of I with the product σ-algebra. The product topology on I^∞ is compact and metrizable (RAP, Theorem 2.2.8, Proposition 2.4.4). Its Borel σ-algebra is the same as the product σ-algebra, by way of the usual base or subbase of the product topology (RAP, Sections 2.1, 2.2). For $t = \{t_n\}_{n \geq 1} \in I^\infty$ let $G(t, x) := \limsup_{n \to \infty} H(t_n, x)$ if the lim sup is finite, otherwise $G(t, x) := 0$. Then G is jointly measurable. Now I^∞, as a Polish space, is Borel-isomorphic to I (RAP, Section 13.1), so we can replace I^∞ by I. Then G is a universal class α function.

If α is not a successor, then there is a sequence $\beta_k \uparrow \alpha$, $\beta_k < \alpha$, meaning that for every $\beta < \alpha$, there is some k with $\beta < \beta_k$. To see this, as $\{\beta : \beta < \alpha\}$ is countable, we can write it as $\{\gamma_i\}_{i=1,2,\dots}$, let $\beta_1 := \gamma_1$, and for each $k \geq 1$, define β_{k+1} as γ_i for the least i such that $\gamma_i > \beta_k$.

For each k, let G_k be a universal class β_k function. Define G on $I^2 \times S$ by $G(s, t, x) = G_k(t, x)$ if $s = 1/k$ and $G(s, t, x) = 0$ otherwise. Then G is jointly measurable. Since I^2 is Borel-isomorphic to I we again have a universal class α function, proving Theorem 5.16. $\square$

With Theorem 5.16, (i) in Theorem 5.15 clearly implies (ii), so (i) through (v) are equivalent.

(iv) implies (vi) directly. If (vi) holds, let Z be a subset of Y on which the map $z \mapsto T(z)(\cdot)$ is one-to-one and onto $\mathcal{F}$. Let $\mathcal{S}_Z$ and T_Z be the restrictions to Z of $\mathcal{S}$ and T respectively. For a function f and set B let $f[B] := \{f(x) : x \in B\}$. Then $\mathcal{F}$ remains image admissible via $(Z, \mathcal{S}_Z, T_Z)$, and $\{T_Z[A] : A \in \mathcal{S}_Z\}$ is an admissible structure for $\mathcal{F}$, giving (iv). $\quad\square$

For $0 \le p < \infty$ and a probability law Q on $(S, \mathcal{B})$ we have the space $\mathcal{L}^p(S, \mathcal{B}, Q)$ of measurable real-valued functions f on S such that $\int |f|^p dQ < \infty$, with the pseudo-metric

$$d_{p,Q}(f, g) := \begin{cases} (\int |f - g|^p dQ)^{1/p}, & 1 \le p < \infty; \\ \int |f - g|^p dQ, & 0 < p < 1; \\ \inf\{\varepsilon > 0 : Q(|f - g| > \varepsilon) < \varepsilon\}, & p = 0. \end{cases}$$

In admissible classes, $d_{p,Q}$-open sets are measurable, as follows:

Theorem 5.17 *Let $(S, \mathcal{B})$ be a separable measurable space, $0 \le p < \infty$, and $\mathcal{F} \subset \mathcal{L}^p(S, \mathcal{B}, Q)$ where $\mathcal{F}$ is admissible. Then if $\mathcal{F}$ is image admissible via $(Y, \mathcal{S}, T)$, $U \subset \mathcal{F}$ and U is relatively $d_{p,Q}$-open in $\mathcal{F}$, we have $T^{-1}(U) \in \mathcal{S}$.*

Proof. Since $(S, \mathcal{B})$ is separable, the pseudo-metric spaces $\mathcal{L}^p(S, \mathcal{B}, Q)$ are separable for $0 \le p < \infty$. To prove this, let $\mathcal{C}$ be a countable set of generators of $\mathcal{B}$, where we can assume $\mathcal{C}$ is an algebra as noted before Theorem 5.14. Let $\mathcal{G}$ be the set of all simple functions $\sum_{i=1}^{n} c_i 1_{A_i}$ for $A_i \in \mathcal{C}$, c_i rational, and $n = 1, 2, \ldots$. Then $\mathcal{G}$ is countable and is dense in $\mathcal{L}^p$ for $0 \le p < \infty$, as is easily checked. So (e.g., RAP, Proposition 2.1.4), U is a countable union of balls $\{f : d_{p,Q}(f, g) < r\}$, $g \in U$, $0 \le r < \infty$. So for $p > 0$ it is enough to show that each function $y \mapsto \int |T(y)(\cdot) - g|^p dQ$ is measurable. For g fixed, $(y, x) \mapsto |T(y)(x) - g(x)|^p$ is jointly measurable. So for $0 < p < \infty$ we can reduce to the case $p = 1$ and $g \equiv 0$ with $T(y)(x) \ge 0$ for all x, y. Now, the Tonelli–Fubini theorem implies the desired measurability. For $p = 0$, given $\varepsilon > 0$, by the Tonelli–Fubini theorem again, the set A_ε of y such that $Q(|T(y)(x) - g(x)| > \varepsilon) < \varepsilon$ is measurable, and $\{y : d_{0,Q}(T(y), g) < r\}$ is the union of A_ε for ε rational, $0 < \varepsilon < r$. $\quad\square$

Corollary 5.18 *If $(S, \mathcal{B})$ is a separable measurable space, $\mathcal{F} \subset \mathcal{L}^1(S, \mathcal{B}, Q)$ and $\mathcal{F}$ is image admissible via $(Y, \mathcal{S}, T)$ then $y \mapsto \int T(y) dQ$ is $\mathcal{S}$-measurable.*

Proof. For any real u, $\{f : \int f dQ > u\}$ is open for $d_{1,Q}$. $\quad\square$

For $1 \le p < \infty$ and $f, g \in \mathcal{L}^p(S, \mathcal{B}, Q)$ let $\rho_{p,Q}(f, g) := d_{p,Q}(f_{0,Q}, g_{0,Q})$ where for $h \in \mathcal{L}^1(S, \mathcal{B}, Q)$, $h_{0,Q} := h - \int h dQ$. Thus for ρ_Q as defined in Section 3.1, $\rho_Q \equiv \rho_{2,Q}$.

Corollary 5.19 *If $(S, \mathcal{B})$ is a separable measurable space, $1 \le p < \infty$, $\mathcal{F} \subset \mathcal{L}^p(S, \mathcal{B}, Q)$, $\mathcal{F}$ is image admissible via $(Y, \mathcal{S}, T)$, $U \subset \mathcal{F}$ and U is $\rho_{p,Q}$-open, then $T^{-1}(U) \in \mathcal{S}$.*

Proof. Let $\mathcal{G}$ be the set of functions $f - \int f dQ$, $f \in \mathcal{F}$. For $y \in Y$ let $W(y)(x) := T(y)(x) - \int T(y) dQ$. Then by Corollary 5.18, W is jointly measurable and $\mathcal{G}$ is image admissible via $(Y, \mathcal{S}, W)$. Now $f \in U$ if and only if $f_{0,Q} \in V$ where V is $d_{p,Q}$-open in $\mathcal{G}$. Then $T^{-1}(U) = W^{-1}(V) \in \mathcal{S}$ by Theorem 5.17.   □

Next, here is a definition extending the definitions of empirical measure and process as given so far.

Definition. A stochastic process $X_{n,m} = X_{n,m}(f, \omega)$ indexed by a class $\mathcal{F}$ of measurable functions on a measurable space $(S, \mathcal{B})$, for ω in Ω, will be called an *empirical-type* process if:

(a) For some integers $n \geq 1$ and $m \geq 0$, the probability space (Ω, Pr) is $\Omega = S^{n+m} \times \Omega_0$, where for some probability measures P and Q on S and Pr_0 on Ω_0, $\mathrm{Pr} = P^n \times Q^m \times \mathrm{Pr}_0$;

(b) The process is of the form, for $\omega_0 \in \Omega_0$,

$$X_{n,m}(f, \omega) = \left(c_0(\omega_0) P + \sum_{i=1}^{n+m} c_i(\omega_0) \delta_{x_i} \right)(f),$$

where x_i are coordinates on S^{n+m}, c_i are real-valued random variables on Ω_0, and for $1 \leq i \leq n+m$, $\mathrm{Pr}(c_i = 0) < 1$.
If $m = 0$, the process will be written as X_n.

Examples of empirical-type processes are: the usual empirical process ν_n, with $m = 0$, $c_0 \equiv -\sqrt{n}$, and $c_j \equiv 1/\sqrt{n}$, $j = 1, \ldots, n$; and the following examples, which all have $c_0 \equiv 0$: the usual empirical measure P_n, with $m = 0$ and $c_j \equiv 1/n$, $j = 1, \ldots, n$; symmetrized empirical processes, with $m = n$, $P = Q$, and $c_{n+j} \equiv -c_j$ for $j = 1, \ldots, n$, to be treated in Lemma 6.5; and the two-sample empirical process $P_n - Q_m$ with $n > 0$ and $m > 0$, or its multiple by $\sqrt{nm/(n+m)}$, to be treated in Section 9.1.

To avoid some pathologies, empirical-type processes are defined via coordinates and product measures on Cartesian product spaces, as opposed to having, for example, $x_1, \ldots, x_n$ i.i.d. P, defined on some arbitrary probability space.

Not all processes to be considered later fit into the above definition: for example, in Poissonized empirical processes, as in Lemma 9.11 and what follows it, n itself is random. These processes, and bootstrapped empirical measures P_n^B and processes $\sqrt{n}(P_n^B - P_n)$ in Section 9.2, will be treated as special cases.

Proposition 5.20 *Let $(S, \mathcal{B})$ be a separable measurable space. Then:*

(a) For $n = 1, 2, \ldots,$ S^n with product σ-algebra $\mathcal{B}^{\otimes n}$ is also separable.

(b) If $\mathcal{F}$ is an image admissible Suslin class of measurable functions on S via $(Y, \mathcal{S}, T)$, and $X_{n,m}$ is an empirical-type process indexed by $\mathcal{F}$, with probability space $S^{n+m} \times \Omega_0$ and $c_0 \equiv 0$, $x = (x_1, \ldots, x_{n+m}) \in S^{n+m}$, and $\omega_0 \in \Omega_0$, then $(y, x, \omega_0) \mapsto X_{n,m}(T(y))$ is jointly measurable.

(c) If possibly $c_0 \neq 0$ and $\mathcal{F} \subset \mathcal{L}^1(S, \mathcal{B}, P)$, then again $(y, x, \omega_0) \mapsto X_{n,m}(T(y))$ is jointly measurable.

Proof. Part (a) is straightforward. For (b), $X_{n,m}(T(y))$ is a Borel measurable function of the $n + m$ jointly measurable functions $(y, x) \mapsto T(y)(x_j)$ and of $c_j(\omega_0)$. For (c) one can combine part (b) and Corollary 5.18 via another Borel measurable function. $\qquad\square$

5.3 Suslin Properties and Selection

Here is another counterexample on measurability, to add to the two at the beginning of the chapter. Let $X = [0, 1]$ with Borel σ-algebra and uniform (Lebesgue) probability measure $P := U[0, 1]$. Let A be a non-Lebesgue measurable subset of $[0, 1]$ (e.g., RAP, Theorem 3.4.4). Let $\mathcal{C} := \{\{x\} : x \in A\}$. Then $\mathcal{C}$ is a collection of disjoint sets, so $S(\mathcal{C}) = 1$ by Theorem 4.10. Also $\mathcal{C}$, being a class of singletons, is admissible, e.g. by Theorem 5.15(ii) with $G(t, s) = 1$ for $t = s$, $G(t, s) = 0$ otherwise. But, $\|P_1\|_{\mathcal{C}}$ is nonmeasurable, being 1 if and only if $X_1 \in A$, and likewise any $\|P_n\|_{\mathcal{C}}$ or $\|P_n - P\|_{\mathcal{C}}$ is nonmeasurable. So some measurability condition beyond admissibility is needed for $\|P_n - P\|_{\mathcal{F}}$ to be measurable. A sufficient condition will be provided by Suslin properties, as follows.

A *Polish space* is a topological space metrizable as a complete separable metric space. A separable measurable space $(Y, \mathcal{S})$ will be called a *Suslin space* iff there is a Polish space X and a Borel measurable map from X onto Y. Recall that by the Borel isomorphism theorem (RAP, Theorem 13.1.1), if A is a Borel set in a Polish space, then there is a 1–1 Borel measurable function with Borel measurable inverse from a Polish space Z onto A, where moreover we can take Z to be either $[0, 1]$, or a converging sequence together with its limit, or a finite set. Thus in the definition of Suslin space, we can equivalently replace the Polish space X by any Borel set in a Polish space.

If $(Y, \mathcal{S})$ is a measurable space, a subset $Z \subset Y$ will be called a *Suslin set* iff it is a Suslin space with the relative σ-algebra $Z \sqcap \mathcal{S}$.

Given a measurable space $(X, \mathcal{B})$ and $M \subset X$, M is called *universally measurable* or *u. m.* iff for every probability law P on $\mathcal{B}$, M is measurable for the completion of P, in other words for some $A, B \in \mathcal{B}$, $A \subset M \subset B$, and $P(A) = P(B)$. In a Polish space, all Suslin sets are universally measurable (RAP, Theorems 13.2.1 and 13.2.6). A function f from X into Z, where $(Z, \mathcal{A})$

is a measurable space, will be called *universally measurable* or *u. m.* iff for each set $B \in \mathcal{A}$, $f^{-1}(B)$ is universally measurable.

If $(\Omega, \mathcal{A})$ is a measurable space and $\mathcal{F}$ a set, then a real-valued function $X : (f, \omega) \mapsto X(f, \omega)$ will be called *image admissible Suslin via* $(Y, \mathcal{S}, T)$ iff $(Y, \mathcal{S})$ is a Suslin measurable space, T is a function from Y onto $\mathcal{F}$, and $(y, \omega) \mapsto X(T(y), \omega)$ is jointly measurable on $Y \times \Omega$. Equivalently, Y could be taken to be Polish with $\mathcal{S}$ its Borel σ-algebra.

As the notation suggests, a main case of interest will be where $\mathcal{F}$ is a set of functions on Ω and $X(f, \omega) \equiv f(\omega)$. Then $\mathcal{F}$ will be called *image admissible Suslin via* $(Y, \mathcal{S}, T)$ if X is.

Recall the notion of separable measurable space defined in the last section. Note that any separable metric space with its Borel σ-algebra is a separable measurable space, as follows from RAP, Proposition 2.1.4. We have:

Theorem 5.21 *A measurable space* $(X, \mathcal{B})$, *where* $\mathcal{B}$ *is countably generated, is separable if and only if it separates the points of* X, *so that for any* $x \neq y$ *in* X, *there is some* $A \in \mathcal{B}$ *containing just one of* x, y.

Proof. If $(X, \mathcal{B})$ is separable, then $\{x\} \in \mathcal{B}$ for each $x \in X$, so $\mathcal{B}$ separates points. The converse direction follows from the proof of Theorem 5.15: (iv) implies (iii). □

Theorem 5.22 (Sainte-Beuve selection theorem) *Let* $(\Omega, \mathcal{A})$ *be any measurable space and let* $X : \mathcal{F} \times \Omega \mapsto \mathbb{R}$ *be image admissible Suslin via* $(Y, \mathcal{S}, T)$. *Then for any Borel set* $B \subset \mathbb{R}$,

$$\Pi_X(B) := \{\omega : X(f, \omega) \in B \text{ for some } f \in \mathcal{F}\}$$

is u. m. in Ω, *and there is a u. m. function* H *from* $\Pi_X(B)$ *into* Y *such that* $X(T(H(\omega)), \omega) \in B$ *for all* $\omega \in \Pi_X(B)$.

Note. Here $(\Omega, \mathcal{A})$ need not be Suslin or even separable.

Proof. As noted above, we can assume Y is Polish and $\mathcal{S}$ is its Borel σ-algebra. For any measurable set V in a product σ-algebra, here $V = \{(y, \omega) : X(T(y), \omega) \in B\} \subset Y \times \Omega$, there are countably many measurable sets $A_n \subset Y$ and $B_n \subset \Omega$ such that V is in the σ-algebra generated by the sets $A_n \times B_n$ (as in the proof of Theorem 5.15, (iv) implies (iii)). Thus $(y, \omega) \mapsto X(T(y), \omega)$ is jointly measurable for the given Suslin σ-algebra $\mathcal{S}$ in Y and a σ-algebra $\mathcal{B}$ in Ω generated by a sequence $\{B_n\}$ of measurable sets.

Define a Marczewski function b, as in the proof of Theorem 5.13, for the sequence $\{B_n\}$ in place of $\{A_n\}$. Then b is a $\mathcal{B}$-measurable function from Ω into the Cantor set $C \subset [0, 1]$ (defined, e.g., in RAP, proof of Proposition 3.4.1). Define $\omega =_\mathcal{B} \omega'$ iff for all n, $\omega \in B_n$ if and only if $\omega' \in B_n$. Equivalently, for all $V \in \mathcal{B}$, $\omega \in V$ if and only if $\omega' \in V$. Then $b(\omega') = b(\omega)$ if and only if $\omega =_{\mathcal{B}} \omega'$.

Let β be a $\mathcal{B}$-measurable function. Then if $b(\omega) = b(\omega')$, we must have $\beta(\omega) = \beta(\omega')$. So $\beta(\omega) \equiv g(b(\omega))$ for some function g. Now, β is measurable for the σ-algebra of all sets $b^{-1}(C)$ where C is a Borel subset of $\mathbb{R}$, since each B_n is such a $b^{-1}(C_n)$ (by the proof of Theorem 5.13). So g can be taken to be Borel measurable (RAP, Theorem 4.2.8). Similarly, $X(T(y), \omega) \equiv F(y, b(\omega))$ for some function F on $Y \times C$.

The next step in the proof of Theorem 5.22 is the following fact:

Lemma 5.23 *Every $\mathcal{S} \otimes \mathcal{B}$ measurable real function ψ on $Y \times \Omega$ can be written as $F(\cdot, b(\cdot))$ where F is $\mathcal{S} \otimes \mathcal{B}_C$ measurable and $\mathcal{B}_C$ is the Borel σ-algebra in C.*

Proof. This is easily seen when ψ is the indicator of a set $S \times B$, $S \in \mathcal{S}$, $B \in \mathcal{B}$, or a finite linear combination of such indicators. A finite union of sets $S_i \times B_i$, $S_i \in \mathcal{S}$, $B_i \in \mathcal{B}$, can be written as a disjoint union (RAP, Proposition 3.2.2). Thus the Lemma holds when ψ is the indicator of any such union. Next, suppose ψ_n are $\mathcal{S} \otimes \mathcal{B}$ measurable, $\psi_n \equiv F_n(\cdot, b(\cdot))$ where F_n are $\mathcal{S} \otimes \mathcal{B}_C$ measurable, and $\psi_n \to \psi$ pointwise on $Y \times \Omega$. For any $y \in Y$ and $\omega, \omega' \in \Omega$ such that $b(\omega) = b(\omega')$, we have $\psi_n(y, b(\omega)) = \psi_n(y, b(\omega'))$ for all n, so $\psi(y, b(\omega)) = \psi(y, b(\omega'))$. Let $F(y, c) := \lim_{n \to \infty} F_n(y, c)$ whenever the sequence converges, as it will if $c = b(\omega)$ for some $\omega \in \Omega$. Otherwise, set $F(y, c) := 0$. Then F is $\mathcal{S} \otimes \mathcal{B}_C$ measurable (RAP, proof of Theorem 4.2.5) and $\psi(y, \omega) = F(y, b(\omega))$ for all $y \in Y$ and $\omega \in \Omega$. Since the finite unions of sets $S_i \times B_i$ form an algebra, the monotone class theorem (RAP, Theorem 4.4.2) proves the Lemma when ψ is the indicator function of any set in the product σ-algebra $\mathcal{S} \otimes \mathcal{B}$. It thus holds when ψ is any simple function (finite linear combination of such indicator functions). Since each measurable function is a pointwise limit of simple functions, the Lemma follows. $\square$

So, $X(T(y), \omega) \equiv F(y, b(\omega))$ for an $\mathcal{S} \otimes \mathcal{B}_C$ measurable function F on $Y \times C$.

Now, $\{(y, c) \in Y \times C : F(y, c) \in B\}$ is a Borel subset of $Y \times C$ (RAP, Proposition 4.1.7) and thus a Suslin set (RAP, Theorem 13.2.1). It follows that $\Pi_F B := \{c \in C : \text{for some } y \in Y, \ F(y, c) \in B\}$ is a Suslin subset of C, and so universally measurable (RAP, Theorem 13.2.6).

By a selection theorem (RAP, Theorem 13.2.7) there is a u. m. function h from $\Pi_F B$ into Y with $F(h(c), c) \in B$ for all $c \in \Pi_F B$. Next we need:

Lemma 5.24 *Let $(D, \mathcal{D})$ and $(G, \mathcal{G})$ be two measurable spaces and g a measurable function from D into G. Let $U \subset G$ be u. m. for $\mathcal{G}$. Then $g^{-1}(U)$ is u. m. for $\mathcal{D}$.*

Proof. Let P be a law on $(D, \mathcal{D})$, so $P \circ g^{-1}$ is a law on $(G, \mathcal{G})$. Take $A, B \in \mathcal{G}$ with $A \subset U \subset B$ and $(P \circ g^{-1})(B \setminus A) = 0$. Then $g^{-1}(A)$ and $g^{-1}(B)$ are in $\mathcal{D}$, $g^{-1}(A) \subset g^{-1}(U) \subset g^{-1}(B)$ and $P(g^{-1}(B) \setminus g^{-1}(A)) = P(g^{-1}(B \setminus A)) = 0$. $\qquad\square$

Remark. The Lemma includes the case where $D \subset G$, g is the identity, and $\mathcal{D}$ includes the relative σ-algebra $\{J \cap D : J \in \mathcal{G}\}$, while possibly $D \notin \mathcal{G}$ and D may not be u. m. for $\mathcal{G}$.

Now to finish the proof of Theorem 5.22, Lemma 5.24 implies that

$$\Pi_X(B) \;=\; b^{-1}\{c \in C : \text{ for some } y \in Y, \; F(y, c) \in B\}$$

is u. m. Let $H := h \circ b$ from Ω into Y. Then for any $E \in \mathcal{S}$, $H^{-1}(E) = b^{-1}(h^{-1}(E))$. Here $h^{-1}(E)$ is a u.m. set in $\Pi_F(B)$, so $b^{-1}(h^{-1}(E))$ is u.m. in Ω by Lemma 5.24, and H is a u. m. function. By choice of h we have $X(T(H(\omega)), \omega) \in B$ for all $\omega \in \Pi_X B$. $\qquad\square$

Some possibilities for the set $B \subset \mathbb{R}$ are the sets $\{x : x > t\}$ or $\{x : |x| > t\}$ for any real t. These choices give:

Corollary 5.25 *Let $(f, \omega) \mapsto X(f, \omega)$ be real-valued and image admissible Suslin via some $(Y, \mathcal{S}, T)$. Then the two functions $\omega \mapsto \sup\{X(f, \omega) : f \in \mathcal{F}\}$ and $\omega \mapsto \sup\{|X(f, \omega)| : f \in \mathcal{F}\}$ are both u. m.*

The image admissible Suslin property is preserved by composing with a measurable function:

Theorem 5.26 *Let $(\Omega, \mathcal{A})$ be a measurable space and $\mathcal{F}_i$ for $i = 1, \ldots, k$ classes of measurable real-valued functions on Ω. Let $X^1, \ldots, X^k$ be image admissible Suslin real-valued functions on $\mathcal{F}_i \times \Omega$, $i = 1, \ldots, k$, via $(Y_i, \mathcal{S}_i, T_i)$, $i = 1, \ldots, k$ respectively. Let g be a Borel measurable function from $\mathbb{R}^k$ into $\mathbb{R}$. Then*

$$(f_1, \ldots, f_k, \omega) \mapsto g(X^1(f_1, \omega), \ldots, X^k(f_k, \omega))$$

is image admissible Suslin via some $(Y, \mathcal{S}, T)$. Specifically, we can let $Y = Y_1 \times \cdots \times Y_k$ with product σ-algebra $\mathcal{S} = \mathcal{S}_1 \otimes \cdots \otimes \mathcal{S}_k$ and let $T(y_1, \ldots, y_k) := (T_1(y_1), \ldots, T_k(y_k))$.

Proof. Clearly $(Y, \mathcal{S})$ is Suslin and the joint measurability holds. $\qquad\square$

Next, here are examples showing that if the Suslin assumption on Y is removed from Theorem 5.22, it may fail. Let $(\Omega, \preceq)$ and the class C of countable initial segments I_x be as in the Example near the beginning of this chapter. We can take Ω to be (in 1–1 correspondence with) a subset of $[0, 1]$, by the axiom of choice (or, by the continuum hypothesis, all of [0, 1]). Here the ordering $\preceq$ has no relation to any usual structure on [0, 1]. Then $\mathcal{C}$ is admissible by Theorem 5.15, since the collection of all countable sets is of bounded Borel class: all finite unions of open intervals (q, r) with rational endpoints are in some $\mathcal{F}_\alpha$, then all finite sets are in $\mathcal{F}_{\alpha+1}$ and all countable sets in $\mathcal{F}_{(\alpha+1)+1} =: \mathcal{F}_{\alpha+2}$.

Let P be a law on Ω which is 0 on countable sets and 1 on sets with countable complement. Such a law, on the σ-algebra generated by singletons, exists on any uncountable set. Under the continuum hypothesis with $\Omega = [0, 1]$ we can take P to be Lebesgue measure or any nonatomic law.

Let $P_1 = \delta_x$, $P_2 = (\delta_x + \delta_y)/2$ and $Q_1 = \delta_z$ where x, y, z are coordinates on Ω^3 with law P^3, so x, y, and z are i.i.d. (P). Then $\sup_{A \in \mathcal{C}}(P_1 - Q_1)(A) = 1$ if and only if $x \prec z$. Thus $\sup_{A \in \mathcal{C}}(P_1 - Q_1)(A)$ is nonmeasurable.

If we let $B := \{(x, y, z) : \sup_{A \in \mathcal{C}} |(P_2 - Q_1)(A)| = 1\}$, it can be seen likewise that B must not be measurable. So "Suslin" cannot simply be removed from Corollary 5.25 or Theorem 5.22.

"Two-sample" empirical processes, which are multiples of $P_m - Q_n$, are of direct interest in statistics, as in testing the hypothesis that P_m and Q_n are sampled from the same distribution $P = Q$. If they are, then $\sqrt{\frac{mn}{m+n}}(P_m - Q_n)$ converges in law to G_P in $\ell^\infty(\mathcal{F})$ if F is a Donsker class for P, as will be seen in Section 9.1. On the other hand, facts can be proved about the one-sample empirical process $\sqrt{n}(P_n - P)$ via symmetrization, subtracting an independent copy of itself to get $\sqrt{n}(P_n - Q_n)$, as will be done in Section 6.1 and applied in Section 6.2 and thereafter.

An admissible structure can be put on spaces of closed sets. Let (X, d) be a separable metric space and $\mathcal{F}_0$ the collection of all nonempty closed subsets of X. Then $\mathcal{F}_0$ is admissible: there is a countable base for the topology of X, so for some α, all finite unions of sets in the base are in $\mathcal{F}_\alpha$, so all open sets are in $\mathcal{F}_{\alpha+1}$. Then all closed sets are in $\mathcal{F}_{\alpha+2}$ since any closed set F is a countable intersection of open sets

$$U_n := \{x : d(x, F) := \inf_{y \in F} d(x, y) < 1/n\}.$$

The topology of X can be metrized by a metric d for which (X, d) is totally bounded (RAP, Theorem 2.8.2). Assume d is such a metric. Let h_d be the Hausdorff metric, $h_d(A, B) := \max\{\sup_{x \in A} d(x, B), \sup_{y \in B} d(y, A)\}$ for any two closed sets A, B.

Since d is totally bounded, it is easily seen that $(\mathcal{F}_0, h_d)$ is separable, since the finite subsets of a countable dense set in X are dense in $\mathcal{F}_0$ for h_d.

The Borel σ-algebra of h_d will be called an *Effros* Borel structure on $\mathcal{F}_0$. (Effros (1965) proved that for d totally bounded this Borel structure is unique.)

Proposition 5.27 *For any separable metric space X with totally bounded metric d, the Effros Borel structure (of h_d) is admissible on $\mathcal{F}_0$. Also, for any law P on the Borel sets of X, $P(\cdot)$ is measurable for the Effros Borel structure.*

Proof. The set $\{(x, F) : x \in F \in \mathcal{F}_0\}$ is closed in $X \times \mathcal{F}_0$: if $x_n \in F_n$ for all n and $(x_n, F_n) \to (x, F)$, then $x_n \to x$ in X and $h_d(F_n, F) \to 0$. Then $d(x_n, F) \to 0$ and so $x \in F$. Since both (X, d) and $(\mathcal{F}_0, h_d)$ are separable, a Borel set in their product is in the product σ-algebra (RAP, Proposition 4.1.7).

For any law P and $c > 0$, the set $\{F \in \mathcal{F}_0 : P(F) \geq c\}$ is closed for h_d, since if $F_n \to F$ for h_d, $P(F_n) \geq c$, and $\varepsilon > 0$, let $F^\varepsilon := \{y : d(x, y) < \varepsilon$ for some $x \in F\}$. Then $F_n \subset F^\varepsilon$ for n large enough, so $P(F^\varepsilon) \geq c$ and letting $\varepsilon = 1/m \downarrow 0$ shows $P(F) \geq c$. So $P(\cdot)$ is upper semicontinuous and so Effros measurable. $\qquad\square$

For families of functions, we have the following:

Theorem 5.28 *(a) Let S be a topological space and $\mathcal{F}$ a family of bounded real functions on S, equicontinuous at each point of S. Then $(f, x) \mapsto f(x)$ is jointly continuous $\mathcal{F} \times S \mapsto \mathbb{R}$, with the supremum norm $\|f\|_\infty := \sup_x |f(x)|$ on $\mathcal{F}$.*

(b) If in addition $\mathcal{F}$ is separable for $\|\cdot\|_\infty$ and S is metrizable as a separable metric space (S, d), then $(f, x) \mapsto f(x)$ is jointly measurable for the Borel σ-algebras of $\|\cdot\|_\infty$ on $\mathcal{F}$ and d on S. Thus $\mathcal{F}$ is admissible. So is $\mathcal{G}$, the collection of all subgraphs $\{(x, y) : 0 \leq y \leq f(x)$ or $f(x) \leq y \leq 0\}$ for $f \in \mathcal{F}$.

(c) If, moreover, $\mathcal{F}$ with $\|\cdot\|_\infty$ distance and its Borel σ-algebra is a Suslin set, then $\mathcal{F}$ and $\mathcal{G}$ are image admissible Suslin.

Proof. Part (a) is immediate. For $\mathcal{F}$, (b) follows from the fact that on a Cartesian product of two separable metric spaces, the Borel σ-algebra of the product topology equals the product σ-algebra of the Borel σ-algebras on each space (RAP, Theorem 4.1.7). For $\mathcal{G}$, the set $\{(y, z) : 0 \leq y \leq z$ or $z \leq y \leq 0\}$ is closed, thus Borel in $\mathbb{R}^2$. Composing its indicator function with $(f, x) \mapsto z := f(x)$ thus gives a measurable function. Then (c) follows directly. $\qquad\square$

Example (Adamski and Gaenssler). Let $H \subset [0, 1]$ be a nonmeasurable set with $\lambda_*(H) = 0$ and $\lambda^*(H) = 1$ where λ is Lebesgue measure (e.g., RAP, Theorem 3.4.4). Then for each Borel set $B \subset [0, 1]$, letting $\mu(B \cap H) := \lambda^*(B \cap H)$ defines a countably additive probability measure on the Borel subsets of H, as a metric space with the usual metric from $\mathbb{R}$ (RAP, Theorem 3.3.6). Likewise, λ^* gives a probability measure ν on the Borel sets of $H^c := [0, 1] \setminus H$. Take the countable product of probability spaces $(\Omega, \rho) := \Pi_{n=1}^\infty (A_n, \mathcal{B}_n, \rho_n)$ where for n odd, $A_n = H$ and $\rho_n = \mu$ while for n even, $A_n = H^c$ and $\rho_n = \nu$. Such a product of probability spaces always exists (e.g., RAP, Theorem 8.2.2). Let X_n be the nth coordinate on Ω, viewed as a map from Ω into $[0, 1]$. Then each X_n is measurable and has law $U[0, 1]$, the uniform distribution on $[0, 1]$. Thus, the X_i are i.i.d. $U[0, 1]$. Let $\mathcal{C}$ be the collection of all finite subsets of H. Let P_n be the empirical measures defined by the given X_i. Then $\|P_n\|_\mathcal{C} \equiv 1/2$ for n even and $\|P_n\|_\mathcal{C} \equiv (n+1)/(2n)$ for n odd. On the other hand if X_i are coordinates on a countable product of copies of $([0, 1], \lambda)$, then $\|P_n\|_\mathcal{C}$ is nonmeasurable. This illustrates that $\mathcal{C}$ is a pathological class of sets, but the pathology can be obscured if one does not define X_j as coordinates on a product space with product probability.

Problems

1. A law Q on $\mathbb{R}^n$ is called *exchangeable* if it is invariant under all permutations f_Π of the coordinates.

(a) Give an example of an exchangeable law which is not of the form P^n for a law P.

(b) Show that the empirical measure P_n is sufficient for any set of exchangeable laws.

2. (continuation) Give an example of a class of non-exchangeable laws on $\mathbb{R}^2$ for which the empirical measure is still sufficient. *Hint*: Consider laws with $X_2 = 2X_1$.

3. An *exponential family* is a set $\{P_\theta\}_{\theta \in \Theta}$ of laws on a measurable space $(X, \mathcal{A})$, all dominated by a σ-finite measure μ, having densities of the form $dP_\theta/d\mu = e^{(f(\theta), g(x))}$ where Θ is a set, f maps Θ into $\mathbb{R}^k$, g maps X into $\mathbb{R}^k$, and $(\cdot, \cdot)$ is the usual inner product in $\mathbb{R}^k$. If $\mathcal{P}$ is such an exponential family and $\mathcal{P}^n := \{P^n : P \in \mathcal{P}\}$ is the corresponding set of laws on X^n, show that $\sum_{j=1}^n g(X_j)$ is a sufficient statistic for $\mathcal{P}^n$.

4. (a) Show that the class $\mathcal{C}$ of initial segments I_x in the example at the beginning of Chapter 5 is admissible. *Hint*: The σ-algebra generated by the sets I_x is not separable, but the set Ω has the smallest possible uncountable cardinality. Thus we can assume Ω is in 1–1 correspondence with a subset of $[0, 1]$ (cf. RAP, Appendix A.3). Use this to show that there exists a separable structure on Ω for which the sets I_x are all measurable. Then, see the discussion after Theorem 5.26.

(b) A measure space $(X, \mathcal{S}, \mu)$ where $\mathcal{S}$ contains all singletons $\{x\}$ is called *nonatomic* if $\mu(\{x\}) = 0$ for all $x \in X$. Let $\mathcal{S}$ be a σ-algebra of subsets of $\mathcal{C}$ from part (a) making it admissible, for some σ-algebra $\mathcal{B}$ of subsets of Ω containing all singletons. Show that there is no nonatomic probability measure on $\mathcal{S}$.

5. (a) Let $(X, \mathcal{E})$ be a measurable space such that $\mathcal{E}$ contains all singletons $\{x\}$ for $x \in X$ and X is uncountable. Suppose that for any nonatomic law P on $(X, \mathcal{E})$, $P(A) = 0$ or 1 for all $A \in \mathcal{E}$, and that there exists at least one nonatomic law P on $(X, \mathcal{E})$. Show that $\mathcal{E}$ cannot be countably generated. *Hint*: Suppose A_k are generators. Let J be the set of all k such that $Q(A_k) = 0$ for every nonatomic law Q on $(X, \mathcal{E})$. Show that we can replace X by the subset $Y := X \setminus \bigcup_{k \in J} A_k$ with the relative σ-algebra $\mathcal{E}_Y := \{A \cap Y : A \in \mathcal{E}\}$ and the assumptions still hold, where $\{B_i\}_{i \geq 1}$ equal to $\{A_k \cap Y : k \notin J\}$ are a nonempty collection and generate $\mathcal{E}_Y$. Thus we can assume that $J = \emptyset$, i.e., for all k, there is a nonatomic law Q_k on $\mathcal{E}$ with $Q_k(A_k) = 1$. Then find a nonatomic law Q on $\mathcal{E}$ with $Q(A_k) = 1$ for all k and consider $\bigcap_k A_k$.

(b) In the Remarks before Theorem 5.13, prove that $\mathcal{T}$ is not countably generated. *Hint*: Use part (a) and the following: for any two sets X and I, let X^I be the product space consisting of all functions from I into X. Suppose $(X, \mathcal{A}, P)$ is a probability space. Let $\mathcal{A}^I$ be the product σ-algebra, the smallest σ-algebra of subsets of X^I for which the coordinate projection $f \mapsto f(i)$ of X^I onto X is measurable for each $i \in I$. Let P^I be the product probability measure on $\mathcal{A}^I$ for which the coordinates are i.i.d. P, which exists (RAP, Theorem 8.2.2). A 1–1 function π from I onto itself is called a *finite permutation* if $\pi(i) = i$ except for at most finitely many values of i. Such a π defines a 1–1 measurable function T_π of X^I onto itself by $T_\pi(\{x_i\}_{i \in I}) := \{x_{\pi(i)}\}_{i \in I}$. A set $B \in \mathcal{A}^I$ is called *invariant* if it is taken onto itself by T_π for every finite permutation π. Apply the Hewitt–Savage 0-1 law (RAP, Theorem 8.4.6), which states that for any invariant set B, $P^I(B) = 0$ or 1.

6. Let (S, d) be a separable metric space. Let $Y_1, Y_2, \ldots$, be Suslin subsets of S, for the Borel σ-algebra on each. Show that $\bigcup_k Y_k$ and $\bigcap_k Y_k$ are also Suslin. *Hints*: Let f_k be Borel functions from Polish spaces S_k onto Y_k. For $\bigcup_k Y_k$ use a disjoint union of copies of S_k, and show it is Polish for a suitable topology or metric. For $\bigcap_k Y_k$, use that the Cartesian product $\prod_k S_k$ with product topology is Polish (RAP, Proposition 2.4.4 gives a metric, which is complete if the metric d_k on S_k is for each k). Find a Borel set in the product which is mapped onto $\bigcap_k Y_k$ by a Borel function, which is sufficient as noted soon after the definition of the Suslin property.

7. Let $(X, \mathcal{A})$ be a measurable space and V a finite-dimensional vector space of measurable real-valued functions on X.

(a) Show that V is image admissible Suslin. *Hint*: See Theorem 5.26.

(b) Show that $\{1_A : A \in \text{pos}(V)\}$ is also image admissible Suslin.

8. Let $(X, \mathcal{A})$ be a measurable space and let $\mathcal{C} \subset \mathcal{A}$ be image admissible Suslin via some $(Y, \mathcal{S}, T)$, i.e. $\{1_B : B \in \mathcal{C}\}$ is. Let $\mathcal{C}^{(k)}$ be the union of all Boolean

algebras generated by k or less sets in $\mathcal{C}$, as in Theorem 4.8. Show that $\mathcal{C}^{(k)}$ is also image admissible Suslin. *Hints*: On any set with the σ-algebra of all its subsets, specifically a finite set, all functions are measurable. Consider the following steps:

(a) Given k, show that the set of all functions $x \mapsto (1_{C_1}(x), \ldots, 1_{C_k}(x)) \in \{0, 1\}^k$ for any $C_1, \ldots, C_k \in \mathcal{C}$ is image admissible Suslin, via Y^k, the Cartesian product of k copies of Y with product σ-algebra, and using $(1_{T(y_1)}, \ldots, 1_{T(y_k)})$.

(b) Define a map from $\{0, 1\}^k$ into $\{0, 1\}^{2^k}$ taking any $(1_{C_1}, \ldots, 1_{C_k})$ into a vector whose components give indicator functions of all atoms of the Boolean algebra generated by $C_1, \ldots, C_k$ and possibly of the empty set.

(c) For any $m = 1, 2, \ldots$, define a function $\mathcal{U}$ from $\{0, 1\}^m = \{b = \{b_j\}_{j=1}^m : b_j = 0 \text{ or } 1 \text{ for all } j\}$ such that for every subset $E \subset \{1, \ldots, m\}$ there is an $r = 1, \ldots, 2^m$ such that $\mathcal{U}(b)_r = \max_{j \in E} b_j$.

Combine steps (a), (b), and (c) and use as the Y space for $\mathcal{C}^{(k)}$ the product space $Y^k \times F_k$ where F_k is the finite set $\{1, 2, \ldots, 2^{2^k}\}$ to get the solution.

9. Let $(X, \mathcal{A})$ be a measurable space such that $\{x\} \in \mathcal{A}$ for all $x \in X$. If P is a law on $\mathcal{A}$, call a set $A \in \mathcal{A}$ a *soft atom* if $P(A) > 0$, for all $B \subset A$ with $B \in \mathcal{A}$ either $P(B) = P(A)$ or $P(B) = 0$, and $P(\{x\}) = 0$ for all $x \in A$. (A "hard atom" would be an ordinary atom, namely, a singleton $\{x\}$ with $P(\{x\}) > 0$.)

(a) If X is a separable metric space with Borel σ-algebra $\mathcal{A}$, show that no law on $\mathcal{A}$ has a soft atom. *Hint*: Show that a soft atom has a sequence of soft atom subsets, decreasing to a singleton.

(b) Give an example of a soft atom. *Hint*: Let X be uncountable and $\mathcal{A}$ consist of countable sets and their complements.

Notes

Notes to Section 5.1. Fisher (1922) invented the idea of sufficiency and Neyman (1935) gave a first form of factorization theorem. Halmos and Savage (1949) proved the factorization theorem, Theorem 5.1, in case $h \in \mathcal{L}^1(S, \mathcal{B}, \mu)$. Vivid *ad hoc* concepts such as "kernel" and "chain" in the proof of Lemma 5.2 are typical of Halmos' style of writing proofs. Bahadur (1954) removed the restriction on h, by the proof given above in the proof of Theorem 5.1.

Theorems 5.7 and 5.8 (sufficiency of P_n) follow from facts given by Neveu (1977, pp. 267–268). I thank Sam Gutmann for showing me the proof actually given in the section and Don Cohn for telling me about Neveu's proof.

I do not have references for Theorem 5.9, Lemma 5.10, or Corollary 5.12. Yet in some sense, it is well known that the empirical distribution function incorporates all the information in an i.i.d. sample (Corollary 5.12).

Notes to Section 5.2. Theorem 5.16, due to Lebesgue, is proved in Natanson (1957, p. 137). Theorem 5.15 is due to Aumann (1961), and the proof here to B. V. Rao (1971), except that the "image admissible" part was added in Dudley (1984). Freedman (1966), Lemma (5), proved that the σ-algebra $\mathcal{T}$ mentioned before Theorem 5.13 is not countably generated.

Notes to Section 5.3. There are several papers on selection theorems. Sainte-Beuve (1974, Theorem 3) gives a selection theorem close to Theorem 5.22, which covers Suslin topological spaces. Sainte-Beuve did not use the terminology "(image) admissible Suslin."

M. Suslin published in 1917 an example of a Borel set in $\mathbb{R}^2$ whose projection into $\mathbb{R}$ is not Borel, disproving a statement in a 1905 paper by Lebesgue (on the history, see the notes to Section 13.2 of RAP). Such projections, or more generally direct images of Borel sets in Polish spaces by Borel functions, are sometimes called analytic sets. They are here called Suslin sets in honor of Suslin's discovery of them, and because analytic sets are generally defined to be subsets of Polish spaces, e.g. in Cohn (1980) and in RAP. Here, Suslin measurable spaces are defined, not necessarily with any metric or even topology.

Darst (1971) showed that even an infinitely differentiable (C^∞) function can take a Borel set onto a non-Borel set in $\mathbb{R}$.

Originally (e.g. Dudley 1984, Section 10.3), the definition of image admissible Suslin required that $(\Omega, \mathcal{A})$ also be a Suslin space. I thank Uwe Einmahl, who pointed out to me around 1988 that the Suslin assumption on Ω was unnecessary. I am grateful to the late Lucien Le Cam for stimulating discussions about this section.

Durst and Dudley (1981) gave the example after Theorem 5.26. Effros (1965) is the original paper on the Effros Borel structure. Strobl (1994, 1995) and Ziegler (1994, 1997a,b) have given special attention to measurability issues for empirical processes. W. Adamski and P. Gaenssler told me in 1992 about the example at the end of the section.

6

Limit Theorems for Vapnik–Červonenkis and Related Classes

6.1 Koltchinskii–Pollard Entropy and Glivenko–Cantelli Theorems

There are some good sufficient conditions for limit theorems (Glivenko–Cantelli or Donsker theorems) over Vapnik–Červonenkis and certain related classes, using the following form of "entropy" or "capacity."

First, let $(X, \mathcal{A})$ be a measurable space and $\mathcal{F} \subset \mathcal{L}^0(X, \mathcal{A})$, the space of all real-valued measurable functions on X. Recall that $F_{\mathcal{F}}(x) := \sup\{|f(x)| : f \in \mathcal{F}\}$ (Section 4.8). Then $F_{\mathcal{F}}(x) \equiv \|\delta_x\|_{\mathcal{F}}$. A measurable function $F \in \mathcal{L}^0(X, \mathcal{A})$ with $F \geq F_{\mathcal{F}}$ is called an *envelope function* for $\mathcal{F}$. If $F_{\mathcal{F}}$ is $\mathcal{A}$-measurable it is called *the* envelope function of $\mathcal{F}$. If a law P is given on $(X, \mathcal{A})$, then $F_{\mathcal{F}}^*$ for P will be called *the* envelope function of $\mathcal{F}$ *for* P, defined up to equality P-a.s.

Let Γ be the set of all laws on X of the form $n^{-1} \sum_{j=1}^{n} \delta_{x(j)}$ for some $x(j) \in X$, $j = 1, \ldots, n$, and $n = 1, 2, \ldots$, where the $x(j)$ need not be distinct. For $\delta > 0, 0 < p < \infty$, and $\gamma \in \Gamma$ recall (Section 4.8) that if F is an envelope function of $\mathcal{F}$,

$$D_F^{(p)}(\delta, \mathcal{F}, \gamma) := \sup \left\{ m : \text{for some } f_1, \ldots, f_m \in \mathcal{F}, \text{ and all } i \neq j, \right. \tag{6.1}$$
$$\left. \int |f_i - f_j|^p \, d\gamma > \delta^p \int F^p d\gamma \right\}.$$

Let $D_F^{(p)}(\delta, \mathcal{F}) := \sup_{\gamma \in \Gamma} D_F^{(p)}(\delta, \mathcal{F}, \gamma)$. Here $D_F^{(p)}(\delta, \mathcal{F}, \gamma)$ is a kind of packing number, involving the envelope function F. The corresponding "capacity" will be the logarithm of $D^{(p)}$. Such logarithms will appear in Section 6.3.

Let $\mathcal{G} := \{f/F : f \in \mathcal{F}\}$, where $0/0$ is replaced by 0. Then $|g(x)| \leq 1$ for all $g \in \mathcal{G}$ and $x \in X$. Given F, p, and $\gamma \in \Gamma$ let $Q(B) := \int_B F^p d\gamma / \gamma (F^p)$, if $\gamma(F^p) > 0$. Then $Q := Q_\gamma$ is a law and for $1 \le p < \infty$, $D_F^{(p)}(\delta, \mathcal{F}, \gamma) = D\left(\delta, \mathcal{G}, d_{p,Q}\right)$, where $d_{p,Q}(f, g) := \left(\int |f - g|^p \, dQ\right)^{1/p}$ (as defined before Theorem 5.17).

For example, if $\mathcal{C}$ is a collection of measurable sets, whose union is all of X, and $\mathcal{F} := \{1_A : A \in \mathcal{C}\}$ the envelope function is $F \equiv 1$ and $D_F^{(p)}(\delta, \mathcal{F}, \gamma) = D(\delta, \mathcal{F}, d_{p,\gamma})$. The next few results will connect $D_F^{(p)}(\delta, \mathcal{F})$ with other ways of measuring the size of certain classes $\mathcal{F}$. First, we have Vapnik–Červonenkis classes $\mathcal{C}$ of sets with $\mathrm{dens}(\mathcal{C}) \le S(\mathcal{C}) < +\infty$ (Corollary 4.4):

Theorem 6.1 *If $\mathcal{C} \subset \mathcal{A}$, $\mathrm{dens}(\mathcal{C}) < +\infty$, $1 \le p < \infty$, $F \in \mathcal{L}^p(X, \mathcal{A}, P)$, $F \ge 0$, and $\mathcal{F} := \{F 1_A : A \in \mathcal{C}\}$, then for any $w > \mathrm{dens}(\mathcal{C})$ there is a $K < \infty$ such that*

$$D_F^{(p)}(\delta, \mathcal{F}) \le K\delta^{-pw}, \quad 0 < \delta \le 1.$$

Proof. Let $\gamma \in \Gamma$ and let G be the smallest set with $\gamma(G) = 1$. We may assume $F(x) > 0$ for some $x \in G$ since otherwise for any $K \ge 1$,

$$D_F^{(p)}(\delta, \mathcal{F}, \gamma) = 1 \le K\delta^{-pw} \text{ for } 0 < \delta \le 1.$$

If $C(1), \ldots, C(m) \in \mathcal{C}$ are such that $[d_{p,\gamma}\left(F 1_{C(i)}, F 1_{C(j)}\right)]^p > \delta^p \int F^p d\gamma$ for $i \ne j$, with m maximal, then for $Q = Q_\gamma$

$$Q(C(i) \triangle C(j)) = \int_{C(i) \triangle C(j)} F^p d\gamma \,/\, \int F^p d\gamma \,>\, \delta^p.$$

Then by maximality and Theorem 4.47, there is a $K(w, \mathcal{C}) < \infty$ such that

$$D_F^{(p)}(\delta, \mathcal{F}, \gamma) \le m \le D\left(\delta^p, \mathcal{C}, d_Q\right) \le K(w, \mathcal{C})\delta^{-pw}. \qquad \square$$

Next, here is a kind of converse to Theorem 6.1:

Proposition 6.2 *Suppose $1 \le p < \infty$, $\mathcal{F} = \{1_B : B \in \mathcal{C}\}$ for some collection $\mathcal{C}$ of sets, $F \equiv 1$, and for some δ with $0 < \delta^p < 1/2$, $D_F^{(p)}(\delta, \mathcal{F}) < \infty$. Then $\mathrm{dens}(\mathcal{C}) \le S(\mathcal{C}) < \infty$.*

Proof. First, $\mathrm{dens}(\mathcal{C}) \le S(\mathcal{C})$ by Corollary 4.4. Next, suppose $\mathcal{C}$ shatters a set G with card $G = n = 2^m$ for some positive integer m. Let γ have mass $1/n$ at each point of G. Then G has m subsets $A(i)$, $i = 1, \ldots, m$, independent for γ, with $\gamma(A(i)) = 1/2$, $i = 1, \ldots, m$ (taking elements of G as strings of m binary digits, let $A(i)$ be the set where the ith digit is 1). Then for $i \ne k$, $\int |1_{A(i)} - 1_{A(k)}|^p \, d\gamma = 1/2$. So for each j in (6.1) we can take $f_j = 1_{A(j)}$. Thus

$$D_F^{(p)}(\delta, \mathcal{F}) \ge D_F^{(p)}(\delta, \mathcal{F}, \gamma) \ge m.$$

For large m this is impossible, so $S(\mathcal{C}) < \infty$. $\qquad \square$

Next, consider families of functions of the form $\mathcal{F} = \{Fg : g \in \mathcal{G}\}$ where $\mathcal{G}$ is a family of functions totally bounded in the supremum norm

$$\|g\|_{\sup} := \sup_{x \in X} |g(x)|,$$

with the associated metric $d_{\sup}(g, h) := \|g - h\|_{\sup}$ and with $\|g\|_{\sup} \le 1$ for all $g \in \mathcal{G}$.

Proposition 6.3 *If* $1 \le p < \infty$ *and* $0 < \varepsilon \le 1$, *then for any such* $\mathcal{F}$,

$$D_F^{(p)}(\varepsilon, \mathcal{F}) \le D\left(\varepsilon, \mathcal{G}, d_{\sup}\right).$$

Proof. If $d_{\sup}(g, h) \le \delta$, then for all x, $|Fg - Fh|^p(x) \le \delta^p F(x)^p$, so the result follows from the definition (6.1). $\quad\square$

Now we come to the Koltchinskii–Pollard method of symmetrization of empirical measures. Given $n = 1, 2, \ldots$, let $x_1, \ldots, x_{2n}$ be coordinates on $\left(X^{2n}, \mathcal{A}^{2n}, P^{2n}\right)$, hence i.i.d. P. Let $\Omega_n := \{0, 1\}^n$ with the uniform probability distribution U_n giving probability $1/2^n$ to each point. Thus the coordinates e_i, $i = 1, \ldots, n$ on Ω_n are i.i.d. and equal 0 or 1 with probability $1/2$ each. Let $\sigma(i) := 2i - e_i$ and $\tau(i) := 2i - 1 + e_i$ for each $i = 1, \ldots, n$. Take the product space $(X^{2n}, P^{2n}) \otimes (\Omega_n, U_n)$, so that $\{x_j\}_{j=1}^{2n}$ is independent of $\{e_j, \sigma(j), \tau(j)\}_{j=1}^n$. Then $x(\sigma(j))$ for $j = 1, \ldots, n$ are i.i.d. P. Let

$$P_n' := n^{-1} \sum_{j=1}^n \delta_{x(\sigma(j))}, \quad P_n'' := n^{-1} \sum_{j=1}^n \delta_{x(\tau(j))},$$

$$\nu_n' := n^{1/2}\left(P_n' - P\right), \quad \nu_n'' := n^{1/2}\left(P_n'' - P\right),$$

$$P_n^0 := P_n' - P_n'', \quad \nu_n^0 := n^{1/2} P_n^0.$$

The variables $\varepsilon_i = 2e_i - 1$ are i.i.d. Rademacher variables, having values ± 1 with probability $1/2$ each. Let E_n be their joint distribution. We have the following alternate representation:

Proposition 6.4 *In* $X^n \times X^n \times \{-1, 1\}^n$,

$$\left\langle \{x_{\sigma(i)}\}_{i=1}^n, \{x_{\tau(i)}\}_{i=1}^n, \{\varepsilon_i\}_{i=1}^n \right\rangle$$

has the same distribution as

$$\left\langle \{x_i\}_{i=1}^n, \{x_{n+i}\}_{i=1}^n, \{\varepsilon_i\}_{i=1}^n \right\rangle,$$

namely, $P^n \times P^n \times E_n$.

Proof. For any given values of $\varepsilon_i = 2e_i - 1$ or of e_i, $\{x_{\sigma(i)}\}_{i=1}^n$ and $\{x_{\tau(i)}\}_{i=1}^n$ are independent, each with distribution P^n. The conclusion follows. $\quad\square$

It follows that v'_n and v''_n are two independent copies of $v_n = \sqrt{n}(P_n - P)$. Here is a symmetrization fact. The Lemma will be used to upper-bound the probability on the right side by $(1 - \zeta^2 \eta^{-2})^{-1}$ times the probability on the left, which is easier to bound by symmetry. It may be compared to the inequalities of P. Lévy, Theorem 1.20, a reflection principle, and that of Ottaviani, Theorem 1.19. The proofs of both those inequalities use a stopping time, the first time that partial sums of independent real random variables reach or cross a given level. Here the class $\mathcal{F}$ has no linear ordering and we cannot choose a "least" f. Instead, measurable selection, from Theorem 5.22, is used in place of a stopping time.

Lemma 6.5 *Let* $\zeta > 0$ *and* $\mathcal{F} \subset \mathcal{L}^2(X, \mathcal{A}, P)$ *with* $\int f^2 dP \le \zeta^2$ *for all* $f \in \mathcal{F}$. *Assume* $\mathcal{F}$ *is image admissible Suslin via some* $(Y, \mathcal{S}, T)$. *Then for any* $\eta > \zeta$

$$\Pr\left\{\|v^0_n\|_{\mathcal{F}} > \eta\right\} \ge \left(1 - \zeta^2 \eta^{-2}\right) \Pr\{\|v_n\|_{\mathcal{F}} > 2\eta\}.$$

Proof. The given events are measurable (up to sets of probability 0) by Proposition 5.20(c) for v_n or (b) for v^0_n, and Corollary 5.25. For $x = \langle x_1, \dots, x_{2n} \rangle$ let $H := \{\langle x, \omega, y \rangle \in X^{2n} \times \Omega_n \times Y : |v''_n(T(y))| > 2\eta\}$. Using the second of the alternate representations in Proposition 6.4, we can take $P''_n = \frac{1}{n} \sum_{j=1}^{n} x_{n+j}$. Let $x' := \{x_{n+j}\}_{j=1}^{n}$ and $\xi = \{x_j\}_{j=1}^{n}$, so that ξ and x' are i.i.d. P^n and v''_n depends only on x', not ξ or $\omega \in \Omega_n$. Let H_1 be the set of $\langle x', y \rangle$ such that $\langle \xi, x', \omega, y \rangle \in H$ for some or equivalently all $\xi \in X^n$ and $\omega \in \Omega_n$. Then by image admissibility, H_1 is a product measurable subset of $X^n \times Y$. The Suslin property implies (Theorem 5.22) that there is a universally measurable selector h such that whenever $\langle \xi, y \rangle \in H_1$ for some $y \in Y$, $h(\xi) \in Y$ and $\langle \xi, y \rangle \in H_1$. Here $h(x')$ is defined if and only if $x' \in J$ for some u.m. set $J \subset X^n$ by Theorem 5.22. On the set where $x' \in J$, since $v^0_n = v'_n - v''_n$,

$$\Pr\left(\|v^0_n\|_{\mathcal{F}} > \eta | x'\right) \ge \Pr\left(\left|v'_n(h_1(x')(\cdot))\right| \le \eta | x'\right).$$

Given x', $T(h(x'))$ is a fixed function $f \in \mathcal{F}$ with $\int f^2 dP \le \zeta^2$. Next since ξ is independent of x', we can apply Chebyshev's inequality to obtain

$$\Pr\left(\left|v'_n(f)\right| \le \eta\right) \ge 1 - (\zeta/\eta)^2.$$

Integrating gives $\Pr\left\{\|v^0_n\|_{\mathcal{F}} > \eta\right\} \ge \left(1 - (\zeta/\eta)^2\right) \Pr\{\|v''_n\|_{\mathcal{F}} > 2\eta\}$, which, since v''_n is a copy of v_n, gives the result. $\square$

Some reversed martingale and submartingale properties of the empirical measures P_n will be proved. Recall that $Q(f) := \int f \, dQ$ for any $f \in \mathcal{L}^1(Q)$, and that in defining empirical measures $P_n := \frac{1}{n} \sum_{j=1}^{n} \delta_{X_j}$, the X_j are always (in this book) taken as coordinates on a product of copies of a probability space $(X, \mathcal{A}, P)$, so that the underlying probability measure for P_n is P^n, a product of n copies of P.

Here are some definitions for reversed (sub)martingales. Let $(\Omega, \mathcal{B}, Q)$ be a probability space. For $n \geq 1$ let $\mathcal{B}_n$ be a sub-σ-algebra of $\mathcal{B}$ such that $\mathcal{B}_n \supset \mathcal{B}_{n+1}$ for all n (the reverse of the usual inclusion for martingales). For each n let $X_n \in \mathcal{L}^1(\Omega, \mathcal{B}_n, P)$. Then $\{X_n\}$ is called a *reversed submartingale* if $E(X_n|\mathcal{B}_{n+1}) \geq X_{n+1}$ almost surely for all n, or a *reversed martingale* if the same holds with $=$ in place of $\geq$.

Remarks. In some presentations (including RAP), the index integers are taken to be negative, so that we have $\mathcal{B}_n \subset \mathcal{B}_{n+1}$ for $n = -2, -3, \ldots$. Here, however, positive integers have been preferred, as in P_n we have n positive.

For each n, let $\mathcal{D}_n$ be the smallest σ-algebra for which all X_m for $m \geq n$ are measurable. Then from the definitions, $\mathcal{B}_n \supset \mathcal{D}_n$ for all n. In some presentations, $\mathcal{B}_n$ is taken to equal $\mathcal{D}_n$. Very often (e.g., in RAP), a conditional expectation $E(X|\mathcal{A})$ is required to be an $\mathcal{A}$-measurable function. Here, however, it may equal such a function almost surely. We need to allow completion measurability (adjoining sets of probability 0) to apply some facts in Chapter 5. As conditional expectations are only defined up to almost sure equality, completion does not make a great difference.

Theorem 6.6 *Let $(X, \mathcal{A}, P)$ be a probability space with $(X, \mathcal{A})$ separable, $\mathcal{F} \subset \mathcal{L}^1(P)$, and P_n empirical measures for P. Let $\mathcal{S}_n$ be the smallest σ-algebra for which $P_k(f)$ are measurable for all $k \geq n$ and all $f \in \mathcal{L}^1(X, \mathcal{A}, P)$. Then:*

(a) *For any $f \in \mathcal{F}$, $\{P_n(f), \mathcal{S}_n\}_{n \geq 1}$ is a reversed martingale; in other words*

$$E\left(P_{n-1}(f)|\mathcal{S}_n\right) = P_n(f) \text{ a.s., if } n \geq 2.$$

(b) *(F. Strobl) Suppose $\mathcal{F}$ has an envelope function $F \in \mathcal{L}^1(X, \mathcal{A}, P)$ and that for each n, $\|P_n - P\|_{\mathcal{F}}$ is measurable for the completion of P^n. Then $(\|P_n - P\|_{\mathcal{F}}, \mathcal{S}_n)_{n \geq 1}$ is a reversed submartingale; in other words we have $\|P_{n+1} - P\|_{\mathcal{F}} \leq E(\|P_n - P\|_{\mathcal{F}}|\mathcal{S}_{n+1})$ a.s. for all n.*

Remark. $\|P_n - P\|_{\mathcal{F}}$ will be completion measurable if $\mathcal{F}$ is image admissible Suslin, by Proposition 5.20(c) and Corollary 5.25.

Proof. For each n, any set in $\mathcal{S}_n$ is invariant under permutations of the first n coordinates X_i, where $P_n := \frac{1}{n}\left(\delta_{X_1} + \cdots + \delta_{X_n}\right)$. So if $1 \leq i < j \leq n$, $E\left(f(X_i)|\mathcal{S}_n\right) = E\left(f(X_j)|\mathcal{S}_n\right)$. Summing over $i = 1, \ldots, k$ and dividing by k gives $E\left(P_k(f)|\mathcal{S}_n\right) = E\left(f(X_1)|\mathcal{S}_n\right)$ for $k = 1, \ldots, n$. Letting $k = n - 1$ and n proves (a).

Proof of (b): clearly $\|P\|_{\mathcal{F}} \leq PF < \infty$ and $\|P_n\|_{\mathcal{F}} \leq P_nF$. Since by assumption $\|P_n - P\|_{\mathcal{F}}$ is completion measurable, we have $E\|P_n - P\|_{\mathcal{F}} \leq 2PF < \infty$.

Next, it will be shown that $\|P_n - P\|_{\mathcal{F}}$ is measurable for the completion of $\mathcal{S}_n$. It is enough to show that for each rational $q \geq 0$, the set $A_q := A_{q,n}$ where $\|P_n - P\|_{\mathcal{F}} > q$ is measurable for the completion of $\mathcal{S}_n$. Let

$$X^\infty := \{\{x_j\}_{j \geq 1} : x_j \in X\} = X^n \times X^{(n)}$$

where $X^{(n)} = \{\{x_j\}_{j > n} : x_j \in X\}$. The set A_q itself is invariant under all transformations in the set Π_n of permutations of the first n coordinates. Here Π_n may be viewed as acting either on X^∞ or on X^n. Let $\mathcal{A}^{[n]} := \{B \times X^{(n)} : B \in \mathcal{A}^n\}$ and $P^n(B \times X^{(n)}) := P^n(B)$. By assumption there are sets C_q, D_q in $\mathcal{A}^{[n]}$ with $C_q \subset A_q \subset D_q$ and $P^n(D_q \setminus C_q) = 0$. Each image of C_q by an element of Π_n is also in $\mathcal{A}^{[n]}$ and included in A_q. The union U_q of these $n!$ images, being in $\mathcal{A}^{[n]}$ and symmetric under all $n!$ permutations of the indices $1, \ldots, n$, is $\mathcal{S}_n$-measurable by Theorem 5.8, and included in A_q. Likewise, the intersection F_q of all images of D_q by transformations in Π_n is $\mathcal{S}_n$-measurable and includes A_q. Clearly $P^n(F_q \setminus U_q) = 0$, so A_q and $\|P_n - P\|_{\mathcal{F}}$ are measurable for the completion of $\mathcal{S}_n$ as desired.

For $i = 1, \ldots, n+1$ let $P_{n,i} := \frac{1}{n} \sum \{\delta_{X_j} : j = 1, \ldots, n+1, \ j \neq i\}$. Then since the X_j are coordinates on a product space, $P_{n,i}$ has the same properties as P_n for each i, and $P_{n,n+1} \equiv P_n$. Thus $\|P_{n,i} - P\|_{\mathcal{F}}$ is completion measurable with respect to $\mathcal{A}^{[n+1]}$.

Now, the conditional expectations $E\left(\|P_{n,i} - P\|_{\mathcal{F}} | \mathcal{S}_{n+1}\right)$ are all the same (equal to each other almost surely) for $i = 1, \ldots, n+1$, so all are a.s. equal to $E(\|P_n - P\|_{\mathcal{F}} | \mathcal{S}_{n+1})$, because for each i, a transformation defined by a permutation of the first $n+1$ coordinates takes $P_{n,i}$ into P_n while leaving all sets in the σ-algebra $\mathcal{S}_{n+1}$ invariant. Then for any $n = 1, 2, \ldots$, we have

$$\|P_{n+1} - P\|_{\mathcal{F}} = \frac{1}{n+1} \left\| \sum_{i=1}^{n+1} (P_{n,i} - P) \right\|_{\mathcal{F}} \leq \frac{1}{n+1} \sum_{i=1}^{n+1} \|P_{n,i} - P\|_{\mathcal{F}}, \text{ and}$$

$$\|P_{n+1} - P\|_{\mathcal{F}} = E(\|P_{n+1} - P\|_{\mathcal{F}} | \mathcal{S}_{n+1})$$

$$\leq \frac{1}{n+1} \sum_{i=1}^{n+1} E\left(\|P_{n,i} - P\|_{\mathcal{F}} | \mathcal{S}_{n+1}\right) = E(\|P_n - P\|_{\mathcal{F}} | \mathcal{S}_{n+1})$$

a.s., finishing the proof. $\square$

Recall $D_{\mathcal{F}}^{(p)}(\delta, \mathcal{F})$ as defined after (6.1). Here is a law of large numbers (generalized Glivenko–Cantelli theorem):

Theorem 6.7 Let $(X, \mathcal{A}, P)$ be a probability space such that $(X.\mathcal{A})$ is separable, $F \in \mathcal{L}^1(X, \mathcal{A}, P)$, and $\mathcal{F}$ a collection of measurable functions on X having F as an envelope function. Suppose $\mathcal{F}$ is image admissible Suslin via $(Y, \mathcal{S}, T)$.

Assume that

$$D_F^{(1)}(\delta, \mathcal{F}) < \infty \text{ for all } \delta > 0. \tag{6.2}$$

Then $\lim_{n \to \infty} \| P_n - P \|_{\mathcal{F}} = 0$ *a.s.*

Proof. By Theorem 6.6, $\{\| P_n - P \|_{\mathcal{F}}, \mathcal{S}_n\}_{n \geq 1}$ is a reversed submartingale. Being nonnegative, it converges almost surely and in $\mathcal{L}^1$ (RAP, Theorem 10.6.4). It will now be enough to show $\| P_n - P \|_{\mathcal{F}} \to 0$ in probability.

Given $\varepsilon > 0$, take $M \geq 1$ large enough so that $P(F1_{F>M}) < \varepsilon/4$. Then

$$\| (P_n - P) 1_{F>M} \|_{\mathcal{F}} \leq \| P_n 1_{F>M} \|_{\mathcal{F}} + P(F1_{F>M})$$

$$\leq (P_n + P)(F1_{F>M}) \to 2P(F1_{F>M}) < \varepsilon/2$$

almost surely as $n \to \infty$. Replacing each $f \in \mathcal{F}$ by $f 1_{F \leq M}$ takes $\mathcal{F}$ onto another class of functions which is still image admissible Suslin since $\langle x, y \rangle \to T(y)(x) 1_{F(x) \leq M}$ is jointly measurable while the Suslin space $(Y, \mathcal{S})$ is unchanged. So we may assume $F \leq M$.

Next apply Lemma 6.5 with $\zeta = M$ and $\eta = n^{1/2}\varepsilon/4 > 2M$ for n large. Then it is enough to show that $\| P_n^0 \|_{\mathcal{F}} = \| P_n' - P_n'' \|_{\mathcal{F}} \to 0$ in probability.

For any $f \in \mathcal{F}$, we can write

$$P_n^0(f) = \frac{1}{n} \sum_{j=1}^{n} \varepsilon_j [f(x_{2j}) - f(x_{2j-1})] \tag{6.3}$$

where ε_j are Rademacher variables independent of $x = \{x_j\}_{j=1}^{2n}$ as in Proposition 6.4. Let $\omega := \{\varepsilon_j\}_{j=1}^{n} \in \mathcal{E} := \{-1, 1\}^n$ with distribution E_n. The function

$$(x, \omega, y) \mapsto |P_n^0(T(y))| = \frac{1}{n} \left| \left(\sum_{j=1}^{n} \varepsilon_j \left(\delta_{x_{2j}} - \delta_{x_{2j-1}} \right) \right) (T(y)) \right|$$

is jointly measurable by image admissibility (Proposition 5.20). Then by Corollary 5.25, $\| P_n^0 \|_{\mathcal{F}}$ is jointly completion measurable in (x, ω).

For $n \geq n_0$ large enough, $\delta P_{2n}(F) < \varepsilon/4$ on a set X' of x with $P^{2n}(X') > 1 - \varepsilon/4$. Let $\delta := \varepsilon/(9M)$ and $K(\delta) := D_F^{(1)}(\delta, \mathcal{F})$. Let $A := A_\varepsilon$ be the event that $\| P_n^0 \|_{\mathcal{F}} > \varepsilon/2$. Then by the Tonelli–Fubini theorem

$$\Pr(A_\varepsilon) \leq \frac{\varepsilon}{4} + \int_{X'} \int_{\mathcal{E}} 1_A(x, \omega) dE_n(\omega) dP^{2n}(x). \tag{6.4}$$

For each fixed $x \in X'$, an upper bound will be given for the inner integral. There exist $m = m(x) \leq K(\delta)$ functions $f_1, \ldots, f_m \in \mathcal{F}$ such that for any $f \in \mathcal{F}$, for some $j \leq m$,

$$P_{2n}(|f - f_j|) \leq \delta P_{2n}(F) < \varepsilon/4,$$

by the definitions of $K^{(1)}(\delta)$ and X'. Then

$$\left|P_n^0(f) - P_n^0(f_j)\right| = \left|P_n'(f - f_j) - P_n''(f - f_j)\right|$$
$$\leq (P_n' + P_n'')(|f - f_j|) = 2P_{2n}(|f - f_j|) \leq 2\delta P_{2n}(F) < \varepsilon/4.$$

From the Rademacher distribution we have $\int P_n^0(f)dE_n(\omega) = 0$ for all x. We also have $|f(x_{2j}) - f(x_{2j-1})| \leq 2M$ for each $f \in \mathcal{F}$ and each j. It follows that for any $x \in X'$, the variance of $P_n^0(f)$ with respect to $E_n(\omega)$ is at most $4M^2/n$. Thus, denoting probability with respect to $E_n(\omega)$ for fixed x as $\Pr_{n,x}$, we have that

$$\Pr_{n,x}\left\{\max_{j\leq m}\left|P_n^0(f_j)\right| > \varepsilon/4\right\} \leq$$

$$K(\delta)\sup\left\{\Pr_{n,x}\left\{\left|P_n^0(f)\right| > \varepsilon/4\right\} : |f| \leq M\right\} \leq 4K(\delta)M^2(4/\varepsilon)^2/n < \varepsilon/4$$

for n large using Chebyshev's inequality. So, integrating with respect to the distribution of $x \in X'$, we have for n large enough the unconditional probability

$$\Pr\left\{\left\|P_n^0\right\|_{\mathcal{F}} > \varepsilon/2\right\} < \frac{\varepsilon}{4} + \frac{\varepsilon}{4} = \varepsilon/2,$$

proving Theorem 6.7.   □

Vapnik and Červonenkis first proved the following (under a different measurability assumption).

Corollary 6.8 (Vapnik and Červonenkis) *Let $\mathcal{C} \subset \mathcal{A}$ where $(X, \mathcal{A})$ is a separable measurable space, $S(\mathcal{C}) < \infty$, and $\mathcal{C}$ is image admissible Suslin. Then for any probability law P on $\mathcal{A}$, we have*

$$\lim_{n\to\infty}\sup_{A\in\mathcal{C}}|(P_n - P)(A)| = 0 \text{ a.s.}$$

Proof. In Theorem 6.7, take $F \equiv 1$ and apply Theorem 6.1 with $p = 1$.   □

Corollary 6.9 *Let $(X, \mathcal{A}, P)$ be a probability space, and $\mathcal{F}$ a VC subgraph class of measurable functions on X with an envelope function $F \in \mathcal{L}^1(X, \mathcal{A}, P)$. Suppose $\mathcal{F}$ is image admissible Suslin via $(Y, \mathcal{S}, T)$.*
    Then $\lim_{n\to\infty}\|P_n - P\|_{\mathcal{F}} = 0$ a.s.

Proof. By Theorem 4.52 with $p = 1$, the hypothesis (6.2) of Theorem 6.7 holds. As the other hypotheses are assumed, the conclusion holds.   □

6.2 Glivenko–Cantelli Properties for Given P

The laws of large numbers (Glivenko–Cantelli theorems) uniformly over a class $\mathcal{C}$ of sets, for a given P, to be proved in this section, are essentially due to Vapnik and Červonenkis (without assuming that $\mathcal{C}$ is a VC class, a case already treated in Corollary 6.8) and to J. M. Steele; only the measurability assumptions here differ from theirs.

Let $(X, \mathcal{A}, P)$ be a probability space and $\mathcal{C} \subset \mathcal{A}$. Let $\{x_n\}_{n\geq 1}$ be coordinates in $(X^\infty, \mathcal{A}^\infty, P^\infty)$ so that x_j are i.i.d. P. For certain classes $\mathcal{C}$ with $S(\mathcal{C}) = +\infty$ one will have $\Delta^\mathcal{C}(\{x_1, \ldots, x_n\}) < 2^n$ with P^n-probability converging to 1. For such classes, with sufficient measurability properties, a law of large numbers will still hold. Here a main result is as follows:

Theorem 6.10 *If $(X, \mathcal{A}, P)$ is any probability space with $(X, \mathcal{A})$ separable, $\mathcal{C} \subset \mathcal{A}$, and $\mathcal{C}$ is image admissible Suslin, then the following are equivalent:*

(a) $\|P_n - P\|_\mathcal{C} \to 0$ *a.s. as $n \to \infty$;*
(b) $\|P_n - P\|_\mathcal{C} \to 0$ *in probability as $n \to \infty$;*
(c) $\lim_{n\to\infty} n^{-1} E \log \Delta^\mathcal{C}(\{x_1, \ldots, x_n\}) = 0$.

For a finite set $F \subset X$ and collection $\mathcal{C} \subset 2^X$, let $k^\mathcal{C}(F) := S(\mathcal{C} \sqcap F)$.

Lemma 6.11 *Under the hypotheses of Theorem 6.10, both $\Delta^\mathcal{C}(x_1, \ldots, x_n)$ and $k^\mathcal{C}(\{x_1, \ldots, x_n\})$ are universally measurable.*

Proof. Let $\mathcal{C}$ be image admissible Suslin via $(Y, \mathcal{Y}, T)$. For any decomposition D of $\{1, \ldots, n\}$ into subsets, let X_D^n be the set of all ordered n-tuples $\langle x_1, \ldots, x_n \rangle$ such that $x_i = x_j$ if and only if i and j belong to the same subset in D. Then X_D^n is $\mathcal{A}^n$ measurable as follows: by Theorem 5.13, up to measurable isomorphism, we can take $X \subset [0, 1]$ with $\mathcal{A}$ as the Borel σ-algebra of X as a metric space (or equivalently, the intersections with X of Borel sets in $[0, 1]$). In that case, each set $\{x : x_i = x_j\}$ for $i \neq j$ is clearly measurable. It is enough to prove the Lemma on each X_D^n; specifically, where D is the decomposition into singletons $\{j\}$, the set on which the x_i are all different.

For each set $J \subset \{1, \ldots, n\}$, and $x := \langle x_1, \ldots, x_n \rangle$, $U(J) := \{\langle x, y \rangle : x_j \in T(y)$ if and only if $j \in J\}$ is product measurable by image admissibility. Thus its projection $\amalg U(J)$ into X^n is universally measurable by Theorem 5.22. Now

$$\Delta^\mathcal{C}(\{x_1, \ldots, x_n\}) = \sum_J 1_{\amalg U(J)}(x) \quad \text{and}$$
$$\{k^\mathcal{C}(\{x_1, \ldots, x_n\}) \geq m\} = \bigcup_{G \subset \{1,\ldots,n\}, \; |G|=m} \bigcap_{H \subset G} \amalg U(H). \tag{6.5}$$

$\square$

The main step in Steele's proof uses Kingman's subadditive ergodic theorem (the equivalent superadditive ergodic theorem is RAP, Theorem 10.7.1). To state it, here is some terminology. Let $\mathbb{N}$ denote the set of nonnegative integers. A *subadditive process* is a doubly indexed array $\{x_{mn}\}_{0 \le m < n < \infty}$ of real random variables, $m, n \in \mathbb{N}$, $m < n$, such that

$$x_{kn} \le x_{km} + x_{mn} \quad \text{whenever} \quad k < m < n. \tag{6.6}$$

Let $x_{nn} := 0$ for all $n \in \mathbb{N}$. If instead of (6.6),

$$x_{kn} \ge x_{km} + x_{mn}, \quad k < m < n, \tag{6.7}$$

then $\{x_{mn}\}_{0 \le m < n}$ is called *superadditive*.

A process which is both subadditive and superadditive is called *additive* and can clearly be written as $x_{kn} = \sum_{k < j \le n} x_j$, where $x_j := x_{j-1,j}$; i.e., one has just partial sums of a sequence of random variables.

A subadditive process $\{x_{mn}\}_{0 \le m < n}$ defined on a probability space $(\Omega, \mathcal{B}, \text{Pr})$ will be called *stationary* if there is a measure-preserving transformation V of Ω onto itself such that for any integers $0 \le m < n$, $x_{mn}(V(\omega)) = x_{m+1,n+1}(\omega)$. Recall that $(f \circ g)(x) := f(g(x))$. Let $V^k := V \circ (V \circ (\cdots \circ V) \cdots)$ to k terms. Then for $k = 1, 2, \ldots$, $x_{mn} \circ V^k = x_{m+k,n+k}$. Let $\mathcal{S}$ be the σ-algebra of all $B \in \mathcal{B}$ such that $V^{-1}(B) = B$.

Another useful hypothesis for subadditive processes is:

$$\text{For each } n \in \mathbb{N}, \ E|x_{0n}| < +\infty, \text{ and } \kappa := \inf_{n \ge 1} Ex_{0n}/n > -\infty. \tag{6.8}$$

A σ-algebra $\mathcal{D} \subset \mathcal{B}$ will be called *degenerate* if $\Pr(D) = 0$ or 1 for all $D \in \mathcal{D}$. Here is Kingman's subadditive ergodic theorem.

Theorem 6.12 (Kingman) *Let* $\{x_{mn}\}_{0 \le m < n}$ *be a stationary subadditive process satisfying (6.8). Then as* $n \to \infty$, x_{0n}/n *converges a.s. and in* $\mathcal{L}^1$ *to a random variable* $y := \inf_{n \ge 1} n^{-1} E(x_{0n}|\mathcal{S})$ *with* $Ey = \kappa$. *If* $\mathcal{S}$ *is degenerate,* $y = \kappa$ *a.s.*

Proof. RAP, Theorem 10.7.1 applies with f_n there defined as $-x_{0n}$. From near the end of the proof we have $Ey \le Ex_{0n}/n$ for all n and $Ex_{0n}/n \to Ey$ as $n \to \infty$, so $Ey = \kappa$. $\qquad\square$

To apply Theorem 6.12 in proving Theorem 6.10 we have:

Theorem 6.13 *Let* $(\Omega, \mathcal{T}, \Pr)$ *be a probability space,* $(X, \mathcal{A})$ *a measurable space,* X_1 *a measurable function from* Ω *into* X, *and* V *a measure-preserving transformation of* Ω *onto itself. Let* $X_j := X_1 \circ V^{j-1}$ *for* $j = 2, 3, \ldots$. *Let* $\mathcal{C}$ *be image admissible Suslin and* $\emptyset \ne \mathcal{C} \subset \mathcal{A}$. *Then each of the following is a stationary subadditive process satisfying (6.8):*

(a) $D_{mn}^{\mathcal{C}} := \sup_{A \in \mathcal{C}} \left| \sum_{m < i \le n} (1_A(X_i) - P(A)) \right|$;

(b) $\log \Delta_{mn}^{\mathcal{C}} := \log \Delta^{\mathcal{C}}(\{X_{m+1}, \ldots, X_n\})$;

(c) $k_{mn}^{\mathcal{C}} := k^{\mathcal{C}}(\{X_{m+1}, \ldots, X_n\})$.

Proof. We have measurability by Corollary 5.25 in (a) and Lemma 6.11 in (b) and (c). Stationarity clearly holds for the same V in each case. Subadditivity is clear in (a) and not difficult for (b) and (c). All three processes are nonnegative: in (b), C nonempty implies $\Delta^C \geq 1$, so (6.8) holds. $\square$

Now Theorem 6.10 will be proved. The quantities $\|P_n - P\|_C$ are completion measurable by Corollary 5.25. The probability space, as usual, is a countable Cartesian product of copies of $(X, \mathcal{A}, P)$, with coordinates $x_1, x_2, \ldots$. The measure-preserving transformation will be $V(\{x_j\}_{j\geq 1}) := \{x_{j+1}\}_{j\geq 1}$. By the Kolmogorov 0-1 law (RAP, Theorem 8.4.5), $\mathcal{S}$ is degenerate. Then by Theorems 6.12 and 6.13, we have almost sure limits

$$c_1 := \lim_{n\to\infty} D^C_{0n}/n, \quad c_2 := \lim_{n\to\infty} \left(\log \Delta^C_{0n}\right)/n, \quad c_3 := \lim_{n\to\infty} k^C_{0n}/n \quad (6.9)$$

for some constants c_i. Thus in Theorem 6.10, (a) and (b) are equivalent (as also shown in the proof of Theorem 6.7 via the reversed submartingale property). It remains to prove (b) equivalent to (c).

To show (c) implies (b), given $0 < \varepsilon < 1$, first, apply the symmetrization Lemma 6.5 with $\zeta = 1$ and $\eta = \varepsilon n^{1/2}/2$ for n large enough so that $\eta \geq 2$, giving

$$\Pr\{\|P_n - P\|_C > \varepsilon\} \leq 2\Pr\left\{\|P'_n - P''_n\|_C > \varepsilon/2\right\}.$$

Thus it suffices to show $\|P'_n - P''_n\|_C \to 0$ in probability (these variables are universally measurable as usual by Proposition 5.20 and Corollary 5.25). For any fixed set $A \in \mathcal{A}$, the conditional probability

$$\Pr_{A,\varepsilon,n} := \Pr\left\{\left|P^0_n(A)\right| > \varepsilon|\{x_j\}^{2n}_{j=1}\right\}$$

$$= \Pr\left\{\left|\sum_{j=1}^{n} \varepsilon_j \left(\delta_{x_{2j}} - \delta_{x_{2j-1}}\right)(A)\right| > n\varepsilon \Big| \{x_i\}^{2n}_{i=1}\right\}$$

where $\varepsilon_j := 2 1_{\{\sigma(j)=2j\}} - 1$ are Rademacher variables independent of the x_i as in (6.3). Thus by Hoeffding's inequality, Proposition 1.12, with $a_j = -1, 0$ or 1, $\Pr_{A,\varepsilon,n} \leq 2\exp\left(-n\varepsilon^2/2\right)$.

For some n_0, $E \log \Delta^C_{0n} < n\varepsilon^4$ for $n \geq n_0$. Then by Markov's inequality,

$$\Pr\left(\log \Delta^C_{0n} > n\varepsilon^3\right) < \varepsilon.$$

Given $\{x_j\}^{2n}_{j=1}$, the event $\left|P^0_n(A)\right| > \varepsilon$ is the same for any two sets A having the same intersection with $\{x_1, \ldots, x_{2n}\}$. Thus

$$\Pr\left\{\|P^0_n\|_C > \varepsilon|\{x_j\}^{2n}_{j=1}\right\} \leq 2\Delta^C_{0,2n} \exp\left(-n\varepsilon^2/2\right).$$

Hence, for $\varepsilon < 1/8$ and $n \geq n_0$ large enough so that $\exp\left(-n\varepsilon^2/4\right) < \varepsilon$, we have

$$\Pr\left(\left\|P_n^0\right\|_{\mathcal{C}} > \varepsilon\right) < \varepsilon + 2\exp\left(2n\varepsilon^3 - n\varepsilon^2/2\right)$$
$$< \varepsilon + 2\exp\left(-n\varepsilon^2/4\right) < 3\varepsilon,$$

so (c) implies (b).

Now it will be shown that (b) implies (c). By (6.9) we have $1 \geq n^{-1}\log\Delta_{0n}^{\mathcal{C}} \to c$ a.s. for some constant $c := c_2 \geq 0$. Thus $n^{-1}E\log\Delta_{0n}^{\mathcal{C}} \to c$ and we want to prove $c = 0$. Suppose $c > 0$. Given $\varepsilon > 0$, for n large enough

$$\Pr\left\{(2n)^{-1}\log\Delta_{0,2n}^{\mathcal{C}} > c/2\right\} = \Pr\left\{\Delta_{0,2n}^{\mathcal{C}} > e^{nc}\right\} > 1 - \varepsilon.$$

Next, to symmetrize,

$$\Pr\left\{\left\|P_n' - P_n''\right\|_{\mathcal{C}} > 2\varepsilon\right\} \leq \Pr\left\{\left\|P_n' - P\right\|_{\mathcal{C}} > \varepsilon\right\} + \Pr\left\{\left\|P_n'' - P\right\|_{\mathcal{C}} > \varepsilon\right\}$$
$$= 2\Pr\left\{\left\|P_n - P\right\|_{\mathcal{C}} > \varepsilon\right\}.$$

So it will suffice to prove $\left\|P_n' - P_n''\right\|_{\mathcal{C}} \nrightarrow 0$ in probability. If $2 \leq k := [\alpha n]$ where $[x]$ is the largest integer $\leq x$ and $0 < \alpha < 1/2$, then $\alpha n \leq k + 1 \leq 3k/2$, so by Proposition 4.3 and Stirling's formula (Theorem 1.17) we have

$$_{2n}C_{\leq k} \leq (2ne/k)^k \leq (3e/\alpha)^{\alpha n}.$$

As $\alpha \downarrow 0$, $(3e/\alpha)^{\alpha} \to 1$. Thus for α small enough, $(3e/\alpha)^{\alpha} < e^c$. Choose and fix such an $\alpha > 0$. Then for n large enough,

$$n \geq 2/\alpha \quad \text{and} \quad (3e/\alpha)^{\alpha n} < e^{nc}. \tag{6.10}$$

Hence by Sauer's theorem 4.2, if $\Delta_{0,2n}^{\mathcal{C}} > e^{nc}$, then $k_{0,2n}^{\mathcal{C}} \geq [\alpha n]$. Fix n satisfying (6.10). Let $k := [\alpha n]$.

Now on an event U with $\Pr(U) > 1 - \varepsilon$, there is a subset T of the indices $\{1, \ldots, 2n\}$ such that $\operatorname{card} T = k$, $\mathcal{C}$ shatters $\{x_i : i \in T\}$, and $x_i \neq x_j$ for $i \neq j$ in T. If there is more than one such T, select each of the possible T's with equal probability, using for this a random variable Y independent of x_j and σ_j, $1 \leq j \leq 2n$. Then since x_j are i.i.d., T is uniformly distributed over its $\binom{2n}{k}$ possible values. For any distinct $j_i \in \{1, \ldots, 2n\}$, $N := 2n$, we have, where the following equations are conditional on U,

$$\Pr(j_1 \in T) = k/N,$$

$$\Pr(j_1, j_2 \in T) = k(k-1)/N(N-1), \tag{6.11}$$

$$\Pr(j_i \in T, \quad i = 1, 2, 3, 4) = \binom{k}{4} \bigg/ \binom{N}{4}.$$

Let M_n be the number of values of $j \leq n$ such that both $2j - 1$ and $2j$ are in T. Then from (6.11),

$$EM_n = k(k - 1)/2(N - 1) = \alpha^2 n/4 + \mathcal{O}(1), \quad n \to \infty;$$

$$EM_n^2 = k(k - 1)/2(N - 1) + n(n - 1)\binom{k}{4} \Big/ \binom{N}{4},$$

and a bit of algebra gives $\sigma^2(M_n) = EM_n^2 - (EM_n)^2 = \mathcal{O}(n)$ as $n \to \infty$. Thus for $0 < \delta < \alpha^2/4$, by Chebyshev's inequality, $\Pr(M_n \geq \delta n) > 1 - 2\varepsilon$ for n large. On the event $U \cap \{M_n \geq \delta n\}$, let's make a measurable selection, Theorem 5.22, of a sequence J of $[\delta n]$ values of i such that $J' := \bigcup_{i \in J}\{2i - 1, 2i\} \subset T$. Here M_n and J are independent of the $\sigma(j)$. Now measurably select, by Theorem 5.22 again with $y(\cdot) = H(\cdot)$, a set $A = A(\omega) = T(y(\omega)) \in \mathcal{C}$ such that $\{j \in J' : x_j \in A\} = \{\sigma(i) : i \in J\}$. Then $\sum_{i \in J}\left(\delta_{x_{\sigma(i)}} - \delta_{x_{\tau(i)}}\right)(A) = [n\delta]$. Here $y(\cdot)$ is measurable for the σ-algebra $\mathcal{B}_J$ generated by all the x_j, by Y, and by $\sigma(i)$ for $i \in J$. Conditional on $\mathcal{B}_J$,

$$\sum_{i \notin J}\left(\delta_{x_{\sigma(i)}} - \delta_{x_{\tau(i)}}\right)(A) = \sum_{i \notin J} s_i a_i$$

where a_i are $\mathcal{B}_J$-measurable functions with values -1, 0, and 1, and s_i have values ± 1 with probability $1/2$ each, independently of each other and of $\mathcal{B}_J$. Thus by Chebyshev's inequality

$$\Pr\left\{\sum_{i \notin J} s_i a_i > n\delta/3 \,\middle|\, \mathcal{B}_J\right\} \leq 9/\left(n\delta^2\right)$$

on the event where J is defined. Thus for n large

$$\Pr\left(\left(P_n' - P_n''\right)(A(\omega)) > \delta/3\right) > 1 - 3\varepsilon$$

and $\left\|P_n' - P_n''\right\|_{\mathcal{C}} \not\to 0$ in probability. So Theorem 6.10 is proved. $\qquad\square$

Theorem 6.14 *In (6.9), if any c_i is 0, all three are 0.*

Proof. In the last proof we saw that $c_1 = 0$ if and only if $c_2 = 0$ and just after (6.10) that if $c_2 > 0$ then $c_3 > 0$. On the other hand if $c_3 > 0$, then for some $\delta > 0$, $k_{0n}^{\mathcal{C}} \geq \delta n$ for n large enough a.s., and then $\Delta_{0n}^{\mathcal{C}} \geq 2^{\delta n}$. Thus $c_2 \geq c_3 \log 2 > 0$. $\qquad\square$

6.3 Pollard's Central Limit Theorem

By way of the Koltchinskii–Pollard kind of entropy and law of large numbers (Section 6.1 above) the following will be proved:

Theorem 6.15 (Pollard) *Let $(X, \mathcal{A}, P)$ be a probability space and let $\mathcal{F} \subset L^2(X, \mathcal{A}, P)$. Let $\mathcal{F}$ be image admissible Suslin via $(Y, \mathcal{S}, T)$ and have an envelope function $F \in L^2(X, \mathcal{A}, P)$. Suppose that*

$$\int_0^1 \left(\log D_F^{(2)}(x, \mathcal{F})\right)^{1/2} dx < \infty. \tag{6.12}$$

Then $\mathcal{F}$ is a Donsker class for P.

Before the proof of the theorem, here is a consequence:

Theorem 6.16 (Jain and Marcus) *Let (K, d) be a compact metric space. Let $C(K)$ be the space of continuous real functions on K with supremum norm. Let $X_1, X_2, \ldots$ be i.i.d. random variables in $C(K)$. Suppose $E X_1(t) = 0$ and $E X_1(t)^2 < \infty$ for all $t \in K$. Assume that for some random variable M with $E M^2 < \infty$,*

$$|X_1(s) - X_1(t)| \, (\omega) \leq M(\omega) d(s, t) \quad \text{for all} \quad \omega \quad \text{and} \quad s, t \in K.$$

Suppose that

$$\int_0^1 (\log D(\varepsilon, K, d))^{1/2} d\varepsilon < \infty. \tag{6.13}$$

Then the central limit theorem holds, i.e., in $C(K)$, $\mathcal{L}\left(n^{-1/2}(X_1 + \cdots + X_n)\right)$ converges to some Gaussian law.

Proof. For a real-valued function h on K recall (as in Section 3.6)

$$\|h\|_L := \sup_{s \neq t} |h(s) - h(t)|/d(s, t), \quad \|h\|_{\sup} := \sup_t |h(t)|,$$

$$\|h\|_{BL} := \|h\|_L + \|h\|_{\sup}, \quad BL(K) := \{h \in C(K) : \|h\|_{BL} < \infty\}.$$

To apply Theorem 6.15, take as probability space $X = BL(K)$, $\mathcal{A} = \sigma$-algebra induced by the Borel sets of $\| \cdot \|_{\sup}$ (or equivalently evaluations at points of K). Let $P = \mathcal{L}(X_1)$, $F(h) := \|h\|_{BL}, h \in X$. Then for any $s \in K$,

$$E \|X_1\|_{\sup}^2 \leq 2E \left(X_1(s)^2 + M(\omega)^2 \sup_t d(s, t)^2\right) < \infty$$

and $E \|X_1\|_L^2 \leq E M(\omega)^2 < \infty$, so $E F^2 < \infty$ (note that F is measurable, $F \in L^2(X, \mathcal{A}, P)$).

Let $\mathcal{G}$ be the collection of functions $\delta_t/F : h \mapsto h(t)/F(h), h \in X$, where we replace $h(t)/F(h)$ by 0 if $h \equiv 0$ in $BL(K)$, and t runs through K. Then $|g(h)| \leq 1$ for all $g \in \mathcal{G}$ and $h \in X$. Let

$$\mathcal{F} := \{Fg : g \in \mathcal{G}\} = \{\delta_t := (h \to h(t)) : t \in K\}.$$

For any $s, t \in K$ and $h \in X$, $|(\delta_s/F - \delta_t/F)(h)| \le d(s, t)$. Then by Proposition 6.3, for $0 < x \le 1$,

$$D_F^{(2)}(x, \mathcal{F}) \le D(x, \mathcal{G}, d_{\sup}) \le D(x, K, d).$$

Thus (6.12) holds and Theorem 6.15 applies to give that $\mathcal{F}$ is a Donsker class. Since $\mathcal{F}$ is the set of evaluations at points of K, uniform convergence over $\mathcal{F}$ (as in the definition of Donsker class) implies uniform convergence of functions on K. Since $BL(K) \subset C(K)$, which is complete for uniform convergence, the limiting Gaussian process G_P for our Donsker class must also have sample functions in $C(K)$ (almost surely). Since $C(K)$ is separable, the laws $\mathcal{L}\left(n^{-1/2}(X_1 + \cdots + X_n)\right)$ are defined on all Borel sets of $C(K)$ and converge to $\mathcal{L}(G_P)$. $\qquad\square$

Remark. In the situation of Theorem 6.16, K may be given originally with a metric e. The metric d may be chosen, perhaps as a function $d = f(e)$, where $f(x)$ may approach 0 slowly as $x \downarrow 0$, e.g., $f(x) = x^\varepsilon$ for $\varepsilon > 0$ or $f(x) = 1/\max(|\log x|, 2)$. Thus one can increase the possibilities for obtaining the Lipschitz property of X_1 with respect to d, so long as (6.13) holds for d.

Now to prove Theorem 6.15 we first have:

Lemma 6.17 *Let $(X, \mathcal{A}, P)$ be a probability space, $F \in \mathcal{L}^2(X, \mathcal{A}, P)$ and $\mathcal{F} \subset \mathcal{L}^2(X, \mathcal{A}, P)$ having F as an envelope function. Let $H := 4F^2$ and $\mathcal{H} := \{(f - g)^2 : f, g \in \mathcal{F}\}$. Then $0 \le \varphi(x) \le H(x)$ for all $\varphi \in \mathcal{H}$ and $x \in X$, and for any $\delta > 0$,*

$$D_H^{(1)}(4\delta, \mathcal{H}) \le D_F^{(2)}(\delta, \mathcal{F})^2.$$

Proof. Clearly $0 \le \varphi \le H$ for $\varphi \in \mathcal{H}$. Given any $\gamma \in \Gamma$, choose $m \le D_F^{(2)}(\delta, \mathcal{F})$ and $f_1, \dots, f_m \in \mathcal{F}$ such that (6.1) holds with $p = 2$. For any $f, g \in \mathcal{F}$, take i and j such that

$$\max\left(\gamma((f - f_i)^2), \gamma((g - f_j)^2)\right) \le \delta^2 \gamma(F^2).$$

Then by the Cauchy–Bunyakovsky–Schwarz inequality,

$$
\begin{aligned}
\gamma\left((f - g)^2 - (f_i - f_j)^2\right) &= \gamma\left((f - g - f_i + f_j)(f - g + f_i - f_j)\right) \\
&\le \gamma\left((f - f_i - (g - f_j))^2\right)^{1/2} 4\gamma\left(F^2\right)^{1/2} \\
&\le 8\delta\gamma\left(F^2\right) = 2\delta\gamma(H).
\end{aligned}
$$

Thus letting $h_{k(i,j)} := f_i - f_j$ where $k(i, j) := mi - m + j$, $i, j = 1, \dots, m$, we get an approximation of all functions in $\mathcal{H}$, in the $\mathcal{L}^1(\gamma)$ norm, within $2\delta\gamma(H)$, by functions h_k^2, $k = 1, \dots, m^2$, which implies the Lemma. $\qquad\square$

Lemma 6.17 gives in particular that if $D_F^{(2)}(\delta, \mathcal{F}) < \infty$ for all $\delta > 0$, then $D_H^{(1)}(\varepsilon, \mathcal{H}) < \infty$ for all $\varepsilon > 0$. Thus hypothesis (6.12) lets us apply Theorem 6.7, with $\mathcal{F}$ there $= \mathcal{H}$.

Proposition 6.18 *Let* $(X, \mathcal{A}, P)$ *be a probability space where* $(X, \mathcal{A})$ *is separable and* $\mathcal{F}$ *an image admissible Suslin class of measurable real-valued functions on* X *with an envelope function* $F \in \mathcal{L}^2(X, \mathcal{A}, P)$. *If* $D_F^{(2)}(\beta, \mathcal{F}) < \infty$ *for all* $\beta > 0$, *then* $\mathcal{F}$ *is totally bounded in* $\mathcal{L}^2(X, \mathcal{A}, P)$.

Proof. We may assume $\int F^2 dP > 0$, as otherwise $\mathcal{F}$ is quite totally bounded. The class $\{(f - g)^2 : f, g \in \mathcal{F}\}$ is image admissible Suslin by Theorem 5.26. By Lemma 6.17 and Theorem 6.7 we have

$$\sup\left\{\left|(P_n - P)\left((f - g)^2\right)\right| : f, g \in \mathcal{F}\right\} \to 0 \quad \text{a.s., } n \to \infty. \quad (6.14)$$

Also, as $n \to \infty$, $\int F^2 dP_{2n} \to \int F^2 dP$ a.s. Given $\varepsilon > 0$ take n_1 large enough and a value of P_{2n}, $n \geq n_1$, such that $\int F^2 dP_{2n} < 2\int F^2 dP$ and

$$\sup\left\{\left|(P_{2n} - P)\left((f - g)^2\right)\right| : f, g \in \mathcal{F}\right\} < \varepsilon/2.$$

Take $0 < \beta < \left(\varepsilon/(4P(F^2))\right)^{1/2}$ and choose $f_1, \ldots, f_m \in \mathcal{F}$ to satisfy (6.1) for $\delta = \beta$, $p = 2$, and $\gamma = P_{2n}$. Then for each $f \in \mathcal{F}$ we have for some j

$$\int (f - f_j)^2 dP < \frac{\varepsilon}{2} + \int (f - f_j)^2 dP_{2n} \leq \frac{\varepsilon}{2} + \beta^2 \int F^2 dP_{2n} < \varepsilon. \quad \square$$

To continue the proof of Theorem 6.15, assume as we may that $P(F^2) > 0$. Consider the $\mathcal{L}^2$ pseudometric $e_P(f, g) := (P((f - g)^2))^{1/2}$ for $f, g \in \mathcal{L}^2(P)$. In Theorem 3.34(III) for $\tau = e_P$, we have proved the total boundedness. It remains to check the asymptotic equicontinuity condition. Let $0 < \varepsilon < 1$. For any δ with $0 < \delta < \varepsilon/2$, let

$$\mathcal{F}(\delta) := \{f - g : f, g \in \mathcal{F}, e_P(f, g) < \delta\}.$$

Then $\mathcal{F}(\delta)$ is image admissible Suslin for each $\delta > 0$ via some $(\mathcal{Y}, \beta, \tau)$ by Theorem 5.26 and Corollary 5.18. It will be enough to show that for δ small enough and n_0 large enough, and any $n \geq n_0$,

$$\Pr(\|\nu_n\|_{\mathcal{F}(\delta)} > 6\varepsilon) < 5\varepsilon. \quad (6.15)$$

By the symmetrization lemma 6.5 applied to $\mathcal{F}(\delta)$, $\eta = 3\varepsilon$, and $\zeta = \delta \leq \varepsilon/2$, we have

$$\Pr(\|\nu_n\|_{\mathcal{F}(\delta)} > 6\varepsilon) \leq \frac{36}{35}\Pr(\|\nu_n^0\|_{\mathcal{F}(\delta)} > 3\varepsilon). \quad (6.16)$$

Thus to prove (6.15), given $\varepsilon > 0$, it will suffice to find a $\delta > 0$ small enough so that for $n \geq n_0$ large enough,

$$\Pr(\|\nu_n^0\|_{\mathcal{F}(\delta)} > 3\varepsilon) \leq 4\varepsilon. \quad (6.17)$$

A suitable value of δ will be found only late in the proof, in (6.30). For $x := \{x_j\}_{j=1}^{2n} \in X^{2n}$, $\omega := \{\varepsilon_j\}_{j=1}^{n} \in \mathcal{E} := \{-1, 1\}^n$, and $y \in \mathcal{Y}$, $(x, \omega, y) \mapsto v_n^0(\tau(y))$ is jointly measurable by Proposition 5.20. It follows that $(x, \omega) \mapsto \|v_n^0\|_{\mathcal{F}(\delta)}$ is jointly completion measurable by Corollary 5.25. Thus the event $A = A_{n,\delta,\varepsilon}$ in (6.17) is jointly completion measurable. For any real-valued function g on X let $\|g\|_{2n} := \left(P_{2n}(g^2)\right)^{1/2}$. Let $B_n := \left\{\|F\|_{2n}^2 \le 2P(F^2)\right\}$. Then

$$\Pr(B_n) > 1 - \varepsilon \ \text{ for } \ n \ge n_2 = n_2(\varepsilon) \tag{6.18}$$

for some n_2 not depending on δ. By (6.14), which applies, we have given $\delta > 0$ and $\varepsilon > 0$ for $n \ge n_1(\delta, \varepsilon)$ large enough

$$\sup\left\{\left|(P_n - P)\left((f - g)^2\right)\right| : f, g \in \mathcal{F}\right\} < \delta^2 \tag{6.19}$$

on an event $C_n = C_n(\delta, \varepsilon)$ with $\Pr(C_n) > 1 - \varepsilon$. On C_n, we will have for all $f, g \in \mathcal{F}$ with $f - g \in \mathcal{F}_\delta$,

$$P_{2n}((f - g)^2) < 2\delta^2. \tag{6.20}$$

Let $X' := X'_n := B_n \cap C_n$. Similarly as in (6.4), we have

$$\Pr(A_{n,\delta,\varepsilon}) \le 2\varepsilon + \int_{X'_n} \int_{\mathcal{E}} 1_A(x, \omega) d E_n(\omega) d P^{2n}(x). \tag{6.21}$$

Again an upper bound for the inner integral (in fact, a finite sum over 2^n points)

$$\int_{\mathcal{E}} 1_A(x, \omega) d E_n(\omega) \tag{6.22}$$

will be sought. In doing this, we have a fixed $x \in X^{2n}$. Some choices depending on x will be made. These need not be made as measurable functions of x, as long as the eventual upper bound for (6.22) is measurable in x. Let $\delta_i := 2^{-i}$, $i = 1, 2, \ldots$. For our fixed x, choose finite subsets $\mathcal{F}(1, x), \mathcal{F}(2, x), \ldots$ of $\mathcal{F}$ such that for all i, and $f \in \mathcal{F}$,

$$\min\{\|f - g\|_{2n} : g \in \mathcal{F}(i, x)\} \le \delta_i\|F\|_{2n}, \tag{6.23}$$

with $k(i, x) := \mathrm{card}(\mathcal{F}(i, x)) \le D_F^{(2)}(\delta_i, \mathcal{F})$. We can write

$$\mathcal{F}(i, x) = \left\{g_{i,1}^{(x)}, \ldots, g_{i,k(i,x)}^{(x)}\right\}.$$

For each $f \in \mathcal{F}$, let $f_i := g := g_{im} \in \mathcal{F}(i, x)$ achieve the minimum in (6.23), with m minimal in case of a tie. Now $\|f_i - f\|_{2n} \to 0$ as $i \to \infty$ by (6.23), and for any fixed r, $f - f_r = \sum_{r<j<\infty} f_j - f_{j-1}$ pointwise on $S = \{x_1, \ldots, x_{2n}\}$.

Let $H_j := \log D_F^{(2)}(2^{-j}, \mathcal{F})$, $j = 1, 2, \ldots$. The integral condition (6.12) is equivalent to

$$\sum_{j=1}^{\infty} 2^{-j} H_j^{1/2} < \infty. \tag{6.24}$$

For all x and j, card $\mathcal{F}(j, x) \leq \exp(H_j)$. Let

$$\eta_j := \max \left(j\delta_j, (576 P(F^2)\delta_j^2 H_j)^{1/2} \right) > 0.$$

Then

$$\sum_{j \geq 1} \eta_j < \infty, \tag{6.25}$$

$$\eta_j^2 \geq 576 P(F^2)\delta_j^2 H_j, \quad \text{and} \tag{6.26}$$

$$\sum_{j \geq 1} \exp\left(-\eta_j^2/(288\delta_j^2 P(F^2))\right) \leq \sum_{j \geq 1} \exp\left(-j^2/(288 P(F^2))\right) < \infty. \tag{6.27}$$

It follows that, with Pr_x denoting probability with respect to $E_n(\omega)$ for our fixed $x \in X_n' \subset X^{2n}$,

$$\mathrm{Pr}_x \left\{ \sup_{f \in \mathcal{F}} |v_n^0(f - f_r)| > \sum_{j > r} \eta_j \right\}$$

$$\leq \sum_{j > r} \mathrm{Pr}_x \left\{ \sup_{f \in \mathcal{F}} |v_n^0(f_j - f_{j-1})| > \eta_j \right\}$$

$$\leq \sum_{j > r} \exp(H_j)\exp(H_{j-1}) \sup_{f \in \mathcal{F}} \mathrm{Pr}_x \left\{ |v_n^0(f_j - f_{j-1})| > \eta_j \right\}, \tag{6.28}$$

since there are $\exp(H_i)$ possibilities for f_i, $i = j - 1, j$. For a fixed j and f let

$$z_i := (f_j - f_{j-1})(x_{2i}) - (f_j - f_{j-1})(x_{2i-1}).$$

Then by (6.3),

$$v_n^0(f_j - f_{j-1}) = n^{-1/2} \sum_{i=1}^{n} \varepsilon_i z_i$$

for the Rademacher variables ε_i. By an inequality of Hoeffding (Proposition 1.12 above)

$$\mathrm{Pr}_x \left\{ n^{-1/2} \left| \sum_{i=1}^{n} \varepsilon_i z_i \right| > \eta_j \right\} \leq 2\exp\left(-\tfrac{1}{2}n\eta_j^2 / \sum_{i=1}^{n} z_i^2\right).$$

On B_n we have

$$\sum_{i=1}^{n} z_i^2 \leq 4n \int (f_j - f_{j-1})^2 dP_{2n} \leq 4n \left(\|f - f_j\|_{2n} + \|f - f_{j-1}\|_{2n} \right)^2$$

$$\leq 4n \|F\|_{2n}^2 (\delta_j + \delta_{j-1})^2 \leq 72n\delta_j^2 P(F^2)$$

by (6.23) and the few lines after it. Then the last sum in (6.28) is less than

$$\sum_{j>r} \exp(2H_j) 2 \exp\left(-\eta_j^2/(144\delta_j^2 P(F^2))\right)$$

$$\leq 2 \sum_{j>r} \exp\left(-\eta_j^2/(288\delta_j^2 P(F^2))\right) \tag{6.29}$$

by (6.26). There will be four conditions for r to be sufficiently large, expressed as $r \geq R_j$ for positive integers R_j. None of these conditions will depend on n. The first is that $r \geq R_1$ large enough so that the expression in (6.29) is $< \varepsilon$, which exists by (6.27). The second is that $r \geq R_2 > R_1$ where R_2 is large enough so that $\sum_{j>r} \eta_j < \varepsilon$, which exists by (6.25). The third is that $r \geq R_3 > R_2$ for R_3 large enough so that $\varepsilon^2 \geq (256H_r + 1)\delta_r^2 P(F^2)$, which exists since $H_r\delta_r^2 \to 0$ as $r \to \infty$ by (6.24). The fourth is that $r \geq R_4 \geq R_3$ large enough so that $2\exp\left(-\varepsilon^2/(128\delta_r^2 P(F^2))\right) < \varepsilon$. Now let for any $r \geq R_4$, specifically $r = R_4$,

$$\delta := \delta_r P(F^2)^{1/2}/2. \tag{6.30}$$

Then $r \geq R_3$ implies that $\delta^2 = \delta_r^2 P(F^2)/4 \leq \varepsilon^2/4$ and so $\delta \leq \varepsilon/2$ as needed before (6.16). Take any $n \geq \max(n_1(\delta, \varepsilon), n_2(\varepsilon))$ for $n_1(\delta, \varepsilon)$ as just before (6.19) and $n_2(\varepsilon)$ as in (6.18). We will have the events B_n and $C_n = C_n(\delta, \varepsilon)$ such that $\Pr(B_n) > 1 - \varepsilon$ and $\Pr(C_n) > 1 - \varepsilon$. Thus defining $X_n' := B_n \cap C_n$ we have

$$\Pr(X_n') > 1 - 2\varepsilon. \tag{6.31}$$

We obtain almost surely on X_n' that, by (6.28) and (6.29) and since $r > R_2 > R_1$,

$$\Pr_x\left\{\sup_{f \in \mathcal{F}} \left|\nu_n^0(f - f_r)\right| > \varepsilon\right\} < \varepsilon. \tag{6.32}$$

Next, almost surely on X_n', by (6.20) we have for all $f, g \in \mathcal{F}$ with $f - g \in \mathcal{F}(\delta)$, $\|f - g\|_{2n}^2 < 2\delta^2 = \delta_r^2 P(F^2)/2$, so

$$\|f - g\|_{2n} < \delta_r\sqrt{P(F^2)/2} < \delta_r\sqrt{P(F^2)}.$$

It follows that on X_n', using (6.23)

$$\begin{aligned}
\|f_r - g_r\|_{2n} &\leq \|f_r - f\|_{2n} + \|f - g\|_{2n} + \|g - g_r\|_{2n} \\
&< \delta_r P(F^2)^{1/2} + 2\delta_r\|F\|_{2n} \\
&\leq 4\delta_r(P(F^2))^{1/2}. \tag{6.33}
\end{aligned}$$

Then by the same Hoeffding inequality Proposition 1.12, on X'_n,

$$\Pr_x \left\{ \sup \left\{ |\nu_n^0(f_r - g_r)| : f, g \in \mathcal{F}, f - g \in \mathcal{F}(\delta) \right\} > \varepsilon \right\}$$
$$\leq (\text{card } \mathcal{F}(r, x))^2 \cdot 2 \exp \left(-\varepsilon^2 / [64\delta_r^2 P(F^2)] \right)$$
$$\leq 2 \exp \left(2H_r - \varepsilon^2 / (64\delta_r^2 P(F^2)) \right) \tag{6.34}$$
$$\leq 2 \exp \left(-\varepsilon^2 / (128\delta_r^2 P(F^2)) \right) \quad \text{with} \quad \varepsilon^2 \geq 256 H_r \delta_r^2 P(F^2)$$
$$< \varepsilon$$

since $r \geq R_4 \geq R_3$. We have

$$|\nu_n^0(f - g)| \leq |\nu_n^0(f - f_r)| + |\nu_n^0(f_r - g_r)| + |\nu_n^0(g - g_r)|.$$

Combining (6.32) and (6.34) we get that for $x \in X'_n$,

$$\Pr_x(\sup\{|\nu_n^0(h)| : h \in \mathcal{F}(\delta)\} > 3\varepsilon) < 2\varepsilon.$$

So for $x \in X'_n$, the integral in (6.22) is bounded above by 2ε. Thus in (6.21), $\Pr(A_{n,\delta,\varepsilon}) \leq 4\varepsilon$, i.e. (6.17), and so (6.15) holds, proving Theorem 6.15. $\square$

Corollary 6.19 (Pollard) *Let $(X, \mathcal{A}, P)$ be a probability space, and let $\mathcal{F}$ be an image admissible Suslin Vapnik–Červonenkis subgraph class of functions with envelope $F \in \mathcal{L}^2(X, \mathcal{A}, P)$. Then $\mathcal{F}$ is a Donsker class for P.*

Proof. This follows from Theorem 6.15 and Theorem 4.52 for $p = 2$. $\square$

Corollary 6.20 *Let $(X, \mathcal{A}, P)$ be a probability space, $F \in \mathcal{L}^2(X, \mathcal{A}, P)$, and $\mathcal{F} = \{F1_C : C \in \mathcal{C}\}$ where $\mathcal{C}$ is an image admissible Suslin Vapnik–Červonenkis class of sets. Then $\mathcal{F}$ is a Donsker class for P.*

Proof. Since F is measurable, the image admissible Suslin property of $\mathcal{F}$ follows from that of $\mathcal{C}$. By Theorem 6.1 for $p = 2$, (6.12) holds and Theorem 6.15 applies. $\square$

For VC major and hull classes we have the following:

Theorem 6.21 *Let $(X, \mathcal{A})$ be a measurable space and $\mathcal{C} \subset \mathcal{A}$ a VC class of sets. Assume $\mathcal{C}$ is image admissible Suslin. Let $\mathcal{F}$ be a VC hull class of functions for $\mathcal{C}$, such as in particular a uniformly bounded VC major class for $\mathcal{C}$. Then $\mathcal{F}$ is a Donsker class for every probability measure P on $(X, \mathcal{A})$.*

Proof. A uniformly bounded VC major class is VC hull by Theorem 4.51. The class $\mathcal{C}$, or in other words $\mathcal{F} = \{1_A : A \in \mathcal{C}\}$, is Donsker for every P by Corollary 6.20 with $F \equiv 1$. Then each $\overline{H}_s(\mathcal{F}, M)$ is also Donsker for every P by Theorem 3.41, which gives the result. $\square$

One might ask whether in Corollary 6.19, "VC subgraph" can be replaced by "VC major" for any $F \in \mathcal{L}^2(P)$, as Corollary 6.20 might suggest. It cannot. A somewhat stronger condition on the envelope F is needed, as the following will show.

Proposition 6.22 *If a probability space X has a decomposition into disjoint measurable sets A_k such that*

$$\sum_{k \geq 1} 2^k P(A_k)^{1/2} = +\infty, \tag{6.35}$$

then there exists a VC major class $\mathcal{F}$ with envelope $F := \sum_{k \geq 1} 2^k 1_{A_k}$ which is not P-pregaussian, even though F may be in $\mathcal{L}^2(P)$.

Proof. If $P(A_k) = C/(4^k k^\alpha)$ where $1 < \alpha \leq 2$ and $C = C_\alpha$ is the suitable normalizing constant, then it is easy to check that $F \in \mathcal{L}^2(P)$ but (6.35) holds. Consider

$$\mathcal{F} := \left\{ \sum_{k=1}^{M} \left(1 - \frac{\sigma_k}{3}\right) 2^k 1_{A_k} : \sigma_k = 0 \text{ or } 1 \text{ for } k \geq 1, \ M = 1, 2, \dots \right\}.$$

Since $4/3 > 1$, the sets $\{x : f(x) > c\}$, $f \in \mathcal{F}$, $c \in \mathbb{R}$, are all of the form $B_{N,M} := \bigcup_{k=N}^{M} A_k$. Thus they form a VC class $\mathcal{C}$ with $S(\mathcal{C}) \leq 2$, since if $x \in A_i$, $y \in A_j$ and $z \in A_k$ with $i \leq j \leq k$, then any set $B_{N,M}$ containing x and z must also contain y. So $\mathcal{F}$ is a VC major class.

For any function f let $f_+ := \max(f, 0)$, $f_- := -\min(f, 0)$. Let

$$T_N := \sup_{\sigma_k = 0, 1} \sum_{k=1}^{N} (1 - \sigma_k/3) 2^k G_P(A_k)$$

and $T := \sup_N T_N$. Then $E|T_1| < \infty$ (note also that $ET_1 > 0$) implies that ET is well-defined (possibly $+\infty$), and for each N, $ET_N \leq ET \leq E\|G_P\|_{\mathcal{F}}$. Now

$$
\begin{aligned}
ET_N &= \sum_{k=1}^{N} 2^k E G_P(A_k)_+ - \frac{2}{3} \sum_{k=1}^{N} 2^k E G_P(A_k)_- \\
&= \frac{1}{2} \sum_{k=1}^{N} 2^k E|G_P(A_k)| - \frac{1}{3} \sum_{k=1}^{N} 2^k E|G_P(A_k)| = \frac{1}{6} \sum_{k=1}^{N} 2^k E|G_P(A_k)| \\
&= \frac{1}{6} \sum_{k=1}^{N} 2^k P^{1/2}(A_k)(1 - P(A_k))^{1/2} (2/\pi)^{1/2} \to \infty
\end{aligned}
$$

as $N \to \infty$ since the A_k are disjoint, so $1 - P(A_k) \to 1$. So $\mathcal{F}$ is not P-pregaussian, by Theorem 2.47(a) or by Lemma 2.10, since the set of all real sequences $\{x_j\}_{j=1}^{\infty}$ with the smallest σ-algebra making each x_j measurable is a measurable vector space. $\qquad\square$

Let F be a measurable, finite-valued function on $(X, \mathcal{A})$ with $F \geq 1$. Denote by M_F the set of all functions of the form $\phi \circ F$, where $\phi : \mathbb{R}^+ \to \mathbb{R}^+$ is an arbitrary nondecreasing function such that $\phi(u) \leq u$ for all $u \in \mathbb{R}^+$. Then M_F is a VC major class for the class $\mathcal{C}$ of sets $\{x : F(x) \geq c\}$ or $\{x : F(x) > c\}$ for $1 \leq c < \infty$, which are linearly ordered by inclusion, and so $S(\mathcal{C}) = 1$. Clearly $\mathcal{F}$ has envelope F.

The Lorentz space $\mathcal{L}_{p,q}$ is defined (e.g., Ledoux and Talagrand 1986) as the space of real random variables η such that $\int_0^\infty (t^p \Pr(|\eta| > t))^{q/p} dt/t < \infty$. Thus $\mathcal{L}_{2,1}$ is the space of the η such that $\int_0^\infty \Pr(|\eta| > t)^{1/2} dt < \infty$. Dudley and Koltchinskii (1994) proved the following, stated here without proof.

Proposition 6.23 *For any image admissible Suslin VC major class $\mathcal{F}$ with envelope F, including M_F, $\mathcal{F}$ is a P-Donsker class if and only if $F \in \mathcal{L}_{2,1}(P)$.*

In the proof it is shown that if $F \notin \mathcal{L}_{2,1}(P)$, then

$$\sum_{k=1}^\infty 2^{k-1} P^{1/2}(A_k) = +\infty$$

where $A_k := \{2^{k-1} \leq F < 2^k\}$, $k \geq 1$.

6.4 Necessary Conditions for Limit Theorems

Theorems 6.1 and 6.15 imply that every class $\mathcal{C} \subset \mathcal{A}$ with $S(\mathcal{C}) < +\infty$, and which is image admissible Suslin, is a Donsker class, for an *arbitrary law* P on $\mathcal{A}$. In this section it will be shown that to obtain, for all P, such a central limit theorem (or even the pregaussian property), for a class $\mathcal{C}$ of sets, the condition $S(\mathcal{C}) < +\infty$ is necessary. Then it will be noted that some measurability, beyond that of $\| P_n - P \|_\mathcal{C}$, is needed to obtain even a law of large numbers for $S(\mathcal{C}) < +\infty$ (Corollary 6.8). Lastly, it will be shown that $S(\mathcal{C}) < \infty$ is necessary so that $\| P_n - P \|_\mathcal{C} \to 0$ in outer probability as $n \to \infty$, uniformly in P.

Theorem 6.24 *Let $(X, \mathcal{A})$ be a measurable space and $\mathcal{C} \subset \mathcal{A}$. Then $S(\mathcal{C}) < +\infty$ if and only if for all laws P on $\mathcal{A}$, $\{1_A : A \in \mathcal{C}\}$ is a pregaussian class (as defined in Section 3.1).*

Proof. "Only if" was proved in Corollary 4.48. To prove "if," suppose $S(\mathcal{C}) = +\infty$. Then for each $n \geq 1$, $\mathcal{C}$ shatters some set F_n with $\text{card}(F_n) = 4^n$. Let $G_n := F_n \setminus \bigcup_{j<n} F_j$. Then the sets G_n are disjoint, $\text{card}(G_n) > 2^n$, and $\mathcal{C}$ shatters G_n. Take $E_n \subset G_n$ with $\text{card}(E_n) = 2^n$. Then the E_n are disjoint and shattered by $\mathcal{C}$. Some countable subset $\mathcal{D} \subset \mathcal{C}$ shatters every E_n. We have

$$\sum_{n=1}^\infty \frac{1}{n(n+1)} = \sum_{n=1}^\infty \frac{1}{n} - \frac{1}{n+1} = 1.$$

Let P be the law on $\bigcup_{n=1}^{\infty} E_n$ with $P(\{x\}) = 1/[2^n n(n+1))]$ for each $x \in E_n$.

Given n, for the isonormal process W_P on $\mathcal{L}^2(P)$, for each $C \in \mathcal{D}$, $W_P(C) = W_P(C \cap E_n) + W_P(C \setminus E_n)$. For $0 < K < \infty$, define the events

$$\mathcal{E}_1 := \{|W_P(B)| \leq 2K \text{ for all } B \subset E_n\},$$

$$\mathcal{E}_2 := \{|W_P(B)| > 2K \text{ for some } B \subset E_n, \text{ and for all such } B$$

$$\text{and all } C \in \mathcal{D} \text{ with } C \cap E_n = B, \ |W_P(C \setminus E_n)| > K\}.$$

Then $\{|W_P(C)| < K \text{ for all } C \in \mathcal{D}\} \subset \mathcal{E}_1 \cup \mathcal{E}_2$. Let $S_n := \sum_{x \in E_n} |W_P(\{x\})|$. Then

$$\sup\{|W_P(B)| : \ B \subset E_n\} \geq S_n/2,$$

so $\mathcal{E}_1 \subset \{S_n \leq 4K\}$. For each $x \in E_n$, $W_P(\{x\})$ has a Gaussian law with mean 0 and variance $1/(n(n+1)2^n) =: \sigma_n^2$, so $E|W_P(\{x\})| = (2/\pi)^{1/2}\sigma_n$ and $\text{Var}(|W_P(\{x\})|) = \sigma_n^2(1 - \frac{2}{\pi})$. Thus $ES_n = 2^n(2/\pi)^{1/2}\sigma_n$. Since W_P has independent values on disjoint sets, $\text{Var}(S_n) = (1 - \frac{2}{\pi})/[n(n+1)]$. For n large, $ES_n \geq 4K$, and then by Chebyshev's inequality

$$\Pr\{S_n \leq 4K\} \leq \Pr\{|S_n - ES_n| \geq ES_n - 4K\} < \frac{1}{n^2(ES_n - 4K)^2}$$

$$\leq 1 \bigg/ \left(\left(2^n \cdot \frac{2n}{\pi(n+1)}\right)^{1/2} - 4Kn\right)^2 =: f(n, K) \to 0 \text{ as } n \to \infty.$$

Turning to $\mathcal{E}_2$, let $t(n) := 2^{2^n}$. Let the subsets of E_n be $B_1, \ldots, B_{t(n)}$. Let $M_0 := \emptyset$ and recursively for $j \geq 1$, $M_j := \{|W_P(B_j)| > 2K\} \setminus \bigcup_{0 \leq i < j} M_i$. Let $D_j := M_j \cap A_{j,\mathcal{D},K}$ where $A_{j,\mathcal{D},K}$ is the event that for all $C \in \mathcal{D}$ such that $C \cap E_n = B_j$, we have $|W_P(C \setminus E_n)| > K$.

For any set A, $W_P(A)$ has a Gaussian distribution with mean 0 and variance $P(A)$. Let $\Phi(x) := (2\pi)^{-1/2} \int_{-\infty}^{x} \exp(-t^2/2)dt$. Then since W_P has independent values on disjoint sets,

$$\Pr(D_j) = \Pr(M_j \cap A_{j,\mathcal{D},K})$$

$$\leq \Pr(M_j) \cdot 2\Phi(-K).$$

Now $\mathcal{E}_2 \subset \bigcup_{1 \leq j \leq t(n)} D_j$, so since the sets M_j are disjoint,

$$\Pr(\mathcal{E}_2) \leq \sum_{1 \leq j \leq t(n)} \Pr(M_j) \cdot 2\Phi(-K) \leq 2\Phi(-K).$$

Hence

$$\Pr(|W_P(C)| < K \text{ for all } C \in \mathcal{D}) \leq f(n, K) + 2\Phi(-K).$$

If we let $n \to +\infty$, then $K \to +\infty$, we see that W_P is a.s. unbounded on $\mathcal{D} \subset \mathcal{C}$, and note that $G_P(\cdot)$ can be written as $W_P(\cdot) - P(\cdot)W_P(1)$, so G_P is a.s. unbounded on $\mathcal{C}$.   □

Remark. Now recall the example at the beginning of Chapter 5, where $\mathcal{C}$ is the collection of all countable initial segments of an uncountable well-ordered set $(X, \preceq)$ and P is a continuous law on some σ-algebra $\mathcal{A}$ containing all countable subsets of X. Then $S(\mathcal{C}) = 1$ but $\sup_{A \in \mathcal{C}} |(P_n - P)(A)| \equiv 1$ for all n. Thus the latter random variable is measurable. For this class the weak law of large numbers, hence the strong law and central limit theorem, all fail as badly as possible. This shows that in Theorem 6.7 and Corollary 6.8, the "image admissible Suslin" condition cannot simply be removed, nor replaced by simple measurability of random quantities appearing in the statements of the results. Further, for all $A \in \mathcal{C}$, $1_A = 0$ a.s. P, so vanishing a.s. (P) even with $S(\mathcal{C}) = 1$ does not imply a law of large numbers.

Remark 6.25 *If X is a countably infinite set and $\mathcal{A} = 2^X$, then for an arbitrary law P on $\mathcal{A}$, $\lim_{n \to \infty} \sup_{A \in \mathcal{A}} |(P_n - P)(A)| = 0$ a.s. (see Problem 8 at the end of this chapter). But $S(\mathcal{A}) = +\infty$, so the hypothesis of Theorem 6.24 cannot be weakened to a law of large numbers for all P.*

Next it will be seen that the Vapnik–Červonenkis property is also necessary for a law of large numbers to hold uniformly in P or that there exist an estimator of P based on $X_1, \dots, X_n$ (which might or might not equal P_n) converging to P uniformly over $\mathcal{C}$ and uniformly in P. Here are some definitions.

Let $(X, \mathcal{B})$ be a measurable space. Let its n-fold Cartesian product be $(X^n, \mathcal{B}^n)$. Let $\mathcal{P}$ be the class of all probability measures on $(X, \mathcal{B})$. Let $\mathcal{C} \subset \mathcal{B}$ be any collection of measurable sets. A real-valued function T_n on $X^n \times \mathcal{C}$ will be called a $\mathcal{C}$-*estimator* if it is a stochastic process indexed by $\mathcal{C}$, in other words, for each $A \in \mathcal{C}$, $x \mapsto T_n(x, A)$ is measurable on X^n. A $\mathcal{C}$-estimator T_n will be called an *estimator* if for each x, there is a probability measure μ on $\mathcal{A}$ which equals $T_n(x, \cdot)$ on $\mathcal{C}$.

For any probability measure P on $(X, \mathcal{B})$ and product law P^n on $(X^n, \mathcal{B}^n)$, for $x = (X_1, \dots, X_n)$ so that X_i are i.i.d. (P), we would like T_n to be a good approximation to P with probability $\to 1$ as $n \to \infty$. The goodness of the approximation will be measured by the *loss function* $L(T_n, P) := \|T_n - P\|_{\mathcal{C}}$. From it we get the *risk* $r(T_n, P, \mathcal{C}) := E_P L(T_n, P)^*$ where E_P denotes expectation with respect to P^n. For any class $\mathcal{Q}$ of laws, let $r(T_n, \mathcal{Q}, \mathcal{C}) := \sup\{r(T_n, P, \mathcal{C}) : P \in \mathcal{Q}\}$, and let $r_n(\mathcal{Q}, \mathcal{C})$ be the *minimax risk*, i.e. the infimum of $r(T_n, \mathcal{Q}, \mathcal{C})$ over all $\mathcal{C}$-estimators T_n.

In finding minimax risks for C-estimators we can assume T_n takes values in $[0, 1]$ since $\max(0, \min(T_n, 1))$ will clearly have risks no larger than those of T_n.

For any C and Q, clearly $0 \leq r_n(Q, C) \leq 1$ and r_n is nonincreasing in n. The following theorem holds for C-estimators and so *a fortiori* for estimators, whose values are probability measures. The following fact is mainly due to P. Assouad (see the Notes to this section).

Theorem 6.26 *Let* $\mathcal{P}$ *be the class of all probability measures on a sample space* $(X, \mathcal{B})$ *and* $C \subset \mathcal{B}$. *If the minimax risk* $r_n(\mathcal{P}, C) < 1/2$ *for some* n, *then* C *is a Vapnik–Červonenkis class.*

Proof. Suppose $S(C) = +\infty$. Given n, for any $m > n$ let F be a set with $2m$ elements, shattered by C. Let $\mathcal{D} \subset C$ be a collection of 2^{2m} sets which shatters F. Then $\|\cdot\|_C^* \geq \|\cdot\|_C \geq \|\cdot\|_\mathcal{D}$, and since $\mathcal{D}$ is finite, $\|T_n - P\|_\mathcal{D}$ will be measurable for any C-estimator T_n and any P. Let T_n be a C-estimator defined on X^n. Let π be a ("prior") probability distribution on the set (finite-dimensional simplex) of all probability laws on F, with mass $1/\binom{2m}{m}$ at each law uniformly distributed over an m-element subset of F. Then the maximum risk is bounded below by the risk for π and $\mathcal{D}$,

$$\sup_P E_P \|T_n - P\|_C^* \geq E_\pi E_P \|T_n - P\|_\mathcal{D}.$$

For each n-tuple $x = (x_1, \ldots, x_n) \in F^n$, let ran x denote the range of x, ran $x = \{x_1, \ldots, x_n\}$, and let $k = k(x)$ be the number of distinct x_i, which is the cardinality of ran x.

Let μ be the distribution of x in F^n averaged with respect to π, $\mu := \int P^n d\pi(P)$. Then for each $x \in F^n$, let $\pi(x) := \pi_x$ be the posterior distribution defined by π given x, namely, the law π_x giving mass $1/\binom{2m-k}{m-k}$ to each law uniform on a set of m elements of F including ran x. Then for any C-estimator T_n,

$$E_\pi E_P \|T_n - P\|_\mathcal{D} = E_\mu E_{\pi(x)} \|T_n - P\|_\mathcal{D}.$$

Let $\mathcal{A} := 2^F$. For each $A \in \mathcal{A}$, there is a unique $D_A \in \mathcal{D}$ with $D_A \cap F = A$. For any P in the support of π, $P(D_A) = P(A)$. Let $V_n(x, A) := T_n(x, D_A)$, an $\mathcal{A}$-estimator. Then $\|T_n - P\|_\mathcal{D} \geq \|V_n - P\|_\mathcal{A}$, and we want to find a lower bound for $E_{\pi(x)} \|V - P\|_\mathcal{A}$ for each P and x fixed where $V(A) := V_n(x, A)$. Let $\gamma(k) := \pi_x$. If P_0 is any fixed law in the support of $\gamma(k)$, and if τ is uniformly distributed over the group G of all $(2m - k)!$ permutations of F which equal the identity on ran x (and thus permute the other $2m - k$ elements of F), then $P_0 \circ \tau^{-1}$ will have distribution $\gamma(k)$. Let $\tau[A] := \{\tau(y) : y \in A\}$ for any set A. Each permutation τ defines a 1–1 transformation $A \mapsto \tau[A]$ of

$\mathcal{A}$ onto itself. Let $(V \circ \tau)(A) := V(\tau[A])$. Then we have

$$E_{\gamma(k)} \|V - P\|_{\mathcal{A}} = \frac{1}{(2m - k)!} \sum_{\tau \in G} \|V - P_0 \circ \tau^{-1}\|_{\mathcal{A}}$$

$$= \frac{1}{(2m - k)!} \sum_{\tau \in G} \|V \circ \tau - P_0\|_{\mathcal{A}},$$

which by the triangle inequality is bounded below by

$$\|[(2m - k)!^{-1} \sum_{\tau \in G} V \circ \tau] - P_0\|_{\mathcal{A}} =: \|\bar{V} - P_0\|_{\mathcal{A}}.$$

Now let $\mathcal{A}(m) := \{A \in \mathcal{A} : \operatorname{ran} x \subset A \text{ and } |A| = m\}$. For any $A, B \in \mathcal{A}(m)$ there is a $\tau \in G$ with $\tau[A] = B$. It follows that $\bar{V}$, restricted to $\mathcal{A}(m)$, is a constant. The support of P_0 is in $\mathcal{A}(m)$. For another set $C \in \mathcal{A}(m)$, $P_0(C) = k/m$. Thus

$$\|\bar{V} - P_0\|_{\mathcal{A}(m)} \geq \max\left(\left|1 - \bar{V}\right|, \left|\frac{k}{m} - \bar{V}\right|\right) \geq \frac{1}{2} - \frac{k}{2m} \geq \frac{1}{2} - \frac{n}{2m}.$$

This last bound, uniform in k, holds after integration with respect to μ. For a given n, letting $m \to \infty$, we see that the minimax risk for $\|\cdot\|_{\mathcal{C}}$ is at least $1/2$. $\qquad\square$

A corollary of Theorem 6.26, taking T_n as the empirical measure P_n, is:

Theorem 6.27 (P. Assouad) *If $(X, \mathcal{B})$ is a measurable space and $\mathcal{C} \subset \mathcal{B}$ is a uniform Glivenko–Cantelli class of sets, that is, $\sup_P E_P \|P_n - P\|_{\mathcal{C}}^* \to 0$ as $n \to \infty$, where the supremum is over all probability laws on $(X, \mathcal{B})$, then $\mathcal{C}$ is a Vapnik–Červonenkis class.*

Now it will be shown that the constant $1/2$ in Theorem 6.26 is sharp:

Proposition 6.28 *For the class $\mathcal{P}$ of all probability measures on a sample space $(X, \mathcal{B})$ there is always a $\mathcal{B}$-estimator T, not depending on n or x, with $r(T, \mathcal{P}, \mathcal{B}) \leq 1/2$, so that for any $\mathcal{Q} \subset \mathcal{P}$, $\mathcal{C} \subset \mathcal{B}$, and n we have $r_n(\mathcal{Q}, \mathcal{C}) \leq 1/2$. Moreover when $(X, \mathcal{B})$ is the unit interval $[0, 1]$ with the Borel σ-algebra, there is a class $\mathcal{C}$ which is not a Vapnik–Červonenkis class, and for which T on $\mathcal{C}$ is given by a probability measure, so T is an estimator, not only a $\mathcal{C}$-estimator.*

Proof. A $\mathcal{C}$-estimator T (not depending on n or x) is defined by $T(x, A) \equiv 1/2$ for all $A \in \mathcal{B}$. Then $r(T, \mathcal{Q}, \mathcal{C}) \leq 1/2$ for any $\mathcal{Q}$ and $\mathcal{C}$, so $r_n(\mathcal{Q}, \mathcal{C}) \leq 1/2$ for all n.

For Lebesgue measure λ on $[0, 1]$ let $\mathcal{C} := \{C_k\}_{k \geq 1}$ be an infinite sequence of sets independent with probability $1/2$; e.g., C_k is the set where the kth binary digit is 1. Recall that sets C_i, $i = 1, \ldots, k$, are called "independent in X" if for any set $J \subset \{1, \ldots, k\}$, the intersection of all the C_i for $i \in J$ and of the complements of the C_i for $i \notin J$ is nonempty. In other words, the (Boolean) algebra generated by the C_i has the maximum number, 2^k, of possible atoms. Clearly C_i are independent in $[0, 1]$. For any $m = 1, 2, \ldots$, a collection of 2^m sets independent in a set X shatters some subset with m elements by Theorem 4.49. So $\mathcal{C}$ is not a Vapnik–Červonenkis class. $\qquad \square$

Next, it will be shown that one of the hypotheses of Theorem 6.7, existence of an integrable envelope function, is essentially necessary for a law of large numbers (Glivenko–Cantelli property). This is related to the fact that that for i.i.d. real random variables $Y_1, \ldots, Y_n, \ldots, (Y_1 + \cdots Y_n)/n$ converges a.s. to a finite limit if and only if $E|Y_1| < \infty$ (RAP, Theorem 8.3.5).

Theorem 6.29 *If $(S, \mathcal{B}, P)$ is a probability space, $\mathcal{F} \subset \mathcal{L}^1(S, \mathcal{B}, P)$, $\|P\|_{\mathcal{F}} < \infty$, and if $\mathcal{F}$ is a strong Glivenko–Cantelli class for P, i.e., $\|P_n - P\|_{\mathcal{F}}^* \to 0$ a.s., then $\mathcal{F}$ has an integrable envelope function: $E\|P_1\|_{\mathcal{F}}^* < \infty$.*

Proof. We have as $n \to \infty$ $\|P_{n-1} - P\|_{\mathcal{F}}^* \to 0$ a.s., $\|\frac{n-1}{n}(P_{n-1} - P)\|_{\mathcal{F}}^* \to 0$ a.s., and $\|P/n\|_{\mathcal{F}} \to 0$, so $\|\delta_{X_n}/n\|_{\mathcal{F}}^* \to 0$ a.s. Let $F(x) := \|\delta_x\|_{\mathcal{F}} := \sup_{f \in \mathcal{F}} |f(x)|$ for $x \in S$. Let S_j be copies of S whose Cartesian product S^∞ is taken in the standard model. For each n write $S^\infty = S_n \times \Pi_{j \neq n} S_j$. Then Lemma 3.6 implies that $\|\delta_{X_n}/n\|_{\mathcal{F}}^*$, where $\|\cdot\|^*$ is with respect to P^∞ on S^∞, equals $F^*(X_n)/n$. The random variables $F^*(X_n)$ are i.i.d. Thus by the Borel–Cantelli Lemma $\sum_{n=1}^\infty P^\infty(F^*(X_n) > n) < \infty$, so $\sum_{n=1}^\infty P(F^* > n) < \infty$, which implies $EF^*(X_1) < \infty$ (RAP, Lemma 8.3.6). $\qquad \square$

Problems

1. On $[0, 1]$ with Lebesgue (uniform) law $P = U[0, 1]$, for any $\alpha > 0$ let $\mathcal{F}(\alpha) := \{k^\alpha 1_{[0,1/k]}\}_{k=1}^\infty$.

(a) Show that for $\alpha < 1$, $\|P_n - P\|_{\mathcal{F}(\alpha)}$ is measurable and $\to 0$ a.s. as $n \to \infty$. *Hint*: This follows from the Chebyshev inequality only for some values of α. Instead, find the envelope function F of $\mathcal{F}(\alpha)$ and show that it is integrable for P. Use the VC subgraph property. All subgraphs, by definition, include the set $X \times \{0\}$, in this case the x axis. Except for that, the subgraphs are rectangles with sides parallel to the axes and lower left vertex at the origin. Show that the subgraphs form a VC class of sets. (There is a relation between width and height of the rectangles, but that need not be used in showing the VC property.)

(b) Show that for $\alpha \geq 1$, $\|P_n - P\|_{\mathcal{F}(\alpha)}$ does not approach 0 in probability. (Prove this directly, without using facts from Chapter 6. Let P_n be based on

$X_1, \ldots, X_n$ i.i.d. $U[0, 1]$ of which the smallest is $X_{(1)} > 0$. What happens when $k > 1/X_{(1)}$?)

2. If C is the collection of all intersections of four closed half-spaces in $\mathbb{R}^3$ (which includes all tetrahedra), and P is any law on the Borel sets of $\mathbb{R}^3$, show that $\|P_n - P\|_C$ is measurable and goes to 0 a.s. as $n \to \infty$. *Hint*: You can assume the result of Chapter 5, Problem 8. (Any polyhedron with 4 faces in $\mathbb{R}^3$ is convex and is an intersection of four half-spaces.)

Let $Co(\mathbb{R}^k)$ be the class of all closed convex sets in $\mathbb{R}^k$.

3. Let $C := Co(\mathbb{R}^2)$ and $F := \{(0, 0), (1, 1), (2, 4), (3, 1), (4, 0)\}$. Evaluate $\Delta^C(F)$ and $k^C(F)$. *Hint*: The subsets $G \subset F$ not in $C \sqcap F$ are those whose convex hull contains a point of F not in G. What are these sets?

4. For $C = Co(\mathbb{R}^k)$ show that for any law P on $\mathbb{R}^k$ (on the Borel σ-algebra, as usual) $\sup_{A \in C}(P_n - P)(A)$ is measurable.

5. Let C be the collection of all unions of three intervals in $\mathbb{R}$. Show that $\Delta^C(F)$ and $k^C(F)$ depend only on the cardinality $m = |F|$ and evaluate each of them as a function of $m = 1, 2, \ldots$.

6. Let $\mathcal{F} := \{t^{-1/3} 1_{[0,t]} : 0 < t \leq 1\}$. Show that $\mathcal{F}$ is a Donsker class for $P = U[0, 1]$. *Hint*: What is the envelope function F of $\mathcal{F}$? As in Problem 1, use the VC subgraph property. Apply Corollary 6.19.

7. For a law P on $\mathbb{R}$ suppose $F \in \mathcal{L}^2(\mathbb{R}, P)$. Let $\mathcal{G}$ be the class of functions g on $\mathbb{R}$ with total variation ≤ 1 and $\|g\|_{\sup} \leq 1$. Show that $\{Fg : g \in \mathcal{G}\}$ is a Donsker class for P. *Hints*: Recall that a function of bounded variation is a difference of two bounded nondecreasing functions. You can assume the results of Problem 6 of Chapter 3. Look also at Theorems 3.36 and 4.51(b). The class of nondecreasing functions with a given bound (such as 1 or 2) is not a VC subgraph class (problem 4.10), but it is a VC major class. Apply Theorem 6.21.

8. Show that if X is countable and $\mathcal{A} = 2^X$ then $\mathcal{A}$ is a universal Glivenko–Cantelli class (i.e., prove the first statement in Remark 6.25). *Hint*: Given any P on X and $\varepsilon > 0$, show there is a finite set F with $P(F) > 1 - \varepsilon/k$ for a given k (such as 2 or 4). For each $x \in F$ and n large enough, $P_n(\{x\})$ is as close as we want to $P(\{x\})$. Use this to show that $P_n(X \setminus F)$ becomes less than some multiple of ε.

9. Let $P(\{k\}) := p_k := ck^{-5/4}$ for $k = 1, 2, \ldots$, with c such that $\sum_{k=1}^{\infty} p_k = 1$. Let $\mathbb{N}^+$ be the set of all positive integers and C the class of all subsets of $\mathbb{N}^+$. Show that for some $\alpha > 0$, $\|P_n - P\|_C \geq \alpha n^{-1/4}$ for all possible values of P_n, and so, C is not Donsker for P. *Hint*: For each n and possible value of P_n there is a set $A_n \subset \{1, 2, \ldots, 2n\}$ with at least n members and $P_n(A_n) = 0$. Find a

lower bound on the possible values of $P(A_n)$. A sum can be bounded below by an integral.

10. In Problem 9, if p_k are replaced by c/k^3 for the appropriate constant c, $\mathcal{C}$ is a Donsker class for P, as will follow easily from Theorem 7.9, but does this follow from Theorem 6.15? *Hints*: What is the envelope function F? Consider Proposition 6.2. Does it help to consider larger envelope functions $G > F$?

11. Let $X = H$ be a separable Hilbert space with an orthonormal basis $\{e_j\}_{j=1}^{\infty}$. Let $k_j := 1/(1 + \log j)$ and $P(k_j e_j) := P(-k_j e_j) := 3/(\pi^2 j^2)$ for $j = 1, 2, \dots$. Then P is a law on H (Chapter 2, Problem 19). Let $\mathcal{F} := \{y \in H : \|y\| \le 1\}$ with $y(x) := (x, y)$. Show that $\mathcal{F}$ is a weak Glivenko–Cantelli class, but not a Glivenko–Cantelli class. *Hint*: $\mathcal{F}$ is not order-bounded.

12. Let $\mathcal{F}$ be a universal Glivenko–Cantelli class.

(a) If $f \in \mathcal{F}$, show that f is bounded. *Hint*: Otherwise, it is not in $\mathcal{L}^1(P)$ for some law P.

(b) Show that $\mathcal{F}_0 := \{f - \inf f : f \in \mathcal{F}\}$ is uniformly bounded.

13. Let $(X, \mathcal{A})$ be a measurable space and $X = \cup_{j=1}^{\infty} A_j$ for some disjoint sets $A_j \in \mathcal{A}$. Let $\mathcal{C} \subset \mathcal{A}$ be image admissible Suslin and such that for each j, $S(\mathcal{C} \sqcap A_j) < \infty$ (but may grow arbitrarily fast with j). Show that $\mathcal{C}$ is a universal Glivenko–Cantelli class.

Notes

Notes to Section 6.1. Koltchinskii (1981) defined $D_F^{(p)}(\delta, \mathcal{F}, \gamma)$ when $F \equiv 1$ and $\gamma = P_n$. Pollard (1982) independently defined it for general F, for $p = 2$, and for distinct $x(j)$. Both, in fact, used the minimal cardinality of an ε-net (see Section 1.2 above). Theorem 6.1 is due to Pollard (1982, proof of Theorem 9). I do not know references for Propositions 6.2 or 6.3. The symmetrization method is also due independently to Koltchinskii (1981) and Pollard (1982). Lemma 6.5 is adapted from Pollard (1982, Lemma 11), who proves it for a countable class $\mathcal{F}$, thus avoiding measurability difficulties. From the countable case one can infer the result if there is a countable subset $\mathcal{H} \subset \mathcal{F}$ such that $\|v_n\|_{\mathcal{H}} = \|v_n\|_{\mathcal{F}}$ almost surely. Thus it suffices for v_n on $\mathcal{F}$ to be separable in the sense of Doob. For most classes $\mathcal{F}$ arising in applications the empirical process is naturally separable. But the collection $\mathcal{C}$ of all singletons in $[0, 1]$, for $P =$ Lebesgue measure, appears as a rather regular collection for which the empirical process is not separable. In such a case it may appear unnatural to use a separable modification of the process, so that, e.g., $P_1(\{x\}) = 0$ for all x, even $x = x_1$! Strobl (1995) proved Theorem 6.6(b) without supplementary measurability assumptions.

Theorem 6.7 extends Pollard (1982, Theorem 12), and Wolfowitz (1954) for half-spaces in $\mathbb{R}^d$.

Vapnik and Červonenkis, with their different measurability assumption, announced Corollary 6.8 in 1968, in the first publication to define the classes now called VC classes. This was the main theorem they announced. In 1971 they published a longer paper giving proofs.

Notes to Section 6.2. In Theorem 6.10, Vapnik and Červonenkis (1971, Theorem 4) proved equivalence of (b) and (c). The proof here was first given in Dudley (1984). The Koltchinskii–Pollard techniques allowed some simplification of the proof of Vapnik and Červonenkis. Steele (1978b) proved equivalence of (a) and (b) using Kingman's theorem and proved Theorem 6.13 and (6.9). However, Steele's assumption that $\| P_n - P \|_\mathcal{C}$ is measurable has been strengthened to "image admissible Suslin." Some such strengthening is needed in the proof.

Notes to Section 6.3. Pollard (1982) proved Theorem 6.15 when the empirical processes ν_n are stochastically separable in the sense of Doob, as is usually true *ab initio* in cases of interest and can always be obtained by modifications of the process, which may, occasionally, appear unnatural (see the Notes to Section 6.1). The "image admissible Suslin" formulation was given in Dudley (1984). The proof, based on Pollard's, is somewhat different, not only as regards measurability, but notably in that Proposition 6.18 extends a more specific result of Pollard.

The implication from Pollard's theorem 6.15 to Theorem 6.16, of Jain and Marcus (1975), was first given in Dudley (1984).

Notes to Section 6.4. Theorem 6.24 and the example before Example 6.25 are from Durst and Dudley (1981). Assouad (1982, Proposition C) proved a form of Theorem 6.26 with $1/(8e^2)$ in place of $1/2$, and so proved Theorem 6.27. I thank David Pollard for a remark leading to Proposition 6.28. Theorem 6.26 and Proposition 6.28 were given in Assouad and Dudley (1990), which, like Assouad (1982), was not published. Assouad (1985) is a related work.

7

Metric Entropy, with Inclusion and Bracketing

7.1 Definitions and the Blum–DeHardt Law of Large Numbers

Definitions. Given a measurable space $(A, \mathcal{A})$, recall that $\mathcal{L}^0(A, \mathcal{A})$ is the set of all real-valued $\mathcal{A}$-measurable functions on A. Given $f, g \in \mathcal{L}^0(A, \mathcal{A})$ with $f \le g$, i.e. $f(x) \le g(x)$ for all $x \in A$, let $[f, g] := \{h \in \mathcal{L}^0(A, \mathcal{A}) : f \le h \le g\}$. A set $[f, g]$ will be called a *bracket*. Given a probability space $(A, \mathcal{A}, P)$, $1 \le q \le \infty$, $\mathcal{F} \subset \mathcal{L}^q(A, \mathcal{A}, P)$ with usual seminorm $\|\cdot\|_q$, and $\varepsilon > 0$, let $N_{[]}^{(q)}(\varepsilon, \mathcal{F}, P)$ denote the smallest m such that for some $f_1, \dots, f_m$ and $g_1, \dots, g_m$ in $\mathcal{L}^q(A, \mathcal{A}, P)$, with $\|g_i - f_i\|_q \le \varepsilon$ for $i = 1, \dots, m$,

$$\mathcal{F} \subset \bigcup_{i=1}^{m} [f_i, g_i]. \tag{7.1}$$

Here $\log N_{[]}^{(q)}(\varepsilon, \mathcal{F}, P)$ will be called a *metric entropy with bracketing*.

Note that the f_j and g_j are not required to be in $\mathcal{F}$. For example, if $\mathcal{F}$ is the set of indicators of half-planes in $\mathbb{R}^2$, then $f \le h \le g$ for f, g, h in $\mathcal{F}$ would require the boundary lines of all three half-planes to be parallel. If instead we let f be the indicator of an intersection of two half-planes and g that of a union, then there can be a nondegenerate set of $h \in \mathcal{F}$ with $f \le h \le g$.

Also note that an individual bracket $[f, g]$ has $\max(-f, g) = \max(|f|, |g|)$ as envelope function, and so if (7.1) holds, for some ε, then $\mathcal{F}$ has an envelope function $\max_{1 \le j \le m} \max(-f_j, g_j) \in \mathcal{L}^q(A, \mathcal{A}, P)$. So, in this chapter, unlike the last, separate assumptions about envelope functions are not needed.

If $\mathcal{F} \subset \mathcal{L}^r$, then for $q \le r \le \infty$, $\mathcal{F} \subset \mathcal{L}^q$ and

$$N_{[]}^{(q)}(\varepsilon, \mathcal{F}, P) \le N_{[]}^{(r)}(\varepsilon, \mathcal{F}, P) \quad \text{for all} \quad \varepsilon > 0. \tag{7.2}$$

Recall that for a bounded real-valued function f, $\|f\|_{\sup} := \sup_x |f(x)|$. Let $d_{\sup}(f, g) := \|f - g\|_{\sup}$. It is easily seen using brackets $[f_j - \varepsilon, f_j + \varepsilon]$ that

269

for any law P,

$$N_{[\,]}^{(\infty)}\,(2\varepsilon, \mathcal{F}, P) \le D\left(\varepsilon, \mathcal{F}, d_{\sup}\right). \tag{7.3}$$

Thus, for example, a set $\mathcal{F}$ of continuous functions, totally bounded for $d_{\sup}$ with given bounds $D\left(\varepsilon, \mathcal{F}, d_{\sup}\right)$, will have the same bounds on all $N_{[\,]}^{(q)}\,(2\varepsilon, \mathcal{F}, P)$, $1 \le q \le \infty$.

If $\mathcal{F}$ consists of indicator functions of measurable sets, then in finding brackets $[f_i, g_i]$ to cover $\mathcal{F}$, it is no loss to assume $0 \le f_i \le g_i \le 1$ for all i. Next, if $C(i) := \{x : f_i(x) > 0\}$, $D(i) := \{x : g_i(x) = 1\}$, and $f_i \le 1_C \le g_i$ then

$$f_i \le 1_{C(i)} \le 1_C \le 1_{D(i)} \le g_i.$$

So, $[f_i, g_i]$ can be replaced by $[1_{C(i)}, 1_{D(i)}]$. If $\mathcal{C}$ is a collection of measurable sets and $\varepsilon > 0$, let

$$N_I(\varepsilon, \mathcal{C}, P) := \inf\{m : \text{ for some } C_1, \ldots, C_m \text{ and } D_1, \ldots, D_m \text{ in } \mathcal{A},$$

$$\text{for all } C \in \mathcal{C} \text{ there is an } i \text{ with } C_i \subset C \subset D_i \text{ and } P\,(D_i \setminus C_i) \le \varepsilon\}.$$

Here the I in N_I indicates "inclusion." Then it follows that

$$N_I\,(\varepsilon, \mathcal{C}, P) = N_{[\,]}^{(1)}\,(\varepsilon, \mathcal{F}, P) \quad \text{where} \quad \mathcal{F} = \{1_C : C \in \mathcal{C}\}. \tag{7.4}$$

We have the following law of large numbers:

Theorem 7.1 (Blum–DeHardt) *Suppose* $\mathcal{F} \subset \mathcal{L}^1(A, \mathcal{A}, P)$ *and for all* $\varepsilon > 0$, $N_{[\,]}^{(1)}(\varepsilon, \mathcal{F}, P) < \infty$. *Then* $\mathcal{F}$ *is a strong Glivenko–Cantelli class, that is,*

$$\lim_{n \to \infty} \| P_n - P \|_{\mathcal{F}}^* = 0 \text{ a.s.}$$

Proof. Given $\varepsilon > 0$ take $f_1, \ldots, f_m$ and $g_1, \ldots, g_m$ in $\mathcal{L}^1$, $m < \infty$, to satisfy (7.1) for $q = 1$. Let $f_{m+j} := g_j$ for $j = 1, \ldots, m$. By the ordinary strong law of large numbers there is an N such that

$$\Pr\left\{\sup_{n \ge N} \max_{j \le 2m} \left|(P_n - P)\,(f_j)\right| > \varepsilon\right\} < \varepsilon.$$

For each $f \in \mathcal{F}$ let $f_i \le f \le g_i$ with $P\,(g_i - f_i) \le \varepsilon$. Then if

$$\max_{j \le 2m} \left|(P_n - P)\,(f_j)\right| \le \varepsilon,$$

we have

$$|(P_n - P)(f)| \le |(P_n - P)(f_i)| + |(P_n - P)(f - f_i)|$$

$$\le \varepsilon + (P_n + P)(g_i - f_i)$$

$$\le 3\varepsilon + (P_n - P)(g_i - f_i) \le 5\varepsilon.$$

Thus

$$\Pr^* \left\{ \sup_{n \geq N} \| P_n - P \|_{\mathcal{F}} > 5\varepsilon \right\} < \varepsilon.$$

One can then apply Lemma 3.8, saying that for any real-valued function X and real t, $\Pr^*(X > t) = \Pr(X^* > t)$, and the easily checked fact that for any sequence ψ_n of real-valued functions, $(\sup_n \psi_n)^* = \sup_n \psi_n^* \leq +\infty$ almost surely. $\qquad\square$

Remark. No measurability assumption such as image admissible Suslin was needed in Theorem 7.1, as it was not needed in the proof. We used the law of large numbers over a finite set of f_i, g_i, then we used bracketing to control $(P_n - P)(f)$ for $f_i \leq f \leq g_i$. We did have a star in the statement, which is not needed if $\mathcal{F}$ is image admissible Suslin. For similar reasons, measurability assumptions are not needed in the statement or proof of bracketing central limit theorems such as Theorem 7.6.

The sufficient condition in Theorem 7.1 is not necessary, as the next Proposition will show. For a class $\mathcal{C}$ of sets, we saw in Remark 6.25 that the Glivenko–Cantelli property holds for all P for the family $\mathcal{C} = 2^{\mathbb{N}}$ of all subsets of $\mathbb{N}$, which is not a VC class. Conditions on a class of sets equivalent to the Glivenko–Cantelli property for a given P were given in Section 6.2.

Proposition 7.2 *There is a probability space* $(A, \mathcal{A}, P)$ *and a strong Glivenko–Cantelli class* $\mathcal{F} := \{1_C : C \subset \mathcal{C}\}$ *for* P, *where* $\mathcal{C} \subset \mathcal{A}$ *is such that for all* $\varepsilon < 1/2$, *we have* $N_{[\,]}^{(1)}(\varepsilon, \mathcal{F}, P) = +\infty$.

Proof. Let $A = [0, 1]$ with $P = U[0, 1]$ the uniform (Lebesgue) law. Let $C_m := C(m)$ be independent sets with $P(C_m) = 1/m$. Then to show

$$\lim_{n \to \infty} \sup_m |(P_n - P)(C_m)| = 0 \quad \text{a.s.,} \tag{7.5}$$

let $0 < \varepsilon < 1$. For $m < 3/\varepsilon$ we have by Bernstein's inequality (Theorem 1.11)

$$\Pr\left(|(P_n - P)(C_m)| > \varepsilon\right) \leq 2\exp\left(-n\varepsilon^2 \Big/ \left(\frac{2}{m} + \frac{2\varepsilon}{3}\right)\right)$$

$$\leq 2\exp\left(-mn\varepsilon^2/4\right).$$

For $m \geq 3/\varepsilon$, inequality (1.7) gives

$$\Pr\left(|(P_n - P)(C_m)| > \varepsilon\right) \leq E(n\varepsilon, n, 1/m) \leq (e/(m\varepsilon))^{n\varepsilon - 1}.$$

We have for $r = 1$ or 2

$$\sum_{n \geq 2/\varepsilon} \left(\frac{e}{m\varepsilon}\right)^{n\varepsilon - 1} \leq \left(\frac{e}{m\varepsilon}\right)^r \Big/ \left[1 - \left(\frac{e}{m\varepsilon}\right)^\varepsilon\right]$$

$$\leq \left(\frac{e}{m\varepsilon}\right)^r \Big/ \left[1 - \left(\frac{e}{3}\right)^\varepsilon\right].$$

Thus $\sum_{n \geq 2/\varepsilon} \sum_{m \geq 1} \Pr\left(|(P_n - P)(C_m)| > \varepsilon\right) < \infty$, and by the Borel–Cantelli Lemma, we get (7.5).

Now, given functions $f_1, \ldots, f_k$ and $g_1, \ldots, g_k$ in $\mathcal{L}^1$ with $P(g_i - f_i) < 1/2$ for each i suppose

$$\{C_m\}_{m \geq 1} \subset \bigcup_{i=1}^k [f_i, g_i].$$

We may assume $0 \leq f_i \leq g_i \leq 1$ for all i. For each i with $P(g_i - f_i) < 1/2$, we have $\sum \{P(C_m) : f_i \leq 1_{C(m)} \leq g_i\} < +\infty$ since if the series diverges, then for a subsequence $C_{m(r)}$ we have $\sum_r P\left(C_{m(r)}\right) = +\infty$ and for $C := \bigcup_r C_{m(r)}$, we have by the Borel–Cantelli lemma $P(C) = 1$, so $f_i \leq 1_C \leq g_i$ implies $P(g_i) = 1$ and $P(f_i) \geq 1 - P(g_i - f_i) > 1 - \frac{1}{2} = \frac{1}{2} > 0$, but then $f_i \leq 1_{C(m)}$ for only finitely many m, a contradiction. Thus

$$N_{[]}^{(1)} (\varepsilon, \{C_m\}, P) = +\infty \text{ for every } \varepsilon < 1/2,$$

which finishes the proof.   □

On the other hand, let $\mathcal{C}$ be the collection of all finite subsets of $[0, 1]$ with Lebesgue law P. Then $\|P_n - P\|_{\mathcal{C}} \equiv 1 \not\to 0$ although $1_A = 0$ a.s. for all $A \in \mathcal{C}$. This shows that in Theorem 7.1, $N_I < \infty$ cannot be replaced by $N\left(\varepsilon, \mathcal{F}, d_p\right) \equiv 1$ for any $\mathcal{L}^p$ distance d_p.

A Banach space $(S, \|\cdot\|)$ has a dual space $(S', \|\cdot\|')$ of continuous linear forms $f : S \mapsto \mathbb{R}$ with $\|f\|' := \sup\{|f(x)| : x \in S, \|x\| \leq 1\} < \infty$ (RAP, Section 6.1). One way to apply Theorem 7.1 is via the following:

Proposition 7.3 *Let $(S, \|\cdot\|)$ be a separable Banach space and P a law on the Borel sets of S such that $\int \|x\| \, dP(x) < \infty$. Let $\mathcal{F}$ be the unit ball of the dual space S', $\mathcal{F} := \{f \in S' : \|f\|' \leq 1\}$. Then for every $\varepsilon > 0$, $N_{[]}^{(1)} (\varepsilon, \mathcal{F}, P) < \infty$.*

Proof. By Ulam's theorem (RAP, Theorem 7.1.4) take a compact $K \subset S$ such that $\int_{S \setminus K} \|x\| \, dP(x) < \varepsilon/4$. The elements of $\mathcal{F}$, restricted to K, form a uniformly bounded, equicontinuous family, hence totally bounded for the sup norm $\|\cdot\|_K$ on K by the Arzelà–Ascoli theorem. Take $f_1, \ldots, f_m \in \mathcal{F}, m < \infty$, such that for all $f \in \mathcal{F}$, $\|f - f_j\|_K < \varepsilon/4$ for some j. Let $g_j := f_j - \varepsilon/4$ on

$K, g_j(x) := - \|x\|, x \notin K; h_j := f_j + \varepsilon/4$ on $K, h_j(x) := \|x\|, x \notin K$, all for $j = 1, \ldots, m$. Then for any $f \in \mathcal{F}$, if $\|f - f_j\|_K \le \varepsilon/4$, then $g_j \le f \le h_j$ and $P(h_j - g_j) < \varepsilon$, so $N_{[\,]}^{(1)}(\varepsilon, \mathcal{F}, P) \le m < \infty$. $\qquad \square$

Corollary 7.4 (Mourier) *Let* $(S, \|\cdot\|)$ *be a separable Banach space,* P *a law on* S *such that* $\int \|x\| \, dP(x) < \infty$, *and* $X_1, X_2, \ldots$ *i.i.d.* P. *Let* $S_n := X_1 + \cdots + X_n$. *Then* S_n/n *converges a.s. in* $(S, \|\cdot\|)$ *to some* $x_0 \in S$.

Proof. By Theorem 7.1 and Proposition 7.3, S_n/n is a Cauchy sequence a.s. for $\|\cdot\|$, hence converges a.s. to some random variable $Y \in S$. For each $f \in \mathcal{F}$, $f(Y) = P(f)$ a.s., i.e., $Y \in f^{-1}(\{P(f)\})$ a.s. Let $\{f_m\}_{m \ge 1} \subset \mathcal{F}$ be a countable total set: if $f_m(x) = 0$ for all m, then $x = 0$. Such f_m exist by the Hahn–Banach theorem (RAP, Corollary 6.1.5). Let $D := \bigcap_m f_m^{-1}(\{P(f_m)\})$. Then $Y \in D$ a.s., so D is nonempty. But if $y, z \in D$ then $\|y - z\| = \sup_m |f_m(y - z)| = 0$, so $D = \{x_0\}$ for some x_0. $\qquad \square$

Direct proof. Given $\varepsilon > 0$, there is a Borel measurable function g from S into a finite subset of itself such that $P(\|x - g(x)\|) < \varepsilon$. To show this, let $\{x_i\}_{i=1}^{\infty}$ be dense in S, with $x_0 := 0$. For $k = 1, 2, \ldots$, let $g_k(x) := x_i$ for the smallest i such that $\|x - x_i\| = \min_{r \le k} \|x - x_r\|$. Then g_k is Borel measurable, $\|x - g_k(x)\| \le \|x\|$ for all k and x, and $\|x - g_k(x)\| \downarrow 0$ as $k \to \infty$ for all x, so $P(\|x - g_k(x)\|) \to 0$ by dominated or monotone convergence. So choose $k = k(\varepsilon)$ such that $P(\|x - g_k(x)\|) < \varepsilon$ and let $g := g_k$.

The strong law of large numbers holds for the finite-dimensional variables $g(X_j)$, with some limit x_ε, and

$$n^{-1} \left\| \sum_{j=1}^{n} X_j - g(X_j) \right\| \le n^{-1} \sum_{j=1}^{n} \|X_j - g(X_j)\|$$

$$\to E \|X_1 - g(X_1)\| < \varepsilon \text{ as } n \to \infty$$

by the one-dimensional strong law. Thus

$$\lim_{n \to \infty} \|x_\varepsilon - S_n/n\| < \varepsilon \text{ a.s.}$$

Letting $\varepsilon \downarrow 0$ through some sequence, we get S_n/n converging a.s. to some x_0. $\qquad \square$

Corollary 7.5 *If* $\mathcal{F} \subset \mathcal{L}^1(A, \mathcal{A}, P)$ *and* $\{\delta_x : x \in A\}$ *is separable for* $\|\cdot\|_{\mathcal{F}}$, *then* $\|P_n - P\|_{\mathcal{F}} = \|P_n - P\|_{\mathcal{F}}^* \to 0$ *a.s.*

Proof. This follows from Corollary 7.4, since finite linear combinations of δ_x, $x \in A$, with rational coefficients, are dense in their completion for $\|\cdot\|_{\mathcal{F}}$, a Banach space. $\qquad \square$

The proof of Proposition 7.3 and Corollary 7.4 together from Theorem 7.1 is no shorter than the direct proof. On the other hand, if $\mathcal{F} = \{1_{[0,t]} : 0 < t < 1\}$ and P is Lebesgue measure on $[0, 1]$, then Theorem 7.1 applies but Corollary 7.5 does not.

7.2 Central Limit Theorems with Bracketing

In this section the bracketing will be in L^2. A bracket $[f, h]$ will be called a δ-bracket if $(\int (h - f)^2 dP)^{1/2} \leq \delta$. The following main theorem will be proved. Then, Corollary 7.8 gives a hypothesis on $N_{[]}^{(1)}$ for uniformly bounded classes of functions.

Theorem 7.6 (M. Ossiander) *Let $(X, \mathcal{A}, P)$ be a probability space and let $\mathcal{F} \subset \mathcal{L}^2(X, \mathcal{A}, P)$ be such that*

$$\int_0^1 \left(\log N_{[]}^{(2)}(x, \mathcal{F}, P) \right)^{1/2} dx < \infty.$$

Then $\mathcal{F}$ is a P-Donsker class.

Proof (Arcones and Giné). Throughout the proof, $P(f)$ or $E(f)$ will be used interchangeably to mean $\int f \, dP$ if it is defined.

First, a lemma will help. Recall that an envelope G for a class $\mathcal{G}$ of functions is a measurable function such that $|g(x)| \leq G(x)$ for all $g \in \mathcal{G}$ and all x.

Lemma 7.7 *Let $(X, \mathcal{A}, P)$ be a probability space, and $\mathcal{G}$ a set of real-valued measurable functions on X, with an envelope G. Let $B \in \mathcal{A}$ and $\delta > 0$. Suppose $P(G1_B) < \delta/2$, and that for each $g \in \mathcal{G}$, $A(g)$ is a measurable set with $A(g) \subset B$. Then*

$$\Pr^*\{\|(P_n - P)(g1_{A(g)})\|_{\mathcal{G}} > 2\delta\} \leq \Pr\{|(P_n - P)(G1_B)| > \delta\}.$$

Proof. $|P(g1_{A(g)})| \leq P(G1_B) < \delta/2$ for all $g \in \mathcal{G}$, so

$$\Pr^*\{\|(P_n - P)(g1_{A(g)})\|_{\mathcal{G}} > 2\delta\} \leq \Pr^*\{\|P_n(g1_{A(g)})\|_{\mathcal{G}} > 3\delta/2\}$$

$$\leq \Pr\{P_n(G1_B) > 3\delta/2\} \leq \Pr\{|(P_n - P)(G1_B)| > \delta\}. \qquad \square$$

Now to prove the theorem, let $N_k := N_{[]}^{(2)}(2^{-k}, \mathcal{F}, P)$, $k = 1, 2, \ldots$. Let $\gamma_k := (\log(kN_1 \ldots N_k))^{1/2}$. Then γ_k is increasing in k. By the integral test,

$\sum_{k=1}^{\infty} (\log N_k)^{1/2} / 2^{k+1} < \infty$, so

$$\sum_{k=1}^{\infty} 2^{-k} \gamma_k \leq \sum_{k=1}^{\infty} 2^{-k} \left[(\log k)^{1/2} + \sum_{j=1}^{k} (\log N_j)^{1/2} \right]$$

$$\leq \sum_{k=1}^{\infty} (\log k)^{1/2} / 2^k + \sum_{j=1}^{\infty} (\log N_j)^{1/2} \sum_{k=j}^{\infty} 2^{-k} < \infty.$$

Let $\beta_k := \sum_{j=k}^{\infty} \gamma_j / 2^j$. Then $\beta_k \to 0$ as $k \to \infty$. Let $S_{ki} := [f_{ki}, h_{ki}]$, $i = 1, \ldots, N_k$, be a set of 2^{-k}-brackets covering $\mathcal{F}$. Let $T_{ki} := S_{ki} \setminus \bigcup_{s<i} S_{ks}$, so that the sets T_{ki} are disjoint and each is included in a 2^{-k}-bracket. If $s_j \in \{1, \ldots, N_j\}$ for $j = 1, \ldots, k$, let $A_{k,s_1,\ldots,s_k} := \bigcap_{j=1}^{k} T_{j,s_j}$. For each nonempty set $A_{k,s}$, $s = (s_1, \ldots, s_k)$, choose $f_{k,s} \in A_{k,s}$. For each $f \in \mathcal{F}$ let $A_k(f) := A_{k,s}$ and $\pi_k f := f_{k,s}$ for the unique $s := s(f) := s(f, k)$ such that $f \in A_{k,s}$. Let $\Delta_k f := \Delta_k(f) := \sup\{|g - h| : g, h \in A_k(f)\}$. Then for any $f \in \mathcal{F}$,

$$\Delta_k(f) \text{ is nonincreasing in } k \text{ and } E\Delta_k^2(f) \leq 2^{-2k}. \tag{7.6}$$

Note that $\Delta_k f$ depends on f only through $s(f)$. For $k \geq 1$, $n \geq 1$ and $f \in \mathcal{F}$ let

$$B_k := B(k) := B(k, f, n) := \left\{ x \in X : \Delta_k f > n^{1/2} / \left(2^{k+1} \gamma_{k+1} \right) \right\}. \tag{7.7}$$

For any fixed j and n and $x \in X$ let

$$\tau f := \tau_{j,n}(f, x) := \min\{k \geq j : x \in B(k, f, n)\},$$

where $\min \emptyset := +\infty$. Then $\{\tau f = j\} = B(j, f, n)$, $\{\tau f > j\} = \{\Delta_j f \leq n^{1/2} / (2^{j+1} \gamma_{j+1})\}$, and for $k > j$,

$$\{\tau f \geq k\} \subset X \setminus B_{k-1} \subset \left\{ \Delta_k f \leq \Delta_{k-1} f \leq n^{1/2} / \left(2^k \gamma_k \right) \right\} \tag{7.8}$$

by (7.6), and $\{\tau f = k\} \subset B_k \setminus B_{k-1}$.

For any $f, g \in \mathcal{F}$ let $\rho_{[]}(f, g) := 1/2^K$ for the largest K such that for some s, f and g are both in $A_{K,s}$, or $\rho_{[]}(f, g) = 0$ if this holds for arbitrarily large K. Then $\mathcal{F}$ is totally bounded for $\rho_{[]}$. So by Theorem 3.34 it will be enough to prove the asymptotic equicontinuity condition for $\rho_{[]}$, in other words that for every $\alpha > 0$,

$$\lim_{j \to \infty} \limsup_{n \to \infty} \Pr^* \{ n^{1/2} \| (P_n - P)(f - \pi_j f) \|_{\mathcal{F}} > \alpha \} = 0. \tag{7.9}$$

We can assume $0 < \alpha < 1$. Then for any positive integers $j < r$, $f - \pi_j f$ will be decomposed as follows:

$$f - \pi_j f = (f - \pi_j f)1_{\tau f = j} + (f - \pi_r f)1_{\tau f \geq r} + \sum_{k=j+1}^{r-1} (f - \pi_k f)1_{\tau f = k}$$

$$+ \sum_{k=j+1}^{r} (\pi_k f - \pi_{k-1} f)1_{\tau f \geq k}; \qquad (7.10)$$

this is easily seen for $r = j + 1$ and then by induction on r. The decomposition (7.10) will give a bound for the outer probability in (7.9) by a sum of four terms to be labeled (I), (II), (III), and (IV) respectively below.

Let $\varepsilon := \alpha/8$. Fix $j = j(\varepsilon)$ large enough so that

$$\beta_j < \varepsilon/24 \quad \text{and} \quad \sum_{k>j} k^{-12} < 2\varepsilon. \qquad (7.11)$$

Then, choose $r > j$ large enough so that, since γ_r increases with r,

$$n^{1/2} 2^{-r} < \varepsilon/4 \quad \text{and} \quad 2 \cdot \exp(-\gamma_r 2^{r-1} \varepsilon) < \varepsilon. \qquad (7.12)$$

Lemma 7.7 will be applied to classes of functions

$$\mathcal{G} := \mathcal{G}(k, s) = \mathcal{G}_{k,s} := \{f - \pi_k f : f \in \mathcal{F}, \ \pi_k f = f_{k,s}\}$$

with envelope $\leq G := G_{k,s} := \Delta_k f_{k,s}$.

About (I): for any function $\psi \geq 0$ and $t > 0$, $1_{\psi > t} \leq \psi/t$. So by (7.7),

$$n^{1/2} E(1_{B_j} \Delta_j f) \leq 2^{j+1} \gamma_{j+1} E((\Delta_j f)^2)$$

$$\leq 2^{1-j} \gamma_{j+1} \leq 4\beta_j < \varepsilon/4$$

for all $f \in \mathcal{F}$. Then since $\{\tau f = j\} = B_j$,

$$n^{1/2} \left| P\left((f - \pi_j f)1_{\tau f = j}\right)\right| \leq n^{1/2} P\left(1_{B_j} \Delta_j f\right) \leq \varepsilon/4.$$

Apply Lemma 7.7 to $\mathcal{G}_{j,s}$ for each s with $B := B_j := B(j, f_{j,s}, n)$, $\delta := \varepsilon/n^{1/2}$, and $A(f - \pi_j f) = B_j = \{\tau f = j\}$ in this case. Then

$$\mathrm{Pr}^* \left\{ \left\| n^{1/2}(P_n - P)(g)1_{B_j} \right\|_{\mathcal{G}(j,s)} > 2\varepsilon \right\}$$

$$\leq \mathrm{Pr} \left\{ n^{1/2} \left| (P_n - P)\left((\Delta_j f_{j,s})1_{B_j}\right)\right| > \varepsilon \right\}.$$

Then summing over s,

$$\text{(I)} := \mathrm{Pr}^* \left\{ \left\| n^{1/2}(P_n - P)((f - \pi_j f)1_{\tau f = j}) \right\|_{\mathcal{F}} > 2\varepsilon \right\}$$

$$\leq \exp\left(\gamma_j^2\right) \max_s \mathrm{Pr} \left\{ n^{1/2} | (P_n - P)(\Delta_j f_{j,s} 1_{B_j}) | > \varepsilon \right\}$$

$$\leq \exp\left(\gamma_j^2\right) \varepsilon^{-2} \max_s \mathrm{Var}(\Delta_j f_{j,s} 1_{B_j}).$$

As $n \to \infty$, for fixed j and ε, by (7.6) and (7.7), since for each j and s, $P(B_j) = P(B(j, f_{j,s}, n)) \to 0$ we have (I) converging to 0, so (I) is less than ε for n large enough.

About (II): we have by (7.6) and (7.12)

$$
\begin{aligned}
n^{1/2} E(\Delta_r f 1_{\{\Delta_r f \le n^{1/2}/(2^r \gamma_r)\}}) &\le n^{1/2} (E(\Delta_r f)^2)^{1/2} \\
&\le n^{1/2} 2^{-r} < \varepsilon/4,
\end{aligned} \tag{7.13}
$$

and

$$
\begin{aligned}
E((\Delta_r f)^2 1_{\{\Delta_r f \le n^{1/2}/(2^r \gamma_r)\}}) &\le [\varepsilon/(4n^{1/2})][n^{1/2}/(2^r \gamma_r)] \\
&= \varepsilon/(2^{r+2} \gamma_r).
\end{aligned} \tag{7.14}
$$

Now by (7.8) for $k = r$, and since $|f - \pi_r f| \le \Delta_r f$, we have by (7.13), $n^{1/2}|E((f - \pi_r f)1_{\tau f \ge r})| < \varepsilon/4$ for all $f \in \mathcal{F}$. Apply Lemma 7.7 for each s with $A(f - \pi_r f) := \{\tau f \ge r\}$ and $B := \{\Delta_r f_{r,s} \le n^{1/2}/(2^r \gamma_r)\}$, noting that $\Delta_r f \le \Delta_{r-1} f$ for all f. Thus by (7.8) again,

$$
\begin{aligned}
(\text{II}) &:= \Pr^* \{ \| n^{1/2}(P_n - P)((f - \pi_r f)1_{\tau f \ge r}) \|_{\mathcal{F}} > 2\varepsilon \} \\
&\le \Pr\{ n^{1/2} \|(P_n - P)(\Delta_r f \cdot 1_{\{\Delta_r f \le n^{1/2}/(2^r \gamma_r)\}} \|_{\mathcal{F}} > \varepsilon \}.
\end{aligned}
$$

Then by Bernstein's inequality (Theorem 1.11) and (7.8),

$$
\begin{aligned}
(\text{II}) &\le 2 \cdot \exp\{\gamma_r^2 - \frac{\varepsilon^2}{2\varepsilon/(2^{r+2}\gamma_r) + 2\varepsilon/(3 \cdot 2^r \gamma_r)}\} \\
&= 2 \cdot \exp(\gamma_r^2 - 2^r \gamma_r \varepsilon/(7/6)) \\
&\le 2 \cdot \exp(-\gamma_r 2^{r-1} \varepsilon)
\end{aligned}
$$

since $2^{-r} \gamma_r \le \beta_r < \beta_j < \varepsilon/24$ by definition of β_r and (7.11). Then (II) $< \varepsilon$ by (7.12).

About (III): for $k = j+1, \ldots, r-1$, by (7.7) and (7.6),

$$
n^{1/2} E((\Delta_k f)1_{\tau f = k}) \le 2^{k+1} \gamma_{k+1} E((\Delta_k f)^2) \le 2^{1-k} \gamma_{k+1}. \tag{7.15}
$$

Then

$$
\begin{aligned}
(\text{III}) &:= \Pr^* \{ n^{1/2} \|(P_n - P)(\sum_{k=j+1}^{r-1} (f - \pi_k f)1_{\tau f = k}) \|_{\mathcal{F}} > 2\varepsilon \} \\
&\le \sum_{k=j+1}^{r-1} \Pr^* \{ n^{1/2} \|(P_n - P)(f - \pi_k f)1_{\tau f = k}) \|_{\mathcal{F}} > 2^{-k} \varepsilon \gamma_{k+1}/\beta_j \},
\end{aligned}
$$

since $\sum_{k=j+1}^{r-1} 2^{-k} \varepsilon \gamma_{k+1}/\beta_j \le 2\varepsilon$. To apply Lemma 7.7 for each $k = j+1, \ldots, r-1$ and for each s, let $\delta := 2^{-k-1} \varepsilon \gamma_{k+1}/\beta_j$, and $A(f) := B := \{\tau f = k\}$ for $f - \pi_k f \in \mathcal{G}_{k,s}$. The hypothesis of the Lemma holds since $\beta_j < \varepsilon/24$

(7.11) implies $2^{1-k}\gamma_{k+1} < 2^{-k-2}\gamma_{k+1}\varepsilon/\beta_j$, and by (7.8). So

$$(\text{III}) \leq \sum_{k=j+1}^{r-1} \Pr\{n^{1/2}\|(P_n - P)((\Delta_k f)1_{\tau f=k})\|_{\mathcal{F}} > 2^{-k-1}\varepsilon\gamma_{k+1}/\beta_j\}.$$

Then by Bernstein's inequality again, (7.6) and (7.8),

$$(\text{III}) \leq \sum_{k=j+1}^{r-1} 2 \cdot \exp\left(\gamma_k^2 - \frac{\varepsilon^2\gamma_{k+1}^2/(2^{2k+2}\beta_j^2)}{2^{1-2k} + 2^{-k}\varepsilon\gamma_{k+1}/(3 \cdot 2^k\gamma_k\beta_j)}\right)$$

$$= \sum_{k=j=1}^{r-1} 2 \cdot \exp\left(\gamma_k^2 - \frac{\varepsilon^2\gamma_{k+1}^2}{4\beta_j^2(2 + \varepsilon \cdot \gamma_{k+1}/(3\gamma_k\beta_j))}\right).$$

Now since γ_k is increasing with k,

$$-\varepsilon^2\gamma_{k+1}^2 \Big/ \left(4\beta_j^2\left[2 + \frac{\varepsilon}{3} \cdot \frac{\gamma_{k+1}}{\gamma_k\beta_j}\right]\right)$$

$$= -\varepsilon^2\gamma_{k+1} \Big/ \left(4\beta_j^2\left[\frac{2}{\gamma_{k+1}} + \frac{\varepsilon}{3\gamma_k\beta_j}\right]\right)$$

$$\leq -\varepsilon^2\gamma_k^2 \Big/ \left(4\beta_j^2\left[2 + \frac{\varepsilon}{3\beta_j}\right]\right).$$

The latter expression, since $\beta_j < \varepsilon/24$ by (7.11) and so $2 < \varepsilon/(12\beta_j)$, is bounded above by

$$-\varepsilon^2\gamma_k^2/4\beta_j^2\left(\frac{\varepsilon}{\beta_j}\right) \cdot \frac{5}{12} = -3\varepsilon\gamma_k^2/(5\beta_j).$$

It follows that

$$(\text{III}) \leq \sum_{k=j+1}^{r-1} 2 \cdot \exp(-\varepsilon\gamma_k^2/(2\beta_j)) \leq \sum_{k=j+1}^{r-1} 2 \cdot \exp(-12\gamma_k^2)$$

$$= 2 \cdot \sum_{k=j+1}^{r-1} 1/(kN_1 \cdots N_k)^{12} \leq 2 \cdot \sum_{k>j} k^{-12} < 4\varepsilon$$

by (7.11).

About (IV): if $k = j+1, \ldots, r$ and $g \in A_k(f)$ then $\pi_k(g) = \pi_k(f)$, $\pi_{k-1}(g) = \pi_{k-1}(f)$, and $\Delta_s f = \Delta_s g$ for all $s = 1, \ldots, k$, so $\{\tau f \geq k\} = \{\tau g \geq k\}$. Thus the number of distinct functions $(\pi_k f - \pi_{k-1}f)1_{\tau f \geq k}$ is at most $\exp(\gamma_k^2)$. Also, $\pi_k f \in A_{k-1}(f)$ and so by (7.6) $E((\pi_k f - \pi_{k-1}f)^2) \leq 2^{2-2k}$. Now

$$(\text{IV}) := \Pr^*\left\{n^{1/2}\left\|(P_n - P)\left(\sum_{k=j+1}^{r}(\pi_k f - \pi_{k-1}f)1_{\tau f \geq k}\right)\right\|_{\mathcal{F}} > \varepsilon\right\}$$

$$\leq \sum_{k=j+1}^{r} \Pr\left\{n^{1/2} \left\|(P_n - P)\left((\pi_k f - \pi_{k-1} f)1_{\tau f \geq k}\right)\right\|_{\mathcal{F}} > 2^{-k} \varepsilon \gamma_k / \beta_j\right\}$$

from the definition of β_j. Bernstein's inequality and (7.8) give

$$(\text{IV}) \leq \sum_{k=j+1}^{r} 2 \cdot \exp\left(\gamma_k^2 - \frac{\varepsilon^2 2^{-2k} \gamma_k^2 \beta_j^{-2}}{2^{3-2k} + \frac{2}{3}\varepsilon 2^{-k} \beta_j^{-1} 2^{-k}}\right).$$

Now since $\beta_j < \varepsilon/24$, $8 + \frac{2}{3}\varepsilon\beta_j^{-1} < \varepsilon/\beta_j$ and

$$(\text{IV}) \leq 2 \sum_{k=j+1}^{r} \exp(\gamma_k^2(1 - \varepsilon/\beta_j)) \leq 2 \sum_{k=j+1}^{r} \exp(-23\gamma_k^2) < 4\varepsilon$$

as in (III). Thus the expression in (7.9) is less than α. Letting $\alpha \downarrow 0$, $j \to \infty$, and $n \to \infty$, the proof of Theorem 7.6 is complete. $\qquad\square$

Theorem 7.6 implies the following for L^1 entropy with bracketing:

Corollary 7.8 *Let $(X, \mathcal{A}, P)$ be a probability space and $\mathcal{F}$ a uniformly bounded set of measurable functions on X. Suppose that*

$$\int_0^1 \left(\log N_{[\,]}^{(1)}(x^2, \mathcal{F}, P)\right)^{1/2} dx < \infty.$$

Then $\mathcal{F}$ is a Donsker class for P.

Proof. Suppose $|f(x)| \leq M < \infty$ for all $f \in \mathcal{F}$ and $x \in X$. Since multiplication by a constant preserves the Donsker property (by Theorem 3.34), we can assume $M = 1/2$. Then for any $f, g \in \mathcal{F}$ and $\varepsilon > 0$, $|f - g| \leq 1$ everywhere. So if $\int |f - g| dP \leq \varepsilon^2$, then $(\int |f - g|^2 dP)^{1/2} \leq \varepsilon$. So $N_{[\,]}^{(2)}(\varepsilon, \mathcal{F}, P) \leq N_{[\,]}^{(1)}(\varepsilon^2, \mathcal{F}, P)$, and the result follows from Theorem 7.6. $\qquad\square$

It will be seen in the next section that Corollary 7.8, and thus Theorem 7.6, are best possible (provide a characterization of the Donsker property) in some cases.

7.3 The Power Set of a Countable Set: Borisov–Durst Theorem

Let P be a law on the set $\mathbb{N}$ of nonnegative integers. The next theorem gives a criterion for the Donsker property of the collection $2^{\mathbb{N}}$ of all subsets of $\mathbb{N}$, for P, in terms of the numbers $p_m := P(\{m\})$ for $m \geq 0$. We also find that the sufficient condition given in Corollary 7.8 is necessary for $2^{\mathbb{N}}$. Recall N_I as defined above Theorem 7.1.

Theorem 7.9 (Borisov–Durst) *The following are equivalent:*

(a) $2^{\mathbb{N}}$ *is a Donsker class for P;*

(b) $\sum_m p_m^{1/2} < \infty;$

(c) $\int_0^1 \left(\log N_I \left(x^2, 2^{\mathbb{N}}, P\right)\right)^{1/2} dx < \infty.$

Proof. We have (c) $\Rightarrow$ (a) by Corollary 7.8. Next, to prove (a) $\Rightarrow$ (b), suppose $\sum p_m^{1/2} = \infty$. The random variables $W(m) := W_P \left(1_{\{m\}}\right)$ (for the isonormal W_P on $L^2(P)$ as defined in Section 2.7) are independent and Gaussian with mean 0 and variances p_m. We can write $G_P(f) = W_P(f) - P(f)W_P(1)$ since the right side is Gaussian and has mean 0 and the covariances of G_P. Then

$$\sum_m E|W_P(\{m\})| = \sum_m (2/\pi)^{1/2} p_m^{1/2}$$

diverges, while $\sum_m \text{Var}(|W_P(\{m\})|) \leq \sum_m p_m < \infty$. Thus for any $M < \infty$, by Chebyshev's inequality,

$$\lim_{m\to\infty} P\left(\sum_{j=1}^m |W_P(\{j\})| > M\right) = 1.$$

So $\sum_j |W_P(\{j\})| = +\infty$ almost surely. Now $\sum_m p_m |W_P(1_{\mathbb{N}})| < \infty$ a.s., so $\sum |G_P(\{m\})| = +\infty$ a.s. Hence $\sup_{A \subset \mathbb{N}} G_P(1_A) = +\infty$ a.s. and $2^{\mathbb{N}}$ is not a pregaussian class, so *a fortiori* not a Donsker class. Thus (a) $\Rightarrow$ (b).

Next, to prove (b) $\Rightarrow$ (c). Equivalently, let us prove

$$\sum_{k=1}^{\infty} 2^{-k} \left(\log N_I \left(4^{-k}, 2^{\mathbb{N}}, P\right)\right)^{1/2} < \infty.$$

We can assume $p_m \geq p_r > 0$ for $m \leq r$. For $j = 0, 1, 2, \ldots$, let r_j be the number of values of m such that $4^{-j-1} < p_m^{1/2} \leq 4^{-j}$ and let $C_j := r_j/4^j$. Then $\sum_j C_j < \infty$. For $k \geq k_0$ large enough there is a unique $j(k)$ such that

$$\sum_{j>j(k)} C_j/4^j \leq 4^{-k} < \sum_{j\geq j(k)} C_j/4^j. \qquad (7.16)$$

Let $m(k) := m_k := \sum_{j=0}^{j(k)} r_j$. Then

$$\sum_{m>m(k)} p_m \leq \sum_{j>j(k)} r_j/4^{2j} \leq 4^{-k}.$$

Let A_i run over all subsets of $\{1, \ldots, m(k)\}$ where $i = 1, \ldots, 2^{m(k)}$. Let $B_i := A_i \cup \{m \in \mathbb{N} : m > m(k)\}$. Then for any $C \subset \mathbb{N}$, $C \cap \{1, \ldots, m(k)\} = A_i$ for some i. Then $A_i \subset C \subset B_i$ and $P(B_i \setminus A_i) \leq 4^{-k}$. So $N_I \left(4^{-k}, 2^{\mathbb{N}}, P\right) \leq 2^{m(k)+1}$. Thus it will be enough to prove

$$\sum_k m_k^{1/2}/2^k < \infty, \qquad (7.17)$$

with $\sum_k$ restricted to $k \geq k_0$. We have

$$\sum_k m_k^{1/2}/2^k = \sum_k \left(\sum_{j=0}^{j(k)} 4^j C_j\right)^{1/2} \Big/ 2^k$$

$$\leq \sum_k \sum_{j=0}^{j(k)} 2^{j-k} C_j^{1/2} = \sum_{j=0}^{\infty} C_j^{1/2} \sum_{k:\, j\leq j(k)} 2^{j-k}.$$

To prove this converges, since $\sum_j C_j < \infty$, it is enough by Cauchy's inequality to prove $\sum_j \left(\sum_{k:\, j\leq j(k)} 2^{j-k}\right)^2 < \infty$. Let $k(j)$ be the smallest k such that $j(k) \geq j$. Then

$$\sum_{k:\, j\leq j(k)} 2^{j-k} \leq 2^{j+1-k(j)}.$$

To prove that $\sum_j 4^{j-k(j)} < \infty$, setting $j(k_0 - 1) := 0$ we have

$$\sum_{j\geq 1} 4^{j-k(j)} \leq \sum_k \sum_{j:\, j(k-1)<j\leq j(k)} 4^{j-k} \leq \sum_k 4^{1+j(k)-k}.$$

For each k, let $\kappa(k)$ be the smallest κ such that $j(\kappa) = j(k)$. Then from (7.16) for κ, letting $\mathcal{K}$ denote the range of $\kappa(\cdot)$, $4^{-\kappa(k)} < \sum_{j\geq j(k)} C_j/4^j$, so

$$\sum_k 4^{j(k)-k} \leq \sum_k 4^{j(k)-k+\kappa(k)} \sum_{j\geq j(k)} C_j/4^j$$

$$= \sum_j C_j 4^{-j} \sum_{\kappa\in\mathcal{K},\, j(\kappa)\leq j} 4^{j(\kappa)+\kappa} \sum_{k:\, \kappa(k)=\kappa} 4^{-k}$$

$$< 2 \sum_j C_j 4^{-j} \sum_{\kappa\in\mathcal{K},\, j(\kappa)\leq j} 4^{j(\kappa)}.$$

Since $j(\cdot)$ is one-to-one on $\mathcal{K}$, the sum is at most $4 \sum_j C_j < \infty$. $\qquad \square$

Recall that if $\mathcal{C} = 2^{\mathbb{N}}$ then $\sup_{A\in\mathcal{C}} |(P_n - P)(A)| \to 0$ a.s. as $n \to \infty$ for any law P on $\mathcal{C}$ (Remark 6.25).

Problems

1. Let (K, d) be a compact metric space. For $0 < \alpha < 1$ let $L_\alpha(K, d)$ be the set of all real-valued functions f on K such that $|f(x) - f(y)| \leq d(x, y)^\alpha$ for all $x, y \in K$. Show that for any law P on K, $L_\alpha(K, d)$ is a strong Glivenko–Cantelli class for P.

2. Show that $\{f \in C[0, 1] : \|f\|_{\sup} \leq 1\}$ is not a weak Glivenko–Cantelli class for $P = U[0, 1]$.

3. Let $(X, \mathcal{A}, P)$ be a probability space, such that $\{x\} \in \mathcal{A}$ for all $x \in X$ and P has no soft atoms, as defined in Problem 9 of Chapter 5. Let $\mathcal{C} \subset \mathcal{A}$ be such that $\mathcal{C}$ is linearly ordered by inclusion and generates $\mathcal{A}$. Give an upper bound on $N_I(\varepsilon, \mathcal{C}, P)$ for $0 < \varepsilon \le 1$.

4. If $X = \mathbb{N}$, $\mathcal{A} = 2^{\mathbb{N}}$ and $p_n := P(\{n\}) := 1/2^{n+1}$ for $n = 0, 1, 2, \ldots,$ evaluate $N_I(\varepsilon, \mathcal{A}, P)$ for $0 < \varepsilon < 1$.

5. If $X = \mathbb{N}$, $\mathcal{A} = 2^{\mathbb{N}}$ and for some $\alpha \in \mathbb{R}$, $p_n := P(\{n\}) :=$ $c_\alpha/[(n+1)^2(\log(n+2))^\alpha]$ for all $n \in \mathbb{N}$, where c_α is such that $\sum_n p_n = 1$, for what values of α is $\mathcal{A}$ a Donsker class for P?

6. Let $(X, \mathcal{A}, P)$ be a probability space. For $i = 1, \ldots, k$ let $\mathcal{F}_i \subset \mathcal{L}^0(X, \mathcal{A})$ each satisfy the hypothesis of Theorem 7.6. Then show that the hypothesis also holds for:

(a) $\mathcal{F} := \bigcup_{i=1}^k \mathcal{F}_i$;

(b) $\mathcal{F} := \{\sum_{j=1}^k f_j : f_j \in \mathcal{F}_j \text{ for each } j\}$.

Let $\mathcal{C}_i$ be a collection of sets, $\mathcal{F}_i := \{1_A : A \in \mathcal{C}_i\}$. Show that the hypothesis of Theorem 7.6 also holds for $\mathcal{F} := \{1_A : A \in \mathcal{C}\}$ where

(c) $\mathcal{C} := \{\bigcup_{j=1}^k A_j : A_j \in \mathcal{C}_j \text{ for each } j\}$.

7. Let $\mathcal{C} := \{[0, a] \times [0, b] : 0 \le a \le 1, 0 \le b \le 1\}$. Let P be the uniform distribution on the unit square $[0, 1] \times [0, 1]$. Give an upper bound for $N_I(\varepsilon, \mathcal{C}, P)$ for $0 < \varepsilon \le 1$.

8. In the unit cube $X := [0, 1]^d$ in $\mathbb{R}^d$, with uniform (Lebesgue) law P, for some $K < \infty$ and $c > 0$, let $\mathcal{C} := \mathcal{C}(K, d)$ be the collection of all subsets A of X such that whenever X is decomposed as a union of n^d cubes of side $1/n$, each with vertices having coordinates i/n, $i = 0, 1, \ldots, n$, the boundary of A intersects at most Kn^{d-c} of these cubes. Show that $N_I(\varepsilon, \mathcal{C}, P) < \infty$ for all $\varepsilon > 0$. *Hints:* (a) Given $\varepsilon > 0$, show that for $n \ge n_0(\varepsilon)$ large enough, any union U of at most Kn^{d-c} of the n^d cubes at the nth stage has $P(U) < \varepsilon$.

(b) Let the sets $C(i)$ and $D(i)$ in the definition of N_I for the given ε each be unions of cubes in the nth decomposition of $[0, 1]^d$. How many such unions are there? (The number may be large, but you need to show that it is finite. Give an explicit bound.)

(c) Give an upper bound for the number of pairs of such unions. (You need not restrict to pairs such that $P(D(i) \setminus C(i)) < \varepsilon$, since again, you only need a finite upper bound.)

9. Show that the hypothesis of the previous problem holds if $\mathcal{C}$ is the collection of convex subsets of the unit square in $\mathbb{R}^2$.

10. Prove the same for the unit cube in $\mathbb{R}^d$ for any finite dimension d.

11. If $\mathcal{F}$ is a class of indicators of measurable sets and $\varepsilon > 0$, show that $N_{[]}^{(2)}(\varepsilon, \mathcal{F}, P) = N_{[]}^{(1)}(\varepsilon^2, \mathcal{F}, P)$ (compare Corollary 7.8 and its proof).

12. Prove Corollary 7.8 directly by Bernstein's inequality (Theorem 1.11).

Notes

Notes to Section 7.1. Theorem 7.1 is due to Blum (1955, Lemma 1) for families of (indicators of) sets and to DeHardt (1971, Lemma 1) for uniformly bounded families of functions. Mourier (1951, 1953 pp. 195–196) proved the law of large numbers in general separable Banach spaces, Corollary 7.4. I do not know a reference for Propositions 7.2 or 7.3. In the proof of 7.4 (Bochner or Pettis) integrals of Banach-valued functions (defined in Appendix E) were not assumed, so they had to be, in part, reconstructed.

Notes to Section 7.2. Theorem 7.6 is due to M. Ossiander (1987). The shorter proof presented here is an expanded version of that of Arcones and Giné (1993, Theorem 4.10) and applies a technique from Andersen et al. (1988), who proved an extended bracketing central limit theorem. Corollary 7.8 was proved earlier, first for classes of sets in Dudley (1978), then for classes of functions in Dudley (1984).

Notes to Section 7.3. In Theorem 7.9, it is not hard to prove that (b) $\Rightarrow$ (a). Durst and Dudley (1981) proved the equivalence of (a) and (b). I. S. Borisov (1981) discovered and announced the more difficult implication (b) $\Rightarrow$ (c). I have not seen his proof.

8

Approximation of Functions and Sets

8.1 Introduction: The Hausdorff Metric

In this chapter upper and lower bounds will be shown for the metric entropies or capacities of various concrete classes of functions on Euclidean spaces and sets in such spaces. Some metric entropies with bracketing are treated, and some without. Metrics for functions are in $\mathcal{L}^p$, $1 \leq p \leq \infty$. For sets we use d_P metrics $d_P(B, C) := P(B \triangle C)$ or the Hausdorff metric, defined as follows.

For any metric space (S, d), $x \in S$, and a nonempty $B \subset S$, let

$$d(x, B) := \inf\{d(x, y) : y \in B\}.$$

Then $d(x, B) = 0$ if and only if x is in the closure of B. If $A \subset B$, then clearly $d(x, B) \leq d(x, A)$. (In order to preserve this when (S, d) is unbounded and $A = \emptyset$ we need to set $d(x, \emptyset) := +\infty$ and so we will do that in all cases.) For nonempty bounded sets $B, C \subset S$ the Hausdorff pseudo-metric is defined by

$$h(B, C) := \max \left(\sup_{x \in B} d(x, C), \sup_{y \in C} d(y, B) \right).$$

To check that this is a pseudo-metric, clearly $h(B, C) = h(C, B) \geq 0$. To show that the triangle inequality $h(B, D) \leq h(B, C) + h(C, D)$ holds for any nonempty, bounded sets B, C, and D, first, for any $\delta > 0$ and $x \in B$ there is some $y \in C$ with $d(x, y) < h(B, C) + \delta$. Then there is some $z \in D$ with $d(y, z) < h(C, D) + \delta$, so $d(x, z) < h(B, C) + h(C, D) + 2\delta$. Letting δ decrease to 0 we get $d(x, D) \leq h(B, C) + h(C, D)$. A symmetric argument starting with a point of D gives the triangle inequality. Thus h is a pseudometric, as claimed.

Then h is a metric, called the *Hausdorff metric*, on the collection of bounded, closed, nonempty subsets of S, since if B and C are two distinct such sets, clearly $h(B, C) > 0$.

284

On $\mathbb{R}^d$ we have the usual Euclidean metric $d(x, y) := |x - y|$ where $|u| := \left(u_1^2 + \cdots + u_d^2\right)^{1/2}, u \in \mathbb{R}^d$. If $d \geq 2$, for any set $H \subset \mathbb{R}^{d-1}$ and function f from H into $[0, \infty]$ let

$$J_f := J(f) := \left\{x \in \mathbb{R}^d : 0 \leq x_d \leq f(x_{(d)}), \ x_{(d)} \in H\right\}$$

where $x_{(d)} := (x_1, \ldots, x_{d-1})$. Then $J(f)$ is the subgraph of f. For any two bounded functions $f \geq 0$ and $g \geq 0$ on H, clearly $h(J_f, J_g) \leq d_{\sup}(f, g)$. Thus for any collection $\mathcal{F}$ of bounded real functions ≥ 0 on H, and any $\varepsilon > 0$,

$$D\left(\varepsilon, \{J_f : f \in \mathcal{F}\}, h\right) \leq D\left(\varepsilon, \mathcal{F}, d_{\sup}\right). \tag{8.1}$$

From here on, assume that H is a bounded Borel set in $\mathbb{R}^{d-1}$ with nonempty interior, whose Lebesgue measure therefore satisfies $0 < \lambda^{d-1}(H) < +\infty$. In one case to be considered, H will be the $(d-1)$-dimensional unit cube I^{d-1} where $I^d := \{x \in \mathbb{R}^d : 0 \leq x_j \leq 1, \ j = 1, \ldots, d\}$.

For two nonnegative functions f and g on H, the symmetric difference of J_f and J_g is given by

$$J_f \Delta J_g = \{(u, v) : u \in H, \ f(u) < v \leq g(u) \ \text{or} \ g(u) < v \leq f(u)\}.$$

If $\varepsilon > 0$ and $d_{\sup}(f, g) \leq \varepsilon$, let $F := \max(f - \varepsilon, 0)$. Then $0 \leq F \leq g \leq f + \varepsilon$, so $J_F \subset J_g \subset J_{f+\varepsilon}$. If f and g are also measurable functions it follows from this and the Tonelli–Fubini theorem that we have the following bounds for Lebesgue d-dimensional measures:

$$\lambda^d(J_f \Delta J_g) \leq \varepsilon \lambda^{d-1}(H), \quad \lambda^d(J_{f+\varepsilon} \setminus J_F) \leq 2\varepsilon \lambda^{d-1}(H).$$

Let $\mathcal{F}$ be a class of bounded nonnegative continuous functions on H, totally bounded for $d_{\sup}$. Given $\varepsilon > 0$, let $m := D(\varepsilon, \mathcal{F}, d_{\sup})$ and let $f_1, \ldots, f_m$ in $\mathcal{F}$ be such that $d_{\sup}(f_i, f_j) > \varepsilon$ for $1 \leq i < j \leq m$. Then for each $g \in \mathcal{F}$, $d_{\sup}(g, f_j) \leq \varepsilon$ for some $j = 1, \ldots, m$ by definition of $D(\varepsilon, \cdot, \cdot)$. Thus for $F_j := \max(0, f_j - \varepsilon)$ for $j = 1, \ldots, m$ we have $F_j \leq f_j$ for each j, $f_j + \varepsilon - F_j \leq 2\varepsilon$, and $\mathcal{F} \subset \bigcup_{j=1}^m [F_j, f_j + \varepsilon]$. Moreover, $d_{\sup}(f_j, g) \leq \varepsilon$ implies that $J_{F_j} \subset J_g \subset J_{f_j+\varepsilon}$. From the last two facts it follows that we have the bracketing covering number bounds, related to (7.3) in light of (7.2):

(i) if P is any law on H, then for any r with $1 \leq r \leq \infty$,

$$N_{[]}^{(r)}(2\varepsilon, \mathcal{F}, P) \leq D(\varepsilon, \mathcal{F}, d_{\sup}); \tag{8.2}$$

(ii) if Q is any law on $H \times [0, \infty)$ having a density q with respect to Lebesgue measure on $\mathbb{R}^d$ with $q(x) \leq M < \infty$ for all x, then by the Tonelli–Fubini theorem again

$$N_I(2M\varepsilon\lambda^{d-1}(H), \{J_f : f \in \mathcal{F}\}, Q) \leq D(\varepsilon, \mathcal{F}, d_{\sup}). \tag{8.3}$$

In the converse direction here is a lower bound for the Hausdorff distance of subgraphs of Lipschitz functions. Recall that

$$\|f\|_L := \sup_{x \neq y} |f(x) - f(y)| / |x - y|.$$

Then we have:

Lemma 8.1 *Let H be a bounded nonempty subset of $\mathbb{R}^{d-1}$. Let $0 < K < \infty$ and let f and g be functions from H into $[0, +\infty)$ such that $\|f\|_L \leq K$ and $\|g\|_L \leq K$. Then $h(J_f, J_g) \geq d_{\mathrm{sup}}(f, g)/\sqrt{1 + K^2}$.*

Proof. If $t := d_{\mathrm{sup}}(f, g) = 0$, $f \equiv g$ and there is no problem, so assume that $t > 0$. Take δ such that $0 < \delta < t$ and take $u \in H$ such that $|(f - g)(u)| \geq t - \delta$. By symmetry we can assume that $f(u) > g(u)$. To find a lower bound for $d((u, f(u)), J_g)$, let $G(w) := g(u) + K d(w, u)$ for all $w \in H$. Then since $\|g\|_L \leq K$ we have $g \leq G$ and $J_g \subset J_G$. So $d((u, f(u)), J_g) \geq d((u, f(u)), J_G)$. Consider half-lines on which $w = u + s(w_0 - u)$ for fixed $w_0 \neq u$ and $s \geq 0$. Then the graph of points $(w, G(w))$ on such a half-line is itself a half-line L in $\mathbb{R}^d$. Since $(u, f(u))$ is not in J_G, a line segment joining $(u, f(u))$ to a point of J_G must pass through the boundary of J_G, which is the graph of G. The line extending L forms an angle $\theta = \tan^{-1} K$ with the subspace $\{x \in \mathbb{R}^d : x_{(d)} = 0\}$. The closest point p to $(u, f(u))$ in L is such that $(u, f(u)) - p$ is perpendicular to L, and so the three points $(u, g(u)), (u, f(u))$, and p form a triangle whose angle at $(u, f(u))$ is also θ. It follows that

$$d((u, f(u)), J_G) \geq (f - g)(u) \cos \theta = (f - g)(u)/\sqrt{1 + K^2}$$
$$\geq (d_{\mathrm{sup}}(f, g) - \delta)/\sqrt{1 + K^2}.$$

Letting $\delta \downarrow 0$, the conclusion follows. $\square$

Recall that a bounded number of Boolean operations preserve the Vapnik–Červonenkis property (Theorem 4.8). The same holds for classes of sets satisfying bounds on metric entropy (with inclusion). For any families $\mathcal{C}_j$ of subsets of a set X, extending the notation in Section 4.5, let

$$\sqcap_{j=1}^k \mathcal{C}_j := \{\cap_{j=1}^k A_j : A_j \in \mathcal{C}_j \text{ for all } j\},$$
$$\sqcup_{j=1}^k \mathcal{C}_j := \{\cup_{j=1}^k A_j : A_j \in \mathcal{C}_j \text{ for all } j\}.$$

Theorem 8.2 *Let $(X, \mathcal{A}, P)$ be a probability space and $\mathcal{C}_j \subset \mathcal{A}$ for $j = 1, \ldots, k$. Let $f_{1j}(\varepsilon) := \log N_I(\varepsilon, \mathcal{C}_j, P)$, $f_{2j}(\varepsilon) := \log D(\varepsilon, \mathcal{C}_j, d_P)$ for $j = 1, \ldots, k$. Let $\mathcal{C}_0 := \sqcap_{j=1}^k \mathcal{C}_j$. If for $i = 1$ or 2, there are a $\gamma > 0$ and constants $M_1, \ldots, M_k$ such that $f_{ij}(\varepsilon) \leq M_j \varepsilon^{-\gamma}$ for $0 < \varepsilon < 1$ and $j = 1, \ldots, k$, then the same holds for $j = 0$. The statements also hold for $i = 1$ or 2 for $\mathcal{C}_0 := \sqcup_{j=1}^k \mathcal{C}_j$.*

Proof. For N_I, given $0 < \varepsilon < 1$, for each $j = 1, \ldots, k$, take m_j brackets $[A_{jr}, B_{jr}]$ covering C_j with $P(B_{jr} \setminus A_{jr}) \le \varepsilon/k$ for $r = 1, \ldots, m_j$ and $m_j \le \exp(M_j(k/\varepsilon)^\gamma)$. If $A_{jr(j)} \subset C_{jr(j)} \subset B_{jr(j)}$ for $j = 1, \ldots, k$ and some $r(j)$, then

$$A_{(r)} := \cap_{j=1}^k A_{jr(j)} \subset C_{(r)} := \cap_{j=1}^k C_{jr(j)} \subset B_{(r)} := \cap_{j=1}^k B_{jr(j)}$$

and $P(B_{(r)} \setminus A_{(r)}) \le k(\varepsilon/k) = \varepsilon$. The result for N_I then follows with $M_0 := k^\gamma (M_1 + M_2 + \cdots + M_k)$. Without inclusions, and/or for $\sqcup$ instead of $\cap$, the proof is similar. $\square$ It follows that if each C_j for $1 \le j \le k$ satisfies an inclusion

(bracketing) condition sufficient for the Glivenko–Cantelli or Donsker property respectively, then so does $C_0 := \cap_{j=1}^k C_j$ or $\sqcup_{j=1}^k C_j$:

Corollary 8.3 *Under the assumptions of Theorem 8.2 for $i = 1$,*

(a) For any γ with $0 < \gamma < +\infty$, C_0 is a Glivenko–Cantelli class for P.

(b) If $0 < \gamma < 1$, then C_0 is a Donsker class for P.

Proof. This follows from Theorem 8.2 and for (a), from the Blum–DeHardt theorem 7.1. For (b) it follows from Corollary 7.8. $\square$

8.2 Spaces of Differentiable Functions and Sets with Differentiable Boundaries

For any $\alpha > 0$, spaces of functions will be defined having "bounded derivatives through order α." If β is the largest integer $< \alpha$, the functions will have partial derivatives through order β bounded, and the derivatives of order β will satisfy a uniform Hölder condition of order $\alpha - \beta$. Still more specifically: for $x := (x_1, \ldots, x_d) \in \mathbb{R}^d$ and $p = (p_1, \ldots, p_d) \in \mathbb{N}^d$ (where $\mathbb{N}$ is the set of nonnegative integers) let $[p] := p_1 + \cdots + p_d$ and

$$D^p := \partial^{[p]}/\partial x_1^{p_1} \cdots \partial x_d^{p_d}.$$

For a real-valued function f on an open set $U \subset \mathbb{R}^d$ having all partial derivatives $D^p f$ of orders $[p] \le \beta$ defined everywhere on U, let

$$\|f\|_\alpha := \|f\|_{\alpha, U} := \max_{0 \le [p] \le \beta} \sup \{|D^p f(x)| : x \in U\}$$

$$+ \max_{[p]=\beta} \sup_{x \ne y, \, x, y \in U} \{|D^p f(x) - D^p f(y)| / |x - y|^{\alpha - \beta}\}.$$

Here $D^0 f := D^{(0,0,\ldots,0)} f := f$. Let I^d denote the unit cube $\{x \in \mathbb{R}^d : 0 \le x_j \le 1, j = 1, \ldots, d\}$, and recall that $x_{(d)} = (x_1, \ldots, x_{d-1})$, $x \in \mathbb{R}^d$. Let $F \subset \mathbb{R}^d$ be a closed set which is the closure of its interior U. Let $\mathcal{F}_{\alpha, K}(F)$ denote the set of all continuous $f : F \to \mathbb{R}$ such that when f is restricted to

U, $\|f\|_{\alpha,U} \leq K$. For $\alpha = 1$, $\mathcal{F}_{1,K}(F)$ is the set of bounded Lipschitz functions f on F with $\|f\|_{\sup} + \|f\|_{L} \leq K$. A real-valued function g on any metric space (S, e), e.g., $\mathbb{R}^d$ with usual metric, is said to be *Hölder of order* α if $0 < \alpha \leq 1$ and $|g(x) - g(y)| \leq e(x, y)^\alpha$ for all $x, y \in S$.

Recall the bounded Lipschitz norm $\|f\|_{BL} := \|f\|_L + \|f\|_{\sup}$. It will be seen how the Lipschitz property is equivalent to having bounded derivatives of order 1 (partial derivatives, in dimension $d \geq 2$). In dimension $d = 1$, a Lipschitz function f is absolutely continuous, and so by a classical theorem of Lebesgue, it has a derivative $f'(x)$ for almost all x. Clearly $|f'(x)| \leq \|f\|_L$ whenever $f'(x)$ exists. On the other hand, if g is any bounded measurable function with $\|g\|_\infty \leq M$ and $f(x) := \int_0^x g(t)dt$, then f is Lipschitz with $\|f\|_L \leq M$ and $f'(x) = g(x)$ for Lebesgue almost all x (e.g., RAP, Theorem 7.2.1). Here f' need not be continuous, e.g. if $f(x) \equiv |x|$, at 0.

A real-valued function f on an open set $U \subset \mathbb{R}^d$ is said to be *Fréchet differentiable* at $x \in U$ if there is a vector $v := (Df)(x)$ such that

$$f(y) = f(x) + v \cdot (y - x) + o(|y - x|) \text{ as } y \to x.$$

Clearly, this implies that the partial derivatives $\partial f(x)/\partial x_j$ for $j = 1, \ldots, d$ exist and are the components of $(Df)(x)$. From Lebesgue's 1-dimensional theorem and the Tonelli–Fubini theorem, it is clear that if f is Lipschitz, then these partial derivatives exist for λ^d-almost all $x \in U$. H. Rademacher proved that moreover, f is Fréchet differentiable at λ^d-almost all x (see the Notes).

Next some related families of sets will be defined. Let $\mathcal{G}_{\alpha,K,d} := \mathcal{F}_{\alpha,K}(I^d)$. For $d \geq 2$ let $\mathcal{C}(\alpha, K, d)$ be the collection of all sets

$$J_f = J(f) = \left\{ x \in I^d : 0 \leq x_d \leq f(x_{(d)}) \right\}, \quad f \in \mathcal{G}_{\alpha,K,d-1}, \quad f \geq 0.$$

If g and h and two functions defined for (small enough) $y > 0$, recall that $g \asymp h$ (as $y \downarrow 0$) means that

$$0 < \liminf_{y \downarrow 0} (g/h)(y) \leq \limsup_{y \downarrow 0} (g/h)(y) < +\infty. \tag{8.4}$$

Clearly, if $f \in \mathcal{G}_{\alpha,K,d}$ and $[p] < \alpha$, then $D^p f \in \mathcal{G}_{\alpha-[p],K,d}$. Let $B_d := \{x \in \mathbb{R}^d : |x| < 1\}$ and let $\overline{B}_d = \{x \in \mathbb{R}^d : |x| \leq 1\}$ be the open and closed unit balls respectively in $\mathbb{R}^d$. Here are some bounds on metric entropies, of which Kolmogorov proved the main, first conclusion (a).

Theorem 8.4 (Kolmogorov) *Let* $0 < K < \infty$, $0 < \alpha < \infty$ *and* $d \geq 1$.

(a) As $\varepsilon \downarrow 0$

$$\log D(\varepsilon, \mathcal{G}_{\alpha,K,d}, d_{\sup}) \asymp \varepsilon^{-d/\alpha}.$$

(b) For some $T := T(\alpha, K, d)$, *any law* P *on* I^d, $1 \leq r \leq \infty$ *and* $0 < \varepsilon < 1$,

$$\log N_{[\,]}^{(r)}(\varepsilon, \mathcal{G}_{\alpha,K,d}, P) \leq T\varepsilon^{-d/\alpha}.$$

(c) For the Hausdorff metric we also have for such a T

$$\log D(\varepsilon, \mathcal{C}(\alpha, K, d+1), h) \le T\varepsilon^{-d/\alpha}.$$

(d) If Q is a law on I^{d+1} having a density with respect to Lebesgue measure bounded by M, then for some $M_1 = M_1(M, d, K, \alpha)$,

$$\log N_I(\varepsilon, \mathcal{C}(\alpha, K, d+1), Q) \le M_1\varepsilon^{-d/\alpha}, \quad 0 < \varepsilon \le 1.$$

(e) Parts (a) and (b) hold for $\overline{B}_d$ in place of I^d and so $\mathcal{F}_{\alpha,K}(\overline{B}_d)$ in place of $\mathcal{G}_{\alpha,K,d}$, with a possibly larger constant T.

Corollary 8.5 *Let $0 < K < \infty$. For any dimension d:*

(a) $\mathcal{G}_{\alpha,K,d}$ is a Glivenko–Cantelli class for any $\alpha > 0$ and law P on I^d.

(b) $\mathcal{G}_{\alpha,K,d}$ is a Donsker class for any $\alpha > d/2$ and law P on I^d.

(c) If $\alpha > d$, then $\mathcal{C}(\alpha, K, d+1)$ is a Donsker class for any law Q on I^{d+1} with a bounded density.

Proof. Parts (a) and (b) follow from Theorem 8.4(b) for $r = 1$ by the Blum–DeHardt Theorem 7.1, and for $r = 2$ and Ossiander's Theorem 7.6, respectively. Part (c) follows from Theorem 8.4(d) and Corollary 7.8. □

Side Remark. The need for differentiability of order larger than $d/2$ for some conclusion, as for the Donsker property in Corollary 8.5(b), also occurs in the very important Sobolev embedding, relating to Schwartz (–Sobolev) distribution theory and partial differential equations.

For $1 \le q < \infty$ let $\mathcal{L}_{\text{loc}}^q(\mathbb{R}^d)$ be the space of measurable real-valued functions f such that $\int_U |f(x)|^q \, dx < \infty$ for every bounded open set U. If $f \in \mathcal{L}_{\text{loc}}^1(\mathbb{R}^d)$, then a Schwartz distribution (generalized function) $[f]$ is defined by $[f](\phi) = \int_{\mathbb{R}^d} f(x)\phi(x)dx$ for every $\phi \in \mathcal{D}$, the space of C^∞ functions with compact support. For a multi-index $p = (p_1, \ldots, p_d)$, $p_j \in \mathbb{N}$, one writes $D^p[f] = [g]$ if also $g \in \mathcal{L}_{\text{loc}}^1$ and $(-1)^{[p]}[f](D^p\phi) = [g](\phi)$ for all $\phi \in \mathcal{D}$. (If f is C^∞, then one can take $g = D^p f$ in the classical sense via integration by parts $[p]$ times.) One form of Sobolev embedding (e.g., Hörmander, 1983, Theorem 4.5.13(ii)) states:

Theorem. Let $1 \le q < \infty$, let $m > 0$ be an integer, and let $u \in \mathcal{L}_{\text{loc}}^1(\mathbb{R}^d)$. Suppose for each p with $[p] = m$, $D^p[u] = [u_p]$ for some $u_p \in \mathcal{L}_{\text{loc}}^q(\mathbb{R}^d)$. If $mq > d$, then $[u] = [v]$ ($u = v$ almost everywhere for Lebesgue measure) where v is continuous and moreover Hölder of order γ if $0 < \gamma < 1$ and $\gamma \le m - (d/q)$, meaning that for any compact $K \subset \mathbb{R}^d$, $\sup_{x,y\in K, x\ne y} |v(x) - v(y)|/|x - y|^\gamma < \infty$.

Often $q = 2$ is taken, as Hilbert space properties of $\mathcal{L}^2$ are convenient. In that case the condition on m is $m > d/2$, as in Corollary 8.5(b). An example shows sharpness: let $d = 2$, $m = 1$, $q = 2$, and for $r = \sqrt{x^2 + y^2}$ on $\mathbb{R}^2$, let $f(x, y) = \log(\log(1/r))$ for $r \leq 1/e$, defined otherwise elsewhere to make it a C^∞ function except at $(0, 0)$, approaching which it is (very slowly) unbounded and discontinuous. The first partial derivatives of f are in $\mathcal{L}^q_{\mathrm{loc}}$ for $q = 2$ but not for any $q > 2$ (which would contradict the embedding theorem).

Note. For $\alpha \geq 1$, in Theorem 8.4(c) about h, the order $\varepsilon^{-d/\alpha}$ is precise; see Corollary 8.10.

Proof of Theorem 8.4. If part (a) holds, then part (b) follows using (7.2) and (7.3). Then parts (c) and (d) follow by (8.1) and (8.3). To begin the proof of part (a), for each $f \in \mathcal{G}_{\alpha, K, d}$, $x \in I^d$ and $x + h \in I^d$ write the Taylor series with remainder

$$f(x + h) - \sum_{k=0}^{\beta} Q_k(x, h) = R(x, h) \tag{8.5}$$

where for each x, $Q_k(x, \cdot)$ is a homogeneous polynomial of degree k in h and by the mean value theorem

$$|R(x, h)| \leq C|h|^\alpha \tag{8.6}$$

for some constant $C = C(d, K, \alpha)$. Then $C \geq 1$ will be taken large enough so that whenever $[p] \leq \beta$ we also have

$$\begin{aligned} |R_p(x, h)| &:= \left| D^p f(x + h) - \sum_{k=0}^{\beta - [p]} Q_{k,p}(x, h) \right| \\ &\leq C|h|^{\alpha - [p]} \end{aligned} \tag{8.7}$$

where for each k, p and x, $Q_{k,p}(x, \cdot)$ is also a homogeneous polynomial of degree k.

To prove one half of the first conclusion in Theorem 8.4 it needs to be shown that

$$\limsup_{\varepsilon \downarrow 0} \left[\log D(\varepsilon, \mathcal{G}_{\alpha, K, d}, d_{\mathrm{sup}}) \right] \varepsilon^{d/\alpha} < \infty. \tag{8.8}$$

Given $0 < \varepsilon < 1$, let $\Delta := (\varepsilon/(4C))^{1/\alpha} < 1$. Let $x_{(1)}, \ldots, x_{(s)}$ be a $\Delta/2$-net in I^d, i.e., $\sup \left\{ \inf_{j \leq s} |x - x_{(j)}| : x \in I^d \right\} \leq \Delta/2$. Here we can take $s \leq M_2 \varepsilon^{-d/\alpha}$ for some constant $M_2 = M_2(d, C)$; specifically, one can choose the $x_{(j)}$ as centers of cubes of a decomposition of I^d into cubes of side $1/m$ where m is the least integer $\geq d^{1/2}/\Delta$. For each multi-index p with $[p] = k \leq \beta$ let $p! := \prod_{j=1}^{d} p_j!$ and for $h \in \mathbb{R}^d$, $h^p := \prod_{j=1}^{d} h_j^{p_j}$. Then for $f \in \mathcal{G}_{\alpha, K, d}$ let

$Q^{(p)}(x,h) := (D^p f)(x)h^p/p!$. Thus in (8.5)

$$Q_k(x,h) = \sum_{[p]=k} Q^{(p)}(x,h). \tag{8.9}$$

Let $\varepsilon_k := \varepsilon/(2\Delta^k e^d)$, $k = 0, 1, \ldots,$

$$A_{i,p} := A_{i,p}(f) := \left[D^p f(x_{(i)})/\varepsilon_k \right], \quad i = 1, \ldots, s, \quad [p] = k \le \beta$$

where $[x]$ denotes the largest integer $\le x$, $-\infty < x < \infty$.

Given some $A := \{A_{i,p} : i \le s, [p] \le \beta\}$ let

$$\mathcal{G}_{\alpha,K,d}(A) := \left\{ f \in \mathcal{G}_{\alpha,K,d} : A_{i,p}(f) = A_{i,p} \text{ for all } i \le s, [p] \le \beta \right\}.$$

Lemma 8.6 *If $f, g \in \mathcal{G}_{\alpha,K,d}(A)$ for some A, then*

$$\sup \left\{ |(f-g)(x)| : x \in I^d \right\} \le \varepsilon.$$

Proof. Let $F := f - g$. Whenever $[p] \le \beta$ and $i \le s$, we will have $|(D^p F)(x_{(i)})| \le \varepsilon_{[p]}$. Also, by (8.7),

$$|D^p F(x+h) - D^p F(x)| \le 2C|h|^{\alpha-\beta} \text{ if } [p] = \beta.$$

For each $y \in I^d$ take an $x_{(i)}$ with $|y - x_{(i)}| \le \Delta/2$. Then from (8.5), (8.6), and (8.9) with $h := y - x_{(i)}$,

$$|F(y)| \le 2C|h|^\alpha + \sum_{k=0}^\beta \varepsilon_k \sum_{[p]=k} |h^p|/p!$$

$$\le 2C\Delta^\alpha + \sum_{k=0}^\beta \varepsilon_k \Delta^k \left(\sum_{[p]=k} 1/p! \right)$$

$$\le e^d \max_{k \le \beta} \varepsilon_k \Delta^k + \varepsilon/2 \le \varepsilon,$$

proving the Lemma. $\square$

Now continuing the proof of Theorem 8.4, it follows that

$$D(\varepsilon, \mathcal{G}_{\alpha,K,d}, d_{\text{sup}}) \le N_{\alpha,K,d}$$

where $N_{\alpha,K,d}$ is the number of distinct nonempty sets $\mathcal{G}_{\alpha,K,d}(A)$. Let the $x_{(j)}$ be ordered so that for $1 < j \le s$, $|x_{(i)} - x_{(j)}| \le \Delta$ for some $i < j$. Such an ordering clearly exists for $d = 1$. Then by induction on d, we enumerate subcubes beginning and ending with subcubes at vertices of I^d, where we first enumerate cubes on one face I^{d-1}, then on the adjoining level, etc.

Now suppose $f \in \mathcal{G}_{\alpha,K,d}(A)$ (so that this set is nonempty) and suppose given the values $A_{i,r}$ for $i < j$ for some $j \le s$. Choose $i < j$ such that $|x_{(i)} - x_{(j)}| \le$

Δ. Take the Taylor expansion as in (8.5) with $x = x_{(i)}$, $h = x_{(j)} - x_{(i)}$. By (8.7) and (8.9),

$$\left| D^p f(x_{(j)}) - \sum_{k=0}^{\beta - [p]} \sum_{[q]=k} D^{p+q} f\left(x_{(i)}\right) h^q / q! \right| \le C |h|^{\alpha - [p]}.$$

Now $|D^{p+q} f(x_{(i)}) - A_{i,p+q}\varepsilon_{k+[p]}| \le \varepsilon_{[p+q]}$ for $[q] = k = 0, 1, \ldots, \beta - [p]$. Let

$$D(j, f, p) := \left[D^p f(x_{(j)}) - \sum_{k=0}^{\beta - [p]} \varepsilon_{k+[p]} \sum_{[q]=k} A_{i,p+q} h^q / q! \right] / \varepsilon_{[p]}.$$

Then by the latter two inequalities,

$$|D(f, j, p)| \le \left(C \Delta^{\alpha - [p]} + \sum_{k=0}^{\beta - [p]} \varepsilon_{[p]+k} \Delta^k \sum_{[q]=k} 1/q! \right) / \varepsilon_{[p]}$$

$$\le 2Ce^d / 4C + e^d \le 3e^d / 2.$$

So for $f \in \mathcal{G}_{\alpha, K, d}$ and given the $A_{i,r}(f)$ for $i < j$ there are at most $3e^d + 2 < 4e^d$ possible values of $A_{j,p}$ for a given p. The number of different $p \in \mathbb{N}^d$ with $[p] \le \beta$ is bounded above by $(\beta + 1)^d$. Thus the number of possible sets of values $\{A_{j,p}\}_{[p] \le \beta}$ for the given $j \ge 2$ is at most $\exp((d + \log 4)(\beta + 1)^d)$. The number of possible values of the vectors $\{A_{1,p}\}_{[p] \le \beta}$ is at most

$$(2(2K + 1)e^d / \varepsilon)^{(\beta + 1)^d}.$$

Thus by Lemma 8.6

$$\begin{aligned} N_{\alpha, K, d} &\le \left(2e^d (2K + 1)(4e^d)^s / \varepsilon \right)^{(\beta + 1)^d} \\ &\le \exp((\beta + 1)^d \{ (d + \log 4) M_2 \varepsilon^{-d/\alpha} + \log \left(e^d (4K + 2)/\varepsilon \right) \}) \\ &\le \exp(J \varepsilon^{-d/\alpha}) \end{aligned}$$

for some $J = J(d, \alpha, K)$ not depending on ε, proving (8.8).

In the other direction we need to prove

$$\liminf_{\varepsilon \downarrow 0} \log D(\varepsilon, \mathcal{G}_{\alpha, K, d}, d_{\sup}) \varepsilon^{d/\alpha} > 0. \tag{8.10}$$

Let f be a C^∞ function on $\mathbb{R}^d$, 0 outside I^d and positive on its interior, such as

$$f(x) := \prod_{j=1}^d g(x_j), \quad \text{where}$$

$$g(t) := \begin{cases} \exp(-1/t)\exp(-1/(1 - t)), & 0 < t < 1 \\ 0 & \text{elsewhere.} \end{cases} \tag{8.11}$$

For $m = 1, 2, \ldots$, decompose I^d into m^d subcubes A_{mi} of side $1/m$, $i = 1, \ldots, m^d$. Let $x_{(i)}$ be the vertex of A_{mi} closest to the origin, in other words, the vertex of A_{mi} at which all coordinates are smallest. Given $\alpha > 0$ set $f_i(x) := m^{-\alpha} f\left(m\left(x - x_{(i)}\right)\right)$. Note that $x \mapsto m(x - x_{(i)})$ is a 1–1 affine map of $\mathbb{R}^d$ onto itself, taking the interior of A_{mi} onto that of I^d. Here an *affine transformation* is a function A from $\mathbb{R}^d$ onto itself of the form $A(x) = Bx + v$ where B is a linear transformation, defined by a $d \times d$ matrix which in this case will need to be nonsingular, and v is any fixed element of $\mathbb{R}^d$. Thus $f_i(x) > 0$ if and only if x is in the interior of A_{mi}. Let $s := \sup_x f(x) (= e^{-4d}$ for our f). For any $S \subset \{1, \ldots, m^d\}$ let $f_S := \sum_{i \in S} f_i$. Then $\| f_S \|_\alpha \le \| f \|_\alpha =: B$ while for $S \ne T$, $\sup_x |(f_S - f_T)(x)| = m^{-\alpha} s$. Thus for any $\varepsilon_m := Km^{-\alpha}s/(3B)$ we have $D\left(\varepsilon_m, \mathcal{G}_{\alpha,K,d}, d_{\sup}\right) \ge 2^{m^d} \ge \exp\left(C\varepsilon_m^{-d/\alpha}\right)$ for some $C = C(K, d, \alpha, s)$ not depending on m. Since $\varepsilon_{m+1}/\varepsilon_m \to 1$ as $m \to \infty$ this is enough to prove (8.10), and so finish the proof of Theorem 8.4(a).

Now it will be shown how to adapt the proof to part (e) for the ball $\overline{B}_d$ in place of I^d. In $S^{d-1} := \{x \in \mathbb{R}^d : |x| = 1\}$, for $0 < \varepsilon \le 1$, take $D(\varepsilon, S^{d-1}, e)$ points at distances $> \varepsilon$ apart where e is the Euclidean metric on $\mathbb{R}^d$. The balls of radius $\varepsilon/2$ with centers at these points are disjoint. Thus by volumes, then the mean value theorem,

$$D(\varepsilon, S^{d-1}, e) \left(\frac{\varepsilon}{2}\right)^d \le \left(1 + \frac{\varepsilon}{2}\right)^d - \left(1 - \frac{\varepsilon}{2}\right)^d$$

$$\le d\varepsilon \left(1 + \frac{\varepsilon}{2}\right)^{d-1} \le d\varepsilon \left(\frac{3}{2}\right)^{d-1},$$

and so $D(\varepsilon, S^{d-1}, e) \le 2d(3/\varepsilon)^{d-1}$. Take a maximal set S in S^{d-1} of points at distances $> \varepsilon/2$ apart, with $|S| \le 2d(6/\varepsilon)^{d-1}$. Let W consist of 0 and all points tx for $x \in S$ and $t = j\varepsilon/2$, $j = 1, 2, \ldots, \lfloor 2/\varepsilon \rfloor$. Then W is ε-dense in $\overline{B}_d$ and

$$|W| \le 1 + 2d(2/\varepsilon)(6/\varepsilon)^{d-1} \le d(6/\varepsilon)^d.$$

Then, starting at 0 and moving outward along each segment tx, $x \in S$, $0 \le t \le 1$, through points in W, one can do the same proof as for (8.8) in I^d, except for larger constants J, M_1. For a lower bound of the form (8.10), note that $\overline{B}_d$ includes a cube of side $d^{-1/2}$ centered at 0. This finishes the proof of Theorem 8.4. $\qquad\square$

Next, some lower bounds for covering or packing numbers will be given. For a collection $\mathcal{F} \subset \mathcal{L}^1(A, \mathcal{A}, P)$, we have the $\mathcal{L}^1$ distance $d_{1,P}(f, g) := P(|f - g|)$.

Theorem 8.7 *Let P be a law on I^d having a density with respect to Lebesgue measure bounded below by $\gamma > 0$. Then for some $C = C(\gamma, \alpha, K, d) > 0$, for*

$\varepsilon > 0$ *small enough, and* $1 \le r \le \infty$,

$$N_{[]}^{(r)}(\varepsilon, \mathcal{G}_{\alpha, K, d}, P) \ge N_{[]}^{(1)}(\varepsilon, \mathcal{G}_{\alpha, K, d}, P) \ge D(\varepsilon, \mathcal{G}_{\alpha, K, d}, d_{1, P})$$
$$\ge \exp\left(C\varepsilon^{-d/\alpha}\right).$$

If $d \ge 2$, *for small enough* $\varepsilon > 0$, *and* $M := C(\gamma, \alpha, K, d - 1)$,

$$N_I(\varepsilon, \mathcal{C}(\alpha, K, d), P) \ge D(\varepsilon, \mathcal{C}(\alpha, K, d), d_P) \ge \exp\left(M\varepsilon^{-(d-1)/\alpha}\right).$$

Proof. The following combinatorial fact will be used:

Lemma 8.8 *Let B be a set with n elements, $n = 0, 1, \ldots$. Then there exist subsets $E_i \subset B$, $i = 1, \ldots, k$, where $k \ge e^{n/6}$, such that for $i \ne j$, the symmetric difference $E_i \Delta E_j$ has at least $n/5$ elements.*

Proof. For any set $E \subset B$, the number of sets $F \subset B$ such that $\mathrm{card}(E \Delta F) \le n/5$ is $2^n B(n/5, n, 1/2)$, where binomial probabilities $B(k, n, p)$ are as defined before the Chernoff inequality (Theorem 1.15). If S_n is the sum of n independent Rademacher variables X_i taking values ± 1 with probability $1/2$ each, then by one of Hoeffding's inequalities (Proposition 1.12), defining "success" as $X_i = -1$,

$$B(n/5, n, 1/2) = \Pr\left(S_n \ge 3n/5\right) \le \exp(-9n/50) < e^{-n/6}.$$

Let $d(E, F) := \mathrm{card}(E \Delta F)$. Thus for any E, $\mathrm{card}(\{F : d(E, F) \le n/5\}) \le 2^n e^{-n/6}$. Recursively, choose a set E_1, say $\emptyset$. Given $E_1, \ldots, E_m$ such that $d(E_i, E_j) > n/5$ for $i \ne j, i, j \le m$, we have

$$\mathrm{card}\left(\bigcup_{j=1}^{m}\{F : d(E_j, F) \le n/5\}\right) \le m 2^n e^{-n/6} < 2^n$$

if $m < e^{n/6}$. Then we can choose E_{m+1} such that $d(E_{m+1}, E_j) > n/5$ for all $j = 1, \ldots, m$ and continue until $m = k \ge e^{n/6}$ as stated. □

Now to prove Theorem 8.7, the first inequality follows from (7.2), and the second is also straightforward. For the lower bound on D, let us use again the construction in the proof of (8.10). Let λ denote Lebesgue measure on I^d and $\delta := \gamma \int f \, d\lambda$ for the f in (8.11). Then for each i, $\int f_i \, dP \ge \delta m^{-\alpha - d}$. Applying Lemma 8.8 and obtaining sets S with $\mathrm{card}(S) \ge m^d/5$ gives $\int f_S \, dP > \delta m^{-\alpha}/6$, and $\int |f_S - f_T| \, dP = \int f_{S \Delta T} \, dP$. So

$$D\left(\delta m^{-\alpha}/6, \mathcal{G}_{\alpha, K, d}, d_{1, P}\right) \ge \exp\left(m^d/6\right).$$

Thus if $0 < \varepsilon \le \delta/6$, since $[x] \ge x/2$ for $x \ge 1$,

$$D\left(\varepsilon, \mathcal{G}_{\alpha, K, d}, d_{1, P}\right) \ge \exp\left(\left[(\delta/(6\varepsilon))^{1/\alpha}\right]^d / 6\right)$$
$$\ge \exp\left(2^{-d}(\delta/(6\varepsilon))^{d/\alpha} / 6\right),$$

proving the statement about $\mathcal{G}_{\alpha,K,d}$.

If $C = J(f)$ and $D = J(g)$, then

$$d_P(C, D) := P(C \triangle D) \geq \gamma \lambda(C \triangle D) = \gamma \int |f - g| d\lambda, \text{ so for } \varepsilon > 0$$

$$N_I(\varepsilon, C(\alpha, K, d), P) \geq D(\varepsilon, C(\alpha, K, d), d_P) \geq D\left(\varepsilon/\gamma, \mathcal{G}_{\alpha,K,d-1}, d_{1,\lambda}\right),$$

which finishes the proof of Theorem 8.7. □

To get lower bounds for the Hausdorff metric, the following will help:

Lemma 8.9 *If* $\alpha \geq 1$ *and* $f, g \in \mathcal{G}_{\alpha,K,d}$, *then*

$$h\left(J_f, J_g\right) \geq d_{\sup}(f, g)/(2 \max(1, Kd)).$$

Proof. Note that for $\alpha \geq 1$, any $g \in \mathcal{G}_{\alpha,K,d}$ is Lipschitz in each coordinate by the mean value theorem with $|g(x) - g(y)| \leq K|x - y|$ if $x_j = y_j$ for all but one value of j. In the cube, one can go from a general x to a general y by changing one coordinate at a time, so g is Lipschitz with $\|g\|_L \leq Kd$ and Lemma 8.1 applies. □

Corollary 8.10 *If* $\alpha \geq 1$ *and* $d = 1, 2, \ldots$, *then as* $\varepsilon \downarrow 0$,

$$\log D(\varepsilon, C(\alpha, K, d+1), h) \asymp \varepsilon^{-d/\alpha}.$$

Proof. This follows from Lemma 8.9 and Theorem 8.4. □

Remark 8.11 For $m = 1, 2, \ldots$, let I^d be decomposed into a grid of m^d subcubes of side $1/m$. Let E be the set of centers of the cubes. For any $A \subset I^d$ let $B \subset E$ be the set of centers of the cubes in the grid that A intersects. Then $h(A, B) \leq d^{1/2}/(2m)$, which includes the possibility that $A = B = \emptyset$. For $0 < \varepsilon < 1$ there is a least $m = 1, 2, \ldots$ such that $d^{1/2}/m \leq \varepsilon$, namely, $m = \lceil d^{1/2}/\varepsilon \rceil$. It follows that

$$D\left(\varepsilon, 2^{I^d}, h\right) \leq 2^{\left(1 + d^{1/2}/\varepsilon\right)^d}.$$

Hence for $\alpha < d/(d+1)$, Corollary 8.10 cannot hold, nor can the upper bound for h in Theorem 8.4 be sharp.

The classes $C(\alpha, K, d)$ considered so far contain sets with flat faces except for one curved face. There are at least two ways to form more general classes of sets with piecewise differentiable boundaries, still satisfying the bounds in Theorem 8.4. One is to take a bounded number of Boolean operations. Let $v_1, \ldots, v_k$ be nonzero vectors in $\mathbb{R}^d$ where $d \geq 2$. For constants $c_1, \ldots, c_k$ let $H_j := \{x \in \mathbb{R}^d : (x, v_j) = c_j\}$, a hyperplane. Let π_j map each $x \in \mathbb{R}^d$ to its nearest point in H_j, $\pi_j(x) := x - ((x, v_j) - c_j)v_j/|v_j|^2$. Let T_j be a cube in

H_j and α, $K > 0$. Let f_j be an affine transformation taking T_j onto I^{d-1}. For $g \in \mathcal{G}_{\alpha,K,d-1}$ with $g \geq 0$, let

$$J_j(g) := \{x \in \mathbb{R}^d : \pi_j(x) \in T_j, \ c_j \leq (v_j, x) \leq c_j + g(f_j(\pi_j(x)))\}.$$

Let

$$\mathcal{C}_j := \mathcal{C}_j(\alpha, K, v_j, c_j) := \{J_j(g) : g \in \mathcal{G}_{\alpha,K,d-1}\}.$$

Then Theorem 8.4 implies that if C is a cube in $\mathbb{R}^d$ including all sets in $\mathcal{C}_j$, and P is a law on C having bounded density with respect to Lebesgue measure λ, then for some $M_j < \infty$,

$$\log D(\varepsilon, \mathcal{C}_j, d_P) \ \leq \ \log N_I(\varepsilon, \mathcal{C}_j, P) \ \leq \ M_j \varepsilon^{(1-d)/\alpha}.$$

We then have by Theorem 8.2 the following:

Theorem 8.12 Let $d \geq 2$ and let $\mathcal{C}_0 := \cap_{j=1}^k \mathcal{C}_j$ or $\mathcal{C}_0 := \sqcup_{j=1}^k \mathcal{C}_j$, for $\mathcal{C}_j$ as just defined. Then for some $M < \infty$,

$$\log D(\varepsilon, \mathcal{C}_0, d_P) \ \leq \ \log N_I(\varepsilon, \mathcal{C}_0, P) \ \leq \ M \varepsilon^{(1-d)/\alpha}.$$

By intersections or unions of k sets in classes $\mathcal{C}_j$ (with k depending on d), one can obtain sets with smooth boundaries (through order α) such as ellipsoids; see Problem 5. Unions work more easily. One can also get more general sets, since, e.g. for $\alpha > 1$, the minimum or maximum of two functions in $\mathcal{G}_{\alpha,K,d}$ need not have first derivatives everywhere and then will not be in $\mathcal{G}_{\gamma,\kappa,d}$ for any $\gamma > 1$ and $\kappa < \infty$.

Recall that a C^∞ real-valued function on an open set on $\mathbb{R}^d$ is one such that the partial derivatives $D^p f$ exist for all $p \in \mathbb{N}^d$ and are continuous. For functions $f := (f_1, \ldots, f_k)$ into $\mathbb{R}^k$, for f to be C^∞ means that each f_j is. Another way to generate sets with boundaries differentiable of order α is as follows. The unit sphere $S^{d-1} := \{x \in \mathbb{R}^d : |x| = 1\}$ is a C^∞ manifold, specifically as follows. S^{d-1} is the union of two sets $A := \{x \in S^{d-1} : x_1 > -1/2\}$ and $C := \{x \in S^{d-1} : x_1 < 1/2\}$. There is a 1–1, C^∞ function ψ from $\{x \in \mathbb{R}^{d-1} : |x| < 9/8\}$ into $\mathbb{R}^d$, with derivative matrix $\{\partial \psi_i / \partial x_j\}_{i=1,j=1}^{d,d-1}$ of maximum rank $d - 1$ everywhere, such that ψ takes $B_{d-1} := \{x \in \mathbb{R}^{d-1} : |x| < 1\}$ onto A. Let $\eta(y) := (-\psi_1(y), \psi_2(y), \ldots, \psi_d(y))$. Then the above statements for ψ and A also hold for η and C.

For $0 < \alpha$, $K < \infty$ let $\mathcal{F}_{\alpha,K}(S^{d-1})$ be the set of functions $h : S^{d-1} \to \mathbb{R}$ such that for $\overline{B}_{d-1} := \{x \in \mathbb{R}^{d-1} : |x| \leq 1\}$, $h \circ \psi$ and $h \circ \eta \in \mathcal{F}_{\alpha,K}(\overline{B}_{d-1})$, recalling that $f \circ g(y) := f(g(y))$. Let $\mathcal{F}_{\alpha,K}^{(d)}(S^{d-1})$ be the set of functions $h = (h_1, \ldots, h_d)$ such that $h_j \in \mathcal{F}_{\alpha,K}(S^{d-1})$ for each $j = 1, \ldots, d$.

Two continuous functions F, G from one topological space X to another, Y, are called *homotopic* iff there exists a jointly continuous function H from

$X \times [0, 1]$ into Y such that $H(\cdot, 0) \equiv F$ and $H(\cdot, 1) \equiv G$. H is then called a *homotopy* of F and G. Let $I(F)$ be the set of all $y \in Y$, not in the range of F, such that among mappings of X into $Y \setminus \{y\}$, F is not homotopic to any constant map $G(x) \equiv z \neq y$.

For a function F let $R(F) := \mathrm{ran}(F) := \mathrm{range}(F)$ and $C(F) := I(F) \cup R(F)$.

For example, if F is the identity from S^{d-1} onto itself in $\mathbb{R}^d$, then $I(F) = \{y : |y| < 1\}$ by well-known facts in algebraic topology, e.g., Eilenberg and Steenrod (1952, Chapter 11, Theorem 3.1).

Let $I(d, \alpha, K) := \{I(F) : F \in \mathcal{F}^{(d)}_{\alpha, K}(S^{d-1})\}$ and $\mathcal{K}(d, \alpha, K) := \{C(F) : F \in \mathcal{F}^{(d)}_{\alpha, K}(S^{d-1})\}$. Then $I(d, \alpha, K)$ is a collection of open sets and $\mathcal{K}(d, \alpha, K)$ of compact sets, each of which, in a sense, have boundaries differentiable of order α. (For functions F that are not one-to-one, the boundaries may not be differentiable in some other senses.) For $\mathcal{K}(d, \alpha, K)$ and to some extent for $I(d, \alpha, K)$ there are bounds as for other classes of sets with α times differentiable boundaries (Theorem 8.12):

Theorem 8.13 *For each* $d = 2, 3, \ldots, K \geq 1$ *and* $\alpha \geq 1$,

(a) there is a constant $H_{d,\alpha,K} < \infty$ *such that for* $0 < \varepsilon \leq 1$, *and the Hausdorff metric* h,

$$\log D(\varepsilon, \mathcal{K}(d, \alpha, K), h) \leq H_{d,\alpha,K} / \varepsilon^{(d-1)/\alpha}.$$

(b) For any $\zeta < \infty$ *there is a there is a constant* $A_{d,\alpha,K,\zeta} < \infty$ *such that for any law* P *on* $\mathbb{R}^d$ *having density with respect to* λ^d *bounded above by* ζ, *for* $0 < \varepsilon \leq 1$,

$$\max(\log N_I(\varepsilon, \mathcal{K}(d, \alpha, K), P), \log N_I(\varepsilon, I(d, \alpha, K), P))$$
$$\leq A_{d,\alpha,K,\zeta} / \varepsilon^{(d-1)/\alpha}.$$

The proof will follow from a sequence of lemmas.

Lemma 8.14 *If* H *is a homotopy of* F *and* G, *then* $I(F) \Delta I(G) \subset R(H)$.

Proof. Suppose $y \in I(F) \setminus I(G)$ and y is not in the range of H. Then F and G are homotopic as maps into $Y \setminus \{y\}$. Homotopy is clearly a transitive relation. Since G is homotopic to a constant map into $Y \setminus \{y\}$, so is F, a contradiction. The Lemma is proved.   □

Lemma 8.15 *If* F *is a continuous map from a compact Hausdorff topological space* K *into a Hausdorff space* Y, *then* $C(F)$ *is closed.*

Proof. If $y \notin C(F)$, then there is a homotopy H of F to a constant map $G(x) \equiv z$ into $Y \setminus \{y\}$. Then $R(H)$ is compact, thus closed (RAP, Theorem

2.2.3 and Proposition 2.2.9). Clearly $I(G) = \emptyset$, so by the previous lemma, $I(F) \subset R(H)$. Then the open complement $Y \setminus R(H) \subset Y \setminus C(F)$, so $Y \setminus C(F)$ is open and $C(F)$ is closed. $\qquad \square$

Lemma 8.16 *If F is a continuous map from a compact Hausdorff topological space into some $\mathbb{R}^d$, then $I(F)$ is open, and its boundary is included in $R(F)$.*

Proof. Let $x \in I(F)$. Then for some $\delta > 0$, $d(x, y) < 2\delta$ implies $y \notin R(F)$. Let $d(x, y) < \delta$. Then there is a homeomorphism g of $\mathbb{R}^d$ which leaves $\{u : |u - x| \geq 2\delta\}$, and so the values of F, fixed and takes x to y. To define such a g we can assume $x = 0$ and $\delta = 1/2$, let $g(u) := u$ for $|u| \geq 1$ and $g(u) := u + y(1 - |u|)$ for $|u| \leq 1$. Then g is the identity for $|u| \geq 1$ and is continuous, with $g(0) = y$. Also, g is 1–1 since $|g(u)| < 1$ for $|u| < 1$, and if $g(u) = g(v)$ with $|u|, |v| < 1$, then $u - v = y(|u| - |v|)$ and $|u - v| \leq |(|u| - |v|)|/2 \leq |u - v|/2$, so $u = v$. Thus $y \in I(F)$, so $I(F)$ is open. Since $C(F)$ is closed by Lemma 8.15, it follows that the boundary of $I(F)$ is included in $R(F)$. $\qquad \square$

Recall that for a metric space (S, d), set $A \subset S$ and $\delta > 0$, the δ-interior of A is defined by $_\delta A := \{x : d(x, y) < \delta \text{ implies } y \in A\}$, and the δ-neighborhood by $A^\delta := \{y : d(x, y) < \delta \text{ for some } x \in A\}$.

Lemma 8.17 *For continuous functions F, G from S^{d-1} into $\mathbb{R}^d$, if $d_{\sup}(F, G)$ $:= \sup\{|F(u) - G(u)| : u \in S^{d-1}\} < \delta$, then*

$$_\delta I(F) \subset I(G) \subset C(G) \subset C(F)^\delta.$$

Proof. If $x \in _\delta I(F)$ and $x \in R(G)$, then $d(x, y) < \delta$ for some $y \in R(F)$, so $y \notin I(F)$, a contradiction. So $x \notin R(G)$. For $0 \leq t \leq 1$ and $u \in S^{d-1}$, let $H(u, t) := (1 - t)F(u) + tG(u)$. Then H is a homotopy of F and G, and $R(H) \subset R(F)^\delta$, but $I(F) \cap R(F) = \emptyset$, so $x \notin R(H)$. Thus by Lemma 8.14, $x \in I(G)$.

Next, let $y \in C(G)$. If $y \in R(G)$ then $y \in R(F)^\delta \subset C(F)^\delta$. Otherwise $y \in I(G)$. Then $y \in I(F)$ or by Lemma 8.14, $y \in R(H) \subset R(F)^\delta$. $\qquad \square$

Lemma 8.18 *For any continuous function F from a compact Hausdorff space K into $\mathbb{R}^d$, $C(F)^\delta \setminus {}_\delta I(F) \subset R(F)^\delta$.*

Proof. Let $x \in C(F)^\delta \setminus {}_\delta I(F)$. Suppose $d(x, R(F)) \geq \delta$. Then $|x - y| < \delta$ for some $y \in I(F)$. Since the boundary of $I(F)$ is included in $R(F)$ by Lemma 8.16, the line segment $\{tx + (1 - t)y : 0 \leq t \leq 1\} \subset I(F)$. It follows that $x \in I(F)$ and then likewise that $z \in I(F)$ whenever $|x - z| < \delta$. Thus $x \in {}_\delta I(F)$, a contradiction. $\qquad \square$

The Lipschitz seminorm $\|F\|_L$ is defined for functions with values in $\mathbb{R}^d$ just as for real-valued functions. Let v_k be the Lebesgue volume of the unit ball in $\mathbb{R}^k$.

Lemma 8.19 *For $k = 1, 2, \ldots$, if (T, d) is a metric space, $\delta > 0$, for some $M < \infty$, $D(\delta, T, d) \leq M\delta^{1-k}$, and F is Lipschitz from T into $\mathbb{R}^k$, with $\|F\|_L \leq \kappa$, then*

$$\lambda^d(C(F)^\delta \setminus {}_\delta I(F)) \leq v_k M(\kappa + 2)^k \delta.$$

Proof. For the usual metric e on $\mathbb{R}^k$, we have $D(\kappa\delta, R(F), e) \leq M\delta^{1-k}$. It follows that $D((\kappa + 2)\delta, R(F)^\delta, e) \leq M\delta^{1-k}$. Lemma 8.18 gives the conclusion. $\qquad\square$

Proof of Theorem 8.13. By the definitions, a function $F \in \mathcal{F}_{\alpha,K}^{(d)}(S^{d-1})$ is given by a pair $F_{(1)}, F_{(2)}$ of functions $F_{(j)} := (F_{(j)1}, \ldots, F_{(j)d})$ where each $F_{(j)i} \in \mathcal{F}_{\alpha,K}(\overline{B}_{d-1})$ and $\overline{B}_j$ is the closed unit ball in $\mathbb{R}^j$. Since $\alpha \geq 1$, each $F_{(j)i}$ is Lipschitz with $\|F_{(j)i}\|_L \leq K$, so each $F_{(j)}$ is Lipschitz with $\|F_{(j)}\|_L \leq dK$. Let $T := T_1 \cup T_2$ be a union of two disjoint copies T_i of $\overline{B}_{d-1}$, with the Euclidean metric e on each and $e(x, y) := 2$ for $x \in T_i$, $y \in T_j$, $i \neq j$. Letting $G := F_{(j)}$ on T_j, $j = 1, 2$, gives a function $G := G_F$ on T with

$$\|G\|_L \leq \max_j \|F_{(j)}\|_L \leq dK. \tag{8.12}$$

Let $0 < \varepsilon \leq 1$. Then there are $D(\varepsilon, \overline{B}_j, e)$ disjoint balls of radius $\varepsilon/2$, included in a ball of radius $1 + (\varepsilon/2)$. It follows by volumes that

$$D(\varepsilon, \overline{B}_j, e) \leq [(2+\varepsilon)/2]^j (2/\varepsilon)^j \leq (3/\varepsilon)^j. \tag{8.13}$$

By Theorem 8.4 for the ball case, for any $K \geq 1$, $d \geq 2$ and $\alpha \geq 1$, there is a $C = C(K, d, \alpha) < \infty$ such that for $0 < \delta \leq 1$,

$$\log D(\delta, \mathcal{F}_{\alpha,K}(\overline{B}_{d-1}), d_{\sup}) \leq C/\delta^{(d-1)/\alpha}.$$

It follows from the definitions with $T := T_1 \cup T_2$ that

$$\log D(\delta, \mathcal{F}_{\alpha,K}(S^{d-1}), d_{\sup}) \leq 2C/\delta^{(d-1)/\alpha}$$

and thus that for $0 < \delta \leq 1$,

$$\log D(\delta, \mathcal{F}_{\alpha,K}^{(d)}(S^{d-1}), d_{\sup}) \leq 2C(d/\delta)^{(d-1)/\alpha}.$$

Given $0 < \delta \leq 1$ take a set of functions $f_1, \ldots, f_m \in \mathcal{F}_{\alpha,K}^{(d)}(S^{d-1})$ such that for $i \neq j$, $d_{\sup}(f_i, f_j) > \delta/2$ and maximal m where $m \leq \exp(\beta_d/\delta^{(d-1)/\alpha})$ and $\beta_d := 2C \cdot (2d)^{(d-1)/\alpha}$. The brackets $[{}_\delta I(f_j), C(f_j)^\delta]$ for $j = 1, \ldots, m$ cover $I(\alpha, K, d)$ and $\mathcal{K}(\alpha, K, d)$ by Lemma 8.17 (some sets ${}_\delta I(f_j)$ may be empty). If $d_{\sup}(g, f_j) \leq \delta/2$, then by Lemma 8.17, $C(g) \subset C(f_j)^\delta$ and $C(f_j) \subset C(g)^\delta$, so $h(C(g), C(f_j)) < \delta$ and part (a) of Theorem 8.13 follows. Then Lemma

8.19 applies with $M = 2 \cdot 3^{d-1}$ by (8.13) and $\kappa = Kd$ by (8.12), and holds for P in place of λ^d with an additional factor of ζ. Theorem 8.13(b) then follows with $\delta := \varepsilon/[2v_d\zeta \cdot 3^{d-1}(Kd+2)^d]$ and $A_{d,\alpha,K,\zeta} := \beta_d[2v_d\zeta \cdot 3^{d-1}(Kd+2)^d]^{(d-1)/\alpha}$. $\qquad\square$

Corollary 8.20 *For any law P on $\mathbb{R}^d$, $d \geq 2$, having bounded density with respect to Lebesgue measure, and $K < \infty$,*

(a) (Tze-Gong Sun) $I(d, \alpha, K)$ is a Donsker class for P if $\alpha > d - 1$.

(b) $I(d, \alpha, K)$ is a Glivenko–Cantelli class for P whenever $\alpha \geq 1$.

Proof. Apply 8.13(b) and, for part (a), Corollary 7.8; for part (b), the Blum–DeHardt theorem 7.1. $\qquad\square$

8.3 Lower Layers

A set $B \subset \mathbb{R}^d$ is called a *lower layer* if and only if for all $x = (x_1, \ldots, x_d) \in B$ and $y = (y_1, \ldots, y_d)$ with $y_j \leq x_j$ for $j = 1, \ldots, d$, we have $y \in B$. Let $\mathcal{LL}_d$ denote the collection of all nonempty lower layers in $\mathbb{R}^d$ with nonempty complement. Recall that $\emptyset$ is the empty set and let

$$\mathcal{LL}_{d,1} := \{L \cap I^d : L \in \mathcal{LL}_d, \ L \cap I^d \neq \emptyset\}.$$

Let $\lambda := \lambda_I^d$ denote Lebesgue measure on I^d. Thus d_λ is defined for any two Lebesgue measurable sets in $\mathbb{R}^d$ by $d_\lambda(A, B) := \lambda^d((A \triangle B) \cap I^d)$. The size of $\mathcal{LL}_{d,1}$ will be bounded first when $d = 1$ and 2. Let $\lceil x \rceil$ be the smallest integer $\geq x$.

Theorem 8.21 *For $d = 1$,*

$$D\left(\varepsilon, \mathcal{LL}_{1,1}, h\right) = D\left(\varepsilon, \mathcal{LL}_1, d_\lambda\right) = N_I\left(\varepsilon, \mathcal{LL}_{1,1}, \lambda\right) = \lceil 1/\varepsilon \rceil.$$

For $d = 2$ and any $m = 1, 2, \ldots,$ we have

$$N_I\left(2/m, \mathcal{LL}_{2,1}, \lambda_I^2\right) \leq \binom{2m-2}{m-1} \leq 2^{2m-2} = 4^{m-1}.$$

For $0 < \varepsilon \leq 1$, $N_I\left(\varepsilon, \mathcal{LL}_{2,1}, \lambda_I^2\right) \leq 4^{\lceil 2/\varepsilon \rceil}$. Lastly, for $0 < s < 1/m$,

$$D(\sqrt{2}/m, \mathcal{LL}_{2,1}, h) \leq \binom{2m}{m} - 1 \leq D(s, \mathcal{LL}_{2,1}, h)$$

and $D\left(\varepsilon, \mathcal{LL}_{2,1}, h\right) \leq 4^{1+\sqrt{2}/\varepsilon}$.

Proof. For $d = 1$, sets in $\mathcal{LL}_{1,1}$ are intervals $[0, t), 0 < t \leq 1$, or $[0, t], 0 \leq t \leq 1$. For any ε with $0 < \varepsilon < 1$, let $m := m(\varepsilon) := \lceil 1/\varepsilon \rceil$. Then the collection of m brackets $[[0, (k-1)\varepsilon], [0, k\varepsilon]]$, $k = 1, \ldots, m$, covers $\mathcal{LL}_{1,1}$ with minimal m for ε, showing that $N_I(\varepsilon, \mathcal{LL}_{1,1}, \lambda) = m$. For any $\delta > 0$, the points $k(\varepsilon + \delta)$

for $k = 0, 1, \ldots, [1/(\varepsilon + \delta)]$, are at distances at least $\varepsilon + \delta$ apart. For $0 \le x \le y \le 1$, $h([0, x], [0, y]) = d_\lambda([0, x], [0, y]) = y - x$, so

$$D(\varepsilon, \mathcal{LL}_{1,1}, h) = D(\varepsilon, \mathcal{LL}_1, d_\lambda) = D(\varepsilon, [0, 1], d) =: D(\varepsilon)$$

for the usual metric d. Letting $\delta \downarrow 0$ gives $D(\varepsilon) = \lceil 1/\varepsilon \rceil = m$ (cf. Chapter 1, Problem 5), finishing the proof for $d = 1$.

For $d = 2$, decompose the unit square I^2 into a union of m^2 squares $S_{ij} := [(i - 1)/m, i/m) \times [(j - 1)/m, j/m)$, $i, j = 1, \ldots, m - 1$, but for $i = m$ or $j = m$, replace "i/m" or "j/m" respectively by "1." For any $L \in \mathcal{LL}_{2,1}$, let $_mL$ be the union of the squares in the grid included in L and L_m the union of the squares which intersect L. Then $_mL \subset L \subset L_m$, and both $_mL$ and L_m are in $\mathcal{LL}_{2,1} \cup \{\emptyset\}$, with $L_m \ne \emptyset$.

For each m and each function f from $\{2, 3, \ldots, 2m - 1\}$ into $\{0, 1\}$ taking the value 1 exactly $m - 1$ times, define a sequence $S(f)(k), k = 1, \ldots, 2m - 1$ of squares in the grid as follows. Let $S(f)(1)$ be the upper left square S_{1m}. Given $S(f)(k - 1) = S_{ij}$, let $S(f)(k)$ be the square $S_{i+1,j}$ just to its right if $f(k) = 1$, otherwise the square $S_{i,j-1}$ just below it, for $k = 2, \ldots, 2m - 1$. Then $S(f)(2m - 1)$ is always the lower right square S_{m1}. Let $B_m(f) := \bigcup_{k=1}^{2m-1} S(f)(k)$. Let $A_m(f)$ be the union of the squares not in $B_m(f)$, below and to the left of it, and $C_m(f) := A_m(f) \cup B_m(f)$. Here $A_m(f)$ and $C_m(f)$ belong to $\mathcal{LL}_{2,1} \cup \{\emptyset\}$ with $C_m(f) \ne \emptyset$. Also, if $f \ne g$, then $h(C_m(f), C_m(g)) \ge 1/m$.

Let $L \in \mathcal{LL}_{2,1} \cup \emptyset$. Let $\overline{L}$ be its closure and

$$M := M_L := \overline{L} \cup \{(0, y) : 0 \le y \le 1\} \cup \{(x, 0) : 0 \le x \le 1\}.$$

Then $M \subset I^2$ is compact and is in $\mathcal{LL}_{2,1}$. The range of $x - y$ on M (or on I^2) is $[-1, 1]$. For each $t \in [-1, 1]$ there is a unique $(x, y) := (x(t), y(t)) \in M$ with $x - y = t$ such that $x + y$ is maximized. It follows from the lower layer properties that $x(\cdot)$ is nondecreasing and $y(\cdot)$ is nonincreasing. Let $g(t) := (x(t), y(t))$ and $G := \{g(t) : -1 \le t \le 1\}$. For $-1 \le s \le t \le 1$ we have

$$x(t) = y(t) + t \le y(s) + t = x(s) + t - s,$$

so $x(\cdot)$ is a Lipschitz function with $\|x(\cdot)\|_L \le 1$. Likewise $\|y(\cdot)\|_L \le 1$. In particular $x(\cdot)$ and $y(\cdot)$ are continuous, and the curve G is connected. We have $x(-1) = 0$, $y(-1) = 1$, $x(1) = 1$, and $y(1) = 0$. For t near -1, $g(t) \in S_{1m}$. If G intersects a square S_{ij} for $i < m$ or $j > 1$, then it next intersects, as t increases, one of the squares $S_{i+1,j}$ or $S_{i,j-1}$. (It cannot go directly to $S_{i+1,j-1}$ since the upper left vertex of that square is in $S_{i+1,j}$.) For t near 1, $g(t) \in S_{m1}$. Thus for some f as above, there is a sequence of squares $S(f)(1) = S_{1m}, \ldots, S(f)(2m - 1) = S_{m1}$ intersected by G, and no other squares S_{ij} are. It follows that

$$A_m(f) \subset {}_mL \subset L \subset L_m \subset C_m(f),$$

so that $L_m \setminus {}_mL \subset B_m(f)$, and $\lambda_I^2 (B_m(f)) = (2m - 1)/m^2 < 2/m$. Since the number of functions f is $\binom{2m-2}{m-1} \leq 4^{m-1}$, the first sentence of the Theorem for $d = 2$ is proved. For $0 < \varepsilon \leq 1$, let $m = \lceil 1/\varepsilon \rceil$. Then $m - 1 < 1/\varepsilon$, and the second sentence follows.

For the statements about the Hausdorff metric h, let $M \in \mathcal{LL}_2$ be such that $L := M \cap I^2 \neq \emptyset$, let ∂M be the boundary of M and $\partial_M L := \partial M \cap I^2$. Possibly $\partial_M L = \emptyset$, if I^2 is included in the interior of M. Otherwise $\partial_M L$ equals the graph of a function $(x(t), y(t))$ defined for $a \leq t \leq b$ where $-1 \leq a \leq b \leq 1$, with $x(a) = 0$ or $y(a) = 1$ or both, and $x(b) = 1$ or $y(b) = 0$ or both. Without loss of generality, we can assume that $(-1/m, 1) \in \partial M$ and $(1, -1/m) \in \partial M$.

There is a largest $j = j(L)$ with $1 \leq j \leq m$ such that S_{1j} intersects L, and a largest $i = i(L)$ with $1 \leq i \leq m$ such that S_{i1} intersects L. Given j and i, for each function f from $\{2, \dots, i + j - 1\}$ into $\{0, 1\}$ taking the value 1 exactly $i - 1$ times, we define $B_m(f)$, $A_m(f)$ and $C_m(f)$ as before, replacing $A_m(f)$ if it is empty by $\{0, 0\}$. Then for each L there is an f such that $A_m(f) \subset L \subset C_m(f)$. If L and L' have the same f, then $h(L, L') \leq \sqrt{2}/m$ and $h(L, C_m(f)) \leq \sqrt{2}/m$. The total number of possible functions f is

$$\sum_{i=1}^m \sum_{j=1}^m \binom{i + j - 2}{i - 1} = \sum_{i=0}^{m-1} \sum_{j=0}^{m-1} \binom{i + j}{i} = \binom{2m}{m} - 1.$$

To see this, consider an $(m + 1) \times (m + 1)$ grid of squares in $[-1/m, 1] \times [-1/m, 1]$, giving $\binom{2m}{m}$ possible $C_{m+1}(f)$, all but one of which intersect $S_{1,1}$. Or, to see the equality of the first and last expressions in the display, consider the numbers of strings of m 0's and m 1's, beginning with $m - i$ 0's and ending with $m - j$ 1's, which summed over i and j from 1 to m give all such strings except the one with first m 0's, then m 1's. It follows that $D(\sqrt{2}/m, \mathcal{LL}_{2,1}, h) \leq \binom{2m}{m} - 1$. Conversely, each set $C_m(f) \neq \emptyset$ is in $\mathcal{L}_{2,1}$, and two distinct such sets are at distance at least $1/m$ apart for h, so the right-hand inequality in the last display of the theorem follows. For $0 < \varepsilon \leq 1$ let $m := \lceil \sqrt{2}/\varepsilon \rceil$. Then $\sqrt{2}/m \leq \varepsilon$ and $\binom{2m}{m} - 1 < 4^m \leq 4^{1+\sqrt{2}/\varepsilon}$, proving the last statement. $\qquad\square$

Recalling again the definition (8.4) of $\asymp$, for dimension ≥ 2 we then have:

Theorem 8.22 *For each $d \geq 2$, as $\varepsilon \downarrow 0$,*

$$\log D(\varepsilon, \mathcal{LL}_{d,1}, h) \asymp \log D(\varepsilon, \mathcal{LL}_d, d_\lambda) = \log D(\varepsilon, \mathcal{LL}_{d,1}, d_\lambda)$$

$$\asymp \log N_I \left(\varepsilon, \mathcal{LL}_{d,1}, \lambda \right) \asymp \varepsilon^{1-d}.$$

Proof. First, for the Hausdorff metric h, it will be shown that for some constants c_d with $1 \leq c_d < \infty$,

$$\log D(\varepsilon, \mathcal{LL}_{d,1}, h) \leq c_d \varepsilon^{1-d} \qquad (8.14)$$

for $0 < \varepsilon < 1$. For $d = 2$ this holds by the previous theorem. It will be proved for $d \geq 2$ by induction on d. Suppose it holds for $d - 1$, for $d \geq 3$. Given $0 < \varepsilon < 1$, take a maximal number of sets $L_1, \ldots, L_m \in \mathcal{LL}_{d-1,1}$ such that $h(L_i, L_j) > \varepsilon/4$ for $i \neq j$, where $m \leq \exp(c_{d-1}(4/\varepsilon)^{d-2})$. Let $k := \lceil 3/\varepsilon \rceil$ and $A \in \mathcal{LL}_{d,1}$. For $j = 0, 1, \ldots, k$ let $A_j := \{x \in I^{d-1} : \langle x, j/k \rangle \in A\}$ and $A_{(j)} := A_j \times \{j/k\} \subset A$. Then $A_j = \emptyset$ or $A_j \in \mathcal{LL}_{d-1,1}$. In the latter case we can choose $i := i(j, A)$ such that $h(A_j, L_i) \leq \varepsilon/4$. Let $L_0 := \emptyset$ and $i := i(j, A) := 0$ if $A_j = \emptyset$, so $h(A_j, L_i) = 0 \leq \varepsilon/4$ in that case also.

Let $A, B \in \mathcal{LL}_{d,1}$ and suppose that $i(j, A) = i(j, B)$ for $j = 0, 1, \ldots, k - 1$. It will be shown that $h(A, B) \leq \varepsilon$. Let $x \in A$. There is a $j = 0, 1, \ldots, k - 1$ such that $j/k \leq x_d \leq (j+1)/k$. Let $y := (x_1, \ldots, x_{d-1}, j/k) \in A_{(j)}$. Then $A_j \neq \emptyset$ and $h(A_j, B_j) \leq \varepsilon/2$, so for some $z \in B_{(j)} \subset B$, we have $|y - z| < 2\varepsilon/3$ and $|x - z| < k^{-1} + 2\varepsilon/3 < \varepsilon$. So $d(x, B) \leq \varepsilon$, and by symmetry, $h(A, B) \leq \varepsilon$. Thus

$$D(\varepsilon, \mathcal{LL}_{d,1}, h) \leq (m+1)^k \leq [\exp(2c_{d-1}(4/\varepsilon)^{d-2})]^{4/\varepsilon} \leq \exp(c_d/\varepsilon^{d-1})$$

for $c_d := 2 \cdot 4^{d-1} c_{d-1}$, so (8.14) is proved.

For the metrics in terms of λ we have the following:

Lemma 8.23 Let $\delta > 0$ and let $A, B \in \mathcal{LL}_{d,1}$ with $h(A, B) \leq \delta$. Then $\lambda^d(A \triangle B) \leq d^{d/2}\delta$.

Proof. Let U be a rotation of $\mathbb{R}^d$ which takes $v := (1, 1, \ldots, 1)$ into $(0, 0, \ldots, d^{1/2})$. Let $\pi_d(y) := (y_1, \ldots, y_{d-1}, 0)$. Let C be the cube $C := U[I^d] := \{U(x) : x \in I^d\}$. Each point of I^d or C is within $d^{1/2}/2$ of its respective center. Thus each point $z \in H := \pi_d[C]$ is within $d^{1/2}/2$ of 0. Also, for any $z \in \mathbb{R}^{d-1}$, $\{t \in \mathbb{R} : \langle z, t \rangle \in C\}$ is empty or a closed interval $h(z) \leq t \leq j(z)$. Let $C_z := \{\langle w, t \rangle \in C : w = z\}$, a line segment. The intersections of $U[A]$ and $U[B]$ with C_z are each either empty or line segments with the same lower endpoint $\langle z, h(z) \rangle$ as C_z, so the two sets are linearly ordered by inclusion. Thus the intersection of $U[A] \triangle U[B] = U[A \triangle B]$ with C_z is some line segment $S_{A,B,z}$. It will be shown that $S_{A,B,z}$ has length $\leq d^{1/2}\delta$. Suppose not. Then by symmetry we can assume that there is some $\zeta > \delta$ and a point $x \in B \setminus A$ such that $v := x + \zeta(1, 1, \ldots, 1) \in B$. The orthant $\Theta := \{y : y_j > x_j \text{ for all } j = 1, \ldots, d\}$ is disjoint from A. But, the open ball of radius ζ and center v is included in Θ, contradicting $h(A, B) \leq \delta$. Now, H is included in a cube of side $d^{1/2}$ in $\mathbb{R}^{d-1}$ with center at 0, so by the Tonelli–Fubini theorem

$$\lambda^d(A \triangle B) \leq (\delta d^{1/2})(d^{1/2})^{d-1} \leq \delta d^{d/2},$$

proving the Lemma. $\square$

Returning to the proof of Theorem 8.22, from the last Lemma and (8.14) it follows that for each $d = 2, 3, \ldots$ and some $C_d < \infty$, for $0 < \varepsilon \leq 1$,

$$\log D(\varepsilon, \mathcal{LL}_d, d_\lambda) \; = \; \log D(\varepsilon, \mathcal{LL}_{d,1}, d_\lambda) \; \leq \; C_d \varepsilon^{1-d}.$$

Next, consider the remaining upper bound statement, for N_I. The angle between $v := (1, 1, \ldots, 1)$ and each hyperplane $x_j = 0$ is

$$\cos^{-1}\left(((d-1)/d)^{1/2}\right) = \sin^{-1} d^{-1/2} = \tan^{-1}((d-1)^{-1/2}).$$

Thus for any nonempty lower layer $B \neq \mathbb{R}^d$ and point p on its boundary, $U[B]$ includes the cone

$$\left\{ x : \; \left| x_{(d)} - q_{(d)} \right| < (q_d - x_d)(d-1)^{-1/2} \right\},$$

where $q := U(p)$ and recalling that $x_{(d)} = (x_1, \ldots, x_{d-1})$. Hence the boundary of $U[B]$ is the graph of a function $f : \mathbb{R}^{d-1} \to \mathbb{R}$ where for any $s, t \in \mathbb{R}^{d-1}$,

$$f(s) \geq f(t) - K|s - t|, \quad K := (d-1)^{1/2}.$$

Hence, interchanging s and t, $|f(s) - f(t)| \leq K|s - t|$. So $\|f\|_L \leq K$. Let $\mathcal{J}(f) := \{x : -\infty < x_d \leq f(x_{(d)})\}$. Thus for each $B \in \mathcal{LL}_{d,1}$ we have $U[B] = \mathcal{J}(f) \cap U[I^d]$ for a function $f = f_B$ on $\mathbb{R}^{d-1}$ with $\|f\|_L \leq K$. We can restrict the functions f to a cube T of side $d^{1/2}$ centered at the origin in $\mathbb{R}^{d-1}$ parallel to the axes, which includes the projection of $U[I^d]$. We can also assume that $\|f_B\|_{\sup} \leq d^{1/2}$ for each B, since replacing f by $\max(-d^{1/2}, (\min(f, d^{1/2}))$ does not change $\mathcal{J}(f) \cap U[I^d]$, nor does it increase $\|f\|_L$ (RAP, Proposition 11.2.2(a), since $\|g\|_L = 0$ if g is constant). Now, apply Theorem 8.4 for $\alpha = 1$ and $d - 1$ in place of d, where by a fixed affine transformation we have a correspondence between I^d and the cube T. In this case, I^{d-1} means the set of points $(x_1, \ldots, x_{d-1}, 0)$ of $\mathbb{R}^d$ such that $0 \leq x_j \leq 1$ for $j = 1, \ldots, d - 1$.

Since $f \leq g$ implies $\mathcal{J}(f) \subset \mathcal{J}(g)$ and $\mathcal{J}(f) \cap U[I^d] \subset \mathcal{J}(g) \cap U[I^d]$, the bracketing parts of Theorem 8.4 imply the desired upper bound for $\log N_I(\varepsilon, \mathcal{LL}_{d,1}, \lambda)$ with $-(d-1)/\alpha = 1 - d$. This finishes the proof for upper bounds.

Now for lower bounds, it will be enough to prove them for $D(\varepsilon, \mathcal{L}_d, d_\lambda)$ in light of Lemma 8.23 and since $N_I(\varepsilon, \ldots) \geq D(\varepsilon, \ldots)$. The angle between $v = (1, 1, \ldots, 1)$ and each coordinate axis is

$$\theta_d := \cos^{-1} d^{-1/2} = \sin^{-1}\left(((d-1)/d)^{1/2}\right) = \tan^{-1}((d-1)^{1/2}).$$

Thus if $f : \mathbb{R}^{d-1} \to \mathbb{R}$ satisfies $\|f\|_L \leq (d-1)^{-1/2}$, then $L := U^{-1}(\mathcal{J}(f))$ is a lower layer: if not, then for some $x \in L$ and $y \notin L$, $x_i = y_i$ for all i except that $x_j < y_j$ for some j. Then U transforms the line through x, y to a line ℓ forming an angle θ_d with the dth coordinate axis. Writing ℓ as $t_d = h(t_{(d)})$ we have $\|h\|_L = \cot \theta_d = (d-1)^{-1/2}$, which yields a contradiction.

Recall (Section 8.2) that for $\delta > 0$ and a metric space Q,

$$\mathcal{F}_{1,\delta}(Q) := \{f : Q \to \mathbb{R}, \ \max(\|f\|_L, \|f\|_{\sup}) \leq \delta\}.$$

Let $\delta := (d-1)^{-1/2}d^{-1}$. Let Q be a small enough cube in $\mathbb{R}^{d-1}$ with center at 0. Then for each $f \in \mathcal{F}_{1,\delta}(Q)$, we have $\frac{1}{2}v + U^{-1}(J(f)) \subset I^d$. For such f, $J(f) \leftrightarrow \frac{1}{2}v + U^{-1}(J(f))$ is an isometry for h and for d_λ and preserves inclusion. Each f can be extended to $\mathbb{R}^{d-1}$, preserving $\|f\|_L \leq \delta$ (RAP, Theorem 6.1.1). So the lower bound with $d_{\sup}$ in Theorem 8.4 gives, via Lemma 8.1, the lower bound with h in Theorem 8.22. Theorem 8.7, likewise adapted from I^{d-1} to Q, gives the lower bounds with λ, proving Theorem 8.22. $\square$

Corollary 8.24 *For any law P on $\mathbb{R}^d$ having a bounded density with respect to Lebesgue measure, $\mathcal{LL}_{d,1}$ is a Glivenko–Cantelli class.*

Proof. This follows from the statements about N_I in Theorem 8.22 (and in the degenerate case $d = 1$, in Theorem 8.21) and the Blum–DeHardt theorem 7.1. $\square$

For what d does the Donsker property hold under the same hypotheses? For $d = 1$, it does easily. For $d \geq 2$, the hypothesis of Corollary 7.8, because of the x^2 in it, does not follow from Theorem 8.22. In fact, the Donsker property fails for $d = 2$ as will be shown in Theorem 11.10.

8.4 Metric Entropy of Classes of Convex Sets

A C^2 function f on $\mathbb{R}^d$ is convex if and only if its Hessian matrix $\partial^2 f / \partial x_i \partial x_j$ is everywhere nonnegative definite. For a general convex function, these derivatives need not exist everywhere, although the Hessian in the generalized sense of Schwartz distributions exists as a nonnegative definite matrix-valued measure (Bakel'man 1965, Reshetnyak 1968; cf. Dudley 1980). Thus convex functions are comparable to functions differentiable just of order 2, not necessarily for any $\alpha > 2$. It will be seen that metric entropy or capacity of convex subsets of a given bounded open subset of $\mathbb{R}^d$ for $d \geq 2$ is of the same order as that of subsets with boundaries given by twice differentiable functions.

Let $\mathcal{C}_d$ denote the class of all nonempty closed convex subsets of the open unit ball $B(0,1) := \{x : |x| < 1\}$ in $\mathbb{R}^d$. Let λ be the uniform Lebesgue measure on $\mathbb{R}^d$. Upper and lower bounds will be given for the metric entropy of $\mathcal{C}_d$ for the metric d_λ and for the Hausdorff metric h.

Theorem 8.25 (E. M. Bronštein) *For each $d \geq 2$ we have*

$$\log D(\varepsilon, \mathcal{C}_d, d_\lambda) \asymp \log D(\varepsilon, \mathcal{C}_d, h) \asymp \varepsilon^{(1-d)/2} \quad as \quad \varepsilon \downarrow 0.$$

Remark. For $d = 1$, C_1 is just the class of subintervals of the open interval $(-1, 1)$. Then it is rather easy to see that

$$D(\varepsilon, C_1, d_\lambda) \asymp D(\varepsilon, C_1, h) \asymp \varepsilon^{-2}.$$

Before proving the theorem, let us prove from it:

Corollary 8.26 *In $\mathbb{R}^d$, for $d \geq 2$, if P is a law whose restriction to $B(0, 1)$ has a bounded density f with respect to Lebesgue measure, then*

(a) $\log N_I(\varepsilon, C_d, P) = O(\varepsilon^{(1-d)/2})$ *as $\varepsilon \downarrow 0$.*

(b) (E. Bolthausen) For $d = 2$, C_2 is a Donsker class for P.

(c) For any d, C_d is a Glivenko–Cantelli class.

(d) If also $f \geq v$ on $B(0, 1)$ for some constant $v > 0$, then

$$\log N_I(\varepsilon, C_d, P) \asymp \varepsilon^{(1-d)/2} \quad \text{as } \varepsilon \downarrow 0.$$

Note. For $d = 3$, the class C_3 of convex sets is not a Donsker class for, e.g., the uniform distribution on the unit cube, as will be shown in Theorem 11.10.

Proof. Let $0 \leq f(x) \leq V < \infty$ for all x. For $B \subset \mathbb{R}^d$ and $\delta > 0$ recall that $_\delta B := \{x : y \in B \text{ whenever } |x - y| < \delta\}$ and $B^\delta := \{x : d(x, B) < \delta\}$. For a closed set B, we also have $_\delta B = \{x : y \in B \text{ whenever } |x - y| \leq \delta\}$. Then B^δ is always open and $_\delta B$ is always closed. We have $_\delta B \subset B \subset B^\delta$. For any two sets B, C, if $h(B, C) < \delta$, then $C \subset B^\delta$. It will be shown next that if B and C are closed and convex, then also $_\delta B \subset C$: if not, let $x \in {}_\delta B \setminus C$. Take a closed half-space $J \supset C$ with $x \notin J$ (RAP, Theorem 6.2.9). On the line through x perpendicular to the hyperplane bounding J, on the side opposite J, there are points $y \in B \setminus C^\delta$, a contradiction. So $_\delta B \subset C \subset B^\delta$. For d_λ, the following will help. For any set B, let ∂B denote the boundary of B.

Lemma 8.27 *For any $d = 1, 2, \ldots$, there is a $K = K(d) < \infty$ such that for any $B \in C_d$ and $0 < \delta \leq 1$, $\lambda \left(B^\delta \setminus {}_\delta B \right) \leq K\delta$.*

Proof. If B is convex and has empty interior, then it is included in some hyperplane (RAP, Theorem 6.2.6). Then $\lambda \left(B^\delta \setminus {}_\delta B \right) = \lambda \left(B^\delta \right) \leq K_1 \delta$ where K_1 is twice the $(d - 1)$-dimensional volume of a ball of radius 2 in $\mathbb{R}^{d-1}$. For a convex set B and $\varepsilon > 0$, the set B^ε is the vector sum $B + \varepsilon B(0, 1)$. The volume $\lambda(B^\varepsilon)$ can be written as a polynomial in ε,

$$\lambda \left(B^\varepsilon \right) = \lambda(B) + C_1(B)\varepsilon + C_2(B)\varepsilon^2 + \cdots + C_d\varepsilon^d$$

where the coefficients C_i are known as mixed volumes of B and $B(0, 1)$ (e.g., Eggleston 1958, pp. 82–89; Bonnesen and Fenchel 1934, pp. 38, 46–47). Here $C_1(B) = \lim_{\varepsilon \downarrow 0}(\lambda(B^\varepsilon) - \lambda(B))/\varepsilon$ is the $(d - 1)$-dimensional surface area of ∂B (Eggleston 1958, p. 88), and $C_d = \lambda(B(0, 1))$.

All the mixed volumes $C_i(B)$ are nondecreasing functions of the convex set B (Eggleston 1958, Theorem 42 p. 86; Bonnesen and Fenchel 1934, p. 41). Thus, all the $C_i(B)$ are maximized for $B \subset B(0, 1)$ when $B = B(0, 1)$. It follows that the derivative $d\lambda(B^\varepsilon)/d\varepsilon$ is bounded above uniformly for $B \in \mathcal{C}_d$ and $0 < \varepsilon \le 1$ by a constant $K_2 = K_2(d)$, so that $\lambda(B^\varepsilon \setminus B) \le K_2\varepsilon$ for all $B \in \mathcal{C}_d$ and $0 < \varepsilon \le 1$.

Now suppose B has an interior. Then $B^\delta \setminus {}_\delta B = (\partial B)^\delta$. For a set of points on ∂B at distance more than δ apart, the balls of radius $\delta/2$ with centers at the points are disjoint, and the outer halves of these balls cut by support hyperplanes to B at the points are outside of B. Thus for $0 < \delta < 1$

$$(\delta/2)K_2 \ge \lambda\left(B^{\delta/2} \setminus B\right) \ge \tfrac{1}{2}C_d\,(\delta/2)^d\,D(\delta, \partial B, \rho)$$

where ρ is the Euclidean distance. Thus

$$D(\delta, \partial B, \rho) \le 2^d K_2 \delta^{1-d}/C_d.$$

Then for $0 < \delta < 1$

$$\lambda((\partial B)^\delta) \le D(\delta, \partial B, \rho)C_d(2\delta)^d \le 4^d K_2\delta,$$

and the Lemma follows. $\qquad\square$

Now continuing the proof of Corollary 8.26, given $0 < \delta \le 1$, let $\mathcal{N}_d(\delta)$ be a δ-net in $\mathcal{C}_d$ for h with cardinality at most $\exp(A_d\delta^{(1-d)/2})$ where A_d is a large enough constant. The brackets $[{}_\delta B, B^\delta]$ for $B \in \mathcal{N}_d(\delta)$ cover $\mathcal{C}_d$: in other words, for any $C \in \mathcal{C}_d$ there is such a B with ${}_\delta B \subset C \subset B^\delta$, as seen just before Lemma 8.27, and $\lambda(B^\delta \setminus {}_\delta B) \le K\delta$, so $P\left(B^\delta \setminus {}_\delta B\right) \le KV\delta$, and

$$N_I(\varepsilon, \mathcal{C}_d, P) \le D(\epsilon/(KV), \mathcal{C}_d, h) \le \exp\left(A_d\left(\varepsilon/(KV)\right)^{(1-d)/2}\right)$$

for ε small enough, proving part (a).

Given part (a), part (b) follows from Corollary 7.8 and part (c) from the Blum–DeHardt theorem 7.1.

For part (d), we have

$$\begin{aligned} N_I\left(\varepsilon, \mathcal{C}_d, P\right) &\ge\ D\left(\varepsilon, \mathcal{C}_d, d_P\right) \\ &\ge\ D\left(\varepsilon/v, \mathcal{C}_d, d_\lambda\right) \ge \exp\left(c_d(\varepsilon/v)^{(1-d)/2}\right) \end{aligned}$$

for some $c_d > 0$ and all ε small enough, which finishes the proof of Corollary 8.26 from Theorem 8.25. $\qquad\square$

Now Theorem 8.25 will be proved. Let $0 < \varepsilon < 1$. For any set $C \subset \mathbb{R}^d$ and $r \ge 0$ let $C^{r]} := \{x \in \mathbb{R}^d : d(x, C) \le r\}$. Then the open set C^r is included in the closed set $C^{r]}$, $\partial C^r = \partial C^{r]}$, a closed set, and $h(C^r, C^{r]}) = 0$.

Lemma 8.28 *For any $C, D \in \mathcal{C}_d$ and $r \ge 0$, $h(C^{r]}, D^{r]}) = h(C, D)$, in other words $\phi_r : E \to E^{r]}$ is an isometry for h.*

Proof. Let $s > 0$. It will be shown that $D \subset C^s$ if and only if $D^{r]} \subset \left(C^{r]}\right)^s = C^{r+s}$. "Only if" is straightforward. To prove "if," suppose not. Let $a \in D \setminus C^s$. There is a unique point of q of $C^{s]}$ closest to a: there is a nearest point q since $C^{s]}$ is compact, and if b is another nearest point, then $(q + b)/2 \in C^{s]}$ since $C^{s]}$ is convex and $(q + b)/2$ is nearer to a, a contradiction. (Possibly $q = a$.) Now $q \in \partial(C^{s]})$. If $q = a$, take a support hyperplane H to $C^{s]}$ at q (RAP, Theorem 6.2.7). If $q \neq a$, then the hyperplane H through q perpendicular to the line segment aq is a support hyperplane to $C^{s]}$ at q (if there were a point c of $C^{s]}$ on the same side of H as a, then on the line segment cq there would be a point of $C^{s]}$ closer to a than q is, a contradiction). Let p be a point at distance r from a in the direction perpendicular to H and heading away from $C^{s]}$. Then $p \in D^{r]}$, but $p \notin (C^{s]})^r = C^{r+s}$, a contradiction. So "if" is proved. Since C and D can be interchanged, the Lemma follows.   □

For $r > 0$, ϕ_r is a useful smoothing, as it takes a convex set D, which may have a sharply curved boundary (vertices, edges, etc.) to a convex set $D^{r]}$ whose boundary is no more curved than a sphere of radius r, and so will be easier to approximate.

Now, for a given $C \in \mathcal{C}_d$ and $\varepsilon > 0$ let

$$\mathcal{N}_\varepsilon(C) := \{D \in \mathcal{C}_d : h(C, D) \leq \varepsilon\},$$

$$\tilde{\mathcal{N}}_\varepsilon(C) := \{D \in \mathcal{N}_\varepsilon(C) : C \subset D\}.$$

For any convex C, $\phi_{2+\varepsilon}$ is an isometry from $\mathcal{N}_\varepsilon(C)$ into $\tilde{\mathcal{N}}_{2\varepsilon}(C^2)$.

A sequence of lemmas will be proved. Here $|\cdot|$ will denote the usual Euclidean norm on $\mathbb{R}^d$. Let $\mathcal{C}_{d,1,3}$ denote the class of all closed convex sets C in $\mathbb{R}^d$ such that $B(0, 1) \subset C \subset B(0, 3)$. Note that if $C \in \mathcal{C}_d$, then clearly $C^{2]} \in \mathcal{C}_{d,1,3}$.

Lemma 8.29 *Let $E \in \mathcal{C}_{d,1,3}$. Let $x \in \partial E$ and let H be a support hyperplane to E at x. Let p be the point of H closest to 0. Then $|p| \geq 1$ and $\angle 0xp \geq \sin^{-1}(1/3)$.*

Proof. Existence of support hyperplanes is proved in RAP (Theorem 6.2.7). Clearly H is disjoint from $B(0, 1)$, so $|p| \geq 1$. Since $|x| \leq 3$ and $\angle 0px = \pi/2$, the Lemma follows.   □

Lemma 8.30 *If $E \in \mathcal{C}_{d,1,3}$, $r > 0$, and z is a point such that $z \in E^r \setminus E$, let $x(z)$ be the point on the half-line from 0 to z and in ∂E. Then $|z - x(z)| \leq 3r$.*

Proof. Note that $x = x(z)$ is uniquely determined since $E \supset B(0, 1)$. Apply Lemma 8.29. Let z_1 be the point of H closest to z. Then $|z - z_1| \leq r$. The vectors $z - z_1$ and p are parallel, so $\phi := \angle p0z = \angle 0zz_1$, and $|z - x| = |z - z_1||x|/|p| \leq 3r$.   □

Lemma 8.31 *Suppose* $C \in \mathcal{C}_d$, $y \in \partial C$, $x \in \partial C^2$, *and* $|x - y| = 2$. *Then for any two-dimensional subspace* V *containing* x, $V \cap B(y, 2)$ *is a disk containing* 0 *of radius at least* $2/3$.

Proof. Clearly $0 \in B(y, 2) \subset C^2$. Apply Lemma 8.29 again. Then H is also a support hyperplane to $B(y, 2)$ at x, so $x - y$ is orthogonal to H and in the same direction as p. Let q be the point on the segment $[0, x]$ closest to y. Then $\theta := \angle 0xp = \angle qyx$, $\sin \theta \geq 1/3$, and so $|x - q| \geq 2/3$. It follows that $V \cap B(y, 2) \supset V \cap B(q, 2/3)$. $\qquad\square$

Now polyhedra to approximate convex sets will be constructed. Let W_d be the cube centered at 0 in $\mathbb{R}^d$ of side $2/d^{1/2}$, parallel to the axes, so that the coordinates of the vertices are $\pm 1/d^{1/2}$. Recall that $f \sim g$ means $f/g \to 1$. Given $\varepsilon > 0$, decompose the $2d$ faces of W_d into equal $(d - 1)$-cubes of side $s_d := s_d(\varepsilon)$ where $s_d \sim c(\varepsilon/d)^{1/2}$ as $\varepsilon \downarrow 0$ and $c := 10^{-4}/(d^{1/2}(d - 1))$, so $s_d \sim \varepsilon^{1/2}10^{-4}/(d(d - 1))$. Specifically, let $s_d := 2/(d^{1/2}k_d)$ where $k_d := k_d(\varepsilon)$ is the smallest positive integer such that $s_d \leq \varepsilon^{1/2}10^{-4}/(d(d - 1))$. Then for $0 < \varepsilon < 1$,

$$\varepsilon^{1/2}10^{-4}/d^2 \leq s_d \leq \varepsilon^{1/2}10^{-4}/(d(d - 1)).$$

Let $\mathcal{L}_d$ be the set of all $(d - 1)$-cubes thus formed.

The diameter of each cube in $\mathcal{L}_d$ is $d^{1/2}s_d \leq \varepsilon^{1/2}10^{-4}/\left(d^{1/2}(d - 1)\right)$. The next fact follows directly, by the law of sines, since $0 < \varepsilon \leq 1$ and $d \geq 2$, and $\sin^{-1} x \leq 1.1x$ for $0 \leq x \leq 10^{-4}$.

Lemma 8.32 *(a) For any cube in* $\mathcal{L}_d$ *and any two vertices* p *and* q *of the cube,*

$$\angle p0q \leq \sin^{-1}(10^{-4}\varepsilon^{1/2}/(d - 1)) \leq (1.1)10^{-4}\varepsilon^{1/2}/(d - 1).$$

(b) The total number of vertices of all the cubes in $\mathcal{L}_d$ *is less than*

$$2d \cdot (k_d(\varepsilon) + 1)^{d-1} \leq \kappa_d \varepsilon^{(1-d)/2}$$

where $\kappa_d := 2d\left((2 \cdot 10^4 + 1)\, d^{3/2}\right)^{d-1}$.

Next, there is a triangulation of each cube in $\mathcal{L}_d$, in other words, a decomposition of the cube into $(d - 1)$-simplices, with disjoint interiors, where each simplex is a convex hull of some d of the 2^{d-1} vertices of the cube. That such a triangulation (without additional vertices) exists (is well known to algebraic topologists and) can be seen as follows. By induction, it will be enough to treat $S \times [0, 1]$ where S is a simplex with vertices $v_0, \ldots, v_p$. Let $a_i := \langle v_i, 0\rangle$, $b_i := \langle v_i, 1\rangle$. Then for each $i = 0, 1, \ldots, p$, the points $a_0, \ldots, a_i, b_i, \ldots, b_p$ are vertices of a $(p + 1)$-dimensional simplex S_i. To see that these S_i give the desired decomposition of $S \times [0, 1]$, note first that each point of a simplex is a unique convex combination of the vertices. For each point z of $S \times [0, 1]$, $z = \left\langle \sum_i \lambda_i v_i, x\right\rangle$ for some unique $x \in [0, 1]$

and $\lambda_i \geq 0$ with $\sum_i \lambda_i = 1$. Then $z = \sum_{i \leq j} \mu_i a_i + \sum_{i \geq j} \rho_i b_i$, where $\mu_i \geq 0$, $\rho_i \geq 0$, and $\sum_{i \leq j} \mu_i + \sum_{k \geq j} \rho_k = 1$, if and only if $\mu_i = \lambda_i$ for $i < j$ $\rho_k = \lambda_k$ for $k > j$, $\lambda_j = \mu_j + \rho_j$, and $x = \sum_{k \geq j} \rho_k$. Thus z is in S_j if and only if $\sum_{i > j} \lambda_i \leq x \leq \sum_{i \geq j} \lambda_i$. If both inequalities are strict, then j is unique. Every point of $S \times [0, 1]$ is in some S_j, and a point in more than one S_j is on the boundary of both.

Let K be a convex set including a neighborhood of 0. For each vertex p_i of a cube in $\mathcal{L}_d$, let H_i be the half-line starting at 0 passing through p_i and let v_i be the unique point at which H_i passes through the boundary of K. For each simplex S_j in the triangulation of the cubes in $\mathcal{L}_d$, let T_j be the corresponding simplex with vertices v_i in place of p_i. Let $\pi_\varepsilon(K)$ be the polyhedron with faces T_j, in other words the union of the d-dimensional simplices which are convex hulls of $T_j \cup \{0\}$. For $d \geq 3$, $\pi_\varepsilon(K)$ is not necessarily convex.

Lemma 8.33 *Let $E \in \mathcal{C}_{d,1,3}$, $\delta > 0$ and $\varepsilon > 0$. For $i = 1, 2$ let z_i be points such that $z_i \in E^{16\varepsilon} \setminus E$. Assume that $\angle z_1 0 z_2 \leq \delta \leq \frac{1}{15}$. Then $|z_1 - z_2| \leq 5\delta + 96\varepsilon$.*

Proof. Let $x_i := x(z_i)$ be the point where the line segment from 0 to z_i intersects ∂E, $i = 1, 2$. By Lemma 8.30, $|z_i - x_i| \leq 48\varepsilon$, $i = 1, 2$. We have $|z_i| \geq 1$ and $|x_i| \geq 1$ for all i.

By symmetry we can assume $|x_1| \geq |x_2|$. To get a bound for $|x_1 - x_2|$ we can assume x_1 and x_2 are not on the same line through 0, or they would be equal. Take a half-line L starting at x_1 which is tangent to the unit circle $\partial B(0, 1)$ at a point v and crosses the half-line from 0 through z_2 at a point y.

If v is between x_1 and y, then $|x_1 - v| \leq \tan \delta \leq 2\delta$ since $\delta \leq \frac{1}{15} < \pi/4$. So $1 \leq |x_2| \leq |x_1| \leq (1 + 4\delta^2)^{1/2} \leq 1 + 2\delta^2 \leq 1 + \delta$, and likewise $1 \leq |y| \leq 1 + \delta$, so

$$|x_2 - v| \leq |x_2 - y| + |y - v| < \delta + 2\delta = 3\delta$$

and $|x_1 - x_2| < 5\delta$.

Otherwise, y is between x_1 and v. Then $|x_1 - x_2|$ is maximized when $|x_2| = |x_1|$ or $x_2 = y$, since x_2 must not be in the convex hull of $\{x_1\} \cup B(0, 1)$. If $|x_2| = |x_1|$, then

$$|x_1 - x_2| \leq 2 \left(2 \sin \tfrac{\delta}{2}\right) \leq 2\delta.$$

To bound $|x_1 - y|$ let $\zeta := \angle 0 x_1 v$. Then $\sin \zeta \geq 1/2$, and

$$|x_1 - y| \leq \tan \left(\tfrac{\pi}{2} - \zeta\right) - \tan \left(\tfrac{\pi}{2} - \zeta - \delta\right)$$
$$\leq \qquad \delta \sec^2 \left(\tfrac{\pi}{2} - \zeta\right) \qquad \leq 4\delta.$$

So $|x_1 - x_2| \leq 5\delta$ in all cases, and $|z_1 - z_2| \leq 5\delta + 96\varepsilon$. $\qquad \square$

Lemma 8.34 *Let $E \in \mathcal{C}_{2,1,3}$. Suppose $2 \leq K < \infty$ and $0 < \varepsilon \leq 10^{-12}/(K - 1)^2$. Let $c := (1.1)10^{-4}/(K^{1/2}(K - 1))$. Let $r \geq 2/3$ and let a, b, α be points* of $\mathbb{R}^2$ such that $a \notin E$, $B(\alpha, r) \subset E$, b is on the boundary of E and of $B(\alpha, r)$, $d(a, E) \leq 16\varepsilon$, and $\beta := \angle a0b \leq c(K\varepsilon)^{1/2}$. Then $d(a, B(\alpha, r)) \leq 50\varepsilon$. Also, $|a - b| \leq .0008\varepsilon^{1/2}/(K - 1)$ and $\angle a\alpha b \leq D(K)\varepsilon^{1/2}$ where $D(K) := .002/(K - 1)$.

Proof. Apply Lemma 8.29 at $x = b$. The tangent line L to E at b is also tangent to the circle $\partial B(\alpha, r)$.

Let $\gamma := \angle 0bp$. Then $\gamma \geq \sin^{-1}(1/3) > 1/3$. We have

$$\beta \leq c(K\varepsilon)^{1/2} = (1.1)10^{-4}\varepsilon^{1/2}/(K - 1) \leq (1.1)10^{-10}. \qquad (8.15)$$

Let H be the half-plane including E bounded by L.

First suppose $a \notin H$. Then $d(a, H) \leq 16\varepsilon$. Let the line from 0 to a intersect L at a point η. If η is between p and b, then $\gamma + \beta \leq \pi/2$ and

$$|\eta - b| = |p|(\cot \gamma - \cot(\gamma + \beta)) \leq 3\beta \csc^2 \gamma \leq 27\beta.$$

If p is between η and b, then

$$\begin{aligned}|\eta - b| &= |p| \left(\cot \gamma + \tan \left(\gamma + \beta - \tfrac{\pi}{2}\right)\right) \\ &= |p|(\cot \gamma - \cot(\gamma + \beta)),\end{aligned}$$

where now $\pi/2 < \gamma + \beta < \pi$. Thus

$$\begin{aligned}|\eta - b| &\leq |p|\beta \max \left(\csc^2 \gamma, \csc^2(\gamma + \beta)\right) \\ &\leq 3\beta \max \left(\csc^2 \gamma, \csc^2 \left(\tfrac{\pi}{2} + 10^{-9}\right)\right) \\ &\leq 27\beta \leq 27c(K\varepsilon)^{1/2}.\end{aligned}$$

The other possibility is that b is between η and p. Then

$$|\eta - b| = |p|(\cot(\gamma - \beta) - \cot \gamma).$$

Now $\beta \leq (1.1)10^{-9}$ and $\gamma \geq \sin^{-1}(1/3)$ imply $\sin(\gamma - \beta) \geq 0.333$, so

$$|\eta - b| \leq 3\beta/(.333)^2 \leq 28\beta \leq 28c(K\varepsilon)^{1/2}.$$

For any ordering of b, p and η, and under the same condition on ε,

$$|a - \eta| \leq 16\varepsilon/\sin(\gamma - \beta) \leq 49\varepsilon.$$

Next, let x be the distance from a varying point ζ on L to b. Then the distance y from ζ to the circle $\partial B(\alpha, r)$ satisfies $y = \left(r^2 + x^2\right)^{1/2} - r$. Now $\left(r^2 + t\right)^{1/2} \leq r + t$ for $t \geq 0$ and $r \geq 2/3$, so $0 \leq y \leq x^2$ for all x. So, the distance from a to $B(\alpha, r)$ is at most $C\varepsilon$ for $C = 49 + 28^2 c^2 K < 50$, giving the first conclusion for $a \notin H$.

A line W through α, orthogonal to the line V through 0 and b, meets V at a point ξ. Then $\angle b\alpha\xi = \gamma > \sin^{-1}(1/3)$, so $|\alpha - \xi| < r\left(1 - \tfrac{1}{9}\right)^{1/2}$. Let q be the point on the circle $|q - \alpha| = r$ and the line W, on the same side of α as ξ is. Then $|q - \xi| \geq \tfrac{2}{3}\left(1 - \left(\tfrac{8}{9}\right)^{1/2}\right) \geq .03$. By (8.15), $\beta < \tan^{-1}(.01)$, so the line A through 0, η and a must intersect $B(\alpha, r)$. Then since E is convex, $a \notin E$, $0 \in E$, and $B(\alpha, r) \subset E$, for $a \in H$, $d(a, B(\alpha, r))$ is maximized when $a = \eta$ on L, and $d(a, B(\alpha, r)) \le C\varepsilon$ for the same $C < 50$ as before. So the first conclusion is proved.

Lemma 8.33 with $\delta = \beta$ gives

$$|a - b| \le 5\beta + 96\varepsilon \le .0008\varepsilon^{1/2}/(K - 1),$$

so $\angle a\alpha b \le \sin^{-1}(|a - b|/(2/3)) \le .002\varepsilon^{1/2}/(K - 1)$. So Lemma 8.34 is proved. $\qquad\square$

Lemma 8.35 *Let $\alpha \in \mathbb{R}^2$, $B := B(\alpha, r) \subset \mathbb{R}^2$, where $r \ge 2/3$ and $0 \in B$, so $|\alpha| < r$. Let $2 \le K < \infty$ and $0 < \varepsilon \le 10^{-12}/(K - 1)^2$. Let a_i, $i = 1, 2$, be points not in B with $d(a_i, B) \le 50\varepsilon$, $|a_1 - a_2| \le D(K)\varepsilon^{1/2}$, and $\angle a_1\alpha a_2 \le 2D(K)\varepsilon^{1/2}$, where again $D(K) = .002/(K - 1)$. Let $S \supset B$ be a bounded convex set with $a_i \in \partial S$, $i = 1, 2$. Let S_1 be the triangle with vertices 0, a_1, a_2. Let W be the convex wedge with vertex 0 bounded by the half-lines from 0 through a_1 and a_2. Then*

$$h(S \cap W, S_1) \le \frac{\varepsilon}{9(K - 1)}.$$

Proof. Since $S_1 \subset S \cap W$, it will be enough to show that for $z_0 \in \partial S \cap W$, $d(z_0, [a_1, a_2]) \le \varepsilon/(9(K - 1))$ where $[a_1, a_2]$ is the line segment joining a_1 to a_2.

It is easy to see that a_1, a_2, and α are not all on a line, unless $a_1 = a_2$, when the result clearly holds, so assume they are not on a line. Let L_i be a tangent line to B, at a point b_i, through a_i, where of the two such tangent lines, L_i is the one for which $\angle b_i\alpha a_i < \angle b_i\alpha a_{3-i}$, $i = 1, 2$, and b_i is not in W.

Now $|a_i - \alpha| \le r + 50\varepsilon$, so

$$\angle b_i\alpha a_i \le \cos^{-1}(r/(r + 50\varepsilon)) \le \cos^{-1}(1 - 75\varepsilon).$$

By derivatives, $\cos x \le 1 - \frac{1}{2}x^2 + \frac{1}{24}x^4$ for all x, so $\cos x \le 1 - 11x^2/24$, $0 \le x \le 1$. Thus $\cos^{-1}\left(1 - 11x^2/24\right) \le x$, and $\cos^{-1}(1 - 75\varepsilon) \le (164\varepsilon)^{1/2}$. Now $2D(K)\varepsilon^{1/2} < 10^{-8}$, and $(164\varepsilon)^{1/2} < 10^{-4}$, so the angles $\angle b_i\alpha a_i$, $i = 1, 2$, and $\angle a_1\alpha a_2$ add up to

$$\angle b_1\alpha b_2 \le (2(164)^{1/2} + 2D(K))\varepsilon^{1/2} < .001 < \pi. \qquad (8.16)$$

So the lines L_i intersect, in a unique point m.

Now z_0 is in the triangle $a_1 a_2 m$ because: $z_0 \in W$ and $a_1, a_2 \in \partial S$, $b_1, b_2 \in S$, so z_0 cannot be in the triangle $\alpha a_1 a_2$ unless it is on $[a_1, a_2]$. Also if z_0 were on the side of L_i away from α, then $a_i \notin \partial S$, a contradiction, $i = 1, 2$.

Next, $\angle m a_1 a_2 + \angle m a_2 a_1 = \pi - \angle b_1 m b_2 = \left(\frac{\pi}{2} - \angle \alpha m b_1\right) + \left(\frac{\pi}{2} - \angle \alpha m b_2\right) = \angle b_1\alpha b_2$.

We have $d(z_0, [a_1, a_2]) \le d(m, [a_1, a_2])$. Then $d(m, [a_1, a_2]) \le |a_1 - a_2|$ $\tan \angle ma_1a_2$, and by (8.16)

$$\tan \angle ma_1a_2 \le 2\angle ma_1a_2 \le (52 + 4D(K))\varepsilon^{1/2},$$

and $|a_1 - a_2| \le D(K)\varepsilon^{1/2}$ by assumption. So

$$d(m, [a_1, a_2]) \le (52D(K) + 4D(K)^2)\varepsilon \le \varepsilon/(9(K - 1)). \qquad \square$$

Lemma 8.36 *Let* $d \ge 2$, $C \in \mathcal{C}_d$, $0 < \varepsilon \le 10^{-12}/(d - 1)^2$, *and* $N' \in \tilde{\mathcal{N}}_{4\varepsilon}\left(C^2\right)$. *Let* N *be the polyhedron* $\pi_\varepsilon(N')$. *Then* $h(N, N') \le \varepsilon/9$.

Proof. For each j, the simplex S_j and the origin span a convex cone W_j, the union of all half-lines with endpoint 0 passing through S_j. It suffices to show that for each j,

$$h\left(N' \cap W_j, N \cap W_j\right) \le \varepsilon/9.$$

Fix such a cone $W = W_j$ and simplex $S = S_j$. For $s = 1, \ldots, d$, let $W^{(s)}$ be the union of the s-dimensional faces of W, so that $W^{(d)} = W$, $W^{(1)}$ is the union of half-lines through 0 and vertices of S, and so on. It will be shown by induction on s that if $z \in W^{(s)} \cap N'$, then

$$d(z, N) \le \varepsilon(s - 1)/(9(d - 1)). \tag{8.17}$$

For $s = d$ this will give the desired conclusion.

For $s = 1$, (8.17) holds by definition of N. Suppose it holds for a given s. Take any $z \in N' \cap W^{(s+1)}$. Let $x = x(z)$ be the point at which the half-line from 0 through z intersects ∂C^2. There is a point $\beta \in \partial C$ such that $|x - \beta| = 2$: to see this let H be a support hyperplane to C^2 at x. Let H_1 be a hyperplane parallel to H at distance 2 in the direction toward C. Then it is easily seen that H_1 is a support hyperplane to C at a point β, the nearest point in H to x. Clearly $B(\beta, 2) \subset C^2$.

Let F_{s+1} be an $(s + 1)$-dimensional face of W containing z. Let π_2 be a two-dimensional subspace with $z \in \pi_2 \cap W \subset F_{s+1}$. Then the two rays at the edges of the wedge $\pi_2 \cap W$ are included in $W^{(s)}$. Let these rays intersect $\partial N'$ at points u, v.

By Lemma 8.31, the disk $\pi_2 \cap B(\beta, 2)$ contains 0 and has radius at least $2/3$. So it is a disk $\pi_2 \cap B(\alpha, r)$, $\alpha \in \pi_2$, with $|x - \alpha| = r \ge 2/3$. Then $u, v \in \partial N'$ implies u, v are not in C^2 and so not in $B(\alpha, r)$. Now apply Lemma 8.34 to $E = C^2 \cap \pi_2$, with $K = d$, first for $(a, b) = (u, x)$, then for $(a, b) = (v, x)$. To justify the application we need the following:

Claim $d(u, E) \le 16\varepsilon$ *and* $d(v, E) \le 16\varepsilon$.

To prove the claim, there is a point $r \in C^2$ with $|u - r| < 4\varepsilon$. Let U be the two-dimensional subspace spanned by u and r (if u and r are on a line through

0 then $r \in \pi_2$ and $d(u, E) \leq 4\varepsilon$). Let M be the line in U through r which crosses the half-line D from 0 through u at a point t and is tangent to the unit disk $B(0, 1)$ in U at a point w. Let z be the closest point to r on D. Then $|r - z| \leq 4\varepsilon$. Let $\zeta = \angle 0tw = \angle ztr$. Since $t \in C^2$, $|t| \leq 3$ so $\sin \zeta \geq 1/3$ and $|r - t| \leq 12\varepsilon$ so $|t - u| \leq |t - r| + |r - u| \leq 16\varepsilon$, proving the Claim for u and by symmetry for v.

It will be shown that:

(a) if u and v are points of a simplex with vertices w_i, then

$$\angle u0v \leq \max_{i,j} \angle w_i 0 w_j.$$

Here (a) will follow from

(b) if u is fixed and v is in a simplex with vertices w_i, then $\angle u0v \leq \max_i \angle u0w_i$, since (b) could be applied in stages to prove (a). Further, it will be enough to prove (b) for a simplex reducing to a line segment from w_1 to w_2, since then, the case of general v would reduce in stages to cases where v is on a boundary face of the simplex, then a lower-dimensional face, and so on until v is a vertex. So it will be enough to show that

(c) For all u, x, y in $\mathbb{R}^3$ and $0 < \lambda < 1$,

$$\angle u0(\lambda x + (1 - \lambda)y) \leq \max(\angle u0x, \angle u0y).$$

Expressing (c) in terms of cosines, and then of scalar products and lengths, we can reduce to the case where $|x| = |y| = |u| = 1$, and then check the condition.

Now it is easily seen using also Lemma 8.32 that all the hypotheses of Lemma 8.34 hold, hence so do its conclusions. Next, Lemma 8.35 will be applied with again $K = d$, and with $a_1 = u$, $a_2 = v$, and $S := N' \cap \pi_2$. It will be checked that the hypotheses of Lemma 8.35 hold. We have $\angle a_1 \alpha a_2 \leq 0.004\varepsilon^{1/2}/(d - 1)$. The angles $\angle u\alpha x$ and $\angle x\alpha v$ both have the upper bound $0.002\varepsilon^{1/2}/(d - 1)$. The other hypotheses of Lemma 8.35 follow from Lemma 8.34, so Lemma 8.35 does apply. Its conclusion gives $d(z, [u, v]) \leq \varepsilon/(9(d - 1))$. By induction hypothesis, $d(y, N) \leq \varepsilon(s - 1)/(9(d - 1))$ for $y = u, v$, and then by convexity for any $y \in [u, v]$. It follows that $d(z, N) \leq \varepsilon s/(9(d - 1))$, completing the induction and the proof of Lemma 8.36. □

Lemma 8.37 *Let* $C \in \mathcal{C}_d$. *Then for any* $\varepsilon > 0$ *with* $\varepsilon \leq 10^{-12}/(d - 1)^2$, *there is an* $\varepsilon/2$-*net for* $\mathcal{N}_{2\varepsilon}(C)$ *containing at most* $g_d(\varepsilon)$ *points where* $g_d(\varepsilon) :=$ $\exp\left(a(d)\varepsilon^{(1-d)/2}\right)$ *and* $a(d) = \kappa_d \log 24$ *with* κ_d *as in Lemma 8.32.*

Proof. Recall that $\phi_{2+2\varepsilon}$ gives an isometry for h of $\mathcal{N}_{2\varepsilon}(C)$ into $\tilde{\mathcal{N}}_{4\varepsilon}(C^2)$. Let $N' \in \tilde{\mathcal{N}}_{4\varepsilon}(C^2)$ and let the vertices of the polyhedron $N := \pi_\varepsilon(N')$ be y_i, $i = 1, 2, \ldots$. By Lemma 8.36, $h(N, N') \leq \varepsilon/9$.

For each i, on the half-line from 0 through y_i, take the interval $I_i :=$ $[y_i, y_i + 12\varepsilon y_i / |y_i|]$ of length 12ε starting at y_i. Then I_i has an $\varepsilon/4$-net J_i containing 24 points (midpoints of subintervals of length $\varepsilon/2$).

By Lemma 8.30, for every $M \in \widetilde{\mathcal{N}}_{4\varepsilon}(C^2)$, $\pi_\varepsilon(M)$ has a vertex v_i in I_i, which is within $\varepsilon/4$ of some $u_i \in J_i$. The $(d-1)$-simplex with vertices v_i is within $\varepsilon/4$ for h of the one with vertices u_i. The same is true of the d-simplices where the vertex 0 is adjoined to both. Thus if $\mathcal{L}_\varepsilon$ is the set of all polyhedra with one vertex in J_i for each i, defined as in the definition of π_ε, then $\mathcal{L}_\varepsilon$ is $\varepsilon/2$-dense in $\widetilde{\mathcal{N}}_{4\varepsilon}(C^2)$. The number of values of i is at most $\kappa_d \varepsilon^{(1-d)/2}$ by Lemma 8.32, and Lemma 8.37 follows. $\qquad\square$

Now it will be shown that there are constants K_d, $L_d < \infty$ such that, for $0 < \varepsilon < 1$, there is an ε-net for $\mathcal{C}_d$ containing at most $f_d(\varepsilon)$ sets where

$$f_d(\varepsilon) := L_d \exp(K_d \varepsilon^{(1-d)/2}). \qquad (8.18)$$

This will clearly imply the upper bound for h in Theorem 8.25.

It will be enough (possibly changing K_d and L_d) to prove (8.18) for ε small enough, specifically for $0 < \varepsilon \leq \varepsilon_0 := 10^{-12}/(d-1)^2$. For such ε, and $a(d)$ from Lemma 8.37, let $K_d := a(d)/(1 - 2^{(1-d)/2})$. ($L_d$ will be specified below.) We have the decomposition

$$(0, \varepsilon_0] = \bigcup_{k=1}^\infty I_k, \quad I_k := (\varepsilon_0/2^k, \varepsilon_0/2^{k-1}].$$

Before getting more precise bounds, it will help to see that $N(\varepsilon, \mathcal{C}_d) < \infty$ for all $\varepsilon > 0$. The cube $[-1, 1]^d$ can be written as a finite union $\cup_i C_i$ of cubes of side $< \varepsilon/d^{1/2}$ and so of diameter $< \varepsilon$. For each nonempty $C \in \mathcal{C}_d$, let $B(C)$ be the union of all C_i which intersect C. Then $C \subset B(C)$ and $h(B(C), C) < \varepsilon$. It follows that $N(\varepsilon, \mathcal{C}_d) < \infty$.

So taking $\varepsilon = \varepsilon_0/2$, L_d can be and hereby is chosen so that (8.18) holds for $\varepsilon \in I_1$. Then (8.18) will be proved for $\varepsilon \in I_k$ by induction on k. Suppose it holds for all $\varepsilon \in I_k$. Let $\delta \in I_{k+1}$. Then $\varepsilon := 2\delta \in I_k$. By induction, there is an ε-net $\{C_i\}$ for $\mathcal{C}_d$ with at most $f_d(\varepsilon)$ values of i. By Lemma 8.37, there exist at most $g_d(\varepsilon/2)$ sets K_j, not necessarily convex, such that each $A \in \mathcal{N}_\varepsilon(C_i)$ is within $\varepsilon/4$ of some K_j. Thus there is a δ-net for $\mathcal{N}_\varepsilon(C_i)$ containing at most $g_d(\delta)$ sets, and a δ-net for $\mathcal{C}_d$ containing at most $g_d(\delta) f_d(2\delta)$ sets. Now $a(d) + 2^{(1-d)/2} K_d \leq K_d$ by choice of K_d, so $g_d(\delta) f_d(2\delta) \leq f_d(\delta)$, and the upper bound in Theorem 8.25 is proved.

Now to prove the lower bound, let ρ be the Euclidean metric on $\mathbb{R}^d$ and recall that S^{d-1} denotes the unit sphere $\{x \in \mathbb{R}^d : |x| = 1\}$. Let $d \geq 2$. If v_d is the volume of $B(0, 1) \subset \mathbb{R}^d$, then

$$\lambda\left(\left(S^{d-1}\right)^\varepsilon\right) \geq v_d[(1+\varepsilon)^d - 1] \geq d v_d \varepsilon.$$

Then by the left side of the last displayed inequality in the proof of Lemma 8.27, there is an $a_d > 0$ such that

$$D\left(\varepsilon, S^{d-1}, \rho\right) \geq a_d \varepsilon^{1-d} \quad \text{for} \quad 0 < \varepsilon \leq 1.$$

Given ε, take a set $\{x_i\}_{i=1}^m$ of points of S^{d-1} more than 2ε apart, of maximal cardinality $m := D(2\varepsilon) := D\left(2\varepsilon, S^{d-1}, \rho\right)$. As above let $\angle abc$ denote the angle at b in the triangle abc. Then for $i \neq j$,

$$\theta := \angle x_i 0 x_j > 2 \sin \tfrac{\theta}{2} > 2\varepsilon.$$

Let K_i be the half-line from 0 through x_i. Let C_i be the spherical cap cut from the unit ball $\overline{B}_d$ by a hyperplane orthogonal to K_i at a distance $\cos \varepsilon$ from 0. Then the caps C_i are disjoint.

For any set $I \subset \{1, \ldots, m\}$ let $D_I := \overline{B}_d \setminus \cup_{i \in I} C_i$. Then each D_I is convex. Let λ^d be d-dimensional Lebesgue measure (volume). Then for all i, $\lambda^d(C_i) > b_d \varepsilon^{d+1}$ for some constant $b_d > 0$. By Lemma 8.8 there are at least $e^{m/6}$ sets $I(j)$ such that for all $j \neq k$ the symmetric difference $I(j) \Delta I(k)$ contains at least $m/5$ elements. Then

$$\lambda^d(D_{I(j)} \Delta D_{I(k)}) \geq c_d \varepsilon^{1-d+d+1} = c_d \varepsilon^2$$

for a constant $c_d := a_d b_d 2^{1-d}/5$. So for some $\beta_d > 0$,

$$D\left(\delta, C_d, d_\lambda\right) \geq \exp\left(\beta_d \delta^{(1-d)/2}\right) \quad \text{for} \quad 0 < \delta \leq 1.$$

This finishes the proof of the lower bound for d_λ and so also for h, so Theorem 8.25 is proved.   □

Problems

1. Let (K, d) be a compact metric space. Show that the collection of all nonempty closed subsets of K, with the Hausdorff metric, is also compact. *Hints*: Find an ϵ-net in the collection of compact nonempty subsets of K by taking an ϵ-net F in K, then taking all nonempty subsets of F. If H_n is a Cauchy sequence of closed sets for the Hausdorff metric, let H be the set of x such that for some x_n in H_n there is a subsequence x_{n_k} converging to x. Show that H is a nonempty compact set and that H_n converges to it for the Hausdorff metric.

2. In the proof of Theorem 8.4, just after Lemma 8.6, show that the cubes can be ordered so that $i = j - 1$.

3. If Proposition 1.16 is used instead of Proposition 1.12 in the proof of Lemma 8.8, what is the result?

4. Show that in Theorem 8.13(a), $\mathcal{K}(d, \alpha, K)$ cannot be replaced by $I(d, \alpha, K)$, or the collection of nonempty subsets of $I(d, \alpha, K)$, if $d = 2$ and $\alpha > 1$.

Hint: Let $f((\cos\theta, \sin\theta)) := (1 - \cos\theta)/2$, so f takes S^1 onto $[0, 1]$. Then $I((f, 0)) = \emptyset$ where $(f, 0)(u, v) := (f(u, v), 0)$. For any interval $(a, b) \subset \mathbb{R}$ there is a C^∞ function $g_{(a,b)} \geq 0$ with $g > 0$ just on (a, b). Show that for functions $\psi = (f, \sum_{i=1}^k \delta_i g_{(a_i, b_i)}(f))$ for disjoint (a, i, b_i) and small enough $\delta_i > 0$, depending on k, $\|\psi\|_\alpha$ can remain bounded as k increases while $I(\psi)$ can approximate any finite subset of $[0, 1] \times \{0\}$ for h.

5. Show that the unit disk $\{(x, y) : x^2 + y^2 \leq 1\}$ belongs to a class $\mathcal{C}_0$ of unions in Theorem 8.12 for $k = 4$, any $\alpha \in (0, \infty)$ and some $K = K(\alpha) < \infty$. *Hint:* The function $g(x) := (1 - x^2)^{1/2}$ is smooth on intervals $|x| \leq \zeta$ for $\zeta < 1$.

6. Find a constant $c > 0$ such that $\liminf_{\varepsilon \downarrow 0} \varepsilon \log N_I \left(\varepsilon, \mathcal{LL}_{2,1}, \lambda_I^2\right) \geq c$, and likewise for $D(\varepsilon, \mathcal{LL}_{2,1}, d_\lambda)$. *Hint:* Consider squares along a decreasing diagonal, $S_j := S_{j,m+1-j}$, $j = 1, \ldots, m - 1$, for the grid defined in the proof of Theorem 8.21. The union of $\cup_{i < m+1-j} S_{ji}$ and an arbitrary set of S_j gives a set in $\mathcal{LL}_{2,1}$. Apply Lemma 8.8.

7. (a) The proof of Theorem 8.21 used the inequality $\binom{2k}{k} \leq 4^k$. Show that, conversely, for all $k \geq 1$, $\binom{2k}{k} \geq 4^k k^{-1/2}/3$. *Hint:* Use Stirling's formula (Theorem 1.17).

(b) Use this to give a lower bound for $\liminf_{\varepsilon \downarrow 0} \varepsilon \log D \left(\varepsilon, \mathcal{LL}_{2,1}, h\right)$.

8. In the proof of Lemma 8.22, after Lemma 8.23, a function f is defined with $\|f\|_L \leq (d - 1)^{1/2}$. For $d = 2$, deduce this from monotonicity of $x(\cdot)$, $y(\cdot)$ in the proof of Theorem 8.21.

9. Show that the lower layers in $\mathbb{R}^d$ form a Glivenko–Cantelli class for any law having a bounded support and bounded density with respect to Lebesgue measure.

Notes

Notes to Section 8.1. Hausdorff (1914, Section 28) defined his metric between closed, bounded, nonempty subsets of a metric space.

Notes to Section 8.2. Rademacher (1919) proved his theorem on almost everywhere Fréchet differentiability of Lipschitz functions on $\mathbb{R}^d$. Ziemer (1989) has an exposition.

Kolmogorov (1955) gave the first statement in Theorem 8.4. The proof of that part as given is essentially that of Kolmogorov and Tikhomirov (1959). Lorentz (1966, p. 920) sketches another proof. Theorem 8.7 is essentially due to Clements (1963, Theorem 3); the proof here is adapted from Dudley (1974, Lemmas 3.5, 3.6). Remark 8.11, for which I am very grateful to Joseph Fu, shows that there is an error in Dudley (1974, (3.2)), even with the correction (1979), which is proved only for $\alpha \geq 1$. Theorem 8.13 and its proof by a sequence of lemmas are newly corrected and extended versions of results and proofs of Dudley (1974). Tze-Gong Sun (1976) first proved Corollary 8.20(a). The statement was apparently first published in Dudley (1978, Theorem 5.12) and attributed to Sun. There is a related technical report by Sun and Pyke (1982).

Notes to Section 8.3. For Theorem 8.22 and its proof for $d \geq 3$ I am grateful to Lucien Birgé (1982, personal communication) for the idea of the transformation U. For Theorem 8.21 I am much indebted to earlier conversations with Mike Steele. Any errors, however, are mine.

A lower bound for empirical processes on lower layers in the plane will be given in Section 11.4 below, where P is uniform on the unit square. For other, previous results, e.g. laws of large numbers uniformly over $\mathcal{LL}_d$ for more general P, and on the statistical interest of lower layers (monotone regression), see Wright (1981) and references given there.

Notes to Section 8.4. Bolthausen (1978) proved the Donsker property of the class of convex sets in the plane for the uniform law on I^2 (cf. Corollary 8.26(b)).

Theorem 8.25, as mentioned, is due to Bronšteĭn (1976). Specifically, the smoothing $C \mapsto C^r$ and the set of lemmas used follows mainly, though not entirely, his original proof. Perhaps most notably, it appears that the polyhedra used to approximate convex sets in Bronšteĭn's construction need not be convex themselves, and this required some adjustments in the proof. A more minor point is that if $C \in \mathcal{C}_d$, then C^1 does not necessarily include $B(0, 1)$, although C^2 does. So Bronšteĭn's Lemma 1 seems incorrect as stated but could be repaired by changing various constants.

In an earlier result of Dudley (1974, Theorem 4.1), for $d \geq 2$, in the upper bound $\varepsilon^{(1-d)/2}$ was multiplied by $|\log \varepsilon|$. This bound, weaker than Bronšteĭn's, is easier to prove and suffices to give Corollary 8.26(b) and (c). The lower bound in Dudley (1974) was reproduced here. I thank James Munkres for telling me about the triangulation method used after Lemma 8.32.

Gruber (1983) surveys other aspects of approximation of convex sets.

9

The Two-Sample Case, the Bootstrap, and Confidence Sets

9.1 The Two-Sample Case

Let $X_1, \ldots, X_m, \ldots, Y_1, \ldots, Y_n, \ldots$, be some random variables taking values in a set A where $(A, \mathcal{A})$ is a measurable space. Thus $(X_1, \ldots, X_m)$ and $(Y_1, \ldots, Y_n)$ are "samples," of which we have two. Let

$$P_m := \frac{1}{m} \sum_{i=1}^{m} \delta_{X_i}, \quad Q_n := \frac{1}{n} \sum_{j=1}^{n} \delta_{Y_j}.$$

The object of two-sample tests in statistics is to decide whether P_m and Q_n are empirical measures from the same, but unknown, law (probability measure) P on $(A, \mathcal{A})$. Since P is unknown, we cannot directly compare P_m or Q_n to it by forming $m^{1/2}(P_m - P)$ or $n^{1/2}(Q_n - P)$. Instead, P_m and Q_n can be compared to each other, setting

$$\nu_{m,n} := \left(\frac{mn}{m+n} \right)^{1/2} (P_m - Q_n).$$

The *basic hypothesis* will be that there are two laws P, Q on $(A, \mathcal{A})$ and a product of two countable products of copies of $(A, \mathcal{A})$ with factor laws P and Q respectively, namely,

$$(\Omega, \mathcal{D}, \Pr) = (\Omega_1, \mathcal{B}_1, \Pr_1) \times (\Omega_2, \mathcal{B}_2, \Pr_2)$$

where

$$(\Omega_1, \mathcal{B}_1, \Pr_1) = \prod_{i=1}^{\infty} (A_i, \mathcal{A}, P), \quad (\Omega_2, \mathcal{B}_2, \Pr_2) = \prod_{j=1}^{\infty} (B_j, \mathcal{A}, Q),$$

and each A_i and B_j is a copy of A. On these products let X_i be the A_i coordinate and Y_j the B_j coordinate. If $\mathcal{P}$ is a class of laws on $(A, \mathcal{A})$, the $(\mathcal{P})$ *null hypothesis* is that in addition, $P = Q \in \mathcal{P}$. A class $\mathcal{F}$ of measurable functions

319

on $(A, \mathcal{A})$ will be called a $\mathcal{P}$-*universal Donsker class* if it is a P-Donsker class for every $P \in \mathcal{P}$.

Theorem 9.1 *Let $\mathcal{F}$ be a $\mathcal{P}$-universal Donsker class of functions on $(A, \mathcal{A})$. Then for each $P \in \mathcal{P}$, under the (P) null hypothesis, $\nu_{m,n} \Rightarrow G_P$ as $m, n \to \infty$.*

Proof. Let $P \in \mathcal{P}$. By Theorem 3.30, it suffices to show that $\beta(\nu_{m,n}, G_P) \to 0$ as $m, n \to \infty$. Since $\mathcal{F}$ is P-Donsker, it is P-pregaussian, so that by Theorem 3.2 there exists a coherent G_P process on $\mathcal{F}$. Since $\mathcal{F}$ is ρ_P-totally bounded, the functions $f \mapsto G_P(f)(\omega)$ on $\mathcal{F}$ belong to the space $S_0 := \mathrm{UC}(\mathcal{F}, \rho_P)$ of all ρ_P-uniformly continuous and thus bounded functions on $\mathcal{F}$. Here $\mathrm{UC}(\mathcal{F}, \rho_P)$ is a separable subspace of the Banach space $(\ell^\infty(\mathcal{F}), \| \cdot \|_\mathcal{F})$, because uniformly continuous functions on a totally bounded metric space extend uniquely to functions in the space $C(K)$ of continuous functions on the compact completion K, and $C(K)$ is separable in the sup norm for every compact metric space K: RAP Corollary 11.2.5. The map $\omega \mapsto G_P(\cdot)(\omega)$ is Borel measurable from the underlying probability space into S_0: to see this, it is enough by second-countability (e.g., RAP, Proposition 2.1.4) to show that the inverse image of an open ball $\{f : \|f - f_0\|_\mathcal{F} < r\}$ is measurable, which is true since on $\mathrm{UC}(\mathcal{F}, \rho_P)$ the supremum can be taken over a countable ρ_P-dense set in $\mathcal{F}$. Thus G_P has a law (Borel probability measure) μ_0 on $\mathrm{UC}(\mathcal{F}, \rho_P)$, as in Theorem 2.32. It is easily seen that every coherent G_P process on $\mathcal{F}$ has the same law on $\mathrm{UC}(\mathcal{F}, \rho_P)$ (consider finite subsets increasing up to a countable ρ_P-dense subset of $\mathcal{F}$).

Since $\mathcal{F}$ is a P-Donsker class, by Theorem 3.24 there exist probability spaces $(V_i, \mathcal{S}_i, \tau_i)$, $i = 1, 2$, perfect measurable functions g_{ni} from V_i into Ω_i such that $\tau_i \circ g_{ni}^{-1} = \mathrm{Pr}_i$ for all n and $i = 1, 2$, and coherent G_P processes $G_P^{(i)}$ on V_i for $i = 1, 2$ such that $m^{1/2}(P_m - P) \circ g_{m1} \to G_P^{(1)}$ almost uniformly for $\| \cdot \|_\mathcal{F}$ as $m \to \infty$ and $n^{1/2}(Q_n - P) \circ g_{n2} \to G_P^{(2)}$ almost uniformly for $\| \cdot \|_\mathcal{F}$ as $n \to \infty$. Form the product $(V, \mathcal{S}, \tau) := (V_1, \mathcal{S}_1, \tau_1) \times (V_2, \mathcal{S}_2, \tau_2)$ and define $h_{mn} : V \mapsto \Omega_1 \times \Omega_2$ by $h_{mn}(u, v) := (g_{m1}(u), g_{n2}(v))$. Then $\tau \circ h_{mn}^{-1} = \mathrm{Pr}_1 \times \mathrm{Pr}_2$ for all m and n. To show that h_{mn} is perfect, the proof of Theorem 3.24, to be called the "1-sample" proof, will be adapted to the 2-sample case.

Apply Theorem 3.24 with (A^n, P^n) in place of (X_n, Q_n) and with the notation τ instead of Q. Take the Cartesian product of two copies of the probability space Ω in the statement, which will be $(V_1, \tau_1) \times (V_2, \tau_2)$. Apply the constructions in the 1-sample proof in defining the two spaces, specifically the definitions $T_n = A^n \times I_n$ for $n \geq 1$ and $T_0 = S_0 \times I_0$ where each I_j is a copy of $[0, 1]$.

In the last passage of the 1-sample proof, showing that g_n are perfect, for given m and n we mainly want to show that h_{mn} is perfect for $m \geq 1$ and $n \geq 1$. The case where just one of m or n is 0 is not needed.

Take a set $B \subset V_1 \times V_2$ with $(\tau_1 \times \tau_2)(B) > 0$. Write points of $V_i = \prod_{n \geq 0} T_n$ as $(x, \xi) \in V_1$ and $(y, \eta) \in V_2$ where $x, y \in T_0$ and $\xi, \eta \in T := \prod_{n \geq 1} T_n$. Then

$$0 < \int \int \int \int 1_B(x, \xi, y, \eta) d\rho_x(\xi) d\rho_y(\eta) d\mu_0(x) d\mu_0(y)$$

for ρ_x and μ_0 as in the 1-sample proof. Choose and fix an x and y such that

$$0 < \int \int 1_B(x, \xi, y, \eta) d\rho_x(\xi) d\rho_y(\eta).$$

In the one-sample proof write $\rho_x = \alpha_{mx} \times \beta_{mjx}$ rather than $P_{mx} \times Q_{mx}$. Then the latter double integral equals

$$\int \int \int \int 1_B(x, \xi_m, \xi^m, y, \eta_m, \eta^m) d\alpha_{mx}(\xi_m) d\beta_{mx}(\xi^m) d\alpha_{ny}(\eta_n) d\beta_{ny}(\eta^n)$$

where ξ_m ranges over $T_m = A^m \times I_m$, ξ^m over $\prod_{1 \leq k \neq m} T_k$, and likewise for η_n and η^n. Choose and fix ξ^m and η^n such that

$$0 < \int \int 1_B(x, \xi_m, \xi^m, y, \eta_m, \eta^m) d\alpha_{mx}(\xi_m) d\alpha_{ny}(\eta_n).$$

Let $\xi_m = (u, v), u \in A^m, \eta_n = (w, z), w \in A^n, z \in I_n$. Recalling that Q_n in the 1-sample proof is here set equal to P^n, we get as in that proof

$$0 < \int \int \int \int 1_B(x, u, v, \xi^m, y, \eta_m, \eta^n) dP^m(u) dv dP^n(w) dz.$$

Choose and fix v and z so that

$$0 < \int \int 1_B(x, u, v, \xi^m, y, \eta_m, \eta^n) dP^m(u) dP^n(w).$$

Let $C := \{(u, w) \in A^m \times A^n : (x, u, v, \xi^m, y, w, z, \eta^n) \in B\}$. Then $(P^m \times P^n)(C) > 0$ and $h_{mn}[B] \supset C$, showing that h_{mn} is perfect by Theorem 3.18(b). We have

$$v_{mn} = \frac{(mn)^{1/2}}{(m+n)^{1/2}}(P_m - P - (Q_n - P))$$

$$= \left(\frac{n}{m+n}\right)^{1/2} m^{1/2}(P_m - P) - \left(\frac{m}{m+n}\right)^{1/2} n^{1/2}(Q_n - P).$$

Let $G_P^{(m,n)} := \left(n/(m+n)\right)^{1/2} G_P^{(1)} - \left(m/(m+n)\right)^{1/2} G_P^{(2)}$. Now, if Y, Z are independent coherent G_P processes on $\mathcal{F}$ and a, b are any real numbers with $a^2 + b^2 = 1$, then $aY + bZ$ is a coherent G_P process on $\mathcal{F}$. To see this, note that $aY + bZ$ is a Gaussian process indexed by $\mathcal{F}$, has the covariances of G_P, and is coherent. So each $G_P^{(m,n)}$ is a coherent G_P process. We have

$$\| \nu_{m,n} \circ h_{mn} - G_P^{(m,n)} \|_{\mathcal{F}} \leq \left(\frac{n}{m+n} \right)^{1/2} \| m^{1/2}(P_m - P) \circ g_{m1} - G_P^{(1)} \|_{\mathcal{F}}$$

$$+ \left(\frac{m}{m+n} \right)^{1/2} \| n^{1/2}(Q_n - P) \circ g_{n2} - G_P^{(2)} \|_{\mathcal{F}} \to 0$$

almost uniformly as $m, n \to \infty$.

Let H be a function on $\ell^\infty(\mathcal{F})$ with $\|H\|_{BL} \leq 1$, so that $\|H\|_{\sup} \leq 1$ and $|H(f) - H(g)| \leq \|f - g\|_{\mathcal{F}}$ for all $f, g \in \ell^\infty(\mathcal{F})$. Given $\varepsilon > 0$, there are $m_0 < \infty$ and $n_0 < \infty$ such that for $m \geq m_0$ and $n \geq n_0$, $\| \nu_{m,n} \circ h_{m,n} - G_P^{(m,n)} \|_{\mathcal{F}} < \varepsilon$ on a set $V_\varepsilon \subset V$ with $\mu(V_\varepsilon) > 1 - \varepsilon$. It follows that

$$d_{m,n} := |H(\nu_{m,n} \circ h_{mn}) - H(G_P^{(m,n)})| < \varepsilon$$

on V_ε, and $d_{m,n} \leq 2$ everywhere. Since $G_P^{(m,n)}$ has a law defined on the Borel sets of $UC(\mathcal{F}, \rho_P)$, thus of $\ell^\infty(\mathcal{F})$, for $\| \cdot \|_{\mathcal{F}}$, $H(G_P^{(m,n)})$ is measurable and

$$\left| \int^* H(\nu_{m,n} \circ h_{mn}) d\mu - EH(G_P^{(m,n)}) \right| \leq 3\varepsilon.$$

It follows that $\beta(\nu_{m,n} \circ h_{mn}, G_P) \to 0$ as $m, n \to \infty$. Since each h_{mn} is perfect and measure-preserving, it follows that $\nu_{m,n} \Rightarrow G_P$ for $m, n \to \infty$ as in Theorem 3.20. $\qquad\square$

The classical two-sample situation is the special case where $A = \mathbb{R}$, $\mathcal{A}$ is the Borel σ-algebra, $\mathcal{F}$ is the set of all indicator functions of half-lines $(-\infty, x]$, and $\mathcal{P}$ is either the set of all laws on $\mathbb{R}$, or the set of all continuous (nonatomic) laws. Thus $m^{1/2}(P_m - P)((-\infty, x]) = m^{1/2}(F_m - F)(x)$ where F is the distribution function of P and F_m an empirical distribution function. Here $\mathcal{F}$ is a universal Donsker class by any of several previous results, for example, Corollary 6.20, and $G_P(1_{(-\infty,x]}) = y_{F(x)}$ where y is the Brownian bridge (as in Chapter 1). Actually, $\mathcal{F}$ is a uniform Donsker class as defined in Section 10.4. Without using any machinery, one can see directly that the convergence to $y_{F(x)}$ is as fast for any distribution function F as it is for the $U[0, 1]$ distribution function. If F is continuous, it takes all values in the open interval $(0, 1)$, $y_0 \equiv y_1 \equiv 0$, and $y_t \to 0$ as $t \downarrow 0$ or $t \uparrow 1$. Thus the distributions of $\sup_x y_{F(x)}$ and $\sup_x |y_{F(x)}|$ and the joint distribution of $(\inf_u y_{F(u)}, \sup_x y_{F(x)})$ do not depend on F, for F continuous. Let F_m and G_n be independent empirical distribution functions for the same F. Let $H_{mn} := (mn/(m+n))^{1/2}(F_m - G_n)$. By Theorem 3.31 and Proposition 3.32, which extend straightforwardly to limits as $m, n \to \infty$, the distributions of the supremum, supremum of absolute value, and supremum minus infimum of H_{mn} converge to those of the same functionals for y_t. Thus we get for $u > 0$ (about $u = 0$, see Problem 3):

Corollary 9.2 *If F_m and G_n are independent empirical distribution functions for a continuous distribution on $\mathbb{R}$, then for any $u > 0$,*

(a) $\lim_{m,n\to\infty} \Pr(\sup_x H_{mn}(x) > u) = \exp(-2u^2)$,

(b) $\lim_{m,n\to\infty} \Pr(\sup_x |H_{mn}(x)| > u) = 2\sum_{k=1}^{\infty}(-1)^{k-1}\exp(-2k^2u^2)$,

(c) $\lim_{m,n\to\infty} \Pr(\sup_x H_{mn}(x) - \inf_y H_{mn}(y) > u)$
$= 2\sum_{k=1}^{\infty}(4k^2u^2 - 1)\exp(-2k^2u^2)$.

Proof. The distributions of the given functionals of the Brownian bridge y_t are given in RAP, Propositions 12.3.3, 12.3.4, and 12.3.6. All three are continuous in u for $u > 0$. Thus convergence follows from RAP, Theorem 9.3.6, adapted to limits as $m, n \to \infty$. $\qquad\square$

The quantity $\sup_x H_{mn}(x) - \inf_y H_{mn}(y)$ on the left in (c) is called a Kuiper statistic. Suppose we have samples on the unit circle in the plane. The unit circle could be written as the set of $(\cos\theta, \sin\theta)$ for $0 \le \theta < 2\pi$ or for $-\pi \le \theta < \pi$. Using, for example, the Kolmogorov–Smirnov statistic as in (b), the value of the statistic would depend on the choice of range for θ, whereas the Kuiper statistic is rotationally invariant (this is left as Problem 1 at the end of the chapter).

For multidimensional sample spaces, one in general does not know specific statistics which, like those in Corollary 9.2, have given asymptotic distributions not depending on P in a large class, such as all nonatomic laws.

9.2 A Bootstrap Central Limit Theorem in Probability

Iterating the operation by which we get an empirical measure P_n from a law P, we form the bootstrap empirical measure P_n^B by sampling n independent points whose distribution is the empirical measure P_n. The bootstrap was first introduced in nonparametric statistics, where the law P is unknown and we want to make inferences about it from the observed P_n. This can be done by way of bootstrap central limit theorems, which say that under some conditions, including n large enough, $n^{1/2}(P_n^B - P_n)$ behaves like $n^{1/2}(P_n - P)$ and both behave like G_P.

Let $(S, \mathcal{S}, P)$ be a probability space and $\mathcal{F}$ a class of real-valued measurable functions on S. Let as usual $X_1, X_2, \ldots$, be coordinates on a countable product of copies of $(S, \mathcal{S}, P)$. Then let $X_{n1}^B, \ldots, X_{nn}^B$ be independent with distribution P_n. Let

$$P_n^B := \frac{1}{n}\sum_{j=1}^{n}\delta_{X_{nj}^B}.$$

Then P_n^B will be called a *bootstrap empirical measure*. More precisely, take the product of (S^∞, P^∞) with a probability space $(\Omega_B, \mathrm{Pr}^B)$ on which for all $n = 1, 2, \ldots$, i.i.d. random variables $i_{n1}, \ldots, i_{nn}$ are defined with uniform distribution on $\{1, \ldots, n\}$. Then set $X_{nj}^B := X_{i_{nj}(\omega)}$.

A statistician has a data set, represented by a fixed P_n, and estimates the distribution of P_n^B by repeated resampling from the same P_n. So we are interested not so much in the unconditional distribution of P_n^B as P_n varies, but rather in the conditional distribution of P_n^B given P_n or $(X_1, \ldots, X_n)$. Let $\nu_n^B := n^{1/2}(P_n^B - P_n)$.

The limit theorems will be stated in terms of the dual-bounded-Lipschitz "metric" β of Section 3.6, which metrizes convergence in distribution for not necessarily measurable random elements of a possibly nonseparable metric space (S, d), to a limit which is a measurable random variable with separable range. Let $\beta_{\mathcal{F}}$ be the β distance where d is the metric defined by the norm $\| \cdot \|_{\mathcal{F}}$.

Definition. Let $(S, \mathcal{S}, P)$ be a probability space and $\mathcal{F}$ a class of measurable real-valued functions on S. Then the *bootstrap central limit theorem holds in probability* (respectively, *almost surely*) for P and $\mathcal{F}$ if and only if $\mathcal{F}$ is pregaussian for P and $\beta_{\mathcal{F}}(\nu_n^B, G_P)$, conditional on $X_1, \ldots, X_n$, converges to 0 in outer probability (resp., almost uniformly) as $n \to \infty$.

To make the conditioning more explicit, given $X_1, \ldots, X_n$, there are finitely many possible values for ν_n^B, each on a measurable set in Ω_B. For any bounded Lipschitz function H on $\ell^\infty(\mathcal{F})$, $E(H(\nu_n^B)|X_1, \ldots, X_n)$ is defined, and $EH(G_P)$ is some number. Thus

$$\sup\{|E(H(\nu_n^B)|X_1, \ldots, X_n) - EH(G_P)| : \|H\|_{BL} \le 1\}$$

is a function of $X_1, \ldots, X_n$, not necessarily measurable, which according to the definition converges to 0 in one sense or the other.

Thus, the bootstrap central limit theorem holds in probability for $\mathcal{F}$ if and only if for any $\varepsilon > 0$, $E^*(\mathrm{Pr}(\beta_{\mathcal{F}}(\nu_n^B, G_P) > \varepsilon|\{X_j\}_{j=1}^n)) \to 0$ as $n \to \infty$, where the outer expectation is taken with respect to the distribution of $X_1, \ldots, X_n$.

A main bootstrap limit theorem in probability will be stated. It will be proved in the rest of this section, via a number of lemmas and auxiliary theorems.

Theorem 9.3 (Giné and Zinn) *Let $(X, \mathcal{A}, P)$ be any probability space and $\mathcal{F}$ a Donsker class for P. Then the bootstrap central limit theorem holds in probability for P and $\mathcal{F}$.*

Remarks. Giné and Zinn (1990), see also Giné (1997), also proved "only if" under a measurability condition and proved a corresponding almost sure form of the theorem where $\mathcal{F}$ has an $\mathcal{L}^2$ envelope up to additive constants.

Proof. E^B, $\Pr^B$, and $\mathcal{L}^B$ will denote the conditional expectation, probability, and law respectively given the sample $X^{(n)} := (X_1, \ldots, X_n)$. Given the sample, ν_n^B has only finitely many possible values.

First, a finite-dimensional bootstrap central limit theorem is needed. $\qquad\square$

Theorem 9.4 *Let $X_1, X_2, \ldots$ be i.i.d. random variables with values in $\mathbb{R}^d$ and let $X_{n,i}^B$, $i = 1, \ldots, n$, be i.i.d. (P_n), where $P_n := \frac{1}{n} \sum_{j=1}^n \delta_{X_j}$. Let $\overline{X}_n := \frac{1}{n} \sum_{j=1}^n X_j$. Assume that $E|X_1|^2 < \infty$. Let C be the covariance matrix of X_1, $C_{rs} := E(X_{1r} X_{1s}) - E(X_{1r})E(X_{1s})$, $r, s = 1 \ldots, d$. Then for the usual convergence of laws in $\mathbb{R}^d$, almost surely as $n \to \infty$,*

$$\mathcal{L}^B \left(n^{-1/2} \sum_{j=1}^n (X_{n,j}^B - \overline{X}_n) \right) \to N(0, C). \qquad (9.1)$$

Proof. Note: $N(0, 0)$ is defined as a point mass at 0. Suppose the theorem holds for $d = 1$. Then for each $t \in \mathbb{R}^d$, the theorem holds for the variables (t, X_j) and thus $(t, X_{n,j}^B)$, and so on. This implies that the characteristic functions of the laws on the left in (9.1) converge pointwise to that of $N(0, C)$, and thus that the laws converge by the Lévy continuity theorem (RAP, Theorem 9.8.2). (This last argument is sometimes called the "Cramér–Wold" device.) So, we can assume $d = 1$, and on the right in (9.1) we have a law $N(0, \sigma^2)$ where σ^2 is the variance of X_1. If $\sigma^2 = 0$ there is no problem, so assume $\sigma^2 > 0$. We can assume that $EX_1 = 0$ since subtracting EX_1 from all X_j does not change the expression on the left in (9.1). Let

$$s_n'^2 := \frac{1}{n} \sum_{j=1}^n (X_j - \overline{X}_n)^2, \quad s_n' := (s_n'^2)^{1/2}.$$

Then by the strong law of large numbers for the X_j and the X_j^2, $s_n'^2$ converges a.s. as $n \to \infty$ to σ^2.

For a given n and sample $X^{(n)}$, $\overline{X}_n$ and s_n' are fixed and the variables $Y_{ni} := (X_{n,i}^B - \overline{X}_n)/n^{1/2}$ for $i = 1, \ldots, n$ are i.i.d. We have a triangular array and will apply the Lindeberg theorem (RAP, Theorem 9.6.1). For each n and i, $E^B Y_{n,i} = 0$. Each Y_{ni} has variance $\text{Var}^B(Y_{ni}) = s_n'^2/n$. The sum from $i = 1$ to n of these variances is $s_n'^2$. Let $Z_{ni} := Y_{ni}/s_n'$. Then Z_{ni} remain i.i.d. given P_n, and the sum of their variances is 1. We would like to show by the Lindeberg theorem that $\mathcal{L}^B(\sum_{i=1}^n Z_{ni})$ converges to $N(0, 1)$ as $n \to \infty$, for almost all $X_1, X_2, \ldots$. We have $E^B(Z_{ni}) = 0$. Let $1\{\ldots\} := 1_{\{\ldots\}}$ for any event $\{\ldots\}$. For $\varepsilon > 0$ let $E_{nj\varepsilon} := E^B(Z_{nj}^2 1\{|Z_{nj}| > \varepsilon\})$. It remains to show that $\sum_{j=1}^{n} E_{nj\varepsilon} = n E_{n1\varepsilon} \to 0$ as $n \to \infty$, almost surely. Now

$$A_{n1} := \{|Z_{n1}| > \varepsilon\} = \{|X_{n,1}^B - \overline{X}_n| > n^{1/2} s_n' \varepsilon\},$$

which for n large enough, almost surely, is included in the set

$$C(n) := C_n := \{|X_{n,1}^B| > n^{1/2}\sigma\varepsilon/2\}.$$

Also, $Z_{n1}^2 = (X_{n,1}^B - \overline{X}_n)^2/(ns_n'^2) \le 2[(X_{n,1}^B)^2 + \overline{X}_n^2]/(ns_n'^2)$. Then $\overline{X}_n^2/s_n'^2$, which is constant given the sample, approaches 0 a.s. as $n \to \infty$, and so does $E^B(1_{C(n)}\overline{X}_n^2/s_n'^2) \le \overline{X}_n^2/s_n'^2$. For the previous term,

$$E^B((X_{n,1}^B)^2 1_{C(n)})/s_n'^2 = n^{-1} \sum_{i=1}^{n} X_i^2 1\{|X_i| > n^{1/2}\sigma\varepsilon/2\}/s_n'^2,$$

which goes to 0 a.s. as $n \to \infty$, by the strong law of large numbers for the variables $X_i^2 1\{|X_i| > K\}$ for any fixed K, then letting $K \to \infty$. Theorem 9.4 is proved. $\qquad\square$

Here is what is called a desymmetrization fact:

Lemma 9.5 *Let T be a set and for any real-valued function f on T let $\|f\|_T := \sup_{t\in T} |f(t)|$. Let X and Y be two stochastic processes indexed by $t \in T$ defined on a probability space $(\Omega \times \Omega', \mathcal{S} \otimes \mathcal{S}', P \times P')$, where $X(t)(\omega, \omega')$ depends only on $\omega \in \Omega$ and $Y(t)(\omega, \omega')$ only on $\omega' \in \Omega'$. Then*

(a) For any $s > 0$ and any $u > 0$ such that $\sup_{t\in T} \Pr\{|Y(t)| \ge u\} < 1$, we have

$$P^*(\|X\|_T > s) \le \Pr^*\{\|X - Y\|_T > s - u\}/[1 - \sup_{t\in T} P'(|Y(t)| \ge u)].$$

(b) If $\theta > \sup_{t\in T} E(Y(t)^2)$, then for any $s > 0$,

$$P^*(\|X\|_T > s) \le 2\Pr^*(\|X - Y\|_T > s - (2\theta)^{1/2}).$$

Proof. Let $X(\omega)(t) := X(t)(\omega)$ and $A(s) := \{\omega : \|X(\omega)\|_T > s\}$. For part (a) we have by (3.3) in Theorem 3.9

$$\Pr^*(\|X - Y\|_T > s - u) \ge E_P^*(P')^*(\|X - Y\|_T > s - u)$$

$$\ge P^*(\|X\|_T > s) \inf_{\omega \in A(s)} (P')^*(\|X(\omega) - Y\|_T > s - u)$$

$$\ge P^*(\|X\|_T > s) \inf_{t\in T} P'(|Y(t)| < u),$$

since if $\|X(\omega)\|_T > s$, then for some $t \in T$, $|X(\omega)(t)| > s$. This proves part (a). Then part (b) follows by Chebyshev's inequality. $\qquad\square$

Recall that ε_i are called Rademacher variables if $\Pr(\varepsilon_i = 1) = \Pr(\varepsilon_i = -1) = 1/2$. Some hypotheses will be given for later reference. Let $(S, \mathcal{S}, P)$ be a probability space and $(S^n, \mathcal{S}^n, P^n)$ a Cartesian product of n copies of

$(S, \mathcal{S}, P)$. Let $\mathcal{F} \subset \mathcal{L}^2(S, \mathcal{S}, P)$. Let $\varepsilon_1, \ldots, \varepsilon_n$ be i.i.d. Rademacher variables defined on a probability space $(\Omega', \mathcal{A}, P')$. Then take the probability space

$$(S^n \times \Omega', \mathcal{S}^n \otimes \mathcal{A}, P^n \times P'). \tag{9.2}$$

References to (9.2) will also be to the preceding paragraph.

Here is another fact on symmetrization and desymmetrization.

Lemma 9.6 *Under (9.2), for any $t > 0$ and $n = 1, 2, \ldots,$*

(a) $\mathrm{Pr}^* \left(\| \sum_{i=1}^n \varepsilon_i f(X_i) \|_{\mathcal{F}} > t \right) \leq 2 \max_{k \leq n} \mathrm{Pr}^* \left(\| \sum_{i=1}^k f(X_i) \|_{\mathcal{F}} > t/2 \right).$

(b) Suppose that $\alpha^2 := \sup_{f \in \mathcal{F}} \int (f - Pf)^2 dP < \infty$. Then for $t > 2^{1/2} \alpha n^{1/2}$ and all $n = 1, 2, \ldots,$

$$C_t := \mathrm{Pr}^* \left(\left\| \sum_{i=1}^n (f(X_i) - Pf) \right\|_{\mathcal{F}} > t \right)$$

$$\leq 4 \mathrm{Pr}^* \left\{ \left\| \sum_{i=1}^n \varepsilon_i f(X_i) \right\|_{\mathcal{F}} > (t - (2n)^{1/2} \alpha)/2 \right\}.$$

Proof. For part (a), let $E(n) := \{ \tau := \{ \tau_i \}_{i=1}^n : \tau_i = \pm 1 \text{ for each } i \}$. Let $\mathcal{E}_n := \{ \varepsilon_i \}_{i=1}^n$. Then

$$P_t := \mathrm{Pr}^* \left(\left\| \sum_{i=1}^n \varepsilon_i f(X_i) \right\|_{\mathcal{F}} > t \right)$$

$$\leq \sum_{\tau \in E(n)} \mathrm{Pr}^* \left[\{ \mathcal{E}_n = \tau \} \times \left\{ \left\| \sum_{\tau_i = 1} f(X_i) - \sum_{\tau_i = -1} f(X_i) \right\|_{\mathcal{F}} > t \right\} \right].$$

By Lemma 3.6 and Theorem 3.3, for laws μ, ν and sets A, B,

$$(\mu \times \nu)^*(A \times B) = \mu^*(A) \nu^*(B) = \mu(A) \nu^*(B)$$

if A is measurable, so

$$P_t \leq 2 \sum_{\tau \in E(n)} 2^{-n} \max_{k \leq n} (P^n)^* (\| \sum_{i=1}^k f(X_i) \|_{\mathcal{F}} > t/2)$$

$$= 2 \max_{k \leq n} \mathrm{Pr}^* (\| \sum_{i=1}^k f(X_i) \|_{\mathcal{F}} > t/2).$$

This proves (a). For (b), take a further product, so that we have (9.2) with $2n$ in place of n. Apply Lemma 9.5(b) to the process $f \mapsto \sum_{i=1}^n f(X_i)$, letting

$Y_i := X_{n+i}$ for $i = 1, \ldots, n$. We get for any $\tau \in E(n)$,

$$C_t \leq 2\mathrm{Pr}^*(\| \sum_{i=1}^{n} f(X_i) - f(Y_i)\|_{\mathcal{F}} > t - (2n)^{1/2}\alpha)$$

$$= 2(P^{2n})^*(\| \sum_{i=1}^{n} \tau_i(f(X_i) - f(Y_i))\|_{\mathcal{F}} > t - (2n)^{1/2}\alpha).$$

Then by Theorem 3.9,

$$C_t \leq 2E_Q^*(P^n)^*(\| \sum_{i=1}^{n} \varepsilon_i(f(X_i) - f(Y_i))\|_{\mathcal{F}} > t - (2n)^{1/2}\alpha)$$

$$\leq 4\mathrm{Pr}^*(\| \sum_{i=1}^{n} \varepsilon_i f(X_i)\|_{\mathcal{F}} > [t - (2n)^{1/2}\alpha]/2),$$

finishing the proof.   □

Next, we have some consequences or forms of Jensen's inequality.

Lemma 9.7 *(a) Let $(S, \mathcal{S}, P)$ be a probability space and $\mathcal{F} \subset \mathcal{L}^1(S, \mathcal{S}, P)$. Then $\|Ef\|_{\mathcal{F}} \leq E\|f\|_{\mathcal{F}}^*$.*

(b) Let T be a set and $(\Omega', \mathcal{S}', Q')$ and $(\Omega'', \mathcal{S}'', Q'')$ two probability spaces. Take the product probability space $(\Omega' \times \Omega'', \mathcal{S}' \otimes \mathcal{S}'', Q' \times Q'')$. Let $g(t, \omega')$ be a real-valued stochastic process indexed by $t \in T$ with $\omega' \in \Omega'$, and let $h(t, \omega'')$ another such process with $\omega'' \in \Omega''$. For any real-valued function f on T let $\|f\|_T := \sup_{t \in T} |f(t)|$. Assume that $E_Q(h(t, \cdot)) = 0$ for all $t \in T$. Then for $1 \leq p < \infty$

$$E^*\|g\|_T^p \leq E^*\|g + h\|_T^p. \tag{9.3}$$

Proof. (a) For each $g \in \mathcal{F}$, by Jensen's inequality (e.g., RAP, 10.2.6), $|Eg| \leq E|g|$, while $E|g| \leq E\|f\|_{\mathcal{F}}^*$. Then, taking the supremum over $g \in \mathcal{F}$ proves (a).

For (b), recall that for any function of ω', E^* for Q' and $Q' \times Q''$ are the same by Lemma 3.6 and Theorem 3.3. For any values of t and ω', $z \mapsto |g(t, \omega') + z|^p$ is a convex function of z. By Jensen's inequality again,

$$|g(t, \omega')|^p = |g(t, \omega') + Eh(t, \cdot)|^p \leq \int |g(t, \omega') + h(t, \omega'')|^p dQ''(\omega'')$$

$$\leq \sup_{s \in T} \int |g(s, \omega') + h(s, \omega'')|^p dQ''(\omega''),$$

and so

$$\|g(\cdot, \omega')\|_T^p \leq \sup_{s \in T} \int |g(s, \omega') + h(s, \omega'')|^p dQ''(\omega'').$$

For all $s \in T$, $|g(s, \omega') + h(s, \omega'')|^p \leq \|g(\cdot, \omega') + h(\cdot, \omega'')\|_T^{*p}$, so

$$E_P^* \|g\|_T^p \leq E_P^* E_Q \|g + h\|_T^{*p} = E^* \|g + h\|_T^p$$

by the Tonelli–Fubini theorem since $\|g + h\|_T^*$ is measurable and by Theorem 3.3 ($E^* f = E f^* \leq +\infty$, $f \geq 0$). $\qquad\square$

The following lemma and theorem are known as Hoffmann-Jørgensen inequalities.

Lemma 9.8 *Let* $(S^n, \mathcal{S}^n, \Pi_{j=1}^n P_j)$ *be a product probability space with coordinates* $X_1, \ldots, X_n$. *Let* $\mathcal{F}$ *be a class of measurable real-valued functions on* $(S, \mathcal{S})$. *Let* $S_k(f) := \sum_{j=1}^k f(X_j)$ *for* $k = 1, \ldots, n$. *Then for any* $s > 0$ *and* $t > 0$,

$$\begin{aligned}
&\Pr\left(\max_{k \leq n} \|S_k(f)\|_{\mathcal{F}}^* > 3t + s\right) \\
&\leq \left(\Pr\left\{\max_{k \leq n} \|S_k\|_{\mathcal{F}}^* > t\right\}\right)^2 + \Pr\left(\max_{j \leq n} \|X_j\|_{\mathcal{F}}^* > s\right).
\end{aligned} \tag{9.4}$$

Proof. Let $\|\cdot\| := \|\cdot\|_{\mathcal{F}}$. Let $\tau := \min\{j \leq n : \|S_j\|^* > t\}$, or $\tau := n + 1$ if there is no such $j \leq n$. Then for $j = 1, \ldots, n$, $\{\tau = j\}$ is by Lemma 3.6 a measurable function of $X_1, \ldots, X_j$ and $\{\max_{k \leq n} \|S_k\|^* > t\}$ is the disjoint union $\cup_{j=1}^n \{\tau = j\}$. On $\{\tau = j\}$, we have $\|S_k\|^* \leq t$ if $k < j$ and when $k \geq j$, $\|S_k\|^* \leq t + \|X_j\|^* + \|S_k - S_j\|^*$. Thus in either case,

$$\max_{k \leq n} \|S_k\|^* \leq t + \max_{i \leq n} \|X_i\|^* + \max_{\tau < k \leq n} \|S_k - S_j\|^*.$$

For $j < k \leq n$, by Lemma 3.6, $\|S_k - S_j\|^*$ is (a.s. equal to) a measurable function of $X_{j+1}, \ldots, X_k$ and thus is independent of $\{\tau = j\}$. So

$$\begin{aligned}
&\Pr(\tau = j, \max_{k \leq n} \|S_k\|^* > 3t + s) \\
&\leq \Pr(\tau = j, \max_{i \leq n} \|X_i\|^* > s) + \Pr(\tau = j)\Pr(\max_{j < k \leq n} \|S_k - S_j\|^* > 2t).
\end{aligned}$$

Since $\max_{j < k \leq n} \|S_k - S_j\|^* \leq 2 \max_{k \leq n} \|S_k\|^*$, summing over $j = 1, \ldots, n$ gives (9.4), and the Lemma is proved. $\qquad\square$

Theorem 9.9 *Let* $0 < p < \infty$, $n = 1, 2, \ldots$, *let* $X_1, \ldots, X_n$ *be coordinates on a product probability space* $(S^n, \mathcal{S}^n, \Pi_{j=1}^n P_j)$. *Let* $\mathcal{F}$ *be a class of measurable real-valued functions on* $(S, \mathcal{S})$ *such that for* $i = 1, \ldots, n$, $E(\|f(X_i)\|_{\mathcal{F}}^{*p}) < \infty$. *Let*

$$u := \inf\left\{t > 0 : \Pr\left[\max_{k \leq n} \left\|\sum_{i=1}^k f(X_i)\right\|_{\mathcal{F}}^* > t\right] \leq 1/(2 \cdot 4^p)\right\}.$$

Then

$$E \max_{k \leq n}\left(\left\|\sum_{i=1}^k f(X_i)\right\|_{\mathcal{F}}^{*p}\right) \leq 2 \cdot 4^p E(\max_{j \leq n}(\|f(X_j)\|_{\mathcal{F}}^*)^p) + 2(4u)^p.$$

Proof. Recall that for any random variable $Y \geq 0$ with law P,

$$\int_0^\infty P(Y > t)dt = \int_0^\infty \int_{t+}^\infty dP(y)dt = \int_0^\infty \int_0^y dt\, dP(y)$$

$$= \int_0^\infty y\, dP(y) = EY.$$

Let $\|\cdot\| := \|\cdot\|_\mathcal{F}$. Then

$$E(\max_{k \leq n} \| S_k \|^{*p}) = 4^p \int_0^\infty \Pr(\max_{k \leq n} \| S_k \|^* > 4t)dt^p$$

$$= 4^p \left(\int_0^u + \int_u^\infty \right) P \left(\max_{k \leq n} \| S_k \|^* > 4t \right) dt^p,$$

which by (9.4) with $s = t$ is less than or equal to

$$(4u)^p + 4^p \int_u^\infty (\Pr(\max_{k \leq n} \| S_k \|^* > t))^2 dt^p + 4^p \int_u^\infty \Pr(\max_{i \leq n} \| X_i \|^* > t)dt^p$$

$$\leq (4u)^p + 4^p \Pr(\max_{k \leq n} \| S_k \|^* > u) \int_0^\infty \Pr(\max_{k \leq n} \| S_k \|^* > t)dt^p$$

$$+ 4^p E \max_{i \leq n} \| X_i \|^{*p}$$

$$\leq (4u)^p + \frac{1}{2} E \left(\max_{k \leq n} \| S_k \|^{*p} \right) + 4^p E \left(\max_{i \leq n} \| X_i \|^{*p} \right),$$

since $4^p \Pr(\max_{k \leq n} \| S_k \|^* > u) \leq 1/2$ by choice of u. Then since $E(\max_{k \leq n} \| S_k \|^{*p}) < \infty$, we can subtract the term with factor $1/2$ from both sides, then multiply by 2 and get

$$E(\max_{k \leq n} \| S_k \|^{*p}) \leq 2(4u)^p + 2 \cdot 4^p E(\max_{i \leq n} \| X_i \|^{*p}).$$

The theorem is proved. $\square$

Next is another symmetrization-desymmetrization inequality:

Lemma 9.10 *Under* (9.2),

$$\tfrac{1}{2} E^* \| \textstyle\sum_{j=1}^n \varepsilon_j (f(X_j) - Pf) \|_\mathcal{F} \leq E^* \| \textstyle\sum_{j=1}^n (f(X_j) - Pf) \|_\mathcal{F}$$

$$\tag{9.5}$$

$$\leq 2 E^* \| \textstyle\sum_{j=1}^n \varepsilon_j (f(X_j) - Pf) \|_\mathcal{F},$$

which also holds if the Pf in the last expression is deleted.

Proof. Replacing $\mathcal{F}$ by $\{f - Pf : f \in \mathcal{F}\}$, we can assume $Pf = 0$ for all $f \in \mathcal{F}$. Then by (3.4),

$$E^* \| \sum_{j=1}^{n} \varepsilon_j f(X_j) \|_{\mathcal{F}} = E_\varepsilon E_X^* \| \sum_{i: \varepsilon_i = 1, i \leq n} \varepsilon_i f(X_i) + \sum_{i: \varepsilon_i = -1, i \leq n} \varepsilon_i f(X_i) \|_{\mathcal{F}}$$

$$\leq E_\varepsilon E_X^* \| \sum_{i: \varepsilon_i = 1, i \leq n} \varepsilon_i f(X_i) \|_{\mathcal{F}} + E_\varepsilon E_X^* \| \sum_{i: \varepsilon_i = -1, i \leq n} \varepsilon_i f(X_i) \|_{\mathcal{F}}$$

$$\leq 2 E_\varepsilon E_X^* \| \sum_{j=1}^{n} f(X_j) \|_{\mathcal{F}} = 2E^* \| \sum_{j=1}^{n} f(X_j) \|_{\mathcal{F}}$$

by (9.3). The left side of (9.5) follows. For the right side, extend (9.2) by taking the Cartesian product with another copy of $(S^n, \mathcal{S}^n, P^n)$ and let the new coordinate functions be $\xi_1, \ldots, \xi_n$. Then if $\tau_i = \pm 1$ for each i,

$$E^* \| \sum_{j=1}^{n} f(X_j) \|_{\mathcal{F}} \leq E^* \| \sum_{j=1}^{n} (f(X_j) - f(\xi_j)) \|_{\mathcal{F}}$$

$$= E^* \| \sum_{j=1}^{n} \tau_j (f(X_j) - f(\xi_j)) \|_{\mathcal{F}},$$

since (9.3) holds as well if $+$ is replaced by $-$, and by symmetry of $f(X_j) - f(\xi_j)$. So by (3.3),

$$E^* \| \sum_{i=1}^{n} f(X_i) \|_{\mathcal{F}} \leq E_\varepsilon E_X^* \| \sum_{i=1}^{n} \varepsilon_i (f(X_i) - f(\xi_i)) \|_{\mathcal{F}}$$

$$\leq 2 E_\varepsilon E_X^* \| \sum_{i=1}^{n} \varepsilon_i f(X_i) \|_{\mathcal{F}} = 2E^* \| \sum_{i=1}^{n} \varepsilon_i f(X_i) \|_{\mathcal{F}},$$

which finishes the proof of the Lemma. $\qquad\square$

Next will be some Poissonization facts. A stochastic process $\xi(t, \omega)$ with values in $\mathbb{R}^n$ is called *centered* if $E\xi(t) = 0$ for all $t \in T$. Recall that Y has a Poisson distribution with parameter $\lambda > 0$ if and only if $\Pr(Y = k) = e^{-\lambda} \lambda^k / k!$ for $k = 0, 1, 2, \ldots$.

Lemma 9.11 *Let T be any set. Let $\xi := \{X_{i,1}\}_{i=1}^{n}(t, \omega_1)$, $t \in T$, be a centered stochastic process with values in $\mathbb{R}^n$ with $\omega_1 \in \Omega_1$ for a probability space $(\Omega_1, \mathcal{S}_1, Q_1)$. For each $j \geq 1$ let $(\Omega_j, \mathcal{S}_j, Q_j)$ be a copy of $(\Omega_1, \mathcal{S}_1, Q_1)$. Take the product $(\Omega, \mathcal{S}, Q) = \prod_{j=1}^{\infty}(\Omega_j, \mathcal{S}_j, Q_j)$. Let $\pi_j(\omega) := \omega_j$ be the coordinate projection from Ω onto Ω_j. Define stochastic processes $\{X_{i,j}\}_{i=1}^{n}$ for $j \geq 2$ by $X_{i,j}(\omega) := X_{i,1}(\pi_j(\omega))$ for $i = 1, \ldots, n$ and all $j \geq 1$.*

Let $N_i := N(i)$, $i = 1, 2, \ldots$, be i.i.d. Poisson variables with parameter $\lambda = 1$, defined on a probability space (Ω', P'), and take the product of this space with Ω, so that $N_1, N_2, \ldots$ are independent of $X_{i,j}$. Let $\|f\|_T := \sup_{t \in T} |f(t)|$ and $x \wedge y := \min(x, y)$. Then

$$
E^* \left\| \sum_{i=1}^{n} X_i \right\|_T \leq \frac{e}{e-1} E^* \left\| \sum_{i=1}^{n} \sum_{j=1}^{N(i)} X_{i,j} \right\|_T. \tag{9.6}
$$

Proof. By Jensen's inequality, Lemma 9.7(a), letting E_N be expectation with respect to the Poisson variables N_i, we have

$$
M_n := \frac{e-1}{e} E^* \left\| \sum_{i=1}^{n} X_{i,1} \right\|_T
$$

$$
= E_X^* \left\| \sum_{i=1}^{n} E(N_i \wedge 1) X_{i,1} \right\|_T \leq E_X^* E_N \left\| \sum_{i=1}^{n} (N_i \wedge 1) X_{i,1} \right\|_T.
$$

Then by (3.4), $M_n \leq E_N E_X^* \| \sum_{i=1}^{n} (N_i \wedge 1) X_{i,1} \|_T$. For each i and given $N_i \equiv N(i)$,

$$
\sum_{j=1}^{N(i)} X_{i,j} - (N_i \wedge 1) X_{i,1} = \sum_{j=2}^{N(i)} X_{i,j}
$$

$(= 0$ for $N_i \leq 1)$, which is a centered process independent of $(N_i \wedge 1) X_{i,1}$. For any fixed values of $N_1, \ldots, N_n$, apply another form of Jensen's inequality, Lemma 9.7(b), with $p = 1$, $\Omega' := \Omega_1$, $\Omega'' := \prod_{j \geq 2} \Omega_j$, $g(t, \omega') := (N_i \wedge 1) X_{i,1}(t)$, and $h(t, \omega'') := \sum_{j=2}^{N(i)} X_{ij}(t)$. We get $M_n \leq E_N E_X^*$ $\| \sum_{i=1}^{n} \sum_{j=1}^{N(i)} X_{i,j} \|_T$. Then by (3.4), $E_N E_X^*$ can be replaced by E^*, and the Lemma is proved. $\square$

For any two finite signed measures μ and ν on the Borel sets of a separable Banach space B recall that the convolution $\mu * \nu$ is defined by $(\mu * \nu)(A) := \int \mu(A - x) d\nu(x)$ for any Borel set A. Here convolution is commutative and associative. For any finite signed measure μ on B and $k = 1, 2, \ldots$, let μ^k be the kth convolution power $\mu * \cdots * \mu$ to k factors. Let $e^\mu := \exp(\mu) := \sum_{k=0}^{\infty} \mu^k / k!$. Let $\mu^0 := \delta_0$. If $\mu \geq 0$ let $\mathrm{Pois}(\mu) := e^{-\mu(B)} e^\mu$. If μ and ν are two finite measures on B, it is straightforward to check that $\mathrm{Pois}(\mu + \nu) = \mathrm{Pois}(\mu) * \mathrm{Pois}(\nu)$. If X is a measurable function from a probability space $(\Omega, \mathcal{S}, P)$ into a measurable space $(S, \mathcal{A})$, recall that the *law* of X is the image measure $\mathcal{L}(X) := P \circ X^{-1}$ on $\mathcal{A}$. For any $c > 0$ and $x \in B$, $\mathrm{Pois}(c\delta_x) = \mathcal{L}(N_c x)$ where N_c is a Poisson random variable with parameter c.

If $\mathcal{L}(X_{i,j}) = \mu_i$ for $j = 1, 2, \ldots$ and $N_i := N(i)$ are Poisson with parameter 1, where all $X_{i,j}$ and N_r are jointly independent, $i, r = 1, \ldots, k$,

$j = 1, 2, \ldots$, then by induction on k,

$$\mathcal{L}\left(\sum_{i=1}^{k}\sum_{j=1}^{N(i)} X_{i,j}\right) = \text{Pois}\left(\sum_{r=1}^{k}\mu_r\right).$$

If X_i are independent random variables with values in a separable Banach space with $EX_i = 0$ for each $i = 1, \ldots, n$, then the depoissonization inequality (9.6) gives

$$E\left\|\sum_{j=1}^{n} X_j\right\| \leq \frac{e}{e-1}\int \|x\|\, d\text{Pois}\left(\sum_{j=1}^{n}\mathcal{L}(X_j)\right). \tag{9.7}$$

Here is another depoissonization inequality.

Lemma 9.12 *Let $(B, \|\cdot\|)$ be a normed space. For each $n = 1, 2, \ldots$, let $v_1, \ldots, v_n$ be n distinct points of B and $v := (v_1 + \cdots v_n)/n$. Let $V_1, \ldots, V_n$ be i.i.d. B-valued random variables with $\Pr(V_i = v_j) = 1/n$ for $i, j = 1, \ldots, n$. Let $N_1, \ldots, N_n$ be Poisson variables with parameter 1. Let all V_i and N_j be jointly independent. Then*

$$E\left\|\sum_{j=1}^{n}(V_j - v)\right\| \leq \frac{e}{e-1}E\left\|\sum_{j=1}^{n}(N_j - 1)(v_j - v)\right\|. \tag{9.8}$$

Proof. The variables V_j and N_i are discrete, so the norms in (9.8) are both measurable random variables. Next, with $v(j) := v_j$,

$$\sum_{j=1}^{n}\mathcal{L}(V_j - v) = n\mathcal{L}(V_1 - v) = \sum_{j=1}^{n}\delta_{v(j)-v}.$$

It follows from the above properties of Poissonization that

$$\begin{aligned}\text{Pois}\left(\sum_{j=1}^{n}\mathcal{L}(V_j - v)\right) &= \text{Pois}(\delta_{v(1)-v}) * \cdots * \text{Pois}(\delta_{v(n)-v}) \\ &= \mathcal{L}\left(\sum_{j=1}^{n} N_j(v_j - v)\right) = \mathcal{L}\left(\sum_{j=1}^{n}(N_j - 1)(v_j - v)\right).\end{aligned} \tag{9.9}$$

Then by (9.7) for $X_j = V_j - v$, the Lemma is proved. $\qquad\square$

The next proof will use the fact that for any real-valued function f on a measure space, $1_{\{f^* > t\}} = (1_{\{f > t\}})^*$. This is true since $1_{\{f > t\}} \leq 1_{\{f^* > t\}}$, where the latter is measurable, and Lemma 3.8(a) applies. The next fact is about triangular arrays with i.i.d. summands.

Theorem 9.13 *Let (T, d) be a totally bounded pseudo-metric space. For each $n = 1, 2, \ldots$, suppose given a product of n copies of a probability space, $\Omega_n^n := (\Omega_n, \mathcal{A}_n, P^{(n)})^n$. For each n, let Y_n be be a real-valued stochastic process indexed by T on Ω_n. For $\omega := \{\omega_j\}_{j=1}^{n} \in \Omega^n$ let $X_{n,j}(\omega) := Y_n(\omega_j)$.*

Thus $X_{n,j}$ are i.i.d. copies of Y_n. Suppose that each Y_n has bounded sample paths a.s. Assume that

(i) For all t in a dense subset $D \subset T$ for d and for all $\beta > 0$,

$$\lim_{n \to \infty} n\Pr^*\{|X_{n,1}(t)| > \beta n^{1/2}\} = 0, \tag{9.10}$$

(ii) For any $\delta > 0$, $\sup_{t \in T} \Pr\{|Y_n(t)| > \delta n^{1/2}\} \to 0$ as $n \to \infty$, and
(iii) for all $\varepsilon > 0$, as $\delta \downarrow 0$,

$$\limsup_{n \to \infty} \Pr^*\left\{ n^{-1/2} \sup_{d(s,t) \le \delta} \left| \sum_{i=1}^{n} X_{n,i}(t) - EX_{n,i}(t) \right.\right.$$

$$\left.\left. -X_{n,i}(s) + EX_{n,i}(s) \right| > \varepsilon \right\} \to 0.$$

Then for any $\gamma > 0$,

$$\lim_{n \to \infty} n\Pr^*\left\{ \|X_{n,1}\|_T > \gamma n^{1/2} \right\} = 0. \tag{9.11}$$

If the hypotheses hold only along a subsequence n_k, then the conclusion also holds along the subsequence.

Proof. Taking more Cartesian products, let $\{V_{n,j}\}$ be an independent copy of $\{X_{n,j}\}$, $n = 1, 2, \ldots$, $j = 1, \ldots, n$. For any $m = 1, 2, \ldots$ and $u > 0$ we have easily

$$\Pr^*\left\{ \|X_{n,1} - V_{n,1}\| > u \right\} \le 2\Pr^*\left\{ \|X_{n,1}\| > u/2 \right\}. \tag{9.12}$$

Then, condition (iii) implies that as $\delta \downarrow 0$,

$$\limsup_{n \to \infty} \Pr^*\left\{ n^{-1/2} \sup_{d(s,t) \le \delta} \left| \sum_{i=1}^{n} X_{n,i}(t) - V_{n,i}(t) \right.\right.$$

$$\left.\left. -X_{n,i}(s) + V_{n,i}(s) \right| > \varepsilon \right\} \to 0. \tag{9.13}$$

Let $U_{n,j} := X_{n,j} - V_{n,j}$. For any $\tau > 0$, let $\{s_1, \ldots, s_{N(\tau)}\}$ be a maximal subset of T with $d(s_i, s_j) > \tau/3$ for $i \neq j$. Let $B_i := \{t \in T : d(t, s_i) \le \tau/3\}$ and $C_i := C_{i,\tau} := B_i \setminus \cup_{r<i} B_r$. Then C_i for $i = 1, \ldots, N(\tau)$ are disjoint sets of diameter $< \tau$ whose union covers T and such that $C_{i,\tau} \cap D \neq \emptyset$ for each i. Choose $t_i \in C_{i,\tau}$ for each i and let $U_{n,j,\tau}(t) := U_{n,j}(t_i)$ for each $t \in C_{i,\tau}$. Then

$$\Pr^*\left\{ \|U_{n,1}\|_T > 2\gamma\sqrt{n} \right\} \le \Pr^*\left\{ \|U_{n,1,\tau}\|_T > \gamma\sqrt{n} \right\}$$

$$+ \Pr^*\left\{ \|U_{n,1} - U_{n,1,\tau}\|_T > \gamma\sqrt{n} \right\}. \tag{9.14}$$

Then, as $n \to \infty$,

$$n\Pr^*\left\{ \|U_{n,1,\tau}\|_T > \gamma\sqrt{n} \right\}$$

$$\le \sum_{i=1}^{N(\tau)} n\Pr\left\{ |U_{n,1}(t_i)| > \gamma\sqrt{n} \right\} \to 0 \tag{9.15}$$

by hypothesis (i) and (9.12). For an arbitrary set A in a probability space, let A^* be a measurable cover, so that A^* is a measurable set and $(1_A)^* = 1_{A^*}$. Then it is easily checked that for any sets $A_1, \ldots, A_k$, $(\bigcup_{i=1}^k A_i)^* = \bigcup_{i=1}^k A_i^*$ up to sets of probability zero. Take $A_i := \{\|U_{n,i} - U_{n,i,\tau}\|_T > \gamma n^{1/2}\}$. Then by Theorem 3.3 and Lemma 3.6,

$$1 - \exp(-n\Pr^*(A_1)) \leq 1 - (1 - \Pr^*(A_1))^n = \Pr(\cup_{j=1}^n A_j^*)$$

$$= \Pr^* \left\{ \max_{1 \leq i \leq n} \|U_{n,i} - U_{n,i,\tau}\|_T > \gamma \sqrt{n} \right\}.$$

Let $S_k := \sum_{i=1}^k U_{n,i} - U_{n,i,\tau}$. Then the last expression in the preceding display is

$$\leq \Pr^* \left(\max_{1 \leq k \leq n} \|S_k\|_T > \gamma \sqrt{n}/2 \right),$$

which, by Lévy's inequality with stars (3.12(b)), is

$$\leq 2\Pr^* \left\{ \left\| \sum_{i=1}^n (U_{n,i} - U_{n,i,\tau}) \right\|_T > \gamma \sqrt{n}/2 \right\}$$

$$\leq 2\Pr^* \left\{ n^{-1/2} \sup_{d(s,t) \leq \tau} \left| \sum_{i=1}^n (X_{n,i}(t) - V_{n,i}(t) - X_{n,i}(s) + V_{n,i}(s)) \right| > \gamma/2 \right\}.$$

By (9.13) the latter becomes less than any given $\zeta > 0$ for $0 < \tau \leq \tau_0(\zeta)$ small enough and $n \geq n_0(\zeta)$ large enough. Take $\zeta \leq 1/2$. Then $1 - \zeta \geq e^{-2\zeta}$, and so $n \Pr^*(A_1) \leq 2\zeta$. Thus as $\tau \downarrow 0$ and $n \to \infty$,

$$n\Pr^*(\|U_{n,1} - U_{n,1,\tau}\|_T > \gamma \sqrt{n}) \to 0.$$

Thus, applying (9.14) and (9.15) gives

$$\lim_{\tau \to 0}(\lim \sup)_{n \to \infty} n\Pr^* \left\{ \|U_{n,1} - U_{n,1,\tau}\|_T > \gamma \sqrt{n} \right\}$$
$$= \lim_{n \to \infty} n\Pr^* \left\{ \|U_{n,1}\|_T > \gamma \sqrt{n} \right\} = 0. \tag{9.16}$$

Next, by Lemma 3.6 and its proof, $\|X_{n,i}\|_T^*$ and $\|V_{n,i}\|_T^*$ are independent. Applying also Lemma 3.8, and recalling that $X_{n,1}$ and $V_{n,1}$ are copies of Y_n, we get

$$n\Pr^*\{\|U_{n,1}\|_t > \gamma \sqrt{n}\} \geq n\Pr\left\{\|Y_n\|_T^* > 2\gamma \sqrt{n}\right\} \Pr\left\{\|Y_n\|_T^* \leq \gamma \sqrt{n}\right\}.$$

By Lemma 9.5(a),

$$n\Pr\{\|Y_n\|_T^* > \gamma n^{1/2}\} \leq \frac{n\Pr^*\{\|U_{n,1}\|_T > \gamma n^{1/2}/2\}}{1 - \sup_{t \in T} \Pr\{|Y_n(t)| > \gamma n^{1/2}/2\}}.$$

Here the numerator $\to 0$ as $n \to \infty$ by (9.16) and $\sup_{t \in T} \Pr\{|Y_n(t) > \gamma n^{1/2}/2\} \to 0$ as $n \to \infty$ by hypothesis (ii). So the conclusion (9.11) holds. The proof for subsequences is the same. $\square$

Next will be a characterization of Donsker classes, in which the asymptotic equicontinuity condition (Theorem 3.34) is put into symmetrized forms. For a class $\mathcal{F}$ of functions let

$$\mathcal{F}'_\delta := \mathcal{F}_\delta := \{f - g : f, g \in \mathcal{F}, \ \rho_P(f, g) < \delta\}.$$

Let $\| \cdot \|_{\delta,\mathcal{F}} := \| \cdot \|_{\mathcal{G}}$ where $\mathcal{G} := \mathcal{F}'_\delta$.

Theorem 9.14 *Let* $(X, \mathcal{A}, P)$ *be a probability space and* $\mathcal{F}$ *a class of functions with* $\mathcal{F} \subset \mathcal{L}^2(X, \mathcal{A}, P)$. *Assume (9.2), so that* ε_i *are i.i.d. Rademacher functions, independent of* $X_1, X_2, \ldots$. *Suppose that for each* $x \in X$,

$$F_c(x) := \sup_{f \in \mathcal{F}} |f(x) - Pf| < \infty. \tag{9.17}$$

Then the following are equivalent:

(a) $\mathcal{F}$ *is a Donsker class for* P;

(b) $\mathcal{F}$ *is totally bounded for* ρ_P *and for any* $\varepsilon > 0$, *as* $\delta \to 0$,

$$\limsup_{n \to \infty} \mathrm{Pr}^* \left\{ n^{-1/2} \left\| \sum_{i=1}^n \varepsilon_i (f(X_i) - Pf) \right\|_{\delta,\mathcal{F}} > \varepsilon \right\} \to 0;$$

(c) $(\mathcal{F}, \rho_P)$ *is totally bounded and as* $\delta \to 0$,

$$\limsup_{n \to \infty} n^{-1/2} E^* \left\| \sum_{i=1}^n \varepsilon_i (f(X_i) - Pf) \right\|_{\delta,\mathcal{F}} \to 0;$$

(d) $(\mathcal{F}, \rho_P)$ *is totally bounded and for* $\nu_n := n^{1/2}(P_n - P)$, *as* $\delta \to 0$, $\limsup_{n \to \infty} E^* \|\nu_n\|_{\delta,\mathcal{F}} \to 0$.

To prove the theorem, we first show that (a) implies (b). Applying Lemma 9.6(a) to the class $\mathcal{F}'_\delta$, the outer probability in (b) is bounded above by

$$2 \max_{k \leq n} \mathrm{Pr}^* \left\{ n^{-1/2} \left\| \sum_{i=1}^k f(X_i) - Pf \right\|_{\delta,\mathcal{F}} > \varepsilon/2 \right\}.$$

For any fixed N, we can write for $n > N$ that

$$\max_{k \leq n} \equiv \max(\max_{k \leq N}, \max_{N < k \leq n}).$$

By the asymptotic equicontinuity condition (Theorem 3.34(b)), there are an $N < \infty$ and a $\delta > 0$ such that for $N \leq k \leq n$

$$\mathrm{Pr}^* \left\{ n^{-1/2} \left\| \sum_{i=1}^k (f(X_i) - Pf) \right\|_{\delta,\mathcal{F}} > \varepsilon/4 \right\} < \varepsilon/4, \tag{9.18}$$

since $n^{-1/2} \le k^{-1/2}$. Then for $1 \le j < N$, note that

$$\sum_{i=1}^{j} = \sum_{i=1}^{N+j} - \sum_{i=j+1}^{N+j}$$

and inequalities for $\sum_{i=1}^{N}$ also hold for $\sum_{i=j+1}^{N+j}$. Thus (9.18) holds also, with $\varepsilon/4$ replaced by $\varepsilon/2$, for $1 \le k < N$, and (b) holds.

Now to prove (b) implies (c), Theorem 9.13 will be applied with $D := T :=$ $\mathcal{F}$ and $X_{n,j}(f) := \varepsilon_j(f(X_j) - Pf)$. Then $EX_{n,j}(f) = 0$ for all $f \in \mathcal{F}$, and (b) gives hypothesis (iii). For hypothesis (i) (9.10), again letting $1\{\dots\} := 1_{\{\dots\}}$, we have

$$n\Pr\{|f(X_j) - Pf| > \beta n^{1/2}\}$$
$$\le \beta^{-2} E(|f(X_j) - Pf|^2 1\{|f(X_1) - Pf| > \beta n^{1/2}\}) \to 0$$

as $n \to \infty$ by dominated convergence. Hypothesis (ii) holds since $\mathcal{F}$ is totally bounded for ρ_P. So all three hypotheses of Theorem 9.13 and its conclusion (9.11) hold for the given $X_{n,j}$. In proving (c) the following will be helpful.

Lemma 9.15 *Let ξ_i, $i = 1, 2, \dots$, be i.i.d. nonnegative random variables such that*

$$M := \sup_{t>0} t^2 \Pr\{\xi_i > t\} < \infty. \tag{9.19}$$

Then, for all r such that $0 < r < 2$,

$$\sup_{n} n^{-r/2} E \max_{1 \le i \le n} \xi_i^r < \infty. \tag{9.20}$$

Proof. We have, as in the proof of Theorem 9.9, for each n,

$$n^{-r/2} E \max_{1 \le i \le n} \xi_i^r = rn^{-r/2} \int_0^\infty t^{r-1} \Pr\{\max_{1 \le i \le n} \xi_i > t\}dt$$
$$\le 1 + rn^{1-r/2} \int_{\sqrt{n}}^\infty t^{r-1} \Pr\{\xi_1 > t\}dt$$
$$\le 1 + Mrn^{1-r/2} \int_{\sqrt{n}}^\infty t^{r-3}dt = 1 + M\frac{r}{2-r},$$

proving (9.20). $\qquad\square$

Now continuing the proof that (b) implies (c), for $\xi_i = \|f(X_i) - Pf\|_{\mathcal{F}}^*$, condition (9.19) follows from equation (9.11). This implies, e.g. for $r = 3/2$, that the sequence $\{\max_{i \le n} \xi_i/\sqrt{n}\}_{n=1}^\infty$ is uniformly integrable. But, again by (9.11), we also have, letting $\gamma \to 0$, that $\max_{i \le n} \xi_i/\sqrt{n} \to 0$ in probability. It follows (RAP, Theorem 10.3.6) that

$$E \max_{i \le n} \xi_i/\sqrt{n} \to 0. \tag{9.21}$$

Next, Theorem 9.9 will be applied, where the original sample space S as in (9.2) is replaced by $S \times \{-1, 1\}$, the random variables are (X_j, ε_j), and in place of $\mathcal{F}$, we take the class $\mathcal{G}$ of functions $g := g_h$ where for each $h \in \mathcal{F}'_\delta$, $g(x, s) := s(h(x) - Ph)$ for $s = \pm 1$, so that $g(X_j, \varepsilon_j) = \varepsilon_j(h(X_j) - Ph)$. Also let $p := 1$. Then by (9.21), the hypothesis of Theorem 9.9 holds, and in the conclusion, the first term on the right is $o(n^{1/2})$ as $n \to \infty$ and $\delta \to 0$. In the definition of u, by the Lévy inequality (3.12), we have $u \le 2t_1$ if

$$\Pr\left(\left\| \sum_{i=1}^{n} g(X_i, \varepsilon_i) \right\|_{\mathcal{G}}^* > t_1 \right) \le 4^{-p-1},$$

and $t_1 = o(n^{1/2})$ also by (b). So (c) follows.

(c) implies (d) by desymmetrization, Lemma 9.10.

(d) implies (a) by Markov's inequality and the usual asymptotic equicontinuity condition, Theorem 3.34.

The proof of Theorem 9.14 is complete. $\qquad\qquad\square$

Next, Theorem 9.14 will be extended to multipliers other than Rademacher variables. Suppose we have a probability space $(S, \mathcal{S}, P)$, a countable product Ω' of copies of $(S, \mathcal{S}, P)$ with coordinates X_j, and a probability space Ω'', another product space with coordinates ε_j which are i.i.d. Rademacher variables. Let Q be a Borel probability measure on $[0, \infty)$ with $Q(\{0\}) < 1$ and let Ω''' be a countable product of copies of $[0, \infty)$ with coordinates v_j, each with Borel σ-algebra and probability Q.

For any real random variable Y let

$$\Lambda_{2,1}(Y) := \int_0^\infty [\Pr(|Y| > t)]^{1/2} dt.$$

The following is known. One reference, Stein and Weiss (1971, Section V.3 on Lorentz "$L(p, q)$" spaces), is related but considers function spaces on $[0, +\infty)$ with Lebesgue measure. I will give an elementary direct proof.

Lemma 9.16 *For a real random variable Y,*

(a) $\Lambda_{2,1}(Y) < \infty$ implies $E(Y^2) < \infty$.

(b) For any $\delta > 0$, $E|Y|^{2+\delta} < \infty$ implies $\Lambda_{2,1}(Y) < \infty$.

Proof. Let F be the distribution function of $|Y|$. (a): given $\int_0^\infty \sqrt{1 - F(t)}dt < +\infty$, we want to show $\int_0^\infty t^2 dF(t) < +\infty$, or equivalently $\int_0^\infty t^2 d(1 - F)(t) > -\infty$. Integrating by parts, we get $t^2(1 - F)(t)|_0^{+\infty} - 2\int_0^\infty t(1 - F(t))dt > -\infty$. The boundary term at 0 is clearly 0. Let $g(t) := \sqrt{1 - F(t)}$ for $t \ge 0$. We need to show that for a nonincreasing function $g \ge 0$ with $\int_0^\infty g(t)dt < +\infty$ that $tg(t) \to 0$ as $t \to +\infty$, as is surely well known, but suppose not. Choose $t_k \to +\infty$ with $t_k g(t_k) \ge \varepsilon > 0$ for all k. We can and do assume $t_k \geq 2t_{k-1}$ for all k. Then for each $k \geq 2$,

$$\int_{t_{k-1}}^{t_k} g(t)dt \geq \left(1 - \frac{t_{k-1}}{t_k}\right)\varepsilon \geq \frac{\varepsilon}{2}.$$

Summing over k gives a contradiction. So the upper boundary term is also 0 and we need to show $\int_0^\infty tg^2(t)dt < +\infty$. This integral is bounded above by

$$\sum_{n=0}^\infty (n+1)g(n)^2 = \sum_{n=0}^\infty \sum_{j=0}^n 1 \cdot g(n)^2 = \sum_{j=0}^\infty \sum_{n=j}^\infty g(n)^2$$

$$\leq \sum_{j=1}^\infty g(j) \sum_{n=j}^\infty g(n) \leq \left[\int_0^\infty g(t)dt\right]^2 < +\infty,$$

which proves (a).

(b): Suppose $E|Y|^{2+\delta} < \infty$. Then $+\infty > g(y) := \int_y^\infty t^{2+\delta}dF(t) \to 0$ as $y \to \infty$ and $g(y) \geq y^{2+\delta}(1 - F(y))$ for $y > 0$. Thus $g(y) = -\int_y^\infty t^{2+\delta}d(1 - F(t))$ can be integrated by parts and equals $\int_y^\infty (1 - F(t))(2 + \delta)t^{1+\delta}dt$. By the Cauchy–Bunyakovsky–Schwarz inequality this implies

$$\int_1^\infty \sqrt{1 - F(t)}dt = \int_1^\infty \sqrt{1 - F(t)}\frac{t^{(1+\delta)/2}}{t^{(1+\delta)/2}}dt$$

$$\leq \left(\int_1^\infty (1 - F(t))t^{1+\delta}dt\right)^{1/2} \left(\int_1^\infty t^{-1-\delta}dt\right)^{1/2} < +\infty,$$

proving (b) and the Lemma.   □

Remark. Part (a) shows that part (b) does not hold for any $\delta < 0$ and will be used later in this chapter.

Lemma 9.17 *Let $\Omega = \Omega' \times \Omega'' \times \Omega'''$ be the product of the three probability spaces defined in the last paragraph, with coordinates X_j, ε_j, and v_j. Let $\mathcal{F} \subset \mathcal{L}^2(S, \mathcal{S}, P)$ and $F_{\mathcal{F}}(x) := \sup_{f \in \mathcal{F}} |f(x)|$. Assume that $E^* F_{\mathcal{F}} < +\infty$.*

(a) Let $\xi_j := \varepsilon_j v_j$ for each j, so that $v_j \equiv |\xi_j|$ and ξ_j are i.i.d. real symmetric random variables. Then for integers $0 \leq m < n < \infty$ we have

$$n^{-1/2}(Ev_1)E^*\|\textstyle\sum_{i=1}^n \varepsilon_i f(X_i)\|_{\mathcal{F}} \leq n^{-1/2}E^*\|\textstyle\sum_{i=1}^n \xi_i f(X_i)\|_{\mathcal{F}}$$

$$\leq mn^{-1/2}(E^*\|f(X_1)\|_{\mathcal{F}})E(\max_{i \leq n} v_i) \tag{9.22}$$

$$+ \Lambda_{2,1}(v_1)\max_{m<k\leq n} k^{-1/2}E^*\|\textstyle\sum_{i=1}^k \varepsilon_i f(X_i)\|_{\mathcal{F}}.$$

(b) If Ω''' is replaced by a countable product of copies of $\mathbb{R}$ and Q by a Borel probability measure with $\int_{-\infty}^\infty x\,dQ(x) = 0$, so that the coordinates v_i

are centered ($Ev_i = 0$), *then we have for* $0 \le m < n$

$$n^{-1/2}(E|v_1 - v_2|/2)E^* \| \textstyle\sum_{i=1}^{n} \varepsilon_i f(X_i) \|_{\mathcal{F}} \le n^{-1/2} E^* \| \textstyle\sum_{i=1}^{n} \varepsilon_i v_i f(X_i) \|_{\mathcal{F}}$$
$$\le 2mn^{-1/2}(E^* \| f(X_1) \|_{\mathcal{F}}) E(\max_{i \le n} |v_i|)$$
$$+ 3\Lambda_{2,1}(v_1) \max_{m < k \le n} k^{-1/2} E^* \| \textstyle\sum_{i=1}^{k} \varepsilon_i f(X_i) \|_{\mathcal{F}}. \tag{9.23}$$

Proof. Part (a): writing $\| \cdot \| := \| \cdot \|_{\mathcal{F}}$, we have by Theorem 3.9 (the one-sided Tonelli–Fubini theorem)

$$E^* \left\| \sum_{i=1}^{n} \xi_i f(X_i) \right\| \ge E_\varepsilon E_X^* E_v^* \left\| \sum_{i=1}^{n} \varepsilon_i v_i f(X_i) \right\| \ge E^* \left\| \sum_{i=1}^{n} \varepsilon_i E v_i f(X_i) \right\|$$

by Jensen's inequality, Lemma 9.7(a), applied to E_v. The first inequality in (9.22) follows. For the second we have that

$$E(n) := E^* \left\| \sum_{i=1}^{n} \xi_i f(X_i) \right\| = E^* \left\| \sum_{i=1}^{n} \varepsilon_i v_i f(X_i) \right\|$$

$$= E^* \left\| \sum_{i=1}^{n} \left(\int_0^\infty 1_{\{t \le v_i\}} dt \right) \varepsilon_i f(X_i) \right\|$$

$$= E^* \left\| \int_0^\infty \left(\sum_{i=1}^{n} 1_{\{t \le v_i\}} \varepsilon_i f(X_i) \right) dt \right\|.$$

Let $F_t(X, \varepsilon, v) := n^{-1/2} \| \sum_{i=1}^{n} 1_{\{t \le v_i\}} \varepsilon_i f(X_i) \|$ and $v_{\max} := \max_{j \le n} v_j$. For any t and $\omega = (\omega', \omega'', \omega''')$ we have

$$F_t(X, \varepsilon, v) \le G_t(X, v) := n^{-1/2} 1_{\{t \le 1 + v_{\max}\}} \sum_{i=1}^{n} F_{\mathcal{F}}(X_i),$$

which is finite almost surely. For almost all ω, $F_t(X, \varepsilon, v)$ is a finite-valued step function of t, equaling 0 for $t > v_{\max}$. Thus $\int_0^\infty F_t(X, \varepsilon, v) dt$ is defined and finite. Also, $\int_0^{1/k} F_t(X, \varepsilon, v) dt \to 0$ as $k \to \infty$ almost surely, by dominated convergence. For each $j, k = 1, 2, \ldots$ and $t \ge 1/k$ let $H_{k,t}(\omega) := H_{k,t}(X, \varepsilon, v) := F_{j/k}(X, \varepsilon, v)$ for $j/k \le t < (j+1)/k$. Then $H_{k,t}(X, \varepsilon, v) \to F_t(X, \varepsilon, v)$ for all $t > 0$ as $k \to \infty$ since $t \mapsto F_t$ is left-continuous. We have $H_{k,t}(\omega) \le G_t(x, V)$ since if $H_{k,t}(\omega) \ne 0$, then $H_{k,t}(\omega) = F_{j/k}(X, \varepsilon, v)$, where $j/k \le v_{\max}$ and $j/k \le t < (j+1)/k$ imply $t \le 1 + v_{\max}$. We have

$$\int_0^\infty G_t(X, v) dt \le n^{-1/2}(1 + v_{\max}) \sum_{i=1}^{n} F_{\mathcal{F}}(X_i) < +\infty$$

for almost all ω. Therefore, by dominated convergence,

$$\int_0^{+\infty} F_t(X, \varepsilon, v)dt = \lim_{k \to \infty} \int_{1/k}^{\infty} H_{k,t}(X, \varepsilon, v)dt = \lim_{k \to \infty} \frac{1}{k} \sum_{j=1}^{\infty} F_{j/k}(X, \varepsilon, v)$$

almost surely. Then

$$n^{-1/2} E^* \left\| \int_0^{\infty} \left(\sum_{i=1}^{n} 1_{t \le v_i} \varepsilon_i f(X_i) \right) dt \right\|$$

$$\le E^* \int_0^{\infty} F_t(X, \varepsilon, v)dt = E^* \lim_{k \to \infty} \sum_{j=1}^{\infty} F_{j/k}(X, \varepsilon, v)/k$$

$$\le E^* \left\{ \liminf_{k \to \infty} \sum_{j=1}^{\infty} F^*_{j/k}(X, \varepsilon, v)/k \right\}$$

$$\le \liminf_{k \to \infty} \sum_{j=1}^{\infty} E^* F_{j/k}(X, \varepsilon, v)/k$$

by Fatou's lemma with stars, Theorem 3.11.

For each set $G \subset \{1, 2, \dots, n\}$ and $t \ge 0$ let

$$H(G, t) := \{v := \{v_i\}_{i=1}^{n} \in [0, +\infty)^n : t \le v_i \text{ if and only if } i \in G\}.$$

It is easily seen that for any measurable set A and functions $f, g \ge 0$ with $f = g$ on A, $(f 1_A)^* = 1_A f^* = 1_A g^*$ a.s. Given t, the sets $H(G, t)$ are disjoint and Borel measurable with union $[0, \infty)^n$. For $v \in H(G, t)$, we have $F_t(X, \varepsilon, v) = n^{-1/2} \| \sum_{i \in G} f(X_i) \|$. So by Lemma 3.6(c),

$$F^*_t(X, \varepsilon, v) = n^{-1/2} \sum_G 1_{H(G,t)}(v) \left\| \sum_{i \in G} \varepsilon_i f(X_i) \right\|^*.$$

Each term of the finite sum is a function of v times a function of (X, ε). As we have a Cartesian product probability space, it follows by Theorem 3.3 and then the ordinary Tonelli–Fubini theorem that

$$E^* F_t = E(F^*_t) = n^{-1/2} \sum_G \Pr(v \in H(G, t)) E^* \left\| \sum_{i \in G} \varepsilon_i f(X_i) \right\|.$$

The $\sum_G$ can be written as $\sum_{k=0}^{n} \sum_{G: |G|=k}$. For $|G| = k = 0$, the $\sum_{i \in G}$ and the envelope of its norm are 0. For each $k = 1, \dots, n$,

$$\sum_{G: |G|=k} \Pr\{v \in H(G, t)\} = \Pr \left\{ \sum_{i=1}^{n} 1_{v_i \ge t} = k \right\}.$$

For any G with $|G| = k$,

$$E^* \left\| \sum_{i \in G} \varepsilon_i f(X_i) \right\| = E^* \left\| \sum_{i=1}^{k} \varepsilon_i f(X_i) \right\|$$

since probabilities for $\{(\varepsilon_i, X_i)\}_{i=1}^{n}$ are preserved by permutations of the indices i. It follows that

$$E^* F_t(X, \varepsilon, v) = n^{-1/2} \sum_{k=1}^{n} \Pr\left(\sum_{i=1}^{n} 1_{\{v_i \geq t\}} = k \right) \cdot E^* \left\| \sum_{i=1}^{k} \varepsilon_i f(X_i) \right\|.$$

Thus $E(n)$ is bounded above by

$$\int_0^{\infty} \left(\sum_{k=1}^{n} \Pr\left\{ \sum_{i=1}^{n} 1_{\{v_i \geq t\}} = k \right\} E^* \left\| \sum_{i=1}^{k} \varepsilon_i f(X_i) \right\| \right) dt$$

$$\leq \left(\int_0^{\infty} \Pr\left\{ \sum_{i=1}^{n} 1_{\{v_i \geq t\}} > 0 \right\} dt \right) \max_{k \leq m} E^* \left\| \sum_{i=1}^{k} \varepsilon_i f(X_i) \right\|$$

$$+ \left(\int_0^{\infty} \sum_{k=m+1}^{n} \sqrt{k} \Pr\left\{ \sum_{i=1}^{n} 1_{\{v_i \geq t\}} = k \right\} dt \right)$$

$$\times \max_{m < k \leq n} E^* \left\| k^{-1/2} \sum_{i=1}^{k} \varepsilon_i f(X_i) \right\|.$$

Now,

$$\sum_{k > m} \sqrt{k} \Pr\left\{ \sum_{i=1}^{n} 1_{\{v_i \geq t\}} = k \right\} = E\left(\left(\sum_{i=1}^{n} 1_{\{v_i \geq t\}} \right)^{1/2} \right)$$

$$\leq \left(E \sum_{i=1}^{n} 1_{\{v_i \geq t\}} \right)^{1/2} = n^{1/2} \Pr(v_1 \geq t)^{1/2}.$$

Thus by subadditivity of E^* of norms (Lemma 3.5)

$$E(n) \leq m \left(\int_0^{\infty} \Pr\{v_{\max} \geq t\} dt \right) E^* \| f(X_1) \|$$

$$+ n^{1/2} \Lambda_{2,1}(v_1) \max_{m < k \leq n} E^* \left\| k^{-1/2} \sum_{i=1}^{k} \varepsilon_i f(X_i) \right\|$$

$$= m E^* \| f(X_1) \| E(v_{\max})$$

$$+ n^{1/2} \Lambda_{2,1}(v_1) \max_{m < k \leq n} E^* \left\| k^{-1/2} \sum_{i=1}^{k} \varepsilon_i f(X_i) \right\|,$$

finishing the proof for part (a).

For part (b), let $\zeta_j := v_j - v_{n+j}$, $j = 1, \ldots, n$. We have

$$E^* \left\| \sum_{i=1}^{n} v_i f(X_i) \right\| \leq E^* \left\| \sum_{i=1}^{n} \zeta_i f(X_i) \right\|$$

by Jensen's inequality (9.3) applied to functions $g(X, v) := vf(X)$. Part (a) applies for ζ_i in place of ξ_i. Clearly, $E(\max_{i \leq n} |\zeta_i|) \leq 2E(\max_{i \leq n} |v_i|)$. We also have

$$\Lambda_{2,1}(\zeta_1) = \int_0^\infty \Pr(|\zeta_1| > t)^{1/2} dt$$

$$\leq \int_0^\infty (2 \Pr(|v_1| > t/2))^{1/2} dt \leq 3 \Lambda_{2,1}(v_1).$$

For the lower bound, it is easily seen that

$$E^* \left\| \sum_{i=1}^{n} \zeta_i f(X_i) \right\| \leq 2E^* \left\| \sum_{i=1}^{n} v_i f(X_i) \right\|,$$

and the conclusion follows. $\quad\square$

Next, here is a characterization of Donsker classes in terms of multipliers ξ_i.

Theorem 9.18 *Let $\mathcal{F}$ be a class such that hypotheses (9.2) hold and $\{f - Pf : f \in \mathcal{F}\}$ has a finite envelope function (9.17). Let ξ_i be i.i.d. centered real random variables, independent of $X_1, X_2, \ldots$, specifically, defined on a different factor of a product probability space, such that $E|\xi_1| > 0$ and $\Lambda_{2,1}(\xi_1) < \infty$. Then $\mathcal{F}$ is Donsker for P if and only if both $(\mathcal{F}, \rho_P)$ is totally bounded and*

$$\lim_{\delta \downarrow 0} \limsup_{n \to \infty} E^* \left\{ n^{-1/2} \left\| \sum_{i=1}^{n} \xi_i (f(X_i) - Pf) \right\|_{\delta, \mathcal{F}} \right\} = 0. \quad (9.24)$$

Proof. We can assume that each $f \in \mathcal{F}$ is centered, replacing f by $f - Pf$. Recall that (a) implies (c) in Theorem 9.14. Also, we have $\sup_{t>0} t(\Pr(|\xi_1| > t))^{1/2} \leq \Lambda_{2,1}(\xi_1) < \infty$, so (9.19) holds for $|\xi_1|$. Thus by (9.22), it follows that (9.24) holds. Conversely, assume that $\mathcal{F}$ is ρ_P-totally bounded and (9.24) holds. Then by the first displayed inequality in Lemma 9.17, applied to $\mathcal{F}'_\delta$ for each $\delta > 0$, and since $E|\xi_1| > 0$, it follows that (c) in Theorem 9.14 holds, so by that theorem, $\mathcal{F}$ is Donsker for P. $\quad\square$

Now, the proof of the bootstrap central limit theorem in probability, Theorem 9.3, will be finished. Let $\mathcal{F}$ be P-Donsker. We can, again, assume that $Pf = 0$ for all $f \in \mathcal{F}$ since each $f \in \mathcal{F}$ can be replaced by $f - Pf$ without changing $\nu_n(f)$ or $\nu_n^B(f)$. Recall that the conclusion is equivalent to a statement in terms of the metric $\beta_{\mathcal{F}}$. Since $\mathcal{F}$ is totally bounded for ρ_P by Theorem 3.34, for any

$\tau > 0$ there is an $N(\tau) < \infty$ and a map π_τ from $\mathcal{F}$ into a subset having $N(\tau)$ elements such that $\rho_P(\pi_\tau f, f) < \tau$ for all $f \in \mathcal{F}$. Let $v_{n,\tau}^B(f) := v_n^B(\pi_\tau f)$ and $G_{P,\tau}(f) := G_P(\pi_\tau(f))$. Then for any bounded Lipschitz function H on $\ell^\infty(\mathcal{F})$ for $\|\cdot\|_\mathcal{F}$,

$$|E^B H(v_n^B) - EH(G_P)| \leq |E^B H(v_n^B) - E^B H(v_{n,\tau}^B)|$$

$$+|E^B H(v_{n,\tau}^B) - EH(G_{P,\tau})| + |EH(G_{P,\tau}) - EH(G_P)|$$

$$=: A_{n,\tau}(H) + B_{n,\tau}(H) + C_\tau(H).$$

Let $BL(1)$ denote the set of all functions H on $\ell^\infty(\mathcal{F})$ with bounded Lipschitz norm ≤ 1. Then $\sup_{H \in BL(1)} C_\tau(H) \to 0$ as $\tau \to 0$ since by definition of Donsker class, G_P is a.s. uniformly continuous with respect to ρ_P, so $G_{P,\tau} \to G_P$ uniformly on $\mathcal{F}$ as $\tau \downarrow 0$. For fixed $\tau > 0$, the supremum for $H \in BL(1)$ of $B_{n,\tau}(H)$ is measurable and converges to 0 a.s. by Theorem 9.4, the finite-dimensional bootstrap central limit theorem, in $\mathbb{R}^{N(\tau)}$. Recall the definition of $\|\cdot\|_{\delta,\mathcal{F}}$ from before Theorem 9.14. For the $A_{n,\tau}$ term, we have $\sup_{H \in BL(1)} A_{n,\tau}(H) \leq 2E^B \|v_n^B\|_{\tau,\mathcal{F}}$, so it will be enough to show that for all $\varepsilon > 0$, as $\delta \downarrow 0$,

$$\limsup_{n \to \infty} \text{Pr}^* \left\{ E^B \|v_n^B\|_{\delta,\mathcal{F}} > \varepsilon \right\} \to 0. \tag{9.25}$$

Apply (9.8) with $B := \ell^\infty(\mathcal{F}_\delta')$ and $v_i := \delta_{X_i(\omega)}$ for each ω, and the starred Tonelli–Fubini theorem where one coordinate is discrete (3.4). We get

$$E^* E^B \|v_n^B\|_{\delta,\mathcal{F}} = n^{-1/2} E^* E^B \left\| \sum_{i=1}^n (\delta_{X_{n,i}^B} - P_n) \right\|_{\delta,\mathcal{F}}$$

$$\leq n^{-1/2} \frac{e}{e-1} E^* \left\| \sum_{i=1}^n (N_i - 1)(\delta_{X_i} - P_n) \right\|_{\delta,\mathcal{F}}$$

$$\leq n^{-1/2} \frac{e}{e-1} E^* \left\| \sum_{i=1}^n (N_i - 1)\delta_{X_i} \right\|_{\delta,\mathcal{F}}$$

$$+ n^{-1/2} \frac{e}{e-1} E^* \left(\left| \sum_{i=1}^n (N_i - 1) \right| \|P_n\|_{\delta,\mathcal{F}} \right) =: S + T.$$

Since $\mathcal{F}$ is Donsker for P, the limit as $\delta \downarrow 0$ of $\limsup_{n \to \infty} S$ is 0 by Theorem 9.18. For T, recalling (3.4), we have $E| \sum_{i=1}^n (N_i - 1)| \leq n^{1/2}$, so $T \leq (e/(e-1))E^* \|P_n\|_{\delta,\mathcal{F}}$. Recalling that each $f \in \mathcal{F}$ is taken to be centered, we have $E^* \|P_n\|_{\delta,\mathcal{F}} = E^* \|n^{-1/2} v_n + P\|_{\delta,\mathcal{F}} \to 0$ as $n \to \infty$ and $\delta \to 0$ by Theorem 9.14(d). Thus Theorem 9.3 is proved. $\qquad\square$

9.3 Other Aspects of the Bootstrap

B. Efron (1979) invented the bootstrap, and by now there is a very large literature about it. This section will address some aspects of the application of the Giné–Zinn theorems. These do not cover the entire field by any means. For example, some statistics of interest, such as $\max(X_1, \ldots, X_n)$, are not averages $\frac{1}{n}(f(X_1) + \cdots + f(X_n))$ as f ranges over a class $\mathcal{F}$.

Some bootstrap limit theorems are stated in probability, and others for almost sure convergence. To compare their usefulness, first note that almost sure convergence is not always preferable to convergence in probability:

Example. Let X_n be a sequence of real-valued random variables converging to some X_0 in probability but not almost surely. Then some subsequences X_{n_k} converge to X_0 almost surely. Suppose this occurs whenever $n_k \geq k^2$ for all k. Let $Y_n := X_{2^k}$ for $2^k \leq n < 2^{k+1}$ where $k = 0, 1, \ldots$. Then $Y_n \to X_0$ almost surely, but in a sense, $X_n \to X_0$ faster although it only converges in probability.

Another point is that almost sure convergence is applicable in statistics when inferences will be made from data sets with increasing values of n, in other words, in the part of statistics called *sequential analysis*. But suppose one has a fixed value of the sample size n, as has generally been the case with the bootstrap. Then the probability of an error of a given size, for a given n, which relates to convergence in probability, may be more relevant than the question of what *would* happen for values of $n \to \infty$, as in almost sure convergence.

The rest of this section will be devoted to confidence sets. A basic example of a confidence set is a confidence interval. As an example, suppose $X_1, \ldots, X_n$ are i.i.d. with distribution $N(\mu, \sigma^2)$ where σ^2 is known but μ is not. Then $\overline{X} := (X_1 + \cdots + X_n)/n$ has a distribution $N(\mu, \sigma^2/n)$. Thus

$$\Pr(\overline{X} \leq \mu - 1.96\sigma/n^{1/2}) \doteq .025 \doteq \Pr(\overline{X} \geq \mu + 1.96\sigma/n^{1/2}).$$

So we have 95% confidence that the unknown μ belongs to the interval $[\overline{X} - 1.96\sigma/n^{1/2}, \overline{X} + 1.96\sigma/n^{1/2}]$, which is then called a 95% confidence interval for μ.

Next, suppose $X_1, \ldots, X_n$ are i.i.d. in $\mathbb{R}^k$ with a normal (Gaussian) distribution $N(\mu, \sigma^2 I)$ where I is the identity matrix. Suppose $\alpha > 0$ and $M_\alpha = M_\alpha(k)$ is such that $N(0, I)\{x : |x| \geq M_\alpha\} = \alpha$. Then $n^{1/2}(\overline{X} - \mu)/\sigma$ has distribution $N(0, I)$ so $\Pr(|\overline{X} - \mu| \geq M_\alpha \sigma/n^{1/2}) = \alpha$. Thus, the ball with center $\overline{X}$ and radius $M_\alpha \sigma/n^{1/2}$ is called a $100(1 - \alpha)\%$ confidence set for the unknown μ.

When the distribution of the X_i is not necessarily normal, but has finite variance, then the distribution of $\overline{X}$ will be approximately normal by the central limit theorem for n large, so we get some approximate confidence sets.

Now to extend these ideas to the bootstrap, let $X_1, \ldots, X_n$ be i.i.d. from an otherwise unknown distribution P. Let P_n be the empirical measure formed from $X_1, \ldots, X_n$. Let $\mathcal{F}$ be a universal Donsker class (Section 10.2). Then we know from the Giné–Zinn theorem in the last section that v_n^B and v_n have asymptotically the same distribution on $\mathcal{F}$. By repeated resampling, given a small $\alpha > 0$ such as $\alpha = .05$, one can find $M = M(\alpha)$ such that approximately $\Pr(\|v_n^B\|_{\mathcal{F}} > M \mid P_n) \doteq \alpha$. Then

$$\{Q : \|Q - P_n\|_{\mathcal{F}} \le M/n^{1/2}\}$$

is an approximate $100(1 - \alpha)\%$ confidence set for P.

Problems

1. If the sample space is the unit circle $S^1 := \{(\cos\theta, \sin\theta) : 0 \le \theta < 2\pi\}$, and distribution functions are defined by evaluating laws on arcs $\{(\cos\theta, \sin\theta) : 0 \le \theta \le x\}$, show that the quantity in Corollary 9.2(c), $\sup_x H_{mn}(x) - \inf_y H_{mn}(y)\}$, is invariant under rotations of the circle.

2. (Continuation) For $m = n = 20$, find the approximation given by Corollary 9.2(c) to the probability that there exists an arc $A := \{(\cos\theta, \sin\theta) : a \le \theta \le b\}$ such that $X_j \in A$ and $Y_j \notin A$ for $j = 1, \ldots, 20$, if all 40 variables are i.i.d. from the same continuous distribution on the circle. *Hint:* Not many terms of the series should be needed.

3. In Corollary 9.2,

   (i) Show that the series in (b) and (c) diverge when $u = 0$.

   (ii) Show that for $u = 0$, the probabilities in (b) and (c) equal 1 for any $m \ge 1$ and $n \ge 1$ and for y_t.

   (iii) Show that the three expressions on the right are all less than 1 for $u > 0$ and converge to 1 as $u \downarrow 0$, so that the corresponding distribution functions are continuous everywhere. *Hint:* For (b) and (c) use the expressions in terms of Brownian bridge limits.

4. Let $X_1, X_2, \ldots$, be i.i.d. in $\mathbb{R}$ with a continuous distribution. Given n, arrange $X_1, \ldots, X_n$ in order as $X_{(1)} < X_{(2)} < \cdots < X_{(n)}$. Let X_{nj}^B, $j = 1, \ldots, n$, be i.i.d. (P_n), a bootstrap sample. Thus $\Pr(X_{nj}^B = X_{(i)}) = 1/n$ for $i, j = 1, \ldots, n$. Let $X_{(1)}^B := \min_{1 \le j \le n} X_{nj}^B$.

   (a) Find $p_{ni} := \Pr(X_{(1)}^B = X_{(i)})$ for $i = 1, \ldots, n$.

   (b) What does p_{ni} converge to as $n \to \infty$ for each $i = 1, 2, \ldots$?

In the next two problems, let $(X, \mathcal{A}, P)$ be a probability space and suppose that $\mathcal{F} \subset \mathcal{L}^2(X, \mathcal{A}, P)$ is a Donsker class for P. Let $X_1, X_2, \ldots$, be i.i.d. (P). Convergence in law, conditional on P_n or P_{2n}, in outer probability as $n \to \infty$, is defined just as in the definition of bootstrap Donsker class early in Section 9.2, except for some interchanges of n and $2n$.

5. Given P_n formed from $X_1, \ldots, X_n$, let $2n$ points $X(j, \beta) := X_j^\beta$, $j = 1, \ldots, 2n$, be i.i.d. (P_n), and $P_{2n}^\beta := \frac{1}{2n} \sum_{j=1}^{2n} \delta_{X(j,\beta)}$. (The β superscript is analogous to the B superscript for the bootstrap, but here the bootstrap sample size is $2n$.) Show that $(2n)^{1/2}(P_{2n}^\beta - P_n)$, conditional on P_n, converges in law to G_P, in outer probability.

6. Given P_{2n} formed from $X_1, \ldots, X_{2n}$, let n points $X(j, \gamma) := X_j^\gamma$, $j = 1, \ldots, n$, be i.i.d. (P_{2n}), and $P_n^\gamma := \frac{1}{n} \sum_{j=1}^n \delta_{X(j,\gamma)}$. Show that $n^{1/2}(P_n^\gamma - P_{2n})$, conditional on P_{2n}, converges in law to G_P, in outer probability.

7. Prove the statements before Lemma 9.17, that $\Lambda_{2,1}(Y) < \infty$ implies $E(Y^2) < \infty$, and for any $\delta > 0$, $E|Y|^{2+\delta} < \infty$ implies $\Lambda_{2,1}(Y) < \infty$.

Notes

Notes to Section 9.1. I do not have a reference for Theorem 9.1. A form of the theorem for classes of sets was given in Dudley (1978). Corollary 9.2 is classical and gives asymptotic distributions of so-called Kolmogorov–Smirnov (parts (a), (b)) and Kuiper (part (c)) statistics. See the notes to RAP, Section 12.3.

Notes to Section 9.2. The section is based mainly on Giné (1997), an update (as regards measurability) of the fundamental theorems of Giné and Zinn (1990). Lemma 9.8 and Theorem 9.9 are based on analogous results of Hoffmann-Jørgensen (1974). The proofs also include some new elements. Giné (1997) gives a lot of references, not reproduced here, on different parts of the proof. Some special cases of the Giné–Zinn theorems were published earlier by Bickel and Freedman (1981), Singh (1981), and Gaenssler (1986) among others. These papers also give some results other than special cases of the Giné–Zinn theorems, on bootstrap asymptotics for sample quantiles and other functionals.

Notes to Section 9.3. Efron (1979) discovered the bootstrap. Three books on it are Hall (1992), Efron and Tibshirani (1993), and Shao and Tu (1995).

10

Uniform and Universal Limit Theorems

In this chapter we look at cases where a class $\mathcal{F}$ of measurable functions is a Glivenko–Cantelli or Donsker class for all probability laws P on the underlying space and ask if so, whether the convergence is uniform in P.

10.1 Uniform Glivenko–Cantelli Classes

Let $(X, \mathcal{A})$ be a measurable space. Let $\mathcal{P}(X) = \mathcal{P}(X, \mathcal{A})$ be the set of all probability measures on $(X, \mathcal{A})$. Let $\mathcal{F}$ be a class of measurable real-valued functions on X. Then $\mathcal{F}$ is called a *strong uniform Glivenko–Cantelli class* if for every $\varepsilon > 0$ there is an n_0 such that

$$\sup_{P \in \mathcal{P}(X, \mathcal{A})} \Pr(\|P_n - P\|_{\mathcal{F}}^* > \varepsilon \text{ for some } n \geq n_0) < \varepsilon.$$

A class $\mathcal{F}$ will be called a *uniform Glivenko–Cantelli class in probability* if for every $\varepsilon > 0$ there is an n_0 such that

$$\Pr(\|P_n - P\|_{\mathcal{F}}^* > \varepsilon) < \varepsilon$$

for all $n \geq n_0$ and all $P \in \mathcal{P}(X, \mathcal{A})$.

Proposition 10.1 *Let $(\Omega, \mathcal{B}, P)$ be a probability space and $(X, \|\cdot\|)$ a normed space. Let $X_1, X_2, \ldots,$ be X-valued functions on Ω such that for each n and $s_j = 1$ or -1 for each $j = 1, \ldots, n$, $\|\sum_{j=1}^n s_j X_j\|$ and $\|X_n\|$ are measurable, the joint distribution of $\|\sum_{i=1}^j s_i X_i\|$ for $j = 1, \ldots, n$ does not depend on $s_1, \ldots, s_n$, and $\|X_j\|$ for $j = 1, \ldots, n$ are i.i.d. Let $S_n := \sum_{j=1}^n X_j$. If $\|S_n\|/n \to 0$ in probability as $n \to \infty$, then $n \Pr(\|X_1\| > n) \to 0$.*

348

Proof. Let $S_0 := 0$. For any n and $j = 1, 2, \ldots, n$, if $\|X_j\| > n$, then we cannot have $\|S_i\| \le n/2$ for both $i = j - 1$ and j. Thus

$$\Pr\left(\max_{1 \le j \le n} \|X_j\| > n\right) \le \Pr\left(\max_{j \le n} \|S_j\|/n > \frac{1}{2}\right). \tag{10.1}$$

Lemma 3.12 is a form of P. Lévy inequality. Its hypotheses per se do not apply, but the conclusion holds with about the same proof. We do not need stars since the norms here are measurable. It gives that the right side of (10.1) is $\le 2\Pr(\|S_n\|/n > 1/2) \to 0$ as $n \to \infty$. Then by (10.1), for $p_n := \Pr(\|X_1\| > n)$, $(1 - p_n)^n \to 1$, which implies $np_n \to 0$. $\square$

Proposition 10.2 *For any measurable space $(X, \mathcal{A})$ and class $\mathcal{F}$ of real-valued measurable functions on it, if $\mathcal{F}$ is a Glivenko–Cantelli class for every P on $(X, \mathcal{A})$, then $\sup_{f \in \mathcal{F}}(\sup f - \inf f) < +\infty$.*

Proof. Suppose not. Then there are $f_k \in \mathcal{F}$ and points x_k and y_k such that $f_k(x_k) - f_k(y_k) > 8^k$ for all positive integers k. Let $P := \sum_{k=1}^{\infty}(\delta_{x_k} + \delta_{y_k})/2^{k+1}$. Then P is a probability measure (law) defined on all subsets of X and so on $\mathcal{A}$, and $\mathcal{F}$ is Glivenko–Cantelli for P. The countable subset $\mathcal{G} = \{f_k\}_{k \ge 1}$ is also Glivenko–Cantelli for P. Norms $\|\cdot\|_{\mathcal{G}}$ in what follows will all be measurable because $\mathcal{G}$ is countable. Let P'_n be an independent copy of P_n. Then $\|P_n - P'_n\|_{\mathcal{G}} \to 0$ in probability. Here $P_n - P'_n = \frac{1}{n}\sum_{i=1}^{n} V_i$ where $V_i = \delta_{X(i)} - \delta_{Y(i)}$ where $X(1), \ldots, X(n), Y(1), \ldots, Y(n)$ are i.i.d. (P). Thus V_i are symmetric, i.e., any V_i can be interchanged with $-V_i$ without changing the joint distribution of the $\|\cdot\|_{\mathcal{G}}$ seminorms of any partial sums. It follows from Proposition 10.1 that $n\Pr(\|V_1\|_{\mathcal{G}} > n) \to 0$ as $n \to \infty$. Then by definition of P, for each k, and $n = 8^k$,

$$n\Pr(\|V_1\|_{\mathcal{G}} > n) \ge n\Pr(X(1) = x_k \text{ and } Y(1) = y_k) \ge 8^k/4^{k+1} \to \infty$$

as $k \to \infty$, a contradiction, proving the proposition. $\square$

If $\mathcal{F}$ is a class of bounded functions, let $\mathcal{F}_0 := \{f - \inf f : f \in \mathcal{F}\}$. The following is immediate:

Proposition 10.3 *Let $(X, \mathcal{A})$ be a measurable space and $\mathcal{F}$ be a family of bounded measurable real-valued functions on X. For any $f \in \mathcal{F}$, constant $c = c_f$, in particular $c_f = \inf f$, and any two probability measures P and Q on $\mathcal{A}$, $(P - Q)(f) = (P - Q)(f - c_f)$. Thus $\|P - Q\|_{\mathcal{F}} = \|P - Q\|_{\mathcal{F}_0}$. This holds in particular if Q is any P_n.*

We have the following equivalence:

Theorem 10.4 *For any separable measurable space $(X, \mathcal{A})$ and class $\mathcal{F}$ of measurable real-valued functions on X such that $\mathcal{F}_0$ is image admissible Suslin,*

$\mathcal{F}$ is a uniform Glivenko–Cantelli class in probability if and only if it is a strong uniform Glivenko–Cantelli class. If $\mathcal{F}$ is such a class, then

$$\lim_{n \to \infty} \sup_P E \| P_n - P \|_{\mathcal{F}} = 0. \tag{10.2}$$

Proof. "If" is immediate. For "only if," by Proposition 10.2, $\mathcal{F}_0$ is uniformly bounded. By Proposition 10.3, we can assume that $\mathcal{F} = \mathcal{F}_0$ and that $\| f \|_{\sup} \le 1$ for all $f \in \mathcal{F}$. Then always $\| P_n - P \|_{\mathcal{F}} \le 1$. By Theorem 6.6(b), for any P on $(X, \mathcal{A})$, $(\| P_n - P \|_{\mathcal{F}}, \mathcal{S}_n)$ is a reversed submartingale with $\mathcal{S}_n$ the smallest σ-algebra for which all $P_k(f)$ for $k \ge n$ and $f \in \mathcal{L}^1(X, \mathcal{A}, P)$ are measurable. By assumption, for any $\delta > 0$ there is an $n_0 = n_0(\delta)$ not depending on P such that for all $n \ge n_0$, $\Pr(\| P_n - P \|_{\mathcal{F}} > \delta) < \delta$. Given $\varepsilon > 0$, let $\delta = \varepsilon^2/2$. For any $N > n_0(\delta)$ let $Y_j = \| P_{N-j} - P \|_{\mathcal{F}}$ for $j = 1, \ldots, N - n_0$. Then Y_j form a submartingale for the σ-algebras $\mathcal{S}_{N-j}$. Let $A(\varepsilon, k) = \{\max_{1 \le j \le k} Y_j \ge \varepsilon\}$. By Doob's maximal inequality (e.g., RAP, Theorem 10.4.2) with $k = N - n_0$,

$$\varepsilon \Pr(A(\varepsilon, k)) \le E \| P_{n_0} - P \|_{\mathcal{F}} < 2\delta = \varepsilon^2. \tag{10.3}$$

Thus $\Pr(A(\varepsilon, k) < \varepsilon$, and this holds for all $N > n_0$. Letting $N \to \infty$ we have

$$\Pr\left(\sup_{n \ge n_0} \| P_n - P \|_{\mathcal{F}} > \varepsilon\right) \le \varepsilon,$$

showing that $\mathcal{F}$ is a strong uniform Glivenko–Cantelli class, and by (10.3) for any $n \ge n_0(\delta)$ in place of n_0, (10.2) also holds. $\qquad \square$

Proposition 10.5 *Let $(X, \mathcal{A})$ be a measurable space and $\mathcal{F}$ a uniformly bounded class $\mathcal{F}$ of measurable functions, totally bounded for $d_{\sup}$. Then $\mathcal{F}$ is a strong uniform Glivenko–Cantelli class.*

Proof. Given $\varepsilon > 0$, suppose that functions $f_1, \ldots, f_m$ form an $\varepsilon/4$-net in $\mathcal{F}$ for $d_{\sup}$. For each $f \in \mathcal{F}$ there is an f_j such that $d_{\sup}(f, f_j) < \varepsilon/4$. It follows that for any n, $|(P_n - P)(f - f_j)| < \varepsilon/2$. Thus if we can find an n_0 such that $|(P_n - P)(f_j)| < \varepsilon/2$ for all $n \ge n_0$ then $|(P_n - P)(f)| < \varepsilon$ for all $n \ge n_0$. So it suffices to show that a finite set of bounded measurable functions is a strong uniform Glivenko–Cantelli class, as it clearly is by the strong law of large numbers. $\qquad \square$

If a class $\mathcal{C}$ of measurable sets is a uniform Glivenko–Cantelli class, it must be a VC class by Assouad's theorem 6.27. Conversely, uniformly bounded classes $\mathcal{F}$ of functions satisfying conditions of Vapnik–Červonenkis type and suitable measurability conditions will be shown to be uniformly Glivenko–Cantelli.

Let $(X, \mathcal{A})$ be a measurable space and $\mathcal{F}$ a family of bounded measurable functions on X. For $x = (x_1, \ldots, x_n), n = 1, 2, \ldots, 1 \le p < \infty$, and any

$f, g \in \mathcal{F}_0$, define the pseudometric

$$e_{x,p}(f, g) := \left[\frac{1}{n} \sum_{j=1}^{n} |f(x_j) - g(x_j)|^p \right]^{1/p}.$$

Then $D(\varepsilon, \mathcal{F}_0, e_{x,p}) \leq +\infty$ is defined. Let

$$H_{n,p}(\varepsilon, \mathcal{F}_0) := \sup_{x \in X^n} \log D(\varepsilon, \mathcal{F}_0, e_{x,p}).$$

We have:

Theorem 10.6 *Let $\mathcal{F}$ be family of bounded functions on X for a separable measurable space $(X, \mathcal{A})$ such that $\mathcal{F}_0$ is image admissible Suslin. Then the following are equivalent:*

(a) $\mathcal{F}$ is a uniform Glivenko–Cantelli class in probability;

(b) $\mathcal{F}$ is a strong uniform Glivenko–Cantelli class;

For $1 \leq p < \infty$,

(c_p) $\mathcal{F}_0$ is uniformly bounded, and for all $\varepsilon > 0$, $\lim_{n \to \infty} H_{n,p}(\varepsilon, \mathcal{F}_0)/n = 0$.

Proof. Statements (a) and (b) are equivalent by Theorem 10.4, and (a) implies that $\mathcal{F}_0$ is uniformly bounded by Proposition 10.2. By Proposition 10.3, each of (a) and (b) holds for $\mathcal{F}_0$ if and only if it holds for $\mathcal{F}$. So we can and will assume $\mathcal{F}$ is uniformly bounded in the rest of the proof.

Next it will be shown that (c_1) implies (b). Let P be any probability measure on $(X, \mathcal{A})$. Let $\{\varepsilon_k\}_{k \geq 1}$ be i.i.d. Rademacher variables independent of $\{X_j\}_{j \geq 1}$. As in the definitions before Proposition 6.4, assume given as there random variables $\sigma(i)$ and $\tau(i)$ independent of each other and the X_j. Specifically, take a countable product X^∞ of copies of $(X, \mathcal{A}, P)$ on which X_j are coordinates, take another probability space Ω' on which the $\sigma(i)$ and $\tau(i)$ are defined, and take yet another Ω_ε on which i.i.d. ε_k are defined. Take the products $\Omega := X^\infty \times \Omega_\varepsilon$ and $\Omega'' := \Omega \times \Omega'$ with product probabilities. For P_n' and P_n'' as defined before Proposition 6.4 we have $P_n' = \frac{1}{n} \sum_{j=1}^{n} \delta_{X_j'}$ for $X_j' = X_{\sigma(j)}$ and likewise $P_n'' = \frac{1}{n} \sum_{j=1}^{n} \delta_{Y_j'}$ for $Y_j' = X_{\tau(j)}$. Thus

$$P_n' - P_n'' = \frac{1}{n} \sum_{j=1}^{n} (\delta_{X_j'} - \delta_{Y_j'}). \tag{10.4}$$

Now $\|P_n - P\|_\mathcal{F}$ and $\|P_n' - P_n''\|_\mathcal{F}$ are universally measurable since $\mathcal{F} = \mathcal{F}_0$ is image admissible Suslin, by Proposition 5.20 and Corollary 5.25. The expression in (10.4) has all the properties of

$$V_n := \frac{1}{n} \sum_{j=1}^{n} \varepsilon_j (\delta_{X_j'} - \delta_{Y_j'}), \tag{10.5}$$

specifically, their $\|\cdot\|_{\mathcal{F}}$ norms are equal in distribution, because replacing any particular ε_j by $-\varepsilon_j$ is equivalent to interchanging $\sigma(j)$ and $\tau(j)$. Both operations preserve probabilities.

Given $\varepsilon > 0$, let $n \geq 16/\varepsilon^2$. In Lemma 6.5 take $\zeta := 1$ and $\eta = \sqrt{n}\varepsilon/2 > 2$. Then the Lemma gives

$$\Pr(\|P_n - P\|_{\mathcal{F}} > \varepsilon) \leq 2\Pr(\|P'_n - P''_n\|_{\mathcal{F}} > \varepsilon/2) = 2\Pr(\|V_n\|_{\mathcal{F}} > \varepsilon/2. \quad (10.6)$$

For $\xi \in X^{\infty}$ let $x := x(\xi) := x_n(\xi) := (X_1(\xi), \dots, X_{2n}(\xi))$. By definition of $D(\varepsilon, \mathcal{F}_0, e_{x,1})$, for each $\xi \in X^{\infty}$ there is a map $\pi_n = \pi_n^{\xi}$ from $\mathcal{F}_0$ onto a subset $\mathcal{G}_{\xi}$ of cardinality at most $D(\varepsilon/8, \mathcal{F}_0, e_{x_n(\xi),1})$ such that for all $f \in \mathcal{F}_0$, $e_{x_n(\xi),1}(f, \pi_n f) \leq \varepsilon/8$.

Let $\mathcal{F}_0$ be image admissible Suslin via $(S, \mathcal{S}, T)$. To show that the choice of $\mathcal{G}_{\varepsilon}$ can be made measurably, we have that $(x, s, t) \mapsto e_{x,1}(T(s), T(t))$ is jointly measurable. To construct $\mathcal{G}_{\xi} := \{f_j^{\xi}\}_{j=1}^{k(\xi)}$ recursively, let f_1^{ξ} be any fixed $f_1 \in \mathcal{F}_0$. Suppose given $f_1^{\xi}, \dots, f_r^{\xi}$ with $e_{x,1}(f_i^{\xi}, f_j^{\xi}) > \varepsilon/8$ for $1 \leq i < j \leq r$. Let $A_r^{\xi} := \{f \in \mathcal{F}_0 : e_{x,1}(f, f_i) > \varepsilon/8\}$ for $1 \leq i \leq r$. If $A_r^{\xi} = \emptyset$, then $k(\xi) = r \leq D(\varepsilon/8, \mathcal{F}_0, e_{x,1})$, and the recursion is finished. If $A_r^{\xi} \neq \emptyset$, then $\{(\xi, s) \in X^{\infty} \times S : f = T(s) \in A_r^{\xi}\}$ is an $\mathcal{A}^{\infty} \times \mathcal{S}$ measurable subset of $X^{\infty} \times S$. Let $B_r := \{\xi : A_r^{\xi} \neq \emptyset\}$. Then by measurable selection (Theorem 5.22), B_r is universally measurable, and there is a universally measurable map ζ_r from B_r into S such that $f_{r+1}^{\xi} := T(\zeta_r(\xi)) \in A_r(\xi)$ for all $\xi \in B_r$.

We have

$$\frac{1}{2}\left| \int (f - \pi_n(f))dV_n \right| \leq \frac{1}{2}\int |f - \pi_n f| d(P'_n + P''_n) \leq \varepsilon/8,$$

and therefore

$$\Pr(\|V_n\|_{\mathcal{F}} > \varepsilon/2) \leq \Pr(\|V_n(\pi_n(f))\|_{\mathcal{F}} > \varepsilon/4).$$

The latter norm is a supremum over a finite set of functions for fixed ξ and then is a measurable function of $\omega_{\varepsilon} := \{\varepsilon_k\}_{k \geq 1}$. Let $\Pr_{\varepsilon}$ denote probability with respect to ω_{ε} for ξ fixed. Then by the Hoeffding inequality for linear combinations of Rademacher functions, Proposition 1.12, for each ξ,

$$\Pr_{\varepsilon}(\|V_n(\pi_n(f))\|_{\mathcal{F}} > \varepsilon/4) \leq 2D(\varepsilon/8, \mathcal{F}, e_{x_n(\xi),1})\exp(-n\varepsilon^2/(32)).$$

By (c_1), for all n large enough, $D(\varepsilon/8, \mathcal{F}, e_{x,1}) \leq \exp(n\varepsilon^2/(64))$ for all possible $x = x_n(\xi)$. For such n, taking the expectation with respect to ξ, where we have joint measurability in $(\xi, \omega_{\varepsilon})$ by the image admissible Suslin condition and measurable choice of f_i^{ξ},

$$\Pr(\|V_n\|_{\mathcal{F}} > \varepsilon/2) \leq 2\exp(-n\varepsilon^2/(64)).$$

Then by (10.6), summing a geometric series gives that for each $\varepsilon > 0$ there is an n_ε and a $C = C(\varepsilon) = 4/[1 - \exp(-\varepsilon^2/(64))]$ such that

$$\sup_P \sum_{k \geq n_\varepsilon} \Pr(\|P_k - P\|_{\mathcal{F}} > \varepsilon) \leq C \exp(-n_\varepsilon \varepsilon^2/(64)) < \varepsilon$$

where the supremum is over all probability measures P on $(X, \mathcal{A})$, which implies (b).

Next it will be shown that (a) implies (c_2). The following is used:

Lemma 10.7 *If $\mathcal{F}$ satisfies (a) and P is such that $\mathcal{F} \subset L^p(P)$, $p \geq 1$, and $Pf = 0$ for all $f \in \mathcal{F}$, and ε_i are i.i.d Rademacher variables independent of the variables X_j (in the stronger product space sense), then, for any functionals $a_i = a_i(f)$,*

$$2^{-p} E^* \left\| \sum \varepsilon_i \delta_{X_i} \right\|_{\mathcal{F}}^p \leq E^* \left\| \sum \delta_{X_i} \right\|_{\mathcal{F}}^p \leq 2^p E^* \left\| \sum \varepsilon_i (\delta_{X_i} + a_i) \right\|_{\mathcal{F}}^p. \quad (10.7)$$

Proof. The subscript $\mathcal{F}$ will be dropped from the norms for simplicity. By Proposition 10.2, $\mathcal{F}_0$ is uniformly bounded. For each $f \in \mathcal{F}$, $Pf = 0$ implies that $\sup f \geq 0$ and $\inf f \leq 0$. Thus $\mathcal{F}$ is uniformly bounded. Let A and B be disjoint sets of indices, let E_A denote integration with respect to X_i, $i \in A$, and likewise for E_B. In the following display, the first two equal E^*'s can be written as E_A^* by Lemma 3.6. The first inequality holds by the Jensen inequality in the form of Lemma 9.7(b), and the last by the 1-sided Tonelli–Fubini theorem, Theorem 3.9, so we have

$$E^* \left\| \sum_{i \in A} f(X_i) \right\|^p = E^* \left\| \sum_{i \in A} f(X_i) + E \sum_{i \in B} f(X_i) \right\|^p$$

$$\leq E_A^* E_B^* \left\| \sum_{i \in A \cup B} f(X_i) \right\|^p \leq E^* \left\| \sum_{i \in A \cup B} f(X_i) \right\|^p. \quad (10.8)$$

If E_X denotes E with respect to $\{X_i\}$ for $\{\varepsilon_i\}$ fixed, and E_ε expectation with respect to $\{\varepsilon_i\}$, then

$$E^* \left\| \sum \varepsilon_i f(X_i) \right\|^p = E_\varepsilon E_X^* \left\| \sum_{i:\varepsilon_i=1} f(X_i) - \sum_{i:\varepsilon_i=-1} f(X_i) \right\|^p \quad (10.9)$$

by Theorem 3.9 again, now using discreteness of $\{\varepsilon_i\}$. If U and V are random elements of a normed space $(Y, \|\cdot\|)$ and $1 \leq p < \infty$, then by convexity

$$\left\| \frac{U - V}{2} \right\|^p \leq \frac{1}{2} \left(\|U\|^p + \|V\|^p \right).$$

Let A and B be the two sets of i's on the right side of (10.9). Let $U := \sum_{i \in A} f(X_i)$ and $V := \sum_{i \in B} f(X_i)$. Using (10.8), also with A and B

interchanged, gives using again the discreteness of $\{\varepsilon_i\}$

$$E^* \|U-V\|^p \leq 2^p E^* \|U+V\|^p \leq 2^p E_\varepsilon E_X^* \left\| \sum f(X_i) \right\|^p = 2^p E^* \left\| \sum f(X_i) \right\|^p,$$

proving the first inequality in (10.7). For the second inequality, similarly, again applying the Jensen inequality Lemma 9.7,

$$E^* \left\| \sum_{i=1}^n f(X_i) \right\|^p = E^* \left\| \sum_{i=1}^n (f(X_i) - Ef(X_{n+i})) \right\|^p$$

$$\leq E^* \left\| \sum_{i=1}^n (f(X_i) + a_i(f)) - \sum_{i=1}^n (f(X_{n+i}) + a_i(f)) \right\|^p,$$

which for any $\{\varepsilon_i\}$, because $P^{\mathbb{N}}$ is invariant under permutations of the coordinates, equals

$$E_X^* \left\| \sum_{i=1}^n \varepsilon_i (f(X_i) + a_i(f) - f(X_{n+i}) - a_i(f)) \right\|^p$$

$$\leq 2^p E^* \left\| \sum_{i=1}^n \varepsilon_i (f(X_i) + a_i(f)) \right\|^p,$$

as in the first half of the proof.   □

Now to continue the proof that (a) implies (c_2), recall $\Lambda_{2,1}$ and Lemma 9.16 about it. In (10.10) below $\Lambda_{2,1}$ of a random variable appears in an upper bound, which is useless when it is $+\infty$. Thus Lemma 9.16(b) gives a sufficient condition, namely $E(|\xi_1|^r) < \infty$ for some $r > 2$, for the bound to be finite.

Theorem 10.8 *Let $\mathcal{F}$ be an image admissible Suslin class of P-integrable functions on X for a separable measurable space $(X, \mathcal{A})$. Let X_i be X-valued random variables, and let $\varepsilon_i, \xi_i, i \in \mathbb{N}$, be respectively a Rademacher sequence and a sequence of symmetric i.i.d. real random variables, independent of each other, all coordinates on a product probability space $X^\infty \times \{-1, 1\}^\infty \times \mathbb{R}^\infty$, in particular, all independent. Then, for every $0 \leq n_0 < \infty$ and $n_0 < n \in \mathbb{N}$, we have*

$$(E|\xi_1|)E \left\| \frac{1}{\sqrt{n}} \sum_{i=1}^n \varepsilon_i f(X_i) \right\|_{\mathcal{F}} \leq E \left\| \frac{1}{\sqrt{n}} \sum_{i=1}^n \xi_i f(X_i) \right\|_{\mathcal{F}}$$

$$\leq n_0 (E\|f(X_1)\|_{\mathcal{F}}) E \left[\frac{1}{\sqrt{n}} \max_{i \leq n} |\xi_i| \right] \qquad (10.10)$$

$$+ \Lambda_{2,1}(\xi_1) \max_{n_0 < k \leq n} E \left\| \frac{1}{\sqrt{k}} \sum_{i=n_0+1}^k \varepsilon_i f(X_i) \right\|_{\mathcal{F}}.$$

Proof. To begin, the first inequality in (10.10) will be proved. For any $n = 1, 2, \ldots$, the class $\mathcal{F}_n$ of functions

$$\{(X_i, \varepsilon_i, \xi_i)\}_{i=1}^n \mapsto \sum_{i=1}^n \varepsilon_i |\xi_i| f(X_i)$$

for $f \in \mathcal{F}$ is image admissible Suslin on $X^n \times \mathbb{R}^n \times \mathbb{R}^n$ (by Theorem 5.26), and likewise if each $\varepsilon_i |\xi_i|$ is replaced by ξ_i. By symmetry, the joint distribution of the ξ_i equals that of the $\varepsilon_i |\xi_i|$, so that

$$E \left\| \frac{1}{\sqrt{n}} \sum_{i=1}^n \xi_i f(X_i) \right\|_{\mathcal{F}} = E \left\| \frac{1}{\sqrt{n}} \sum_{i=1}^n \varepsilon_i |\xi_i| f(X_i) \right\|_{\mathcal{F}} \qquad (10.11)$$

where the norms are universally measurable by Corollary 5.25, so the expectations exist. For fixed $\{\varepsilon_i\}$, $\{X_i\}$, and f, $\sum_{i=1}^n \varepsilon_i |\xi_i| f(X_i)$ is a linear function of $\eta_n := \{|\xi_i|\}_{i=1}^n$ and so its absolute value is a convex function of η_n. Thus $\| \sum_{i=1}^n \varepsilon_i |\xi_i| f(X_i) \|_{\mathcal{F}}$, as the supremum of a family of convex functions, is a convex function of η_n. It follows then by Jensen's inequality that the equal expressions in (10.11) are

$$\geq E \left\| \frac{1}{\sqrt{n}} \sum_{i=1}^n \varepsilon_i \left(E|\xi_i| \right) f(X_i) \right\|_{\mathcal{F}},$$

which proves the first inequality. Now consider the second. From here on $\mathcal{F}$ is omitted from the norm signs. Let $N_t := \#\{i \leq n : |\xi_i| \geq t\}$. The expressions in (10.11) are also equal to

$$E \left\| n^{-1/2} \sum_{i=1}^n \left(\int_0^\infty 1_{t \leq |\xi_i|} dt \right) \varepsilon_i f(X_i) \right\|$$

$$= E \left\| n^{-1/2} \int_0^\infty \left(\sum_{i=1}^n 1_{t \leq |\xi_i|} \varepsilon_i f(X_i) \right) dt \right\|$$

$$\leq \int_0^\infty E \left\| n^{-1/2} \sum_{i=1}^n 1_{t \leq |\xi_i|} \varepsilon_i f(X_i) \right\| dt$$

$$= \int_0^\infty E \left\| n^{-1/2} \sum_{j=1}^{N_t} \varepsilon_j f(X_j) \right\| dt$$

$$\leq \int_0^\infty \left(\sum_{k=1}^n \Pr\{N_t = k\} E \left\| \frac{1}{\sqrt{n}} \sum_{i=1}^k \varepsilon_i f(X_i) \right\| \right) dt \leq T + UV$$

where

$$T := \left(\int_0^\infty \Pr\{N_t > 0\}dt\right) \max_{k \le n_0} E \left\| n^{-1/2} \sum_{i=1}^k \varepsilon_i f(X_i) \right\|,$$

$$U := \frac{1}{\sqrt{n}} \int_0^\infty \sum_{k=n_0+1}^{\infty} \sqrt{k} \, \Pr\{N_t = k\} \, dt,$$

$$V := \max_{n_0 < k \le n} E \left\| \frac{1}{\sqrt{k}} \sum_{i=n_0+1}^{k} \varepsilon_i f(X_i) \right\|.$$

We have $T \le \left(\int_0^\infty \Pr\left\{\max_{i \le n} |\xi_i| \ge t\right\} dt\right) n_0 E \|f(X_1)/\sqrt{n}\|$ and

$$U \le \frac{1}{\sqrt{n}} \int_0^\infty \sum_{k=1}^{n} \sqrt{k} \, \Pr(N_t = k)dt.$$

Let $|\xi_1|, \ldots, |\xi_n|$ in order (order statistics) be $|\xi|_{(1)} \le |\xi|_{(2)} \le \cdots \le |\xi|_{(n)}$, with $|\xi|_{(0)} := 0$. Then N_t is the number of values of $i \le n$ with $|\xi|_{(i)} \ge t$, and $N_t = k$ if and only if $|\xi|_{(n-k+1)} \ge t > |\xi|_{(n-k)}$. Now

$$\int_0^\infty \Pr(|\xi|_{(n-k)} < t \le |\xi|_{(n-k+1)})dt = \int_0^\infty E 1_{|\xi|_{(n-k)} < t \le |\xi|_{(n-k+1)}} dt$$

$$= E \int_{|\xi|_{(n-k)}}^{|\xi|_{(n-k+1)}} dt.$$

Next,

$$E \sum_{k=1}^{n} \int_{|\xi|_{(n-k)}}^{|\xi|_{(n-k+1)}} \sqrt{k} \, dt = E \int_0^{|\xi|_{(n)}} \sqrt{N_t} \, dt,$$

and $E\sqrt{N_t} \le \sqrt{EN_t} = \left(E \sum_{i=1}^{n} 1_{|\xi_i| \ge t}\right)^{1/2} = (n \Pr(|\xi_1| \ge t))^{1/2}$, so $U \le \int_0^\infty \Pr(|\xi_1| \ge t)^{1/2} dt = \Lambda_{2,1}(\xi_1)$. Theorem 10.8 now follows. $\qquad \square$

To continue further the proof that (a) implies (c_2), for a given P, and $\mathcal{F}$ and $\mathcal{F}_0$ as in Theorem 10.6, let $\mathcal{G} := \{f - Pf : f \in \mathcal{F}\}$. Since clearly $Pf \in [\inf f, \sup f]$ for each $f \in \mathcal{F}$, by Proposition 10.2, $\mathcal{G}$ is uniformly bounded. We have clearly $\mathcal{G}_0 = \mathcal{F}_0$, so $\mathcal{G}_0$ is image admissible Suslin. By Proposition 10.3, $\mathcal{G}_0$ and $\mathcal{G}$ are uniform Glivenko–Cantelli classes, so (a) holds for $\mathcal{G}$. For the given P, clearly $\mathcal{G} \subset \mathcal{L}^1(P)$, and $Pg = 0$ for all $g \in \mathcal{G}$, so the hypothesis of Lemma 10.7 holds for $\mathcal{G}$ in place of $\mathcal{F}$ and so also the conclusion. For $p = 1$

this gives for $m = 1, 2, \ldots,$

$$\frac{1}{2} E^* \left\| \sum_{i=1}^{m} \varepsilon_i \delta_{X_i} \right\|_{\mathcal{G}} \leq E^* \left\| \sum_{i=1}^{m} \delta_{X_i} \right\|_{\mathcal{G}}$$

$$= E^* \| m P_m \|_{\mathcal{G}} = m E^* \| P_m - P \|_{\mathcal{F}}$$

$$= m E \| P_m - P \|_{\mathcal{F}_0} = o(m) \tag{10.12}$$

as $m \to \infty$, uniformly in $P \in \mathcal{P}(X, \mathcal{A})$, by Theorem 10.4.

To continue the proof that (a) implies (c_2), the following Gaussianization lemma will be used:

Lemma 10.9 *Under the hypotheses of Theorem 10.6 and (a), if $g_1, g_2, \ldots$ are i.i.d. $N(0, 1)$, then*

$$\lim_{n \to \infty} \sup_{P \in \mathcal{P}(X, \mathcal{A})} E \left\| \frac{1}{n} \sum_{j=1}^{n} g_j \delta_{X_j} \right\|_{\mathcal{F}_0} = 0. \tag{10.13}$$

Proof. Apply Theorem 10.8 in case $\xi_i = g_i$, for which clearly $\Lambda_{2,1}(g_1) < \infty$. Of the three expressions in the two inequalities (10.10), we want to show that the middle expression $E \| \sum_{i=1}^{n} g_i f(X_i) \|_{\mathcal{F}} / \sqrt{n}$, divided by a further $\sqrt{n}$, is $o_p(1)$ uniformly in $P \in \mathcal{P}(X, \mathcal{A})$. Let the two terms added in the last expression in (10.10) with $\xi_i = g_i$ be $T_1 + T_2$. So we want to show that $(T_1 + T_2)/\sqrt{n} \to 0$. In T_1, since $\mathcal{F}$ is uniformly bounded by some M, $E \| f(X_1) \|_{\mathcal{F}} \leq M$. By Proposition 2.5, for any $n \geq 2$ and $C > 0$,

$$\Pr(\max_{j \leq n} |g_j| \geq C \sqrt{\log n}) \leq n \exp(-(C^2 \log n)/2) = n^{1 - C^2/2}.$$

It follows that

$$\sup_{n \geq 2} E \left(\max_{j \leq n} |g_j| / \sqrt{\log n} \right) \leq 2 + \sup_{n \geq 2} \sum_{k=2}^{\infty} n^{1 - k^2/2} < \infty,$$

so $E(\max_{j \leq n} |g_j|) = O(\sqrt{\log n})$ as $n \to \infty$. Thus $T_1/\sqrt{n} = O(n_0 \sqrt{\log n}/n)$. To make this $o(1)$, we are free to choose $n_0 = n_0(n) < n$ in (10.10), and we need $n_0 = o(n/\sqrt{\log n})$, so choose n_0 to be asymptotic to $n/(\log n)$.

For T_2, and each value of k with $n_0 < k \leq n$, we have for $m := k - n_0 < k$

$$E \left\| \sum_{i=n_0+1}^{k} \varepsilon_i f(X_i) \right\|_{\mathcal{F}_0} = E \left\| \sum_{i=1}^{m} \varepsilon_i f(X_i) \right\|_{\mathcal{F}_0},$$

which is $o(m)$ and $o(k)$ uniformly in $P \in \mathcal{P}(X, \mathcal{A})$ as $n \to \infty$ by (10.12). When divided by $\sqrt{k}$ it is $o(\sqrt{k}) = o(\sqrt{n})$, so when further divided by $\sqrt{n}$ it is $o(1)$, uniformly in $P \in \mathcal{P}(X, \mathcal{A})$, as desired, and the Lemma is proved. $\square$

Now continuing the proof that (a) implies (c$_2$), let $x = (x_1, \ldots, x_n) \in X^n$ and let P_x be the corresponding empirical measure $P_x := \frac{1}{n} \sum_{i=1}^{n} \delta_{x_i}$. For $j = 1, \ldots, n$ let $m(j)$ be i.i.d. random variables uniformly distributed over $\{1, 2, \ldots, n\}$ and independent of $\{g_i\}_{i=1}^{n}$, defined on another product space factor $\{1, 2, \ldots, n\}^n$. Then $x_{m(j)}$ for $j = 1, \ldots, n$ are i.i.d. P_x (as in bootstrap sampling, Chapter 9). Applying (10.13) to P_x gives

$$\limsup_{n \to \infty} \sup_{x \in X^n} E \left\| \frac{1}{n} \sum_{j=1}^{n} g_j \delta_{x_{m(j)}} \right\| = 0. \tag{10.14}$$

The following claim will be proved:

$$E \left\| \frac{1}{n} \sum_{i=1}^{n} g_i \delta_{x_i} \right\|_{\mathcal{F}_0} \leq (1 - e^{-1})^{-1} E \left\| \frac{1}{n} \sum_{j=1}^{n} g_j \delta_{x_{m(j)}} \right\|_{\mathcal{F}_0}. \tag{10.15}$$

To prove this let $A_{ij} := \{m(j) = i\}$ for $i, j = 1, \ldots, n$. Then for each $j = 1, \ldots, n$, the sets A_{ij} are disjoint for distinct i, each with probability $\Pr(A_{ij}) = 1/n$, and $\bigcup_{i=1}^{n} A_{ij}$ is the whole probability space. Sets $A_{i_1 1}, A_{i_2 2}, \ldots, A_{i_n n}$ are jointly independent for any $i_1, \ldots, i_n$. Let g_{ij} for $i, j = 1, \ldots, n$ be n^2 i.i.d. $N(0, 1)$ random variables. By disjointness and independence, the two $n \times n$ arrays of random variables

$$\{\{g_j 1_{A_{ij}}\}_{i=1}^{n}\}_{j=1}^{n} \quad \text{and} \quad \{\{g_{ij} 1_{A_{ij}}\}_{i=1}^{n}\}_{j=1}^{n}$$

have the same joint distribution on $\mathbb{R}^{n^2}$. It follows that

$$E \left\| \frac{1}{n} \sum_{j=1}^{n} g_j \delta_{x_{m(j)}} \right\|_{\mathcal{F}_0} = E \left\| \frac{1}{n} \sum_{j=1}^{n} g_j \left(\sum_{i=1}^{n} 1_{A_{ij}} \delta_{x_i} \right) \right\|_{\mathcal{F}_0}$$

$$= E \left\| \frac{1}{n} \sum_{i,j=1}^{n} g_{ij} 1_{A_{ij}} \delta_{x_i} \right\|_{\mathcal{F}_0}. \tag{10.16}$$

Conditionally on the events $\{A_{ij}\}$, the random variables $\sum_{j=1}^{n} g_{ij} 1_{A_{ij}}$ for $i = 1, \ldots, n$ are independent with distribution $N(0, \sum_{j=1}^{n} 1_{A_{ij}})$, in other words, equal in distribution to $\left\{ \left(\sum_{j=1}^{n} 1_{A_{ij}} \right)^{1/2} g_i \right\}_{i=1}^{n}$.

Here is a subclaim: for each i, $E\sqrt{S_i} \geq 1 - e^{-1}$ where $S_i := \sum_{j=1}^{n} 1_{A_{ij}}$. To prove the subclaim, the $1_{A_{ij}}$ for fixed i are i.i.d. Bernoulli $(1/n)$ random variables. We have $\sqrt{S_i} \geq 1_{B_i}$ where $B_i := \bigcup_{j=1}^{n} A_{ij}$. Thus

$$
E\sqrt{S_i} \geq \Pr(B_i) = 1 - \Pr\left(\bigcap_{j=1}^{n} A_{ij}^c\right)
$$

$$
= 1 - \left(1 - \frac{1}{n}\right)^n \geq 1 - e^{-1}.
$$

So the subclaim is proved.

We then have

$$
E\left\|\frac{1}{n}\sum_{i,j=1}^{n} g_{ij} 1_{A_{ij}} \delta_{x_i}\right\|_{\mathcal{F}_0} = E\left\|\frac{1}{n}\sum_{i=1}^{n}\left(\sum_{j=1}^{n} 1_{A_{ij}}\right)^{1/2} g_i \delta_{x_i}\right\|_{\mathcal{F}_0}
$$

$$
\geq \left(1 - e^{-1}\right) E\left\|\frac{1}{n}\sum_{i=1}^{n} g_i \delta_{x_i}\right\|_{\mathcal{F}_0},
$$

as follows. The first equality results from writing $E = E_{m(\cdot)}E_{(g)}$, where $E_{m(\cdot)}$ is expectation with respect to the distribution of $m(\cdot)$, a finite sum, and $E_{(g)}$ is conditional expectation given $\{m(j)\}_{j=1}^{n}$, using the conditional distribution given before the subclaim. Then, since $\{g_i\}_{i=1}^{m}$ is independent of $m(\cdot)$, we can reverse the order of integration in the second expectation, where now $E_{(g)}$ is expectation with respect to the unconditional joint distribution of the g_i, namely, i.i.d. $N(0, 1)$. For fixed $\{g_i\}$ we then have by the Jensen inequality, Lemma 9.7(a), that $E_{(m)}\|\dots\| \geq \|E_{(m)}\dots\|$, and then applying the subclaim, we get the last inequality. Then using (10.16), the claim (10.15) is proved. It follows from (10.14) that

$$
\lim_{n\to\infty} \sup_{x\in X^n} E\left\|\frac{1}{n}\sum_{i=1}^{n} g_i \delta_{x_i}\right\|_{\mathcal{F}_0} = 0. \tag{10.17}
$$

For $f \in \mathcal{F}_0$ let $X(f) := \frac{1}{n}\sum_{i=1}^{n} g_i f(x_i)$, a Gaussian process indexed by $\mathcal{F}_0$. For $h \in \mathcal{F}_0$ also we have $e_{x,2}(f, h) \equiv \sqrt{n}d_X(f, h)$. Thus for $\varepsilon > 0$, $D(\varepsilon, \mathcal{F}_0, e_{x,2}) = D(\varepsilon/\sqrt{n}, \mathcal{F}_0, d_X)$. By the Sudakov minoration, in the form of Theorem 2.22(b), we have

$$
E\left\|\frac{1}{n}\sum_{i=1}^{n} g_i \delta_{x_i}\right\| \geq \frac{1}{17} \sup_{\varepsilon>0} \varepsilon(\log D(\varepsilon, \mathcal{F}_0, e_{x,2}))^{1/2}/n^{1/2}.
$$

Then (c$_2$) follows from this and (10.17).

Lastly, it will be shown that the conditions (c$_p$) for $1 \leq p < \infty$ are all equivalent. Take $M < \infty$ such that $\|f\|_{\sup} \leq M$ for all $f \in \mathcal{F}_0$. For any $\{x_j\}_{j=1}^{n}$, for the probability measure $P_n := \frac{1}{n}\sum_{j=1}^{n} \delta_{x_j}$ and $1 \leq r < \infty$, $e_{x,r}(f, g) = \|f - g\|_{r,n}$, the L^r norm for P_n, which is nondecreasing in r by Hölder's inequality.

Let $1 \le p < q < \infty$. Then for any $\varepsilon > 0$, $H_{n,p}(\varepsilon, \mathcal{F}_0) \le H_{n,q}(\varepsilon, \mathcal{F}_0)$, so (c_q) implies (c_p). Conversely, $e_{x,q}(f, g) \le e_{x,p}(f, g)^{p/q}(2M)^{(q-p)/q}$, which implies $H_{n,q}(\varepsilon, \mathcal{F}_0) \le H_{n,p}(\varepsilon^{q/p}/(2M)^{(q-p)/p}, \mathcal{F}_0)$. So (c_p) implies (c_q).   □

10.2 Universal Donsker Classes

Let X be a set and $\mathcal{A}$ a σ-algebra of subsets of X. Then a class $\mathcal{F}$ of measurable functions on X will be called a *universal Donsker class* if it is a P-Donsker class for every probability measure P on $(X, \mathcal{A})$.

Recall that every universal Donsker class of sets is a Vapnik–Červonenkis class (Theorem 6.24), and the converse holds (Corollary 6.20) under the usual image admissible Suslin measurability condition.

For a real-valued function f let $\operatorname{diam}(f) := \sup f - \inf f$. The following shows that a universal Donsker class is uniformly bounded up to additive constants:

Proposition 10.10 *If $\mathcal{F}$ is a universal Donsker class, then $\sup_{f \in \mathcal{F}} \operatorname{diam}(f) < \infty$.*

Proof. Suppose not. Then take $x_k \in X$, $y_k \in X$ and $f_k \in \mathcal{F}$ so that $f_k(x_k) - f_k(y_k) > 2^k$ for $k = 1, 2, \ldots$. Let

$$P := \sum_{k=1}^{\infty} (\delta_{x_k} + \delta_{y_k})/2^{k+1}.$$

Then P is a probability measure defined on all subsets of X and so on $\mathcal{A}$. For P,

$$E(f_k - Ef_k)^2 \ge 2^{-k-1} \inf_{c \in \mathbb{R}} \{(f_k(x_k) - c)^2 + (f_k(y_k) - c)^2\}.$$

The infimum is attained when $c = (f_k(x_k) + f_k(y_k))/2$, so

$$E(f_k - Ef_k)^2 \ge 2^{-k-2}(f_k(x_k) - f_k(y_k))^2 > 2^{k-2}$$

for all $k = 1, 2, \ldots$, so $\mathcal{F}$ is unbounded in the ρ_P metric and hence not a Donsker class for P by Theorem 3.34, and not a universal Donsker class.   □

Proposition 10.11 *For any family $\mathcal{F}$ of measurable functions on $(X, \mathcal{A})$, any probability measure P on $(X, \mathcal{A})$, any constants c_f depending on $f \in \mathcal{F}$, $\mathcal{G} := \{f - c_f : f \in \mathcal{F}\}$, and $\mathcal{H} := \mathcal{F} + \mathbb{R} := \{f + c : f \in \mathcal{F}, c \in \mathbb{R}\}$, each of the following properties holds for all three of $\mathcal{F}$, $\mathcal{G}$, and $\mathcal{H}$ if it holds for any one of them:*

(a) Donsker for P;

(b) Universal Donsker;

(c) Glivenko–Cantelli for P;

(d) Glivenko–Cantelli for all P;

(e) Uniform Glivenko–Cantelli.

Proof. Since $P_n - P$ is linear and $(P_n - P)(c) \equiv 0$, (c), (d), and (e) are immediate. In Section 3.1 a G_P process on $\mathcal{F}$ was called *coherent* if each sample function $G_P(\cdot)(\omega)$ is prelinear, bounded, and uniformly continuous on $\mathcal{F}$ with respect to ρ_P. Here $\mathcal{F}$ is P-pregaussian if and only if a coherent G_P process on it exists (Theorem 3.2). For (a), a coherent G_P process has $G_P(0) = 0$ and can be extended to make $G_P(f + c) = G_P(f)$ for all $f \in \mathcal{F}$ and all c, remaining coherent. The total boundedness for ρ_P and asymptotic equicontinuity are equivalent for $\mathcal{F}$, $\mathcal{G}$, and $\mathcal{H}$. It follows by Theorem 3.34 that (a) holds, and (b) follows. $\qquad\square$

Remark. If $\mathcal{F}$ is a class of bounded functions, in Proposition 10.11 we can take $c(f) := \inf f$. Then $\mathcal{G}$ is a class of nonnegative functions. If $\mathcal{F}$ is universal Donsker, then by Proposition 10.10 $\mathcal{G}$ is uniformly bounded.

The Vapnik–Červonenkis properties of classes of functions treated in Sections 4.7 and 4.8 (VC subgraph, VC major, VC hull) all have been (for VC hull and major in Theorem 6.21) or will be seen to imply the universal Donsker property for uniformly bounded classes of functions under some measurability conditions. So the relations among these different VC properties are of interest here. Recall (Section 4.8) that $D^{(p)}(\varepsilon, \mathcal{F}, Q)$ is the largest m such that for some $f_1, \ldots, f_m \in \mathcal{F}$, $\int |f_i - f_j|^p dQ > \varepsilon^p$ for all $i \neq j$. Also, $D^{(p)}(\varepsilon, \mathcal{F})$ is the supremum over all laws Q with finite support of $D^{(p)}(\varepsilon, \mathcal{F}, Q)$.

The following is a continuation of Theorem 4.53.

Proposition 10.12 *There exist uniformly bounded VC major (thus VC hull) classes which do not satisfy (4.12), thus are not VC subgraph classes.*

Proof. A uniformly bounded VC major class is VC hull by Theorem 4.51(b). Let $\mathcal{F}$ be the set of all right-continuous nonincreasing functions f on $\mathbb{R}$ with $0 \leq f \leq 1$. Then since the class $\mathcal{C}$ of open or closed half-lines $(-\infty, x)$ or $(-\infty, x]$ is a VC class (with $S(\mathcal{C}) = 1$), $\mathcal{F}$ is a VC major class. It is rather easy to see (Chapter 4, Problem 10) that $\mathcal{F}$ is not a VC subgraph class. The interest here is in showing that (4.12) fails.

For any f, g (in $\mathcal{F}$) and any law Q, $(\int |f - g|^2 dQ)^{1/2} \geq \int |f - g| dQ$. Thus

$$D^{(2)}(\varepsilon, \mathcal{F}, Q) \geq D^{(1)}(\varepsilon, \mathcal{F}, Q) \quad \text{for any } \varepsilon > 0.$$

Let P be Lebesgue measure on $[0, 1]$. Then $D^{(1)}(\varepsilon, \mathcal{F}, P) = D(\varepsilon, \mathcal{LL}_{2,1}, d_\lambda)$ where $\mathcal{LL}_{2,1}$ is the set of all lower layers (defined in Section 8.3) in the unit square I^2 in $\mathbb{R}^2$, and $d_\lambda(A, B) := \lambda(A \triangle B)$. For some $c > 0$, $D(\varepsilon, \mathcal{LL}_{2,1}, d_\lambda) \geq e^{c/\varepsilon}$ as $\varepsilon \downarrow 0$ by Theorem 8.22. For each $\varepsilon > 0$ small enough, by the law of large numbers, there is a law Q with finite support and $D^{(1)}(\varepsilon, \mathcal{F}, Q) \geq e^{c/\varepsilon} - 1$, so (4.12) fails and by Theorem 4.53, $\mathcal{F}$ is not a VC subgraph class.   □

Let $\mathcal{F}$ be a uniformly bounded class of measurable functions, so that some constant M with $0 < M < \infty$ is an envelope for $\mathcal{F}$. Limit-theorem properties of $\mathcal{F}$ are equivalent to those of $\mathcal{G} := \{f/M : f \in \mathcal{F}\}$, so we can assume that $M = 1$. Then Pollard's entropy condition as in Theorem 6.15 becomes

$$\int_0^1 (\log D^{(2)}(\varepsilon, \mathcal{F}))^{1/2} d\varepsilon < \infty. \tag{10.18}$$

Theorem 10.13 *If $\mathcal{F}$ is a uniformly bounded, image admissible Suslin class of measurable functions and satisfies (10.18), then $\mathcal{F}$ is a universal Donsker class.*

Proof. $\mathcal{F}$ has a finite constant C as an envelope function. For a constant envelope, the hypotheses of Theorem 6.15 do not depend on the law P, so Theorem 10.13 is a corollary of Theorem 6.15. Here are some more details. Let $\mathcal{F}/C := \{f/C : f \in \mathcal{F}\}$, so that $\mathcal{F}/C$ has as an envelope the constant 1. It will be enough to show that $\mathcal{F}/C$ is a universal Donsker class. Then for $\delta > 0$, $D^{(2)}(\delta, \mathcal{F}/C) = D_1^{(2)}(\delta, \mathcal{F}/C)$ as in Theorem 6.15. Make the substitution $\delta = \varepsilon/C$ and note that $D^{(2)}(\varepsilon/C, \mathcal{F}/C) \equiv D^{(2)}(\varepsilon, \mathcal{F})$. It follows that $\mathcal{F}/C$ satisfies (10.18). So Theorem 6.15 applies, and $\mathcal{F}/C$ and $\mathcal{F}$ are universal Donsker classes.   □

Corollary 10.14 *If a VC subgraph class $\mathcal{F}$ of measurable functions is uniformly bounded and image admissible Suslin, then it is a universal Donsker class.*

Proof. This follows from Theorems 4.53 and 10.13.   □

Specializing further, the set of indicators of an image admissible Suslin VC class of sets is a universal Donsker class (Corollary 6.19 for $F = 1$).

For a class $\mathcal{F}$ of real-valued functions on a set X, recall from Section 4.7 the class $H(\mathcal{F}, M)$ which is M times the symmetric convex hull of $\mathcal{F}$, and $\overline{H}_s(\mathcal{F}, M)$ which is the closure of $H(\mathcal{F}, M)$ for sequential pointwise convergence. Note that for any uniformly bounded class $\mathcal{F}$ of measurable functions for a σ-algebra $\mathcal{A}$ and any law Q defined on $\mathcal{A}$, $H(\mathcal{F}, M)$ is dense in $\overline{H}_s(\mathcal{F}, M)$ for the $L^2(Q)$ distance (or any $L^p(Q)$ distance, $1 \leq p < \infty$).

Theorem 10.15 *If $\mathcal{F}$ is a universal Donsker class of measurable real-valued functions on a measurable space $(X, \mathcal{A})$, then for any $M < \infty$, $\overline{H}_s(\mathcal{F}, M)$ is a universal Donsker class.*

Proof. Let $\mathcal{G} := \{f - \inf f : f \in \mathcal{F}\}$, as in the Remark after Proposition 10.11. Then functions in $H(\mathcal{F}, M)$ differ from functions in $H(\mathcal{G}, M)$ by additive constants. If $h_k \in \overline{H}_s(\mathcal{F}, M)$, $h_k \to h$ pointwise, and for all k, $h_k = \phi_k + c_k$ for some $\phi_k \in \overline{H}_s(\mathcal{G}, M)$ and constants c_k, then since ϕ_k are uniformly bounded and h has finite values, c_k are bounded. So, taking a subsequence, we can assume c_k converges to some c. Then ϕ_k converge pointwise to some $\phi \in \overline{H}_s(\mathcal{G}, M)$ and $h = \phi + c$. So all functions in $\overline{H}_s(\mathcal{F}, M)$ differ by additive constants from functions in $\overline{H}_s(\mathcal{G}, M)$ (the converse may not hold). Thus, if $\overline{H}_s(\mathcal{G}, M)$ is a universal Donsker class, so is $\overline{H}_s(\mathcal{F}, M)$ by Proposition 10.11(b). So we can assume $\mathcal{F}$ is uniformly bounded. Then it has an envelope function in $L^2(P)$ for all P, so by Theorem 3.41, $\overline{H}_s(\mathcal{F}, M)$ is a universal Donsker class. $\qquad\square$

Remark. We already saw in Theorem 6.21 that if $\mathcal{F}$ is a uniformly bounded VC major class $\mathcal{F}$ for a VC class $\mathcal{C}$ of sets, such that $\mathcal{C}$ is image admisssible Suslin, then $\mathcal{F}$ is a universal Donsker class.

By Theorem 10.13 above, for any $\delta > 0$, if $\log D^{(2)}(\varepsilon, \mathcal{F}) = O(1/\varepsilon^{2-\delta})$ as $\varepsilon \downarrow 0$, and if $\mathcal{F}$ is image admissible Suslin, then $\mathcal{F}$ is a universal Donsker class. In the converse direction we have:

Theorem 10.16 *For a uniformly bounded class $\mathcal{F}$ to be a universal Donsker class it is necessary that*

$$\log D^{(2)}(\varepsilon, \mathcal{F}) = O(\varepsilon^{-2}) \quad as \ \varepsilon \downarrow 0.$$

Proof. Suppose not. Then there are a universal Donsker class $\mathcal{F}$ and $\varepsilon_k \downarrow 0$ such that $\log D^{(2)}(\varepsilon_k, \mathcal{F}) > k^3/\varepsilon_k^2$ for $k = 1, 2, \ldots$, so there are probability laws P_k with finite support for which $\log D^{(2)}(\varepsilon, \mathcal{F}, P_k) > k^3/\varepsilon_k^2$ for $k = 2, 3, \ldots$. Let P be a law with $P \ge \sum_{k=2}^{\infty} P_k/k^2$. Then for any measurable f and g,

$$(\textstyle\int (f - g)^2 dP)^{1/2} \ge (\int (f - g)^2 dP_k)^{1/2}/k.$$

Let $\delta_k := \varepsilon_k/k$. Then

$$\log D^{(2)}(\delta_k, \mathcal{F}, P) \ge \log D^{(2)}(\varepsilon_k, \mathcal{F}, P_k) > k^3/\varepsilon_k^2 = k/\delta_k^2.$$

So any isonormal process L on $L^2(P)$ is a.s. unbounded on $\mathcal{F}$ by Theorem 2.14. We can write $L(f) \equiv G_P(f) + G \int f\, dP$ where G is a standard normal variable independent of G_P. Since $\mathcal{F}$ is uniformly bounded, G_P is a.s. unbounded on (a countable ρ_P-dense subset of) $\mathcal{F}$, so $\mathcal{F}$ is not P-pregaussian, and so not a P-Donsker class. $\qquad\square$

Theorem 10.16 is optimal, as the following shows:

Proposition 10.17 *There exists a universal Donsker class $\mathcal{E}$ such that*

$$\liminf_{\delta \downarrow 0} \delta^2 \log D^{(2)}(\delta, \mathcal{E}) > 0.$$

Proof. Let $A_j := A(j)$ be disjoint, nonempty measurable sets for $j = 1, 2, \dots$. Let $\| \cdot \|_2$ be the ℓ^2 norm, $\|x\|_2 = (\sum_j x_j^2)^{1/2}$ for $x = \{x_j\}_{j=1}^\infty$. Let

$$\mathcal{E} := \left\{ \sum_j x_j 1_{A(j)} : \|x\|_2 \le 1 \right\}.$$

(So $\mathcal{E}$ is an ellipsoid with center 0 and semiaxes $1_{A(j)}$.) Let P be any probability measure defined on the A_j and let $p_j := P(A_j)$ for $j = 1, 2, \dots$. We can assume that $p_j > 0$ for all j, since if B is the union of all A_j such that $p_j = 0$, then $P(B) = 0$, $G_P(B) = 0$ a.s. and $\nu_n(B) = 0$ a.s. for all n.

Let $\varepsilon > 0$. For any k and n, let $\|\nu_n\|_{2,k} := (\sum_{j=k}^\infty \nu_n(A_j)^2)^{1/2}$. Then for all n,

$$E\|\nu_n\|_{2,k}^2 \le \sum_{j=k}^\infty p_j \to 0 \text{ as } k \to \infty.$$

Take $k = k(\varepsilon)$ large enough so that $\sum_{j=k}^\infty p_j < \varepsilon^3/18$. Then

$$\Pr\{\|\nu_n\|_{2,k} > \varepsilon/3\} < \varepsilon/2. \tag{10.19}$$

If $\|\nu_n\|_{2,k} \le \varepsilon/3$, $\|x\|_2 \le 1$ and $\|y\|_2 \le 1$, then by the Cauchy (–Schwarz) inequality,

$$\left| \nu_n \left(\sum_{j=k}^\infty (x_j - y_j) 1_{A(j)} \right) \right| \le 2\varepsilon/3. \tag{10.20}$$

Also, $E\|\nu_n\|_{2,1} \le (E\|\nu_n\|_{2,1}^2)^{1/2} \le 1$, so

$$\Pr\{\|\nu_n\|_{2,1} > 2/\varepsilon\} < \varepsilon/2. \tag{10.21}$$

Let $\delta := (\min_{j<k} p_j)^{1/2} \varepsilon^2/6 > 0$. Let $f_x := \sum_{j=1}^\infty x_j 1_{A(j)}$ for each x with $\|x\|_2 \le 1$, so $f_x \in \mathcal{E}$. If f_x and $f_y \in \mathcal{E}$ and $e_P(f_x, f_y) := (\int (f_x - f_y)^2 dP)^{1/2} < \delta$, then

$$\left(\sum_{j<k} (x_j - y_j)^2 \right)^{1/2} < \varepsilon^2/6. \tag{10.22}$$

By (10.21) and Cauchy's inequality, $|\sum_{j<k}(x_j - y_j)\nu_n(A_j)| < \varepsilon/3$ for all x and y such that $e_P(f_x, f_y) < \delta$, except on an event with probability $< \varepsilon/2$. Thus by (10.19) and (10.20), there is an event F with $\Pr(F) < \varepsilon$ such that if $\omega \notin F$ then

$$\text{for all } x \text{ and } y \text{ with } e_P(f_x, f_y) < \delta, \ |\nu_n(f_x - f_y)| < \varepsilon. \tag{10.23}$$

Given x with $\sum_j x_j^2 \le 1$, let $y_j = x_j$ for $j \le k$ and $y_j = 0$ for $j > k$. Then $(\int (f_x - f_y)^2 dP)^{1/2} < \varepsilon/4$. Since $x \mapsto f_x$ is continuous from ℓ^2 into $L^2(P)$, the set of all f_x such that $\sum_j x_j^2 \le 1$ and $x_j = 0$ for $j > k$ is compact in $L^2(P)$.

It follows that $\mathcal{E}$ is totally bounded in $L^2(P)$. Thus by (10.23) and Theorem 3.34 for $\tau = e_P$, $\mathcal{E}$ is a Donsker class for P. Since P was arbitrary, $\mathcal{E}$ is a universal Donsker class.

Next, given $0 < \delta < 1/2$, let $m := [1/(4\delta^2)]$, where $[x]$ is the largest integer $\leq x$. Let P be a probability measure with $P(A_j) = 1/m$ for $j = 1, \ldots, m$. Then in $L^2(P)$, $\mathcal{E}$ is an m-dimensional ball with radius $m^{-1/2}$. Let $r := D^{(2)}(\delta, \mathcal{E}, P)$ and let $g_1, \ldots, g_r \in \mathcal{E}$ with $e_P(g_i, g_j) > \delta$ for $1 \leq i < j \leq r$. Let $B_i := \{f_x : e_P(f_x, g_i) \leq 2\delta\}$. Then

$$\bigcup_{i=1}^{r} B_i \supset \left\{ f_x : \left(\sum_{j=1}^{m} x_j^2/m \right)^{1/2} \leq m^{-1/2} + \delta, \ x_j = 0 \text{ for } j > m \right\}.$$

Thus by comparing volumes of balls in $\mathbb{R}^m$, we get by choice of m,

$$D^{(2)}(\delta, \mathcal{E}, P) = r \geq (m^{-1/2} + \delta)^m/(2\delta)^m = \left(\frac{1}{2}[1 + 1/(\delta m^{1/2})] \right)^m \geq (3/2)^m$$

$$\geq \frac{2}{3} \exp \left(\log \left(\frac{3}{2} \right)/(4\delta^2) \right).$$

Letting $\delta \downarrow 0$, Proposition 10.17 follows. $\qquad\square$

Proposition 10.18 *There is a uniformly bounded class $\mathcal{F}$ of measurable functions, which is not a universal Donsker class, such that*

$$\log D^{(2)}(\varepsilon, \mathcal{F}) \leq \frac{2}{\varepsilon^2 \log(1/\varepsilon)} \quad \text{as } \varepsilon \downarrow 0.$$

Proof. Let $B_j := B(j)$ be disjoint nonempty measurable sets. Recall that $Lx := \max(1, \log x)$. Let $\alpha_j := 1/(jLj)^{1/2}$, $j \geq 1$, and

$$\mathcal{F} = \left\{ \sum_{j=1}^{\infty} x_j 1_{B(j)} : x_j = \pm\alpha_j \text{ for all } j \right\}.$$

Take c such that $\sum_{j=1}^{\infty} p_j = 1$, where $p_j := c(\alpha_j/LLj)^2$. Here $\sum_j p_j < \infty$ by the integral test since

$$(d/dx)(1/LLx) = -1/(xLx(LLx)^2) \text{ for } x > e^e.$$

Take a probability measure P with $P(B_j) = p_j$ for all j. Let $\alpha := (1 - p_1)^{1/2}/2 > 0$. Then

$$E\|G_P\|_{\mathcal{F}} = \sum_{j=1}^{\infty} \alpha_j E|G_P(B_j)| = \sum_{j=1}^{\infty} \alpha_j (2/\pi)^{1/2}(p_j(1 - p_j))^{1/2}$$

$$\geq \alpha \sum_{j=1}^{\infty} \alpha_j p_j^{1/2} \geq \alpha c^{1/2} \sum_{j=1}^{\infty} 1/(jLjLLj) = +\infty$$

by the integral test since $(d/dx)(LLLx) = 1/(xLxLLx)$ for x large enough. If $\mathcal{F}$ were P-pregaussian, then since G_P can be treated as an isonormal process (see Section 3.1), by Theorem 2.32 ((a) if and only if (h)) and the material just before it, G_P could be realized on a separable Banach space with norm $\|\cdot\|_{\mathcal{F}}$. Then the norm would have finite expectation by the Landau–Shepp–Marcus–Fernique theorem (Theorem 2.6 above), a contradiction. So $\mathcal{F}$ is not pregaussian for P and so not a universal Donsker class.

For any probability measure Q, $r = 1, 2, \ldots,$

$$f = \sum_j x_j 1_{B(j)} \in \mathcal{F} \text{ and } g = \sum_j y_j 1_{B(j)} \in \mathcal{F},$$

if $x_j = y_j$ for $1 \le j < r$, then

$$e_Q(f, g) = \left(\int \left(\sum_{j=r}^{\infty} (x_j - y_j) 1_{B(j)} \right)^2 dQ \right)^{1/2} \le \left(\sum_{j=r}^{\infty} 4\alpha_j^2 Q(B_j) \right)^{1/2} \le 2\alpha_r.$$

$$(10.24)$$

Given $\varepsilon > 0$, let $r(\varepsilon)$ be the smallest integer $r \ge 1$ such that $\alpha_r < \varepsilon/2$. By (10.24), if $x_j = y_j$ for $1 \le j < r$, then $e_P(f, g) < \varepsilon$. Thus since there are only 2^{r-1} possibilities for $x_j = \pm\alpha_j$, $j < r$, we have $D^{(2)}(\varepsilon, \mathcal{F}, Q) \le 2^{r-1}$, and since r does not depend on Q, $D^{(2)}(\varepsilon, \mathcal{F}) \le 2^{r-1}$ and $\log D^{(2)}(\varepsilon, \mathcal{F}) < r \log 2$. As $\varepsilon \downarrow 0$, we have $\alpha_{r(\varepsilon)} \sim \varepsilon/2$, so

$$\log(1/\varepsilon) \sim \log(2/\varepsilon) \sim \log(1/\alpha_{r(\varepsilon)}) \sim \frac{1}{2} \log(r(\varepsilon)),$$

and $\varepsilon/2 \sim 1/(r(\varepsilon) \cdot 2 \log(1/\varepsilon))^{1/2}$, so $r(\varepsilon) \sim 2/(\varepsilon^2 \log(1/\varepsilon))$. Since

$$\log D^{(2)}(\varepsilon, \mathcal{F}) \le r(\varepsilon) \log 2 < r(\varepsilon),$$

the conclusion follows. $\square$

Theorems 10.13 and 10.16 show that the condition (10.18) comes close to characterizing the universal Donsker property, but Propositions 10.17 and 10.18 show that there is no characterization of the universal Donsker property in terms of $D^{(2)}$.

10.3 Metric Entropy of Convex Hulls in Hilbert Space

Let H be a real Hilbert space and for any subset B of H let $co(B)$ be its convex hull,

$$co(B) := \left\{ \sum_{j=1}^{k} t_j x_j : t_j \ge 0, \sum_{j=1}^{k} t_j = 1, x_j \in B, k = 1, 2, \ldots \right\}.$$

Recall that $D(\varepsilon, B)$ is the maximum number of points in B more than ε apart.

Theorem 10.19 *Suppose that B is an infinite subset of a Hilbert space H, $\|x\| \le 1$ for all $x \in B$ and that for some $K < \infty$ and $0 < \gamma < \infty$, we have $D(\varepsilon, B) \le K\varepsilon^{-\gamma}$ for $0 < \varepsilon \le 1$. Let $s := 2\gamma/(2+\gamma)$. Then for any $t > s$, there are constants C_1 and C_2, which depend only on K, γ and t, such that*

$$D(\varepsilon, co(B)) \le C_1 \exp(C_2 \varepsilon^{-t}) \text{ for } 0 < \varepsilon \le 1.$$

Note. Both van der Vaart and Wellner (1996, Theorem 2.6.9) and Carl (1997) give the sharper bound with $t = s$.

Proof. We may assume $K \ge 1$. Choose any $x_1 \in B$. Let $n \ge 2$ and suppose given $B(n) := \{x_1, \ldots, x_{n-1}\}$. Let $d(x, B(n)) := \min_{y \in B(n)} \|x - y\|$ and $\delta_n := \sup_{x \in B} d(x, B(n))$. Since B is infinite, $\delta_n > 0$ for all n. Choose $x_n \in B$ with $d(x_n, B(n)) > \delta_n/2$. Then for all n, $K(2/\delta_n)^\gamma \ge D(\delta_n/2, B) \ge n$, so $\delta_n \le Mn^{-1/\gamma}$ for all n where $M := 2K^{1/\gamma}$.

Let $0 < \varepsilon \le 1$. Let $N := N(\varepsilon)$ be the next integer larger than $(4M/\varepsilon)^\gamma$. Then $\delta_N < \varepsilon/4$. Let $G := B(N)$. For each $x \in B$ there is an $i = i(x) \le N - 1$ with $\|x - x_i\| \le \delta_N$. For any convex combination $z = \sum_{x \in B} z_x x$ where $z_x \ge 0$ and $\sum_{x \in B} z_x = 1$, with $z_x = 0$ except for finitely many x, let $z_N := \sum_{x \in B} z_x x_{i(x)}$. Then $\|z - z_N\| \le \delta_N < \varepsilon/4$, so

$$D(\varepsilon, co(B)) \le D(\varepsilon/2, co(G)). \tag{10.25}$$

To bound $D(\varepsilon/2, co(G))$, let $m := m(\varepsilon)$ be the largest integer $\le \varepsilon^{-s}$. Note that $\gamma > s$. Then for each i with $m < i \le N$, there is a $j \le m$ such that

$$\|x_i - x_j\| \le \delta_{m+1} \le M\varepsilon^{s/\gamma}. \tag{10.26}$$

Let $j(i)$ be the least such j. Let $\Lambda_m := \{\{\lambda_j\}_{1 \le j \le m} : \lambda_j \ge 0, \sum_{j=1}^{m} \lambda_j = 1\}$. On $\mathbb{R}^m$ we have the ℓ_p metrics

$$\rho_p(\{x_j\}, \{y_j\}) := \left(\sum_{j=1}^{m} |x_j - y_j|^p \right)^{1/p}.$$

By Cauchy's inequality, $\rho_1 \le m^{1/2}\rho_2$. Let $\beta := \varepsilon/6$ and $\delta := \beta/(2m^{1/2})$. The δ-neighborhood of Λ_m for ρ_2 is included in a ball of radius $1 + \delta < 13/12$. We have $D(2\delta, \Lambda_m, \rho_2)$ centers of disjoint balls of radius δ included in the neighborhood. Comparing volumes of balls and recalling that $Lx := \max(1, \log x)$ gives

$$D(\beta, \Lambda_m, \rho_1) \le D(2\delta, \Lambda_m, \rho_2) \le 13^m m^{m/2} \varepsilon^{-m}$$

$$\le \exp(m\{L(1/\varepsilon) + (Lm)/2 + \log(13)\}) \tag{10.27}$$

$$\le \exp(C_3 \varepsilon^{-s} L(1/\varepsilon)) \le \exp(C_4 \varepsilon^{-t}), \quad 0 < \varepsilon \le 1,$$

for some constants C_3, C_4.

For each $j = 1, \ldots, m$, let $A_j := A(j)$ consist of x_j and the set of all x_i, $i = m+1, \ldots, N$, such that $j(i) = j$. Then $G = \cup_{j=1}^m A_j$ and the A_j are disjoint. Take a maximal set $S = S(\varepsilon) \subset \Lambda_m$ with $\rho_1(u, v) > \beta$ for any $u \neq v$ in S. For a given $\lambda = (\lambda_1, \ldots, \lambda_m) \in S$, let

$$F_\lambda := \{x \in \mathrm{co}(G) : x = \sum_{y \in G} \mu_{y,x} y, \quad \text{where } \mu_{y,x} \geq 0 \text{ and}$$

$$\sum_{y \in A(j)} \mu_{y,x} = \lambda_j, \quad \text{for all } j = 1, \ldots, m\}.$$

For any $x \in \mathrm{co}(G)$, let $x = \sum_{y \in G} \mu_{y,x} y$ where $\mu_{y,x} \geq 0$ and for $\tau_j(x) := \sum_{y \in A(j)} \mu_{y,x}$, $\tau(x) := \{\tau_j(x)\}_{j=1}^m \in \Lambda_m$. Take $\lambda(x) \in S$ such that $\sum_j |\lambda_j(x) - \tau_j(x)| \leq \beta$. For $y \in A(j)$, define $\nu_{y,x} := \lambda_j(x)\mu_{y,x}/\tau_j(x)$ if $\tau_j(x) > 0$. If $\tau_j(x) = 0$, choose a $w \in A(j)$ and define $\nu_{w,x} := \lambda_j(x)$ and $\nu_{y,x} := 0$ for $y \neq w$, $y \in A(j)$. If $\tau_j(x) > 0$, then

$$\sum_{y \in A(j)} |\mu_{y,x} - \nu_{y,x}| = \sum_{y \in A(j)} \mu_{y,x}|1 - \lambda_j(x)/\tau_j(x)| = |\tau_j(x) - \lambda_j(x)|.$$

If $\tau_j(x) = 0$, then $\sum_{y \in A(j)} |\mu_{y,x} - \nu_{y,x}| = \nu_{w,x} = \lambda_j(x) = |\tau_j(x) - \lambda_j(x)|$. It follows that $\sum_{y \in G} |\mu_{y,x} - \nu_{y,x}| = \sum_j |\tau_j(x) - \lambda_j(x)| \leq \beta$. Set $z_x := \sum_{y \in G} \nu_{y,x} y$. Then $z_x \in F_{\lambda(x)}$, $\lambda(x) \in S$, and $\|z_x - x\| \leq \beta$. Thus by (10.25) and (10.27),

$$D(\varepsilon, \mathrm{co}(B)) \leq D(\varepsilon/2, \mathrm{co}(G)) \leq D(\beta, \bigcup_{\lambda \in S} F_\lambda)$$

$$\leq \mathrm{card}(S) \max_\lambda D(\beta, F_\lambda) \leq \exp(C_4 \varepsilon^{-t}) \max_\lambda D(\beta, F_\lambda). \tag{10.28}$$

To estimate the latter factor, let $\lambda \in S$. We may assume $\lambda_j > 0$ for all j. For any $j = 1, \ldots, m$ and $x \in F_\lambda$, let $x^{(j)} := \sum_{y \in A(j)} \mu_{y,x} y$. Let Y_j be a random variable with values in $A(j)$ and $P(Y_j = y) = \mu_{y,x}/\lambda_j$ for each $y \in A(j)$. Then $EY_j = x^{(j)}/\lambda_j =: z_j$. Take $Y_1, \ldots, Y_m$ to be independent and let $Y := \sum_{j=1}^m \lambda_j Y_j$. Then $EY = x$ and

$$E\|Y - x\|^2 = E\|\sum_{j=1}^m \lambda_j(Y_j - z_j)\|^2 = \sum_{j=1}^m \lambda_j^2 E\|Y_j - z_j\|^2,$$

since $Y_j - z_j$ are independent and have mean 0, and H is a Hilbert space.

Now the diameter of $A(j)$ is at most $2M\varepsilon^{s/\gamma}$ by (10.26), and z_j is a convex combination of elements of $A(j)$. Thus

$$E\|Y_j - z_j\|^2 = \lambda_j^{-1} \sum_{y \in A(j)} \mu_{y,x} \left\| \lambda_j^{-1} \sum_{z \in A(j)} \mu_{z,x}(y - z) \right\|^2 \leq 4M^2 \varepsilon^{2s/\gamma},$$

and for any set $F \subset \{1, \ldots, m\}$,

$$E \left\| \sum_{j \in F} \lambda_j (Y_j - z_j) \right\|^2 \leq \sum_{j \in F} \lambda_j^2 4M^2 \varepsilon^{2s/\gamma} \leq 4M^2 (\max_{j \in F} \lambda_j) \varepsilon^{2s/\gamma}.$$

Next, an idea of B. Maurey will be applied. For $k = 1, 2, \ldots$, let $Y_{j1}, Y_{j2}, \ldots, Y_{jk}$ be independent with the distribution of Y_j, and with Y_{ji} also independent for different j. Then

$$E \left\| \sum_{j \in F} \lambda_j (k^{-1} \sum_{i=1}^{k} (Y_{ji} - z_j)) \right\|^2 \leq 4M^2 (\max_{j \in F} \lambda_j) \varepsilon^{2s/\gamma} / k,$$

so

$$E \left\| \sum_{j \in F} \lambda_j ((k^{-1} \sum_{i=1}^{k} Y_{ji}) - z_j) \right\| \leq 2M (\max_{j \in F} \lambda_j)^{1/2} \varepsilon^{s/\gamma} / k^{1/2}.$$

Thus, there exist $y_{ji} \in A(j)$, $i = 1, \ldots, k$, $j \in F$, such that

$$\left\| \sum_{j \in F} \lambda_j ((k^{-1} \sum_{i=1}^{k} y_{ji}) - z_j) \right\| \leq 2M (\max_{j \in F} \lambda_j)^{1/2} \varepsilon^{s/\gamma} / k^{1/2}. \quad (10.29)$$

Take $v > 0$ such that $s + v < t$. Let $F(0) := \{j \leq m : \lambda_j \geq \varepsilon^v\}$. Let $k(0)$ be the smallest integer k such that

$$k \geq 6400 M^2 \varepsilon^{-2+2s/\gamma} = 6400 M^2 \varepsilon^{-s}.$$

For $k \geq k(0)$ and $F = F(0)$ the expressions in (10.29) are at most $\varepsilon/4 > 0$.

Let r be the smallest positive integer such that $\varepsilon^v / 4^r \leq (\varepsilon^{1-s/\gamma}/(80M))^2$. For $u = 1, 2, \ldots, r$, let

$$F(u) := \{j \leq m : \varepsilon^v / 4^u \leq \lambda_j < \varepsilon^v / 4^{u-1}\},$$

and let $k = k(u)$ be the smallest integer k such that $2^{2-u} M \varepsilon^{s/\gamma} / k^{1/2} < \varepsilon/(40r)$, i.e., $k > 100 M^2 4^{4-u} r^2 \varepsilon^{-2+2s/\gamma}$. Thus, for some constant C_5, $k(u) \leq 1 + C_5 4^{-u} \varepsilon^{-s} (L(1/\varepsilon))^2$. The y_{ji} for $F = F(u)$ will be called $y_{ji}^{(u)}$ (they also depend on x).

Let $F(r + 1) := \{j \leq m : \lambda_j < \varepsilon^v / 4^r\}$. Let $k(r + 1) := 1$. For $F = F(r + 1)$ and $k = k(r + 1)$, (10.29) is bounded above by $\varepsilon/40$. We have $y_{j1}^{(r+1)}$, a single choice for each j. Let

$$\zeta := \sum_{u=0}^{r+1} \sum_{j \in F(u)} \lambda_j \cdot \frac{1}{k(u)} \sum_{i=1}^{k(u)} y_{ji}^{(u)}.$$

Then by (10.29) and the results for $u = 0$ and $u = r + 1$,

$$\| \zeta - x \| = \left\| \sum_{u=0}^{r+1} \sum_{j \in F(u)} \lambda_j \cdot \frac{1}{k(u)} \sum_{i=1}^{k(u)} (y_{ji}^{(u)} - z_j) \right\|$$

$$\leq \frac{\varepsilon}{40} + \frac{\varepsilon}{40} + \sum_{u=1}^{r} 2M(\varepsilon^v/4^{u-1})^{1/2} \varepsilon^{s/\gamma} / k(u)^{1/2}$$

$$\leq \frac{\varepsilon}{20} + r \cdot \frac{\varepsilon}{40r} = \frac{3\varepsilon}{40} < \frac{\varepsilon}{12}.$$

Here ζ is determined uniquely by the $k(u)$-tuples $(y_{j1}^{(u)}, \ldots, y_{jk(u)}^{(u)})$, $j \in F(u)$, $u = 0, 1, \ldots, r + 1$. Each $A(j)$ has at most N elements, so that for given $u \leq r$ and $j \leq m$, there are at most $N^{k(u)}$ ways of choosing the $y_{ji}^{(u)}$. Now $\mathrm{card}(F(u)) \leq 4^u/\varepsilon^v$, so the number of ways to choose the $y_{ij}^{(u)}$ for given u with $1 \leq u \leq r$ is at most $\exp\{(\log N)(4^u \varepsilon^{-v} + C_5 \varepsilon^{-s-v} L(1/\varepsilon)^2)\}$. There are at most

$$\exp((\log N)[\varepsilon^{-s-v} 6400M^2 + \varepsilon^{-v}])$$

ways to choose the $y_{ij}^{(0)}$. Thus, the total number of ways to choose all the $y_{ij}^{(u)}$ gives

$$D(\varepsilon/6, F_\lambda) \leq \exp(C_6 \{\varepsilon^{-s-v} L(1/\varepsilon)^4 + L(1/\varepsilon)4^r/\varepsilon^v\})$$

for some $C_6 := C_6(K)$. By definition of r, $4^r/\varepsilon^v \leq C_7 \varepsilon^{2s/\gamma-2} = C_7 \varepsilon^{-s}$ for some $C_7 := C_7(K)$. Thus, $D(\varepsilon/6, F_\lambda) \leq \exp(C_8 \varepsilon^{-t})$ for some C_8. Combining with (10.28) completes the proof of Theorem 10.19. $\quad\square$

Example. The exponent $2\gamma/(2+\gamma)$ in Theorem 10.19 is sharp in the following example. Let $\{e_n\}_{n \geq 1}$ be an orthonormal basis of H, and for $0 < \gamma < \infty$ let

$$B := \{n^{-1/\gamma} e_n\}_{n \geq 1} \cup \{-n^{-1/\gamma} e_n\}_{n \geq 1}.$$

For any $\varepsilon > 0$ small enough we have $D(\varepsilon, B) = 2n$ for the least n such that $(n^{-2/\gamma} + (n+1)^{-2/\gamma})^{1/2} \leq \varepsilon$: the points $\pm j^{-1/\gamma} e_j$, $1 \leq j \leq n$, are more than ε apart, so $D(\varepsilon, B) \geq 2n$, while a set of points of B more than ε apart cannot contain $j^{-1/\gamma} e_j$ or $-j^{-1/\gamma} e_j$ for more than one value of $j \geq n$, so $D(\varepsilon, B) \leq 2n$. Thus as $\varepsilon \to 0$, $\varepsilon^2 \sim 2/n^{2/\gamma}$, so for a constant $C = C_\gamma$, $2n \sim C/\varepsilon^\gamma$ and replacing C by a suitable larger K if necessary, the hypothesis of Theorem 10.19 holds.

Let B_n be the intersection of B with the linear span of $e_1, \ldots, e_n$. Let $C_n = \mathrm{co}(B_n)$. Then for any $\varepsilon > 0$, $D(\varepsilon, \mathrm{co}(B)) \geq D(\varepsilon, C_n)$. The n-dimensional volume $v_n(C_n)$ is

$$v_n(C_n) = (n!)^{-1} 2^n \Pi_{j=1}^n j^{-1/\gamma} = 2^n/n!^{1+1/\gamma}.$$

A ball $B(x, r) := \{y : |y - x| \le r\}$ in $\mathbb{R}^n$ has volume $v_n(B(x, r)) = c_n r^n$ where $c_n = \pi^{n/2}/\Gamma((n + 2)/2)$. (This is well known, especially for $n = 1, 2, 3$ since $\Gamma(1/2) = \pi^{1/2}$, and then can be proved by induction from n to $n + 2$). If $x_1, \ldots, x_m$ are points of C_n more than ε apart, for a maximal $m = D(\varepsilon, C_n)$, then the sets $B(x_i, \varepsilon)$ cover C_n, so $m c_n \varepsilon^n \ge v_n(C_n)$. By Stirling's formula (Theorem 1.17), as $n \to \infty$

$$v_n(C_n)/c_n \sim 2^n (e/n)^{n+n/\gamma} (2\pi n)^{-(\gamma+1)/(2\gamma)} \pi^{-n/2} (n/(2e))^{n/2} (\pi n)^{1/2}$$

$$= (e/n)^{n(2+\gamma)/(2\gamma)} (2/\pi)^{n/2} n^{-1/(2\gamma)} D_\gamma$$

for a constant D_γ.

Take any d such that $d > \frac{1}{2} + \frac{1}{\gamma}$. Then for n large enough $v_n(C_n)/c_n \ge n^{-dn}$. Let $g_n(\varepsilon) := n^{-dn}/\varepsilon^n$. Then $m \ge g_n(\varepsilon)$.

The following paragraph is only for motivation. As $\varepsilon \downarrow 0$, $n = n(\varepsilon)$ will be chosen to make $g_n(\varepsilon)$ about as large as possible. We have

$$g_{n+1}(\varepsilon)/g_n(\varepsilon) = \varepsilon^{-1}(n + 1)^{-d(n+1)} n^{dn} \sim \frac{1}{\varepsilon} n^{-d} \left(1 + \frac{1}{n}\right)^{-dn} \sim \frac{1}{\varepsilon}(ne)^{-d}.$$

This sequence is decreasing in n, so to maximize $g_n(\varepsilon)$ for a given ε we want to take n such that the ratio is approximately 1.

At any rate, let $n(\varepsilon)$ be the largest integer $\le f(\varepsilon) := e^{-1}\varepsilon^{-1/d}$. Then for ε small enough

$$g_{n(\varepsilon)}(\varepsilon) \ge f(\varepsilon)^{-df(\varepsilon)} \varepsilon^{1-f(\varepsilon)} = \varepsilon \exp(df(\varepsilon)) = \varepsilon \exp\left(\frac{d}{e}\left(\frac{1}{\varepsilon}\right)^{1/d}\right).$$

Now $\varepsilon \ge \exp(-\varepsilon^{-\delta})$ as $\varepsilon \downarrow 0$ for any $\delta > 0$. Taking $\delta < 1/d$ and letting $d \downarrow (\gamma + 2)/(2\gamma)$ we have $1/d \uparrow (2\gamma)/(2 + \gamma)$, showing that the exponent in Theorem 10.19 is indeed best possible. Recall the definitions of $D^{(2)}$ from Section 4.8 and $\overline{H}_s$ from after Corollary 10.14.

Corollary 10.20 *If $\mathcal{G}$ is a uniformly bounded class of measurable functions and for some $K < \infty$ and $0 < \gamma < \infty$, $D^{(2)}(\varepsilon, \mathcal{G}) \le K\varepsilon^{-\gamma}$ for $0 < \varepsilon < 1$, then for any $t > r := 2\gamma/(2 + \gamma)$, and for the constants $C_i = C_i(2K, \gamma, t)$, $i = 1, 2$, of Theorem 10.19,*

$$D^{(2)}(\varepsilon, \overline{H}_s(\mathcal{G}, 1)) \le C_1 \exp(C_2 \varepsilon^{-t}) \text{ for } 0 < \varepsilon < 1.$$

Proof. We have $D^{(2)}(\varepsilon, \mathcal{G} \cup -\mathcal{G}) \le 2K\varepsilon^{-\gamma}$, $0 < \varepsilon < 1$. Thus for any law Q with finite support, $D^{(2)}(\varepsilon, \mathcal{G} \cup -\mathcal{G}, Q) \le 2K\varepsilon^{-\gamma}$. By Theorem 10.19, $D^{(2)}(\varepsilon, H(\mathcal{G}, 1), Q) \le C_1 \exp(C_2 \varepsilon^{-t})$ for $C_i = C_i(2K, \gamma, t)$. It is easily seen that $H(\mathcal{G}, 1)$ is a dense subset of $\overline{H}_s(\mathcal{G}, 1)$ in $\mathcal{L}^2(Q)$. The conclusion follows. $\qquad\square$

Now, recall the notions of VC subgraph and VC subgraph hull class from Section 4.7.

Corollary 10.21 *If $\mathcal{G}$ is a uniformly bounded VC subgraph class and $M < \infty$, then the VC subgraph hull class $\overline{H}_s(\mathcal{G}, M)$ satisfies (10.18). Also, if $\mathcal{H}$ is a uniformly bounded VC major class, then $\mathcal{H}$ satisfies (10.18).*

Proof. First let $\mathcal{G}$ be VC subgraph. Let $M\mathcal{G} := \{Mg : g \in \mathcal{G}\}$. Then $M\mathcal{G}$ is a uniformly bounded VC subgraph class. By Theorem 4.53(a), $M\mathcal{G}$ satisfies the hypothesis of Corollary 10.20, so $r < 2$ and we can take $t < 2$, and the conclusion holds.

If $\mathcal{H}$ is VC major for a VC class $\mathcal{C}$ of sets, let $\mathcal{F} := \{1_C : C \in \mathcal{C}\}$. Then $\mathcal{F}$ is a VC subgraph class by Theorem 4.51(a) and $\mathcal{H}$ is VC hull, thus VC subgraph hull by Theorem 4.51(b), so the first part of the proof applies with $\mathcal{F}$ in place of $\mathcal{G}$.   □

Remark. It follows by Theorem 6.15 that for a uniformly bounded VC subgraph class $\mathcal{G}$, if $\mathcal{F} \subset \overline{H}_s(\mathcal{G}, M)$, in particular if $\mathcal{F}$ is VC major or (thus) VC hull, and $\mathcal{F}$ is image admissible Suslin, then $\mathcal{F}$ is a universal Donsker class. This also follows from Corollary 10.14 and Theorem 10.15.

Example. Let $\mathcal{C}$ be the set of all intervals $(a, b]$ for $0 \le a \le b \le 1$. Let G be the set of all real functions f on $[0, 1]$ such that $|f(x)| \le 1/2$ for all x, $|f(x) - f(y)| \le |x - y|$ for $0 < x, y < 1$, and $f(x) = 0$ for $x \le 0$ or $x \ge 1$. Each f in G has total variation at most 2 (at most 1 on the open interval $0 < x < 1$ and $1/2$ at each endpoint 0, 1). By the Jordan decomposition we have, for each $f \in G$, $f = g - h$ where g and h are both nondecreasing functions, 0 for $x \le 0$. Then g and h have equal total variations ≤ 1 and $G \subset \overline{H}_s(\mathcal{C}, 2)$ by the proof of Theorem 4.51(b). Let P be Lebesgue measure on $[0, 1]$. By Theorem 8.7, and since $(\int |f|^2 dP)^{1/2} \ge \int |f| dP$, there is a $c > 0$ such that $D^{(2)}(\varepsilon, G) \ge e^{c/\varepsilon}$ as $\varepsilon \downarrow 0$ (consider laws with finite support which approach P). Since $S(\mathcal{C}) = 2$, the exponent γ can be any number larger than 2 by Corollary 4.4 and Theorem 4.47. Letting $\gamma \downarrow 2$, t in Corollary 10.20 can be any number > 1, and we saw above that it cannot be < 1 in this case, so again the exponent is sharp.

10.4 Uniform Donsker Classes

A class $\mathcal{F}$ of measurable functions on a measurable space $(X, \mathcal{A})$ is a uniform Donsker class if it is a universal Donsker class and the convergence in law of ν_n to G_P is also uniform in P. Giné and Zinn (1991) gave a precise formulation of the uniformity in terms of the dual-bounded-Lipschitz distance β as defined just before Theorem 3.28, and gave a characterization of the so defined uniform Donsker property of $\mathcal{F}$, to be stated and proved in this section.

Let $(X, \mathcal{A})$ be a measurable space. Recall that $\mathcal{P}(X) = \mathcal{P}(X, \mathcal{A})$ is the set of all probability measures on $(X, \mathcal{A})$. Let $\mathcal{F}$ be a class of real-valued measurable functions on X. By Proposition 10.10, if $\mathcal{F}$ is universal Donsker, then $\mathcal{F}_0 := \{f - \inf f : f \in \mathcal{F}\}$ is uniformly bounded. By Proposition 10.11, $\mathcal{F}$ is Donsker for a given P if and only if $\mathcal{F}_0$ is, and the same holds for universal Donsker and uniform Glivenko–Cantelli. Once we define uniform Donsker, we will see that it also holds for uniform Donsker.

As in the definition of Donsker class (near the end of Section 3.1) let $\ell^\infty(\mathcal{F})$ be the set of all bounded real-valued functions on $\mathcal{F}$. For any two bounded functions G and H on $\mathcal{F}$, let $d(G, H) := d_{\mathcal{F}}(G, H) := \|G - H\|_{\mathcal{F}}$. In Theorem 3.28 on equivalence of convergence in distribution and convergence for the metric β or ρ, for functions f_m and f from a probability space Ω into $\ell^\infty(\mathcal{F})$, let the metric space S be $\ell^\infty(\mathcal{F})$ and the metric $d = d_{\mathcal{F}}$. Then β will be written as $\beta_{\mathcal{F}}$. Recall that $\beta_{\mathcal{F}}$ is not a metric in the usual sense, in that $\beta_{\mathcal{F}}(f_m, f_0)$ is defined only when f_0 has separable range and is measurable.

Let $\mathcal{P}_f(X)$ be the set of all laws P in $\mathcal{P}(X)$ such that $P(F) = 1$ for some finite F. In the (unusual) case that finite sets are not in the σ-algebra $\mathcal{A}$, we can express this by saying that for some numbers $c_x > 0$ for all $x \in F$ with $\sum_{x \in F} c_x = 1$, $P = \sum_{x \in F} c_x \delta_x$. Such a P is defined on all subsets of X and so in particular on any σ-algebra.

For a pseudometric d on a class $\mathcal{F}$, such as an $\mathcal{L}^2$ distance e_P or ρ_P, and $\delta > 0$, let

$$\mathcal{F}'(\delta, d) := \{f - g : f, g \in \mathcal{F}, \ d(f, g) \leq \delta\}. \tag{10.30}$$

Then for $d = e_P$ or ρ_P, $\mathcal{F}'(\delta, d)$ is image admissible Suslin by Corollary 5.18 and Theorem 5.26.

Definitions. A class $\mathcal{F}$ is *uniformly pregaussian (UPG)* if it is pregaussian for all $P \in \mathcal{P}(X)$, and if, for a coherent version of G_P for each P, we have both

$$\sup_{P \in \mathcal{P}(X)} E \|G_P\|_{\mathcal{F}} < \infty \tag{10.31}$$

and

$$\lim_{\delta \downarrow 0} \sup_{P \in \mathcal{P}(X)} E \|G_P\|_{\mathcal{F}'(\delta, \rho_P)} = 0. \tag{10.32}$$

The class $\mathcal{F}$ is *finitely uniformly pregaussian (UPG$_f$)* if the same holds with $\mathcal{P}_f(X)$ in place of $\mathcal{P}(X)$, namely

$$\sup_{P \in \mathcal{P}_f(X)} E \|G_P\|_{\mathcal{F}} < \infty \tag{10.33}$$

and

$$\lim_{\delta \downarrow 0} \sup_{P \in \mathcal{P}_f(X)} E \|G_P\|_{\mathcal{F}'(\delta, \rho_P)} = 0. \tag{10.34}$$

The class $\mathcal{F}$ is a *uniform Donsker class* if it is uniformly pregaussian and

$$\lim_{n \to \infty} \sup_{P \in \mathcal{P}(X)} \beta_{\mathcal{F}}(\nu_n, G_P) = 0 \tag{10.35}$$

where $\beta_{\mathcal{F}}$ is the dual-bounded-Lipschitz "metric" β based on $\| \cdot \|_{\mathcal{F}}$ as in Theorem 3.28.

The following, due to Giné and Zinn (1991), will be proved:

Theorem 10.22 *Let* $(X, \mathcal{A})$ *be a measurable space and* $\mathcal{F}$ *an image admissible Suslin class of real-valued measurable functions on* X. *Then* $\mathcal{F}$ *is a uniform Donsker class if and only if it is finitely uniformly pregaussian and thus, if and only if it is uniformly pregaussian.*

Remarks. The theorem is a very useful characterization since it is easier to check the finitely uniformly pregaussian property than to check the uniform Donsker property directly.

Giné and Zinn showed that Pollard's entropy condition (10.18), together with uniform boundedness and measurability for $\mathcal{F}$, which imply $\mathcal{F}$ is universal Donsker by Theorem 10.13, actually imply that $\mathcal{F}$ is uniformly Donsker (Theorem 10.26 below). Thus most of the examples of universal Donsker classes treated in Sections 10.2 and 10.3 are uniformly Donsker. An exception is the "ellipsoid" universal Donsker class of Proposition 10.17; see Problem 4 below.

Proof of Theorem 10.22. First, some uniformity in finite dimensional cases will be helpful. Recall the space $BL(\mathbb{R}^d)$ of bounded Lipschitz functions on $\mathbb{R}^d$ and its unit ball

$$BL_1(\mathbb{R}^d) := \{f \in BL(\mathbb{R}^d) : \|f\|_{BL} \le 1\}.$$

Let β_d be the usual bounded Lipschitz distance for laws P, Q on $\mathbb{R}^d$, namely, $\beta_d(P, Q) := \sup\{|\int f d(P - Q)| : f \in BL_1(\mathbb{R}^d)\}$ (RAP, Prop. 11.3.2).

As in Section 8.2, for a multi-index $p = (p_1, \ldots, p_d)$ where p_j are non-negative integers and $[p] := p_1 + \cdots + p_d$, let $x^p := \prod_{j=1}^d x_j^{p_j}$ and if f is a suitably differentiable function, let $D^p f := \partial^{[p]} f / \partial x_1^{p_1} \cdots \partial x_d^{p_d}$. Let C_1^3 be the set of functions f on $\mathbb{R}^d$ such that $D^p f$ exist and are bounded and continuous for $0 \le [p] \le 3$, where $D^0 f := f$, and for which $\sum_{[p] \le 3} \|D^p f\|_{\sup} \le 1$. For probability measures P and Q on $\mathbb{R}^d$ let

$$d_3(P, Q) := \sup\{|\int f d(P - Q)| : f \in C_1^3\}.$$

Let S be a separable Banach space and $\mathcal{P}(S)$ the set of all Borel probability measures on S. Recall that for any $P, Q \in \mathcal{P}(S)$, the convolution is defined by $(P * Q)(A) := \int P(A - x) dQ(x)$ for any Borel set $A \subset S$. Convolution is commutative and associative. The following fact and proof, given here for completeness, are as in Araujo and Giné (1980, p. 37); on p. 67, and in Giné and Zinn (1991), Lindeberg's name is associated with it.

Lemma 10.23 *Let S be a separable Banach space and $\mathcal{F}$ a uniformly bounded, image admissible Suslin family of real-valued Borel functions on S which is translation invariant, i.e., for each $f \in \mathcal{F}$ and $u \in S$, the function $x \mapsto f(x - u)$ is in $\mathcal{F}$. Let P_j and Q_j be in $\mathcal{P}(S)$ for $j = 1, \ldots, n$. Then*

$$\|(P_1 * P_2 * \cdots * P_n) - (Q_1 * Q_2 * \cdots * Q_n)\|_{\mathcal{F}} \leq \sum_{j=1}^{n} \|P_j - Q_j\|_{\mathcal{F}}.$$

Proof. It will suffice to treat $n = 2$, as then one can use induction. We have

$$\| \int f \, d(P_1 * P_2 - Q_1 * Q_2)\|_{\mathcal{F}}$$

$$= \| \int \int f(x + y) dP_1(x) dP_2(y) - dQ_2(x) dQ_1(y)\|_{\mathcal{F}}$$

$$\leq \| \int \int f(x + y) d(P_1 - Q_1)(x) dP_2(y)$$

$$+ \| \int \int f(x + y) d(P_2 - Q_2)(x) dQ_1(y)\|_{\mathcal{F}}$$

$$\leq \int \| \int f d(P_1 - Q_1)\|_{\mathcal{F}} dP_2 + \int \| \int f d(P_2 - Q_2)\|_{\mathcal{F}} dQ_1$$

$$= \|P_1 - Q_1\|_{\mathcal{F}} + \|P_2 - Q_2\|_{\mathcal{F}},$$

proving the Lemma. $\qquad \square$

For $S = \mathbb{R}^d$, $BL_1(\mathbb{R}^d)$ and C_1^3 satisfy the hypotheses on $\mathcal{F}$ in Lemma 10.23.

Lemma 10.24 *For $0 < M < \infty$ and $d = 1, 2, \ldots$, let $\mathcal{P}_M^d$ be the class of Borel probability measures P on $\mathbb{R}^d$ with $P(|x| \leq M) = 1$. For $P \in \mathcal{P}_M^d$ let $\{X_i^P\}_{i \geq 1}$ be i.i.d. (P) and let C_P be the covariance for P. Then we have for a constant $K = K_d$ depending only on the dimension,*

$$d_3 \left[\mathcal{L} \left(\sum_{i=1}^{n} (X_i^P - EX_i^P)/\sqrt{n} \right), N(0, C_P) \right] \leq KM \operatorname{Tr}(C_P)/\sqrt{n} \quad (10.36)$$

and

$$\lim_{n \to \infty} \sup_{P \in \mathcal{P}_M^d} \beta_d \left[\mathcal{L} \left(\frac{1}{\sqrt{n}} \sum_{i=1}^{n} (X_i^P - EX_i^P) \right), N(0, C_P) \right] = 0. \quad (10.37)$$

Proof. For (10.36), first, it is well known that for any covariance matrices C and D, $N(0, C) * N(0, D) = N(0, C + D)$, and this can be iterated to any number of terms. Let Y_i be i.i.d. $N(0, C_P/n)$, $\xi_i := (X_i^P - EX_i^P)/\sqrt{n}$, $i = 1, \ldots, n$, and $\xi := \{\xi_i\}_{i=1}^n$. Thus by Lemma 10.23,

$$d_3\left(\mathcal{L}\left(\sum_{i=1}^n \xi_i\right), N(0, C_P)\right) \leq n d_3\left(\mathcal{L}(\xi_1), N(0, C_P/n)\right). \tag{10.38}$$

If $f \in C_1^3$, then for $p! := \prod_{j=1}^d p_j!$

$$f(\xi_i) = \theta(\xi_i) + \sum_{[p] \leq 2} \frac{D^p f(0)\xi_i^p}{p!} \tag{10.39}$$

where the remainder $\theta(\xi_i) \leq J_d |\xi_i|^3$ for a constant J_d depending only on the dimension d, and similarly for $f(Y_i)$. Here $E\xi_i = EY_i = 0$, $E\xi_1^p = EY_1^p$ whenever $[p] = 2$, and

$$E|\xi|^3 \leq 2M E|\xi|^2/\sqrt{n} = 2M \operatorname{Tr}(C_P)/n^{3/2}. \tag{10.40}$$

Take coordinates in which C_P is diagonalized. For each coordinate $y_j := Y_{1j}$ of Y_1 we have $E(y_j^2) = \sigma_j^2 := E(\xi_j^2)$, and

$$\operatorname{Tr}(C_P) = n\sum_{j=1}^d \sigma_j^2 = n E(|\xi|^2) \leq E(|X_1|^2) \leq M^2.$$

We have

$$E|Y_1|^4 = E((y_1^2 + \cdots + y_d^2)^2) = \sum_{j=1}^d E(y_j^4) + 2\sum_{1 \leq i < j \leq d} \sigma_i^2 \sigma_j^2.$$

For each j, $E(y_j^4) = 3\sigma_j^4 \leq 3M^2\sigma_j^2/n$. It follows that

$$E|Y_1|^4 \leq 3M^2 \operatorname{Tr}(C_P)/n^2 + (\operatorname{Tr}(C_P))^2/n^2 \leq 4M^2 \operatorname{Tr}(C_P)/n^2.$$

Then by the Cauchy–Bunyakovsky–Schwarz inequality

$$E|Y_1|^3 = E(|Y_1|^2|Y_1|) \leq \sqrt{E|Y_1|^4 \operatorname{Tr}(C_P/n)} \leq 2M \operatorname{Tr}(C_P)/n^{3/2}. \tag{10.41}$$

We then have, summing terms in the Taylor series (10.39) for ξ_i and the corresponding ones for Y_i, since $E(\xi^p) = E(Y_1^p)$ for $[p] \leq 2$, that by (10.40) and (10.41),

$$|E[f(\xi_1) - f(Y_1)]| \leq 4M J_d \operatorname{Tr}(C_P)/n^{3/2},$$

which gives (10.36) with $K_d := 4J_d$.

Next, to bound the β_d distance in (10.37), let $f \in BL_1(\mathbb{R}^d)$. Smooth f by convolving it with the $N(0, \varepsilon I)$ density ϕ_ε, namely, for $\phi := \phi_1$, set $f_\varepsilon(x) := \int f(x - \varepsilon y)\phi(y)dy$. Then for some constant $c(d)$

$$\|f - f_\varepsilon\|_{\sup} \leq (2\pi)^{-d/2} \int (\min(2, \varepsilon|y|) \exp(-|y|^2/2)dy \leq c(d)\varepsilon \to 0$$

as $\varepsilon \downarrow 0$. By the changes of variables $u = x - \varepsilon y$, so that $y = (x - u)/\varepsilon$, we get $f_\varepsilon(x) = \int f(u)\phi((x - u)/\varepsilon)du$. We can then differentiate under the integral sign (Appendix A) with respect to coordinates of x. For any multi-index p, $\|D^p f_\varepsilon\|_{\sup} \leq \varepsilon^{-[p]} \int |D^p \phi(y)|dy$. Thus, f can be approximated as well as desired in sup norm by a function having bounded, continuous derivatives up to any given degree, in our case $[p] \leq 3$. Thus (10.37) and and Lemma 10.24 follow. □

Next, here are some definitions. For any nnds (nonnegative definite symmetric) matrix M, there is a unique nnds square root $A = M^{1/2}$, namely, a nnds matrix A with $A^2 = M$. For any $d \times d$ matrix M, define the *Hilbert–Schmidt norm* $\|M\|_2$ as the square root of, for any two orthonormal bases $\{e_j\}_{j=1}^d$ and $\{f_i\}_{i=1}^d$,

$$\|M\|_2^2 := \sum_{j=1}^d |Me_j|^2 = \sum_{j=1}^d \sum_{i=1}^d (Me_j, f_i)^2.$$

This does not depend on the choice of orthonormal basis $\{e_j\}$ since $(Me_j, f_i) \equiv (Mf_i, e_j)$. For (S, d) a separable metric space, $1 \leq p < \infty$, and two Borel probability measures P and Q on S, such that $\int d(x_0, x)^p d\mu(x) < \infty$ for some (any) $x_0 \in S$, the *Wasserstein (p) distance* is here defined by $W_p(P, Q) := \inf\{d(X, Y)^p : \mathcal{L}(X) = P, \mathcal{L}(Y) = Q\}^{1/p}$. (Sometimes in the literature "Wasserstein distance" refers to the special case $p = 1$, and other times to quantities $\sigma(x, y)$ other than pth powers of a given metric.) The *Fréchet distance* between P and Q is $W_2(P, Q)$. We have $W_p(P, Q) \leq W_q(P, Q)$ for $1 \leq p < q \leq \infty$ by the Hölder inequality. Recall $\|f\|_L := \sup_{x \neq y} |f(x) - f(y)|/d(x, y)$. For $p = 1$, $W_1(P, Q) := \sup\{|\int f d(P - Q)| : \|f\|_L \leq 1\}$ by the Kantorovich–Rubinštein theorem (RAP, Theorem 11.8.2). Clearly, the bounded Lipschitz distance $\beta(P, Q) \leq W_1(P, Q)$. We next have:

Lemma 10.25 *Let C and D be two $d \times d$ covariance matrices and $N(0, C)$ and $N(0, D)$ the corresponding normal laws on $\mathbb{R}^d$. Then*

$$\beta_d(N(0, C), N(0, D)) \leq W_1(N(0, C), N(0, D))$$
$$\leq W_2(N(0, C), N(0, D)) \leq \|\sqrt{C} - \sqrt{D}\|_2. \tag{10.42}$$

Proof. Let Z have $N(0, I_d)$ distribution on $\mathbb{R}^d$. Then $\sqrt{C}Z$ has $N(0, C)$ and $\sqrt{D}Z$ has $N(0, D)$. We have

$$W_2(N(0, C), N(0, D)) \leq \left[E(\|(\sqrt{C} - \sqrt{D})(Z)\|^2)\right]^{1/2} = \|\sqrt{C} - \sqrt{D}\|_2.$$

The rest follows from the discussion preceding the Lemma. □

Now, to prove Theorem 10.22, since clearly uniform Donsker implies UPG_f (finitely uniformly pregaussian), it will suffice to prove the converse, which will be done in a series of claims. Let $\mathcal{F} \in UPG_f$ with $\mathcal{F}$ admissible Suslin.

Claim 1. It suffices to consider classes $\mathcal{F}$ uniformly bounded by 1.

Proof of Claim 1. For $x \neq y$ in X let $P(x, y) := \frac{1}{2}(\delta_x + \delta_y)$. Then

$$\sup_{x,y \in X} E\|G_{P(x,y)}\|_{\mathcal{F}} < \infty$$

implies $\sup_{f \in \mathcal{F}}(\operatorname{diam}(f))^2/4 = \sup_{f \in \mathcal{F}} EG^2_{P(x,y)}(f) < \infty$. Thus the class $\mathcal{F}_0 := \{f - \inf f : f \in \mathcal{F}\}$ is uniformly bounded as in Proposition 10.3, so we can replace $\mathcal{F}$ by $\mathcal{F}_0$. For $M := \sup\{\|g\|_{\sup} : g \in \mathcal{F}_0\}$ we can further replace $\mathcal{F}$ by $\{f/M : f \in \mathcal{F}_0\}$ without loss of generality, getting a class bounded by 1, proving Claim 1.

Let $\mathcal{F}^2 := \{fg : f, g \in \mathcal{F}\}$, $\mathcal{F}' := \{f - g : f, g \in \mathcal{F}\}$, and $\mathcal{F}'^2 := \{h^2 : h \in \mathcal{F}'\}$.

Claim 2. Let $\mathcal{G} := \mathcal{F} \cup \mathcal{F}^2 \cup \mathcal{F}' \cup (\mathcal{F}')^2$. Then $\sup_{P \in \mathcal{P}(X)} E_P\|P_n - P\|_{\mathcal{F}} = O(n^{-1/2})$.

Proof of Claim 2. The class $\mathcal{G}$ and each of the given subclasses of it are image admissible Suslin by Theorem 5.26. It suffices to consider $\mathcal{G} = \mathcal{F}'^2$. For $f, g, \phi, \psi \in \mathcal{F}$, by Claim 1, we have

$$\left((f - g)^2 - (\phi - \psi)^2\right)^2 = ((f - g - (\phi - \psi))^2(f - g + \phi - \psi)^2$$
$$\leq 16(f - g - (\phi - \psi))^2,$$

and so

$$E_{P_n}[((f - g)^2 - (\phi - \psi)^2)^2] \leq 16 E_{P_n}[((f - g) - (\phi - \psi))^2]. \quad (10.43)$$

Take a product of our probability space $(X^\infty, \mathcal{A}^{\otimes\infty}, P^\infty)$ with a copy of $[0, 1]$ with its usual σ-algebra and with $U[0, 1]$ distribution. On $[0, 1]$ define a sequence $\{g_i\}_{i \geq 1}$ of i.i.d. $N(0, 1)$ random variables. Define a process for $h \in \mathcal{G}$ by $G(h) := \sum_{i=1}^{n} g_i h(X_i)/\sqrt{n}$. For fixed $\omega \in X^\infty$ and thus $X_1, \ldots, X_n$, and where E_g denotes expectation with respect to the distribution of g_i only, it will be shown that

$$E_g \left\| \sum_{i=1}^{n} g_i \delta_{X_i}/\sqrt{n} \right\|_{(\mathcal{F}')^2} \leq 8 E_g \left\| \sum_{i=1}^{n} g_i \delta_{X_i}/\sqrt{n} \right\|_{\mathcal{F}'}. \quad (10.44)$$

Let $T := \mathcal{F}'$. For $t := h = f - g$ with $f, g \in \mathcal{F}$, let $Y_t := G(h)$ and $s = H = \phi - \psi, \phi, \psi \in \mathcal{F}$. Then

$$E\left((Y_s - Y_t)^2\right) = E\left(\left[\frac{1}{\sqrt{n}}\sum_{i=1}^{n} g_i(h - H)(X_i)\right]^2\right)$$

$$= \frac{1}{n}\sum_{i=1}^{n}(h - H)(X_i)^2 = P_n((h - H)^2).$$

Let $X_t := X_h := \frac{1}{4}Y_{h^2}$. Then we have

$$16E\left((X_s - X_t)^2\right) = E\left(\left[\frac{1}{\sqrt{n}}\sum_{i=1}^{n} g_i\left(h^2 - H^2\right)(X_i)\right]^2\right)$$

$$= \frac{1}{n}\sum_{i=1}^{n}\left(\left(h^2 - H^2\right)(X_i)\right)^2$$

$$= P_n\left(\left[h^2 - H^2\right]^2\right) \leq 16P_n\left((h - H)^2\right)$$

by (10.43). So $E\left((X_s - X_t)^2\right) \leq E\left((Y_s - Y_t)^2\right)$, a main hypothesis of Theorem 2.18. Taking a countable dense subset $\{f_i\}_{i=1}^{\infty}$ in $\mathcal{F}$ with respect to $L^2(P_n)$, $S := \{f_i - f_j\}_{i,j\geq1}$ is countable and dense in $T = \mathcal{F}'$, and $\inf_{i,j}|X_{f_i - f_j}| = 0$ a.s., so by (2.21), which Giné and Zinn call the Slepian–Fernique Lemma, we do get (10.44).

Now suppose, as one may, that beside $\{g_i\}_{i\geq1}$, on $[0, 1]$ we also have defined i.i.d. Rademacher variables $\{\varepsilon_i\}_{i\geq1}$ independent of $\{g_j\}_{j\geq1}$. For any nonnegative random variable Y we have $EY = \int_0^{\infty} \Pr(Y > x)dx$. Thus (the next inequality is well known to hold with a factor of 2 rather than 4, but 4 suffices for present purposes) by desymmetrization (10.6)

$$E\|P_n - P\|_{(\mathcal{F}')^2} \leq 4E\left\|\frac{1}{n}\sum_{i=1}^{n}\varepsilon_i\delta_{X_i}\right\|_{(\mathcal{F}')^2}. \tag{10.45}$$

Next, $\{\varepsilon_i|g_i|\}_{i=1}^{n}$ are equal in distribution to $\{g_i\}_{i=1}^{n}$, and so we have

$$E\left\|\sum_{i=1}^{n}g_i\delta_{X_i}\right\|_{(\mathcal{F}')^2} = E\left\|\sum_{i=1}^{n}\varepsilon_i|g_i|\delta_{X_i}\right\|_{(\mathcal{F}')^2} \geq E|g_1|E\left\|\sum_{i=1}^{n}\varepsilon_i\delta_{X_i}\right\|_{(\mathcal{F}')^2}$$

by Jensen's inequality, for example Lemma 9.7(a), applied to integration E_g with respect to $\{g_i\}$ only for fixed X_i and ε_i. We have joint measurability by the image admissible Suslin property of $(\mathcal{F}')^2$. It follows that the right side of

(10.45) is

$$\leq \frac{4}{E|g_1|} E \left\| \frac{1}{n} \sum_{i=1}^{n} g_i \delta_{X_i} \right\|_{(\mathcal{F}')^2},$$

which then by (10.44), then the triangle inequality, is

$$\leq \frac{32}{\sqrt{n} E|g_1|} E_P E_g \left\| \frac{1}{\sqrt{n}} \sum_{i=1}^{n} g_i \delta_{X_i} \right\|_{\mathcal{F}'}$$

$$\leq \frac{64}{\sqrt{n} E|g_1|} E_P E_g \left\| \frac{1}{\sqrt{n}} \sum_{i=1}^{n} g_i \delta_{X_i} \right\|_{\mathcal{F}}. \tag{10.46}$$

We have $E|g_1| = \sqrt{\pi/2}$, and the Gaussian process W_{P_n} can be written as

$$W_{P_n} = n^{-1/2} \sum_{i=1}^{n} g_i \delta_{X_i}, \tag{10.47}$$

and so the expression in (10.46) is

$$\leq 64\sqrt{\pi/2} n^{-1/2} \sup_{Q \in \mathcal{P}_f(X)} E \|W_Q\|_{\mathcal{F}},$$

which by Claim 1 and the fact that we can write

$$W_P(f) = G_P(f) + Z E_P f \tag{10.48}$$

for a $N(0, 1)$ variable Z independent of G_P, recalling that W_P is the isonormal process on $L^2(P)$, is

$$\leq 64\sqrt{\pi/2} n^{-1/2} \sup_{Q \in \mathcal{P}_f(X)} \left(E \|G_Q\|_{\mathcal{F}} + \sqrt{\frac{2}{\pi}} \right).$$

Combining, since $\mathcal{F} \in \mathrm{UPG}_f$, the left side of (10.45) is $O(n^{-1/2})$ uniformly in $P \in \mathcal{P}(X)$, proving Claim 2.

Recall $\mathcal{F}'(\delta, d)$ as defined in (10.30). The next claim includes a uniform asymptotic equicontinuity condition:

Claim 3. $(\mathcal{F}, e_P)$ is totally bounded for each $P \in \mathcal{P}(X)$ and for all $\varepsilon > 0$,

$$\lim_{\delta \downarrow 0} \limsup_{n \to \infty} \sup_{P \in \mathcal{P}(X)} \mathrm{Pr} \left(\left\| \sqrt{n}(P_n - P) \right\|_{\mathcal{F}'(\delta, e_P)} > \varepsilon \right) = 0. \tag{10.49}$$

Also, $\mathcal{F}$ is universal Donsker.

Proof of Claim 3. Once (10.49) is proved, it will follow by Theorem 3.34 that $\mathcal{F}$ is universal Donsker.

Since $\mathcal{F} \in \mathrm{UPG}_f$, it follows that

$$\sup_{n, P_n} \{ E \left\| G_{P_n} \right\|_{\mathcal{F}} < \infty$$

where the supremum is over all possible empirical measures $P_n := \frac{1}{n} \sum_{i=1}^{n} \delta_{x_i}$ for $x_1, \ldots, x_n$ in X. It follows that for each $\varepsilon > 0$, the supremum over P_n of the packing numbers $D(\varepsilon, \mathcal{F}, \rho_{P_n})$ is bounded by the Sudakov minoration Theorem 2.22(b), specifically, there is a finite c such that for all possible P_n on X and $\varepsilon > 0$,

$$\log D(\varepsilon, \mathcal{F}, e_{P_n}) < c/\varepsilon^2, \tag{10.50}$$

first for ρ_{P_n} in place of e_{P_n}, and then for e_{P_n} with a possibly larger c since $\mathcal{F}$ is uniformly bounded by 1.

Next, for each $P \in \mathcal{P}(X)$, $Y_n := \|P_n - P\|_{(\mathcal{F}')^2}$ is a reversed submartingale by Theorem 6.6. Since $(\mathcal{F}')^2$ is uniformly bounded, Y_n converges a.s. and in L^1 to some limit by theorems of Doob (RAP, Theorem 10.6.4). By Claim 2, the limit in L^1, and so a.s., must be 0,

$$\lim_{n \to \infty} \|P_n - P\|_{(\mathcal{F}')^2} = 0. \tag{10.51}$$

Now

$$|e_{P_n}(f, g)^2 - e_P(f, g)^2| \equiv |P_n((f - g)^2) - P((f - g)^2)|,$$

and so almost surely, for any $P \in \mathcal{P}(X)$,

$$\lim_{n \to \infty} \sup_{f, g \in \mathcal{F}} |e_{P_n}(f, g)^2 - e_P(f, g)^2| = 0. \tag{10.52}$$

From this and (10.50), for all $\varepsilon > 0$,

$$\sup_{P \in \mathcal{P}(X)} \log D(\varepsilon, \mathcal{F}, e_P) \leq c/\varepsilon^2. \tag{10.53}$$

Thus $(\mathcal{F}, e_P)$ is totally bounded, uniformly in P.

Now let, as previously, $\{\xi_j\}_{j \geq 1}$ be a sequence of i.i.d. Rademacher variables defined on the probability space $([0, 1], U[0, 1])$, where we take the product space $X^\infty \times [0, 1]$. For a given $\varepsilon > 0$ and $0 < \delta < \varepsilon$, Lemma 9.6(b) will be applied to $\mathcal{F}'(\delta, e_P)$ in place of $\mathcal{F}$. Then we will have $\alpha \leq \delta$. Thus we can take any $t > \sqrt{2n}\delta$, specifically, $t = 2\sqrt{n}\varepsilon$. Dividing all four expressions in the events whose probabilities are taken in Lemma 9.6(b) by $\sqrt{n}$ we get equivalent events. Let $u = t/\sqrt{n} = 2\varepsilon$. Then we get

$$\Pr\left(\|\nu_n\|_{\mathcal{F}'(\delta, e_P)} > 2\varepsilon\right) \leq 4\Pr\left(\left\|\frac{1}{\sqrt{n}} \sum_{i=1}^{n} \xi_i f(X_i)\right\|_{\mathcal{F}'(\delta, e_P)} > \varepsilon\right).$$

Next, we have

$$\Pr\left(\left\|\frac{1}{\sqrt{n}} \sum_{i=1}^{n} \varepsilon_i \delta_{X_i}\right\|_{\mathcal{F}'(\delta, e_P)} > \varepsilon\right) \leq T_1 + T_2$$

where

$$T_1 := T_1(\delta, \varepsilon) := \Pr\left(\left\| \frac{1}{\sqrt{n}} \sum_{i=1}^{n} \varepsilon_i \delta_{X_i} \right\|_{\mathcal{F}'(\sqrt{2}\delta, e_{P_n})} > \varepsilon \right)$$

and

$$T_2 := T_2(\delta) := \Pr\left(\sup_{f,g \in \mathcal{F}} \left| e_{P_n}(f,g)^2 - e_P(f,g)^2 \right| > \delta^2 \right).$$

Claim 2 implies that $\lim_{n \to \infty} \sup_{P \in \mathcal{P}(X)} T_2(\delta) = 0$ for all $\delta > 0$.

By (10.48) we have for any probability measure Q on $\mathcal{A}$ and $\delta > 0$,

$$E \|W_Q\|_{\mathcal{F}'(\delta, e_Q)} + \sqrt{\frac{2}{\pi}}\delta \leq E \|G_Q\|_{\mathcal{F}'(\delta, \rho_Q)} + \sqrt{\frac{2}{\pi}}\delta.$$

Then Gaussianizing and using Jensen's inequality as in the proof of Claim 2,

$$E \left\| \frac{1}{\sqrt{n}} \sum_{i=1}^{n} \varepsilon_i \delta_{X_i} \right\|_{\mathcal{F}'(\delta, e_{P_n})} \leq \frac{1}{E|g_1|} E_P E_g \left\| \frac{1}{\sqrt{n}} \sum_{i=1}^{n} g_i \delta_{X_i} \right\|_{\mathcal{F}'(\delta, e_{P_n})}$$

$$\leq \frac{1}{E|g_1|} \sup_{Q \in \mathcal{P}_f(X)} E \|Z_Q\|_{\mathcal{F}'(\delta, e_Q)}$$

$$\leq \sqrt{\frac{\pi}{2}} \sup_{Q \in \mathcal{P}_f(X)} E \|G_Q\|_{\mathcal{F}'(\delta, \rho_Q)} + \sqrt{\frac{\pi}{2}}$$

for all $P \in \mathcal{P}(X)$. Since $\mathcal{F} \in \mathrm{UPG}_f$ it follows that for all $\varepsilon > 0$,

$$\lim_{\delta \downarrow 0} \sup_{P \in \mathcal{P}(X)} T_1(\delta, \varepsilon) = 0,$$

which proves (10.49) and so Claim 3.

Claim 4. $\mathcal{F} \in \mathrm{UPG}$.

Proof of Claim 4. For $\mathcal{G}$ as in Claim 2, we have by the same proof as for $(\mathcal{F}')^2$ (10.51) that for any $P \in \mathcal{P}(X)$, almost surely

$$\|P_n - P\|_{\mathcal{G}} \to 0. \tag{10.54}$$

Take an $\omega \in X^\infty$ such that (10.54) holds. Then for any finite sequence $f_1, \ldots, f_r$ of functions in $\mathcal{F}$, we have convergence in distribution

$$\mathcal{L}\left(G_{P_n(\omega)}(f_1), \ldots, G_{P_n(\omega)}(f_r) \right) \to \mathcal{L}(G_P(f_1), \ldots, G_P(f_r)). \tag{10.55}$$

We know that $\mathcal{F}$ is P-Donsker by Claim 3 and thus that it is pregaussian for P, so that G_P can be chosen to have a distribution with separable support in $\ell^\infty(\mathcal{F})$. The same is true of each $G_{P_n(\omega)}$ for our fixed ω. The next aim is to show that $\{G_{P_n(\omega)}\}_{n \geq 1}$ is a Cauchy sequence with respect to the distance $\beta_\mathcal{F}$. Let

$H \in BL_1^{\mathcal{F}} := \{H \in BL(\ell^\infty(\mathcal{F})): \|H\|_{BL} \leq 1\}$. Given $\tau > 0$, since $\rho_P \leq e_P$, by (10.52) and (10.53), there is an $N < \infty$ and $f_1, \ldots, f_N \in \mathcal{F}$ such that for each $f \in \mathcal{F}$ there is an $i \leq N$ with $\rho_{P_m}(f, f_i) \leq \tau$ for all $m \geq n$. Let $i(f, \tau)$ be the least such i and $\pi_\tau f := f_{i(f,\tau)}$. Let $G_{P_n(\omega),\tau}(f) := G_{P_n(\omega)}(\pi_\tau f)$. Then, taking expectations with respect to the distributions on the left in (10.55), still for fixed ω, writing $P_{r,\omega} := P_r(\omega)$, we have

$$\left| EH(G_{P_{n,\omega}}) - EH(G_{P_{m,\omega}}) \right| \leq H_1 + H_2 + H_3 \qquad (10.56)$$

where $H_1 := |EH(G_{P_{n,\omega}}) - EH(G_{P_{n,\omega},\tau})|, H_2 := |EH(G_{P_{n,\omega},\tau}) - EH(G_{P_{m,\omega},\tau})|$, and $H_3 := |EH(G_{P_{m,\omega},\tau}) - EH(G_{P_{m,\omega}})|$. Then

$$H_1 \leq E \left\| G_{P_{n,\omega}} \right\|_{\mathcal{F}'(\tau,\rho_{P_{n,\omega}})}, \qquad H_3 \leq E \left\| G_{P_{m,\omega}} \right\|_{\mathcal{F}'(\tau,\rho_{P_{m,\omega}})},$$

and so for $m \geq n$

$$\max(H_1, H_3) \leq \sup_{Q \in \mathcal{P}_f(X)} E \left\| G_Q \right\|_{\mathcal{F}'(\tau,\rho_Q)}. \qquad (10.57)$$

For a probability measure $Q = P$ or some $P_{r,\omega}$ and $i, j = 1, \ldots, N$, let C_Q be the covariance matrix with respect to Q, with elements $C_{Qij} = \text{Cov}_Q(f_i, f_j)$, and $C_{nij} := C_{P_n ij}$. By (10.54), for the (almost all) ω satisfying it, $C_{nij} \to C_{Pij}$ elementwise as $n \to \infty$. Since the dimension N is fixed, the convergence also holds with respect to any norm on the $N(N + 1)/2$-dimensional vector space S_N of $N \times N$ symmetric matrices. Let H_N be the set of nonnegative definite elements of S_N, with the topology of any norm on S_N. Then H_N is a locally compact metrizable and so Hausdorff space. Any continuous one-to-one mapping of one compact Hausdorff space onto another has a continuous inverse (RAP, Theorem 2.2.11). This holds in suitable locally compact Hausdorff spaces, for example, for the map $A \mapsto A^2$ of H_N onto itself. So, the inverse map $C \mapsto \sqrt{C}$ is continuous $H_N \to H_N$ with respect to any norm(s) on S_N. Thus $\sqrt{C_n} \to \sqrt{C_P}$ as $n \to \infty$, and

$$\lim_{n\to\infty} \sup_{m \geq n} \| \sqrt{C_m} - \sqrt{C_n} \|_2 = 0$$

for the Hilbert–Schmidt norm $\|\cdot\|_2$ used in Lemma 10.25 (or any norm $\|\cdot\|$ on S_N). It follows from that Lemma that G_{P_n} on $\{f_i\}_{i=1}^N$ form a Cauchy sequence with respect to the bounded Lipschitz distance β on $\mathbb{R}^N$. Thus

$$\lim_{n\to\infty} \sup_{m \geq n} \sup_{H \in BL_1^{\mathcal{F}}} H_2 = 0. \qquad (10.58)$$

Since $G_P(f) \equiv W_P(f) - Pf$ for all $f \in \mathcal{L}^2(P)$, and by (10.47), each $G_{P_n(\omega)}$ takes values in the finite-dimensional subspace of $\ell^\infty(\mathcal{F})$ spanned by $\delta_{X_j(\omega)}$, $j = 1, \ldots, n$, and constant functions. Thus, these processes for all n, for the given ω, take values in the separable subspace S_ω of $\ell^\infty(\mathcal{F})$ of functions of the form $c_0 + \sum_{j=1}^\infty c_j \delta_{X_j(\omega)}$ such that $\sum_j |c_j| < \infty$. Since $\mathcal{F}$ is P-pregaussian by

Claim 3, there is a separable subspace T_P of $\ell^\infty(\mathcal{F})$ in which G_P can be defined to take values. There is a separable subspace T_ω of $\ell^\infty(\mathcal{F})$ including both S_ω and T_P. Each $G_{P_n(\omega)}$ has a law, as mentioned in (10.55), which can be taken to be defined on S_ω and thus on T_P. G_P and $G_{P_n(\omega)}$ are measurable random variables with values in T_P. It follows from (10.58) (10.57), and $\mathcal{F} \in \mathrm{UPG}_f$ that $\mathcal{L}(G_{P_n(\omega)})$ on T_P form a Cauchy sequence for $\beta_{\mathcal{F}}$, a metric for laws on T_P.

$\mathcal{F} \in \mathrm{UPG}_f$ implies that

$$\sup_{Q \in \mathcal{P}_f(S)} E\|G_Q\|_{\mathcal{F}}^2 < \infty, \tag{10.59}$$

as follows from a Gaussian concentration inequality, Theorem 2.47. Namely, for a Gaussian process V with mean 0, defined on a parameter space T separable for d_V as all G_Q processes are for $Q \in \mathcal{P}_f(X)$, taking $V(\cdot) \in \mathcal{L}^2(P)$, we have in distribution $V = L \circ V$, so we can write ess. $\sup_{t \in T} |V_t| =_d Y := |L(A)|^*$ where $A := \{V(t) \in \mathcal{L}^2(P) : t \in T\}$. Here A is a GB-set with $EY < \infty$, and Theorem 2.47 and a calculation give $E(Y^2) \le (EY)^2 + K$ where K is an absolute constant depending on the absolute constant C given in Proposition 2.46. Thus $\|G_{P_n(\omega)}\|_{\mathcal{F}}$ are uniformly integrable and $E\|G_{P_n(\omega)}\|_{\mathcal{F}} \to E\|G_P\|_{\mathcal{F}}$ as $n \to \infty$. Likewise, $E\|G_{P_n(\omega)}\|_{\mathcal{F}'(\delta,\rho_P)} \to E\|G_P\|_{\mathcal{F}'(\delta,\rho_P)}$. For n large enough, by (10.54) as before,

$$\|G_{P_n(\omega)}\|_{\mathcal{F}'(\delta,\rho_P)} \le \|G_P\|_{\mathcal{F}'(2\delta,\rho_P)}.$$

Since $\mathcal{F} \in \mathrm{UPG}_f$, we have

$$\lim_{\delta \downarrow 0} \sup_{P \in \mathcal{P}(X)} E\|G_P\|_{\mathcal{F}'(\delta,\rho_P)} = 0.$$

So $\mathcal{F} \in \mathrm{UPG}$ and Claim 4 is proved.

Claim 5. $\mathcal{F}$ is uniform Donsker, i.e., (10.35) holds.

Proof of Claim 5. Claim 4 implies

$$\sup_{P \in \mathcal{P}(X)} E\|G_P\|_{\mathcal{F}} < \infty, \quad \lim_{\delta \downarrow 0} \sup_{P \in \mathcal{P}(X)} E\|G_P\|_{\mathcal{F}'(\delta,\rho_P)} = 0. \tag{10.60}$$

If we wanted to prove $\beta_{\mathcal{F}}(\nu_n^P, G_P) \to 0$ for an individual P, we could prove total boundedness of $\mathcal{F}$ for ρ_P and an asymptotic equicontinuity condition, infer the P-Donsker property of $\mathcal{F}$ from Theorem 3.34, and then get the $\beta_{\mathcal{F}}$ convergence from Theorem 3.28. For uniform Donsker, (10.53) gives that $\mathcal{F}$ is ρ_P- and, in fact, e_P-totally bounded uniformly in $P \in \mathcal{P}(X)$. Then Claim 3, (10.49), gives an asymptotic equicontinuity condition uniformly in $P \in \mathcal{P}(X)$. The conclusion $\beta_{\mathcal{F}}(\nu_n, G_P) \to 0$, uniformly in $P \in \mathcal{P}(X)$, which is the definition of uniform Donsker, will be proved directly.

Let $P \in \mathcal{P}(X)$. Given $\tau > 0$ let $f_1, \ldots, f_{N_P(\tau)}$ be a maximal set of members of $\mathcal{F}$ with $e_P(f_i, f_j) > \tau$ for $1 \le i < j \le N_P(\tau)$. For each $f \in \mathcal{F}$ let $\pi_\tau^P(f) :=$

f_i for the least i such that $e_P(f, f_i) \leq \tau$. For each $f \in \mathcal{F}$ and $j = 1, 2, \ldots$, let $Y_j^P(f) := f(X_j) - Pf$ and $Y_{j,\tau}^P(f) := Y_j^P(\pi_\tau^P f)$. By Theorem 3.2 we can and do take G_P such that for (almost) all ω, G_P is uniformly continuous on $\mathcal{F}$ for ρ_P and thus for e_P. In particular G_P takes values in a separable subspace of $\ell^\infty(\mathcal{F})$. Let $G_{P,\tau}(f) := G_P(\pi_\tau^P f)$ for $f \in \mathcal{F}$. Each $Y_{j,\tau}^P$ and $G_{P,\tau}$ also takes values in a separable subspace. Let $H \in BL_1^{\mathcal{F}}$. Then

$$|E^* H(\nu_n^P) - EH(G_P)| \leq \eta_1 + \eta_2 + \eta_3 \qquad (10.61)$$

where

$$\eta_1 := \left| E^* H\left(\frac{1}{\sqrt{n}} \sum_{i=1}^n Y_i^P \right) - EH\left(\frac{1}{\sqrt{n}} \sum_{i=1}^n Y_{i,\tau}^P \right) \right|,$$

$$\eta_2 := \left| EH\left(\frac{1}{\sqrt{n}} \sum_{i=1}^n Y_{i,\tau}^P \right) - EH(G_{P,\tau}) \right|, \quad \eta_3 := |EH(G_{P,\tau}) - EH(G_P)|.$$

By (10.53), $\sup_{P \in \mathcal{P}(X)} N_P(\tau) < +\infty$. Then by Lemma 10.24 we have for each $\tau > 0$

$$\lim_{n \to \infty} \sup \{\eta_2 : P \in \mathcal{P}(X), H \in BL_1^{\mathcal{F}}\} = 0. \qquad (10.62)$$

For each $\varepsilon > 0$ and $H \in BL_1^{\mathcal{F}}$, since for each $\phi, \psi \in \mathcal{F}$, $|H(\phi) - H(\psi)| \leq \min(2, \|\phi - \psi\|_{\sup})$, we have for each $P \in \mathcal{P}(X)$

$$\eta_1 \leq \varepsilon + 2 \Pr \left(\|\nu_n^P\|_{\mathcal{F}'(\tau, e_P)} > \varepsilon \right),$$

and so by Claim 3 (10.49)

$$\lim_{\tau \downarrow 0} \lim_{n \to \infty} \sup \sup \{\eta_1 : P \in \mathcal{P}(X), H \in BL_1^{\mathcal{F}}\} = 0. \qquad (10.63)$$

Clearly, $\eta_3 \leq \varepsilon + 2 \Pr \left(\|G_P\|_{\mathcal{F}'(\tau, e_P)} > \varepsilon \right)$, and so by (10.60)

$$\lim_{\tau \downarrow 0} \lim_{n \to \infty} \sup \sup \{\eta_3 : P \in \mathcal{P}(X), H \in BL_1^{\mathcal{F}}\} = 0. \qquad (10.64)$$

The displays (10.61) through (10.64) combine to prove Claim 5, i.e. that $\mathcal{F}$ is uniform Donsker (10.35), and so prove Theorem 10.22. □

Under Pollard's entropy condition (10.18), Theorem 10.13 can be strengthened as follows:

Theorem 10.26 *If $(X, \mathcal{A})$ is a measurable space and $\mathcal{F}$ is a uniformly bounded, image admissible Suslin class of functions on $(X, \mathcal{A})$ satisfying Pollard's entropy condition (10.18), then $\mathcal{F}$ is a uniform Donsker class.*

Proof. Pollard's entropy condition in this case can be written as

$$\int_0^\infty \sup_{Q \in \mathcal{P}_f(X)} \left(\log D(\varepsilon, \mathcal{F}, e_Q) \right)^{1/2} d\varepsilon < \infty. \tag{10.65}$$

For $\delta > 0$ let $G(\delta) := \int_0^\delta \sup_{Q \in \mathcal{P}_f(X)} \left(\log D(\varepsilon, \mathcal{F}, e_Q) \right)^{1/2} d\varepsilon$. Thus $G(\delta) \downarrow 0$ as $\delta \downarrow 0$. Using the metric entropy sufficient condition for sample continuity of Gaussian processes, in the modulus of continuity form (Theorem 2.36), more specifically with expectations (2.25), and Theorem 10.22, give that $\mathcal{F}$ is uniform Donsker. $\qquad\square$

Theorem 10.15 showed that for a universal Donsker class $\mathcal{F}$, the symmetric convex hull $H(\mathcal{F}, 1)$ and its closure $\overline{H}_s(\mathcal{F}, 1)$ for sequential pointwise convergence are also universal Donsker classes. Bousquet, Koltchinskii, and Panchenko (2002) proved (or more precisely proved a fact from which it easily follows, as they noted) that the same holds for the uniform Donsker property. They give in detail the sufficient fact, interesting in its own right, in the next theorem.

For a probability space (Ω, P), a pregaussian set $\mathcal{F} \subset \mathcal{L}^2(P)$, and $\delta > 0$ define the modulus of continuity (in expectation) of G_P on $\mathcal{F}$ as

$$\omega(\mathcal{F}, \delta) := E \|G_P\|_{\mathcal{F}'(\delta, \rho_P)}.$$

As $\mathcal{F}$ is pregaussian, by Theorem 3.2, we can take G_P to be coherent, i.e., to have prelinear and ρ_P-uniformly continuous sample functions on $\mathcal{F}$. Moreover by Lemma 2.30, prelinearity on $\mathcal{F}$ implies that each $G_P(\cdot)(\omega)$ has a unique linear extension to the linear span of $\mathcal{F}$ and thus is uniquely defined on the symmetric convex hull of $\mathcal{F}$. On moduli of continuity we then have:

Theorem 10.27 (Bousquet, Koltchinskii, and Panchenko) *For any pregaussian $\mathcal{F}$ and any $\delta > 0$,*

$$\omega(H(\mathcal{F}, 1), \delta) \le \inf_{\varepsilon > 0} \left(4\omega(\mathcal{F}, \varepsilon) + \delta \sqrt{D(\varepsilon, \mathcal{F}, \rho_P)} \right). \tag{10.66}$$

Proof. Let $\varepsilon > 0$ and let $S_\varepsilon \subset \mathcal{F}$ be a maximal set such that $\rho_P(f, g) > \varepsilon$ for $f \ne g$ in S, of cardinality $N := \text{card}(S) = D(\varepsilon, \mathcal{F}, \rho_P)$. Recall that this must be finite since by (Sudakov's) Theorem 2.19 $E_{\mathcal{F}} := E \sup_{f \in \mathcal{F}} G_P(f) < +\infty$ and then from the Sudakov minoration Theorem 2.22(b),

$$E_{\mathcal{F}} \ge \frac{\varepsilon}{17} \sqrt{\log D(\varepsilon, \mathcal{F}, \rho_P)}. \tag{10.67}$$

Consider the Hilbert space $H = L_0^2(P) := \{f \in L^2(P) : E_P f = 0\}$, with the (covariance) inner product. On this H, G_P is the isonormal process, while on the other hand, for any $f \in \mathcal{L}^2(P)$ and constant c, e.g. Pf, $G_P(f - c) \equiv G_P(f)$. Let V be the linear span of S_ε in the Hilbert space H. Let W be the orthogonal complement of V in H, and let π_V and π_W be the orthogonal projections onto V and W respectively. Then for each $f \in \mathcal{F}$, $f = \pi_V f + \pi_W f$, and so

$$\omega(H(\mathcal{F}, 1), \delta) \leq E\left[\sup\{|G_P(\pi_V h)| : h \in H(\mathcal{F}, 1)'(\delta, \rho_P)\}\right]$$
$$+E\left[\sup\{|G_P(\pi_W h)| : h \in H(\mathcal{F}, 1)'(\delta, \rho_P)\}\right].$$

Since any orthogonal projection Π is a contraction, $\|\Pi f - \Pi g\| \leq \|f - g\|$, we have

$$\omega(H(\mathcal{F}, 1), \delta) \leq \omega(\pi_V H(\mathcal{F}, 1), \delta) + \omega(\pi_W H(\mathcal{F}, 1), \delta). \quad (10.68)$$

To bound the first term on the right of (10.68), let d be the dimension of V, for which $d \leq D(\varepsilon, \mathcal{F}, \rho_P)$. We have

$$\omega(\pi_V(H(\mathcal{F}, 1)), \delta) \leq \omega(V, \delta),$$

and since G_P is linear and V is a vector space,

$$\omega(V, \delta) \leq \delta E\|Z\| \leq \delta(E\|Z\|^2)^{1/2}$$

where Z is a standard normal d-dimensional vector, so that

$$\omega(V, \delta) \leq \delta\sqrt{d} \leq \delta\sqrt{D(\varepsilon, \mathcal{F}, \rho_P)}. \quad (10.69)$$

For the second term on the right of (10.68), we have a crude bound, which will be sufficient,

$$\omega(\pi_W H(\mathcal{F}, 1), \delta) \leq 2E\left[\sup\{|G_P(\pi_W f)| : f \in H(\mathcal{F}, 1)\}\right].$$

Since π_W and G_P are linear, the supremum is attained at extreme points of $H(\mathcal{F}, 1)$, which are functions $\pm f$ for $f \in \mathcal{F}$, from which we get

$$\omega(\pi_W H(\mathcal{F}, 1), \delta) \leq 2E\left[\sup\{|G_P(\pi_W f)| : f \in \mathcal{F}\}\right].$$

For $S = \{f_1, \ldots, f_N\}$, given $f \in \mathcal{F}$, let $g = f_j$ be the closest point of S to f, or the one with smallest j in case of a tie. Then $\|f - g\| \leq \varepsilon$ and $g \in V \cap \mathcal{F}$, so $\pi_W g = 0$ and

$$\omega(\pi_W H(\mathcal{F}, 1), \delta)$$
$$\leq 2E\left[\sup\{|G_P(\pi_W f) - G_P(\pi_W g)| : f, g \in \mathcal{F}, \|f - g\| \leq \varepsilon\}\right]$$
$$= 2E\|G_P \circ \pi_W\|_{\mathcal{F}'(\varepsilon, \rho_P)}.$$

Since π_W is a contraction, the last expression can be bounded above using an inequality related to Slepian's, (2.21) in Theorem 2.18, by

$$4E\|G_P\|_{\mathcal{F}'(\varepsilon, \rho_P)} = 4\omega(\mathcal{F}, \varepsilon).$$

Combining this with (10.69) gives (10.66). □

Now the following will be proved:

Theorem 10.28 *Let $(X, \mathcal{S})$ be a measurable space and $\mathcal{F}$ a uniform Donsker class of functions on X which is uniformly bounded and image admissible Suslin. Then the symmetric convex hull $H(\mathcal{F}, 1)$ of $\mathcal{F}$ and its set $\overline{H}_s(\mathcal{F}, 1)$ of limits under sequential pointwise convergence are also uniform Donsker classes.*

Proof. It suffices by Theorem 10.22 to show that if $\mathcal{F}$ is uniformly pregaussian, then so are the other two classes of functions. By (10.31) we have $M := \sup_{P \in \mathcal{P}(X)} E\|G_P\|_{\mathcal{F}} < \infty$ where in (10.67) $E_{\mathcal{F}} \leq M$ for all P. Moreover, $\|G_P\|_{H(\mathcal{F},1)} \equiv \|G_P\|_{\mathcal{F}}$, so that (10.31) holds for $H(\mathcal{F}, 1)$ with the same M.

For each $\varepsilon > 0$, and all $P \in \mathcal{P}(X)$, by (10.67)

$$N := D(\varepsilon, \mathcal{F}, \rho_P) \leq N_0(\varepsilon) := \exp(289M^2/\varepsilon^2).$$

To prove (10.32) for $H(\mathcal{F}, 1)$, as $\delta \downarrow 0$, it will suffice to choose $\varepsilon = \varepsilon(\delta)$ in (10.66) which also converges to 0 and is such that $\delta\sqrt{N_0(\varepsilon)}$ converges to 0, so that the right side of (10.66) will converge to 0. For the latter it will suffice that $N_0(\varepsilon) \leq 1/\sqrt{\delta}$. We can set $\varepsilon = 17\sqrt{2}M/\sqrt{\log(1/\delta)}$ which does converge to 0 as $\delta \downarrow 0$ (although, of course, more slowly than δ does). Sequential pointwise limits do not change any $\|G_P\|_{\mathcal{G}}$ or $\|G_P\|_{\mathcal{G}'(\delta, \rho_P)}$, from $\mathcal{G} = H(\mathcal{F}, 1)$ to $\mathcal{G} = \overline{H}_s(\mathcal{F}, 1)$, so $\overline{H}_s(\mathcal{F}, 1)$ is also uniformly pregaussian and thus uniformly Donsker. $\square$

10.5 Universal Glivenko–Cantelli Classes

Given a measurable space $(X, \mathcal{A})$, a class $\mathcal{F}$ of measurable real-valued functions on X is called a universal Glivenko–Cantelli class if it is Glivenko–Cantelli for each law P on $(X, \mathcal{A})$. The notion of universal Glivenko–Cantelli class seems to be excessively general. For example, if X is a countably infinite set, then the class 2^X of all its subsets is Glivenko–Cantelli for every P defined on it, but it is not uniformly Glivenko–Cantelli (Problem 5); see also Problem 6. Dudley, Giné and Zinn (1991) give various examples of unexpectedly large universal Glivenko–Cantelli classes. Some of them could well be called pathological. It seems that uniform Glivenko–Cantelli classes are of considerable interest in statistics and machine learning (e.g., Alon et al. 1997), where P is unknown, but universal Glivenko–Cantelli classes may not be.

Problems

1. Let V be a finite-dimensional vector space of real-valued functions on a set X. Let W be a subset of V consisting of functions f with $f(x) > 0$ for all

$x \in X$. Show that the set of all quotients f/g for $f \in V$ and $g \in W$ is a VC major class. *Hint*: For any real t, $f/g > t$ if and only if $f - gt > 0$. Apply a fact from Chapter 4.

2. For $d = 1, 2, \ldots$ and $k = 1, 2, \ldots$, let $V_{k,d}$ be the vector space of all polynomials on $\mathbb{R}^d$ of degree at most $2k$. For M with $0 < M < +\infty$, let $V_{k,d,M}$ be the set of all $f \in V_{k,d}$, all of whose coefficients have absolute values $\leq M$. Let $\mathcal{F}$ be the set of all quotients $x \mapsto f(x)/(1 + |x|^2)^k$ for $f \in V_{k,d,M}$. Show that $\mathcal{F}$ is a universal Donsker class. *Hints:* Apply Problem 1 and Corollary 10.21. One needs to show that $\mathcal{F}$ is uniformly bounded. A bound on the dimension of $V_{k,d}$ will help. To prove that $\mathcal{F}$ is image admissible Suslin, use that the set of possible vectors of coefficients is a Polish space and show joint measurability.

3. Show that each class $\mathcal{G}_{\alpha,K,d}$ in Theorem 8.4 for $\alpha > 0$ and $K < \infty$ is a uniform Glivenko–Cantelli class. *Hint*: The sets $\mathcal{G}_{\alpha,K,d}$ are uniformly bounded and totally bounded for $d_{\sup}$, by the Arzelà–Ascoli theorem or more specifically by Theorem 8.4.

4. Show that the "ellipsoid" universal Donsker class of Proposition 10.17 is not uniform Donsker.

5. For a countably infinite set, say the set $\mathbb{N}$ of nonnegative integers, show that the class $\mathcal{C} = 2^{\mathbb{N}}$ of all subsets is a universal Glivenko–Cantelli class as shown in Problem 8 of Chapter 6. Show that it is not a uniform Glivenko–Cantelli class.

6. Let $(X, \mathcal{A})$ be a measurable space and let X be the union of a sequence of disjoint measurable sets A_k. Suppose that for each k, $\mathcal{C}_k$ is an image admissible Suslin VC class of sets, where $S(\mathcal{C}_k)$ may go to ∞ arbitrarily fast as $k \to \infty$. Show that the collection $\mathcal{C}$ of all sets A such that $A \cap A_k \in \mathcal{C}_k$ for each k is a universal Glivenko–Cantelli class. *Hint*: Use the result of the previous problem and the Glivenko–Cantelli property of each VC class $\mathcal{C}_k$.

7. If $\mathcal{F}$ is a universal Glivenko–Cantelli class of functions on $(X, \mathcal{A})$, then $\mathcal{F}$ is uniformly bounded up to additive constants, i.e., the conclusion of Proposition 10.10 holds for it. *Hint*: For each $f \in \mathcal{F}$ and law P on $(X, \mathcal{A})$, $P(|f|)$ must be finite in order for $(P_n - P)(f)$ to make sense. Adapt the first part of the proof of Proposition 10.10, using 8^k in place of 2^k.

Notes

Notes to Section 10.1. The main Theorem 10.6 was proved by Dudley, Giné, and Zinn (1991, Theorem 6). I am very indebted to Evarist Giné for suggestions on the proof given here. Theorem 10.8 is essentially Lemma 2.9 of Giné and Zinn (1984), for which they give credit to G. Pisier and X. Fernique. Later, Alon, Ben-David, Cesa-Bianchi, and Haussler (1997) gave another, interesting characterization of the uniform Glivenko–Cantelli property.

Notes to Section 10.2. This section is based on Dudley (1987).

Notes to Section 10.3. Theorem 10.19 is from Dudley (1987). The argument of B. Maurey used in its proof was published in the proofs of Pisier (1981), Lemma 2, and Carl (1982), Lemma 1. The example showing that $2\lambda/(2 + \lambda)$ is sharp is from Dudley (1967a), Propositions 5.8 and 6.12. van der Vaart and Wellner (1996, Theorem 2.6.9) and Carl (1997) showed that one can take $t = s$ in Theorem 10.19.

Notes to Section 10.4. The section is based on the paper of Giné and Zinn (1991). Lemma 10.24 is their Lemma 2.1, which they say is well known; the bound (10.36) is used in the 1991 proof and given with proof, for $d = 1$ but the proof directly extends to all d, in Araujo and Giné (1980, Theorem 1.3).

In Lemma 10.25 and its proof, I do not claim that among random variables X, Y with distributions $N(0, C)$ and $N(0, D)$, $X = \sqrt{C}Z$ and $Y = \sqrt{D}Z$ with Z having $N(0, I_d)$ achieve the minimum of $E|X - Y|^2$, even requiring (X, Y) to have a Gaussian joint distribution. According to Olkin and Pukelsheim (1982), they do if C and D commute, but not in general.

It appears that Dobrushin (1970) proposed the name "Wasserstein metric," referring to a paper by Vaserštein (1969). Cuesta and Matran (1989) say such metrics are due to Kantorovich (1942). The Kantorovich–Rubinštein theorem about this metric was published in Kantorovich and Rubinštein (1958). Some authors later have used the name "Kantorovich metric." Kantorovich's work on "transportation problems" has become quite widely known.

The main theorem characterizing uniform Donsker classes, Theorem 10.22, is Theorem 2.3 of Giné and Zinn (1991). The fact that Pollard's entropy condition, with boundedness and measurability, implies uniform Donsker, Theorem 10.26, follows from Proposition 3.1 of Giné and Zinn (1991) as stated there, although a different method of providing a sample modulus in expectation was used here.

Bousquet, Koltchinskii, and Panchenko (2002) essentially proved Theorem 10.28 on preservation of the uniform Donsker property by convex hulls, in that they did prove Theorem 10.27 from which it easily follows. (They gave a factor of 2 rather than 4 in (10.66); the 4 is used here to allow a more self-contained proof based on (2.21).)

11

Classes of Sets or Functions Too Large for Central Limit Theorems

11.1 Universal Lower Bounds

This chapter is primarily about asymptotic lower bounds for $\|P_n - P\|_{\mathcal{F}}$ on certain classes $\mathcal{F}$ of functions, as treated in Chapter 8, mainly classes of indicators of sets. Section 11.2 will give some upper bounds which indicate the sharpness of some of the lower bounds. Section 11.4 gives some relatively difficult lower bounds on classes such as the convex sets in $\mathbb{R}^3$ and lower layers in $\mathbb{R}^2$. In preparation for this, Section 11.3 treats Poissonization and random "stopping sets" analogous to stopping times. The present section gives lower bounds in some cases which hold not only with probability converging to 1, but for all possible P_n. Definitions are as in Sections 3.1 and 8.2, with $P := U(I^d) = \lambda^d$ = Lebesgue measure on I^d. Specifically, recall the classes $\mathcal{G}(\alpha, K, d) := \mathcal{G}_{\alpha,K,d}$ of functions on the unit cube $I^d \subset \mathbb{R}^d$ with derivatives through αth order bounded by K, and the related families $\mathcal{C}(\alpha, K, d)$ of sets (subgraphs of functions in $\mathcal{G}(\alpha, K, d-1)$), both defined early in Section 8.2.

Theorem 11.1 (Bakhvalov) *For $P = U(I^d)$, any $d = 1, 2, \ldots$ and $\alpha > 0$, there is a $\gamma = \gamma(d, \alpha) > 0$ such that for all $n = 1, 2, \ldots$, and all possible values of P_n, we have $\|P_n - P\|_{\mathcal{G}(\alpha,1,d)} \geq \gamma n^{-\alpha/d}$.*

Remarks. When $\alpha < d/2$, this shows that $\mathcal{G}(\alpha, K, d)$, $K > 0$, is not a Donsker class. For $\alpha > d/2$, $\mathcal{G}(\alpha, K, d)$ is Donsker by Corollary 8.5(b). The lower bound in Theorem 11.1 is not useful, since it is smaller than the average size of $\|P_n - P\|_{\mathcal{G}(\alpha,1,d)}$, which is at least of order $n^{-1/2}$: even for one function f not constant a.e. P, $E|(P_n - P)(f)| \geq cn^{-1/2}$ for some $c > 0$. For $\alpha = d/2$ and $\mathcal{F} := \mathcal{G}(d/2, 1, d)$, recalling $\nu_n := \sqrt{n}(P_n - P)$, always $\|\nu_n\|_{\mathcal{F}} \geq \gamma > 0$, so $\mathcal{F}$ is not P-Donsker, as follows. Theorem 2.32(b) says that for a pregaussian class $\mathcal{F}$ (a GC-set), $\|G_P\|_{\mathcal{F}} < \gamma$ with positive probability, so ν_n cannot converge to it in law (see Problem 3).

391

Theorem 11.1 gives information about accuracy of possible methods of numerical integration in several dimensions, or "cubature," using the values of a function $f \in \mathcal{G}(\alpha, K, d)$ at just n points chosen in advance (from the proof, it will be seen that one has the same lower bound even if one can use any partial derivatives of f at the n points). It was in this connection that Bakhvalov (1959) proved the theorem.

Proof. Given n let $m := m(n) := \lceil (2n)^{1/d} \rceil$ where $\lceil x \rceil$ is the smallest integer $\geq x$. Decompose the unit cube I^d into m^d cubes C_i of side $1/m$. Then $m^d \geq 2n$. For any P_n let $S := \{i : P_n(C_i) = 0\}$. Then $\mathrm{card}(S) \geq n$. For $x_{(i)}$, f_i and f_S as defined after (8.11), we then have for each i,

$$P(f_i) = m^{-\alpha} \int f(m(x - x_{(i)}))dx = m^{-\alpha} \int_{C_i} f(m(x - x_{(i)}))dx$$

$$= \int_{I^d} f(y)dy/m^{d+\alpha} = m^{-d-\alpha}P(f).$$

Thus $|(P_n - P)(f_S)| = P(f_S) \geq m^{-\alpha-d}nP(f) \geq cn^{-\alpha/d}$ for some constant $c = c(d, \alpha) > 0$, while $\|f_S\|_\alpha \leq \|f\|_\alpha \leq 1$ (dividing the original f by some constant depending on d and α if necessary). $\qquad \square$

Theorem 11.2 For $P = U(I^d)$, any $K > 0$ and $0 < \alpha < d - 1$ there is a $\delta = \delta(\alpha, K, d) > 0$ such that for all $n = 1, 2, \ldots$ and all possible values of P_n, $\|P_n - P\|_{\mathcal{C}(\alpha, K, d)} > \delta n^{-\alpha/(d-1+\alpha)}$.

Remark. Since $\alpha/(d - 1 + \alpha) < 1/2$ for $\alpha < d - 1$, the classes $\mathcal{C}(\alpha, K, d)$ are then not Donsker classes. For $\alpha > d - 1$, $\mathcal{C}(\alpha, K, d)$ is a Donsker class by Theorem 8.4 and Corollary 7.8. For $\alpha = d - 1$, it is not a Donsker class via Theorem 2.32(b) as in the remarks after the preceding theorem. Theorem 11.10 below will show that $\|\nu_n\|_{\mathcal{C}(d-1,K,d)}$ is unbounded in probability (at a logarithmic rate).

Proof. Again, the construction and notation around (8.11) will be applied, now on I^{d-1}. Let $c := K\lambda^{d-1}(f)/\|f\|_\alpha$, $m := m(n) := \lfloor (nc)^{1/(\alpha+d-1)} \rfloor$. Then $1/n \leq cm^{1-d-\alpha}$. Take $n \geq r$ (the result holds for $n < r$) for an r with

$$M := \sup_{n \geq r} nc\, m(n)^{1-d-\alpha} < \infty.$$

Let $\Theta_n := m(n)^{\alpha+d-1}/(nc)$. Then $1/M \leq \Theta_n \leq 1$. Let $g_i := K\Theta_n f_i/(2\|f\|_\alpha)$. Then $\lambda^{d-1}(g_i) = 1/(2n)$ and $\|g_i\|_\alpha \leq K$. Let $B_i := \{x \in I^d : 0 < x_d < g_i(x_{(d)})\}$ where $x_{(d)} := (x_1, \ldots, x_{d-1})$. Then for each i, $\lambda^d(B_i) := P(B_i) = 1/(2n)$ and either $P_n(B_i) = 0$ or $P_n(B_i) \geq 1/n$. Either at least half the B_i have $P_n(B_i) = 0$, or at least half have $P_n(B_i) \geq 1/n$. In either case,

$$\|P_n - P\|_{\mathcal{C}(\alpha, K, d)} \geq m^{d-1}/(4n) \geq cm^{-\alpha}/(4M)$$
$$\geq \delta n^{-\alpha/(\alpha+d-1)}$$

for some $\delta(\alpha, K, d) > 0$. $\qquad\square$

The method of the last proof applies to convex sets in $\mathbb{R}^d$, as follows.

Theorem 11.3 (W. M. Schmidt) *Let $d = 2, 3, \ldots$. For the collection $\mathcal{C}_d$ of closed convex subsets of a bounded nonempty open set U in $\mathbb{R}^d$ there is a constant $b := b(d, U) > 0$ such that for $P = $ Lebesgue measure normalized on U, and all P_n,*

$$\sup\{|(P_n - P)(C)| : C \in \mathcal{C}_d\} \geq bn^{-2/(d+1)}.$$

Proof. We can assume by a Euclidean transformation that U includes the unit ball. Take disjoint spherical caps as at the end of Section 8.4. Let each cap C have volume $P(C) = 1/(2n)$. Then as $n \to \infty$, the angular radius ε_n of such caps is asymptotic to $c_d n^{-1/(d+1)}$ for some constant c_d. Thus the number of such disjoint caps is of the order of $n^{(d-1)/(d+1)}$. Either $(P_n - P)(C) \geq 1/(2n)$ for at least half of such caps, or $(P_n - P)(C) = -1/(2n)$ for at least half of them. Thus for some constant $\eta = \eta(U, d)$, there exist convex sets D, E, differing by a union of caps, such that

$$|(P_n - P)(D) - (P_n - P)(E)| \geq \eta n^{(d-1)/(d+1) - 1},$$

and the result follows with $b = \eta/2$. $\qquad\square$

Thus for $d \geq 4$, $\mathcal{C}_d$ is not a Donsker class for P. If $d = 3$, it is not either, by the same argument as in the remarks after the previous two theorems; cf. Problem 3. $\mathcal{C}_2$ is a Donsker class for λ^2 on I^2 by Theorem 8.25 and Corollary 7.8.

11.2 An Upper Bound

Here, using metric entropy with bracketing N_I as in Section 7.1, is an upper bound for $\|v_n\|_{\mathcal{C}} := \sup_{B \in \mathcal{C}} |v_n(B)|$, which applies in many cases where the hypotheses of Corollary 7.8 fail. Let $(X, \mathcal{A}, Q)$ be a probability space, $v_n := n^{1/2}(Q_n - Q)$, and recall N_I as defined before (7.4).

Theorem 11.4 *Let $\mathcal{C} \subset \mathcal{A}$, $1 \leq \zeta < \infty$, $\eta > 2/(\zeta + 1)$ and $\Theta := (\zeta - 1)/(2\zeta + 2)$. If for some $K < \infty$, $N_I(\varepsilon, \mathcal{C}, Q) \leq \exp(K\varepsilon^{-\zeta})$, $0 < \varepsilon \leq 1$, then*

$$\lim_{n \to \infty} \Pr^* \{\|v_n\|_{\mathcal{C}} > n^{\Theta}(\log n)^{\eta}\} = 0.$$

Remarks. The statement is not interesting if C is a Q-Donsker class, as $\Theta \geq 0$, $\eta > 0$, and then $\Pr^*\{\|\nu_n\|_C > a_n\} \to 0$ for any $a_n \to +\infty$. Nor is the statement useful in proving classes are not Donsker because for that we need lower, not upper, bounds. It is useful in showing that some lower bounds are sharp, with respect to Θ. (With respect to η, it is not sharp as will be seen in Section 11.4.)

The classes $C = C(\alpha, M, d)$ satisfy the hypothesis of Theorem 11.4 for $\zeta = (d-1)/\alpha \geq 1$, i.e. $\alpha \leq d-1$, by the last inequality in Theorem 8.4. Then $\Theta = \frac{1}{2} - \frac{\alpha}{d-1+\alpha}$. Thus Theorem 11.2 shows that the exponent Θ is sharp for $\zeta > 1$. Conversely, Theorem 11.4 shows that the exponent on n in Theorem 11.2 cannot be improved. In Theorem 11.4 we cannot take $\zeta < 1$, for then $\Theta < 0$, which is impossible even for a single set, $C = \{C\}$, with $0 < P(C) < 1$.

Proof. The chaining method will be used as in Section 7.2. Let $\log_2$ be logarithm to the base 2. For each $n \geq 3$ let

$$k(n) := \left[\left(\frac{1}{2} - \Theta\right) \log_2 n - \eta \log_2 \log n\right]. \tag{11.1}$$

Let $N(k) := N_I(2^{-k}, C, Q)$, $k = 1, 2, \ldots$. Then for some A_{ki} and $B_{ki} \in \mathcal{A}$, $i = 1, \ldots, N(k)$, and any $A \in C$, there are i, $j \leq N(k)$ with $A_{ki} \subset A \subset B_{ki}$ and $Q(B_{ki} \backslash A_{ki}) \leq 2^{-k}$. Let $A_{01} := \emptyset$ (the empty set) and $B_{01} := X$. Choose such $i = i(k, A)$ for $k = 0, 1, \ldots$. Then for each $k \geq 0$,

$$Q(A_{ki(k,A)} \triangle A_{k-1,i(k-1,A)}) \leq 2^{2-k}$$

where $\triangle :=$ symmetric difference, $C \triangle D := (C \backslash D) \cup (D \backslash C)$.

For $k \geq 1$ let $\mathcal{B}(k)$ be the collection of sets B, with $Q(B) \leq 2^{2-k}$, of the form $A_{ki} \backslash A_{k-1,j}$ or $A_{k-1,j} \backslash A_{ki}$ or $B_{ki} \backslash A_{ki}$. Then

$$\mathrm{card}(\mathcal{B}(k)) \leq 2N(k-1)N(k) + N(k) \leq 3\exp(2K2^{k\zeta}).$$

For each $B \in \mathcal{B}(k)$, Bernstein's inequality (Theorem 1.11) implies, for any $t > 0$,

$$\Pr\{|\nu_n(B)| > t\} \leq 2\exp(-t^2/(2^{3-k} + tn^{-1/2})). \tag{11.2}$$

Choose $\delta > 0$ such that $\delta < 1$ and $(2 + 2\delta)/(1 + \zeta) < \eta$. Let $c := \delta/(1 + \delta)$ and $t := t_{n,k} := cn^\Theta (\log n)^\eta k^{-1-\delta}$. Then for each $k = 1, \ldots, k(n)$, $2^{3-k} \geq 8n^{\Theta-1/2}(\log n)^\eta \geq t_{n,k}n^{-1/2}$. Hence by (11.2),

$$\Pr\{|\nu_n(B)| > t_{n,k}\} \leq 2\exp(-t_{n,k}^2/2^{4-k}), \quad \text{and}$$

$$P_{nk} := \Pr\left\{\sup_{B \in \mathcal{B}(k)} |\nu_n(B)| > t_{n,k}\right\} \leq 6\exp\left(2K2^{k\zeta} - 2^{k-4}t_{n,k}^2\right)$$

$$= 6\exp\left(2K2^{k\zeta} - 2^{k-4}c^2n^{2\Theta}(\log n)^{2\eta}k^{-2-2\delta}\right).$$

For $k \leq k(n)$ we have $2^k(\log n)^\eta \leq n^{1/2-\Theta}$. Since $\Theta < 1/2$ and $2\Theta/(\frac{1}{2} - \Theta) = \zeta - 1$, we have $n^{2\Theta} \geq 2^{k(\zeta-1)}(\log n)^{\eta(\zeta-1)}$, and

$$P_{nk} \leq 6\exp\left(2K2^{k\zeta} - 2^{k\zeta-4}c^2(\log n)^{\eta(\zeta+1)}k^{-2-2\delta}\right).$$

Let $\gamma := \eta(\zeta + 1) - 2 - 2\delta > 0$ by choice of δ. Since $\frac{1}{2} < \log 2$, $k(n) \leq \log n$ and $P_{nk} \leq 6\exp(2^{k\zeta}(2K - 2^{-4}c^2(\log n)^\gamma))$.

For n large, $(\log n)^\gamma > 64K/c^2$. Then $2K < 2^{-5}c^2(\log n)^\gamma$ and $2^{k\zeta} \geq 1$ for $k \geq 1$, so

$$\sum_{k=1}^{k(n)} P_{nk} \leq 6(\log n)\exp(-2^{-5}c^2(\log n)^\gamma) \to 0 \quad \text{as } n \to \infty.$$

Let $\mathcal{E}_n := \{\omega: \sup_{B\in\mathcal{B}(r)} |v_n(B)| \leq t_{n,r}, \ r = 1, \ldots, k(n)\}$. Then $\lim_{n\to\infty} \Pr(\mathcal{E}_n) = 1$. For any $A \in \mathcal{C}$ and n, let $k := k(n)$ and $i := i(k, A)$. Then for each $\omega \in \mathcal{E}_n$, $|v_n(A_{ki})|$ is bounded above by

$$\sum_{r=1}^{k} |v_n(A_{ri(r,A)} \setminus A_{r-1,i(r-1,A)})| + |v_n(A_{r-1,i(r-1,A)} \setminus A_{r,i(r,A)})|$$

$$\leq 2\sum_{r=1}^{k} t_{n,r} < 2n^\Theta(\log n)^\eta \sum_{r\geq 1} cr^{-1-\delta} \leq 2n^\Theta(\log n)^\eta,$$

$|v_n(B_{ki}\setminus A_{ki})| \leq n^\Theta(\log n)^\eta$, and by (11.1),

$$n^{1/2}Q(B_{ki}\setminus A_{ki}) \leq n^{1/2}/2^k < 2n^\Theta(\log n)^\eta.$$

Hence $n^{1/2}Q_n(B_{ki}\setminus A_{ki}) \leq 3n^\Theta(\log n)^\eta$, and $|v_n(A\setminus A_{ki})| \leq 3n^\Theta(\log n)^\eta$. So on $\mathcal{E}_n$, $|v_n(A)| \leq 5n^\Theta(\log n)^\eta$. As $\eta \downarrow 2/(\zeta + 1)$ the factor of 5 can be dropped. Since $A \in \mathcal{C}$ is arbitrary, the proof is done. $\square$

11.3 Poissonization and Random Sets

Section 11.4 will give some lower bounds $\|v_n\|_\mathcal{C} \geq f(n)$ with probability converging to 1 as $n \to \infty$ where f is a product of powers of logarithms or iterated logarithms. Such an f has the following property. A real-valued function f defined for large enough $x > 0$ is called *slowly varying* (in the sense of Karamata) iff for every $c > 0$, $f(cx)/f(x) \to 1$ as $x \to +\infty$.

Lemma 11.5 *If f is continuous and slowly varying, then for every $\varepsilon > 0$ there is a $\delta = \delta(\varepsilon) > 0$ such that whenever $x > 1/\delta$ and $|1 - \frac{y}{x}| < \delta$ we have* $\left|1 - \frac{f(y)}{f(x)}\right| < \varepsilon$.

Proof. For each $c > 0$ and $\varepsilon > 0$ there is an $x(c, \varepsilon)$ such that for $x \geq x(c, \varepsilon)$, $\left|\frac{f(cx)}{f(x)} - 1\right| \leq \frac{\varepsilon}{4}$. Note that if $x(c, \varepsilon) \leq n$, even for one c, then $f(x) \neq 0$ for all $x \geq n$. By the category theorem (e.g., RAP, Theorem 2.5.2), for fixed $\varepsilon > 0$ there is an $n < \infty$ such that $x(c, \varepsilon) \leq n$ for all c in a set dense in some interval $[a, b]$, where $0 < a < b$, and thus by continuity for all c in $[a, b]$. Then for $c, d \in [a, b]$ and $x \geq n$, $|(f(cx) - f(dx))/f(x)| \leq \varepsilon/2$. Fix $c := (a + b)/2$.

Let $u := cx$. Then for $u \geq nc$, we have

$$\left|\frac{f(ud/c)}{f(u)} - 1\right| \left|\frac{f(u)}{f(u/c)}\right| \leq \frac{\varepsilon}{2}.$$

As $u \to +\infty$, $f(u)/f(u/c) \to 1$. Then there is a $\delta > 0$ with $\delta < (b - a)/(b + a)$ such that for $u \geq 1/\delta$ and all r with $|r - 1| < \delta$ we have $\left|\frac{f(ru)}{f(u)} - 1\right| < \varepsilon$. $\qquad\square$

Recall the Poisson law P_c on $\mathbb{N}$ with parameter $c \geq 0$, so that $P_c(k) := e^{-c} c^k / k!$ for $k = 0, 1, \ldots$. Given a probability space $(X, \mathcal{A}, P)$, let U_c be a Poisson point process on $(X, \mathcal{A})$ with intensity measure cP: for any disjoint $A_1, \ldots, A_m$ in $\mathcal{A}$, $U_c(A_j)$ are jointly independent, $j = 1, \ldots, m$, $U_c(A_1 \cup A_2) = U_c(A_1) + U_c(A_2)$, and for any $A \in \mathcal{A}$, $U_c(A)(\cdot)$ has law $P_{cP(A)}$.

Let $Y_c(A) := (U_c - cP)(A)$, $A \in \mathcal{A}$. Then Y_c has mean 0 on all A and still has independent values on disjoint sets.

Let $x(1), x(2), \ldots$ be coordinates for the product space $(X^\infty, \mathcal{A}^\infty, P^\infty)$. For $c > 0$ let $n(c)$ be a random variable with law P_c, independent of the $x(i)$. Then for $P_n := n^{-1}(\delta_{x(1)} + \cdots + \delta_{x(n)})$, $n \geq 1$, $P_0 := 0$ we have:

Lemma 11.6 *The process $Z_c := n(c) P_{n(c)}$ is a Poisson process with intensity measure cP.*

Proof. We have laws $\mathcal{L}(U_c(X)) = \mathcal{L}(Z_c(X)) = P_c$. If X_i are independent Poisson variables with $\mathcal{L}(X_i) = P_{c(i)}$ and $\sum_{i=1}^m c(i) = c$, then given $n := \sum_{i=1}^m X_i$, the conditional distribution of $\{X_i\}_{i=1}^m$ is multinomial with total n and probabilities $p_i = c(i)/c$. Thus U_c and Z_c have the same conditional distributions on disjoint sets given their values on X. This implies the Lemma. $\qquad\square$

From here on, the version $U_c \equiv Z_c$ will be used. Thus for each ω, $U_c(\cdot)(\omega)$ is a countably additive integer-valued measure of total mass $U_c(X)(\omega) = n(c)(\omega)$. Then

$$Y_c = n(c) P_{n(c)} - cP = n(c)(P_{n(c)} - P) + (n(c) - c)P,$$
$$Y_c / c^{1/2} = (n(c)/c)^{1/2} \nu_{n(c)} + (n(c) - c) c^{-1/2} P. \tag{11.3}$$

The following shows that the empirical process ν_n is asymptotically "as large" as a corresponding Poisson process.

Lemma 11.7 *Let* $(X, \mathcal{A}, P)$ *be a probability space and* $\mathcal{C} \subset \mathcal{A}$. *Assume that for each n and constant t*, $\sup_{A \in \mathcal{C}} |(P_n - tP)(A)|$ *is measurable. Let f be a continuous, slowly varying function such that as* $x \to +\infty$, $f(x) \to +\infty$. *For* $b > 0$ *let*

$$g(b) := \liminf_{x \to +\infty} \Pr\{\sup_{A \in \mathcal{C}} |Y_x(A)| \geq bf(x)x^{1/2}\}.$$

Then for any $a < b$,

$$\liminf_{n \to \infty} \Pr\{\sup_{A \in \mathcal{C}} |\nu_n(A)| \geq af(n)\} \geq g(b).$$

Proof. It follows from Lemma 11.5 that $f(x)/x \to 0$ as $x \to \infty$. From (11.3), $\sup_{A \in \mathcal{C}} |Y_x(A)|$ is measurable. As $x \to +\infty$, $\Pr(n(x) > 0) \to 1$ and $n(x)/x \to 1$ in probability. If the Lemma is false, there is a $\Theta < g(b)$ and a sequence $m_k \to +\infty$ with, for each $m = m_k$,

$$\Pr\{\sup_{A \in \mathcal{C}} |\nu_m(A)| \geq af(m)\} \leq \Theta.$$

Choose $0 < \varepsilon < 1/3$ such that $a(1 + 7\varepsilon) < b$. Then let $0 < \delta < 1/2$ be such that $\delta \leq \delta(\varepsilon)$ in Lemma 11.5 and $(1 + \delta)(1 + 5\varepsilon) < 1 + 6\varepsilon$. We may assume that for all $k = 1, 2, \ldots$, $m = m_k \geq 2/\delta$ and $1 + 2\varepsilon < a\varepsilon f(m)^{1/2}$.

Set $\delta_m := (f(m)/m)^{1/2}$. Then since $f(x)/x \to 0$ we may assume $\delta_m < \delta/2$ for all $m = m_k$. Then for any $m = m_k$, if $(1 - \delta_m)m \leq n \leq m$, then for all $A \in \mathcal{A}$, $mP_m(A) \geq nP_n(A)$, so $m(P_m - P)(A) \geq n(P_n - P)(A) - m\delta_m$. Conversely $nP_n(A) \geq mP_m(A) - m\delta_m$ and

$$n(P_n - P)(A) \geq m(P_m - P)(A) - m\delta_m.$$

Thus

$$\|\nu_m(A)\| \geq (n/m)^{1/2}|\nu_n(A)| - f(m)^{1/2} \geq (1 + \delta)^{-1}|\nu_n(A)| - f(m)^{1/2},$$

so $|\nu_n(A)| \leq (1 + \delta)(|\nu_m(A)| + f(m)^{1/2})$. Next, $|1 - \frac{m}{n}| = \frac{m}{n} - 1 < 2\delta_m < \delta$ implies $\left|\frac{f(m)}{f(n)} - 1\right| < \varepsilon$, so $1/f(n) < (1 + \varepsilon)/f(m)$. Hence

$$|\nu_n(A)|/f(n) < (1 + 2\varepsilon)(|\nu_m(A)|f(m)^{-1} + f(m)^{-1/2}).$$

Thus since $(1 + 2\varepsilon)f(m)^{-1/2} < a\varepsilon$, $\Pr\{\sup_{A \in \mathcal{C}} |\nu_n(A)| \geq af(n)(1 + 3\varepsilon)\} \leq \Theta$.

For each $m = m_k$, set $c = c_m = (1 - \frac{1}{2}\delta_m)m$. Then as $k \to \infty$, since $f(m_k) \to \infty$, by Chebyshev's inequality and since P_c has variance c,

$$\Pr\{(1 - \delta_m)m \leq n(c) \leq m\} \to 1.$$

Then for any y with $\Theta < y < g := g(b)$ and k large enough, since the $x(i)$ are independent of $n(c)$,

$$\Pr\{\sup_{A \in \mathcal{C}} |\nu_{n(c)}(A)| \geq af(n(c))(1 + 3\varepsilon)\} < y.$$

Since $\delta_m \leq \delta(\varepsilon)/2$, for $(1 - \delta_m)m \leq n \leq m$ we have $\left|1 - \frac{f(n)}{f(c)}\right| < \varepsilon$. Thus for k large enough we may assume

$$\Pr\{\sup_{A \in C} |\nu_{n(c)}(A)| \geq af(c)(1 + 5\varepsilon)\} < y.$$

For k large enough, applying Chebyshev's inequality to $n(c) - c$, we have since $c \to \infty$

$$\Pr\left\{\left(\frac{n(c)}{c}\right)^{1/2} > \frac{1 + 6\varepsilon}{1 + 5\varepsilon}\right\} \leq \Pr\left\{\frac{n(c) - c}{c^{1/2}} > \frac{2\varepsilon c^{1/2}}{(1 + 5\varepsilon)^2}\right\} < \frac{g - y}{4},$$

and since $f(c) \to \infty$, $\Pr\{[n(c) - c]/c^{1/2} > af(c)\varepsilon\} < (g - y)/4$. Thus by (11.3),

$$\Pr\left\{\sup_{A \in C} |Y_c(A)| \geq ac^{1/2}f(c)(1 + 7\varepsilon)\right\} < (y + g)/2 < g,$$

a contradiction, proving Lemma 11.7.   □

Next, the Poisson process's independence property on disjoint sets will be extended to suitable random sets. Let $(X, \mathcal{A})$ be a measurable space, and $(\Omega, \mathcal{B}, \Pr)$ a probability space. A collection $\{\mathcal{B}_A : A \in \mathcal{A}\}$ of sub-σ-algebras of $\mathcal{B}$ will be called a *filtration* if $\mathcal{B}_A \subset \mathcal{B}_B$ whenever $A \subset B$ in $\mathcal{A}$. A stochastic process Y indexed by $\mathcal{A}$, $(A, \omega) \mapsto Y(A)(\omega)$, will be called *adapted* to $\{\mathcal{B}_A : A \in \mathcal{A}\}$ if for every $A \in \mathcal{A}$, $Y(A)(\cdot)$ is $\mathcal{B}_A$ measurable. Then the process and filtration will be written $\{Y(A), \mathcal{B}_A\}_{A \in \mathcal{A}}$. A stochastic process $Y \colon \langle A, \omega \rangle \to Y(A)(\omega)$, $A \in \mathcal{A}$, $\omega \in \Omega$, will be said to have *independent pieces* iff for any disjoint $A_1, \ldots, A_m \in \mathcal{A}$, $Y(A_j)$ are independent, $j = 1, \ldots, m$, and $Y(A_1 \cup A_2) = Y(A_1) + Y(A_2)$ almost surely. Clearly each Y_c has independent pieces. If in addition the process is adapted to a filtration $\{\mathcal{B}_A : A \in \mathcal{A}\}$, the process $\{Y(A), \mathcal{B}_A\}_{A \in \mathcal{A}}$ will be said to have *independent pieces* iff for any disjoint sets $A_1, \ldots, A_n$ in $\mathcal{A}$, the random variables $Y(A_2), \ldots, Y(A_n)$ and any random variable measurable for the σ-algebra $\mathcal{B}_{A_1}$ are jointly independent.

For example, for any $C \in \mathcal{A}$ let $\mathcal{B}_C$ be the smallest σ-algebra for which every $Y(A)(\cdot)$ is measurable for $A \subset C$, $A \in \mathcal{A}$. This is clearly a filtration, and the smallest filtration to which Y is adapted.

A function G from Ω into $\mathcal{A}$ will be called a *stopping set* for a filtration $\{\mathcal{B}_A : A \in \mathcal{A}\}$ iff for all $C \in \mathcal{A}$, $\{\omega \colon G(\omega) \subset C\} \in \mathcal{B}_C$. Given a stopping set $G(\cdot)$, let $\mathcal{B}_G$ be the σ-algebra of all sets $B \in \mathcal{B}$ such that for every $C \in \mathcal{A}$, $B \cap \{G \subset C\} \in \mathcal{B}_C$. (Note that if G is not a stopping set, then $\Omega \notin \mathcal{B}_G$, so $\mathcal{B}_G$ would not be a σ-algebra.) If $G(\omega) \equiv H \in \mathcal{A}$, then it is easy to check that G is a stopping set and $\mathcal{B}_G = \mathcal{B}_H$.

Lemma 11.8 *Suppose* $\{Y(A), \mathcal{B}_A\}_{A \in \mathcal{A}}$ *has independent pieces and for all* $\omega \in \Omega$, $G(\omega) \in \mathcal{A}$, $A(\omega) \in \mathcal{A}$ *and* $E(\omega) \in \mathcal{A}$.

Assume that:

(i) $G(\cdot)$ *is a stopping set;*
(ii) *For all ω, $G(\omega)$ is disjoint from $A(\omega)$ and from $E(\omega)$;*
(iii) *Each of $G(\omega)$, $A(\omega)$ and $E(\omega)$ has just countably many possible values $G(j) := G_j \in \mathcal{A}$, $C(i) := C_i \in \mathcal{A}$, and $D(j) := D_j \in \mathcal{A}$ respectively;*
(iv) *For all i, j, $\{A(\cdot) = C_i\} \in \mathcal{B}_G$ and $\{E(\cdot) = D_j\} \in \mathcal{B}_G$.*

Then the conditional probability law (joint distribution) of $Y(A)$ and $Y(E)$ given $\mathcal{B}_G$ satisfies

$$\mathcal{L}\{(Y(A), Y(E))|\mathcal{B}_G\} = \sum_{i,j} 1_{\{A(\cdot)=C(i), E(\cdot)=D(j)\}} \mathcal{L}(Y(C_i), Y(D_j))$$

where $\mathcal{L}(Y(C_i), Y(D_j))$ is the unconditional joint distribution of $Y(C_i)$ and $Y(D_j)$. If this unconditional distribution is the same for all i and j, then $(Y(A), Y(E))$ is independent of $\mathcal{B}_G$.

Proof. The proof will be given when there is only one random set $A(\cdot)$ rather than two, $A(\cdot)$ and $E(\cdot)$. The proof for two is essentially the same. If for some ω, $A(\omega) = C_i$ and $G(\omega) = G_j$, then by (ii), $C_i \cap G_j = \emptyset$ (the empty set). Thus $Y(C_i)$ is independent of $\mathcal{B}_{G(j)}$. Let $B_i := B(i) := \{A(\cdot) = C_i\} \in \mathcal{B}_G$ by (iv). For each j,

$$\{G = G_j\} = \{G \subset G_j\} \setminus \bigcup_i \{\{G \subset G_i\} : G_i \subset G_j, \ G_i \neq G_j\},$$

so by (i)

$$\{G = G_j\} \in \mathcal{B}_{G(j)}. \tag{11.4}$$

Let $H_j := \{G = G_j\}$. For any $D \in \mathcal{A}$, $H_j \cap \{G \subset D\} = \emptyset \in \mathcal{B}_D$ if $G_j \not\subset D$. If $G_j \subset D$, then $H_j \cap \{G \subset D\} = H_j \in \mathcal{B}_{G(j)} \subset \mathcal{B}_D$ by (11.4). Thus

$$H(j) := H_j \in \mathcal{B}_G. \tag{11.5}$$

For any $B \in \mathcal{B}_G$, by (11.4)

$$B \cap H_j = B \cap \{G \subset G_j\} \cap H_j \in \mathcal{B}_{G(j)}. \tag{11.6}$$

We have for any real t almost surely, since $B_i \in \mathcal{B}_G$,

$$\Pr(Y(A) \leq t|\mathcal{B}_G) = \sum_{i,j} \Pr(Y(A) \leq t|\mathcal{B}_G) 1_{B(i)} 1_{H(j)}$$

$$= \sum_{i,j} \Pr(Y(C_i) \leq t|\mathcal{B}_G) 1_{B(i)} 1_{H(j)}.$$

The sums can be restricted to i, j such that $B_i \cap H_j \neq \emptyset$ and therefore $C_i \cap G_j = \emptyset$. Now $\Pr(Y(C_i) \leq t|\mathcal{B}_G)$ is a $\mathcal{B}_G$-measurable function f_i such that for any $\Gamma \in \mathcal{B}_G$,

$$\Pr(\{Y(C_i) \le t\} \cap \Gamma) = \int_\Gamma f_i \, d\Pr.$$

Restricted to H_j, f_i equals a $\mathcal{B}_{G(j)}$-measurable function by (11.6). By (11.5), $\Gamma \cap H_j \in \mathcal{B}_G$, so

$$\begin{aligned}
\int_{\Gamma \cap H(j)} f_i \, d\Pr &= \Pr(\{Y(C_i) \le t\} \cap \Gamma \cap H_j) \\
&= \Pr(Y(C_i) \le t)\Pr(\Gamma \cap H_j)
\end{aligned} \tag{11.7}$$

because by (11.6) $\Gamma \cap H(j) \in \mathcal{B}_{G(j)}$, which is independent of $\{Y(C_i) \le t\}$.

One solution f_i to (11.7) for all j is $f_i \equiv \Pr(Y(C_i) \le t)$. To show that this solution is unique, it will be enough to show that it is unique on H_j for each j. First, for a set $A \subset H_j$ it will be shown that $A \in \mathcal{B}_{G(j)}$ if and only if $A \in \mathcal{B}_G$. By (11.4) and (11.5) H_j itself is in both $\mathcal{B}_G$ and $\mathcal{B}_{G(j)}$. We can write $A = A \cap H(j)$. Then "if" follows from (11.6). To prove "only if" let $A \in \mathcal{B}_{G(j)}$. Then for $C \in \mathcal{A}$, let

$$F := A \cap \{G \subset C\} = A \cap \{G_j \subset C\}.$$

If $G_j \subset C$, then $F = A \in \mathcal{B}_{G(j)} \subset \mathcal{B}_C$. Otherwise, F is empty and in $\mathcal{B}_C$. In either case $F \in \mathcal{B}_C$ so $A \in \mathcal{B}_G$ as desired.

Thus, if g is a function on H_j, measurable for $\mathcal{B}_{G(j)}$ there, and $\int_{D \cap H(j)} g \, d\Pr = 0$ for all $D \in \mathcal{B}_G$, then the same holds for all $D \in \mathcal{B}_{G(j)}$. So, as in the usual proof of uniqueness of conditional expectations,

$$\int_{\{g>0\} \cap H(j)} g \, d\Pr = \int_{\{g<0\} \cap H(j)} g \, d\Pr = 0,$$

so $g = 0$ on $H(j)$ a.s. and $f_i = \Pr(Y(C_i) \le t)$ a.s. on $H(j)$ for all j. Thus $\Pr(Y(A) \le t | \mathcal{B}_G) = \sum_i \Pr(Y(C_i) \le t) 1_{B(i)}$. $\qquad \square$

Here is another fact about stopping sets, which corresponds to a known fact about nonnegative real-valued stopping times or Markov times (e.g., RAP, Lemma 12.2.5):

Lemma 11.9 *If G and H are stopping sets and $G \subset H$, then $\mathcal{B}_G \subset \mathcal{B}_H$.*

Proof. For any measurable set D and $A \in \mathcal{B}_G$, we have

$$A \cap \{H \subset D\} = (A \cap \{G \subset D\}) \cap \{H \subset D\} \in \mathcal{B}_D. \qquad \square$$

11.4 Lower Bounds in Borderline Cases

Recall the classes $\mathcal{C}(\alpha, K, d)$ of subgraphs of functions with bounded derivatives through order α in $\mathbb{R}^d$, defined in Section 8.2. We had lower bounds for $P_n - P$ on $\mathcal{C}(\alpha, K, d)$ in Theorem 11.2, which imply that for $\alpha < d - 1$, $\|\nu_n\|_{\mathcal{C}(\alpha, K, d)} \to \infty$ surely as $n \to \infty$. For $\alpha > d - 1$, $\mathcal{C}(\alpha, K, d)$ is a Donsker class by Theorem 8.4 and Corollary 7.8, so $\|\nu_n\|_{\mathcal{C}(\alpha,K,d)}$ is bounded in probability. Thus $\alpha = d - 1$ is a borderline case. Other such cases are given by the class $\mathcal{LL}_2$ of lower layers in $\mathbb{R}^2$ (Section 8.3) and the class $\mathcal{C}_3$ of convex sets in $\mathbb{R}^3$ (Section 8.4), for λ^d = Lebesgue measure on the unit cube I^d, where $I := [0, 1]$.

Any lower layer A has a closure $\overline{A}$ which is also a lower layer, with $\lambda^d(\overline{A} \setminus A) = 0$, where in the present case $d = 2$. It is easily seen that suprema of our processes over all lower layers are equal to suprema over closed lower layers, so it will be enough to consider closed lower layers. Let $\overline{\mathcal{LL}_2}$ be the class of all closed lower layers in $\mathbb{R}^2$.

Let $P = \lambda^d$ and $c > 0$. Recall the centered Poisson process Y_c from Section 11.3. Let $N_c := U_c - V_c$ where U_c and V_c are independent Poisson processes, each with intensity measure cP. Equivalently, we could take U_c and V_c to be centered. The following lower bound holds for all the above borderline cases:

Theorem 11.10 *For any $K > 0$ and $\delta > 0$ there is a $\gamma = \gamma(d, K, \delta) > 0$ such that*

$$\lim_{x \to +\infty} \Pr\left\{ \|Y_x\|_\mathcal{C} > \gamma(x \log x)^{1/2}(\log \log x)^{-\delta-1/2} \right\} = 1$$

and

$$\lim_{n \to \infty} \Pr\left\{ \|\nu_n\|_\mathcal{C} > \gamma(\log n)^{1/2}(\log \log n)^{-\delta-1/2} \right\} = 1$$

where $\mathcal{C} = \mathcal{C}(d - 1, K, d)$, $d \geq 2$, or $\mathcal{C} = \overline{\mathcal{LL}_2}$, or $\mathcal{C} = \mathcal{C}_3$.

For a proof, see the next section, except that here I will give a larger lower bound with probability close to 1, of order $(\log n)^{3/4}$ in the lower layer case ($\mathcal{C} = \mathcal{LL}_2, d = 2$). Shor (1986) first showed that $E\|Y_x\|_\mathcal{C} > \gamma x^{1/2}(\log x)^{3/4}$ for some $\gamma > 0$ and x large enough. Shor's lower bound also applies to $\mathcal{C}(1, K, 2)$ by a 45° rotation as in Section 8.3. For an upper bound with a 3/4 power of the log also for convex subsets of a fixed bounded open set in $\mathbb{R}^3$, see Talagrand (1994, Theorem 1.6).

To see that the supremum of N_c, Y_c or an empirical process ν_n over $\mathcal{LL}_2$ is measurable, note first for P_n that for each $F \subset \{1, \ldots, n\}$ and each ω, there is a smallest, closed lower layer $L_F(\omega)$ containing the x_j for $j \in F$, with $L_F(\omega) := \emptyset$ for $F = \emptyset$. For any $c > 0$, $\omega \mapsto (P_n - cP)(L_F(\omega))(\omega)$ is measurable. The supremum of $P_n - cP$ over $\mathcal{LL}_2$, as the maximum of these 2^n measurable functions, is measurable. Letting $n = n(c)$ as in Lemma 11.6 and (11.3) then shows $\sup\{Y_c(A)\colon A \in \mathcal{LL}_2\}$ is measurable. Likewise, there is a largest, open lower layer not containing x_j for any $j \in F$, so $\sup\{|Y_c(A)| \colon A \in \mathcal{LL}_2\}$ and $\sup\{|\nu_n(A)| \colon A \in \mathcal{LL}_2\}$ are measurable.

For N_c, taking noncentered Poisson processes U_c and V_c, their numbers of points $m(\omega)$ and $n(\omega)$ are measurable, as are the m-tuple and n-tuple of points occurring in each. For each $i = 0, 1, \ldots, m$ and $k = 0, 1, \ldots, n$, it is a measurable event that there exists a lower layer containing exactly i of the m points and k of the n, and so the supremum of N_c over all lower layers, as a measurable function of the indicators of these finitely many events, is measurable.

Theorem 11.11 (Shor) *For every $\varepsilon > 0$ there is a $\delta > 0$ such that for the uniform distribution P on the unit square I^2, and n large enough,*

$$\Pr\left(\sup\{|\nu_n(A)| : A \in \mathcal{LL}_2\} \geq \delta(\log n)^{3/4}\right) \geq 1 - \varepsilon,$$

and the same holds for $\mathcal{C}(1, 2, 2)$ in place of $\mathcal{LL}_2$. Also, ν_n can be replaced by $N_c/c^{1/2}$ or $Y_c/c^{1/2}$ if $\log n$ is replaced by $\log c$, for c large enough.

Remark. The order $(\log n)^{3/4}$ of the lower bound is best possible, as there is an upper bound in expectation of the same order, not proved here; see Rhee and Talagrand (1988), Leighton and Shor (1989), and Coffman and Shor (1991).

Proof. Let $\mathcal{M}_1$ be the set of all functions f on $[0, 1]$ with $f(0) = f(1) = 1/2$ and $\|f\|_L \leq 1$, i.e., $|f(u) - f(x)| \leq |x - u|$ for $0 \leq x \leq u \leq 1$. Then $0 \leq f(x) \leq 1$ for $0 \leq x \leq 1$. For any $f : [0, 1] \mapsto [0, \infty)$ let S_f be the subgraph of f, $S_f := S(f) := \{(x, y) : 0 \leq x \leq 1, 0 \leq y \leq f(x)\}$. Then for each $f \in \mathcal{M}_1$ we have $S_f \in \mathcal{C}(1, 2, 2)$. Let $\mathcal{S}_1 := \{S_f : f \in \mathcal{M}_1\}$.

Let R be a counterclockwise rotation of $\mathbb{R}^2$ by $45°$, $R = 2^{-1/2}\left(\begin{smallmatrix} 1 & -1 \\ 1 & 1 \end{smallmatrix}\right)$. Then for each $f \in \mathcal{M}_1$, $R^{-1}(S_f) = M \cap R^{-1}(I^2)$ where M is a lower layer, $I^2 := I \times I$, and $I := [0, 1]$. So, it will be enough to prove the theorem for $\mathcal{S}_1 \subset \mathcal{C}(1, 1, 2)$.

Let $\lfloor x \rfloor$ denote the largest integer $\leq x$ and let $0 < \delta \leq 1/3$. For each c large enough so that $\delta^2(\log c) \geq 1$ and ω, functions $f_0 \leq f_1 \leq \cdots \leq f_L \leq g_L \leq \cdots \leq g_1 \leq g_0$ will be defined for $L := \lfloor \delta^2(\log c) \rfloor$, with $f_j := f_j(t) := f_j(\omega, t)$ and likewise g_j for $\omega \in \Omega$ and $0 \leq t \leq 1$.

Each f_i and g_i will be continuous and piecewise linear in t. For each $j = 1, \ldots, 2^i$, one of f_i and g_i will be linear on $I_{ij} := [(j - 1)/2^i, j/2^i]$, and the other will be linear only on each half, $I_{i+1,2j-1}$ and $I_{i+1,2j}$, with $f_i = g_i$ at the endpoints of I_{ij}. Thus over I_{ij}, the region T_{ij} between the graphs of f_i and g_i is a triangle.

Let $f_0(x) := 1/2$ for $0 \leq x \leq 1$. Let $s := 1/(L + 1)$ and $g_0(1/2) := (1 + s)/2$. Given f_i and g_i, to define f_{i+1} and g_{i+1}, we have two cases.

Case 1. Suppose g_i is linear on I_{ij}, as in Figure 11.1. Let $p_k := (x_k, y_k)$ be the point labeled by $k = 1, \ldots, 10$ in Figure 11.1. Then $x_k = (4j + k - 5)/2^{i+2}$ for $k = 1, \ldots, 5$, $x_k = x_{10-k}$ for $k = 6, 7, 8$, $x_9 = x_2 + 1/(3 \cdot 2^{i+2})$, and $x_{10} = x_4 - 1/(3 \cdot 2^{i+2})$. Thus $T := T_{ij}$ is the triangle $p_1 p_3 p_5$. Let $Q := Q_{ij}$ be the quadrilateral $p_3 p_{10} p_7 p_9$, and $V := V_{ij}$ the union $p_2 p_3 p_9 \cup p_3 p_4 p_{10}$ of two

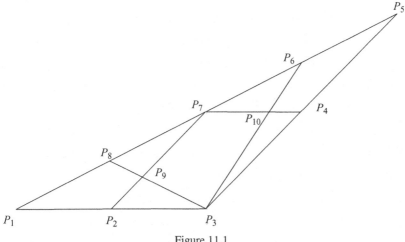

Figure 11.1

triangles. The triangles $p_1 p_2 p_7$, $p_1 p_3 p_8$, $p_3 p_5 p_6$, and $p_4 p_5 p_7$ each have $1/4$ the area of T, Q has $1/3$, and V has $1/6$.

Let ρ_{ij}, $i = 1, 2, \ldots$, $j = 1, \ldots, 2^i$, be i.i.d. Rademacher random variables, so that $P(\rho_{ij} = 1) = P(\rho_{ij} = -1) = 1/2$, independent of the N_c process. If $N_c(Q_{ij}) > 0$, or if $N_c(Q_{ij}) = 0$ and $\rho_{ij} = 1$, say event A_{ij} occurs and set $g_{i+1} \equiv g_i$ on I_{ij}. Then, the graph of f_{i+1} on I_{ij} will consist of line segments joining the points p_1, p_2, p_7, p_4, p_5 in turn, so $T_{i+1,2j-1} = p_1 p_2 p_7$ and $T_{i+1,2j} = p_7 p_4 p_5$.

If A_{ij} does not occur, i.e., $N_c(Q) < 0$ or $N_c(Q) = 0$ and $\rho_{ij} = -1$, set $f_{i+1} := f_i$ on I_{ij} and let the graph of g_{i+1} consist of line segments joining the points p_1, p_8, p_3, p_6, p_5, so that $T_{i+1,2j-1} = p_1 p_8 p_3$ and $T_{i+1,2j} = p_3 p_6 p_5$. This finishes the recursive definition of f_i and g_i in Case 1. The stated properties of f_i and g_i continue to hold.

Case 2. In this case f_i is linear on I_{ij}, see Figure 11.2. Then T_{ij} is the triangle $p_1 p_5 p_7$. The other definitions remain the same as in Case 1.

Lemma 11.12 *For $i = 0, 1, \ldots, L$, all slopes of segments of f_i and g_i have absolute values at most $(i + 1)s < (L + 1)s = 1$, so f_i and g_i are in $\mathcal{G}_{1,2,1}$.*

Proof. For $i = 0$, the slope for f_0 is 0 and those for g_0 are $\pm s$. By induction for $i \geq 1$, the slopes for f_{i-1} are those for $P_1 P_3$ and/or $P_3 P_5$, and those for g_{i-1} are those of $P_1 P_7$ and/or $P_7 P_5$ and by induction assumption are at most is in absolute value. There is an integer $m = m_{ij} = m_{ij}(\omega) \in \mathbb{Z}$ with $|m| \leq i$ such that $p_1 p_5$ has slope ms, $p_1 p_3$ in Figure 11.1 and $p_7 p_5$ in Figure 11.2 have slope $(m - 1)s$, while $p_3 p_5$ in Figure 11.1 and $p_1 p_7$ in Figure 11.2 have slope $(m + 1)s$. We have for each i and j that $m_{i+1,2j-1} - m_{ij}$ and $m_{i+1,2j} -$

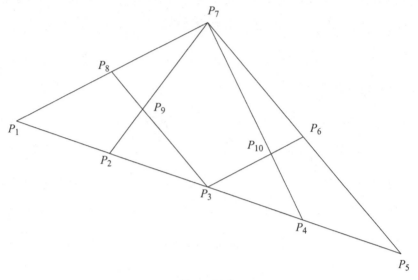

Figure 11.2

m_{ij} have possible values 0 or ± 1. Specifically, on A_{ij} in Case 1 or A_{ij}^c in Case 2, $m_{i+1,2j} = m_{i+1,2j-1} = m_{ij}$. On A_{ij}^c in Case 1, $m_{i+1,2j-1} - m_{ij} = -1$ and $m_{i+1,2j} - m_{ij} = 1$. Or on A_{ij} in Case 2, $m_{i+1,2j-1} - m_{ij} = 1$ and $m_{i+1,2j} - m_{ij} = -1$. Thus in each case, from $i-1$ to i, the maximum absolute value of the slope of a segment increases at most by s, and the conclusions follow.   □

Let (Ω, μ) be a probability space on which N_c and all ρ_{ij} are defined. Take $\Omega_1 := \Omega \times [0, 1]$ with product measure $Pr := \mu \times \lambda$ where λ is the uniform (Lebesgue) law on $[0, 1]$. For almost all $t \in [0, 1]$, t is not a dyadic rational, so for each $i = 1, 2, \ldots$, there is a unique $j := j(i, t)$, $j = 1, \ldots, 2^i$, such that $t \in Int_{ij} := ((j-1)/2^i, j/2^i)$. Thus Int_{ij} is the interior of I_{ij}. The derivatives f_i' and g_i' are step functions, constant on each interval $Int_{i+1,j}$, and possibly undefined at the endpoints.

Some random subsets of I^2 are defined as follows. For $m = 0, 1, 2, \ldots$, let

$$G_m := G(m) := G(m, \omega) := \bigcup_{k=0}^{m} \bigcup_{i=1}^{2^m} Q_{ki}.$$

For $j = 1, \ldots, 2^{m+1}$, let $G(m, j) := G(m) \cup \bigcup_{i=1}^{j} Q_{m+1,i}$.

For each C in the Borel σ-algebra $\mathcal{B} := \mathcal{B}(I^2)$, $k = 0, 1, \ldots$, and $i = 1, \ldots, 2^k$, let $\mathcal{B}_C^{(k,i)}$ be the smallest σ-algebra of subsets of Ω with respect to which all ρ_{mr} for $m = 0, 1, \ldots, k-1$ or $m = k$, $r = 1, \ldots, i$, and all $N_c(A)$, for Borel $A \subset C$, are measurable. It is easily seen that for each fixed k and

i, $\{\mathcal{B}_C^{(k,i)} : C \in \mathcal{B}\}$ is a filtration, and that $\{N_c(A), \mathcal{B}_A^{(k,i)}\}_{A \in \mathcal{B}}$ has independent pieces as defined before Lemma 11.8.

The initial triangle T_{01} is fixed and has area $s/4$. For each $k = 1, 2, \ldots$ and each possible value of the triangle $T_{k-1,i}$, there are two possible values of the pair of triangles $T_{k,2i-1}$, $T_{k,2i}$, each with area $s/4^{k+1}$. This and the area $s/(3 \cdot 4^{k+1})$ of Q_{ki} are nonrandom. For each $m = 0, 1, 2, \ldots$, there are finitely many possibilities for the values of T_{ki} and so Q_{ki} for $k = 0, 1, \ldots m$ and $i = 1, \ldots, 2^k$, each on a measurable event, and for whether A_{ki} holds or not. So $\omega \mapsto N_c(Q_{ij}(\omega))(\omega)$ is measurable, and for each m and j, there are finitely many possible values of $G(m, j)(\omega)$. Next we have:

Lemma 11.13 *For each $k = 0, 1, 2, \ldots$ and $i = 1, \ldots 2^k$,*

(a) $G(k, i)$ is a stopping set for $\{\mathcal{B}_A^{(k,i)}\}_{A \in \mathcal{B}}$;

(b) $A_{ki} \in \mathcal{B}^{k,i} := \mathcal{B}_{G(k,i)}^{(k,i)}$;

(c) If $k \geq 1$, then for each possible value Q of Q_{ki}, $\{Q_{ki} = Q\} \in \mathcal{B}^{(k-1)}$, where $\mathcal{B}^{(r)} := \mathcal{B}_{G(r)}^{(r,2^r)}$ for $r = 0, 1, \ldots$.

Proof. It will be shown by double induction on k and on i for fixed k that (a), (b), and (c) all hold. Let $\mathcal{B}_A^{(k)} := \mathcal{B}_A^{(k,2^k)}$ for any $A \in \mathcal{B}(I^2)$ and $k = 0, 1, \ldots$. For $k = 0$ we have $G(0) \equiv G(0, 1) \equiv Q_{01}$, a fixed set. So as noted just before Lemma 11.8, $G(0)$ is a stopping set, and $\mathcal{B}^{(0)} = \mathcal{B}_{G(0)}^{(0)}$, where $\mathcal{B}_{G(0)}^{(0)}$ is defined as for stopping sets. So (a) holds for $k = 0$.

Since $1_{A_{01}}$ is a measurable function of $N_c(Q_{01})$ and ρ_{01}, we have $A_{01} \in \mathcal{B}^{(0)}$, so (b) holds for $k = 0$.

In (c), each event $\{Q_{ki} = Q\}$ is a Boolean combination of events A_{jr} for $j = 0, \ldots, k - 1$ and $r = 1, \ldots, 2^j$. Thus if (a) and (b) hold for $k = 0, 1, \ldots, m - 1$ for some $m \geq 1$, then (c) holds for $k = m$. So (c) holds for $k = 1$.

To prove (a) for $k = m$, let $J(m, j)$ be the finite set of possible values of $G(m, j)$. If (c) holds for $k = 1, \ldots, m$, then from the definition of $G(m, j)$, for each $G \in J(m, j)$,

$$\{G(m, j) = G\} \in \mathcal{B}^{(m-1)}. \tag{11.8}$$

Thus, for any $D \in \mathcal{B}(I^2)$, since $G(m - 1) \subset G(m, j)$,

$$\{G(m, j) \subset D\} = \{G(m, j) \subset D\} \cap \{G(m - 1) \subset D\}$$

$$= \bigcup_{G \in J(m,j), G \subset D} \{G(m, j) = G\} \cap \{G(m - 1) \subset D\} \in \mathcal{B}_D^{(m,j)},$$

so $G(m, j)$ is a stopping set and (a) holds for $k = m$.

Then to show (b) for $k = m$, for a set $C \in \mathcal{B}(I^2)$ and $i = 1, \ldots, 2^m$ let

$$C_{(m,i)} := \{N_c(C) > 0\} \cup \{N_c(C) = 0, \ \rho_{mi} = 1\}.$$

Then

$$C_{(m,i)} \in \mathcal{B}_C^{(m,i)}. \tag{11.9}$$

If $C(m, i)$ is the finite set of possible values of Q_{mi},

$$A_{mi} = \bigcup_{Q \in C(m,i)} \{Q_{mi} = Q\} \cap Q_{(m,i)}.$$

Thus

$$A_{mi} \cap \{G(m) \subset D\}$$

$$= \{G_m \subset D\} \cap \{G_{m-1} \subset D\} \cap \left[\bigcup_{Q \in C(m,i), Q \subset D} \{Q_{mi} = Q\} \cap Q_{(m,i)} \right].$$

It follows from (11.8) that $\{G(m, i) \subset D\} \in \mathcal{B}^{(m-1)}$ for each m and i. By (c) for $k = m$, each $\{Q_{mi} = Q\} \in \mathcal{B}^{(m-1)}$. Thus

$$\{G(m, i) \subset D\} \cap \{Q_{mi} = Q\} \in \mathcal{B}^{(m-1)},$$

and so

$$\{G(m, i) \subset D\} \cap \{Q_{mi} = Q\} \cap \{G(m-1) \subset D\} \in \mathcal{B}_D^{(m-1)}.$$

For $Q \subset D$, by (11.9), $Q_{(m,i)} \in \mathcal{B}_D^{(m,i)}$. So taking a union of intersections, $A_{mi} \cap \{G(m, i) \subset D\} \in \mathcal{B}_D^{(m,i)}$. Thus $A_{mi} \in \mathcal{B}_{G(m,i)}^{(m,i)}$ and (b) holds for $k = m$. This finishes the proof of the Lemma. □

Next, here is a fact about symmetrized Poisson variables.

Lemma 11.14 *There is a constant $c_0 > 0$ such that if X and Y are independent random variables, each with distribution Poisson with parameter $\lambda \geq 1$, then $E|X - Y| \geq c_0 \lambda^{1/2}$.*

Proof. By a straightforward calculation, $E((X - Y)^4) = 2\lambda + 12\lambda^2 \leq 14\lambda^2$ if $\lambda \geq 1$. By the Cauchy–Bunyakovsky–Schwarz inequality,

$$2\lambda = E((X - Y)^2) \leq (E|X - Y|)^{1/2}(E(|X - Y|^3))^{1/2}$$

and $E(|X - Y|^3) \leq (E(|X - Y|^4))^{3/4}$. The Lemma follows with $c_0 = 4/(14)^{3/4}$. □

Now continuing with the proof of Theorem 11.11, set $G(m, 0) := G_{m-1}$. We apply Lemma 11.13(a) and (c). For $m = 1, 2, \ldots$ and $j = 1, \ldots, 2^m$, the random set Q_{mj} is disjoint from the stopping set $G_{m,j-1}$. The hypotheses of Lemma 11.8 hold for $(G, A) = (G_{m,j-1}, Q_{mj})$ and the process $Y = N_c$ with the filtration $\{\mathcal{B}_A^{(m,j-1)} : A \in \mathcal{B}\}$. Since all possible values of Q_{mj} have the same area, it follows that the random variable $N_c(Q_{mj})$ has the law of $U - V$ for U, V i.i.d. Poisson with parameter

$$cP(Q_{mj}) = cs/(6 \cdot 4^m) \tag{11.10}$$

and is independent of $\mathcal{B}^{m,j-1}$, which was defined in Lemma 11.13(b). Also, $N_c(Q_{ri})$ is $\mathcal{B}^{m,j-1}$ measurable for $r < m$ or for $r = m$ and $i < j$. So, the variables $N_c(Q_{mj})$ for $m = 0, 1, 2, \ldots$ and $j = 1, \ldots, 2^m$ are all jointly independent with the given laws. Also, the i.i.d. Rademacher variables ρ_{mj} are independent of the process N_c.

For $t \in Int_{ij}$, let $h_i(\omega, t)$ be the slope of the longest side of T_{ij}, so $h_i(\omega, t) = g_i'(t)$ in Case 1 and $f_i'(t)$ in Case 2, while $h_0(\omega, t) \equiv 0$. In Case 1, if A_{ij} holds then $h_{i+1} = h_i$ on Int_{ij}. If $\omega \notin A_{ij}$ then $h_{i+1} - h_i = -s$ on $Int_{i+1,2j-1}$ and $+s$ on $Int_{i+1,2j}$.

In Case 2, if $\omega \notin A_{ij}$ then $h_{i+1} = h_i$ on Int_{ij}. If $\omega \in A_{ij}$, then $h_{i+1} - h_i = s$ on $Int_{i+1,2j-1}$ and $-s$ on $Int_{i+1,2j}$.

Let $\zeta_i := h_i - h_{i-1}$ for $i = 1, \ldots, L$. By Lemmas 11.8 and 11.13, and (11.10), any $N_c(Q_{mj})$ or ρ_{mj} for $m \geq k$ is independent of $\mathcal{B}^{(k-1)}$, where $\mathcal{B}^{(-1)}$ is defined as the trivial σ-algebra $\{\emptyset, \Omega_1\}$. So, each of $\zeta_1, \ldots, \zeta_k$ is $\mathcal{B}^{(k-1)}$ measurable while any event A_{kj} and the random function ζ_{k+1} are independent of $\mathcal{B}^{(k-1)}$. It follows that $h_i(\omega, t) = \sum_{r=1}^{i} \zeta_r(\omega, t)$ where ζ_r are i.i.d. variables for $Pr = \mu \times \lambda$ on $\Omega \times [0, 1]$ having distribution $Pr(\zeta_r = 0) = 1/2$, $Pr(\zeta_r = s) = Pr(\zeta_r = -s) = 1/4$. Since ζ_r are independent and symmetric, we have by the P. Lévy inequality (Theorem 1.20) that for any $M > 0$,

$$Pr(|h_r| \geq M \text{ for some } r \leq L) \leq 2Pr(|h_L| \geq M). \tag{11.11}$$

Now, we can write $\zeta_r = \eta_{2r-1} + \eta_{2r}$ where η_j are i.i.d. variables with $Pr(\eta_j = s/2) = Pr(\eta_j = -s/2) = 1/2$. So by one of Hoeffding's inequalities (Proposition 1.12 above), we have $Pr(|h_L| \geq M) \leq \exp(-M^2/(Ls^2))$. For c large, $(1 - 2s)^2 \geq 1/2$. Thus

$$
\begin{aligned}
Pr(|h_r| &> 1 - 2s \text{ for some } r \leq L) \\
&\leq 2\exp(-(1 - 2s)^2/(Ls^2)) \\
&\leq 2\exp(-\log c/(2\delta^2 \log c)) \\
&- 2\exp(-1/(2\delta^2)).
\end{aligned}
\tag{11.12}
$$

Let $\kappa(\omega, t) := \inf\{k : |h_k(\omega, t)| > 1 - 2s\} \leq +\infty$. Then $|f_k'(t)| \leq 1$ for all $k \leq \kappa(\omega, t)$. Let $\Phi_k(\omega, t) := f_{\min(\kappa,k)}(\omega, t)$. Then $\Phi_k \in \mathcal{M}_1$ for each k. If $|h_r(t)| \leq 1 - 2s$ for all $r \leq L$, then $|f_r'(t)| \leq 1$ and $|g_r'(t)| \leq 1$ for all $r \leq L$, $\kappa(\omega, t) > L$, and $\Phi_L(\omega, t) = f_L(\omega, T)$.

Let $\chi_{ki}(\omega) := 1$ if $\kappa(\omega, t) \geq k$ for some, or equivalently all, $t \in Int_{ki}$, otherwise let $\chi_{ki}(\omega) = 0$. For any real x let $x^+ := \max(x, 0)$. For $k =$

$0, 1, \ldots, L - 1$ and $i = 1, \ldots, 2^k$ we have $Q_{ki} \subset S(f_{k+1})$ if and only if A_{ki} holds. So, $Q_{ki} \subset S(\Phi_L(\omega, \cdot))$ if and only if both A_{ki} holds and $\chi_{ki} = 1$. Let $a_{ki} := 1_{A_{ki}}$,

$$S_{\Phi,L} := \sum_{k=0}^{L-1} \sum_{i=1}^{2^k} \chi_{ki} N_c(Q_{ki})^+, \tag{11.13}$$

$$S_{V,L} := \sum_{k=0}^{L-1} \sum_{i=1}^{2^k} V_{ki} \tag{11.14}$$

where $V_{ki} := \chi_{ki} a_{ki} N_c(V_{ki})$. Also, let $U_{ki} := \chi_{ki} N_c(Q_{ki})^+$ and $W := N_c(S(f_0))$ where $S(f_0) = [0, 1] \times [0, 1/2]$. Then $S(f_k)$ is the union, disjoint up to one-dimensional boundary line segments, of $S(f_0)$ and of those Q_{rj} and V_{rj} with $r < k$ such that A_{rj} holds. So

$$N_c(S(\Phi_L)) = S_{\Phi,L} + S_{V,L} + W. \tag{11.15}$$

From the definitions, we can see that the set where $\kappa(\omega, t) \leq k$ is $\mathcal{B}^{(k-1)}$ measurable for each t, and thus so is each χ_{ki}. Also, Q_{ki} is a $\mathcal{B}^{(k-1)}$ measurable random set, disjoint from $G(k-1)$. Since $P(Q_{ki})$ is fixed, by Lemma 11.8, $N_c(Q_{ki})^+$, a function of $N_c(Q_{ki})$, is independent of $\mathcal{B}^{(k-1)}$ and thus of χ_{ki}.

Similarly, a_{ki} is independent of χ_{ki} by Lemma 11.8. Both are measurable with respect to $\mathcal{B}^{k,i} \equiv \mathcal{B}_{G(k,i)}^{(k,i)}$, while V_{ki} is disjoint from $G(k, i)$, and $P(V_{ki})$ is a constant. So by Lemma 11.8 again, $N_c(V_{ki})$ is independent of $\mathcal{B}_{G(k,i)}$, and the three variables $N_c(V_{ki})$, a_{ki}, and χ_{ki} are jointly independent. Since $L \leq (\log c)/9$, in (11.10 the parameters of the Poisson variables for $k \leq L$ are $\geq cs/(6 \cdot 4^L) \geq 1$ for c large enough. By (11.12) and the Tonelli–Fubini theorem we have for $\delta \leq 1/4$ and $k = 0, 1, \ldots, L - 1$ that

$$1/2 \leq 1 - 2\exp(-1/(2\delta^2)) \leq \Pr(\kappa > k) \leq \frac{1}{2^k} \sum_{i=1}^{2^k} E \chi_{ki}. \tag{11.16}$$

Let $X_{ki} := N_c(Q_{ki})^+$ and $p_{ki} := Pr(\chi_{ki} = 0)$. Then by (11.16),

$$\sum_{i=1}^{2^k} p_{ki} \leq 2^{k+1} \exp(-1/(2\delta^2)).$$

For each k and i, $E[(1 - \chi_{ki})X_{ki}] \leq p_{ki}^{1/2}(EX_{ki}^2)^{1/2}$. Thus by (11.10), setting $S_1 := \sum_{k=0}^{L-1} \sum_{i=1}^{2^k}(1 - \chi_{ki})X_{ki}$, we have

$$ES_1 \leq \sum_{k=0}^{L-1} [2^{k+1} \exp(-1/(2\delta^2))]^{1/2} [2^k cs/(6 \cdot 4^k)]^{1/2}$$

$$= 6^{-1/2} L(cs)^{1/2} \exp(-1/(4\delta^2)) \leq (\delta^2 \log c)(cs)^{1/2} \exp(-1/(4\delta^2)),$$

so

$$ES_1 \leq c^{1/2}(\log c)^{3/4}\delta^2 \exp(-1/(4\delta^2)). \tag{11.17}$$

Letting $S_2 := \sum_{k=0}^{L-1}\sum_{i=1}^{2^k} X_{ki}$, we have by Lemma 11.14, and since by symmetry $EN_c(Q_{ki})^+ = E|N_c(Q_{ki})|/2$, that

$$
\begin{aligned}
ES_2 &\geq \sum_{k=0}^{L-1} 2^{k-1}c_0(cs/(6\cdot 4^k))^{1/2} \\
&= 24^{-1/2}c_0 L(cs)^{1/2} \geq c_1\delta^2 c^{1/2}(\log c)^{3/4},
\end{aligned}
$$

where $c_1 := c_0/10$. By (11.17) and Markov's inequality we have for any $\alpha > 0$,

$$Pr(S_1 \geq \alpha\delta^2 c^{1/2}(\log c)^{3/4}) \leq \alpha^{-1}\exp(-1/(4\delta^2)). \tag{11.18}$$

If $j < k$ or $j = k$ and $i < r$, then X_{ji} is measurable for $\mathcal{B}_{G(k,r-1)}$ while by Lemma 11.8, X_{kr} is independent of it. Thus all the variables X_{ji} are independent, and by (11.10),

$$\mathrm{Var}(S_2) = \sum_{j=0}^{L-1}\sum_{i=1}^{2^j}\mathrm{Var}(X_{ji}) \leq \sum_{j=0}^{L-1} 2^j cs/(6\cdot 4^j) < cs = c/(\log c)^{1/2}.$$

Thus S_2 has standard deviation $< c^{1/2}/(\log c)^{1/4}$, and

$$S_2 - ES_2 = o_p(ES_2) \text{ as } c \to \infty \tag{11.19}$$

by Chebyshev's inequality.

To find the covariance $\mathrm{Cov}(\mathcal{V}_{ji}, \mathcal{V}_{kr})$ of two different terms of $S_{V,L}$ in (11.14), we can again assume $j < k$ or $j = k$ and $i < r$. Let $Y_{uv} := N_c(V_{uv})$ for each u and v. Since $E\mathcal{V}_{uv} = 0$, we need to find

$$E_{jikr} := E(\chi_{ji}a_{ji}Y_{ji}\chi_{kr}a_{kr}Y_{kr}) = E(\chi_{ji}N_c(W_{ji})\chi_{kr}a_{kr}Y_{kr})$$

where $W_{ji} := V_{ji}$ on A_{ji} and $W_{ji} := \emptyset$ otherwise. Then W_{ji} and V_{kr} are $\mathcal{B}^{k,r}$-measurable random sets disjoint from the stopping set $G(k, r)$, and each of the three random sets has finitely many possible values. Thus Lemma 11.8 applies to $G = G(k, r)$, $A = W_{ji}$ and $B = V_{kr}$. Also, $A \cap B = \emptyset$. Thus we have

$$
\begin{aligned}
E_{jikr} &= EE(\chi_{ji}N_c(W_{ji})\chi_{kr}a_{kr}Y_{kr}|\mathcal{B}^{k,r}) \\
&= E\left[\chi_{ji}\chi_{kr}a_{kr}E\left(N_c(W_{ji})N_c(V_{kr})|\mathcal{B}^{k,r}\right)\right] \\
&= E\left[\chi_{ji}\chi_{kr}a_{kr}E\left(N_c(W_{ji})|\mathcal{B}^{k,r}\right)E\left(N_c(V_{kr})|\mathcal{B}^{k,r}\right)\right] = 0
\end{aligned}
$$

since the last conditional expectation is 0, and $E(\mathcal{V}_{ji}\mathcal{V}_{kr}) = 0 = E(\mathcal{V}_{ji}) = E(\mathcal{V}_{kr})$. So the terms are orthogonal, and

$$\text{Var}(S_{V,L}) = \sum_{j=0}^{L-1}\sum_{i=1}^{2^j} \text{Var}(\mathcal{V}_{ji}) \le \sum_{j=0}^{L-1} 2^j E((N_c(V_{j1}))^2)$$

$$\le \sum_{j=0}^{L-1} 2^{j+1} cs/(12 \cdot 4^j) < cs.$$

Also, for W in (11.15), $\text{Var}(W) \le c$. Thus by Chebyshev's inequality, $S_{V,L} = o_p(ES_2)$ and $W = o_p(ES_2)$ as $c \to \infty$. Note that we have $N_c(S(\Phi_L)) \equiv S_{\Phi,L} \equiv S_2 - S_1$. Then by (11.15) and (11.19),

$$N_c(S(\Phi_L)) = ES_2 + (S_2 - ES_2) - S_1 + S_{V,L} + W = ES_2 - S_1 + o_p(ES_2).$$

Taking $\alpha := c_1/3$ in (11.18), then $\delta > 0$ small enough so that $(3/c_1)\exp(-1/(4\delta^2)) < \varepsilon/2$, we have for c large enough that

$$Pr\left\{N_c(S(\Phi_L)) > \gamma c^{1/2}(\log c)^{3/4}\right\} > 1 - \varepsilon$$

where $\gamma := c_1\delta^2/3$. Since $\Phi_L \in \mathcal{M}_1$, the conclusion for N_c follows.

Now, suppose the theorem fails for the centered Poisson process Y_c, for some $\varepsilon > 0$. Thus, for arbitrarily small $\delta > 0$, $Pr\left[\|Y_c\|_{\mathcal{C}(1,2,2)} < \delta c^{1/2}(\log c)^{3/4}/2\right] \ge \varepsilon$. Then, subtracting two independent versions of Y_c to get an N_c, we have $Pr\left[\|N_c\|_{\mathcal{C}(1,2,2)} < \delta c^{1/2}(\log c)^{3/4}\right] \ge \varepsilon^2$, a contradiction.

The ν_n case follows from Lemma 11.7.   □

11.5 Proof of Theorem 11.10

Proof. First, there are measurability properties to consider. For $\mathcal{C}(\alpha, K, d)$, let $\alpha = \beta + \gamma$ where $0 < \gamma \le 1$ and β is an integer ≥ 0. The set $\mathcal{G}(\alpha, K, d)$ of functions is compact in the $\|\cdot\|_\beta$ norm: it is totally bounded by the Arzelà–Ascoli Theorem (RAP, Theorem 2.4.7) and closed since uniform (or pointwise) convergence of functions g preserves a Hölder condition $|G(x) - G(y)| \le K|x - y|^\gamma$. Recall that for $x \in \mathbb{R}^d$ we let $x_{(d)} := (x_1, \ldots, x_{d-1})$. The set $\{(x, g) : 0 \le x_d \le g(x_{(d)}), g \in \mathcal{G}(\alpha, K, d)\}$ is compact, for the β norm on g. So the class $\mathcal{C}(\alpha, K, d)$ is image admissible Suslin.

Measurability for the lower layer case $\mathcal{C} = \mathcal{LL}_d$ follows as in the case $d = 2$ treated soon after the statement of Theorem 11.10. For the case $\mathcal{C} = \mathcal{C}_3$ of convex sets in $\mathbb{R}^3$, for any $F \subset \{X_1, \ldots, X_n\}$ there is a smallest convex set including F, its convex hull (a polyhedron). The maximum of $P_n - P$ will be found by maximizing over $2^n - 1$ such sets. Conversely, to maximize $(P - P_n)(C)$, or equivalently $P(C)$, over all closed convex sets C such that $C \cap \{X_1, \ldots, X_n\} = F$, recall that a closed convex set is an intersection of closed half-spaces (RAP, Theorem 6.2.9). If $|F| = k$, it suffices in the present case to take an intersection of no more than $n - k$ closed half-spaces, one to exclude each $X_j \notin F$, $j \leq n$. Thus $\|P_n - P\|_{\mathcal{C}} = \|P_n - P\|_{\mathcal{D}}$ for a finite-dimensional class $\mathcal{D}$ of sets for which an image admissible Suslin condition holds. It follows that the norms $\|Y_x\|_{\mathcal{C}}$ and $\|\nu_n\|_{\mathcal{C}}$ are measurable for the classes $\mathcal{C}$ in the statement.

Theorem 11.10 for Y_x implies it for ν_n by Lemma 11.7.

By assumption $d \geq 2$. Let $J := [0, 1)$, so that $J^{d-1} = \{x \in \mathbb{R}^{d-1} : 0 \leq x_j < 1 \text{ for } j = 1, \ldots, d - 1\}$.

Let f_1 be a C^∞ function on $\mathbb{R}$ such that $f_1(t) = 0$ for $t \leq 0$ or $t \geq 1$, for some κ with $0 < \kappa < 1$, $f(t) = \kappa$ for $1/3 \leq t \leq 2/3$, and $0 < f(t) < \kappa$ for $t \in (0, 1/3) \cup (2/3, 1)$. Specifically, we can let $f_1(t) := (g * h)(t) := \int_{-\infty}^{\infty} g(t - y)h(y)dy$ where $h := 1_{[1/6, 5/6]}$ and g is a C^∞ function with $g(t) > 0$ for $|t| < 1/6$ and $g(t) = 0$ for $|t| \geq 1/6$. For $x \in \mathbb{R}^{d-1}$ let $f(x) := f_1(x_1) \cdot f_1(x_2) \cdots f_1(x_{d-1})$. Then f is a C^∞ function on $\mathbb{R}^{d-1}$ with $f(x) = 0$ outside the unit cube J^{d-1}, $0 < f(x) \leq \gamma$ for x in the interior of J^{d-1}, and $f \equiv \gamma$ on the subcube $[1/3, 2/3]^{d-1}$ where $0 < \gamma := \kappa^{d-1} < 1$. Taking a small enough positive multiple of f_1 we can assume that $\sup_x \sup_{[p] \leq d-1} |D^p f(x)| \leq 1$, possibly with a smaller $\gamma > 0$.

Some indexed families of sets and functions, some of them random, will be defined as follows. For each $j = 1, 2, \ldots$, the unit cube J^{d-1} is decomposed as a union of $3^{j(d-1)}$ disjoint subcubes C_{ji} of side 3^{-j} for $i = 1, 2, \ldots, 3^{j(d-1)}$, where each C_{ji} is also a Cartesian product of left closed, right open intervals. Let B_{ji} be a cube of side 3^{-j-1}, concentric with and parallel to C_{ji}. Let x_{ji} be the point of C_{ji} closest to 0 (a vertex) and y_{ji} the point of B_{ji} closest to 0. For $\delta > 0$ and $j = 1, 2, \ldots$, let $c_j := Cj^{-1}(\log(j + 1))^{-1-2\delta}$, where the constant $C > 0$ is chosen so that $\sum_{j=1}^{\infty} c_j \leq 1$. For $x \in \mathbb{R}^{d-1}$ let

$$f_{ji}(x) := c_j 3^{-j(d-1)} f(3^j(x - x_{ji})),$$

$$g_{ji}(x) := c_j 3^{-(j+1)(d-1)} f(3^{j+1}(x - y_{ji})).$$

Then $f_{ji}(x) > 0$ on the interior of C_{ji} while $f_{ji}(x) = 0$ outside C_{ji}, and likewise for g_{ji} and B_{ji}. Note that

$$f_{ji}(x)/g_{ji}(x) \geq 3^{d-1} > 1 \quad \text{for all } x \in B_{ji}. \tag{11.20}$$

Let $S_0 := 1/2$. Sequences of random variables $s_{ji} = \pm 1$ and random functions on I^{d-1},

$$S_k := S_0 + \sum_{j=1}^{k} \sum_{i=1}^{3^{j(d-1)}} s_{ji}(\omega) f_{ji} \tag{11.21}$$

will be defined recursively. Then since $d \geq 2$, we have $\sum_{j=1}^{\infty} c_j 3^{-j(d-1)} \leq 1/3$, so $0 < S_k < 1$ for all k. Given $j \geq 1$ and S_{j-1}, let

$$D_{ji} := D_{ji}(\omega) := \{x \in J^d : |x_d - S_{j-1}(x_{(d)})(\omega)| < g_{ji}(x_{(d)})\}. \quad (11.22)$$

Let $s_{ji}(\omega) := +1$ if $Y_\lambda(D_{ji}(\omega)) > 0$, otherwise $s_{ji}(\omega) := -1$. This finishes the recursive definition of the s_{ji} and the S_k.

By induction, each D_{ji} has only finitely many possible values, each on a measurable event, so the s_{ji} and S_k are all measurable random variables. Since the cubes C_{ji} do not overlap, and all derivatives $D^p f_{ji}$ are 0 on the boundary of C_{ji}, it follows for any $k \geq 1$ that

$$\begin{aligned}
\sup_{|p| \leq d-1} \sup_x |D^p S_k(x)| &\leq \sum_{j=1}^k \sup_{|p| \leq d-1} |D^p f_{j1}(x)| \\
&\leq \sum_{j=1}^k c_j < 1.
\end{aligned} \quad (11.23)$$

The volume of $D_{ji}(\omega)$ is always

$$P(D_{ji}(\omega)) = 2 \int_{B_{ji}} g_{ji} dx = 2\mu c_j / 9^{(j+1)(d-1)} \quad (11.24)$$

where $0 < \mu := \int_{J^{d-1}} f(x) dx < 1$.

Next, it will be shown that $D_{ji}(\omega)$ for different i, j are disjoint. For $1 \leq j \leq k$ and any ω, i and $x \in B_{ji}$,

$$\begin{aligned}
|S_k(x) - S_{j-1}(x)|(\omega) &\geq \gamma c_j 3^{-j(d-1)} - \sum_{r>j} \gamma c_r 3^{-r(d-1)} \\
&\geq \gamma c_j 3^{-j(d-1)} \left(1 - \frac{1}{2}\right) \\
&\geq \gamma c_j 3^{-j(d-1)}/2 \geq \sup_y (g_{ji} + \sup_r g_{k+1,r})(y).
\end{aligned}$$

Thus if $s_{ji}(\omega) = +1$, then for any r and any $x \in B_{ji}$,

$$S_k(x)(\omega) - g_{k+1,r}(x) \geq S_{j-1}(x)(\omega) + g_{ji}(x).$$

It follows that $D_{ji}(\omega)$ is disjoint from $D_{k+1,r}(\omega)$. They are likewise disjoint if $s_{ji}(\omega) = -1$ by a symmetrical argument. For the same j and different i, the sets $D_{ji}(\omega)$ are disjoint since the projection $x \mapsto x_{(d)}$ from $\mathbb{R}^d$ onto $\mathbb{R}^{d-1}$ takes them into disjoint sets B_{ji}.

Given $\lambda > 0$ let $r = r(\lambda)$ be the largest j, if one exists, such that $2\lambda\mu c_j \geq 9^{(j+1)(d-1)}$. Then

$$r(\lambda) \sim (\log \lambda)/((d-1)\log 9) \quad \text{as} \quad \lambda \to +\infty. \quad (11.25)$$

Let $G_0(\omega) := \emptyset$. For $m = 1, 2, \ldots$, let

$$G(m)(\omega) := G_m(\omega) := \bigcup \{D_{ji}(\omega) : j \leq m\}.$$

Let $H_m(\omega)$ be the subgraph of S_m, $H_m(\omega) := \{x : 0 \le x_d \le S_m(x_{(d)})(\omega)\}$. Then for all ω, by (11.23), $H_m \in \mathcal{C}(d-1, 1, d)$. Let $A_m(\omega) := H_m(\omega) \setminus G_m(\omega)$. From the disjointness proof,

$$H_m(\omega) \cap G_m(\omega) = \bigcup \{D_{ji}(\omega) : j \le m, \ s_{ji} = +1\}. \tag{11.26}$$

For each m, each of the above sets has finitely many possible values, each on a measurable event.

Each G_j is easily seen to be a stopping set, while $A_j(\cdot)$ is $\mathcal{B}_{G(j)}$ measurable and for each ω, $A_j(\omega)$ is disjoint from $G_j(\omega)$. So the hypotheses of Lemma 11.8 hold. Thus, conditional on $\mathcal{B}_{G(j)}$, $X_\lambda(A_j)$ is Poisson with parameter $\lambda P(A_j(\omega))$. Also, for $u^+ := \max(u, 0)$, by (11.26),

$$Y_\lambda((H_m \cap G_m)(\omega)) = \sum_{j=1}^{m} \sum_{i=1}^{3^{j(d-1)}} Y_\lambda(D_{ji})^+. \tag{11.27}$$

Now, $P(D_{ji})$ does not depend on ω nor i, and $D_{ji}(\cdot)$ is $\mathcal{B}_{G(j-1)}$ measurable. Thus by Lemma 11.8, applied to $A := D_{mi}$, replacing G_m by $G_{m-1} \cup \bigcup_{r<i} D_{mr}$, the $Y_\lambda(D_{ji})$ for different i or j are jointly independent, and each has the law of $Y_\lambda(D)$ for a fixed set D with $P(D) := 2 \int g_{j1} dx$.

Taking $m := r(\lambda)$, equation (11.27) gives a sum of independent nonnegative parts of centered Poisson variables with parameters $\lambda P(D_{ji}) \ge 1$, by (11.24). So by Lemma 1.18, for a constant $c > 0$,

$$\lambda^{-1/2} E Y_\lambda((H_m \cap G_m)(\omega)) \ge c \sum_{j=1}^{r(\lambda)} \sum_{i=1}^{3^{j(d-1)}} P(D_{ji})^{1/2}$$

$$= c \sum_{j=1}^{r(\lambda)} 3^{j(d-1)} \left(2\mu c_j / 9^{(j+1)(d-1)}\right)^{1/2} \quad \text{by (11.24)}$$

$$= c 3^{1-d} (2\mu)^{1/2} \sum_{j=1}^{r(\lambda)} (Cj^{-1}(\log(j+1))^{-1-2\delta})^{1/2} \quad \text{(by def. of } c_j\text{)}$$

$$= c(2\mu C)^{1/2} 3^{1-d} \sum_{j=1}^{r(\lambda)} j^{-1/2} ((\log(j+1))^{-0.5-\delta}$$

$$\ge a_d (\log(r(\lambda)+1))^{-0.5-\delta} \sum_{j=1}^{r(\lambda)} j^{-1/2}$$

$$\ge 2a_d (r(\lambda)^{1/2} - 1)(\log(r(\lambda)+1))^{-0.5-\delta}$$

for some constant $a_d > 0$. For λ large, by (11.25), the latter expression is $\ge 3b_d(\log \lambda)^{1/2}(\log \log \lambda)^{-0.5-\delta}$ for some $b_d > 0$.

By independence of the variables $Y_\lambda(D_{ji})$ and (11.24), $Y_\lambda((H_{r(\lambda)} \cap G_{r(\lambda)})(\omega)$ has variance less than

$$\sum_{j=1}^{r(\lambda)} \sum_{i=1}^{3^{j(d-1)}} \lambda P(D_{ji}) = \lambda \sum_{j=1}^{r(\lambda)} 3^{j(d-1)} 2\mu c_j 9^{-(j+1)(d-1)} < \lambda.$$

Thus by Chebyshev's inequality, as $\lambda \to +\infty$,

$$\Pr\{\lambda^{-1/2} Y_\lambda(H_{r(\lambda)} \cap G_{r(\lambda)})(\omega) \geq 2b_d(\log \lambda)^{1/2}(\log\log \lambda)^{-0.5-\delta}\} \to 1.$$

As shown around (11.27), $Y_\lambda(A_{r(\lambda)}(\omega))(\omega)$ has, given $\mathcal{B}_{G(r(\lambda))}$, the conditional distribution of $Y_\lambda(D)$ for $P(D) = P(A_{r(\lambda)}(\omega))$, where since Y_λ is a centered Poisson process, $EY_\lambda(D)^2 \leq \lambda$ for all D. Thus $EY_\lambda(D_{r(\lambda)})^2 \leq \lambda$, $Y_\lambda(A_{r(\lambda)})/\lambda^{1/2}$ is bounded in probability, and

$$\Pr\{Y_\lambda(H_{r(\lambda)}) \geq b_d(\lambda \log \lambda)^{1/2}(\log\log \lambda)^{-0.5-\delta}\} \to 1$$

as $\lambda \to +\infty$. This proves Theorem 11.10 for Y_λ, since given the result for $K = 1$, one can let $\gamma(d, K, \delta) := K\gamma(d, 1, \delta)$. For ν_n, Lemma 11.7 then applies. This completes the proof for classes $\mathcal{C}(d-1, K, d)$. For convex sets in $\mathbb{R}^3$ see Dudley (1982, Theorem 4, proof in Section 5). Lower layers were treated in Section 11.4. $\square$

Problems

1. Find a lower bound for the constant $c(d, \alpha)$ in the proof of Theorem 11.1. *Hint:* $c(d, \alpha) \geq 3^{-d-\alpha} P(f)/\|f\|_\alpha$.

2. In the proof of Theorem 11.2, show that r can be taken as $n+1$ for the largest $n = 0, 1, \ldots$ such that $m(n) = 0$.

3. If $(Y, \|\cdot\|)$ is any Banach space, in particular $(\ell^\infty(\mathcal{F}), \|\cdot\|_\mathcal{F})$ for any $\mathcal{F}$, for any $\gamma > 0$ there is a bounded continuous (in fact, Lipschitz) function $H \geq 0$ on Y such that $H(y) > 0$ if and only if $\|y\| < \gamma$. Use this to prove the statement made in the Remarks after Theorem 11.1 that $\mathcal{G}(d/2, 1, d)$ is not P-Donsker for $P = U(I^d)$. *Hint:* If Y is 1-dimensional, say $Y = \mathbb{R}$, find such a function of the form $h(|y|)$. In general take $h(\|y\|)$.

4. What should replace $n^\Theta(\log n)^\eta$ in Theorem 11.4 if $0 < \zeta < 1$? *Hint:* Any sequence $a_n \to +\infty$.

5. To show that Poisson processes are well-defined, show that if X and Y are independent real random variables with $\mathcal{L}(X) = P_a$ and $\mathcal{L}(Y) = P_b$ then $\mathcal{L}(X + Y) = P_{a+b}$.

6. Let P be a law on $\mathbb{R}$ and U_c the Poisson process with intensity measure cP for some $c > 0$. For a give $y \geq 0$ let X be the least $x \geq 0$ such that $U_c([-x, x]) \geq y$. Show that the interval $[-X, X]$ is a stopping set.

Notes

Notes to Section 11.1. Bakhvalov (1959) proved Theorem 11.1. W. Schmidt (1975) proved Theorem 11.3. Theorem 11.2, proved by the same method as theirs, also is in Dudley (1982, Theorem 1).

Notes to Section 11.2. Walter Philipp (unpublished) proved Theorem 11.4 for $\zeta = 1$, with a refinement ($\eta = 1$ and a power of $\log \log n$ factor). The proof for $\zeta > 1$ follows the same scheme. Theorem 11.4 is Theorem 2 of Dudley (1982).

Notes to Section 11.3. Lemma 11.5 is classical, see e.g. Feller (1971, VIII.8, Lemma 2). Lemma 11.6 and with it the main idea of Poissonization are due to Kac (1949). Pyke (1968), along the line of Lemma 11.7, gives relations between Poissonized and non-Poissonized cases. Evstigneev (1977, Theorem 1) proves a Markov property for random fields indexed by closed subsets of a Euclidean space, somewhat along the line of Lemma 11.8. This entire section is a revision of Section 3 of Dudley (1982), with proofs of Lemmas 11.5 and 11.6 supplied here.

Notes to Section 11.4. Theorem 11.10 was proved in Dudley (1982) for all $d \geq 2$. Another proof was given for lower layers and $d = 2$ in Dudley (1984). Shor (1986) discovered the larger lower bound with $(\log n)^{3/4}$ in expectation, and contributed to a later version of the proof (Coffman and Lueker, 1991, pp. 57–64). Shor gave the definition of f_i and g_i. Theorem 11.11 here shows that the lower bound holds not only in expectation, but with probability going to 1, by the methods of Dudley (1982, 1984). I thank J. Yukich for providing very helpful advice about this section.

Notes to Section 11.5 The proof is from Dudley (1982).

Appendix A
Differentiating under an Integral Sign

The object here is to give some sufficient conditions for the equation

$$\frac{d}{dt}\int f(x,t)d\mu(x) = \int \frac{\partial f(x,t)}{\partial t}d\mu(x). \tag{A.1}$$

Here $x \in X$, where $(X, \mathcal{S}, \mu)$ is a measure space, and t is real-valued. The derivatives with respect to t will be taken at some point $t = t_0$. The function f will be defined for $x \in X$ and t in an interval J containing t_0. Assume for the time being that t_0 is in the interior of J. Let

$$f_2(x, t_0) := \partial f(x, t)/\partial t|_{t=t_0}.$$

Some assumptions are clearly needed for (A.1) even to make sense. A set $\mathcal{F} \subset \mathcal{L}^1(X, \mathcal{S}, \mu)$ is called $\mathcal{L}^1$-*bounded* iff $\sup\{\int |f|d\mu : f \in \mathcal{F}\} < \infty$. Here are some basic assumptions:

$$\text{For some } \delta > 0, \ \{f(\cdot, t) : |t - t_0| < \delta\} \text{ are } \mathcal{L}^1\text{-bounded,}$$

$$f_2(x, t_0) \text{ exists for } \mu\text{-almost all } x, \tag{A.2}$$

$$\text{and } f_2(\cdot, t_0) \in \mathcal{L}^1(X, \mathcal{S}, \mu).$$

Here is an example to show that the conditions (A.2) are not sufficient:

Example. Let (X, μ) be $[0, 1]$ with Lebesgue measure. Let $J = [-1, 1]$, $t_0 = 0$. Let $f(x, t) := 1/t$ if $t > 0$ and $0 \le x \le t$, otherwise let $f(x, t) := 0$. Then for all $x \ne 0$, $f(x, t) = 0$ for t in a neighborhood of 0, so $f_2(x, 0) = 0$ for $x \ne 0$. For each $t > 0$, $\int_0^1 f(x, t)dx = 1$, while $\int_0^1 f(x, 0)dx = 0$. So the function $t \mapsto \int_0^1 f(x, t)dx$ is not even continuous, and so not differentiable, at $t = 0$, so (A.1) fails.

The function f in the last example behaves badly near $(0, 0)$. Another possibility has to do with behavior where x is unbounded, while f may be very regular at finite points. Recall that a function f of two real variables x, t is

417

called C^∞ if the partial derivatives $\partial^{p+q} f/\partial x^p \partial t^q$ exist and are continuous for all nonnegative integers p, q.

Example. For $J = (-1, 1)$, $X = \mathbb{R}$ with Lebesgue measure μ, and $t_0 = 0$, there exists a C^∞ function f on $X \times J$ such that for $0 < t < 1$, $\int_{-\infty}^{\infty} f(x, t)dx = 1$, while for $t_0 = 0$, and any x, $f(x, t) = 0$ for t in a neighborhood of 0 (depending on x), so $f(x, 0) \equiv f_2(x, 0) \equiv 0$. Thus $\int_{-\infty}^{\infty} f_2(x, 0)dx = 0$ while $\int_{-\infty}^{\infty} f(x, t)dx$ is not continuous (and so not differentiable) at 0 and (A.1) fails. To define such an f, let g be a C^∞ function on $\mathbb{R}$ with compact support, specifically $g(x) := c \cdot \exp(-1/(x(1 - x)))$ for $0 < x < 1$ and $g(x) := 0$ elsewhere, where c is chosen to make $\int_{-\infty}^{\infty} g(x)dx = \int_0^1 g(x)dx = 1$. It can be checked (by L'Hospital's rule) that g and all its derivatives approach 0 as $x \downarrow 0$ or $x \uparrow 1$, so g is C^∞. Now let $f(x, t) := g(x - t^{-1})$ for $t > 0$ and $f(x, t) := 0$ otherwise. The stated properties then follow since $f(x, t) = 0$ if $t \leq 0$ or $x \leq 0$ or $0 < t < 1/x$.

For interchanging two integrals, there is a standard theorem, the Tonelli–Fubini theorem (e.g., RAP, Theorem 4.4.5). But for interchanging a derivative and an integral there is apparently not such a handy single theorem yielding (A.1). Some sufficient conditions will be given, beginning with rather general ones and going on to more special but useful conditions.

A set $\mathcal{F}$ of integrable functions on $(X, \mathcal{S}, \mu)$ is called *uniformly integrable* or *u.i.* if for every $\varepsilon > 0$ there exist both a set A with $\mu(A) < \infty$ such that

$$\int_{X \setminus A} |f|d\mu < \varepsilon \quad \text{for all } f \in \mathcal{F}, \tag{A.3}$$

and an $M < \infty$ such that

$$\int_{|f|>M} |f|d\mu < \varepsilon \quad \text{for all } f \in \mathcal{F}. \tag{A.4}$$

In the presence of (A.3), condition (A.4) is equivalent to: $\mathcal{F}$ is L^1-bounded and for any $\varepsilon > 0$ there is a $\delta > 0$ such that

$$\text{for any } C \in \mathcal{S} \text{ with } \mu(C) < \delta \text{ and any } f \in \mathcal{F}, \int_C |f|d\mu < \varepsilon; \tag{A.5}$$

this was proved in RAP (Theorem 10.3.5) for probability measures, where (A.3) is vacuous. The proof extends easily to the general case.

The best-known sufficient condition for uniform integrability is known as domination. The following is straightforward to prove, via (A.4):

Theorem A.1 *If $f \in \mathcal{L}^1(X, \mathcal{S}, \mu)$ and $f \geq 0$, then $\{g \in \mathcal{L}^1(X, \mathcal{S}, \mu) : |g| \leq f\}$ is uniformly integrable.*

Here $|g| \leq f$ means $|g(x)| \leq f(x)$ for all $x \in X$. But domination is not necessary for uniform integrability, as the following shows:

Example. Let $(X, \mathcal{S}, \mu)$ be $[0, 1]$ with Lebesgue measure. Let $g(t, x) := |t - x|^{-1/2}$, $0 \leq t \leq 1$, $0 \leq x \leq 1$. Then the set of all functions $g(t, \cdot)$, $0 \leq t \leq 1$, is uniformly integrable but not dominated.

Uniform integrability provides a general sufficient condition for (A.1). Let

$$\Delta_h f(x, t_0) := f(x, t_0 + h) - f(x, t_0).$$

Theorem A.2 *Assume (A.2) and that for some neighborhood U of 0,*

$$\{\Delta_h f(x, t_0)/h \, : \, 0 \neq h \in U\} \text{ are uniformly integrable.} \qquad \text{(A.6)}$$

Then (A.1) holds and moreover

$$\int |\Delta_h f(x, t_0)/h - f_2(x, t_0)| d\mu(x) \to 0 \text{ as } h \to 0. \qquad \text{(A.7)}$$

Conversely, (A.7) and (A.2) imply (A.6).

Proof. (A.2) implies that the difference-quotients $\Delta_h f(x, t_0)/h$ converge pointwise for almost all x as $h \to 0$ to $\partial f(x, t)/\partial t|_{t=t_0}$. Pointwise convergence and uniform integrability (A.6) imply $\mathcal{L}^1$ convergence (A.7): this was proved in (RAP, Theorem 10.3.6) for a sequence of functions on a probability space. The implication holds in the case here since:

(a) By (A.3), we can assume that up to a difference of at most ε in all the integrals, the functions are defined on a finite measure space, which reduces by a constant multiple to a probability space;

(b) If (A.7) fails, then it fails along some sequence $h = h_n \to 0$, so a proof for sequences is enough. Thus (A.7) follows from the cited fact.

Now suppose (A.7) and (A.2) hold. Then given $\varepsilon > 0$, there is a set A of finite measure with $\int_{X \setminus A} |f_2(x, t_0)| d\mu(x) < \varepsilon/2$, and for h small enough, $\int |\Delta_h f(x, t_0)/h - f_2(x, t_0)| d\mu(x) < \varepsilon/2$, so (A.3) holds for the functions $\Delta_h f/h$, $h \in U$, as desired.

For sequences of integrable functions on a probability space, convergence in $\mathcal{L}^1$ implies uniform integrability (RAP, Theorem 10.3.6). Here again, we can reduce to the case of probability spaces, and if (A.6) fails, for a given ε, then it fails along some sequence, contradicting the cited fact. $\square$

But given (A.2), the uniform integrability condition (A.6), necessary for (A.7), is not necessary for (A.1), as is shown by a modification of a previous example:

Example. Let $(X, \mu) = [-1, 1]$ with Lebesgue measure, $J = [-1, 1]$, and $t_0 = 0$. If $t > 0$, let $f(x, t) = 1/t$ if $0 < x < t$, $f(x, t) = -1/t$ if $-t < x < 0$, and $f(x, t) = 0$ otherwise. Then $\int_{-1}^{1} f(x, t) dx = 0$ for all t and $x \int_{-1}^{1} f_2(x, 0) dx = 0$ since $f_2(x, 0) \equiv 0$, so (A.1) holds.

Domination of the difference-quotients by an integrable function gives the classical "Weierstrass" sufficient condition for (A.1), which follows directly from Theorems A.1 and A.2:

Corollary A.3 *If (A.2) holds and for some neighborhood U of 0, there is a $g \in \mathcal{L}^1(X, \mathcal{S}, \mu)$ such that $|\Delta_h f(x, t_0)/h| \le g(x)$ for $0 \ne h \in U$, then (A.1) holds.*

Corollary A.3 is highly useful, but it is not as directly applicable as is the Tonelli–Fubini theorem on interchanging integrals. In some cases, even when a dominating function $g \in \mathcal{L}^1$ exists, it may not be easy to choose such a g for which integrability can be proved conveniently.

We can take the neighborhoods U to be sets $\{h : |h| < \delta\}$, for $\delta > 0$. For a given $\delta > 0$, the smallest possible dominating function would be

$$g_\delta(x) := \sup\{|\Delta_h f(x, t_0)/h| : 0 < |h| < \delta\}.$$

If $(X, \mathcal{S})$ can be taken to be a complete separable metric space with its σ-algebra of Borel sets, and if f is jointly measurable, then each g_δ is measurable for the completion of μ (RAP, Section 13.2). If so, then the hypothesis of Corollary A.3, beside (A.2), is equivalent to saying that

$$\text{for some } \delta > 0, \ g_\delta \in \mathcal{L}^1(\mu). \tag{A.8}$$

So far, nothing required the partial derivatives $f_2(x, t)$ to exist for $t \ne t_0$, but if they do, we get other sufficient conditions:

Corollary A.4 *Suppose (A.2) holds, $f(\cdot, \cdot)$ is jointly measurable, and there is a neighborhood U of t_0 such that for all $t \in U$, $f_2(x, t)$ exists for almost all x and the functions $f_2(\cdot, t)$ for $t \in U$ are uniformly integrable. Suppose also that for $t \in U$, $f(x, t) - f(x, t_0) = \int_{t_0}^t f_2(x, s)ds$ for μ-almost all x. Then (A.6) and so (A.7) and (A.1) hold.*

Proof. The functions $f_2(\cdot, t)$ for $t \in U$, being uniformly integrable, are $\mathcal{L}^1$-bounded. We can take U to be a bounded interval. Then $\int_U \int_X |f_2(x, t)| d\mu(x) dt < \infty$. Now for $|h|$ small enough,

$$|\Delta_h f(x, t_0)/h| = \left|\frac{1}{h}\int_0^h f_2(x, t_0 + u)du\right| \le \frac{1}{|h|}\int_{-|h|}^{|h|} |f_2(x, t_0 + u)|du < \infty.$$

Thus the difference-quotients $\Delta_h f(\cdot, t_0)/h$ are also L^1-bounded for h in a neighborhood of 0. Condition (A.3) for the functions $f_2(\cdot, t)$ implies it for the functions $\Delta_h f(\cdot, t_0)/h$. Also, condition (A.5) for the functions $f_2(\cdot, t_0 + u)$ directly implies (A.5) and (A.6) for the functions $\Delta_h f(x, t_0)/h$ as desired.   □

Three clear consequences of Corollary A.3 will be given.

Corollary A.5 *Suppose* $f(x, t) \equiv G(x, t)H(x)$ *for measurable functions* $G(\cdot, \cdot)$ *and H where (A.2) holds for f. Suppose that for h in a neighborhood of* $0, |\Delta_h G(x, t_0)/h| \le \alpha(x)$ *for a function* α *such that* $\int |\alpha(x)H(x)|d\mu(x) < \infty$. *Then (A.6), (A.7), and (A.1) hold.*

Corollary A.6 *Let* $G(x, t) \equiv \phi(x - t)$ *where* ϕ *is bounded and has a bounded, continuous first derivative,* $X = \mathbb{R}$, $\mu = $ *Lebesgue measure, and* $H \in \mathcal{L}^1(\mathbb{R})$. *Then the convolution integral* $(\phi * H)(t) := \int_{-\infty}^{\infty} \phi(t - x)H(x)dx$ *has a first derivative with respect to t given by* $(\phi * H)'(t) = \int_{-\infty}^{\infty} \phi'(t - x)H(x)dx$. *If the jth derivative* $\phi^{(j)}$ *is continuous and bounded for* $j = 0, 1, \ldots, n$, *then iterating gives* $(\phi * H)^{(n)}(t) \equiv \int_{-\infty}^{\infty} \phi^{(n)}(t - x)H(x)dx$.

Corollary A.7 *Let* $f(x, t) = e^{it\psi(x)}H(x)$ *where* ψ *and H are measurable real-valued functions on X such that H and* ψH *are integrable. Then (A.6), (A.7), and (A.1) hold.*

Proof. Let $G(x, t) := e^{it\psi(x)}$. For any real t, u, and h, $|e^{i(t+h)u} - e^{itu}| \le |uh|$ (RAP, proof of Theorem 9.4.4). So $|\Delta_h G(x, t)/h| \le |\psi(x)|$. Since

$$\frac{\partial}{\partial t}(e^{it\psi(x)}H(x)) = i\psi(x)e^{it\psi(x)}H(x)$$

and $|i\psi(x)e^{it\psi(x)}| \equiv |\psi(x)|$, conditions (A.2) hold and Corollary A.5 applies. $\qquad \square$

Note. In Corollary A.7, if $X = \mathbb{R}$ and $\psi(x) \equiv x$, we have Fourier transform integrals.

Proposition A.8 *Let* ψ *be a measurable function of x and suppose that* $\eta(t) := \int e^{t\psi(x)}d\mu(x) < \infty$ *for all t in an open interval U. Then* η *is an analytic function having a power series expansion in t in a neighborhood of any* $t_0 \in U$. *Derivatives of* η *of all orders can be found by differentiating under the integral,*

$$\eta^{(n)}(t) = \int \psi(x)^n e^{t\psi(x)}d\mu(x), \quad n = 1, 2, \ldots, \quad \text{for all } t \in U.$$

Proof. Take $t_0 \in U$. Replacing μ by v where $dv(x) = \exp(t_0\psi(x))d\mu(x)$, we can assume $t_0 = 0$. Then for some $\varepsilon > 0$, $\{t : |t| \le \varepsilon\} \subset U$, and for $|t| \le \varepsilon$, $e^{|t\psi(x)|} \le e^{\varepsilon\psi(x)} + e^{-\varepsilon\psi(x)}$, an integrable function. Thus, $\eta(x) = \sum_{n=0}^{\infty} t^n \int \psi(x)^n d\mu(x)/n!$ where the series converges absolutely by dominated convergence since the corresponding series for $e^{|t\psi(x)|}$ does and is a series of positive terms dominating those for $e^{t\psi(x)}$. The derivatives of η at 0 satisfy $\eta^{(n)}(0) = \int \psi(x)^n d\mu(x)$: this holds either by the theorem that for a function η represented by a power series $\sum_{n=0}^{\infty} c_n t^n$ in a neighborhood of 0, we must have $c_n = \eta^{(n)}(0)/n!$ (converse of Taylor's theorem), or by the methods of this Appendix as follows. For any real y, $|e^y - 1| \le |y|(e^y + 1)$ by the mean value theorem since $e^t \leq e^y + 1$ for t between 0 and y. Thus for any real u and h, $|(e^{uh} - 1)/h| \leq |u|(e^{uh} + 1)$. For any $\delta > 0$, there is a constant $C = C_\delta$ such that $|u| \leq C(e^{\delta u} + e^{-\delta u})$ for all real u. If $|h| \leq \delta := \varepsilon/2$, then since $e^{uh} \leq e^{\delta u} + e^{-\delta u}$ and $e^{cu} + e^{-cu}$ is increasing in $u > 0$ for any real c, we have

$$|(e^{uh} - 1)/h| \leq 3C_\delta(e^{\varepsilon u} + e^{-\varepsilon u}).$$

Letting $u = \psi(x)$, we have domination by an integrable function, so Corollary A.3 applies for the first derivative. The proof extends to higher derivatives since $|\psi(x)|^n \leq Ke^{\delta|\psi(x)|}$ for some $K = K_{n,\delta}$. □

Notes to Appendix A. Perhaps the most classical result on interchange of integral and derivative is attributed to Leibniz: if f and $\partial f/\partial y$ are continuous on a finite rectangle $[a, b] \times [c, d]$ and $c < y < d$, then $(d/dy)\int_a^b f(x, y)dx = \int_a^b \partial f(x, y)/\partial y\, dx$.

In the literature of this topic, "uniformly integrable" and even the existence of integrals are sometimes used with different meanings. Let $\mathcal{F}$ be a family of real-valued continuous functions defined on a half-line, say $[1, \infty)$. A function f in $\mathcal{F}$ will be said to have an (improper Riemann) integral if the integrals $\int_1^M f(x)dx$ converge to a finite limit as $M \to \infty$, which may then be called $\int_1^\infty f(x)dx$. This integral is analogous to the (conditional) convergence of the sum of a series, where Lebesgue integrability corresponds to absolute convergence. An example of a non-Lebesgue integrable function having a finite improper Riemann integral is $(\sin x)/x$. The family $\mathcal{F}$ is called uniformly (improper Riemann) integrable if the convergence is uniform for $f \in \mathcal{F}$. For a continuous function f of two real variables having a continuous partial derivative $\partial f/\partial y$, uniform improper Riemann integrability with respect to x, say on a half-line $[a, \infty)$, of $\partial f(x, y)/\partial y$ for y in some interval, and finiteness of $\int_a^\infty f(x, y)dx$, imply that the latter integral can be differentiated with respect to y under the integral sign, e.g., Ilyin and Poznyak (1982, Theorem 9.10).

Ilyin and Poznyak (1982, Theorem 9.8) prove what they call Dini's test: if f is nonnegative and continuous on $[a, \infty) \times [c, d)$, where $-\infty < c < d < +\infty$, and for each $y \in [c, d]$, $I(y) := \int_a^\infty f(x, y)dx < \infty$ and $I(\cdot)$ is continuous, then the $f(\cdot, y)$ for $c \leq y \leq d$ are uniformly (improper Riemann) integrable; in this case they are also uniformly integrable in our sense. Of course, the same holds if f is nonpositive, which would apply to the derivative of a positive, decreasing function.

The Weierstrass domination condition, Corollary A.3, is classical and appears in most of the listed references. Hobson (1926, p. 355) states it (for functions of real variables) and gives references to several works published between 1891 and 1910.

Lang (1983) mentions the convolution case, Corollary A.6. Lang also treats differentiation of integrals $\int_a^b f(t, x)dt$ where $x \in E$, $f \in F$, and E, F are Banach spaces.

Ilyin and Poznyak (1982, Theorem 9.5) and Kartashev and Rozhdestvenskiĭ (1984) treat differentiation of integrals $\int_{a(y)}^{b(y)} f(x, y)dx$.

Brown (1986, Chapter 2) is a reference for Proposition A.16 and some multidimensional extensions, for application to exponential families in statistics.

Appendix B
Multinomial Distributions

Consider an experiment whose outcome is given by a point ω of a set Ω where $(\Omega, \mathcal{A}, P)$ is a probability space. Let A_i, $i = 1, \ldots, m$, be disjoint measurable sets with union Ω. Independent repetition of the experiment n times is represented by taking the Cartesian product Ω^n of n copies of $(\Omega, \mathcal{A}, P)$ (RAP, Theorem 4.4.6). Let $X := \{X_i\}_{i=1}^n \in \Omega^n$ and $P_n(A) := \frac{1}{n} \sum_{i=1}^n \delta_{X_i}(A)$, $A \in \mathcal{A}$, so P_n is an empirical measure for P, and $n P_n(A_i)$ is the number of times the event A_i occurs in the n repetitions.

A random vector $\{n_j\}_{j=1}^m$ is said to have a *multinomial* distribution for n observations, or with *sample size* n, and m *bins* or *categories*, with probabilities $\{p_j\}_{j=1}^m$, if $p_j \geq 0$, $p_1 + \cdots + p_m = 1$, and for any nonnegative integers k_j with $k_1 + \cdots + k_m = n$,

$$\Pr\{n_j = k_j, \ j = 1, \ldots, m\} = \frac{n!}{k_1! k_2! \ldots k_m!} p_1^{k_1} \ldots p_m^{k_m}. \tag{B.1}$$

Otherwise — if k_j are not nonnegative integers, or do not sum to n — the probabilities are 0.

Note that in a multinomial distribution, since the probability is only positive when $k_m = n - \sum_{i<m} k_i$, we have $n_m \equiv n - \sum_{i<m} n_i$. Also, $p_m = 1 - \sum_{i<m} p_i$. So if $p_1, \ldots, p_{m-1}$ are nonnegative and their sum is less than or equal to 1, it also makes sense to say that $\{n_j\}_{j<m}$ have a multinomial distribution for m (not $m - 1$) bins and probabilities $\{p_j\}_{j<m}$ if $\Pr\{n_j = k_j, \ j < m\}$ is given by the right side of (B.1) whenever k_j are nonnegative integers whose sum is at most n, and where k_m and p_m are determined as just described.

Theorem B.1 *For any probability space $(\Omega, \mathcal{A}, P)$ and any disjoint measurable sets A_i, $i = 1, \ldots, m$, with union Ω, $\{n P_n(A_i)\}_{i=1}^m$ have a multinomial distribution for n observations and m categories with probabilities $\{P(A_i)\}_{i=1}^m$. The same holds if we take $i = 1, \ldots, m - 1$.*

Proof. Use induction on m. For $m = 2$, $n P_n(A_1)$ has a binomial distribution: specifically, $n P_n(A_1)$ is the number of successes, and $n P_n(A_2)$ the number of failures, in n independent trials with probability $p = P(A_1)$ of success on each trial. To evaluate $\Pr(n P_n(A_1) = k)$, for $k = 0, \ldots, n$, there are $\binom{n}{k}$ ways to choose k of the n trials. For each way, the probability that just these trials are successes is $p^k(1 - p)^{n-k}$. Then $n P_n(A_2) = n - k$, and the stated result holds.

For the induction step from m to $m + 1$, apply the case of m bins to $A_1 \cup A_2, A_3, \ldots, A_{m+1}$. Then, given that $n P_n(A_1 \cup A_2) = k_1 + k_2$, the conditional probability that $n_1 = k_1$ and $n_2 = k_2$ is binomial. Checking that

$$\frac{(p_1 + p_2)^{k_1+k_2}}{(k_1 + k_2)!} \binom{k_1 + k_2}{k_1} \left(\frac{p_1}{p_1 + p_2}\right)^{k_1} \left(\frac{p_2}{p_1 + p_2}\right)^{k_2} = \frac{p_1^{k_1} p_2^{k_2}}{k_1! k_2!}$$

finishes the proof. $\qquad\qquad\square$

Theorem B.2 *Suppose* $\{n_i\}_{i=1}^m$ *have a multinomial distribution for n observations and m bins with probabilities* $\{p_j\}_{j=1}^m$. *Let T be a subset of* $\{1, \ldots, m\}$. *Then*

(a) $\{n_i\}_{i \in T}$ *have a multinomial distribution for n observations and $\operatorname{card}(T) + 1$ bins, with probabilities* $\{p_i\}_{i \in T}$.

(b) Let S be a subset of $\{1, \ldots, m\}$ *disjoint from T. Then the conditional distribution, or law,* $\mathcal{L}(\{n_i\}_{i \in S} | \{n_j\}_{j \in T})$, *depends on* $\{n_j\}_{j \in T}$ *only through* $\sum_{j \in T} n_j$, *in other words for any integers* $k_j \geq 0$, $j \in T$,

$$\mathcal{L}(\{n_i\}_{i \in S} | n_j = k_j \text{ for all } j \in T) = \mathcal{L}\left(\{n_i\}_{i \in S} \Big| \sum_{j \in T} n_j = \sum_{j \in T} k_j\right).$$

(c) If $S \cup T = \{1, \ldots, m\}$, this distribution is multinomial for $n - \sum_{j \in T} k_j$ observations in $\operatorname{card}(S)$ bins with probabilities $p_j / \sum_{i \in S} p_i$, $j \in S$.

Proof. (a) We can assume that for some r, $T = \{r + 1, \ldots, m\}$. The probability that $n_j = k_j$ for $r < j \leq m$ is a sum of multinomial probabilities,

$$n! \left(\Pi_{j>r} p_j^{k_j} / k_j!\right) \sum \left(\Pi_{j \leq r} p_j^{\kappa_j} / \kappa_j! : \kappa_j \; \kappa := \kappa_1 + \cdots + \kappa_r = n - \sum_{j>r} k_j\right)$$

where κ_j are integers ≥ 0. The sum $\sum(\Pi \ldots)$ equals $(p_1 + \cdots + p_r)^\kappa / \kappa!$ by the multinomial theorem, and (a) follows.

(b) We can assume $S \cup T = \{1, \ldots, m\}$ since the joint distribution of n_i, $i \in S$ is determined by that of n_j, $j \notin T$. Thus $S = \{1, \ldots, r\}$. We need to evaluate conditional probabilities of the form

$$\Pr(n_i = k_i, \; i = 1, \ldots, r | n_j = k_j, \; j > r) = \frac{P(n_i = k_i, \; i = 1, \ldots, m)}{P(n_j = k_j, \; j > r)}.$$

The numerator is a simple multinomial probability as in (B.1). The denominator is a multinomial probability by part (a). Carrying out the division, the $p_j^{k_j}/k_j!$ terms for $j > r$ cancel. The resulting expression, as claimed, depends on the k_j for $j > r$ only through their sum. Also,

(c) It is of the stated multinomial form, by the proof of (a).   □

Appendix C

Measures on Nonseparable Metric Spaces

Let (S, d) be a metric space. Under fairly general conditions, to be given, any probability measure on the Borel sets will be concentrated in a separable subspace.

The problem will be reduced to a problem about discrete spaces. An *open cover* of S will be a family $\{U_\alpha\}_{\alpha \in I}$ of open subsets U_α of S, where I is any set, here called an index set, such that $S = \bigcup_{\alpha \in I} U_\alpha$. An open cover $\{V_\beta\}_{\beta \in J}$ of S, with some index set J, will be called a *refinement* of $\{U_\alpha\}_{\alpha \in I}$ iff for all $\beta \in J$ there exists an $\alpha \in I$ with $V_\beta \subset U_\alpha$. An open cover $\{V_\alpha\}_{\alpha \in I}$ will be called σ-*discrete* if I is the union of a sequence of sets I_n such that for each n and $\alpha \neq \beta$ in I_n, U_α and U_β are disjoint. Recall that the ball $B(x, r)$ is defined as $\{y \in S : d(x, y) < r\}$. For two sets A, B we have $d(A, B) := \inf\{d(x, y) : x \in A,\ y \in B\}$, and $d(y, B) := d(\{y\}, B)$ for a point y.

Theorem C.1 *For any metric space (S, d), any open cover $\{U_\alpha\}_{\alpha \in I}$ of S has an open σ-discrete refinement.*

Proof. For any open set $U \subset S$ let $U_n := \{y : B(y, 2^{-n}) \subset U$. Then $y \in U_n$ if and only if $d(y, S \setminus U_n) \geq 2^{-n}$, and $d(U_n, S \setminus U_{n+1}) \geq 2^{-n} - 2^{-n-1} = 2^{-n-1}$ by the triangle inequality. (The sets U_n are always closed, and not usually open.) Let $U_{\alpha,n} := (U_\alpha)_n$. By well-ordering (e.g., RAP, Section 1.5), we can assume that the index set I is well-ordered by a relation $<$. For each positive integer n and each $\alpha \in I$ let $W_{\alpha,n} := U_{\alpha,n} \setminus \bigcup\{U_{\beta,n+1} : \beta < \alpha\}$. For each $\alpha \neq \beta$ in I and each n, either $W_{\alpha,n} \subset S \setminus U_{\beta,n+1}$ if $\beta < \alpha$ or $W_{\beta,n} \subset S \setminus U_{\alpha,n+1}$ if $\alpha < \beta$. In either case, $d(W_{\alpha,n}, W_{\beta,n}) \geq 2^{-n-1}$. Let $V_{\alpha,n} := \{x : d(x, W_{\alpha,n}) < 2^{-n-3}\}$. Then $d(V_{\alpha,n}, V_{\beta,n}) \geq 2^{-n-2}$. The sets $V_{\alpha,n}$ are all open. For a given n and different values of α, they are disjoint. Let $J := \mathbb{N} \times I$ and define V_γ, $\gamma \in J$, by $V_{(n,\alpha)} := V_{n,\alpha}$. To show that $\{V_\gamma\}_{\gamma \in J}$ is a cover of S, given any $x \in S$, take the least α such that $x \in U_\alpha$. Then for n large enough, $x \in U_{\alpha,n}$, and then

427

$x \in W_{\alpha,n} \subset V_{\alpha,n}$ as desired. So $\{V_\gamma\}_{\gamma \in J}$ is a σ-discrete refinement of $\{U_\alpha\}_{\alpha \in I}$, completing the proof.   □

Cardinal numbers are defined in RAP, in the last part of Appendix A (as smallest ordinals with given cardinality). A cardinal number ζ is said to be *real-valued measurable* if for a set S of cardinality ζ, there exists a probability measure P defined on all subsets of S which is nonatomic, in other words $P(\{x\}) = 0$ for all $x \in S$. If there is no such P, ζ is said to be *of measure* 0.

The continuum hypothesis implies that the cardinality c of the continuum (that is, of $[0, 1]$) is of measure 0 (RAP, Appendix C).

The *separability character* of a metric space is the smallest cardinality of a dense subset. We have:

Theorem C.2 *Let* (S, d) *be a metric space. Let* P *be a probability measure on the* σ-*algebra of Borel sets, generated by the open sets. Then either there is a separable subspace* T *with* $P(T) = 1$, *or the separability character* ζ *of* S *is real-valued measurable.*

Proof. Let $\{x_\alpha\}_{\alpha \in I}$ be dense in S, where I has cardinality ζ. For a fixed positive integer m, the balls $U_\alpha := B(x_\alpha, 1/m)$ form an open cover of S. Take an open, σ-discrete refinement. Then S is the union of countably many open sets V_n, each of which is the union of open sets $V_{\gamma,n}$, $\gamma \in \Gamma_n$, disjoint for different values of γ, and each included in some ball of radius $2/m$. Here each Γ_n has cardinality at most ζ. If a cardinal has measure 0, then so, clearly, do all smaller cardinals. Define a measure μ_n on each Γ_n by $\mu_n(A) := P(\bigcup_{\gamma \in A} V_{\gamma,n})$, where P is defined on all open sets. So if ζ has measure 0, then for each n, there is a countable subset G_n of Γ_n such that $P(V_n) = \sum\{P(V_{\gamma,n}) : \gamma \in G_n\}$. So $P(C_m) = 1$ where C_m is a countable union (over n and members of G_n for each n) of balls of radius $2/m$. Taking an intersection over m, we see that $P(C) = 1$ for some separable subset C.   □

It is consistent with the usual axioms of set theory (including the axiom of choice) that there are no real-valued measurable cardinals, in other words, all cardinals are of measure 0 (e.g., Drake 1974, pp. 67–68, 177–178). So, for practical purposes, a probability measure defined on the Borel sets of a metric space is always concentrated in some separable subspace.

Here is another fact giving separability:

Theorem C.3 *Let* f *be a Borel measurable function from a separable metric space* S *into a metric space* T. *Then, assuming the continuum hypothesis,* f *has separable range.*

Proof. If the range of f is nonseparable, then for some $r > 0$, there is an uncountable set F of values of f any two of which are more than r apart. The continuum hypothesis implies that F has cardinality at least c. All the 2^c or more subsets of F are closed, so all their inverse images under f are distinct Borel sets. On the other hand, in a separable metric space there are at most c Borel sets: the larger collection of analytic sets has cardinality at most c by the Borel isomorphism theorem and universal analytic set theorem (RAP, Theorems 13.1.1 and 13.2.4). So there is a contradiction. $\square$

Notes on Appendix C. Theorem C.1 on σ-discrete refinements is from Kelley (1955, Theorem 4.21 p. 129), who attributes it to A. H. Stone (1948). Theorem C.2 is due to Marczewski and Sikorski (1948), who prove it by a somewhat different method.

Appendix D

An Extension of Lusin's Theorem

Lusin's theorem says that for any measurable real-valued function f, on $[0, 1]$ with Lebesgue measure λ for example, and $\varepsilon > 0$, there is a set A with $\lambda(A) < \varepsilon$ such that restricted to the complement of A, f is continuous. Here $[0, 1]$ can be replaced by any normal topological space and λ by any finite measure μ which is *closed regular*, meaning that for each Borel measurable set B, $\mu(B) = \sup\{\mu(F) : F$ closed, $F \subset B\}$ (RAP, Theorem 7.5.2). Recall that any finite Borel measure on a metric space is closed regular (RAP, Theorem 7.1.3).

Proofs of Lusin's theorem are often based on Egorov's theorem (RAP, Theorem 7.5.1), which says that if measurable functions f_n from a finite measure space to a metric space converge pointwise, then for any $\varepsilon > 0$ there is a set of measure less than ε outside of which the f_n converge uniformly.

Here, the aim will be to extend Lusin's theorem to functions having values in any separable metric space. The proof of Lusin's theorem in RAP, however, also relied on the Tietze–Urysohn extension theorem, which says that a continuous real-valued function on a closed subset of a normal space can be extended to be continuous on the whole space. Such an extension may not exist for some range spaces: for example, the identity from $\{0, 1\}$ onto itself does not extend to a continuous function from $[0, 1]$ onto $\{0, 1\}$; in fact, there is no such function since $[0, 1]$ is connected.

It turns out, however, that the Tietze–Urysohn extension and Egorov's theorem are both unnecessary in proving Lusin's theorem:

Theorem D.1 *Let $(X, \mathcal{T})$ be a topological space and μ a finite, closed regular measure defined on the Borel sets of X. Let f be a Borel measurable function from x into S where (S, d) is a separable metric space. Then for any $\varepsilon > 0$ there is a closed set F with $\mu(X \setminus F) < \varepsilon$ such that f restricted to F is continuous.*

Proof. Let $\{s_n\}_{n \geq 1}$ be a countable dense set in S. For $m = 1, 2, \ldots$, and any $x \in X$, let $f_m(x) = s_n$ for the least n such that $d(f(x), s_n) < 1/m$. Then f_m is measurable and defined on all of X. For each m, let $n(m)$ be large enough so that

$$\mu\{x \,:\, d(f(x), s_n) \geq 1/m \text{ for all } n \leq n(m)\} \leq 1/2^m.$$

For $n = 1, \ldots, n(m)$, take a closed set $F_{mn} \subset f_m^{-1}\{s_n\}$ with

$$\mu(f_m^{-1}\{s_n\} \setminus F_{mn}) < \frac{1}{2^m n(m)}$$

by closed regularity. For each fixed m, the sets F_{mn} are disjoint for different values of n. Let $F_m := \bigcup_{n=1}^{n(m)} F_{mn}$. Then f_m is continuous on F_m. By choice of $n(m)$ and F_{mn}, $\mu(F_m) > 1 - 2/2^m$.

Since $d(f_m, f) < 1/m$ everywhere, clearly $f_m \to f$ uniformly (so Egorov's theorem is not needed). For $r = 1, 2, \ldots$, let $H_r := \bigcap_{m=r}^{\infty} F_m$. Then H_r is closed and $\mu(H_r) \geq 1 - 4/2^r$. Take r large enough so that $4/2^r < \varepsilon$. Then f restricted to H_r is continuous as the uniform limit of continuous functions f_m on $H_r \subset F_m$, $m \geq r$, so we can let $F = H_r$ to finish the proof. $\qquad\square$

Notes on Appendix D. Lusin's theorem for measurable functions with values in any separable metric space, on a space with a closed regular finite measure, was first proved as far as I know by Schaerf (1947), who proved it for f with values in any second-countable topological space (i.e. a space having a countable base for its topology), and for more general domain spaces ("neighborhood spaces"). See also Schaerf (1948) and Zakon (1965) for more extensions.

Appendix E
Bochner and Pettis Integrals

Let $(X, \mathcal{A}, \mu)$ be a measure space and $(S, \|\cdot\|)$ a separable Banach space. A function f from X into S will be called *simple* or *μ-simple* if it is of the form $f = \sum_{i=1}^{k} 1_{A_i} y_i$ for some $y_i \in S$, $k < \infty$ and measurable A_i with $\mu(A_i) < \infty$. For a simple function, the Bochner integral is defined by

$$\int f \, d\mu := \sum_{i=1}^{k} \mu(A_i) y_i \in S.$$

Theorem E.1 *The Bochner integral is well-defined for μ-simple functions, the μ-simple functions form a real vector space, and for any μ-simple functions f, g and real constant c, $\int cf + g \, d\mu = c \int f \, d\mu + \int g \, d\mu$.*

Proof. These facts are proved just as they are for real-valued functions (RAP, Proposition 4.1.4). $\qquad\square$

For any measurable function g from X into S, where measurability is defined for the Borel σ-algebra generated by the open sets of S, $x \mapsto \|g(x)\|$ is a measurable, nonnegative real-valued function on S.

Lemma E.2 *For any two μ-simple functions f and g from X into S, $\| \int f d\mu \| \le \int \|f\| d\mu$ and $\| \int f \, d\mu - \int g \, d\mu \| \le \int \|f - g\| \, d\mu$.*

Proof. Since $f - g$ is μ-simple and Theorem E.1 applies, it will be enough to prove the first statement. Note that $\int f \, d\mu$ is a Bochner integral while $\int \|f\| d\mu$ is an integral of a real-valued function. Let $f = \sum_i 1_{A_i} u_i$, a finite sum. We can assume that the measurable sets A_i are disjoint. Then $\| \int f \, d\mu \| = \| \sum_i \mu(A_i) u_i \| \le \sum_i \mu(A_i) \|u_i\| = \int \|f\| \, d\mu$. $\qquad\square$

Let $\mathcal{L}^1(X, \mathcal{A}, \mu, S) := \mathcal{L}^1(X, \mu, S)$ be the space of all measurable functions f from X into S such that $\int \|f\| \, d\mu < \infty$. By the triangle inequality, it is easily seen that $\mathcal{L}^1(X, \mathcal{A}, \mu, S)$ is a vector space. Also, all μ-simple functions belong to $\mathcal{L}^1(X, \mathcal{A}, \mu, S)$. Define $\|\cdot\|_1$ on $\mathcal{L}^1(X, \mathcal{A}, \mu, S)$ by $\|f\|_1 := \int \|f\| \, d\mu$. It is easily seen that $\|\cdot\|_1$ is a seminorm on $\mathcal{L}^1(X, \mathcal{A}, \mu, S)$. On the vector space of μ-simple functions, the Bochner integral is linear (by Theorem E.1) and continuous (also Lipschitz) for $\|\cdot\|_1$ by Lemma E.2.

Theorem E.3 *For any separable Banach space $(S, \|\cdot\|)$ and any measure space $(X, \mathcal{A}, \mu)$, the μ-simple functions are dense for $\|\cdot\|_1$ in $\mathcal{L}^1(X, \mathcal{A}, \mu, S)$ and the Bochner integral extends uniquely to a linear, real-valued function on the Banach space $\mathcal{L}^1(X, \mathcal{A}, \mu, S)$, continuous for $\|\cdot\|_1$.*

Proof. A first step in the proof will be the following.

Lemma E.4 *For any $f \in \mathcal{L}^1(X, \mathcal{A}, \mu, S)$ there exist μ-simple f_k with $\|f - f_k\|_1 \to 0$ and $\|f - f_k\| \to 0$ almost everywhere for μ as $k \to \infty$ and such that $\|f_k\| \le \|f\|$ on X.*

Proof. Define a measure γ on $(X, \mathcal{A})$ by $\gamma(B) := \int_B \|f(x)\| \, d\mu(x)$. Then γ is a finite measure and has a finite image measure $\gamma \circ f^{-1}$ on S. If $f = 0$ a.e. (μ), so that $\gamma \equiv 0$, set $f_k \equiv 0$. So we can assume $\gamma(S) > 0$. Given $m = 1, 2, \ldots$, by Ulam's theorem (RAP, Theorem 7.1.4), there is a compact $K := K_m \subset S$ with $\gamma(f^{-1}(S \setminus K)) < 1/2^m$. Then take a finite set $F \subset K$ such that each point of K is within $1/m$ of some point of F. For each $y \in F$, let $s(y) := cy$ for a constant with $0 < c < 1$ such that $\|y - cy\| = 1/m$, or $s(y) = 0$ if there is no such c ($\|y\| \le 1/m$). Let $F := \{y_1, \ldots, y_k\}$ for some distinct y_i. We can assume $y_1 := 0 \in F \subset K$. For $y \in K$ let $h_m(y) = s(y_i)$ such that $\|y - y_i\|$ is minimized, choosing the smallest possible index i in case of ties. For $y \notin K$ set $h_m(y) := 0$. Then for each $y \in S$, $\|h_m(y)\| \le \|y\|$, and $\|h_m(y) - y\| \le 2/m$ for all $y \in K$. Define a function g_m by $g_m(x) := h_m(f(x))$. Then g_m and h_m are measurable and have finite range. Let $\delta := \min_{j \ge 2} \|y_j\|$. Then $\delta > 0$. The set where $g_m \ne 0$ and thus $f \in K$ is included in the set where $\|f\| > \delta/2$, which has finite measure for μ. So g_m is a μ-simple function. The set where $\|f - g_m\| > 2/m$ has γ measure at most $1/2^m$. Then, by the Borel–Cantelli Lemma applied to the probability measure $\gamma/\gamma(S)$, it follows that $g_m \to f$ almost everywhere for γ. For any x, if $f(x) = 0$, then $g_m(x) = 0$ for all m. It follows that $\|g_m - f\| \to 0$ almost everywhere for μ. Since $\|g_m\| \le \|f\|$ and $\|g_m - f\| \le 2\|f\|$, it follows by dominated convergence that $\|g_m - f\|_1 \to 0$ as $m \to \infty$, proving the Lemma. $\square$

Then to continue proving the Theorem, the Bochner integral, a linear function, is continuous (and Lipschitz) for the seminorm $\|\cdot\|_1$. Thus it is uniformly continuous on its original domain, the simple functions, and so has a unique continuous extension to the closure of its domain, which is $\mathcal{L}^1(X, \mathcal{A}, \mu, S)$, and the extension is clearly also linear. $\square$

So the Bochner integral is well defined for all functions in $\mathcal{L}^1(X, \mathcal{A}, \mu, S)$, and only for such functions. A function from X into S will be called *Bochner integrable* if and only if it belongs to $\mathcal{L}^1(X, \mathcal{A}, \mu, S)$. The extension of the Bochner integral to $\mathcal{L}^1(X, \mathcal{A}, \mu, S)$ will also be written as $\int \cdot \, d\mu$. Thus Theorem E.3 implies that

$$\int cf + g \, d\mu = c \int f \, d\mu + \int g \, d\mu \qquad (E.1)$$

for any Bochner integrable functions f, g and real constant c. Also, by taking limits in Lemma E.2 it follows that

$$\left\| \int f \, d\mu \right\| \leq \int \| f \| \, d\mu \qquad (E.2)$$

for any Bochner integrable function f.

Although monotone convergence is not defined in general Banach spaces, a form of dominated convergence holds:

Theorem E.5 *Let $(X, \mathcal{A}, \mu)$ be a measure space. Let f_n be measurable functions from X into a Banach space S such that for all n, $\| f_n \| \leq g$ where g is an integrable real-valued function. Suppose f_n converge almost everywhere to a function f. Then f is Bochner integrable and $\| \int f_n - f \, d\mu \| \leq \int \| f_n - f \| \, d\mu \to 0$ as $n \to \infty$.*

Proof. First, f is measurable (RAP, Theorem 4.2.2). Since $\| f \| \leq g$, f is Bochner integrable, and the rest follows from (E.2) for $f_n - f$ and dominated convergence for real-valued functions $\| f_n - f \| \leq 2g$. $\qquad \square$

A Bochner integral $\int g \, d\mu = \int f \, d\mu$ can be defined when g is only defined almost everywhere for μ, f is Bochner integrable and $g = f$ where g is defined, just as for real-valued functions (RAP, Section 4.3). It is easy to check that when $S = \mathbb{R}$, the Bochner integral equals the usual Lebesgue integral.

A Tonelli–Fubini theorem holds for the Bochner integral:

Theorem E.6 *Let $(X, \mathcal{A}, \mu)$ and $(Y, \mathcal{B}, \nu)$ be σ-finite measure spaces. Let f be a measurable function from $X \times Y$ into a Banach space S such that*

$$\int \int \| f(x, y) \| d\mu(x) d\nu(y) < \infty.$$

Then for μ-almost all x, $f(x, \cdot)$ is Bochner integrable from Y into S; for ν-almost all y, $f(\cdot, y)$ is Bochner integrable from X into S, and

$$\int f d(\mu \times \nu) = \int \int f(x, y) d\mu(x) d\nu(y) = \int \int f(x, y) d\nu(y) d\mu(x).$$

Proof. By the usual Tonelli–Fubini theorem for real-valued functions (RAP, 4.4.5), the assumption $\int \| f(x, y) \| d\mu(x) d\mu(y) < \infty$ implies that for ν-almost all y, $\int \|f(x, y)\| d\mu(x) < \infty$ and for μ-almost all x, $\int \|f(x, y)\| d\nu(y) < \infty$. Let TF be the set of all $f \in \mathcal{L}^1(X \times Y, \mu \times \nu, S)$ such that

$$\int f d(\mu \times \nu) = \int \int f(x, y) d\mu(x) d\nu(y) = \int \int f(x, y) d\nu(y) d\mu(x)\}.$$

Then by the usual Tonelli–Fubini theorem, all $(\mu \times \nu)$-simple functions are in TF. For any $f \in \mathcal{L}^1(X \times Y, \mu \times \nu, S)$, by Lemma E.4, take simple f_n with $\|f_n\| \leq \|f\|$ and $\|f_n - f\| \to 0$ almost everywhere for $\mu \times \nu$. Then for ν-almost all y, $f_n(x, y) \to f(x, y)$ for μ-almost all x, and $\|f_n(\cdot, y)\| \leq \|f(\cdot, y)\| \in \mathcal{L}^1(X, \mu)$, so by dominated convergence (Theorem E.5), $\int f_n(x, y) d\mu(x) \to \int f(x, y) d\mu(x)$. Since the functions of y on the left are ν-measurable, so is the function on the right (RAP, Theorem 4.2.2, with a possible adjustment on a set of measure 0). For each y, $\| \int f_n(x, y) d\mu(x)\| \leq \int \|f(x, y)\| d\mu(x)$, which is ν-integrable with respect to y, so by dominated convergence (Theorem E.5) again,

$$\int \int f_n(x, y) d\mu(x) d\nu(y) \to \int \int f(x, y) d\mu(x) d\nu(y) \quad \text{as } n \to \infty.$$

The same holds for the iterated integral in the other order, $\int \int \cdot d\nu d\mu$, and $\int f_n d(\mu \times \nu) \to \int f d(\mu \times \nu)$ by (E.7), so $f \in TF$. $\qquad\square$

Now let $(S, \mathcal{T})$ be any topological vector space, in other words, S is a real vector space, $\mathcal{T}$ is a topology on S, and the operation $(c, f, g) \mapsto cf + g$ is jointly continuous from $\mathbb{R} \times S \times S$ into S. Then the *dual space* S' is the set of all continuous linear functions from S into $\mathbb{R}$. Let $(X, \mathcal{A}, \mu)$ be a measure space. Then a function f from X into S is called *Pettis integrable* with *Pettis integral* $y \in S$ if and only if for every $t \in S'$, $\int t(f) d\mu$ is defined and finite and equals $t(y)$.

The Pettis integral is also due to Gelfand and Dunford and might be called the Gelfand–Dunford–Pettis integral (see the Notes). The Pettis integral may lack interest unless S' separates points of S, as is true for normed linear spaces by the Hahn–Banach theorem (RAP, Corollary 6.1.5).

Theorem E.7 *For any measure space $(X, \mathcal{A}, \mu)$ and separable Banach space $(S, \|\cdot\|)$, each Bochner integrable function f from X into S is also Pettis integrable, and the values of the integrals are the same.*

Proof. The equation $t(\int f \, d\mu) = \int t(f) \, d\mu$ is easily seen to hold for simple functions and then, by Theorem E.3, for Bochner integrable functions. $\qquad\square$

Example. A function can be Pettis integrable without being Bochner integrable. Let H be a separable, infinite-dimensional Hilbert space with orthonormal basis $\{e_n\}$. Let $\mu(f = ne_n) := \mu(f = -ne_n) := p_n := n^{-7/4}$, and $f = 0$

otherwise. Then $\int \|f\| \, d\mu = 2 \sum_n n^{-3/4} = +\infty$, so f is not Bochner integrable. On the other hand for any x_n with $\sum_n x_n^2 < \infty$ we have $|\sum 2n x_n p_n| \leq (\sum x_n^2)^{1/2} 2 (\sum n^{-3/2})^{1/2} < \infty$, and by symmetry the Pettis integral of f is 0.

Example. The Tonelli–Fubini theorem, which holds for the Bochner integral (Theorem E.6), can fail for the Pettis integral. Let H be an infinite-dimensional Hilbert space and let $\xi_{i,j}$ be orthonormal for all positive integers i and j. Let $U(x, y) := 2^i \xi_{i,j}$ for $(j - 1)/2^i \leq x < j/2^i$ and $2^{-i} \leq y < 2^{1-i}$, $j = 1, \ldots, 2^i$, and 0 elsewhere. Then it can be checked that U is Pettis integrable on $[0, 1] \times [0, 1]$ for Lebesgue measure but is not integrable with respect to y for fixed x.

Notes on Appendix E. The treatment of the Bochner integral here is partly based on that of Cohn (1980). The following historical notes are mainly based on those of Dunford and Schwartz (1958).

Graves (1927) defined a Riemann integral for suitable Banach-valued functions. Bochner (1933) defined his integral, a Lebesgue-type integral for Banach-valued functions. The definition of integral given by Pettis (1938) was published at the same time or earlier by Gelfand (1936, 1938) and is equivalent to one of the definitions of Dunford (1936). Gelfand (1936) is one of the first, perhaps the first, of many distinguished publications by I. M. Gelfand. Birkhoff's integral (1935) strictly includes that of Bochner and is strictly included in that of Gelfand–Dunford–Pettis. Price (1940), in a further extension, treats set-valued functions. Dunford and Schwartz (1958, Chapter 3) develop an integral when μ is only finitely additive. Facts such as the dominated convergence theorem still require countable additivity.

Appendix F

Nonexistence of Some Linear Forms

Recall that a real vector space V with a topology $\mathcal{T}$ is called a *topological vector space* if addition is jointly continuous from $V \times V$ to V for $\mathcal{T}$, and scalar multiplication $(c, v) \mapsto cv$ is jointly continuous from $\mathbb{R} \times V$ into V for $\mathcal{T}$ on V and the usual topology on $\mathbb{R}$. If d is a metric on V, then (V, d) is called a *metric linear space* iff it is a topological vector space for the topology of d. First, it will be shown that any measurable linear form on a complete metric linear space is continuous. Then, it will be seen that the only measurable (and thus, continuous) linear form on $L^p([0, 1], \lambda)$ for $0 < p < 1$ is zero. Recall that for a given σ-algebra A of measurable sets, in our case the Borel σ-algebra, a *universally measurable set* is one measurable for the completion of every probability measure on A (RAP, Section 11.5). The universally measurable sets form a σ-algebra, and a function f is called universally measurable if for every Borel set B in its range, $f^{-1}(B)$ is universally measurable.

Theorem F.1 *Let (E, d) be a complete metric real linear space. Let u be a universally measurable linear form: $E \mapsto \mathbb{R}$. Then u is continuous.*

Proof. There exists a metric ρ, metrizing the d topology on E, such that $\rho(x + v, y + v) = \rho(x, y)$ for all $x, y, v \in E$, and $|\lambda| \leq 1$ implies $\rho(0, \lambda x) \leq \rho(0, x)$ for all $x \in E$ (Schaefer 1966, Theorem I.6.1 p. 28). Then we have:

Lemma F.2 *If (E, d) is a complete metric linear space, then it is also complete for ρ.*

Proof. If not, let F be the completion of E for ρ. Because of the translation invariance property of ρ, the topological vector space structure of E extends to F. Then since (E, d) is complete, E is a G_δ (a countable intersection of open sets) in F (RAP, Theorem 2.5.4). So, if E is not complete for ρ, $F \setminus E$ is of first category in F, but $F \setminus E$ includes a translate of E which is carried to E

by a homeomorphism (by translation) of F, so $F \setminus E$ is of second category in F, a contradiction, so the Lemma is proved. $\square$

Now to continue the proof of Theorem F.1, let (E, ρ) be a complete metric linear space where ρ has the stated properties. If u is not continuous, there are $e_n \in E$ with $e_n \to 0$ and $|u(e_n)| \geq 1$ for all n. There are constants $c_n \to \infty$ such that $c_n e_n \to 0$: for each $k = 1, 2, \ldots$, there is a $\delta_k > 0$ such that if $\rho(x, 0) < \delta_k$, then $\rho(kx, 0) < 1/k$. We can assume that δ_k is decreasing as k increases. For some n_k, $\rho(e_k, 0) < \delta_k$ for $n \geq n_k$. We can also assume that n_k increases with k. Let $c_n = 1$ for $n < n_1$ and $c_n = k$ for $n_k \leq n < n_{k+1}$, $k = 1, 2, \ldots$. Then c_n have the claimed properties. So we can assume $e_n \to 0$ and $|u(e_n)| \to \infty$. Taking a subsequence, we can assume $\sum_n \rho(0, e_n) < \infty$. So it will be enough to show that if $\sum_n \rho(0, e_n) < \infty$ then $|u(e_n)|$ are bounded.

Let Q be the Cartesian product $\Pi_{n=1}^{\infty} I_n$ where for each n, I_n is a copy of $[-1, 1]$. (So Q is the closed unit ball of ℓ^{∞}.) For $\lambda := \{\lambda_n\}_{n=1}^{\infty} \in Q$, the series $h(\lambda) := \sum_{n=1}^{\infty} \lambda_n e_n$ converges in E, since by the properties of ρ,

$$\rho\left(0, \sum_{j=k}^{n} \lambda_j e_j\right) \leq \sum_{j=k}^{n} \rho\left(0, \lambda_j e_j\right) \leq \sum_{j=k}^{n} \rho\left(0, e_j\right) \to 0$$

as $n \geq k \to \infty$, and (E, ρ) is complete. Also, h is continuous for the product topology $\mathcal{T}$ on Q. For all n, let μ_n be the uniform law $U[-1, 1]$ (one-half times Lebesgue measure on $[-1, 1]$) and let $\mu := \Pi_{n=1}^{\infty} \mu_n$ on Q. Then μ is a regular Borel measure on the compact metrizable space $(Q, \mathcal{T})$. Thus the image measure $\mu \circ h^{-1}$ is a regular Borel measure on (E, ρ). Since u is universally measurable, it is $\mu \circ h^{-1}$ measurable (that is, measurable for the completion of $\mu \circ h^{-1}$ on the Borel σ-algebra). So $U := u \circ h$ is measurable for the completion of μ on the Borel σ-algebra of $(Q, \mathcal{T})$. For $0 < M < \infty$ let $Q_M := \{x \in Q : |U(x)| \leq M\}$. Then Q_M is μ-measurable and $\mu(Q_M) \uparrow 1$ as $M \uparrow +\infty$. For a given n write $Q = I_n \times Q_{(n)}$ where $Q_{(n)} := \Pi_{j \neq n} I_j$. Let $\mu_{(n)} := \Pi_{j \neq n} \mu_j$. Then $\mu = \mu_n \times \mu_{(n)}$. For $y \in Q_{(n)}$ let $Q_{M,n}(y) := \{t \in I_n : (t, y) \in Q_M\}$. By the Tonelli–Fubini theorem,

$$\mu(Q_M) = \int_{Q_{(n)}} \mu_n(Q_{M,n}(y)) d\mu_{(n)}(y).$$

For $s, t \in Q_{M,n}(y)$, $U(s, y) - U(t, y) = (s - t)u(e_n)$. So $|U(s, y)| \leq M$ and $|U(t, y)| \leq M$ imply $|s - t| \leq 2M/|u(e_n)|$. (We can assume $u(e_n) \neq 0$ for all n.) So $\mu_n(Q_{M,n}(y)) \leq M/|u(e_n)|$ (since μ_n has density $\frac{1}{2} \cdot 1_{[-1,1]}$). So $\mu(Q_M) \leq M/|u(e_n)|$, and $|u(e_n)| \leq M/\mu(Q_M) < \infty$ for all n if $\mu(Q_M) > 0$, which is true for some M large enough. So $|u(e_n)|$ are bounded, proving Theorem F.1. $\square$

For $0 < p < \infty$ let $\mathcal{L}^p([0, 1], \lambda)$ be the space of all Lebesgue measurable real-valued functions f on $[0, 1]$ with $\int_0^1 |f(x)|^p dx < \infty$. Let $L^p :=$ $L^p[0, 1] := L^p([0, 1], \lambda)$ be the space of all equivalence classes of functions in $\mathcal{L}^p([0, 1], \lambda)$ for equality a.e. (λ). For $1 \le p < \infty$, recall that L^p is a Banach space with the norm $\|f\|_p := (\int_0^1 |f(x)|^p dx)^{1/p}$. For $0 < p < 1$, $\| \cdot \|_p$ is not a norm. Instead, let $\rho_p(f, g) := \int_0^1 |f - g|^p(x) dx$. It will be shown that for $0 < p \le 1$, ρ_p defines a metric on L^p. For any $a \ge 0$ and $b \ge 0$, $(a + b)^p \le a^p + b^p$: to see this, let $r := 1/p \ge 1$, $u := a^p$, $v := b^p$, and apply the Minkowski inequality in ℓ^r. The triangle inequality for ρ_p follows, and the other properties of a metric are clear. Also, ρ_p has the properties of the metric ρ in the proof of Theorem F.1. To see that L^p is complete for ρ_p, let $\{f_n\}$ be a Cauchy sequence. As in the proof for $p \ge 1$ (RAP, Theorem 5.2.1) there is a subsequence f_{n_k} converging for almost all x to some $f(x)$, and $\rho_p(f_{n_k}, f) \to 0$, so $\rho_p(f_n, f) \to 0$.

Theorem F.3 *For $0 < p < 1$ there are no nonzero, universally measurable linear forms on $L^p[0, 1]$.*

Proof. Suppose u is such a linear form. Then by Theorem F.1, u is continuous. Now $L^1[0, 1] \subset L^p[0, 1]$ and the injection h from L^1 into L^p is continuous. Thus, $u \circ h$ is a continuous linear form on L^1. So, for some $g \in L^\infty[0, 1]$, $u(h(f)) = \int_0^1 (fg)(x) dx$ for all $f \in L^1[0, 1]$ (e.g., RAP, Theorem 6.4.1). Then we can assume that $g \ge c > 0$ on a set E with $\lambda(E) > 0$ (replacing u by $-u$ if necessary). Let $f_n = n$ on a set $E_n \subset E$ with $\lambda(E_n) = \lambda(E)/n$ and $f_n = 0$ elsewhere. Then $f_n \to 0$ in L^p for $0 < p < 1$ and $u(f_n) = \int_0^1 (f_n g)(x) dx \ge c\lambda(E) > 0$ for all n, a contradiction. $\square$

Note that Theorem F.1 fails if the completeness assumption is omitted: let H be an infinite-dimensional Hilbert space and let h_n be an infinite orthonormal sequence in H. Let E be the set of all finite linear combinations of the h_n. Then there exists a linear form u on E with $u(h_n) = n$ for all n, and u is Borel measurable on E but not continuous.

Notes on Appendix F. The proof of Theorem F.1 is adapted from a proof for the case that E is a Banach space. The proof is due to A. Douady, according to L. Schwartz (1966) who published it. The fact for Banach spaces extends easily to inductive limits of Banach spaces (for definitions see, e.g., Schaefer 1966), or so-called ultrabornologic locally convex topological linear spaces, and from linear forms to linear maps into general locally convex spaces. Among ultrabornologic spaces are the spaces of Schwartz's theory of distributions (generalized functions).

Appendix G
Separation of Analytic Sets; Borel Injections

Recall that a *Polish space* is a topological space S metrizable by a metric for which S is complete and separable. Also, in any topological space X, the σ-algebra of *Borel* sets is generated by the open sets. Two disjoint subsets A, C of X are said to be *separated by Borel sets* if there is a Borel set $B \subset X$ such that $A \subset B$ and $C \subset X \setminus B$. Recall that a set A in a Polish space Y is called *analytic* iff there is another Polish space X, a Borel subset B of X, and a Borel measurable function f from B into Y such that $A = f[B] := \{f(x) : x \in B\}$ (e.g., RAP, Section 13.2). Equivalently, we can take f to be continuous and/or $B = X$ and/or $X =\ni$, where $\mathbb{N}$ is the set of nonnegative integers with discrete topology and $\ni$ the Cartesian product of an infinite sequence of copies of $\mathbb{N}$, with product topology (RAP, Theorem 13.2.1).

Theorem G.1 (Separation theorem for analytic sets) *Let X be a Polish space. Then any disjoint analytic subsets A, C of X can be separated by Borel sets.*

Proof. First, it will be shown that:

Lemma G.2 *If C_j for $j = 1, 2, \ldots$, and D are subsets of X such that for each j, C_j and D can be separated by Borel sets, then $\bigcup_{j=1}^{\infty} C_j$ and D can be separated by Borel sets.*

Proof. For each j, take a Borel set B_j such that $D \subset B_j$ and $C_j \subset X \setminus B_j$. Let $B := \bigcap_{j=1}^{\infty} B_j$. Then $D \subset B$ and $\bigcup_{j=1}^{\infty} C_j \subset X \setminus B$, so the Lemma is proved. $\square$

Next, it will be shown that:

Lemma G.3 *If C_j and D_j for $j = 1, 2, \ldots$, are subsets of X such that for every i and j, C_i and D_j can be separated by Borel sets, then $\bigcup_{i=1}^{\infty} C_i$ and $\bigcup_{j=1}^{\infty} D_j$ can be separated by Borel sets.*

Proof. By Lemma G.2, for each j, $\bigcup_{i=1}^{\infty} C_i$ and D_j can be separated by Borel sets. Applying Lemma G.2 again gives Lemma G.3. $\qquad\square$

To continue the proof of Theorem G.1, we can assume A and C are both nonempty. Then there exist continuous functions f, g from $\ni$ into X such that $f[\ni] = A$ and $g[\ni] = C$.

Suppose A and C cannot be separated by Borel sets. For $k = 1, 2, \ldots$, and $m_i \in \mathbb{N}$, $i = 1, \ldots, k$, let

$$\ni (m_1, \ldots, m_k) := \{n = \{n_i\}_{i=1}^{\infty} \in\ni: n_i = m_i \text{ for } i = 1, \ldots, k\}.$$

By Lemma G.3, there exist m_1 and n_1 such that $f[\ni (m_1)]$ and $g[\ni (n_1)]$ cannot be separated by Borel sets. Inductively, for all $k = 1, 2, \ldots$, there exist m_k and n_k such that $f[\ni (m_1, \ldots, m_k)]$ and $g[\ni (n_1, \ldots, n_k)]$ cannot be separated by Borel sets. Let $m := \{m_i\}_{i=1}^{\infty}$, $n := \{n_j\}_{j=1}^{\infty}$. If $f(m) \neq g(n)$, then there are disjoint open sets U, V with $f(m) \in U$ and $g(n) \in V$. But then by continuity of f and g, for k large enough, $f[\ni (m_1, \ldots, m_k)] \subset U$ and $g[\ni (n_1, \ldots, n_k)] \subset V$, which gives a contradiction. So $f(m) = g(n)$. But $f(m) \in A$, $g(n) \in C$, and $A \cap C = \phi$, another contradiction, which proves Theorem G.1. $\qquad\square$

Lemma G.4 *If X is a Polish space and A_j, $j = 1, 2, \ldots$, are analytic in X, then $\bigcup_{j=1}^{\infty} A_j$ is analytic.*

Proof. For $j = 1, 2, \ldots$, let f_j be a continuous function from $\ni$ onto A_j. For $n := \{n_i\}_{i=1}^{\infty} \in\ni$ define $f(n) := f_{n_1}(\{n_{i+1}\}_{i\geq 1})$. Then f is continuous from $\ni$ onto $\bigcup_{j=1}^{\infty} A_j$. $\qquad\square$

Corollary G.5 *If X is a Polish space and A_j, $j = 1, 2, \ldots$, are disjoint analytic subsets of X, then there exist disjoint Borel sets B_j such that $A_j \subset B_j$ for all j.*

Proof. For each j such that $A_j = \emptyset$ we can take $B_j = \emptyset$. So we can assume all the A_j are nonempty. For each $k = 1, 2, \ldots$, by Lemma G.4, $A^{(k)} := \bigcup_{j \neq k} A_j$ is analytic. Then by Theorem G.1, there is a Borel set C_k with $A_k \subset C_k$ and $A^{(k)} \subset X \setminus C_k$. Let $B_1 := C_1$ and for $j \geq 2$, $B_j := C_j \setminus \bigcup_{k<j} C_k$. Then B_j have the properties stated. $\qquad\square$

Theorem G.6 *Let S be a Polish space, Y a separable metric space, and A a Borel subset of S. Let f be a 1–1, Borel measurable function from A into Y. Then the range $f[A]$ is a Borel subset of Y, and f^{-1} is a Borel measurable function from $f[A]$ onto A.*

Proof. The following fact will be used.

Lemma G.7 *Let S be a Polish space and Y a separable metric space. Let A be a nonempty Borel subset of S and f a 1–1, Borel measurable function from A into Y. Then there exists a Borel measurable function g from Y onto A such that $g(f(x)) = x$ for all x in A.*

Proof. Let d metrize S where (S, d) is complete and separable. Let $\{y_j\}_{j=1}^{\infty}$ be dense in S. Choose $x_0 \in A$. Recall that $B(x, r) := \{y : d(x, y) < r\}$. For each $k = 1, 2, \ldots$ and $j = 1, 2, \ldots$, let $A_{kj} := A \cap B(y_j, 1/k) \setminus \bigcup_{i<j} B(y_i, 1/k)$, where the union is empty for $j = 1$. Then for each k, the A_{kj} for $j = 1, 2, \ldots$ are disjoint Borel sets whose union is A, and for each j, $d(x, y) < 2/k$ for any $x, y \in A_{kj}$. Since f is 1–1, for each k, $f[A_{kj}]$ are disjoint analytic sets in Y. Thus by Corollary G.5 there exist disjoint Borel subsets B_{kj} of Y with $f[A_{kj}] \subset B_{kj}$, and with $B_{kj} = \emptyset$ if $A_{kj} = \emptyset$. For each j such that A_{kj} is nonempty, choose a point $x_{kj} \in A_{kj}$. Define a function g_k on Y by letting $g_k(y) := x_{kj}$ if $y \in B_{kj}$ for any j and $g_k(y) := x_0$ if $y \notin \bigcup_{j=1}^{\infty} B_{kj}$. Then g_k is Borel measurable (e.g. RAP, Lemma 4.2.4).

Since (S, d) is complete, the set G of y in Y such that $g_k(y)$ converges in X is the same as the set on which $\{g_k(y)\}_{k \geq 1}$ is a Cauchy sequence, so

$$G = \bigcap_{m \geq 1} \bigcup_{n \geq 1} \bigcap_{k \geq n} \{y : d(g_k(y), g_n(y)) < 1/m\}.$$

Since S is separable, $d(\cdot, \cdot)$ is product measurable on $S \times S$ (e.g., RAP, Propositions 4.1.7 and 2.1.4). Thus G is a Borel set in Y. On G, let $h(y) := \lim_{k \to \infty} g_k(y)$. Then h is Borel measurable (RAP, Theorem 4.2.2). Let $g(y) := h(y)$ if $y \in G$ and $h(y) \in A$. Otherwise, let $g(y) := x_0$. Then g is a Borel function from Y onto A. If $x \in A$, then for each k, $x \in A_{kj}$ for some j, $f(x) \in B_{kj}$, and $g_k(f(x)) = x_{kj}$, so $d(x, g_k(f(x))) < 2/k$. Thus $g_k(f(x)) \to x \in A$, so $g(f(x)) = h(f(x)) = x$, proving Lemma G.7. □

Now to prove Theorem G.6, we can assume A is nonempty. Take the function g from Lemma G.7. Then if $y \in f[A]$, we have $f(g(y)) = y$, while if $y \notin f[A]$, then $f(g(y)) \neq y$. So $f[A] = \{y \in Y : f(g(y)) = y\}$. If e is a metric for Y, then $f[A] = \{y \in Y : e(f(g(y)), y) = 0\}$. So by the product measurability of e (as for d above), $f[A]$ is a Borel set in Y. Thus for any Borel set $B \subset A$, $(f^{-1})^{-1}(B) = f[B]$ is Borel in Y, so f^{-1} is Borel measurable, and Theorem G.6 is proved. □

Note on Appendix G. This Appendix is entirely based on Cohn (1980), Section 8.3.

Appendix H
Young–Orlicz Spaces

A convex, increasing function g from $[0, \infty)$ onto itself will be called a *Young–Orlicz modulus*. Then g is continuous since it is increasing and onto. Let $(X, \mathcal{S}, \mu)$ be a measure space and g a Young–Orlicz modulus. Let $\mathcal{L}_g(X, \mathcal{S}, \mu)$ be the set of all real-valued measurable functions f on X such that

$$\|f\|_g := \inf\left\{c > 0 : \int g(|f(x)|/c)d\mu(x) \leq 1\right\} < \infty. \quad \text{(H.1)}$$

Let L_g be the set of equivalence classes of functions in $\mathcal{L}_g(X, \mathcal{S}, \mu)$ for equality almost everywhere (μ). By monotone convergence, we have

Proposition H.1 *For any Young–Orlicz modulus g and any $f \in \mathcal{L}_g(X, \mathcal{S}, \mu)$, if $0 < c := \|f\|_g < +\infty$, we have $\int(g(|f|/c))d\mu = 1$, in other words, the infimum in the definition of $\|f\|_g$ is attained. Also, $\|f\|_g = 0$ if and only if $f = 0$ almost everywhere for μ.*

Next, we have

Lemma H.2 *For any Young–Orlicz modulus g, and any measurable functions f_n, if $|f_n| \uparrow |f|$, then $\|f_n\|_g \uparrow \|f\|_g \leq +\infty$.*

Proof. For any $c > 0$, $\int g(|f_n|/c)d\mu \uparrow \int g(|f|/c)d\mu$ by ordinary monotone convergence. Thus $\|f_n\|_g \uparrow t$ for some t. Taking $c = t$ and using Proposition H.1 we get $\|f\|_g \leq t$, while clearly $\|f\|_g \geq t$. $\qquad\square$

Next is a fact showing that (not surprisingly) convergence in $\|\cdot\|_g$ norm implies convergence in measure (or probability):

Lemma H.3 *For any Young–Orlicz modulus g, and any $\varepsilon > 0$, there is a $\delta > 0$ such that if $\|f\|_g \leq \delta$, then $\mu(|f| > \varepsilon) < \varepsilon$.*

Proof. For any $\delta > 0$, $\|f\|_g \leq \delta$ is equivalent to $\int g(|f|/\delta)d\mu \leq 1$ by Proposition H.1. Since g is onto $[0, \infty)$, there is some $M < \infty$ with $g(M) > 1/\varepsilon$ and $M \geq 1$. Thus $\|f\|_g \leq \delta$ implies $\mu(|f| > M\delta) < \varepsilon$. So we can set $\delta := \varepsilon/M$. $\qquad\square$

Theorem H.4 *For any measure space* $(X, \mathcal{S}, \mu)$, $L_g(X, \mathcal{S}, \mu)$ *is a Banach space.*

Proof. For any real number $\lambda \neq 0$ it is clear that a function f is in $\mathcal{L}_g(X, \mathcal{S}, \mu)$ if and only if λf is, with $\|\lambda f\|_g \equiv |\lambda| \|f\|_g$ by definition of $\|\cdot\|_g$. If $f, h \in \mathcal{L}_g(X, \mathcal{S}, \mu)$ and $0 < \lambda < 1$, then since g is convex we have by Jensen's inequality (RAP, 10.2.6) for any $c, d > 0$

$$\int g\left(\lambda\left(\frac{f}{c}\right) + (1 - \lambda)\frac{h}{d}\right) d\mu \leq \lambda \int g\left(\frac{f}{c}\right) d\mu + (1 - \lambda) \int g\left(\frac{h}{d}\right) d\mu,$$
(H.2)

where some or all three integrals may be infinite. Applying this for $c = 1$ and $h = 0$ gives

$$\int g(\lambda f) d\mu \leq \lambda \int g(f) d\mu.$$
(H.3)

Clearly, if the inequality in (H.1) holds for some $c > 0$, it also holds for any larger c. It follows that $\lambda f + (1 - \lambda)h \in \mathcal{L}_g(X, \mathcal{S}, \mu)$. For the triangle inequality, let $c := \|f\|_g$ and $d := \|h\|_g$. If $c = 0$, then $f = 0$ a.e. (μ), so $\|f + h\|_g = \|h\|_g \leq c + d$, and likewise if $d = 0$. If $c > 0$ and $d > 0$, let $\lambda := c/(c + d)$ in (H.6). Applying Proposition H.1 to both terms on the right in (H.6) gives $\int g((f + h)/(c + d))d\mu \leq 1$, and so $\|f + h\|_g \leq c + d$. So the triangle inequality holds and $\|\cdot\|_g$ is a seminorm on $\mathcal{L}_g(X, \mathcal{S}, \mu)$. Clearly it becomes a norm on $L_g(X, \mathcal{S}, \mu)$. To see that the latter space is complete for $\|\cdot\|_g$, let $\{f_k\}$ be a Cauchy sequence. By Lemma H.3, take $\delta_j := \delta$ for $\varepsilon := \varepsilon_j := 1/2^j$. Take a subsequence $f_{k(j)}$ with $\|f_i - f_{k(j)}\|_g \leq \delta_j$ for any $i \geq k(j)$ and $j = 1, 2, \ldots$. Then $f_{k(j)}$ converges μ-almost everywhere, by the proof of the Borel–Cantelli Lemma, to some f. Then $\|f_{k(j)} - f\|_g \to 0$ as $j \to \infty$ by Fatou's Lemma applied to functions $g(|f_{k(j)} - f|/c)$ for $c > 2^{-j}$. It follows that $\|f_i - f\|_g \to 0$ as $i \to \infty$, completing the proof. $\square$

Let Φ be a Young–Orlicz modulus. Then it has one-sided derivatives as follows (RAP, Corollary 6.3.3): $\phi(x) := \Phi'(x+) := \lim_{y \downarrow x}(\Phi(y) - \Phi(x))/(y - x)$ exists for all $x \geq 0$, and $\phi(x-) := \Phi'(x+) := \lim_{y \uparrow x}(\Phi(x) - \Phi(y))/(x - y)$ exists for all $x > 0$. As the notation suggests, for each $x > 0$, $\phi(x-) \equiv \lim_{y \uparrow x} \phi(y)$, and ϕ is a nondecreasing function on $[0, \infty)$. Thus, $\phi(x-) = \phi(x)$ except for at most countably many values of x, where ϕ may have jumps with $\phi(x) > \phi(x-)$. On any bounded interval, where ϕ is bounded, Φ is Lipschitz and so absolutely continuous. Thus since $\Phi(0) = 0$ we have $\Phi(x) = \int_0^x \phi(u)du$ for any $x > 0$ (e.g., Rudin 1974, Theorem 8.18). For any $x > 0$, $\phi(x) > 0$ since Φ is strictly increasing.

If ϕ is unbounded, for $0 \leq y < \infty$ let $\psi(y) := \phi^{\leftarrow}(y) := \inf\{x \geq 0 : \phi(x) \geq y\}$. Then $\psi(0) = 0$ and ψ is nondecreasing. Let $\Psi(y) := \int_0^y \psi(t)dt$.

Then Ψ is convex and $\Psi' = \psi$ except on the at most countable set where ψ has jumps. Thus for each $y > 0$ we have $\psi(y) > 0$ and Ψ is also strictly increasing.

For any nondecreasing function f from $[0, \infty)$ into itself, it is easily seen that for any $x > 0$ and $u > 0$, $f^{\leftarrow}(u) \geq x$ if and only if $f(t) < u$ for all $t < x$. It follows that $(f^{\leftarrow})^{\leftarrow}(x) = f(x-)$ for all $x > 0$. Since a change in ϕ or ψ on a countable set (of its jumps) does not change its indefinite integral Φ or Ψ respectively, the relation between Φ and Ψ is symmetric.

A Young–Orlicz modulus Φ such that ϕ is unbounded and $\phi(x){\downarrow}0$ as $x{\downarrow}0$ will be called an *Orlicz modulus*. Then ψ is also unbounded and $\psi(y) > 0$ for all $y > 0$, so Ψ is also an Orlicz modulus. In that case Φ and Ψ will be called *dual Orlicz moduli*. For such moduli we have a basic inequality due to W. H. Young:

Theorem H.5 (W. H. Young) *Let Φ, Ψ be any two dual Young–Orlicz moduli from $[0, \infty)$ onto itself. Then for any $x, y \geq 0$ we have*

$$xy \leq \Phi(x) + \Psi(y),$$

with equality if $x > 0$ and $y = \phi(x-)$.

Proof. If $x = 0$ or $y = 0$, there is no problem. Let $x > 0$ and $y > 0$. Then $\Phi(x)$ is the area of the region A: $0 \leq u \leq x$, $0 \leq v \leq \phi(u)$ in the (u, v) plane. Likewise, $\Psi(y)$ is the area of the region B: $0 \leq v \leq y$, $0 \leq u < \psi(v)$. By monotonicity and right-continuity of ϕ, $u \geq \psi(v)$ is equivalent to $\phi(u) \geq v$, so $u < \psi(v)$ is equivalent to $\phi(u) < v$, so $A \cap B = \emptyset$. The rectangle $R_{x,y}$: $0 \leq u \leq x$, $0 \leq v \leq y$ is included in $A \cup B \cup C$, where C has zero area, and if $y = \phi(x-)$, then $R_{x,y} = A \cup B$ up to a set of zero area, so the conclusions hold. $\qquad\square$

One of the main uses of inequality H.5 is to prove an extension of the Rogers–Hölder inequality to Young–Orlicz spaces:

Theorem H.6 *Let Φ and Ψ be dual Orlicz moduli, and for a measure space $(X, \mathcal{S}, \mu)$ let $f \in \mathcal{L}_\Phi(X, \mathcal{S}, \mu)$ and $g \in \mathcal{L}_\Psi(X, \mathcal{S}, \mu)$. Then $fg \in \mathcal{L}^1(X, \mathcal{S}, \mu)$ and $\int |fg| d\mu \leq 2\|f\|_\Phi \|g\|_\Psi$.*

Proof. By homogeneity we can assume $\|f\|_\Phi = \|g\|_\Psi = 1$. Then applying Proposition H.1 with $c - 1$ and Theorem H.5, we get $\int |fg| d\mu(x) \leq 2$, and the conclusion follows. $\qquad\square$

Notes to Appendix H. W. H. Young (1912) proved his inequality (Theorem H.5) for smooth functions Φ. Birnbaum and Orlicz (1931) apparently began the theory of "Orlicz spaces," and W. A. J. Luxemburg defined the norms $\|\cdot\|_\Phi$; see Luxemburg and Zaanen (1956). Krasnosel'skiĭ and Rutitskiĭ (1961) wrote a book on the topic.

Appendix I

Modifications and Versions of Isonormal Processes

Let T be any set and $(\Omega, \mathcal{A}, P)$ a probability space. Recall that a real-valued stochastic process indexed by T is a function $(t, \omega) \mapsto X_t(\omega)$ from $T \times \Omega$ into $\mathbb{R}$ such that for each $t \in T$, $X_t(\cdot)$ is measurable from Ω into $\mathbb{R}$. A *modification* of the process is another stochastic process Y_t defined for the same T and Ω such that for each t, we have $P(X_t = Y_t) = 1$. A *version* of the process X_t is a process Z_t, $t \in T$, for the same T but defined on a possibly different probability space $(\Omega_1, \mathcal{B}, Q)$ such that X_t and Z_t have the same laws, i.e., for each finite subset F of T, $\mathcal{L}(\{X_t\}_{t \in F}) = \mathcal{L}(\{Z_t\}_{t \in F})$. Clearly, any modification of a process is also a version of the process, but a version, even if on the same probability space, may not be a modification. For example, for an isonormal process L on a Hilbert space H, the process $M(x) := L(-x)$ is a version, but not a modification, of L.

One may take a version or modification of a process in order to get better properties such as continuity. It turns out that for the isonormal process on subsets of Hilbert space, what can be done with a version can also be done by a modification, as follows.

Theorem I.1 *Let L be an isonormal process restricted to a subset C of Hilbert space. For each of the following properties, if there exists a version M of L with the property, there also is a modification N with the property. For each ω, $x \mapsto M(x)(\omega)$ for $x \in C$ is:*

(a) bounded   (b) uniformly continuous.

Also, if there is a version with (a) and another with (b), then there is a modification $N(\cdot)$ having both properties.

Proof. Let A be a countable dense subset of C. For each $x \in C$, take $x_n \in A$ with $\|x_n - x\| \le 1/n^2$ for all n. Then $L(x_n) \to L(x)$ a.s. Thus if we define $N(x)(\omega) := \limsup_{n \to \infty} L(x_n)(\omega)$, or 0 on the set of probability 0 where the lim sup is infinite, then N is a modification of L. If (a) holds for M, it will also hold for L on A and so for N, and likewise for (b), since a uniformly continuous function $L(\cdot)(\omega)$ on A has a unique uniformly continuous extension to C given by $N(\cdot)(\omega)$. Since $N(\cdot)$ is the same in both cases, the last conclusion follows. $\qquad \square$

Bibliography

Alexander, Kenneth S. (1984). Probability inequalities for empirical processes and a law of the iterated logarithm. *Ann. Probab.* **12**, 1041–1067.

Alexander, K. S. (1987). The central limit theorem for empirical processes on Vapnik-Červonenkis classes. *Ann. Probab.* **15**, 178–203.

Alon, N., Ben-David, S., Cesa-Bianchi, N., and Haussler, D. (1997) Scale-sensitive dimensions, uniform convergence, and learnability. *J. ACM* **44**, 615–663.

Andersen, Niels Trolle (1985a). The central limit theorem for non–separable valued functions. *Z. Wahrsch. verw. Gebiete* **70**, 445–455.

Andersen, N. T. (1985b). The calculus of non-measurable functions and sets. Various Publ. Ser. no. 36, Matematisk Institut, Aarhus Universitet.

Andersen, N. T., and Dobrić, Vladimir (1987). The central limit theorem for stochastic processes. *Ann. Probab.* **15**, 164–177.

Andersen, N. T., and Dobrić, V. (1988). The central limit theorem for stochastic processes II. *J. Theoret. Probab.* **1**, 287–303.

Andersen, N. T., Giné, E., Ossiander, M., and Zinn, J. (1988). The central limit theorem and the law of the iterated logarithm for empirical processes under local conditions. *Probab. Theory Related Fields* **77**, 271–305.

Anderson, Theodore W. (1955). The integral of a symmetric unimodal function over a symmetric convex set and some probability inequalities. *Proc. Amer. Math. Soc.* **6**, 170–176.

Araujo, Aloisio, and Giné, E. (1980). *The Central Limit Theorem for Real and Banach Valued Random Variables.* Wiley, New York.

Arcones, Miguel A., and Giné, E. (1993). Limit theorems for U-processes. *Ann. Probab.* **21**, 1494–1542.

Assouad, Patrice (1981). Sur les classes de Vapnik-Červonenkis. *C. R. Acad. Sci. Paris* Sér. I **292**, 921–924.

Assouad, P. (1982). Classes de Vapnik–Červonenkis et vitesse d'estimation (unpublished manuscript).

Assouad, P. (1983). Densité et dimension. *Ann. Inst. Fourier* (Grenoble) **33**, no. 3, 233–282.

Assouad, P. (1985). Observations sur les classes de Vapnik–Červonenkis et la dimension combinatoire de Blei. In *Séminaire d'analyse harmonique, 1983–84, Publ. Math. Orsay*, Univ. Paris XI, Orsay, 92–112.

Assouad, P., and Dudley, R. M. (1990). Minimax nonparametric estimation over classes of sets (unpublished manuscript).

Aumann, Robert J. (1961). Borel structures for function spaces. *Illinois J. Math.* **5**, 614–630.

Bahadur, Raghu Raj (1954). Sufficiency and statistical decision functions. *Ann. Math. Statist.* **25**, 423–462.

Bakel'man, I. Ya. (1965). *Geometric Methods of Solution of Elliptic Equations* (in Russian). Nauka, Moscow.

Bakhvalov, N. S. (1959). On approximate calculation of multiple integrals. *Vestnik Moskov. Univ. Ser. Mat. Mekh. Astron. Fiz. Khim.* **1959**, No. 4, 3–18.

Bauer, Heinz (1981). *Probability Theory and Elements of Measure Theory*, 2nd ed. Academic Press, London.

Bennett, George W. (1962). Probability inequalities for the sum of bounded random variables. *J. Amer. Statist. Assoc.* **57**, 33–45.

Berkes, István, and Philipp, Walter (1977). An almost sure invariance principle for the empirical distribution function of mixing random variables. *Z. Wahrsch. verw. Gebiete* **41**, 115–137.

Bernštein, Sergei N. (1924). Ob odnom vidoizmenenii neravenstva Chebysheva i o pogreshnosti formuly Laplasa (in Russian). *Uchen. Zapiski Nauchn.-issled. Kafedr Ukrainy, Otdel. Mat.*, vyp. 1, 38–48; reprinted in S. N. Bernštein, *Sobranie Sochineniĭ [Collected Works], Tom IV, Teoriya Veroiatnostei, Matematicheskaya Statistika*, Nauka, Moscow, 1964, pp. 71–79.

*Bernštein, S. N. (1927). *Teoriya Veroiatnostei* (*Theory of Probability*, in Russian). Moscow. 2nd ed., 1934.

Bickel, Peter J., and Doksum, Kjell A. (2001). *Mathematical Statistics*, 2nd ed., vol. **1**. Prentice–Hall, New York.

Bickel, P. J., and Freedman, David A. (1981). Some asymptotic theory for the bootstrap. *Ann. Statist.* **9**, 1196–1217.

Billingsley, Patrick (1968). *Convergence of Probability Measures.* Wiley, New York.

Birkhoff, Garrett (1935). Integration of functions with values in a Banach space. *Trans. Amer. Math. Soc.* **38**, 357–378.

Birnbaum, Zygmunt W., and Orlicz, W. (1931). Über die Verallgemeinerung des Begriffes der zueinander konjugierten Potenzen. *Studia Math.* **3**, 1–67.

Birnbaum, Z. W., and Tingey, F. H. (1951). One-sided confidence contours for probability distribution functions. *Ann. Math. Statist.* **22**, 592–596.

Blum, J. R. (1955). On the convergence of empiric distribution functions. *Ann. Math. Statist.* **26**, 527–529.

Blumberg, Henry (1935). The measurable boundaries of an arbitrary function. *Acta Math.* (Sweden) **65**, 263–282.

Bochner, Salomon (1933). Integration von Funktionen, deren Werte die Elemente eines Vektorraumes sind. *Fund. Math.* **20**, 262–276.

Bolthausen, Erwin (1978). Weak convergence of an empirical process indexed by the closed convex subsets of I^2. *Z. Wahrsch. verw. Gebiete* **43**, 173–181.

Bonnesen, Tommy, and Fenchel, Werner (1934). *Theorie der konvexen Körper.* Berlin, Springer; repr. Chelsea, New York, 1948.

Borell, Christer (1974). Convex measures on locally convex spaces. *Ark. Mat.* **12**, 239–252.

Borell, C. (1975a). Convex set functions in d-space. *Period. Math. Hungar.* **6**, 111–136.

Borell, C. (1975b). The Brunn–Minkowski inequality in Gauss space. *Invent. Math.* **30**, 207–216.

Borisov, Igor S. (1981). Some limit theorems for empirical distributions (in Russian). *Abstracts of Reports, Third Vilnius Conf. Probability Th. Math. Statist.* **1**, 71–72.

Bousquet, O., Koltchinskii, V., and Panchenko, D. (2002). Some local measures of complexity of convex hulls and generalization bounds. COLT [Conference on Learning Theory], *Lecture Notes on Artificial Intelligence* **2375**, 59–73.

Bretagnolle, Jean, and Massart, Pascal (1989). Hungarian constructions from the nonasymptotic viewpoint. *Ann. Probab.* **17**, 239–256.

Bronshtein [Bronštein], E. M. (1976). ε-entropy of convex sets and functions. *Siberian Math. J.* **17**, 393–398, transl. from *Sibirsk. Mat. Zh.* **17**, 508–514.

Brown, Lawrence D. (1986). *Fundamentals of Statistical Exponential Families, Inst. Math. Statist. Lecture Notes–Monograph Ser.* **9**, Hayward, CA.

Cantelli, Francesco Paolo (1933). Sulla determinazione empirica della leggi di probabilità. *Giorn. Ist. Ital. Attuari* **4**, 421–424.

Carl, Bernd (1982). On a characterization of operators from ℓ_q into a Banach space of type p with some applications to eigenvalue problems. *J. Funct. Anal.* **48**, 394–407.

Carl, B. (1997). Metric entropy of convex hulls in Hilbert space. *Bull. London Math. Soc.* **29**, 452–458.

Chernoff, Herman (1952). A measure of asymptotic efficiency for tests of a hypothesis based on the sum of observations. *Ann. Math. Statist.* **23**, 493–507.

Chevet, Simone (1970). Mesures de Radon sur $\mathbb{R}^n$ et mesures cylindriques. *Ann. Fac. Sci. Univ. Clermont* #43 (math., 6° fasc.), 91–158.

Clements, G. F. (1963). Entropies of several sets of real valued functions. *Pacific J. Math* **13**, 1085–1095.

Coffman, E. G., Jr., and Lueker, George S. (1991). *Probabilistic Analysis of Packing and Partitioning Algorithms.* Wiley, New York.

Coffman, E. G., and Shor, Peter W. (1991). A simple proof of the $O(\sqrt{n} \log^{3/4} n)$ upright matching bound. *SIAM J. Discrete Math.* **4**, 48–57.

Cohn, Donald L. (1980). *Measure Theory.* Birkhäuser, Boston. (Second Ed., 2013.)

Cover, Thomas M. (1965). Geometric and statistical properties of systems of linear inequalities with applications to pattern recognition. *IEEE Trans. Elec. Comp.* **EC-14** 326–334.

*Csáki, E. (1974). Investigations concerning the empirical distribution function. *Magyar Tud. Akad. Mat. Fiz. Oszt. Közl.* **23** 239–327; English transl. in *Selected Transl. Math. Statist. Probab.* **15** (1981), 229–317.

Csörgő, Miklós, and Horváth, Lajos (1993). *Weighted Approximations in Probability and Statistics.* Wiley, Chichester.

Csörgő, M., and Révész, Pal (1981). *Strong Approximations in Probability and Statistics.* Academic, New York.

Cuesta, J. A., and Matrán, C. (1989). Notes on the Wasserstein metric in Hilbert spaces. *Ann. Probab.* **17**, 1264–1276.

Danzer, Ludwig, Grünbaum, Branko, and Klee, Victor L. (1963). Helly's theorem and its relatives. *Proc. Symp. Pure Math. (Amer. Math. Soc.)* **7**, 101–180.

Darmois, Georges (1951). Sur une propriété caractéristique de la loi de probabilité de Laplace. *Comptes Rendus Acad. Sci. Paris* **232**, 1999–2000.

Darst, Richard B. (1971). C^∞ functions need not be bimeasurable. *Proc. Amer. Math. Soc.* **27**, 128–132.

Davis, Philip J., and Polonsky, Ivan (1972). Numerical interpolation, differentiation and integration. Chapter 25 in *Handbook of Mathematical Functions*, ed. M. Abramowitz and I. A. Stegun, Dover, New York, 9th printing.

DeHardt, John (1971). Generalizations of the Glivenko–Cantelli theorem. *Ann. Math. Statist.* **42**, 2050–2055.

*Dobrushin, R. L. (1970). Prescribing a system of random variables by conditional distributions. *Theory Probab. Appl.* **15**, 458–486.

Donsker, M. D. (1952). Justification and extension of Doob's heuristic approach to the Kolmogorov–Smirnov theorems. *Ann. Math. Statist.* **23**, 277–281.

Doob, Joseph L. (1949). Heuristic approach to the Kolmogorov–Smirnov theorems. *Ann. Math. Statist.* **20**, 393–403.

Drake, Frank R. (1974). *Set Theory: An Introduction to Large Cardinals*. North-Holland, Amsterdam.

Dudley, R. M. (1966). Weak convergence of probabilities on nonseparable metric spaces and empirical measures on Euclidean spaces. *Illinois J. Math.* **10**, 109–126.

Dudley, R. M. (1967a). The sizes of compact subsets of Hilbert space and continuity of Gaussian processes. *J. Funct. Anal.* **1**, 290–330.

Dudley, R. M. (1967b). Measures on non-separable metric spaces. *Illinois J. Math.* **11**, 449–453.

Dudley, R. M. (1968). Distances of probability measures and random variables. *Ann. Math. Statist.* **39**, 1563–1572.

Dudley, R. M. (1973). Sample functions of the Gaussian process. *Ann. Probab.* **1**, 66–103.

Dudley, R. M. (1974). Metric entropy of some classes of sets with differentiable boundaries. *J. Approx. Theory* **10**, 227–236; Correction **26** (1979), 192–193.

Dudley, R. M. (1978). Central limit theorems for empirical measures. *Ann. Probab.* **6**, 899–929; Correction **7** (1979), 909–911.

Dudley, R. M. (1980). Acknowledgment of priority: Second derivatives of convex functions. *Math. Scand.* **46**, 61.

Dudley, R. M. (1981). Donsker classes of functions. In *Statistics and Related Topics* (Proc. Symp. Ottawa, 1980), North-Holland, New York, 341–352.

Dudley, R. M. (1982). Empirical and Poisson processes on classes of sets or functions too large for central limit theorems. *Z. Wahrsch. verw. Gebiete* **61**, 355–368.

Dudley, R. M. (1984). A course on empirical processes. Ecole d'été de probabilités de St.-Flour, 1982. *Lecture Notes in Math.* (Springer) **1097**, 1–142.

Dudley, R. M. (1985a). An extended Wichura theorem, definitions of Donsker class, and weighted empirical distributions. In *Probability in Banach Spaces V* (Proc. Conf. Medford, 1984), *Lecture Notes in Math.* (Springer) **1153**, 141–178.

Dudley, R. M. (1985b). The structure of some Vapnik–Červonenkis classes. In *Proc. Berkeley Conf. in Honor of J. Neyman and J. Kiefer* **2**, 495–508. Wadsworth, Belmont, CA.

Dudley, R. M. (1987). Universal Donsker classes and metric entropy. *Ann. Probab.* **15**, 1306–1326.

Dudley, R. M. (1990). Nonlinear functionals of empirical measures and the bootstrap. In *Probability in Banach Spaces 7*, Proc. Conf. Oberwolfach, 1988, *Progress in Probability* **21**, Birkhäuser, Boston, 63–82.

Dudley, R. M. (1994). Metric marginal problems for set-valued or non-measurable variables. *Probab. Theory Related Fields* **100**, 175–189.

Dudley, R. M. (2000). *Notes on Empirical Processes. MaPhySto Lecture Notes* **4**, Aarhus, Denmark.

Dudley, R. M. (2002). *Real Analysis and Probability*. 2nd ed., Cambridge University Press, Cambridge.

Dudley, R. M., Giné, E., and Zinn, J. (1991). Uniform and universal Glivenko–Cantelli classes. *J. Theoret. Probab.* **4**, 485–210.

Dudley, R. M., and Koltchinskii, V. I. (1994, 1996). Envelope moment conditions and Donsker classes. *Theory Probab. Math. Statist. No. 51*, 39–48; Ukrainian version, *Teor. Ĭmovĭr. Mat. Stat. No. 51* (1994) 39–49.

Dudley, R. M., Kulkarni, S. R., Richardson, T., and Zeitouni, O. (1994). A metric entropy bound is not sufficient for learnability. *IEEE Trans. Inform. Theory* **40**, 883–885.

Dudley, R. M., and Philipp, Walter (1983). Invariance principles for sums of Banach space valued random elements and empirical processes. *Z. Wahrsch. verw. Gebiete* **62**, 509–552.

Dunford, Nelson (1936). Integration and linear operations. *Trans. Amer. Math. Soc.* **40**, 474–494.

Dunford, N., and Schwartz, Jacob T. (1958). *Linear Operators. Part I: General Theory*. Interscience, New York. Repr. 1988.

Durst, Mark, and Dudley, R. M. (1981). Empirical processes, Vapnik–Chervonenkis classes and Poisson processes. *Probab. Math. Statist.* (Wrocław) **1**, no. 2, 109–115.

Dvoretzky, A., Kiefer, J., and Wolfowitz, J. (1956). Asymptotic minimax character of the sample distribution function and the classical multinomial estimator. *Ann. Math. Statist.* **27**, 642–669.

Eames, W., and May, L. E. (1967). Measurable cover functions. *Canad. Math. Bull.* **10**, 519–523.

Effros, Edward G. (1965). Convergence of closed subsets in a topological space. *Proc. Amer. Math. Soc.* **16**, 929–931.

Efron, Bradley (1979). Bootstrap methods: another look at the jackknife. *Ann. Statist.* **7**, 1–26.

Efron, B., and Tibshirani, Robert J. (1993). *An Introduction to the Bootstrap*. Chapman and Hall, New York.

Eggleston, H. G. (1958). *Convexity*. Cambridge University Press, Cambridge. Reprinted with corrections, 1969.

Eilenberg, Samuel, and Steenrod, Norman (1952). *Foundations of Algebraic Topology*. Princeton University Press, Princeton, NJ.

Eršov, M. P. (1975). The Choquet theorem and stochastic equations. *Analysis Math.* **1**, 259–271.

Evstigneev, I. V. (1977). "Markov times" for random fields. *Theory Probab. Appl.* **22**, 563–569; *Teor. Veroiatnost. i Primenen.* **22**, 575–581.

Feldman, Jacob (1972). Sets of boundedness and continuity for the canonical normal process. *Proc. Sixth Berkeley Symposium Math. Statist. Prob.* **2**, pp. 357–367. University of California Press, Berkeley and Los Angeles.

Feller, William (1968). *An Introduction to Probability Theory and Its Applications*. Vol. 1, 3rd ed. Wiley, New York.

Feller, W. (1971). *An Introduction to Probability Theory and Its Applications*. Vol. 2, 2nd ed. Wiley, New York.

Ferguson, Thomas S. (1967). *Mathematical Statistics: A Decision Theoretic Approach*. Academic Press, New York.

Fernique, Xavier (1964). Continuité des processus gaussiens. *Comptes Rendus Acad. Sci. Paris* **258**, 6058–6060.

Fernique, X. (1970). Intégrabilité des vecteurs gaussiens. *Comptes Rendus Acad. Sci. Paris Sér.* A **270**, 1698–1699.

Fernique, X. (1971). Régularité de processus gaussiens. *Invent. Math.* **12**, 304–320.

Fernique, X. (1975). Régularité des trajectoires des fonctions aléatoires gaussiennes. Ecole d'été de probabilités de St.-Flour, 1974. *Lecture Notes in Math.* (Springer) **480**, 1–96.

Fernique, X. (1985). Sur la convergence étroite des mesures gaussiennes. *Z. Wahrsch. verw. Gebiete* **68**, 331–336.

Fernique, X. (1997). Fonctions aléatoires gaussiennes, vecteurs aléatoires gaussiens. Publications du Centre de Recherches Mathématiques, Montréal.

Fisher, Ronald Aylmer (1922). IX. On the mathematical foundations of theoretical statistics. *Phil. Trans. Roy. Soc. London* Ser. A **222**, 309–368.

Freedman, David A. (1966). On two equivalence relations between measures. *Ann. Math. Statist.* **37**, 686–689.

Gaenssler, Peter (1986). Bootstrapping empirical measures indexed by Vapnik–Chervonenkis classes of sets. In *Probability Theory and Mathematical Statistics* (Vilnius, 1985), Yu. V. Prohorov, V. A. Statulevičius, V. V. Sazonov, and B. Grigelionis, eds., VNU Science Press, Utrecht, 467–481.

Gaenssler, P., and Stute, Winfried (1979). Empirical processes: a survey of results for independent and identically distributed random variables. *Ann. Probab.* **7**, 193–243.

*Gelfand, Izrail' Moiseevich (1936). Sur un lemme de la théorie des espaces linéaires. *Communications de l'Institut des Sciences Math. et Mécaniques de l'Université de Kharkoff et de la Societé Math. de Kharkov (= Zapiski Khark. Mat. Obshchestva)* (Ser. 4) **13**, 35–40.

Gelfand, I. M. (1938). Abstrakte Funktionen und lineare Operatoren. *Mat. Sbornik* (N. S.) **4**, 235–286.

Giné, Evarist (1974). On the central limit theorem for sample continuous processes. *Ann. Probab.* **2**, 629–641.

Giné, E. (1997). Lectures on some aspects of the bootstrap. In *Lectures on Probability Theory and Statistics*, Ecole d'été de probabilités de Saint-Flour (1996), ed. P. Bernard. *Lecture Notes in Math.* (Springer) **1665**, 37–151.

Giné, E., and Zinn, Joel (1984). Some limit theorems for empirical processes. *Ann. Probab.* **12**, 929–989.

Giné, E., and Zinn, J. (1986). Lectures on the central limit theorem for empirical processes. In *Probability and Banach Spaces*, Proc. Conf. Zaragoza, 1985, *Lecture Notes in Math.* **1221**, 50–113. Springer, Berlin.

Giné, E. and Zinn, J. (1990). Bootstrapping general empirical measures. *Ann. Probab.* **18**, 851–869.

Giné, E. and Zinn, J. (1991). Gaussian characterization of uniform Donsker classes of functions. *Ann. Probab.* **19**, 758–782.

Glivenko, V. I. (1933). Sulla determinazione empirica della leggi di probabilità. *Giorn. Ist. Ital. Attuari* **4**, 92–99.

Gnedenko, B. V., and Kolmogorov, A. N. (1949). *Limit Distributions for Sums of Independent Random Variables.* Moscow. Transl. and ed. by K. L. Chung, Addison-Wesley, Reading, Mass., 1954, rev. ed. 1968.

Goffman, C., and Zink, R. E. (1960). Concerning the measurable boundaries of a real function. *Fund. Math.* **48**, 105–111.

Gordon, Yehoram (1985). Some inequalities for Gaussian processes and applications. *Israel J. Math.* **50**, 265–289.

Graves, L. M. (1927). Riemann integration and Taylor's theorem in general analysis. *Trans. Amer. Math. Soc.* **29**, 163–177.

Gross, Leonard (1962). Measurable functions on Hilbert space. *Trans. Amer. Math. Soc.* **105**, 372–390.

Gruber, P. M. (1983). Approximation of convex bodies. In *Convexity and its Applications*, ed. P. M. Gruber and J. M. Wills. Birkhäuser, Basel, pp. 131–162.

Gutmann, Samuel (1981). Unpublished manuscript.

Hall, Peter (1992). *The Bootstrap and Edgeworth Expansion.* Springer, New York.

Halmos, Paul R., and Savage, Leonard Jimmie (1949). Application of the Radon–Nikodym theorem to the theory of sufficient statistics. *Ann. Math. Statist.* **20**, 225–241.

Harary, F. (1969). *Graph Theory.* Addison-Wesley, Reading, MA.

Harding, E. F. (1967). The number of partitions of a set of N points in k dimensions induced by hyperplanes. *Proc. Edinburgh Math. Soc.* (Ser. II) **15**, 285–289.

Hausdorff, Felix (1914). *Mengenlehre*, transl. by J. P. Aumann et al. as *Set Theory*. 3rd English ed. of transl. of 3rd German edition (1937). Chelsea, New York, 1978.

Haussler, D. (1995). Sphere packing mumbers for subsets of the Boolean n-cube with bounded Vapnik–Chervonenkis dimension. *J. Combin. Theory Ser. A* **69**, 217–232.

Hewitt, Edwin, and Ross, Kenneth A. (1979). *Abstract Harmonic Analysis*, vol. 1, 2nd ed. Springer, Berlin.

Hobson, E. W. (1926). *The Theory of Functions of a Real Variable and the Theory of Fourier's Series*, vol. 2, 2nd ed. Repr. Dover, New York, 1957.

Hoeffding, Wassily (1963). Probability inequalities for sums of bounded random variables. *J. Amer. Statist. Assoc.* **58**, 13–30.

Hoffman, Kenneth (1975). *Analysis in Euclidean Space.* Prentice-Hall, Englewood Cliffs, NJ.

Hoffmann-Jørgensen, Jørgen (1974). Sums of independent Banach space valued random elements. *Studia Math.* **52**, 159–186.

Hoffmann-Jørgensen, J. (1984). *Stochastic Processes on Polish Spaces.* Published (1991): *Various Publication Series* no. 39, Matematisk Institut, Aarhus Universitet.

Hoffmann-Jørgensen, J. (1985). The law of large numbers for non-measurable and non-separable random elements. *Astérisque* **131**, 299–356.

Hörmander, L. (1983). *The Analysis of Linear Partial Differential Operators I.* Springer, Berlin.

Hu, Inchi (1985). A uniform bound for the tail probability of Kolmogorov–Smirnov statistics. *Ann. Statist.* **13**, 821–826.

Il'in, V. A., and Pozniak, E. G. (1982). *Fundamentals of Mathematical Analysis.* Transl. from the 4th Russian ed. (1980) by V. Shokurov. Mir, Moscow.

Itô, Kiyosi, and McKean, Henry P., Jr. (1974). *Diffusion Processes and Their Sample Paths.* Springer, Berlin.

Jain, Naresh C., and Marcus, Michael B. (1975). Central limit theorems for $C(S)$-valued random variables. *J. Funct. Anal.* **19**, 216–231.

Kac, M. (1949). On deviations between theoretical and empirical distributions. *Proc. Nat. Acad. Sci. USA* **35**, 252–257.

Kahane, Jean-Pierre (1985). *Some Random Series of Functions*, 2nd ed. Cambridge University Press, Cambridge.

Kahane, J.-P. (1986). Une inégalité du type de Slepian et Gordon sur les processus gaussiens. *Israel J. Math.* **55**, 109–110.

*Kantorovich, L. V. (1942). On the transfer of masses. *Dokl. Akad Nauk. SSSR* **37**, 7–8.

Kantorovich, L. V., and Rubinshtein, G. Sh. (1958). On a space of completely additive functions. *Vestnik Leningrad Univ.* **13** no. 7, *Ser. Math. Astron. Phys.* **2**, 52–59 (in Russian).

Kartashev, A. P., and Rozhdestvenskiĭ, B. L. (1984). *Matematicheskiĭ Analiz* (Mathematical Analysis; in Russian). Nauka, Moscow. French transl.: Kartachev, A., and Rojdestvenski, B. *Analyse mathématique*, transl. by Djilali Embarek. Mir, Moscow, 1988.

Kelley, John L. (1955). *General Topology*. Van Nostrand, Princeton, NJ. Repr. Springer, New York, 1975.

Kolmogorov, A. N. (1933a). Über die Grenzwertsätze der Wahrscheinlichkeitsrechnung. *Izv. Akad. Nauk SSSR (Bull. Acad. Sci. URSS)* (Ser. 7) no. 3, 363–372.

*Kolmogorov, A. N. (1933b). Sulla determinazione empirica di una legge di distribuzione. *Giorn. Ist. Ital. Attuari* **4**, 83–91.

Kolmogorov, A. N. (1955). Bounds for the minimal number of elements of an ε-net in various classes of functions and their applications to the question of representability of functions of several variables by superpositions of functions of fewer variables (in Russian). *Uspekhi Mat. Nauk* **10**, no. 1 (63), 192–194.

Kolmogorov, A. N. (1956). On Skorokhod convergence. *Theory Probab. Appl.* **1**, 215–222.

Kolmogorov, A. N., and Tikhomirov, V. M. (1959). ϵ-entropy and ϵ-capacity of sets in function spaces. *Uspekhi Mat. Nauk* **14**, no. 2, 3–86 = *Amer. Math. Soc. Transl.* (Ser. 2) **17** (1961), 277–364.

Koltchinskii, Vladimir I. (1981). On the central limit theorem for empirical measures. *Theor. Probab. Math. Statist.* **24**, 71–82. Transl. from *Teor. Verojatnost. i Mat. Statist. (1981)* **24**, 63–75.

*Komatsu, Y. (1955). Elementary inequalities for Mills' ratio. *Rep. Statist. Appl. Res. Un. Japan. Sci. Engrs.* **4**, 69–70.

Komlós, János, Major, Péter, and Tusnády, Gábor (1975). An approximation of partial sums of independent RV'-s and the sample DF. I. *Z. Wahrsch. verw. Gebiete* **32**, 111–131.

Krasnosel'skiĭ, M. A., and Rutitskii, Ia. B. (1961). *Convex Functions and Orlicz Spaces.* Transl. from Russian by L. F. Boron. Noordhoff, Groningen.

Landau, H. J., and Shepp, Lawrence A. (1971). On the supremum of a Gaussian process. *Sankhyā* Ser. A **32**, 369–378.

Lang, Serge (1993). *Real and Functional Analysis*, 3rd ed. of *Real Analysis*, Springer, New York.

Ledoux, Michel (1996). Isoperimetry and Gaussian analysis. In *Ecole d'été de probabilités de St.-Flour, 1994, Lecture Notes in Math.* (Springer) **1648**, 165–294.

Ledoux, M. (2001). *The Concentration of Measure Phenomenon*, Math. Surveys and Monographs **89**, American Mathematical Society, Providence, RI.

Ledoux, M., and Talagrand, M. (1991). *Probability in Banach Spaces*. Springer, Berlin.

Lehmann, Erich L. (1986,1991). *Testing Statistical Hypotheses*, 2nd ed. repr. 1997, Springer, New York.

Leighton, Thomas, and Shor, P. (1989). Tight bounds for minimax grid matching, with applications to the average case analysis of algorithms. *Combinatorica* **9**, 161–187.

Lorentz, G. G. (1966). Metric entropy and approximation. *Bull. Amer. Math. Soc.* **72**, 903–937.

Luxemburg, W. A. J., and Zaanen, A. C. (1956). Conjugate spaces of Orlicz spaces. *Akad. Wetensch. Amsterdam Proc.* Ser. A **59** = *Indag. Math.* **18**, 217–228.

Luxemburg, W. A. J., and Zaanen, A. C. (1983). *Riesz Spaces,* vol. 2, North-Holland, Amsterdam.

Marcus, Michael B. (1974). The ε-entropy of some compact subsets of ℓ^p. *J. Approximation Th.* **10**, 304–312.

Marcus, M. B., and Shepp, L. A. (1972). Sample behavior of Gaussian processes. *Proc. Sixth Berkeley Symp. Math. Statist. Prob.* (1970) **2**, 423–441. University of California Press, Berkeley and Los Angeles.

Marczewski, E., and Sikorski, R. (1948). Measures in non-separable metric spaces. *Colloq. Math.* **1**, 133–139.

Mason, David M. (1998). Notes on the the KMT Brownian bridge approximation to the uniform empirical process. Preprint.

Mason, D. M., and van Zwet, Willem (1987). A refinement of the KMT inequality for the uniform empirical process. *Ann. Probab.* **15**, 871–884.

Massart, Pascal (1990). The tight constant in the Dvoretzky–Kiefer–Wolfowitz inequality. *Ann. Probab.* **18**, 1269–1283.

Mourier, Edith (1951), Lois de grands nombres et théorie ergodique. *C. R. Acad. Sci. Paris* **232**, 923–925.

Mourier, E. (1953). Éléments aléatoires dans un espace de Banach. *Ann. Inst. H. Poincaré* **13**, 161–244.

Nachbin, Leopoldo J. (1965). *The Haar Integral.* Transl. from Portuguese by L. Bechtolsheim. Van Nostrand, Princeton, NJ.

Nanjundiah, T. S. (1959). Note on Stirling's formula. *Amer. Math. Monthly* **66**, 701–703.

Natanson, I. P. (1957). *Theory of Functions of a Real Variable*, vol. 2, 2nd ed. Transl. by L. F. Boron, Ungar, New York, 1961.

Neveu, Jacques (1977). Processus ponctuels. *Ecole d'été de probabilités de St.-Flour VI, 1976, Lecture Notes in Math.* (Springer) **598**, 249–445.

*Neyman, Jerzy (1935). Su un teorema concernente le cosidette statistiche sufficienti. *Giorn. Ist. Ital. Attuari* **6**, 320–334.

Okamoto, Masashi (1958). Some inequalities relating to the partial sum of binomial probabilities. *Ann. Inst. Statist. Math.* **10**, 29–35.

Olkin, I., and Pukelsheim, F. (1982). The distance between two random vectors with given dispersion matrices. *Linear Algebra Appl.* **48**, 257–263.

Ossiander, Mina (1987). A central limit theorem under metric entropy with L_2 bracketing. *Ann. Probab.* **15**, 897–919.

Pachl, Jan K. (1979). Two classes of measures. *Colloq. Math.* **42**, 331–340.

Panchenko, Dmitry (2004). 18.465 Topics in Statistics, OpenCourseWare, MIT.

Pettis, Billy Joe (1938). On integration in vector spaces. *Trans. Amer. Math. Soc.* **44**, 277–304.

Pisier, Gilles (1981). Remarques sur un resultat non publié de B. Maurey. *Séminaire d'Analyse Fonctionelle 1980–1981* V.1–V.12. Ecole Polytechnique, Centre de Mathématiques, Palaiseau.

Pisier, G. (1983). Some applications of the metric entropy bound in harmonic analysis. In *Banach Spaces, Harmonic Analysis, and Probability Theory*, Proc. Univ. Connecticut 1980–1981, eds. R. C. Blei and S. J. Sidney, *Lecture Notes in Math.* (Springer) **995**, 123–154.

Pollard, David B. (1982). A central limit theorem for empirical processes. *J. Austral. Math. Soc. Ser. A* **33**, 235–248.

Pollard, D. (1982). A central limit theorem for *k*-means clustering. *Ann. Probab.* **10**, 919–926.

Pollard, D. (1984). *Convergence of Stochastic Processes*. Springer, New York.

Pollard, D. (1985). New ways to prove central limit theorems. *Econometric Theory* **1**, 295–314.

Pollard, D. (1990). *Empirical Processes: Theory and Applications. NSF-CBMS Regional Conference Series in Probab. and Statist.* **2**. Inst. Math. Statist. and Amer. Statist. Assoc.

Posner, Edward C., Rodemich, Eugene R., and Rumsey, Howard, Jr. (1967). Epsilon entropy of stochastic processes. *Ann. Math. Statist.* **38**, 1000–1020.

Posner, E. C., Rodemich, E. R., and Rumsey, H. (1969). Epsilon entropy of Gaussian processes. *Ann. Math. Statist.* **40**, 1272–1296.

Price, G. B. (1940). The theory of integration. *Trans. Amer. Math. Soc.* **47**, 1–50.

Pyke, Ronald (1968). The weak convergence of the empirical process with random sample size. *Proc. Cambridge Philos. Soc.* **64**, 155–160.

Rademacher, Hans (1919). Über partielle und totale Differenzierbarkeit I. *Math. Ann.* **79**, 340–359.

Radon, Johann (1921). Mengen konvexer Körper, die einen gemeinsamen Punkt enthalten. *Math. Ann.* **83**, 113–115.

Rao, B. V. (1971). Borel structures for function spaces. *Colloq. Math.* **23**, 33–38.

Rao, C. Radhakrishna (1973). *Linear Statistical Inference and Its Applications*. 2nd ed. Wiley, New York.

Reshetnyak, Yu. G. (1968). Generalized derivatives and differentiability almost everywhere. *Mat. Sb.* **75**, 323–334 (Russian) = *Math. USSR-Sb.* **4**, 293–302 (English transl.).

Rhee, WanSoo, and Talagrand, M. (1988). Exact bounds for the stochastic upward matching problem. *Trans. Amer. Math. Soc.* **307**, 109–125.

Rudin, Walter (1974). *Real and Complex Analysis*, 2nd ed. McGraw-Hill, New York.

Ryll-Nardzewski, C. (1953). On quasi-compact measures. *Fund. Math.* **40**, 125–130.

Sainte-Beuve, Marie-France (1974). On the extension of von Neumann–Aumann's theorem. *J. Funct. Anal.* **17**, 112–129.

Sauer, Norbert (1972). On the density of families of sets. *J. Combin. Theory* Ser. A **13**, 145–147.

Sazonov, Vyacheslav V. (1962). On perfect measures (in Russian). *Izv. Akad. Nauk SSSR Ser. Mat.* **26**, 391–414.

Schaefer, Helmut H. (1966). *Topological Vector Spaces*. Macmillan, New York. 3rd printing, corrected, Springer, New York, 1971.

Schaerf, Henry M. (1947). On the continuity of measurable functions in neighborhood spaces. *Portugaliae Math.* **6**, 33–44, 66.

Schaerf, H. M. (1948). On the continuity of measurable functions in neighborhood spaces II. *Portugaliae Math.* **7**, 91–92.

Schläfli, Ludwig (1901, posth.). *Theorie der vielfachen Kontinuität*, republ. 1991, Cornell University Library, Ithaca, NY; also in *Gesammelte Math. Abhandlungen* I, Birkhäuser, Basel, 1950.

Schmidt, Wolfgang M. (1975). Irregularities of distribution IX. *Acta Arith.* **27**, 385–396.

Schwartz, Laurent (1966). Sur le théorème du graphe fermé. *C. R. Acad. Sci. Paris Sér.* A **263**, A602–605.

Shao, Jun, and Tu, Dongsheng (1995). *The Jackknife and Bootstrap.* Springer, New York.

Shelah, Saharon (1972). A combinatorial problem: stability and order for models and theories in infinitary languages. *Pacific J. Math.* **41**, 247–261.

Shor, Peter W. (1986). The average-case analysis of some on-line algorithms for bin packing. *Combinatorica* **6**, 179–200.

Shorack, Galen, and Wellner, Jon A. (1986). *Empirical Processes with Applications to Statistics.* Wiley, New York.

Shortt, Rae M. (1983). Universally measurable spaces: an invariance theorem and diverse characterizations. *Fund. Math.* **121**, 169–176.

Shortt, R. M. (1984). Combinatorial methods in the study of marginal problems over separable spaces. *J. Math. Anal. Appl.* **97**, 462–479.

Singh, Kesar (1981). On the asymptotic accuracy of Efron's bootstrap. *Ann. Statist.* **9**, 1187–1195.

Skitovič, V. P. (1954). Linear combinations of independent random variables and the normal distribution law. *Izv. Akad. Nauk SSSR Ser. Mat.* **18**, 185–200 (in Russian).

Skorohod, Anatoli V. (1956). Limit theorems for stochastic processes. *Theory Probab. Appl.* **1**, 261–290.

Skorohod, A. V. (1976). On a representation of random variables. *Theory Probab. Appl.* **21**, 628–632 (English), 645–648 (Russian).

Slepian, David (1962). The one-sided barrier problem for Gaussian noise. *Bell Syst. Tech. J.* **41**, 463–501.

*Smirnov, N. V. (1944). Approximate laws of distributions of random variables from empirical data. *Uspekhi Mat. Nauk* **10**, 179–206 (in Russian).

Smoktunowicz, Agata (1997). A remark on Vapnik–Chervonienkis classes. *Colloq. Math.* **74**, 93–98.

Steele, J. Michael (1975). Combinatorial entropy and uniform limit laws. Ph. D. dissertation, mathematics, Stanford University, Stanford, CA.

Steele, J. M. (1978a). Existence of submatrices with all possible columns. *J. Combin. Theory* Ser. A **24**, 84–88.

Steele, J. M. (1978b). Empirical discrepancies and subadditive processes. *Ann. Probab.* **6**, 118–127.

Stein, Elias M., and Weiss, Guido (1971). *Introduction to Fourier Analysis on Euclidean Spaces.* Princeton University Press, Princeton, NJ.

Steiner, Jakob (1826). Einige Gesetze über die Theilung der Ebene und des Raumes. *J. Reine Angew. Math.* **1**, 349–364.

Stone, Arthur H. (1948). Paracompactness and product spaces. *Bull. Amer. Math. Soc.* **54**, 631–632.

Strobl, Franz (1994). *Zur Theorie empirischer Prozesse.* Dissertation, Mathematik, Universität München.

Strobl, F. (1995). On the reversed sub-martingale property of empirical discrepancies in arbitrary sample spaces. *J. Theoret. Probab.* **8**, 825–831.

Sudakov, Vladimir N. (1969). Gaussian measures, Cauchy measures and ε-entropy. *Soviet Math. Dokl.* **10**, 310–313.

Sudakov, V. N. (1971). Gaussian random processes and measures of solid angles in Hilbert space. *Soviet Math. Dokl.* **12**, 412–415.

Sudakov, V. N. (1973). A remark on the criterion of continuity of Gaussian sample function. In *Proc. Second Japan-USSR Symp. Probab. Theory, Kyoto, 1972, ed. G. Maruyama, Yu. V. Prokhorov; Lecture Notes in Math.* (Springer) **330**, 444–454.

Sun, Tze-Gong (1976). Draft Ph. D. dissertation, Mathematics, University of Washington, Seattle.

Sun, Tze-Gong, and Pyke, R. (1982). Weak convergence of empirical processes. Technical Report, Dept. of Statistics, University of Washington, Seattle.

Talagrand, Michel (1987). Regularity of Gaussian processes. *Acta Math.* (Sweden) **159**, 99–149.

Talagrand, M. (1992). A simple proof of the majorizing measure theorem. *Geom. Funct. Anal.* **2**, 118–125.

Talagrand, M. (1994). Matching theorems and empirical discrepancy computations using majorizing measures. *J. Amer. Math. Soc.* **7**, 455–537.

Talagrand, M. (2005). *The Generic Chaining.* Springer, Berlin.

Topsøe, Flemming (1970). *Topology and Measure. Lecture Notes in Math.* (Springer) **133**.

Uspensky, James V. (1937). *Introduction to Mathematical Probability.* McGraw-Hill, New York.

van der Vaart, Aad [W.] (1996). New Donsker classes. *Ann. Probab.* **24**, 2128–2140.

van der Vaart, A. W., and Wellner, J. A. (1996). *Weak Convergence and Empirical Processes.* Springer, New York.

Vapnik, Vladimir N., and Červonenkis, Alekseĭ Ya. (1968). Uniform convergence of frequencies of occurrence of events to their probabilities. *Dokl. Akad. Nauk SSSR* **181**, 781–783 (Russian) = *Sov. Math. Doklady* **9**, 915–918 (English).

Vapnik, V. N., and Červonenkis, A. Ya. (1971). On the uniform convergence of relative frequencies of events to their probabilities. *Theory Probab. Appl.* **16**, 264–280 = *Teor. Veroiatnost. i Primenen.* **16**, 264–279.

Vapnik, V. N., and Červonenkis, A. Ya. (1974). *Teoriya Raspoznavaniya Obrazov: Statisticheskie problemy obucheniya* [Theory of Pattern Recognition; Statistical problems of learning; in Russian]. Nauka, Moscow. German ed.: *Theorie der Zeichenerkennung*, by W. N. Wapnik and A. J. Tscherwonenkis, transl. by K. G. Stöckel and B. Schneider, ed. S. Unger and K. Fritzsch. Akademie-Verlag, Berlin, 1979 (*Elektronisches Rechnen und Regeln*, Sonderband).

Vapnik, V. N., and Červonenkis, A. Ya. (1981). Necessary and sufficient conditions for the uniform convergence of means to their expectations. *Theory Probab. Appl.* **26**, 532–553.

Vorob'ev, N. N. (1962). Consistent families of measures and their extensions. *Theory Probab. Appl.* **7**, 147–163 (English), 153–169 (Russian).

Vulikh, B. Z. (1961). *Introduction to the Theory of Partially Ordered Spaces* (transl. from Russian by L. F. Boron, 1967). Wolters-Noordhoff, Groningen.

*Wasserstein [Vaserštein], L. N. (1969). Markov processes of spaces describing large families of automata. *Prob. Information Transmiss.* **5**, 47–52.

Watson, D. (1969). On partitions of *n* points. *Proc. Edinburgh Math. Soc.* **16**, 263–264.

Wenocur, Roberta S., and Dudley, R. M. (1981). Some special Vapnik–Červonenkis classes. *Discrete Math.* **33**, 313–318.

Whittaker, E. T., and Watson, G. N. (1927). *Modern Analysis*, 4th ed., Cambridge University Press, Cambridge, repr. 1962.

Wichura, Michael J. (1970). On the construction of almost uniformly convergent random variables with given weakly convergent image laws. *Ann. Math. Statist.* **41**, 284–291.

Wolfowitz, Jacob (1954). Generalization of the theorem of Glivenko–Cantelli. *Ann. Math. Statist.* **25**, 131–138.

Wright, F. T. (1981). The empirical discrepancy over lower layers and a related law of large numbers. *Ann. Probab.* **9**, 323–329.

Young, William Henry (1912). On classes of summable functions and their Fourier series. *Proc. Roy. Soc. London* Ser. A **87**, 225–229.

Zakon, Elias (1965). On "essentially metrizable" spaces and on measurable functions with values in such spaces. *Trans. Amer. Math. Soc.* **119**, 443–453.

Ziegler, Klaus (1994). *On functional central limit theorems and uniform laws of large numbers for sums of independent processes.* Dissertation, Mathematik, Universität München.

*Ziegler, K. (1997a). A maximal inequality and a functional central limit theorem for set-indexed empirical processes. *Results Math.* **31**, 189–194.

*Ziegler, K. (1997b). On Hoffmann-Jørgensen-type inequalities for outer expectations with applications. *Results Math.* **32**, 179–192.

Ziemer, William P. (1989). *Weakly Differentiable Functions.* Springer, New York.

---

* An asterisk indicates a work I have seen discussed in secondary sources but not in the original.

Notation Index

Author Index

Subject Index

Printed in the United States
by Baker & Taylor Publisher Services